Geologic Fracture Mechanics

This lively introduction to geologic fracture mechanics provides a consistent treatment of all common types of geologic structural discontinuities. It explores the formation, growth, and interpretation of fractures and deformation bands, from theoretical, field, and laboratory-based perspectives, bridging the gap between a general textbook treatment and the more advanced research literature. It allows the reader to acquire basic tools to interpret discontinuity origins, geometries, patterns, and implications using many of the leading and contemporary concepts known to specialists in the field. Exercises are provided at the end of each chapter, and worked examples are included within each chapter to illustrate topics and enable self-study. With all common geologic structures including joints, hydrofractures, faults, stylolites, and deformation bands being discussed from a fresh perspective, it will be a useful reference for advanced students, researchers, and industry practitioners interested in structural geology, neotectonics, rock mechanics, planetary geology, and reservoir geomechanics.

Dr. Richard A. Schultz is the owner and principal consultant of Orion Geomechanics LLC, Cypress, Texas. He is a geologist specializing in the geomechanics of faulted overburden and reservoir systems, and in underground natural gas storage. He was previously Senior Research Scientist at The University of Texas at Austin, Principal Geomechanicist with ConocoPhillips, and Foundation Professor of Geological Engineering and Geomechanics (now Emeritus) with the University of Nevada, Reno. He is a member of the Interstate Oil and Gas Compact Commission, a Fellow of the Geological Society of America, a licensed Professional Geologist in the State of Texas, and was an instructor of State oil and gas regulators with TopCorp.

Geologic Fracture Mechanics

Richard A. Schultz

Orion Geomechanics LLC

CAMBRIDGE
UNIVERSITY PRESS

CAMBRIDGE
UNIVERSITY PRESS

University Printing House, Cambridge CB2 8BS, United Kingdom

One Liberty Plaza, 20th Floor, New York, NY 10006, USA

477 Williamstown Road, Port Melbourne, VIC 3207, Australia

314–321, 3rd Floor, Plot 3, Splendor Forum, Jasola District Centre, New Delhi – 110025, India

79 Anson Road, #06–04/06, Singapore 079906

Cambridge University Press is part of the University of Cambridge.

It furthers the University's mission by disseminating knowledge in the pursuit of education, learning, and research at the highest international levels of excellence.

www.cambridge.org
Information on this title: www.cambridge.org/9781107189997
DOI: 10.1017/9781316996737

First published 2019

Printed in the United Kingdom by TJ International Ltd. Padstow Cornwall

A catalogue record for this publication is available from the British Library.

Library of Congress Cataloging-in-Publication Data
Names: Schultz, Richard A., author.
Title: Geologic fracture mechanics / Richard A. Schultz (Orion Geomechanics).
Description: Cambridge ; New York, NY : Cambridge University Press, 2019.
Identifiers: LCCN 2019001105 | ISBN 9781107189997
Subjects: LCSH: Geology, Structural – Textbooks. | Rocks – Fracture – Textbooks. | Rocks – Cleavage – Textbooks. | Rock deformation – Textbooks.
Classification: LCC QE601 .S3285 2019 | DDC 551.8–dc23
LC record available at https://lccn.loc.gov/2019001105

ISBN 978-1-107-18999-7 Hardback

To my Father,
   who taught me initiative;

To my Mother,
   who taught me compassion;

And to my wife and family for their loving support.

Contents

Preface

Structural discontinuities in rock, including all types of fractures, joints, faults, and deformation bands, along with stylolites and fluid-filled cracks such as hydrofractures, veins, and dikes, have been studied extensively using the tools and terminology of geology, geomorphology, geophysics, rock engineering, geomechanics, tectonophysics, hydrology, petroleum geology, mining engineering, quarrying, soil mechanics/geotechnical engineering, crystallography, experimental rock physics, and materials science. Most of what is known about fractures and deformation bands, collectively referred to as geologic structural discontinuities in this book, comes from a synthesis of concepts flowing from these traditionally separate disciplines. Many of the basics of what structural discontinuities are, and how to identify them, can be found in various degrees by using any of a number of good textbooks on particular topics. However, these sources must necessarily cover a wide range of material appropriate for semester-long, comprehensive courses in, for example, structural geology and tectonics. On the other hand, journal articles provide in-depth explorations of these topics for the experienced and knowledgeable reader.

The purpose of this book is to help bridge the gap between general treatments of geologic structural discontinuities found in textbooks or more advanced topical monographs and the published research literature. The book explores **geologic fracture mechanics**—the formation, growth, and interpretation of geologic structural discontinuities—primarily from field-based and theoretical perspectives, with laboratory-based research incorporated as appropriate. In many cases, some large amount (like 70%) of what can be interpreted from a discontinuity set can be gained with only simple tools and observations and without extensive mathematical analysis. Many scientists and engineers working in industry utilize this to balance rigor with deadlines and the limitations of applicable data. Specialists in fracture mechanics (sometimes referred to as "crack aficionados" by other structural geologists who don't live and breathe fracture mechanics to the same extent) can use sophisticated mathematical, geodetic, and computer analysis techniques to pull a greater amount of understanding from the rocks, but for many professional geologists and engineers working in industry, or who may be some years away from their formal academic training, this additional insight may not be necessary. This book will help the reader acquire some basic tools that can help to interpret discontinuity origins, geometries, patterns, and implications using many of the leading or contemporary concepts known to specialists in these areas.

This book is intended to be read and used by geologists and engineers who have had some exposure to the standard principles of structural geology, whether in college-level courses, graduate study, or continuing education. It is not designed as a stand-alone textbook for a class in structural geology or rock mechanics, but it may be used to supplement such a class. Nor should it be regarded as a substitute for formal training at the graduate level for research-grade work in geologic fracture mechanics. Instead, this book is an attempt to translate many of the major findings of the specialized literature on geologic structural discontinuities for the less specialized geoscientist or engineer.

The main emphasis of the book is *field interpretation of geologic structural discontinuities*. It is hoped that the material will provide a useful template for conceptually matching discontinuity sets on the outcrop with the implied stress states, local conditions, or three-dimensional (3-D) structural geometry, insofar as these are known or can be reasonably inferred from inspection. Equations are introduced and included as needed both to illustrate key concepts and to provide simple yet powerful tools for going deeper than just a simple visual scanning of the outcrop.

The literature bearing on geologic structural discontinuities is vast. References to the literature are inserted as called for in the text. The interested reader will quickly find that the rather extensive references provided are but the tip of the iceberg, so that by studying this book an appreciation of the overall scope, tenor, and terminology of the field may be gained that can facilitate a manageable entry into the realms of geologic fracture mechanics.

In preparing this book I have had to make difficult choices as to depth and rigor of treatment. Many of these choices, made to more succinctly or convincingly access and present the material for the nonspecialist, may challenge the purest researchers in fracture mechanics. Specialists may find other topics and applications here beyond their usual focus, while nonspecialists will perhaps appreciate the level of sophistication available about geologic structural discontinuities and the engineering approaches used to understand them.

The book begins with an overview in Chapter 1 of the main concepts, definitions, and terminology that underlie the interpretation and analysis of fractures, deformation bands, and other structures as mechanical discontinuities in an otherwise effectively continuous rock (or rock mass). The astute reader will note that some liberties from a more traditional approach have been taken which could be said to be justifiable extensions and refinements.

Many of the key relationships between elastic *stress and strain* that are helpful in dealing with discontinuities in rock are touched on in Chapter 2. Here *rheological terminology* (such as brittle, ductile, elastic, and plastic) that is important in geologic fracture mechanics is also discussed. The important concept of *stress concentration* due to flaws, cavities, fractures, and other kinds of inhomogeneities is also covered in this chapter. Here lies the basis for fracture mechanics and the redistribution of remote stress states—by structural discontinuities—into spatially variable, inhomogeneous stress states that are so important in fractured rock masses.

Chapter 3 provides a non-traditional summary of *stress analysis, peak-strength concepts, frictional sliding* and the Coulomb criterion, tensile failure and the several Griffith criteria, and concepts of *rock-mass classification systems* and *rock-mass strength criteria*—including the important and versatile Hoek–Brown criterion—that provide an understanding of the mechanics of fracturing and associated rock-mass deformation.

Chapter 4 concentrates on *dilatant cracks*—one of the most common types of brittle geologic structure. Because mechanical and rock engineers deal extensively with these types of structural discontinuities, the understanding of the behavior and analysis of cracks is well advanced; additionally, laboratory simulations can reproduce many of the surface textures and discontinuity patterns that are encountered in geology. As a result, this chapter may bear some resemblance to other reviews and treatises on mode-I cracks that already exist. The chapter ends with an overview of anticracks, stylolites, and pure compaction bands. These structures have the opposite sense of displacement to cracks (dissolution and interpenetration of opposing walls across the surface) yet the same direction (normal to the surface). Pressure solution surfaces (in soluble rocks) and compaction bands (in porous rocks) form patterns suggestive of lateral propagation and interaction, making them a form of stress concentrator and discontinuous deformation mechanism in susceptible rock types.

Some of the common discontinuity patterns are examined in Chapter 5. Here we'll start with two *discovery patterns*—echelon discontinuities and end cracks—that can be interpreted by using the basic principles of geologic fracture mechanics including near-tip stress concentration. Other patterns that reveal key principles will be explored including growth of discontinuities into sets, spacing, fault formation, and the Riedel shear patterns that are so commonly observed in soils and soft porous granular rocks.

The extensive topic of faults and faulting is explored in Chapter 6 from the perspective of geologic fracture mechanics. Rather than relying on traditional and well-known means for determining the sense of fault offsets, such as fault-surface textures or offset stratigraphic markers, the use of stepover and fault-tip kinematics is illustrated as an alternative and additional technique for interpreting fault offsets and fault patterns. The context and growth of faults in various systems including Anderson's classification, three-dimensional (3-D) strain and polygonal fault sets, crustal strength envelopes, the stress polygon, and critically stressed systems is also discussed.

Chapter 7 presents an introduction to the important class of structures known as *deformation bands*. These tabular structures occur preferentially in porous granular rocks such as sandstones and are critically important as a precursor to faults in many sedimentary rocks and sedimentary basins around the world. Certain types of deformation bands can form barriers to fluid flow, defining compartments against fluid migration (water or petroleum) in the subsurface. Deformation bands display a rich variety of kinematics including opening, shearing, closing, and various combinations of these. Because much of the current theoretical or conceptual framework for deformation bands draws from critical-state soil mechanics and Cam cap-type models of pre-peak yielding, an extended but largely non-mathematical synopsis of this approach and its application to bands in porous granular rocks is included in this chapter.

In Chapter 8 an overview of *linear elastic fracture mechanics* (LEFM) is developed. It includes the ideas that are most important and useful in dealing with geologic structural discontinuities, presented in an order that, as will be seen, differs from the traditional engineering sequence. For example, the displacements and stresses generated in the vicinity of geologic structural discontinuities are emphasized early, because these are what are mapped, measured, and worked with most often in the field. Only toward the end is focus placed on the discontinuity tip, since the LEFM characterization of discontinuity tips as bounded by a negligibly small plastic zone describes actual discontinuity tips in rock only under certain sets of conditions.

The final chapter, Chapter 9, introduces some topics of relevance to the rapidly developing field of geologic fracture mechanics. These include subcritical fracture growth, more realistic (linear and "end-zone") models for discontinuity tips, models of discontinuity propagation, displacement–length scaling relations, discontinuity populations, and brittle strain. The issues chosen for inclusion are under active research in contemporary structural geology, and this chapter may provide a readable entry into these exciting topics.

A *learning map* is included at the end of each chapter. Patterned loosely after a dendrite, learning maps are graphical summaries, limited to a single page, that highlight not only the most important points and concepts covered, but also the logical flow and connections between them. They are an exceptionally useful method for synoptically visualizing and remembering an entire chapter and for helping to place the important details into their larger context.

All chapters have *exercise sets* that encourage the reader to work through the material themselves. Many worked examples within the chapters illustrate the topics being discussed, and answers or helpful hints to selected exercises are provided at the publisher's website, www.cambridge.org/schultz. Many of the chapters and the exercises have been tested by undergraduate and graduate students in courses taught by the author, with modifications made as called for. A *glossary* collects definitions of many of the specialized terms used in the book while providing definitions of well-known terms from the perspective of geologic fracture mechanics. Although some of these definitions may challenge traditional ones, their development in the glossary provides an additional approach to exploring the field, assumptions, methods, and utility of geologic fracture mechanics.

I have enjoyed writing this book, and I hope that you will enjoy it too. Learning about geologic fracture mechanics is an odyssey that never ends. It always gets more interesting and more fun—it is never dull or boring. I hope that something in this book spurs you to learn some new material, make new field observations, apply some new concepts, and contribute to this exciting area.

Acknowledgements

In writing this book I am pleased to acknowledge those who have, knowingly and willingly or not, provided many of the philosophical and logical underpinnings for my approach. While a graduate student at Arizona State University, I was introduced to the methods of the "Stanford School of Fracture Mechanics" by Jon Fink, who used Arvid Johnson's excellent book *Physical Processes in Geology* as his text. This innovative book (and Jon's tutelage and encouragement) provided a powerful motivation for me to learn and apply mechanics to geologic problems through unusually clear writing and examples. The philosophy embodied in that book continues to be an inspiration, especially the reflexive maxim that "all good problems begin in the field, and any (kinematic, mathematical, or computer) solution must fit what is seen in the field." Following completion of my MS degree there I moved on to study for my doctorate at Purdue University with Atilla Aydin, who (initially unappreciated by me at the time) was a colleague and classmate of Jon's at Stanford. Atilla taught me that maxim in innumerable and inscrutable ways, and I am grateful to him for it. Many of the concepts in this book have grown from my late-night, classroom, and in-field discussions with Atilla and many others over the years, although the responsibility for anything within these pages remains mine. I also thank my former graduate students at the University of Nevada, Reno—Kathleen Ward, Qizhi Li, Paul Piscoran, Andrea Fori, Edward Wellman, Jason Moore, Will Roadarmel, Paul Caruso, Dr. Scott Wilkins, Dr. Chris Okubo, Clara Balasko, Dr. Cheryl Goudy, Daniel Neuffer, Anjani Polit, Wendy Key, Dr. Amanda Nahm, and Dr. Christian Klimczak, and former postdocs Dr. Daniel Mège and Dr. Roger Soliva—for letting me practise and refine my translations of geologic fracture mechanics.

I have benefited enormously, both professionally and personally, from stimulating and pleasant associations with a number of individuals. At the risk of inadvertent omission, I'd especially like to acknowledge Mike Piburn, Steve Fox, and Roger Hewins (Rutgers University); Uel Clanton and Herb Zook (NASA Johnson Space Center); Pete Schultz (Brown University); Mike Ward (Independence Mining Company); Steve Self (University of

Hawaii); Ron Greeley and Don Ragan (Arizona State University); Mike Malin (Malin Space Science Systems, Inc.); Nick Christensen (University of Wisconsin, Madison), Tom Tharp, Bob Nowack, James Doyle, and C. T. Sun (Purdue University); Jim DeGraff (formerly with ExxonMobil Development Company); Greg Ohlmacher (Black and Veatch); Paul Morgan (Colorado Geological Survey); David Pollard (Stanford University); Steve Martel (University of Hawaii); Terry Engelder (Pennsylvania State University); Herb Frey, Dave Smith, Jim Garvin, and Dave Harding (NASA Goddard Space Flight Center); Cindy Ebinger (Tulane University), Maria Zuber (MIT); Jian Lin (Woods Hole Oceanographic Institution); Greg Hirth (Brown University); Sean Solomon (Lamont-Doherty Earth Observatory); Ken Tanaka and Baerbel Lucchitta (U.S. Geological Survey); Tom Watters, Bob Craddock, and Jim Zimbelman (National Air and Space Museum); Ron Bruhn (University of Utah); Dick Bradt (University of Alabama); John Kemeny (University of Arizona); George McGill and Michele Cooke (University of Massachusetts); Dan Schultz-Ela and Martin Jackson (Texas Bureau of Economic Geology); Daniel Mège (Polish Academy of Sciences); Jacques Angelier and Catherine Homberg (Université Pierre et Marie Curie, Paris VI); Bill Higgs, Chuck Kluth (Colorado School of Mines), and Chuck Sword (Chevron Petroleum Technology Company); Eric Grosfils and Linda Reinen (Pomona College); Philippe Masson, François Costard, Nicolas Mangold, Jean-Pierre Peulvast, and Antonio Benedicto (Université de Paris-Saclay, Orsay); Roger Soliva (Université Montpellier II); Zoe Shipton (University of Strathclyde); Karen Mair (University of Oslo); Teng-fong Wong (formerly with Stony Brook University, now with The Chinese University of Hong Kong); Jim Carr, Bob Watters, Jaak Daemen, Jane Long, John Bell, Steve Wesnousky, John Louie, Raj Siddharthan, and Dhanesh Chandra (University of Nevada, Reno); Peter Hennings, Pete D'Onfro, Bob Krantz, Dave Amendt, Seth Busetti, Pijush Paul, Anastasia Mironova, and Zijun Fang (ConocoPhillips); Jon Olson, Steve Laubach, Maša Prodanović, Matt Balhoff, Larry Lake, Nicolas Espinoza, Hilary Olson, Peter Eichhubl, and Julia Gale (The University of Texas at Austin); Donnie Vereide (Geostock Sandia); Nicolas Bonnier, Louis Londe, and Bruno Paul-Dauphin (Geostock); Russell Bentley (WSP); David Evans (British Geological Survey), Doug Hubbard (Hubbard Decision Research), and Sam Savage (ProbabilityManagement.org and Stanford University).

The material presented here was formulated and initiated while at the University of Nevada, Reno. Much of it was further developed while I was enjoying a productive sabbatical leave at the Woods Hole Oceanographic Institution and the Université Pierre et Marie Curie (Sorbonne University) in Paris; then refined and expanded at ConocoPhillips and at The University of Texas at Austin; the work was completed at Orion Geomechanics LLC in Cypress, Texas. I am grateful to several agencies that sponsored my research financially including the US National Academy of Sciences/National Research Council, NASA, NSF, the Department of Energy, CNRS (France),

Chevron Petroleum Technology Company, Shell International Exploration and Production Company, the American Chemical Society's Petroleum Research Fund, the Norwegian Research Council, UT's Fracture Research and Applications Consortium, and the Texas Center for Integrated Seismicity Research along with various industrial and proprietary contracts.

Early drafts of this book were graciously reviewed by Terry Engelder, Tom Blenkinsop, Chris Wibberley, and Jaak Daemen, who provided vigorous, detailed, and thoughtful critiques that significantly improved its scope, style, and balance. Hunjoo Lee kindly read through Chapter 8. Much of the material including the exercise sets was used in various undergraduate and graduate classes in structural geology and geomechanics, and has been improved through the detailed feedback of these wonderful students.

I'm also happy to acknowledge and thank my wife, Rosemary, for her encouragement and support during my addiction with The Book. Writing a book like this one is not a trivial undertaking, as it impacts both normal scientific productivity and family life. It is a creative exercise that builds for the future while solidifying the present. Her active support was instrumental in producing a quality product. Thank you! Our sons Sebastian and Rainier provided a wonderful counterpoint as this book moved to completion.

1 Introduction to Geologic Structural Discontinuities

1.1 Localized Deformation Structures in Rock

In This Chapter:

♦ *Types and expression of structural discontinuities in geology*
♦ *Displacement modes for geologic structural discontinuities*

Planar breaks in rock are one of the most spectacular, fascinating, and important features in structural geology. Joints control the course of river systems, the extrusion of lava flows and fire fountains, and modulate groundwater flow. Joints and faults are associated with bending of rock strata to form spectacular folds as seen in orogenic belts from British Columbia to Iran, as well as seismogenic deformation of continental and oceanic lithospheres. Anticracks akin to stylolites accommodate significant volumetric strain in the fluid-saturated crust. Deformation bands are pervasive in soft sediments and in porous rocks such as sandstones and carbonates, providing nuclei for fault formation on the continents. Faults also form the boundaries of the large tectonic plates that produce earthquakes—and related phenomena such as mudslides in densely populated regions such as San Francisco, California—in response to tectonic forces and heat transport deep within the Earth. Faults, joints, and deformation bands have been recognized on other planets, satellites, and/or asteroids within our Solar System, attesting to their continuing intrigue and importance to planetary structural geology and tectonics.

Fractures such as joints and faults have long been recognized and described by geologists and engineers as expressions of brittle deformation of rocks (e.g., Price, 1966; Priest and Hudson, 1976; Gudmundsson, 2011; Peacock et al., 2018). They are important geologic structural discontinuities as they reveal types and phases of deformation, and they can be used to constrain paleostrain and paleostress magnitudes. Furthermore, they affect fluid flow in petroleum and groundwater reservoirs in a variety of ways, ranging from highly permeable fracture zones in limestones or crystalline rocks to sealing fault structures in hydrocarbon reservoirs. As a result, they have important practical implications in such fields as structural geology, geo-engineering, landscape geomorphology, hydrogeology, and petroleum geology (e.g., Cook et al., 2007).

On the other hand, deformation bands, identified as thin, tabular zones of cataclasis, pore collapse, and/or grain crushing (e.g., Engelder, 1974; Aydin, 1978), now encompass five kinematic varieties, from opening through shearing to closing senses of displacement that may or may not involve cataclasis (e.g., Aydin et al., 2006). Both classes of discontinuity—fractures and deformation bands—share common attributes, such as approximately planar or gently curved geometries, small displacements relative to their horizontal lengths, echelon or linked geometries, displacement transfer between adjacent segments, variable effect on fluid flow, and systematic variation in displacement magnitude accommodated along them (see Fossen et al., 2007, 2017, for a review and discussion of these attributes for deformation bands). Building on these commonalities, a consistent terminology that encompasses all the variations noted above is available (for example, see Aydin et al., 2006; Schultz and Fossen, 2008) and utilized here.

In this chapter, we review the main types of structural discontinuities in rock. Aspects of rock mechanics and lithology that influence the type of structural discontinuity that can form will be reviewed. The related concept of *modes of displacement* that has proven so useful in studies of joints, faults, and other types of geologic discontinuities will then be outlined in the context of this framework. Subsequent chapters will explore the development, expression, patterns, and geological significance of these fascinating elements of rock deformation.

1.2 What Is a Structural Discontinuity?

In rock engineering, a discontinuity is a general term meant to include a wide range of mechanical defects, flaws, or *planes of weakness*, in a rock mass without regard to consideration of their origins (e.g., Fookes and Parrish, 1969; Attewell and Woodman, 1971; Goodman, 1976, 1989; Priest and Hudson, 1976; International Society for Rock Mechanics, 1978; Bieniawski, 1989; Fig. 1.1). This term includes bedding planes, cracks, faults, schistosity, and other planar surfaces that are characterized by small shear strength, small tensile strength, reduced stiffness, strain softening, and large fluid conductivity relative to the surrounding rock mass (Bell, 1993, p. 37; Brady and Brown, 1993, pp. 52–53; Priest, 1993, p. 1; Hudson and Harrison, 1997, p. 20). By implication, cataclastic deformation bands and solidified igneous dikes or veins in sedimentary host rock, for example, that are stronger and/or stiffer than their surroundings might not be considered as a discontinuity by a rock engineer.

Pollard and Segall (1987) defined a discontinuity as "two opposing surfaces that are bounded in extent, approximately planar compared to the longest dimension, and having a displacement of originally adjacent points on the opposing walls that is both discontinuous and small relative to the longest dimension." According to this definition, cracks, faults,

Fig. 1.1. Examples of **discontinuities in a rock mass** as seen through the eyes of a rock engineer. Several joint planes are visible, including vertical ones exhibiting plumose structure that define the vertical face of the exposure, another vertical set perpendicular to the plane of the exposure that is restricted to certain of the horizontal sandstone layers, and sub-horizontal joints that mark separation along the bedding planes. A few inclined cracks in the uppermost massive bed suggest bedding-plane slip of that layer (top to the right) along the subjacent interface. The section shown is part of the Permian Elephant Canyon Formation exposed in southern Devils Lane graben, Canyonlands National Park, Utah.

veins, igneous dikes, stylolites, pressure solution surfaces, and anticracks would be considered as discontinuities. The term "structural discontinuity" encompasses a wide range of localized geologic structures, over a broad range of scales, strain rates, rock types, formation mechanisms, and tectonic environments.

The most common types of geologic structural discontinuities are listed in Table 1.1. An explanation of the genesis and context of these terms can be found in Schultz and Fossen (2008), with deeper exploration of their utility and significance contained in subsequent chapters.

Structural discontinuities are defined by a pair of surfaces, or fracture walls, that have been displaced (by opening or by shearing) from their original positions. For example, a crack or joint is open between its two walls (Fig. 1.2); these walls join smoothly at a crack or joint tip, defining the maximum horizontal or vertical extent of this structure in the rock. Similarly, a fault (Figs. 1.3–1.4) must be considered to be a pair of planes that are in frictional contact; a single fault plane begs the question of what was sliding against it. These paired fault walls or planes may be separated by gouge or other deformed material, and like cracks, join at a tip (Fig. 1.3) where the displacement magnitude decreases to zero. In the case of deformation bands (Fig. 1.5), the walls can be identified by a comparatively rapid change in displacement gradient or porosity, although they can be more indistinct than the clean sharp breaks typical of joints and fault planes (e.g., Aydin, 1978).

Table 1.1 Geologic structural discontinuities

Discontinuity	Mechanical defect, flaw, or plane of weakness in a rock mass without regard to its origin or kinematics
Structural discontinuity	A localized curviplanar change in strength or stiffness caused by deformation of a rock that is characterized by two opposing surfaces that are bounded in extent, approximately planar compared to the longest dimension, and having a displacement of originally adjacent points on the opposing walls that is small relative to the longest dimension
Sharp discontinuity	A structural discontinuity having a discontinuous change in displacement, strength, or stiffness that occurs between a pair of discrete planar surfaces
Tabular discontinuity	A structural discontinuity having a continuous change in displacement, strength, or stiffness that occurs across a relatively thin band
Fracture	A sharp structural discontinuity having a local reduction in strength and/or stiffness and an associated increase in fluid conductivity between the opposing pair of surfaces
Joint	A sharp structural discontinuity having field evidence for discontinuous and predominantly opening displacements between the opposing walls
Anticrack	A sharp or tabular structural discontinuity having field evidence for predominantly closing displacements between the opposing walls
Deformation band	A tabular structural discontinuity having a continuous change in displacement, strength, or stiffness across a relatively narrow zone in porous rocks
Fault	A sharp structural discontinuity defined by slip planes (surfaces of discontinuous displacement) and related structures including fault core and damage zones that formed at any stage in the evolution of the structure
Fault zone	A set of relatively closely spaced faults having similar strikes
Shear zone	A tabular structural discontinuity having a continuous change in strength or stiffness across a relatively narrow zone of shearing; shear and volumetric strains are continuous across the zone, and large or continuous (linked) slip surfaces are rare or absent
Damage zone	A zone of increased deformation density that is located around a discontinuity that formed at any stage in the evolution of the structure

A brittle fracture implies creation of these surfaces under conditions of relatively small strain and with a decreasing resistance to continued deformation (i.e., a strain-softening response in the post-peak part of the stress–strain curve for the rock; Jaeger, 1969; see also Hudson and Harrison, 1997; Chapter 8). Brittle deformation additionally implies *localization* of strain into one or more discrete planar elements, whereas ductile deformation implies nonlocalized, spatially distributed deformation (e.g., Pollard and Fletcher, 2005, p. 334). The term "fracture" thus implies a local *reduction* in strength and/or stiffness and an associated increase in fluid conductivity between the pair of surfaces.

Fig. 1.2. Several thin, parallel joints from a granitic pluton near Ward Lake, California, are visible to the naked eye. Closer examination reveals that these joints are filled with epidote and other hydrothermal minerals that were present in the fluid that circulated within the joints as they dilated and propagated at mid-crustal depths.

Cracks, joints, and faults are, as a consequence, various types of brittle fractures. A solidified igneous dike that cuts a weaker and/or less stiff sedimentary sequence, however, would not be considered to be a fracture according to this definition, although the *contact* between the igneous rock and the country rock may qualify if it is weaker, or less stiff, than the igneous or country rock.

Fractures are seen to be a pair of surfaces that separate their displaced surroundings from what is between them (e.g., Johnson, 1995). Joints in the subsurface are filled by a variety of liquids including groundwater (e.g., Engelder, 1985; National Academy of Sciences, 1996), natural gas (Lacazette and Engelder, 1992), and petroleum (Dholakia et al., 1998; Engelder et al., 2009). Joints filled by solidified hydrothermal, diagenetic, or magmatic minerals are called **veins** and **igneous dikes** (Figs. 1.6 and 1.7), respectively. Rubin (1995b) calls igneous dikes "magma-filled cracks" to emphasize the mechanical basis for dike dilation and propagation. In all of these examples, it becomes important to determine whether the filling material was there

Fig. 1.3. Normal faults exposed on this dip slope of Upper Cretaceous carbonate rock (Garumnian Formation, Collado de Fumanyá) from the Spanish Pyrenees demonstrate a consistent spacing and clear fault terminations. The shadowed indentations are dinosaur footprints that predate the faults.

Fig. 1.4. A prominent normal fault cuts through a precursory zone of deformation bands in this view of "Aydin's Wall" that formed in Entrada Sandstone east of the San Rafael Swell in eastern Utah. The bright sunlit high-angle planes are large slip surfaces whose corrugations are inherited from the architecture of the deformation band network. A second normal fault at the upper right defines this area as a fault zone having two major subparallel strands.

Fig. 1.5. Deformation bands are an important deformation mechanism in porous rocks such as sandstone, and they define intricate and informative arrays as in this fine example in Navajo Sandstone from southern Utah. The sub-horizontal bedding can be seen to be cut by two oppositely dipping sets of shear-enhanced compaction bands and a sub-vertical set of pure compaction bands. Stratigraphic restriction of these band sets can be observed.

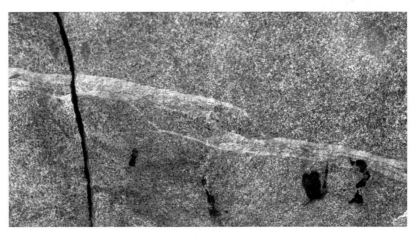

Fig. 1.6. A pair of echelon igneous dikes is shown in granitic rock of the central Sierra Nevada near Donner Lake, California. Each dike segment beyond the overlap region, where the dikes are linked, exhibits a separation of host rock by ~10 cm, with the intervening void within the dike being filled by a sequence of hydrothermal (light-toned) and igneous (dark-toned) fluids that have since crystallized, preserving the extensional strain in the pluton. The later crack in the left-hand part of the image (that cuts the dike) also demonstrates that a fracture must be defined by a **pair** of subparallel walls or surfaces, rather than a single plane.

initially, inside the joint, as it dilated and propagated, or whether the filling material came afterwards and simply occupied the volume within the joint (e.g., Laubach et al., 2010).

Fault walls may be separated by gouge (e.g., Chambon et al., 2006); those of a stylolite, by insoluble residue such as clays (e.g., Fletcher and Pollard, 1981; Engelder and Marshak, 1985; Pollard and Fletcher, 2005, p. 17); shear

Fig. 1.7. The detailed shape of the dike tip is clearly revealed by this view of the stepover between a pair of closely spaced igneous dikes. The dike segment width, normal to its trace, is ~10 cm (camera lens cap for scale). Dike shapes such as these can be predicted very well by representing the dikes as dilatant cracks propagating through a continuous elastic medium.

Discontinuity, part 1: Finite Dimensions

Discontinuity, part 2: Displacement Gradient

zones contain a variety of interesting rocks and fabrics (e.g., Fossen and Cavalcante, 2017). Deformation bands typically have either increased or decreased porosity within them (e.g., Aydin et al., 2006); grain-size reduction may also characterize the interior of a deformation band (e.g., Aydin, 1978; Davis, 1999; Fossen et al., 2007). As a result, cracks, joints, faults, stylolites, and deformation bands all contain various materials between their walls that can differ significantly from the host rock that surrounds them.

The term "**discontinuity**" has two distinct meanings when applied to localized structures. First, joining of the paired crack or fault walls at the fracture tips defines the dimensions of the fracture, such as its length, width, or height. Fractures that have **discrete lengths** are called discontinuous (e.g., Pollard and Segall, 1987) and this is the basis for the field of engineering fracture mechanics (because discontinuous fractures end at the fracture tips). The second, and more recent, use of "discontinuity" refers to the **rate of change of displacement** across the structure (e.g., Johnson, 1995; Borja, 2002). Fractures having a discrete step-wise change in displacement across them, such as cracks and faults (or slip surfaces), are said to have, or be, a displacement discontinuity. In contrast, shear zones and deformation bands may exhibit a continuous displacement across them. The boundary element modeling approach developed by Crouch and Starfield (1983, pp. 208–210) and used by many researchers in geologic fracture mechanics clearly illustrates how these (strong and weak) discontinuities can *both* be easily represented in a fracture model by specifying the properties of the filling material along with the strengths of the enclosing walls. This computational approach parallels the geologic work summarized in this section that motivates the integration of terminology espoused in this book.

The scale of the structure relative to its surroundings is also important to consider. For example, the rock cut by a structure is commonly taken as an "effective continuum" (e.g., Pollard and Segall, 1987, and references therein). A geologic unit is considered to be *continuous* when its properties at the scale of the structure are statistically constant, leading to homogeneous values and behavior (Priest and Hudson, 1981). Rock units with spatially variable properties, such as joint sets, at a given scale are called *discontinuous*. In mechanics, a continuous material is said to be simply-connected, whereas a discontinuous one is multiply-connected (e.g., Nehari, 1952).

An outcrop of columnar-jointed basaltic lava flows (e.g., Fig. 1.8) provides an informative example of relative scale (Schultz, 1996). For a scale of observation of millimeters to centimeters, the rock within a column is intact basaltic rock, and it can be idealized by those properties. At significantly smaller scales, grain size and microcracks become important enough that the assumption of a continuous material may not apply. For a scale of a few meters, however, the rock unit can be described as intact basalt partly separated into irregular blocks by numerous discontinuous fractures (i.e., the columnar joints). Because this scale of observation is comparable to the scale of the fracturing (or equivalently, the block size), the flow must be considered as a discontinuum within which the properties of the rock and fractures both must be considered to understand and represent the behavior of the flow, as in the example shown in Fig. 1.8. An approach called *block theory* (Goodman and Shi, 1985; Priest, 1993, pp. 246–250) would be the method of choice here. At dimensions for which the scale of observation greatly exceeds (e.g., by a factor of 5–10) the block size or fracture spacing, such as would be found on an aerial or satellite image, the lava flow can be described as an *effective continuum* (Priest, 1993) and a *rock mass* (see also Bieniawski, 1989; Schultz, 1993, 1995a,b, 1996; Fig. 1.1). Various methods for "upscaling" rock

Fig. 1.8. View of columnar joints in basalt near Donner Lake, California, looking down the column axes; rock hammer for scale.

properties from laboratory-scale measurements to outcrop or larger scales are routinely used in the oil and gas industry, including geostatistics and geomodels (e.g., Christie, 1996; Rogers et al., 2016).

To summarize, a 10-cm core of the basaltic rock within a basalt column, and a 10-m outcrop of jointed basalt, are both effectively continuous at their respective scales, although the values for strength, deformability, and hydrologic properties will differ in detail for each scale. The lava flow is discontinuous at scales of observation comparable to the grain or block size. Rock engineers have coined the acronyms CHILE (continuous, homogeneous, isotropic, linearly elastic) and DIANE (discontinuous, inhomogeneous, anisotropic, non-elastic) to describe the characteristics of rock units at the particular scale of interest (see Hudson and Harrison, 1997, pp. 164–165). Continuum methods such as fracture mechanics work well with CHILE materials, but processes at the block or grain scale, such as grain reorganization and force chain stability (e.g., Cates et al., 1998; Mandl, 2000, pp. 100–101; Mair et al., 2002; Peters et al., 2005) within a deformation band, for example, require methods such as distinct element or particle flow codes instead to explicitly represent these DIANE materials (e.g., Antonellini and Pollard, 1995; Morgan and Boettcher, 1999). In this book we adopt terminology and analysis techniques for structures consistent with CHILE behavior of their surroundings, while recognizing that the total geologic system of rocks and structural discontinuities functions more like a DIANE system.

To summarize the preceding discussion and paralleling Pollard and Segall (1987):

- A structural discontinuity is a localized curviplanar change in strength or stiffness caused by deformation of a rock that is characterized by two opposing surfaces that are bounded in extent, approximately planar compared to the longest dimension, and having a displacement of originally adjacent points on the opposing walls that is small relative to the longest dimension.

Quantification of "small" displacements of rock on either side of a structural discontinuity can be done by means of a displacement–length diagram (Fig. 1.9), which is a standard tool in geologic fracture mechanics and structural geology (e.g., Cowie and Scholz, 1992a; Dawers et al., 1993; Clark and Cox, 1996; Schlische et al., 1996; Fossen and Hesthammer, 1997; Schultz, 1999; Cowie and Roberts, 2001; Scholz, 2002, pp. 115–117; Olson, 2003; Schultz et al., 2008a, 2013; Fossen, 2010a,b; Fossen and Cavalcante, 2017). As can be seen on Fig. 1.9, longer faults are associated with systematically larger displacements, or shear offsets (D_{max} on the diagram), so that faults accommodate offsets between 0.1% and 10% of their lengths, with smaller variations found for faults in a particular area. Other types of geologic structural discontinuities, such as joints and deformation bands (including the compaction band variety), and shear zones, also show a consistent displacement–length scaling relation (e.g., Vermilye and Scholz, 1995; Olson, 2003; Fossen et al., 2007;

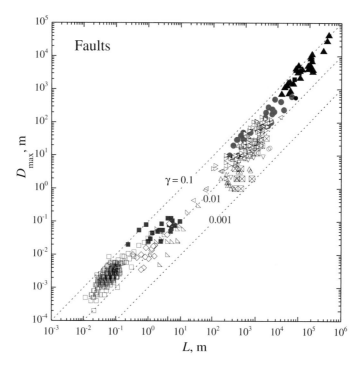

Fig. 1.9. Displacement–length diagram for faults showing that the maximum shearing offsets D_{max} range between 0.1% and 10% of the fault length L ($D_{max}/L = \gamma = 0.001–0.1$) after Schultz et al. (2008a, 2013). More discussion is given in Chapter 9.

Rudnicki, 2007; Schultz et al., 2008a, 2013; Tembe et al., 2008; Schultz, 2009; Schultz and Soliva, 2012; Fossen and Cavalcante, 2017) with the details dependent on the kinematics and mechanics of the structure.

Two general classes of structural discontinuities are now recognized in structural geology and rock mechanics, following Borja and Aydin (2004) and Aydin et al. (2006). These are summarized in Table 1.2.

- Sharp structural discontinuities have a discontinuous change in displacement, strength, or stiffness that occurs between a pair of discrete planar surfaces; sharp discontinuities are associated with a discontinuous (abrupt) change in the displacement distribution across them, leading to the mechanically descriptive terms "displacement discontinuity" (e.g., Crouch and Starfield, 1983, pp. 80–84; Pollard and Segall, 1987; Pollard and Aydin, 1988) and "strong discontinuity" (Borja, 2002; Aydin et al., 2006) in the mechanics literature.
- Tabular structural discontinuities have a continuous change in displacement, strength, or stiffness that occurs across a relatively thin band. These structures show a continuous change in normal or shear strain (i.e., an increase in the displacement gradient) across them and are called "weak discontinuities" by, e.g., Borja (2002) and Aydin et al. (2006).

In these definitions, the terms "sharp" and "tabular" describe the variation in *displacement*, or offset, that is accommodated between the two opposing surfaces, such as opening, closing, or shearing displacements, rather than a tensile, shear, or compressive strength. Specifically,

Table 1.2 Geomechanical classification of structural discontinuities according to the *rate of change of displacement across them* (sharp planes with discontinuous displacement vs. tabular zones with continuous displacement) and by *kinematics* (displacement sense across the structure)

		DISPLACEMENT CONTINUITY	
		Sharp Discontinuity	**Tabular Discontinuity**
DISPLACEMENT SENSE	*Opening*	Crack (Joint, Dike, Sill, Hydrofracture)	Dilation band
	Opening + shear	Mixed-mode crack	Dilational shear band
	Shearing	Fault (Slip patch)	Shear band; also Shear zone
	Shear + closing	None	Compactional shear band Shear-enhanced compaction band
	Closing	Stylolite	Pure compaction band

- A "sharp discontinuity" is one in which the thickness of the structure is very close to zero. This corresponds to two planes in contact or in close proximity.
- A "tabular discontinuity" is one in which the thickness of the structure that accommodates most of the offset is measurable at the scale of interest, such as a thin section, hand sample, outcrop, or a satellite image.

In general, there is a good correlation between the width of a structure and its displacement continuity (e.g., Aydin et al., 2006). Sharp discontinuities typically are associated with discontinuous displacement gradients across them, whereas thicker, tabular ones exhibit more gradual and continuous displacement gradients across them.

Sharp discontinuities (Table 1.2) typically form under peak stress levels in the rock (see Chapter 3), involving cracking and/or shearing processes. These are typically referred to as **fractures** in geology and **discontinuities** in rock engineering, and usually accommodate either shearing (via frictional sliding) or opening displacements.

Tabular discontinuities, on the other hand, commonly referred to as **deformation bands** or **shear zones**, can accommodate the full range of kinematics across their widths, including dilation, shear, compaction, and combinations of these (e.g., Besuelle, 2001a,b; Aydin et al., 2006; Fossen et al., 2007, 2017; Fossen and Cavalcante, 2017). A type of deformation band in which compactional normal strains predominate over shear strains across the band is called a **compaction band** (Mollema and Antonellini, 1996; Sternlof et al., 2005; Aydin et al., 2006; Eichhubl et al., 2010). Two varieties are recognized as *pure compaction bands* (Fig. 1.10), which accommodate essentially compactional

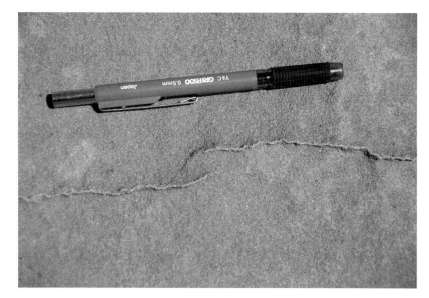

Fig. 1.10. Pure compaction bands in Navajo Sandstone from southern Utah exhibit relays and related characteristics that suggest propagation under compressive stresses oriented normal to the average orientation of the band.

Fig. 1.11. Shear-enhanced compaction bands in Navajo Sandstone, southern Utah, shown here dipping to the right through the cross-bedded strata, attain maximum thicknesses of >1 cm. Sedimentary bedding is slightly offset with a reverse sense along the bands, with shear and compactional strain between the two well-defined band walls.

strain and resemble stylolites in soluble rock, and *shear-enhanced compaction bands*, which accommodate both compactional and shear strains (Fig. 1.11). Depending on the stress state and grain-scale characteristics of the host rock, such as porosity and grain size, the thickness of a tabular discontinuity can attain values of several centimeters.

A sharp discontinuity has a significant displacement over a very narrow zone, leading to a significant or "strong" effect on its surroundings. A tabular discontinuity, on the other hand, exerts a less significant, or weaker, effect on its surroundings. As a result, certain processes, such as mechanical interaction or displacement transfer between closely spaced structural discontinuities, may be more effective, or may operate over larger distances, for sharp discontinuities than for tabular discontinuities.

The term "discontinuity" refers to the *dimensions* parallel to the structure, such as horizontal length or vertical (or down-dip) height, as shown in Fig. 1.11. In general, the width or thickness is a very small fraction of the discontinuity's length or height. With these kept in mind, the term "structural discontinuity" can easily be used to encompass those having either abrupt or continuous changes in displacement across them.

A **fracture** is a subtype of structural discontinuity. It is defined in Table 1.1. Most geologic fractures are *brittle* structures, in the sense (Rutter, 1986; Knipe, 1989) that they break or shear the rock as a result of pressure-dependent deformation mechanisms (e.g., grain cracking, Pollard and Aydin, 1988; frictional sliding, Byerlee, 1978; Scholz, 2002, pp. 53–100). This strain-softening behavior explains why cracks and faults can both be considered to be types of fractures. However, cracks that propagate by temperature-dependent deformation mechanisms, such as void growth ahead of the crack tip, are also known from high-temperature settings; these are referred to as "ductile fractures" (e.g., Eichhubl, 2004).

The definitions of structural discontinuity and fracture given in Table 1.1 require that the rock surrounding the discontinuity be an effective continuum. For example, grain-scale diffusion or cracking implies a different scale than flow in gouge, which is different in scale from fault population statistics. In other words, you have to get far enough out of, for example, the fault gouge to see the fault itself, and then to see that the fault has a width and a length, with fault tips within the host rock. This is a less restrictive requirement than the CHILE (continuous, homogeneous, linearly elastic) or DIANE (discontinuous, inhomogeneous, anisotropic, non-elastic) assumptions that may come into play with more detailed investigations, however. Specifying statistically homogeneous and continuous properties outside the discontinuity (an effective continuum) clearly defines the appropriate scale of observation at which these definitions are meant to apply.

Structural discontinuities that do not rupture the entire rock mass are characteristically discontinuous in length in one or more directions (e.g., Davison, 1994). That is, discontinuities end, or **terminate**, where the displacements across them go to zero (the discontinuity tips, or **tipline**; Figs. 1.3 and 1.12b). However, many discontinuities open or intersect at surfaces such as the ground, excavation faces, or other discontinuities; here, the discontinuity does not have a sharp termination as it would if it ended inside the rock. Discontinuities that completely rupture a rock, causing it to break apart, have no physical terminations.

Discontinuities with lengths less than the dimensions of the rock in which they occur have clearly defined terminations: these are called

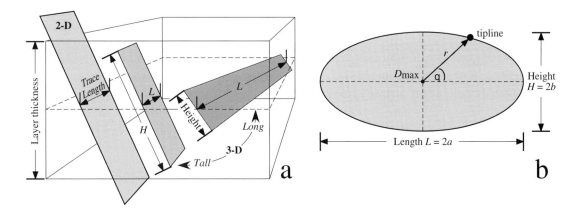

Fig. 1.12. Geometry of dipping surfaces (after Schultz and Fossen, 2002): (a) 3-D planar surfaces have horizontal (trace) lengths L and heights H (measured in their planes), whereas 2-D surfaces only provide measurements of their lengths. (b) Geometric parameters for an elliptical discontinuity in 3-D, with length L being twice the semi-horizontal axis ($L = 2a$) and height H being twice the semi-vertical axis ($H = 2b$) of the dipping plane. A fracture whose height is much greater than its length ($2b \gg 2a$) is called a "tall" or "tunnel" fracture that penetrates deeply into the rock mass relative to its length. A circular, equidimensional fracture ($2a = 2b$) is called a "penny" fracture. A "long" fracture has $2a \gg 2b$. For a 2-D plane fracture, height ($2b$) is indefinitely large.

bounded discontinuities. Sometimes the discontinuity intersects another planar element such as the Earth's surface or another discontinuity. In this case the discontinuity is considered to be **unbounded** at the intersection yet is bounded by the rock elsewhere; this is called a **semi-bounded** discontinuity. Faults that intersect the ground surface, for example, are semi-bounded discontinuities. This distinction of bounded vs. unbounded discontinuities can be important in applying engineering principles, such as those embodied by the Coulomb friction and Mohr circles, as well as stress intensity factors, to naturally and artificially occurring structural discontinuities in the Earth (e.g., Palmer and Rice, 1973; Rudnicki, 1980; Pollard and Segall, 1987; Li, 1987; Rubin, 1995b; Gudmundsson, 2011). This is because discontinuity propagation occurs at its tips, along with most of the close-range mechanical (stress) interaction that governs whether closely spaced discontinuities will link.

A kinematic or **displacement-based classification of structural discontinuities** is constructed by evaluating the sense, and continuity, of offset, or displacement of initially contiguous points, relative to the plane of the structure (e.g., Griggs and Handin, 1960; Pollard and Aydin, 1988; Jamison, 1989; Wojtal, 1989; Johnson, 1995; Aydin, 2000; Aydin et al., 2006). In the field, the geologist or rock engineer would examine and demonstrate the two key parameters: **displacement sense and displacement continuity**, as shown in Table 1.2. Assessing the displacement sense is straightforward—the methods of crosscutting and shear-sense criteria are available in any structural geology text. Assessing the continuity of displacement across the structure reduces in most cases to determining the width of the structure and then examining if the opening, shearing, or closing displacement is smoothly varying across the zone (i.e., a continuous displacement) or if it changes abruptly (i.e., a discontinuous displacement). More often than not, there is a correlation between width of the structure and the displacement continuity (i.e., continuous shearing across a tabular structure 1 mm thick), especially in the early development of a structure.

1.3 The Role of Lithology and Rock Properties

The physical properties of a rock affect its strength, deformability (see Bell, 1993, pp. 165–179, Evans and Kohlstedt, 1995, and Paterson and Wong, 2005, for concise syntheses), and the types of structural discontinuities that form within it (see also Crider and Peacock, 2004). In particular, the **porosity** and **grain size**, along with grain sorting (Cheung et al., 2012), grain angularity, lithology, and degree of diagenesis (cementation), of a rock or soil exert a primary influence on the types of structural discontinuities that form (see reviews and discussions by Wong et al., 1992, 2004; Bürgmann et al., 1994; Aydin, 2000; Pollard and Fletcher, 2005, p. 382; Aydin et al., 2006; Fossen et al., 2007; Laubach et al., 2010; Cheung et al., 2012; Wong and Baud, 2012).

There is a good correlation between the amount of porosity in a rock or soil and the **volumetric changes** it undergoes during deformation (e.g., Evans and Kohlstedt, 1995; Davis and Selvadurai, 2002; Wong et al., 1992, 2004). For example, a "compact" rock (Paterson and Wong, 2005) with negligible porosity, such as a granite or basalt, or a low-porosity soil (called "over-consolidated"), typically expands in volume during shearing, leading to dilatant behavior and localized zones of concentrated shear. In contrast, a rock with high porosity (e.g., $n > 10$–20%), or a high-porosity soil (called "normally consolidated"), generally contracts in volume during shearing, leading to porosity reduction and distributed deformation. In the rock mechanics literature, brittle deformation is associated with localized dilatancy and strain localization, whereas ductile deformation is associated with nonlocalized macroscopic flow (e.g., Evans and Kohlstedt, 1995).

A good correlation also exists between the porosity of a rock and the type of structural discontinuity—sharp or tabular—that forms in it (e.g., Aydin et al., 2006; Eichhubl et al., 2009). Tabular structural discontinuities (deformation bands) form most easily in high-porosity granular rocks, like sandstone, some pyroclastic tuffs, siliceous mudstones, and porous limestones, whereas sharp structural discontinuities (cracks and faults) form in rocks of any value of porosity. Although other variables such as stress state, diagenesis, mineralogy, and pore-water conditions appear to influence the type of structural discontinuities that form in a rock (e.g., Davatzes et al., 2005; Laubach et al., 2010), tabular structural discontinuities are clearly restricted to formation in rocks having non-negligible values of porosity.

This dichotomy in structure type (sharp vs. tabular structural discontinuities) is illustrated in Fig. 1.13 after Schultz and Fossen (2008). Here, the kinematics of the structural discontinuities also must be specified before the structural discontinuities can be named (e.g., Peacock et al., 2018). By specifying the kinematics, a broad class such as deformation bands (see also Table 1.2) can be divided into its five *kinematic* classes (i.e., dilation bands, dilational shear bands, shear bands, compactional shear bands, and

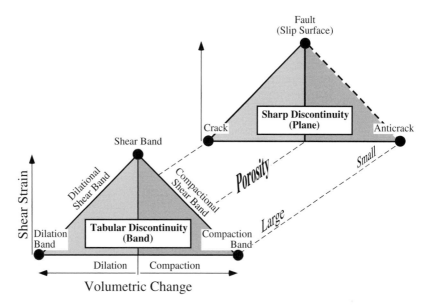

Fig. 1.13. A more detailed classification of structural discontinuities is possible when the porosity of the host rock is considered (after Schultz and Fossen, 2008). The lower-left diagram assumes a porous host rock, such as sandstone, chalk, and many limestones and shales. Tabular discontinuities (deformation bands) form most easily in porous rocks, with the type of band depending on the relative amounts of shear strain and volume change accommodated. The upper-right diagram assumes a low-porosity rock, such as a granite, gneiss, or basalt. Sharp discontinuities (fractures) characteristically form in these rocks. In particular cases, anticracks (stylolites) may form given appropriate conditions of pressure, temperature, and host-rock composition. Although sharp discontinuities can also form in porous rocks, tabular discontinuities rarely if ever occur in the low-porosity rocks.

compaction bands), depending on the relative amounts of volumetric change (dilation or compaction) and shear strain accommodated across the structural discontinuity. Similarly, cracks and faults can be distinguished by the amounts of opening or shear displacements across them. These diagrams indicate the initial types of geologic structural discontinuities (e.g., sharp or tabular) that can occur in a rock, depending on its porosity.

1.4 Displacement Modes

The approach developed in this book for structural discontinuities is based on the deformation of rock or other material in a **local coordinate system** that is defined by the *plane of the structural discontinuity* (Irwin, 1957; Sih et al., 1962; Paris and Sih, 1965; Sih and Liebowitz, 1968; Pollard and Segall, 1987; see Kies et al., 1975, for an historical overview). Analysis of the stresses and displacements associated with structural discontinuities (sharp or tabular) of any displacement mode lays the foundation for the field of **geologic fracture mechanics** that therefore includes all types of geologic structural discontinuities, including fractures and bands.

Suppose we observe a vertical joint that cuts a thin sedimentary layer (as in Fig. 1.14). In order to investigate this joint using the concept of displacement modes and the techniques of fracture mechanics, we can orient this structural discontinuity most conveniently with its plane parallel to the *x*-axis and centered at the origin of this local *xy*-coordinate system, as shown in Fig. 1.15 (see also Crouch and Starfield, 1983, p. 91; Lawn, 1993, p. 3). In this case the horizontal layer containing the discontinuity (as in Fig. 1.14) defines the *xy*-plane and is represented for clarity in Fig. 1.15 as being thin in the vertical, *z*-direction. The fracture's dimension in this *xy*-plane is called its **length** (see Fig. 1.12a).

Stresses acting in the *xy*-plane, such as layer-parallel shear, compression, or tension, are called "**in-plane stresses**" and those acting perpendicular to this plane (having a *z*-component) are called "**out-of-plane**" or "anti-plane" stresses (e.g., Pollard and Segall, 1987, pp. 316–317). The *xy*-plane can be thought of as the Earth's surface, with a vertically or steeply dipping structural discontinuity in map view cutting down into the layer (Figs. 1.14 and 1.15). Although the layer itself is shown as thin in this example, the definition of displacement mode applies regardless of layer thickness. This means we can

Fig. 1.14. Geologists examine a sub-vertical joint exposed in a thin horizontal sedimentary layer in New York State.

Fig. 1.15. Defining the reference plane of interest, such as the horizontal plane shown here, is needed to determine whether the stresses that load a structural discontinuity are (a) in-plane (layer-parallel) or (b) out-of-plane. The structural discontinuity is shown as an ellipse on the surface of the layer. The two-dimensional stress states associated with in-plane loading ("top view") and out-of-plane loading ("cross-strike view" and "along-strike view") are shown along with the axis of rotation implied by the shear stresses for each planar view. Note that $T_{xy} = T_{yx}$, $T_{xz} = T_{zx}$, and $T_{yz} = T_{zy}$ (see Chapter 3).

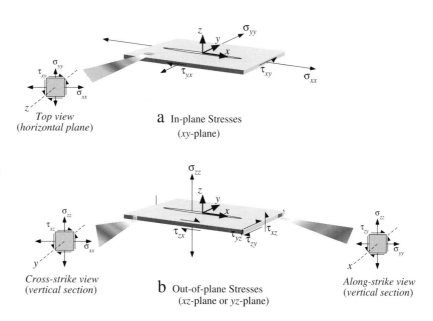

start with the joints that cut a large granitic pluton, such as those shown in Fig. 1.2, or the rotated normal faults as shown in Fig. 1.3, and define our coordinate systems using these structural discontinuities and these layers. In-plane stresses can produce rock rotations about vertical axes, as in strike-slip tectonics, whereas anti-plane stresses can induce rock rotations about horizontal axes, as in normal or thrust faulting.

The three displacement modes (also called loading modes in engineering) are defined in the upper set of diagrams in Fig. 1.16; these are oriented as is typical in engineering treatises (e.g., Paris and Sih, 1965; Sih and Liebowitz, 1968; Kanninen and Popelar, 1985, p. 139; Broek, 1986, p. 8; Atkinson, 1987; Lawn, 1993, p. 24; Anderson, 1995, p. 53; Tada et al., 2000, p. 2; see also Dmowska and Rice, 1986; Pollard and Fletcher, 2005, p. 372). They were defined originally in the engineering context for design of ships, bridges, railroads, and buildings (e.g., Kanninen and Popelar, 1985; Broek, 1986; Lawn, 1993; Anderson, 1995). Structural discontinuities that accommodate more than one principal (or "pure") displacement mode are called "**mixed-mode**." For example, a crack (mode-I) that opens obliquely is called a mixed-mode crack (mode I-II or I-III depending on the sense of shearing along the crack walls; see Chapter 2). The shearing modes (II and III) are commonly introduced in engineering fracture mechanics treatises to explain crack propagation paths (mixed-mode I-II cracks) or breakdown of the crack tip during propagation (mixed-mode I-III) (e.g., Erdogan and Sih, 1963; Sih, 1974; Ingraffea, 1987; Lawn, 1993, p. 23; Lin et al., 2010).

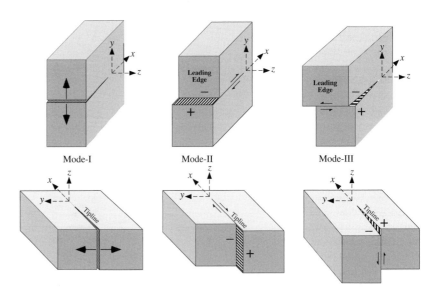

Fig. 1.16. The three displacement modes of a structural discontinuity are identified relative to the reference plane (here, the unshaded xy-plane) and the edge, or tipline, of the discontinuity. The upper diagrams are oriented following standard engineering practice, with mode-II shearing defined as being normal to the leading edge and mode-III shearing being parallel to the leading edge (Tada et al., 2000). The lower diagrams are rotated so that the reference (xy-) plane is horizontal (after Twiss and Moores, 2007, p. 38; Kulander et al., 1979), as is the case for many geologic situations (e.g., Fig. 1.14); modes II and III are defined relative to the tipline.

The displacement modes have long been applied to the analysis and interpretation of **geologic structural discontinuities** (e.g., see Brace and Bombolakis, 1963; Segall and Pollard, 1980; Pollard et al., 1982, 1993; Engelder, 1987; Pollard and Segall, 1987; Aydin, 1988; Pollard and Aydin, 1988; Cowie and Scholz, 1992b; Engelder et al., 1993; Schultz and Balasko, 2003; Pollard and Fletcher, 2005; Aydin et al., 2006; Okubo and Schultz, 2006). For example, a deformation band that accommodates both shearing and volumetric dilation (i.e., a dilational shear band; see Table 1.2) would be a **mixed-mode tabular structural discontinuity**, with the order of the mode numerals indicating the relative amounts of each component (greater magnitude listed first; i.e., mode-II-I or III-I for the case of shear displacement (mode-II or mode-III) exceeding the dilational displacement). A shear-enhanced compaction band is one having subequal amounts of compactional and shear deformation (e.g., Eichhubl et al., 2010) although a separate displacement mode for these tabular structural discontinuities is not used in practice. Although the displacement modes need not imply particular mechanics (such as mode-II faults or deformation bands, which accommodate shear displacement by different physical mechanisms), Chemenda et al. (2011) have drawn such a distinction in their interpretation of plumose structure being related to precursory dilation bands, rather than to joints, in sedimentary rocks.

The displacement mode diagrams can be made more intuitively applicable to geologic situations by rotating them so that the reference xy-plane becomes the horizontal plane; this is shown by the lower set of diagrams in Fig. 1.16. If the discontinuity opens (or dilates) in the in-plane direction, it is called **mode-I** (mode "one," not mode "eye"), analogous to a joint, vein, or dike (e.g., Pollard et al., 1982; Segall and Pollard, 1983a,b; Pollard and Aydin, 1984, 1988; Kulander and Dean, 1985; Delaney et al., 1986; Pollard, 1987; Rubin, 1993b). If the discontinuity accommodates shearing in the xy-plane (with the shear or displacement sense normal to the fault's vertical edge, as in Fig. 1.16, upper-center diagram, or, correspondingly, the xy-plane contains the slip vector), it is **mode-II**, analogous to a strike-slip fault in map view (e.g., Chinnery, 1961, 1963, 1965; Rodgers, 1980; Segall and Pollard, 1980; Rispoli, 1981; Aydin and Page, 1984; Pollard and Segall, 1987; Petit and Barquins, 1988; Davison, 1994). Out-of-plane (i.e., sub-vertical) shearing along the discontinuity (with the shear or displacement sense parallel to the fault's vertical edge, as in Fig. 1.16, upper-right diagram) corresponds to **mode-III**, analogous to a normal or thrust fault seen in map view (e.g., King and Yielding, 1984; Aydin and Nur, 1985; Pollard and Segall, 1987; Davison, 1994; Willemse et al., 1996; Crider and Pollard, 1998; Gupta and Scholz, 2000a). Mode-I structural discontinuities are discussed in detail in Chapter 4; sharp mode-II and mode-III structural discontinuities (faults) are considered in Chapter 6.

A sharp or tabular structural discontinuity can thus be categorized as:

The three loading or displacement modes

- Mode-I: opening along a structural discontinuity (called "opening mode").
- Mode-II: in-plane shear along a structural discontinuity (called "shearing mode").
- Mode-III: out-of-plane shear along a structural discontinuity (called "tearing mode").

The mode-I case with a horizontal reference plane (Fig. 1.16, lower-left diagram) was used by DeGraff and Aydin (1987) and Pollard and Aydin (1988) to illustrate crack propagation paths for a vertically dipping joint. Segall and Pollard (1980), Pollard and Segall (1987, their figure 8.18), Li (1987), and others note that a strike-slip fault corresponds to a mode-II structural discontinuity in map view and a mode-III structural discontinuity exposed in cross-sectional view. The reoriented mode-II diagram in Fig. 1.16 (lower center) follows those of Marone (1998b, his figure 15) and Pollard and Segall (1987, their figure 8.16a). Similarly, a normal or thrust fault corresponds to a dipping (i.e., nonvertical) mode-III structural discontinuity in map view and a mode-II structural discontinuity exposed in cross-sectional view (e.g., Aydin,

1988). Determination of the displacement mode in relation to the shear sense is also important in the study of deformation bands (e.g., Aydin, 1978; Aydin and Johnson, 1978, 1983; Schultz and Balasko, 2003; Aydin et al., 2006; Okubo and Schultz, 2006).

Although many geologic structural discontinuities can have a component of displacement in the other sense, in most cases *one mode predominates*. For example, a fault may have a width of perhaps a meter, with comparable amounts of opening or dilation, along with strike-parallel displacement ("offset") of many kilometers. Because the parallel component is so much larger than the normal component, one can neglect the small normal component because, in nature, the parallel component will dominate the changes in stress and displacement associated with the fault. Similarly, the normal component of displacement across dilatant cracks is considerably greater than the parallel component. Even mixed-mode cracks, with oblique dilation, are seen to work much like pure mode-I cracks (see Chapters 4 and 9), with the oblique component serving as a modifying factor in the stress and displacement fields about the crack.

Another variety of structural discontinuity can be defined by noting that material displacements perpendicular to the discontinuity surface result from a volume loss across the structure. Pressure solution surfaces, or *stylolites* (Twiss and Moores, 1992), have been likened to *anticracks* (Fletcher and Pollard, 1981; Fueten and Robin, 1992; Zhou and Aydin, 2010, 2012) that are common in soluble rocks such as carbonates when subjected to large compressive stresses. These structures are included here as the **anticrack class of structural discontinuity** (e.g., Jamison, 1989; Wojtal, 1989) even though they physically represent a very different process than bond breakage in a rock. However, their growth and interaction are similar in many respects to other types of structural discontinuities (Dunne and Hancock, 1994; Benedicto and Schultz, 2009; Zhou and Aydin, 2010).

Anticracks have the opposite kinematic significance to cracks, and their stress and displacement fields can sometimes be represented by using a mode-I crack with a compactional sense of strain across it (e.g., Fletcher and Pollard, 1981; Rudnicki and Sternlof, 2005; Sternlof et al., 2005; Rudnicki, 2007; Tembe et al., 2008; Meng and Pollard, 2014), resulting in closing (or interpenetrating) discontinuity walls. Mollema and Antonellini (1996) suggested using "anti-mode-I" for their anticracks, as did Green et al. (1990); the term "mode –I" (minus one) has also been used informally to describe these structures (Scholz, 2002, p. 331). Introducing a negative sign before "mode-I" elegantly conveys the kinematic significance of anticracks but it may also blur the important physical distinctions between cracks and the several varieties of anticrack-like structures (e.g., Fletcher and Pollard, 1981; Knipe, 1989) such as pressure-solution seams (stylolites) and pure compaction bands (Eichhubl et al., 2010). As shown in Table 1.2, anticracks can be either sharp or tabular, permitting both stylolites and pure compaction bands to be investigated kinematically and, to a degree, mechanically, as anticrack structural discontinuities.

1.5 Using Geologic Fracture Mechanics in Research and Practice

Throughout this book and your study of fracture mechanics applied to geologic structural discontinuities, you will repeatedly encounter a number of major concepts, such as stress concentration or displacement–length scaling. Sometimes these concepts or principles that are based on mechanics are readily apparent, as for example when working out the growth history of a set of tectonic joints or in reconstructing the sequence of fault segment linkage by measuring the displacement profiles along the faults. In other situations, however, you might not think that fracture mechanics would apply. For example, the lateral margins of regional-scale detachment structures in Tertiary-age extensional terranes, of thrust sheets in a Paleozoic contractional orogen, and of landslides that scallop out a mountainside, all likely nucleated, and continued to function, as particular types of fault arrays. Many problems in finite strain started as small, infinitesimal strains and developed from there (for example, see papers by Chinnery (1961), Aydin and Nur (1982), Aydin and Page (1984), ten Brink et al. (1996), and Kattenhorn and Marshall (2006) on strike-slip tectonics; Cowie and Scholz (1992b) and Schultz et al. (2006, 2008a, 2013) on fault displacement–length scaling; Willemse (1997) and Soliva et al. (2008) on fault growth and linkage; Niño et al. (1998) and Cooke and Kameda (2002) on fold-and-thrust belts; Martel (2004) on slope failure and landslide nucleation; and Fossen and Cavalcante (2017) on shear zones). These phenomena obey the laws of fracture mechanics even though the strains involved may ultimately attain large, finite values (such as several tens of percent or more). Once the rules that govern how fractures operate are identified, the sequence of deformation can become clear regardless of how complicated a structural suite has eventually become.

Mechanics is to structural geology as thermodynamics is to petrology

Acquire new tools

The material in this book will also provide a number of new tools and techniques that should become part of a geologist's arsenal in the field. For example, we will explore how to decipher what the shape of the termination of a dike or fault means mechanically, and what that says about how the fracture grew. Several reliable rules of thumb for relating the amount of offset or opening displacement to the length of a structural discontinuity will be developed that can be used, in real time, by a field geologist to produce better interpretations, maps, and cross sections. In sum, a clearer appreciation can be gained from this book of research being done at the cutting edge of geologic fracture mechanics in structural geology and tectonics.

1.6 Review of Important Concepts

1. Fractures are a class of geologic structural discontinuity that involve discontinuous displacements across their surfaces. Fractures tend to be weaker, and less stiff, than the surrounding host rock, making them important geologic flaws and stress concentrators as well as conduits for enhanced fluid flow.

2. Deformation bands are a class of geologic structural discontinuity that accommodate a continuous, smoothly varying displacement across their widths. The type of deformation band depends on its kinematics, leading to five kinematic varieties, and on its micromechanics (such as cataclasis within a compactional shear band).

3. Structural discontinuities are identified in the field by using a *displacement-based classification* scheme. The two main (engineering) end-members of displacement sense are normal (mode-I) and parallel (mode-II, III) to the discontinuity surface, leading to opening-mode and shearing-mode structures, respectively. Closing-mode structures are also encountered in rock.

4. Structural discontinuities whose walls have been displaced *normal* (perpendicular) to the discontinuity surface are *cracks* (having opening displacements) and *anticracks* (stylolites and pure compaction bands, both having primarily closing displacements). Fractures or deformation bands whose walls have been displaced parallel to the surface are called *faults* (if the displacement is discontinuous across the fracture) or *shearing deformation bands* (having continuous displacements across the deformation band), respectively.

5. Strike-slip faults are mode-II structures in map view, and mode-III structures in cross-sectional view. Conversely, dip-slip faults (normal and thrust) are dipping, nonvertical mode-III structures in map view, and mode-II structures in cross-sectional view. These mechanical idealizations of fault surfaces are helpful to understand the patterns of faults and the strains that develop with these types of structures.

6. Deformation bands form in porous granular geomaterials such as soils and porous rock, but are not observed in compact rocks. Conversely, fractures can form in compact or porous rocks depending on factors such as stress state, mineralogy, and pore-water conditions.

1.7 Exercises

1. Define the following terms: fracture, crack, joint, dike, fault, fault zone, anticrack, and deformation band, giving an example of each of these, that you have encountered in the field or in a similar first-hand situation, that best illustrates the components of the definitions developed in this chapter.

2. What is the definition of a geologic structural discontinuity? How can you explain or defend each of the parts of the definition given your experience?

3. Describe the relationship between the physical properties of a host rock and the initial type of structure formed. How do these physical properties relate to lithology?

4. Summarize the engineering classification of fractures and deformation bands based on displacement mode. Place the main types of fractures, including dip-slip and strike-slip faults, into this scheme while accounting for the orientation of the viewing plane (e.g., map view or cross-sectional view).

5. What is the difference between a tabular and a sharp geologic discontinuity? Provide examples of each along with your justification.

6. Describe the difference between the plane-stress and a plane-strain assumptions and give an example of a geologic situation, such as a cross section or deformation of a sedimentary basin, that clearly illustrates each.

7. Give the displacement mode for each of the structures shown in map view: (a) strike-slip fault; (b) thrust fault; (c) normal fault; (d) stylolite. How would the displacement mode for each change if the structures were viewed in cross-section instead?

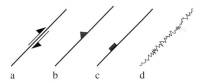

a   b   c   d

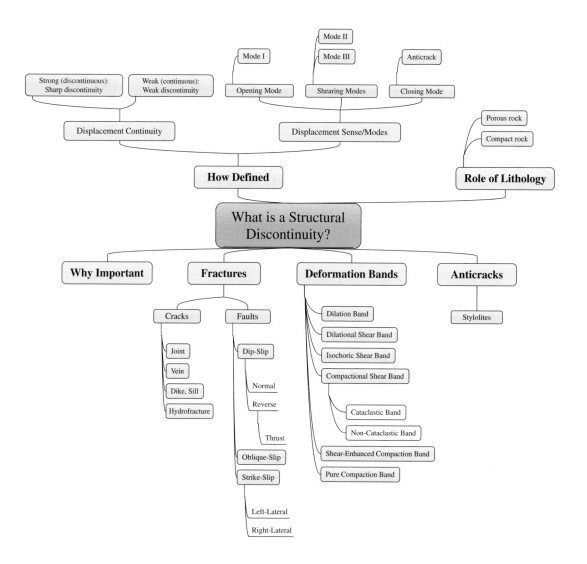

2

Elastic Rock Rheology and Stress Concentration

2.1 Introduction

In This Chapter:

♦ *Rock testing*
♦ *Terminology*
♦ *Hooke's Law relationships*
♦ *Elastic parameters*
♦ *Complete Stress–Strain Curve for Rock*
♦ *Stress concentration*

Rheology is the study of flow or, more generally, the response of a material like rock to imposed stresses or strains (e.g., Johnson, 1970, pp. 13–22; Weijermars, 1997, p. 13; Karato, 2008). In this chapter we first review some aspects of experimental rock deformation that are relevant to the simplest and perhaps most widely used rheologic model for rocks, that of an **elastic** material. We'll then examine the terminology of deformation and strain that flows from the corpus of laboratory studies of rock deformation.

The basic equations for calculating the stresses and strains in an ideal elastic material are then presented. Hooke's Law is introduced with the familiar one-dimensional (1-D) form, then the equations are given for Hooke's Law for the more realistic and appropriate two-dimensional (2-D) and three-dimensional (3-D) situations. These equations are useful if the stresses acting in a rock mass, such as a slope, are known and we need to estimate the elastic strains that would be predicted in the rock mass. Conversely, sometimes the strains are known (for example, if they are measured from a field exposure or seismic section); then from these equations you can back-calculate the stresses generated in the rock mass associated with the strains. Stress–strain curves are presented and used to illustrate the concept of the Complete Stress–Strain for Rock, which finds application to small laboratory-scale rock samples as well as to large-scale outcrops, particularly when the concept of the **rock mass** (Chapter 3) is considered.

Many geologic elements can act as stress concentrators. For example, joints in sedimentary sequences can often be observed to originate at fossils and loadcasts that concentrated the far-field tectonic stresses sufficiently to lead to joint nucleation there (e.g., Pollard and Aydin, 1988; Chapter 4). Shear displacements along faults can begin as local areas of frictional slip along pre-existing joints or bedding planes (e.g., Rice, 1979; Rudnicki, 1980; Martel and Pollard, 1989; Crider and Peacock, 2004). These and other structural discontinuities in a rock mass can themselves also act as flaws, or **stress**

concentrators, in much the same way that rivet holes do in engineering materials such as steel. Because geologic structural discontinuities both concentrate and redistribute stress at their terminations as well as along their lengths, such discontinuities contribute a significant, if not dominant, influence (e.g., Hoek and Brown, 1980; Hoek, 1983; Bieniawski, 1974, 1989; Lajtai, 1991) in *weakening* a large-scale, outcrop-sized rock mass relative to the strength and deformability of small, intact, unfractured hand samples or core.

Accordingly, we'll then step through some examples that illustrate how structural elements such as loadcasts and fractures weaken a material by concentrating stress. First we'll look at the ideal or theoretical strength of a rock, such as might be calculated by a physicist. Next we'll calculate the state of stress in an intact material, such as a simple sheet or plate of rock. Then we'll imagine producing a hole in the plate and looking at the resulting change in stress in the plate near the hole. Lastly we'll use a simple but powerful equation, now more than a century old, to assess the amount of stress concentration, or "strength," of a structural discontinuity such as a joint.

2.2 Laboratory Testing of Rocks

Experimental rock deformation is a vast, fascinating, and intricate field of research, with applications to structural geology, seismology, petroleum production and storage, coal and precious metals extraction, hydrology, rock mechanics, and rock engineering to name just a few (e.g., Brady and Brown, 1993; Lockner, 1995; Hudson and Harrison, 1997; Marone, 1998a,b; Scholz, 1998, 2002; Karner, 2006). Rather than summarize all this material here, a brief overview of perhaps the most common and important type of test—the **compressive strength test**—will be given in this section. Given that the state of stress underground in the Earth's crust is, in general, compressive (e.g., McGarr and Gay, 1978; Zoback et al., 2003), compressive strength tests are appropriate for understanding the deformation of rocks under common geologic conditions (Fig. 2.1). Additionally, because many of the key physical properties of rocks can be related to the rock's compressive strength, such as its tensile strength and modulus (e.g., Jaeger and Cook, 1979; Hoek and Brown, 1980; Farmer, 1983, p. 65; Bell, 1993, p. 171; Lockner, 1995), the basic results of this set of tests will be presented.

There are three main types of compressive strength tests, as illustrated in Fig. 2.2; these are the uniaxial (or unconfined), triaxial (strictly, an axisymmetric triaxial test; Paterson and Wong, 2005, p. 6), and polyaxial tests (e.g., Brady and Brown, 1993, pp. 101–106). In these tests, compression is taken to be positive (as are shortening or contractional normal strain; see Chapter 3), with $\sigma_1 > \sigma_2 > \sigma_3$. The seldom-used biaxial compression test, performed on a cubic sample (Brady and Brown, 1993, pp. 101–102), is a subset of the polyaxial test in which $\sigma_1 \geq \sigma_2$, $\sigma_3 = 0$. Jaeger and Cook (1979), Brady and

Fig. 2.1. A large-capacity computerized testing frame, such as the one shown here, deforms rocks under uniaxial or triaxial compression to provide the basic data for generating the Complete Stress–Strain Curve for Rock and the associated strength and deformability parameters such as peak strength and Young's modulus.

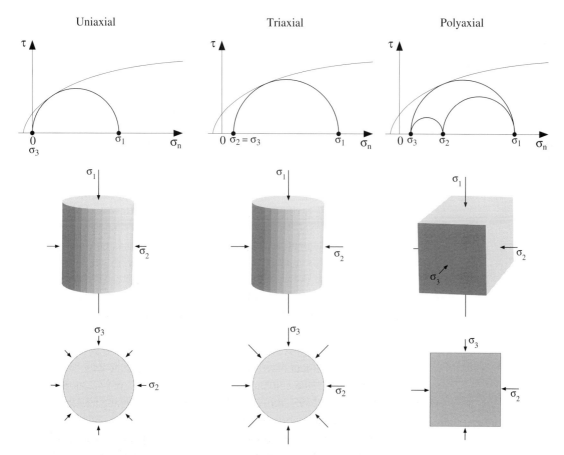

Fig. 2.2. Mohr diagrams (upper panels) and sample geometries (middle and lower panels) for the uniaxial, triaxial, and polyaxial compression tests.

Fig. 2.3. A fine-grained tuff from Yucca Mountain, Nevada, showing a fault inclined to the axis of the ~7.5 cm diameter sample formed during or following peak stress during a uniaxial compression test (sample courtesy of Jaak Daemen, University of Nevada).

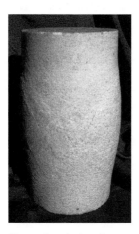

Fig. 2.4. Bow-tie shaped pattern of distributed shear (Lüders') bands in a limestone deformed under triaxial compression (see Farmer, 1983, p. 87; sample diameter at ends is ~7.5 cm).

Brown (1993), and Paterson and Wong (2005) provide concise introductions to the various tests, sample preparation, and interpretation; the detailed standards for rock testing are contained in a set of reports prepared under the auspices of the International Society for Rock Mechanics (available at www.isrm.net).

In the **uniaxial test** (e.g., Goodman, 1989, pp. 60–61; $\sigma_1 > \sigma_2 = \sigma_3 = 0$; Fig. 2.2, left-hand column), the rock cylinder compresses as the load is increased along the axis of the sample, eventually leading to failure at what is known as the peak stress value (the rock's *peak strength*) and also its **unconfined compressive strength** (UCS, or σ_c). This value of strength is commonly used as a simple and repeatable measure of a rock's baseline strength, to which other properties such as tensile strength are commonly scaled. **Axial splitting** is often the failure mode for rocks that are heterogeneous and coarse-grained at the scale of the test sample (e.g., Wawersik and Fairhurst, 1970; Holzhausen and Johnson, 1979; Brady and Brown, 1993, p. 97), whereas a fault may form near peak stress in these tests (as implied by the Mohr diagram in Fig. 2.2), either when the rock is fine-grained or when the testing machine is softer than the rock (i.e., one that is loaded by hydraulic oil), leading to uncontrolled and explosive failure (Wawersik and Fairhurst, 1970). Faulting may also be promoted in coarse- or fine-grained samples late in the test due to tensile stresses generated in samples whose ends are restrained by rigid end-caps (called platens) such as steel (e.g., Hawkes and Mellor, 1970), as in the sample shown in Fig. 2.3.

The influence of confining pressure in rock deformation is evaluated by using a **triaxial test** (e.g., Griggs, 1936; Jaeger and Cook, 1979, pp. 147–157; Farmer, 1983, pp. 95–111; Goodman, 1989, pp. 61–65; Brady and Brown, 1993, pp. 102–105; Lockner, 1995; Paterson and Wong, 2005, pp. 5–14; $\sigma_1 > \sigma_2 = \sigma_3 > 0$; Fig. 2.2, center column); this test can simulate how depth below the Earth's surface affects the rock's response to loading. Other factors that can be investigated include temperature and strain rate (e.g., Griggs et al., 1960; Paterson, 1978; Paterson and Wong, 2005). The cylinder is compressed axially with confinement being applied to its vertical sides by a fluid, such as oil (the sample is normally isolated from the fluid by an impermeable membrane). A rock deformed in a triaxial test typically develops a barrel shape (e.g., Farmer, 1983), in part due to yielding in the central part of the sample (Olsson, 2000) and in part due to end effects (e.g., Brady and Brown, 1993, pp. 92–93). A nice example is shown in Fig. 2.4 in which a limestone deformed to large strains without localized failure under elevated confining pressure. A halite sample, tested under triaxial conditions in support of underground product storage, shown in Fig. 2.5 demonstrates the large amount of axial shortening and barreling that can occur for this rock.

Polyaxial tests ($\sigma_1 > \sigma_2 > \sigma_3 > 0$; Fig. 2.2, right-hand column) make it possible to vary each of the three principal stresses, along with their rates of loading, independently (e.g., Jaeger and Cook, 1979, pp. 158–160; Brady and

Fig. 2.5. A West Texas halite (NaCl) in its original sample size (left, diameter ~ 10 cm) and deformed under triaxial conditions (right; samples courtesy of Jaak Daemen, University of Nevada). Note the increase in diameter at the sample's ends, indicating less restraint during the test than the limestone shown in Fig. 2.4.

Brown, 1993, pp. 105–106). The samples are cubes rather than right-circular cylinders, and this type of compression test can best simulate the state of stress in the Earth's crust and predict the number and orientations of faults that may form in the three-dimensional stress states (e.g., Reches, 1978, 1983; Krantz, 1988, 1989) that are generally developed within the Earth (e.g., Aydin and Reches, 1982; Reches and Dieterich, 1983; Krantz, 1989).

2.3 Terminology from Rock Mechanics

Many of the terms used in structural geology and tectonics to describe deformation and structures come from experimental rock mechanics. Some of these, such as *Young's modulus*, have already been covered. Some of the most common and important terms are listed in Table 2.1 and discussed in this section. We'll follow the approach and conventions advanced by Rutter (1986), Rutter and Brodie (1991), Kohlstedt et al. (1995), Lockner (1995), and Paterson and Wong (2005) among others.

Process: What happens
Mechanism: How it happens

    The terms from rock mechanics motivate a distinction between *phenomenology, or process* (a description or classification of what you see in the field, or **what happens**) and the *physical deformation mechanisms* responsible for deforming the rock mass (**how it happens**). For example, folding (the process) can occur by either brittle mechanisms (with *cataclasis* in the hinges) or

Table 2.1 Terminology of rock deformation

1. Ideal Material	Characteristics	Example
Elastic	Stress related to recoverable strain	Damp sponge
Linearly Elastic	Stress *linearly* related to recoverable strain	Rock
Viscous	Stress related to strain rate	Fluid
Linearly Viscous	Stress *linearly* related to strain rate	Water
Plastic	Stress related to permanent strain	Friction
Bingham plastic	Stress related to (permanent) strain-rate	Mud, lava
Ductile	Capacity for large strain without failure	Folds, stylolites

2. Deformation Mechanism	Key Control	Example
Brittle	Pressure sensitive	Cracking, faulting
Plastic	Temperature sensitive	Twinning, recrystallization
Cataclasis	Pressure sensitive	Fragmentation

3. Useful Combinations	Characteristics	Example
Elastic–inelastic	Recoverable-permanent strain, type unspecified	Complete stress-strain curve
Brittle–plastic	Pressure- to temperature-sensitive deformation mechanisms	Strength envelopes
Elastic–ductile	Small to large strains, type unspecified	Boudinage
Brittle–ductile transition	Change from localized to distributed deformation	Slip surface to cataclasis
Brittle–plastic transition	Change from pressure- to temperature-sensitive deformation mechanism	Seismogenic depth

4. Poor Combinations	Characteristics	Example
Brittle–ductile	Combines deformation mechanism at small strains with unspecified mechanism for large strains	None
Elastic–plastic	Combines unspecified deformation mechanism at small strains with temperature-sensitive mechanism	None

by plastic mechanisms (with *flow* at higher temperatures). In both cases, the folds are considered to be *ductile structures* because they accommodate large contractional strains. The observation that strata are folded is not sufficient in itself, however, to demonstrate the physical factors implicated in folding such as (in this example) low strain rates or elevated temperatures. Keeping process and mechanism separated is one of the challenges in defining a consistent and workable set of terms.

Elastic material – an ideal rheology where the strains caused by stresses are fully recoverable when the load is removed; a special case is a linearly elastic material where the relationship between stress and resulting strain is linear, with a constant value of Young's or shear modulus, and the strain is also fully recoverable; no provision for failure is implied in elastic behavior.

Viscous material – an ideal rheology in which an applied stress leads to a particular strain rate and flow; a special case is a linear viscous material with a constant value of viscosity. Usually synonymous with "viscous fluid."

Plastic material – an ideal rheology involving two-fold behavior under stress: no deformation if the stress is less than a specified level, the "yield strength," and permanent deformation for greater values of stress. Important subtypes include elastic–plastic and visco-plastic materials.

Plastic rheology – a deformation mechanism involving temperature-dependent processes such as dislocation movement, twinning, and creep.

Brittle – a pressure-dependent deformation mechanism usually involving nucleation, growth, and coalescence of dilatant cracks. It is also used to describe the behavior of rock near a discontinuity tip for which rock strength greatly exceeds stress magnitudes, promoting small-scale yielding conditions.

Brittle–ductile transition – a widely used term to describe a change from faulting to flow in the crust, sometimes considered to be a planar mappable interface; because ductile deformation can occur using brittle mechanisms (cataclasis), the term has largely been replaced by "brittle–plastic transition" in tectonic applications. In rock deformation the brittle–ductile transition separates localized deformation (such as faulting) from distributed deformation (such as flow).

Brittle–plastic transition – a change from brittle, pressure-dependent deformation mechanisms, such as cracking and frictional sliding (faulting), to temperature-dependent deformation mechanisms such as creep; the transition is often gradual and dependent on several factors such as grain size and mineralogy, so that regimes marked by simultaneous brittle and plastic deformation are not uncommon.

Cataclastic deformation – fracturing and grain-size reduction within a zone or throughout the volume of a rock mass; can accommodate large ("ductile") strains by brittle mechanisms such as cracking and frictional sliding.

Ductile – the capacity for a rock to sustain distributed flow or large deformations; the specific deformation mechanism through which this occurs (brittle cataclasis or plastic creep) is not specified in the term.

Starting with a given load or stress state, a rock or stratigraphic sequence will deform in various ways depending on its rheologic behavior. In general, deformation is characterized by **two parts**: *rigid-body deformation* and *internal deformation* (Fig. 2.6). **Rigid-body deformation** measures the movement of the rock by translating (linear motion in a certain direction, such as "up") and/or rotating (spinning), without any changes to the rock's internal characteristics. It is sometimes convenient to consider these two parts separately, although to fully describe rock-mass deformation, both parts may be needed. For example, how do we interpret the paleomagnetic signal of an accreted terrane? The paleomagnetic inclination and declination can differ from the present ones not just by the internal rotations due to shear strains, but also by the rigid-body rotation and translation of the terrane across the Earth's surface. Similarly, the movement of thrust sheets at convergent plate margins involves both internal deformation and translation of the deforming mass.

Internal deformation is the component that is observed when we visit an outcrop. Here there are two parts as well: *distortion*—the change in shape of the rock—and *volume change* (sometimes referred to as the "dilatation" (see Oertel, 1996, p. 65), and the "dilation" (see Suppe, 1985, p. 80; Price and Cosgrove, 1990, p. 8), or the volumetric strain (Jaeger and Cook, 1979, p. 40; Davis and Selvadurai, 1996, pp. 13–14). We'll choose to use the term "volume change" here so it doesn't conflict with *dilation*, associated with the

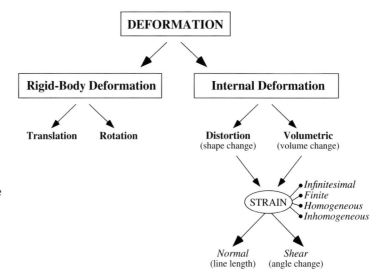

Fig. 2.6. Flowchart showing the different classes of deformation that may be encountered in rocks and rock-masses subjected to displacement or stress loading.

displacement sense for opening-mode cracks (Chapters 4 and 8), or *dilatancy*, associated with either: (a) pressure-dependent crack growth seen macroscopically for rocks and rock-masses loaded in compression (e.g., Jaeger and Cook, 1979, p. 92; Scholz, 1990, pp. 21–38) or (b) frictional sliding across rough surfaces under small confining pressure (e.g., Byerlee, 1978; Scholz, 1990, pp. 56–60). Volumetric reductions can occur due to burial and lithification (e.g., Cartwright and Lonergan, 1996) or to mass loss associated with pressure solution and growth of stylolites (e.g., Price and Cosgrove, 1990, p. 8; Bayly, 1992, pp. 140–143).

Strain is a measure of the internal deformation of a rock mass, and the normal and shear strains can result from either *distortion* or *volumetric changes* of the rock mass (or both). Strain is typically defined as either the change in length (normal strain) or change in angle (shear strain) within a rock due to its deformation. With progressive deformation, rotations and finite strains (Jaeger et al., 2007, pp. 60–64) can lead to non-coaxial strains, whose principal directions are no longer parallel to the original far-field principal stress directions (e.g., Weijermars, 1997, pp. 192–194). Excellent treatments of infinitesimal stress–strain relations are given by Timoshenko and Goodier (1970, pp. 219–234; Jaeger et al., 2007, pp. 41–60), and of relationships between strain and more general deformation by Suppe (1985, pp. 76–109). With the wide availability of thorough treatises on strain, as well as how to measure and calculate the displacements and strains in rock that are well known to structural geologists, no attempt will be made to summarize or reproduce that material here. Instead, we will continue to focus on those aspects of rock deformation that are germane to geologic fracture mechanics.

2.4 Hooke's Law

The properties of an elastic material are given by the relationship between *stress and strain*. In contrast, a viscous material (like corn syrup or water) is defined by the relationship between stress and strain rate. Elastic materials are the simplest kind of *rheology*, or idealized material behavior, that can be applied to rock. Other types of material behaviors, including viscous, plastic, and combinations of these basic rheologies are elaborated upon by Johnson (1970), Turcotte and Schubert (1982), Price and Cosgrove (1990), and Fossen (2010a,b).

A widely used special case is that of a **linearly elastic** material. Here the stress–strain curve is a straight line having a constant slope (Fig 2.7b). Because the slope is constant, rather than varying along each point for the general elastic material (Fig. 2.7a), the relationship between stress and strain is much simpler. We call the slope of the stress–strain curve the *Young's* (or

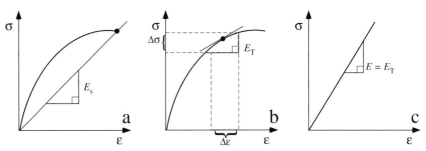

Fig. 2.7. Schematic stress–strain curves for (a, b) a general elastic material, and (c) a linearly elastic material. The local slope of the curve is the tangent Young's modulus, E_T and the secant Young's modulus is E_s

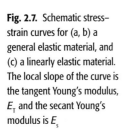

elastic) *modulus* of the material; it has the symbol E (for Elastic modulus). The equations listed in this section all assume linear elasticity—we find that many rocks in nature obey this relationship until they are stressed to their elastic limits, justifying the use of these rules and equations for fracture problems in rock.

An important aspect about the elastic material is that there is **no provision for failure**. This means that any time you use elastic stress equations, you are only completing half the task, because the Hooke's Law equations are not appropriate to describing the yielding or failure of a material. For example, although values of 1000 MPa tension can be calculated, they would not be physically meaningful. In such cases we need to superimpose a suitable failure criterion that can tell us how much stress is considered necessary for the rock to break, slide, or fold (e.g., Lockner, 1995). This is the purpose of the Coulomb criterion for frictional sliding (Chapter 3) and the fracture toughness for crack propagation (Chapter 8). Without explicitly comparing stress to rock strength, a stress analysis is lacking in its applicability to rock deformation.

As long as the stresses to which a rock is subjected do not lead to significant damage within the rock, the strains will be linearly related to the stresses, and they will be transient, or recoverable. The rock can then be represented by its region B behavior (see the Complete Stress–Strain Curve for Rock, Section 2.6, and Fig. 2.12)—the *linearly elastic*, or Hookean, response. The basic equations for elasticity were developed by Robert Hooke (1635–1703) and have been useful in engineering design ever since.

2.4.1 Hooke's Law for 1-D

The simplest form of Hooke's Law can be written for only one dimension (1-D). For 1-D linear elastic behavior, the rock is considered to be a cylindrical or rectangular slab, like a long, negligibly thin wafer, with the long dimension being the direction of both the normal stress and the normal strain. This is illustrated in Fig. 2.8. The elastic equation that relates the amount of stress acting on the rock to the amount of strain that results is (e.g., Jaeger et al., 2007, p. 81):

Hooke's Law—1-D

$$\sigma_x = E\varepsilon_x \qquad (2.1)$$

The proportionality between stress and strain (in 1-D) is the slope of the stress–strain curve—the Young's modulus, E. The equation can easily be solved for the normal strain (contraction or extension) to give

Hooke's Law—1-D, alternative form

$$\varepsilon_x = \frac{\sigma_x}{E} \qquad (2.2)$$

Because these equations (2.1 and 2.2) are defined only for use in one direction, they are not very useful in typical geologic situations.

For a particular stress–strain curve, two Young's moduli can be defined (e.g., Brady and Brown, 1993, p. 91; Jaeger et al., 2007, p. 82). A nonlinear elastic material, having a non-uniform slope, can be measured by taking either the complete values of stress and strain at a point (Fig. 2.7a), referred to as the **secant Young's modulus** E_s, or the local slope at a particular point, as noted in Section 2.6 and commonly at half the peak strength of the rock, referred to as the **tangent Young's modulus**, E_T (Fig. 2.7b). The relations are given by

Fig. 2.8. Conceptual setup for Hooke's Law in 1-D.

Secant Young's modulus

$$E_s = \frac{\sigma}{\varepsilon} \qquad (2.3)$$

and

Tangent Young's modulus

$$E_T = \frac{\Delta\sigma}{\Delta\varepsilon} \qquad (2.4)$$

For the equations discussed in this chapter for 2-D and 3-D Hooke's Law, linear elastic behavior is assumed, so that the rock or outcrop of interest deforms as Region B in Fig. 2.12; in this case, the Young's modulus $E = E_T = E_s$ (Fig. 2.7c) so only E is used in the following equations for brevity.

2.4.2 Hooke's Law for 3-D

Let's now examine the full results for an elastic material in three dimensions (3-D). This is done before going to 2-D for several reasons. First, the 2-D equations are simplifications of the 3-D ones, so by seeing the 3-D results first you'll gain a clearer understanding of the 2-D equations, which are the most useful of all the Hooke's Law results. Second, you can see that the individual terms in 3-D stress or strain are additive; the way that the 2-D equations are often shown can obscure this principle.

The physical situation for 3-D stress and strain is shown in Fig. 2.9. Now we have a cube of rock that is loaded by three principal normal stresses (upper panel), each one acting along one of the coordinate axes. As before, the normal strains (lower panel) associated with each stress will also be aligned

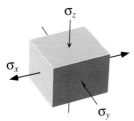

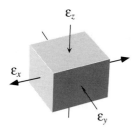

Fig. 2.9. Geometry for problems using 3-D Hooke's Law. Upper panel shows stresses, lower panel shows strains.

with the coordinate axes. This situation, in which the stresses and strains are aligned during the deformation, is called *coaxial strain*. The stresses and elastic strains for 3-D linearly elastic materials are given by (Timoshenko and Goodier, 1970, p. 11)

$$\sigma_x = \frac{vE}{(1+v)(1-2v)}(\varepsilon_x + \varepsilon_y + \varepsilon_z) + \frac{E}{(1+v)}\varepsilon_x$$

$$\sigma_y = \frac{vE}{(1+v)(1-2v)}(\varepsilon_x + \varepsilon_y + \varepsilon_z) + \frac{E}{(1+v)}\varepsilon_y \qquad (2.5)$$

$$\sigma_z = \frac{vE}{(1+v)(1-2v)}(\varepsilon_x + \varepsilon_y + \varepsilon_z) + \frac{E}{(1+v)}\varepsilon_z$$

As you can see, the stress in the *x*-direction, for example, depends upon the strains in all three directions (*x, y,* and *z*), the elastic properties of the rock (here, the Young's modulus, *E*), and a second elastic parameter, *v*, which is called *Poisson's ratio*. It can be defined as the ratio between two strains like this:

$$v = \left| \frac{\text{lateral strain}}{\text{axial strain}} \right| \qquad (2.6)$$

Poisson's ratio

or, using the coordinate directions to be consistent with our 2-D and 3-D equations (Jaeger et al., 2007, p. 108),

$$v = -\frac{\varepsilon_y}{\varepsilon_x} = -\frac{\varepsilon_z}{\varepsilon_x} \qquad (2.7)$$

Poisson's ratio for elastic materials ranges from 0 to 0.5 (Beer et al., 2004, pp. 66–67; Gercek, 2007). A Poisson's ratio of zero implies that a strain in one (say, the *x*) direction does not produce any strain, whether extension or contraction, in the other direction. An example would be a piece of foam or perhaps very soft rock such as diatomite or nonwelded tuff: squeezing it between your fingers leads to crushing between them yet little or no strain in the other directions. The maximum value for *v*, 0.5, requires that all the strain that occurs in one direction is perfectly transferred into strain in the other directions. This is an *incompressible* material, like water. Rocks deep in the Earth that flow in response to elevated temperatures and pressures are sometimes taken to be incompressible. Most intact rocks have values of *v* of about 0.25, with rock masses being somewhat higher at *v* = 0.3 or so (see Farmer, 1983, pp. 50–51). Although Poisson's ratio is defined here for perfectly elastic materials, values of *v* greater than 0.5 can occur due to internal microcracking, dilatancy, and volume increase during deformation (Farmer, 1983, p. 51; Schultz and Li, 1995; Gercek, 2007). Some specialized materials may have an inelastic Poisson's ratio that is negative.

Recall that the units of stress (in SI) are given in MPa (mega pascals), or Pa × 10⁶. Strain is a dimensionless quantity, as you can see from equation (2.2) where MPa (for stress) is divided by GPa (for Young's modulus). In engineering, the term "microstrain" is commonly used as an informal unit: for example, 12 μstrains. Often in geology, though, with strains being large (finite) compared to the typically tiny (infinitesimal) strains in engineering materials, strain is multiplied by 100 to get values in percent. Poisson's ratio is also a dimensionless quantity as it is the ratio between two strains.

In linear elasticity, only two elastic parameters (modulus and Poisson's ratio) are required to characterize how the material deforms in response to stress (except for the uncommon 1-D case, where there is only the one parameter, E). In parallel with Young's modulus being the slope of the normal-stress–normal-strain curve, the shear modulus, G, is the slope of the shear-stress vs. shear strain curve. In any practical problem using 2-D or 3-D isotropic elasticity, as considered in this chapter, either E or G may be available, or more convenient, for calculation. Because the material is elastic and cannot deform inelastically, however, all three of these parameters are related by a simple relationship (Timoshenko and Goodier, 1970, p. 10):

Shear modulus, G

$$G = \frac{E}{2(1+v)} \tag{2.8}$$

In practice, E and G are interchangeable because of the relationship given in equation (2.8).

Several other elastic parameters are useful. The first is called the *bulk modulus* (or *incompressibility*), K (or the "modulus of volume expansion": Timoshenko and Goodier, 1970, p. 11; Jaeger et al., 2007, p. 108) and is given by

Bulk modulus or incompressibility, K

$$K = \frac{E}{3(1-2v)} \tag{2.9}$$

This parameter (Jaeger and Cook, 1979, p. 111) represents how much volumetric strain is experienced in the rock due to a uniform, all-around, hydrostatic pressure (all three normal stresses being equal). Its reciprocal, $1/K$, is called the *compressibility* of the rock. Although the bulk modulus traditionally is denoted by the symbol K, it is different than the stress intensity factor (discussed in Chapter 8, and usually though not always accompanied by a subscript related to the displacement mode, $i = 1, 2,$ or 3) or the related dynamic fracture toughness, K_c.

Equations (2.10)–(2.12) for 3-D that follow are called *Generalized Hooke's Law for Multiaxial Loading*. The *normal strains* are given in their simplest form by (Beer et al., 2004, p. 64)

$$\varepsilon_x = \frac{\sigma_x}{E} - v\frac{\sigma_y}{E} - v\frac{\sigma_z}{E}$$

$$\varepsilon_y = v\frac{\sigma_x}{E} - \frac{\sigma_y}{E} - v\frac{\sigma_z}{E}$$

$$\varepsilon_z = v\frac{\sigma_x}{E} - v\frac{\sigma_y}{E} - \frac{\sigma_z}{E} \qquad (2.10)$$

Notice that the strain in the x-direction, ε_x, is the sum of the Hooke's Law terms for each of the three coordinate directions. The strain depends on the stresses in the x-, y-, and z-directions along with Young's modulus and Poisson's ratio. The strain equations can be recast into more familiar forms by collecting terms to obtain (Timoshenko and Goodier, 1970, p. 8)

Normal strains—3-D

$$\varepsilon_x = \frac{1}{E}[\sigma_x - v(\sigma_y + \sigma_z)]$$

$$\varepsilon_y = \frac{1}{E}[\sigma_y - v(\sigma_x + \sigma_z)] \qquad (2.11)$$

$$\varepsilon_z = \frac{1}{E}[\sigma_z - v(\sigma_x + \sigma_y)]$$

Here the strain in, for example, the x-direction depends on the two elastic parameters, E and v, and on the stresses in each of the three coordinate directions. You could also write these equations using the shear modulus G by substituting equation (2.8) into (2.10) and (2.11).

Up until now we've only considered normal stresses and normal strains. The stress transformation equations describe how normal stresses are associated with shear stresses within the rock. Whereas normal strains represent either contraction or extension of an originally straight line in the rock, shear strains cause the line to rotate or twist without changing its length. The total strain in a rock is the sum or combination of the normal and shear strains at a particular point.

The elastic *shear strains* in 3-D are given by (Beer et al., 2004, p. 69)

Shear strains—3-D

$$\gamma_{xy} = \frac{\tau_{xy}}{G}$$

$$\gamma_{yz} = \frac{\tau_{yz}}{G} \qquad (2.12)$$

$$\gamma_{xz} = \frac{\tau_{xz}}{G}$$

in which γ_{xy} are the shear strains, τ_{xy} are the shear stresses referenced to the xyz coordinate system used before, and G is the shear modulus. Again, although it is customary to use Young's modulus for normal strains and shear modulus for shear strains, they are interchangeable depending on convenience.

2.4.3 Hooke's Law for 2-D

Many physical problems in structural geology and rock mechanics that involve 3-D can be represented in only two dimensions (2-D) while still retaining enough of the information and processes to make an investigation worthwhile. This means that we can simplify the 3-D equations for stress and strain by eliminating one of the coordinate directions (usually the *z*-direction).

There are two common ways to approximate the 2-D behavior of a 3-D material: **plane stress** and **plane strain**. These are not exact solutions but are accurate enough in many instances. The two ways represent end-members of 3-D geometries that are frequently encountered in engineering as well as geologic situations.

Imagine an expanse of rock that is shaped like a plate: long in the *x*- and *y*-directions and thin in the *z*-direction (Fig. 1.14). Called a "thin plate," this geometry can provide a realistic representation of stresses and strains within a crustal section that is, for example, 100 km by 100 km in map view yet only extends down 10 km or less. Plate tectonics assumes that the tectonic *plates* are shaped this way (neglecting their curvature along the Earth's surface). *A thin plate like this uses the plane-stress approach.*

Conversely, imagine a railroad tunnel that was blasted through a mountain. The tunnel may be 10 m in diameter by 10 m high in cross-sectional view and 1000 m long. Here the length in (what we'll call) the *z*-direction is much greater than the dimensions in the other directions (10 m). This geometry is referred to as a "thick plate," with the thickness again being oriented in, or assigned to, the *z*-direction. *A thick plate uses the plane-strain approximation.*

Plane stress = Thin plate

Physically, these two 2-D situations are very different. If you squeeze the **thin plate** in the *x*- and *y*-directions (that is, along the edges of the plate: its in-plane directions), the plate is thin enough that it may buckle, flex, or otherwise deform upward and downward, in the *z*-direction. Think of squeezing a sheet of paper, or even a thin book, along its edges to make it flex and you get the idea. Here, you impose stresses in the *x*- and *y*-directions and get a strain in the *x*-, *y*-, and *z*-directions. Yet the paper is so thin (in the *z*-direction) that there isn't any significant variation in the *z*-component of stress in the *z*-direction. For this **plane-stress** situation, the stress in the vertical, *z*-direction is set to equal zero in the 3-D equations, yet the strain in the *z*-direction is nonzero because the plate must be free to flex and deform in this direction. An easy mnemonic to remember this is: for plane stress, stress (in the *z*-direction) is zero, and strain (in the *z*-direction) is nonzero.

Plane stress:
Stress (z) = 0
Strain (z) ≠ 0

Now let's look at **plane strain**. If the crustal stress state compresses the railroad tunnel along its diameter or height (*x*- or *y*-directions), the rock along the tunnel's length is unable to deform significantly in the long (*z*-) direction. Instead, stress builds up in the *z*-direction. So for a **thick plate**, in plane strain, strain in the *z*-direction is taken to be zero while stress in the *z*-direction is definitely nonzero. Setting the strain component equal to zero again eliminates some terms in the 3-D stress–strain equations, but they are different terms than those eliminated in the plane-stress approximation.

Plane strain = Thick plate

Plane strain:
Strain (z) = 0
Stress (z) ≠ 0

The equations for stress using Hooke's Law for 2-D, **plane stress**, are (Timoshenko and Goodier, 1970, p. 75)

Normal stresses: Plane stress

$$\sigma_x = \left[\frac{E}{(1-v^2)}\right](\varepsilon_x + v\varepsilon_y)$$

$$\sigma_y = \left[\frac{E}{(1-v^2)}\right](\varepsilon_y + v\varepsilon_x) \qquad (2.13)$$

$$\sigma_z = 0$$

and the strains are given by (Timoshenko and Goodier, 1970, p. 29)

Normal and shear strains: Plane stress

$$\varepsilon_x = \frac{1}{E}\left(\sigma_x - v\sigma_y\right)$$

$$\varepsilon_y = \frac{1}{E}\left(\sigma_y - v\sigma_x\right) \qquad (2.14)$$

$$\varepsilon_z = -\left(\frac{v}{1-v}\right)(\varepsilon_x + \varepsilon_y)$$

$$\gamma_{xy} = \frac{1}{G}\tau_{xy} = \frac{2(1+v)}{E}\tau_{xy}$$

For **plane-strain** conditions, the 2-D stresses are given by (Timoshenko and Goodier, 1970, pp. 75, 30)

Normal stresses: Plane strain

$$\sigma_x = \frac{E}{(1+v)(1-2v)}\left[\varepsilon_x + \left(\frac{v}{1-v}\right)\varepsilon_y\right]$$

$$\sigma_y = \frac{E}{(1+v)(1-2v)}\left[\varepsilon_y + \left(\frac{v}{1-v}\right)\varepsilon_x\right] \qquad (2.15)$$

$$\sigma_z = v\left(\sigma_x + \sigma_y\right)$$

and the strains are given by (Timoshenko and Goodier, 1970, p. 30)

Normal and shear strains: Plane strain

$$\varepsilon_x = \frac{1}{E}\left[(1-v^2)\sigma_x - v(1+v)\sigma_y\right]$$

$$\varepsilon_y = \frac{1}{E}\left[(1-v^2)\sigma_y - v(1+v)\sigma_x\right] \qquad (2.16)$$

$$\varepsilon_z = 0$$

$$\gamma_{xy} = \frac{2(1+v)}{E}\tau_{xy}$$

2.4.4 Related Parameters

There are several useful parameters that help to characterize a deforming rock. Some of the more common and important ones for geologic fracture mechanics are described in this section.

In fracture mechanics it is useful to talk about the deformability or *stiffness* of the rock surrounding a discontinuity, often in connection to how it will adjust to the deformations generated in it by displacements along them (e.g., Rubin, 1992, 1993a). The expression for rock stiffness by Rubin (1993a) is

Rock stiffness

$$S = \frac{G}{1-v} = \frac{E}{2(1-v^2)} \tag{2.17}$$

This parameter increases with Poisson's ratio and may be found in the expressions for fracture-surface displacements (e.g., Chapter 8). It can be thought of as a modification of the modulus by Poisson's ratio to take into account the 2-D and 3-D nature of Hooke's Law. The shear modulus, bulk modulus, compressibility, and stiffness are plotted in Fig. 2.10 to show how they vary with Poisson's ratio and how they relate to Young's modulus.

In a rock mechanics experiment, we apply a load to a rock, record the resulting strains, and then measure the slope of the stress–strain curve to obtain a value of tangent modulus—E for normal loads and G for shear loads (and shear strains; see also Brady and Brown, 1993, pp. 35–37). Written using tensors (e.g., Oertel, 1996, p. 85),

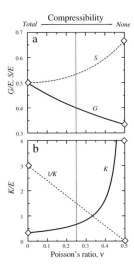

$$\sigma_{ij} = c_{ijkl}\varepsilon_{kl} \tag{2.18}$$

in which σ_{ij} is the stress tensor, c_{ijkl} is the **stiffness matrix** (notice that the expression uses a "c" rather than an "E"), and ε_{kl} is the strain tensor (e.g., Oertel, 1996, p. 85). But in the laboratory we measure strains, not the stresses (which are imposed during the test), so that we have (Jaeger and Cook, 1979, p. 138; Oertel, 1996, p. 85)

$$\varepsilon_{ij} = s_{ijkl}\sigma_{kl} \tag{2.19}$$

Fig. 2.10. Plots showing how shear modulus G, the simple stiffness S, and the bulk modulus K depend on Poisson's ratio v. To find the value of G, S, or K for a practical situation, choose v then multiply the result by E to scale the result to the appropriate value. Note how the compressibility of the rock, $1/K$, goes to zero for $v = 0.5$.

in which s_{ijkl} is the **compliance matrix** (compliance with an "s") and because strain is a function of stress and the material properties (here, the compliances). As clearly stated by Jaeger and Cook (1979, p. 138), the stiffnesses c are called moduli by most or all workers, but since we measure s in (2.19), the moduli are the inverse of stiffness. The desire for precision in terminology led Rubin (1992, 1993a) to define rock stiffness in (2.17) as proportional to modulus, because this is consistent with the current use of these terms—specifically, following (2.18). A rock with larger Young's (or shear) modulus has a greater resistance to deformation and strain, so it makes sense to say it is more "stiff" than a rock with a smaller value of modulus. Correspondingly, if we put the modulus in the denominator of an expression, such as that for fracture surface displacements (see Chapter 8), then the inverse of a stiff rock (with E or G in the denominator) is called more **compliant**.

Compliance = 1/modulus
Less Stiff = More
Compliant

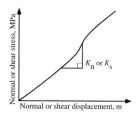

Fig. 2.11. Schematic plot of displacement vs. load (normal or shear stress) showing definition of normal or shear stiffness as the local slope of the curve.

There are other terms that are also referred to as stiffness—both normal and shear—that differ from that in (2.17). In rock mechanics, *stiffness is defined as the slope of the stress–displacement curve* (Fig. 2.11), rather than the slope of the stress–strain curve. Strain is dimensionless, so (stress)/(strain) has units of stress (Pa). The units imply that stiffness defined in this second way, as the local slope of the stress–*displacement* curve, has units of Pa m^{-1}, which are different from those implied by (2.17), or GPa.

These normal and shear stiffnesses are frequently used to characterize the resistance to opening, shear, or closure of joints and other types of fractures. Using the equations for the stress–displacement curve (such as that shown schematically in Fig. 2.11) and denoting the normal and shear displacements as D_n and D_s, respectively,

$$\sigma_n = K_n D_n$$
$$\sigma_s = \tau_{xy} = K_s D_s \qquad (2.20)$$

The normal and shear stiffnesses, K_n and K_s, respectively, are given by (e.g., Crouch and Starfield, 1983, pp. 208–210; Goodman, 1989, p. 196; Brady and Brown, 1993, p. 130)

Normal Stiffness, K_n
Shear Stiffness, K_s

$$K_n = \frac{\sigma_n}{D_n}$$
$$K_s = \frac{\sigma_s}{D_s} \qquad (2.21)$$

Suppose you have an open joint. If you apply a compressive load normal to the joint, the fracture walls will move toward each other, and perhaps eventually close, at a rate (its "stiffness") that depends on rock properties and the joint geometry and roughness (e.g., Goodman, 1989, p. 164). Using a joint width (between its open walls) of T, then (2.20) becomes

$$\sigma_n = E\varepsilon_n = E\left(\frac{D_n}{T}\right) = \left(\frac{E}{T}\right)D_n = K_n D_n$$
$$\sigma_s = G\varepsilon_s = G\left(\frac{D_s}{T}\right) = \left(\frac{G}{T}\right)D_s = K_s D_s \qquad (2.22)$$

in which K_n is the normal stiffness (see also Crouch and Starfield, 1983, p. 210). The same approach can be used to obtain a shear stiffness (K_s) from the second of (2.22), and this is used in studies of shearing structural discontinuities, such as faults and deformation bands (Schultz and Balasko, 2003) as well as closed fractures such as bedding planes at depth. These stiffnesses, such as K_n, can be distinguished from other rock properties, such as stress intensity factors (e.g., K_I) and bulk modulus K (equation (2.9)), by their subscripts and units.

The next useful quantity is the *volumetric strain, e*:

Volumetric strain, e

$$e = \left(\varepsilon_x + \varepsilon_y + \varepsilon_z \right) \tag{2.23}$$

The volumetric strain is the sum of the normal strains (Davis and Selvadurai, 1996, pp. 13–14; Jaeger et al., 2007, p. 60). This parameter quantifies the total change in volume (expansion, more commonly called **dilatancy**, or contraction, sometimes called the **compactancy**) that occurs in the rock due to the applied stresses. For a perfectly incompressible elastic material, $v = 0.5$, $G/E = 0.33$, $K = \infty$, $1/K = 0$, and $e = 0$ (no volume change during strain). Volumetric changes in rock during deformation are one of the key properties to observe and measure as they provide insights into localization of strain, such as into fractures and deformation bands, along with the transition from localized brittle deformation (faulting) to distributed ductile deformation (flow; Paterson and Wong, 2005, pp. 68–76).

Closely related to volumetric changes in a rock is its inherent *porosity, n*, which is a measure of the ratio of open space in a rock to its total volume (e.g., Farmer, 1983, p. 18; Hook, 2003; Jaeger et al., 2007, p. 169). Porosity is given by

Porosity

$$n = \frac{V_{\text{pores}}}{V_{\text{pores}} + V_{\text{matrix}}} \tag{2.24}$$

and can range in values $0 \leq n < 1$. In practice the porosity of a rock is commonly expressed as a percentage, so that a sandstone, for example, has a porosity of $n = 0.2$ or $n = 20\%$. A related quantity that is more commonly used in soil mechanics and geotechnical engineering is the *void ratio, e_v*; this is given by

Void ratio

$$e_v = \frac{V_{\text{pores}}}{V_{\text{matrix}}} \tag{2.25}$$

where the subscript has been added here to distinguish void ratio, that usually is given the symbol e (e.g., Farmer, 1983, pp. 18–19; Jaeger et al., 2007, p. 170) from volumetric strain e just discussed. These useful parameters are related by

$$n = \frac{e_v}{1 + e_v} \tag{2.26}$$

In rocks, as in soils, the pore space may be filled by a gas (such as air), a liquid (such as water or petroleum) or, in many lithified sedimentary and high-grade metamorphic rocks, a solid (cement or crystallized mineral phases). In addition, the degree of connectivity of the pores determines how well the rock or soil will conduct fluids, such as groundwater or petroleum; further information on geo-hydrologic properties such as hydraulic conductivity and permeability can be found in Freeze and Cherry (1979), Jaeger et al. (2007), pp. 186–187, and Aadnøy and Looyeh (2010). A fracture porosity can also be considered in strata that are cut by fracture networks.

Porosity is such an important rock property that it leads to a *two-fold classification of rocks* as far as their brittle deformation characteristics are concerned (Paterson and Wong, 2005, p. 2). A **compact rock** is one having very small porosity, less than a few percent, with its strength and stiffness properties influenced by tiny grain-boundary flaws. A **porous rock** has substantial porosity (commonly well in excess of 5%) so that its strength and stiffness properties depend on the solid grains, the pore-filling material (solid, liquid, or gas), and the grain interfaces. Porous rocks can be further separated into two subclasses (Paterson and Wong, 2005, pp. 134–137). A **strongly cohesive porous rock** is characterized by grain boundaries that are strong enough to resist deformation themselves, so that the rock behaves like a somewhat weaker compact rock. In contrast, a **weakly cohesive porous rock** has grain boundaries that are weak enough for the rock to behave more as a porous granular aggregate consisting of particles that can slip, roll, and crack (forming intra-granular cracks); this subclass commonly localizes strain to form *deformation bands* rather than fractures as would be more common in a strongly cohesive porous rock or in a compact rock (see Fig. 1.12 and Chapter 7).

The *work* done on a rock is the product of an applied load (or force), P, and the amount of displacement caused by the force, D. The work is also known as the *strain energy, u* (Beer et al., 2004, p. 472), which is the area under the force–displacement curve and given for a linear one by

Compact rock

Porous rock

Strongly cohesive porous rock

Weakly cohesive porous rock

Strain energy, u, or work done by the applied forces

$$u = \frac{1}{2}PD \qquad (2.27)$$

Work and strain energy have units of N m, or equivalently, J (joules). By analogy (Timoshenko and Goodier, 1970, p. 244), compression of a gas produces a temperature increase, with the change in temperature being the work done by compression of the gas. For an elastic material, the stresses produce a quantity of energy that is stored in it, ready to be released by, for example, strain localization; this stored elastic energy is the strain energy.

The area under the stress–strain curve (like the ones shown in Fig. 2.7) is called the *strain energy density, U*, with units of either MPa or MJ m^{-3} (Timoshenko and Goodier, 1970, p. 246; Beer et al., 2004, pp. 474–478; Jaeger et al., 2007, p. 130). Strain energy density is the strain energy per unit volume of the rock.

For a full 3-D state of stress, the strain energy density is given by

Strain energy density, U, or the area under the stress–strain curve in 3-D

$$U = \frac{1}{2E}\left[\sigma_x^2 + \sigma_y^2 + \sigma_z^2 - 2v(\sigma_x\sigma_y + \sigma_x\sigma_z + \sigma_y\sigma_z)\right]$$
$$+ \frac{1}{2G}\left(\tau_{xy}^2 + \tau_{yz}^2 + \tau_{xz}^2\right) \tag{2.28}$$

and for loading by *two remote principal stresses*, the strain energy density is

Strain energy density with two remote principal stresses

$$U = \frac{1}{2E}\left[\sigma_1^2 + \sigma_2^2 - 2v(\sigma_1\sigma_2)\right] \tag{2.29}$$

For the simple but rarely utilized 1-D case,

Strain energy density under uniaxial loading

$$U = \frac{\sigma_1^2}{2E} \tag{2.30}$$

The "modulus of resilience" is given in equations (2.28)–(2.30) in which the stress has been set to the yield stress (peak stress in Fig. 2.12). The "modulus of toughness" is the area under the entire stress–strain curve for any expression for strain energy density (Beer et al., 2004, p. 473).

Strain energy density is commonly separated into two components: the first for volumetric changes in the rock due to the applied stresses, called the *volumetric strain energy density* U_v, and the second for shearing-related changes in the rock, called the *distortional strain energy density*, U_d; these expressions are given by (Timoshenko and Goodier, 1970, pp. 247–248; Jaeger et al., 2007, p. 131)

Volumetric strain energy density

Distortional strain energy density

$$U_v = \frac{1-2v}{6E}\left(\sigma_x + \sigma_y + \sigma_z\right)$$
$$U_d = \frac{1+v}{6E}\left[\left(\sigma_x - \sigma_y\right)^2 + \left(\sigma_y - \sigma_x\right)^2\right] \tag{2.31}$$
$$+ \frac{1}{2G}\left(\tau_{xy}^2 + \tau_{yz}^2 + \tau_{xz}^2\right)$$

Volumetric strain energy density has been used extensively as a criterion for propagation of cracks (mode-I and mixed-mode) in engineering (Sih, 1973, 1974; 1991; Ingraffea, 1987) and, more recently, for the nucleation of deformation bands (Okubo and Schultz, 2005); critical values of U_v can be defined at which a crack or band will grow (Sih and Macdonald, 1974; Okubo and Schultz, 2005). **Distortional strain energy density** has been used as a propagation criterion for faults (Du and Aydin, 1993, 1995) and deformation bands (Schultz and Balasko, 2003; Okubo and Schultz, 2005, 2006) that propagate under pre-peak strength conditions.

Propagation of geologic structural discontinuties may also be investigated by calculating the strain energy release rate, G, following Griffith's (1921, 1924) pioneering work that set the stage for engineering fracture mechanics. As developed in Chapter 8, G can be related to stress intensity factors for sharp geologic discontinuities like cracks or faults, but it can also be obtained for deformation bands that are weak discontinuities (e.g., Palmer and Rice, 1973; Rudnicki and Sternlof, 2005; Rudnicki, 2007) that lack a stress singularity at their tips by using the J-integral method (Kanninen and Popelar, 1985, p. 167; Rudnicki and Sternlof, 2005; Schultz and Soliva, 2012; Chapter 9). In general, $J = G$, which reduces to an equivalence between J, G, and the stress intensity factor for sharp discontinuities (Kanninen and Popelar, 1985, p. 166). A particularly useful form for sharp or tabular geologic discontinuities is given as

*Strain energy release rate
from J-integral*

$$J_{band} = \sigma_{yield} \delta \tag{2.32}$$

in which σ_{yield} is the yield stress or, equivalently, the driving stress (Li, 1987) that led to discontinuity growth, and δ is the opening, closing, or shear displacement across the discontinuity. Units of strain energy release rate are J m^{-2}.

2.5 The Complete Stress–Strain Curve for Rock

Testing of rocks in the laboratory provides the basis for understanding how rocks deform under stress. As a result, a grasp of experimental rock mechanics is fundamental to accurately interpreting geologic structures that we observe and map in the field.

Results from several decades of careful test results demonstrate that a wide range of natural geologic conditions, and deformation mechanisms, can be reproduced with a high degree of confidence in the laboratory. We also know that the behavior of rock under certain sets of conditions is well described by simple rheologic models, or idealizations; these are very useful in engineering design and in structural geology.

This section presents key aspects of rock deformation under shallow crustal conditions (upper few kilometers) that are appropriate in the study and interpretation of structural discontinuities such as joints and faults. You can find more comprehensive and detailed treatments of this topic, and others in experimental rock mechanics, in Jaeger and Cook (1979), Farmer (1983), Goodman (1989), Brady and Brown (1993), Paterson and Wong (2005), and in scientific journals such as the *International Journal of Rock Mechanics and Mining Sciences*, along with many annual and topical conference proceedings volumes.

If we take a cylindrical core of rock, apply a compressive load to its ends in a testing machine (Fig. 2.1), and record the resulting deformation, we can map out a consistent sequence of behaviors that is common to many rocks under many types of conditions. This sequence is called the **Complete Stress–Strain Curve for Rock** (see Wawersik and Fairhurst, 1970; Wawersik and Brace, 1971; Hudson et al., 1971; Farmer, 1983, p. 77; Goodman, 1989, pp. 69–70; Paterson and Wong, 2005, pp. 111–114, for clear and detailed treatments). Shown in Fig. 2.12, this curve represents a compact way to understand the progressive deformation of a rock sample, from the onset of the smallest of loads through the formation of continuous, through-going, fault surfaces. It also applies, conceptually if not semi-quantitatively, to a **field-scale exposure** of rock mass as it deforms tectonically.

A typical **rock mass** (Hoek and Brown, 1980; Chapter 3) that is jointed or otherwise deformed exhibits a stress–strain response that is analogous to that found for small intact rock samples (e.g., Farmer, 1983, pp. 184–189; Bieniawski, 1978, 1989). In a rock mass, the slope of the stress–strain curve for loading conditions is called the **deformation modulus** E^* (e.g., Bieniawski, 1978; Goodman, 1989, p. 184; equation (3.35)); this value is invariably smaller than the corresponding tangent Young's modulus for the intact rock material. Values for the tensile strength, peak shear strength, unconfined compressive strength, friction angle or coefficient, and cohesion for the rock mass can be obtained by using standard relationships that are based largely on the Rock Mass Rating (RMR) of the exposure (e.g., Hoek, 1983; Schultz, 1996; Chapter 3). These quantities are important to engineering design of many practical projects including coal and salt mine pillars, slopes of open-pit gold mines, and underground chambers for powerhouses.

As the load is applied to the sample such as that shown in Fig. 2.1, the sample begins to compress. **Region A** is characterized (Fig. 2.12) by *closure*

Complete Stress–Strain Curve for Rock

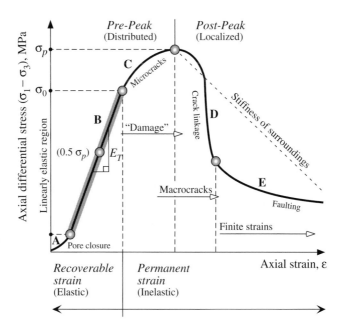

Fig. 2.12. The Complete Stress–Strain Curve for Rock drawn for axial loading. The important regions and parameters are identified in the figure and discussed in the text.

of pore volume and any microcracks in the rock—it is also called the "seating" region where the sample becomes firmly coupled to the testing frame. In a porous rock like sandstone, the pore spaces close up until only the quartz grains are left. In a compact crystalline rock like granite, there is very little pore volume to close, so this region may be very small on the diagram.

Region B is one of *continuous elastic compression* of the rock's constituent mineral grains. This region (shaded in Fig. 2.12) may be large enough to dominate the stress-strain curve of a rock. Here, an increase in the load (moving vertically up on the diagram) produces a proportional contractional strain, and the relationship is *linear*. If the load is removed while the rock is in either region A or region B, the rock uncompresses back to its initial shape, without sustaining any damage from the test. This recoverable strain is called *elastic strain*; the linearly related stress and strain in region B is called the *linearly elastic region*.

The point in region B, denoted 0.5 σ_p, is the point corresponding to 50% of the peak strength, σ_p, or 50% of the greatest value of load that the rock can withstand. This point is used to systematically calculate the slope of the *linearly elastic part* of the stress–strain curve. An elastic material is one that deforms in response to an applied load, then recovers its initial shape completely when the load is removed. Any strain or deformation in the elastic material is temporary and recoverable, not permanent. The *constant slope* in region B is calculated and called the rock's Young's modulus, E (specifically, the tangent modulus E_T; see Brady and Brown, 1993, p. 91). As can be seen from Fig. 2.12, a slope could also be calculated for other parts of the curve, and these may have special meanings depending on the region from which they were calculated.

Measurements of many intact rock samples demonstrate a correlation between the tangent Young's modulus and the rock's peak strength (Deere and Miller, 1966; Bell, 1993; Sonmez et al., 2006). This relationship, shown in Fig. 2.13, provides a means for obtaining a value of peak strength, for example, using a value of Young's modulus for the same rock, measured under the same test conditions. It also underlies the common observation that stronger rocks are also stiffer than weaker ones (e.g., Farmer, 1983, p. 65); this is illustrated in Fig. 2.14.

As the load increases above 0.5 σ_p, the sample begins to become damaged, with some grain cracking and opening of new void space. When this damage becomes *sufficiently pronounced to affect the average properties of the sample*, the rock enters **region C**, where the stress–strain curve becomes concave downward, with a corresponding decrease in its slope and the associated Young's modulus. This region is characterized by stable microcrack growth (Horii and Nemat-Nasser, 1985; Kemeny and Cook, 1986, 1987; Kemeny, 1991), **strain hardening** (Desai and Siriwardane, 1984, p. 228), and progressive damage to the rock. The increased (open) volume due to cracks is the reason why the modulus and slope are reduced in this region (Segall, 1984b; Kemeny and Cook, 1986; Kachanov, 1992). The crack growth is

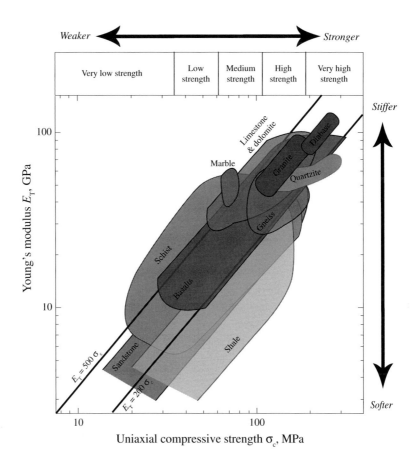

Fig. 2.13. Correlation of strength and deformability for rocks, after Deere and Miller (1966) and Bell (1993).

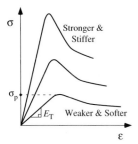

Fig. 2.14. Schematic relationship between strength and deformability of rocks illustrated by using their stress–strain curves.

stable because the rate of crack growth is greater than, or equal to, the rate of load increase—that is, the growth of new cracks can keep pace with the rate of deformation applied to the rock. Coalescence of microcracks and sound generation (acoustic emissions) can typify this region.

The onset of sample damage in region C is noted on the stress axis by the point σ_0, called variously the "maximum stress for elastic behavior," "elastic limit," or the **yield stress**. This point commonly occurs around 66% of the peak stress, but it may be much closer (for example, >90%) depending on the rock type and rate of loading vs. rate of microcrack growth (Segall, 1984b; Kemeny, 1991).

Beginning with region C, the damage to the sample is permanent. This means that, if you were to stop the experiment somewhere in this region, the sample would not return to its original shape, but would retain some degree of flattening. Once a rock enters region C, it *ceases to be the rock it was* during the early stages of the test and systematically becomes a new rock, with a different (greater) fracture porosity and average grain size. The internal structure of the damaged rock is markedly different—region C marks the transition between an intact rock sample and a faulted, broken-up rock.

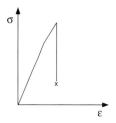

Fig. 2.15. A stress–strain curve obtained from some uniaxial compression tests in which the loading rate is much greater than the sample can accommodate in region C (Fig. 2.12), or if the testing frame is softer than the rock (hydraulically driven machines fit this description). The sample fails explosively and catastrophically somewhere near the peak stress.

Region D is initiated at the **peak stress**, which is also known as the ultimate stress. Although sometimes referred to as the "failure stress," this notion is based on results from some early testing machines (Fig. 2.15) that were less stiff than the rock (see Hudson et al., 1972; Brady and Brown, 1993, pp. 94–98). If a rock is supported adequately by its surroundings (either the testing machine in the laboratory or the adjacent rock in the field), the peak stress marks, instead, the beginning of an important phase in the rock's journey to its final faulted state (e.g., Wawersik and Fairhurst, 1970; Wawersik and Brace, 1971). Region D is characterized by **strain softening** due to the lateral *growth, or propagation, and linkage* of microcrack arrays to form larger macrocracks that are now visible on the scale of the sample. These macrocracks are analogous, in a conceptual way, to joint sets seen in outcrops (Nemat-Nasser and Hori, 1982; Schultz and Zuber, 1994) although the stress conditions for cracks in rock samples and joints in the crust are very different.

Once the rock reaches the peak stress, and if deformation continues, the rock enters what is known as the **post-peak region**, where the rock response is a *compromise* between the energy going in and the energy going out (dissipated by fracture growth, frictional wear, and the like). We'll discuss this compromise later in this section; key terms and concepts that describe what a rock or outcrop is doing in the post-peak region are illustrated in Fig. 2.16.

The final stage in the progressive evolution of the rock sample is the *accumulation of large displacements* along (and parallel to) the coalescing macrocracks. Frictional sliding along these crack surfaces characterizes **region E**, where the stress–strain curve flattens to a more horizontal, constant slope. The slope is called the *residual strength* and is analogous to the coefficient of friction along a well-developed, mature fault.

The Complete Stress–Strain Curve for Rock is a useful tool for understanding the behavior of more than just small laboratory samples—it summarizes how **larger rock exposures** can deform too, as noted above. As long as the stresses acting on a large-scale rock exposure remain small enough that the rock remains in its region B, then the laboratory-scale properties of Young's

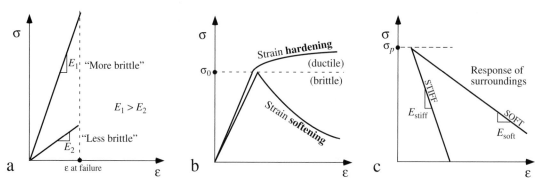

Fig. 2.16. Some important relationships from the stress–strain curve. (a) Strain at failure vs. modulus; (b) departures from linear behavior on approach of peak stress; (c) unloading modulus for stiff and soft rocks after peak stress.

modulus (and Poisson's ratio) will apply to the larger-scale exposure. Once the load begins to damage the rock, though, then the laboratory-derived properties will need to be re-evaluated.

A rock having a steeper slope in region B—corresponding to a larger value of Young's (or the shear) modulus—is said to be stiffer, or sometimes "less competent," than one with a smaller slope and modulus (Fig. 2.16a). However, if you choose a particular value of strain, then the rock that fails (with the stress then corresponding to the rock's peak strength; Fig. 2.12) is considered to be "more brittle" than the one that has not failed at that value of strain. The unfailed one can carry the load without unstable deformation and is called "ductile" (see Figs. 2.4 and 2.5).

Once the elastic (yield) strength σ_0 of a rock is achieved, the macroscopic (or average) response of the rock can define two basic paths (Fig. 2.16b). Some rocks can sustain a larger amount of stress as they deform to larger values of strain; this is called *strain* (or work) *hardening*. Here, deformation within the rock does not localize, but becomes widely distributed throughout, promoting *stable straining* (Jaeger and Cook, 1979, p. 157; region C in Fig. 2.12). The larger, nonlocalized strains in this strain-hardening regime are called "**ductile**" and can occur in rocks such as marble, limestone, salt, and others as temperature increases and non-brittle deformation mechanisms take over. On the other hand, a rock that crumbles and fractures into little pieces in the *post-peak* region (regions D and E in Fig. 2.12) traces a shallower unloading path with a progressive degradation in strength as increasing strain (Fig. 2.16b). Many compact igneous rocks show this *strain-softening* behavior as dilatant cracking and faulting (*strain localization*) occur. Strain-softening curves demonstrate "**brittle**" behavior and may respond unstably to further straining, depending on the rate of unloading relative to that of the surroundings (see below). The "**brittle–ductile transition**" is defined by the right combination of pressure–temperature–strain rate–composition conditions to produce a *horizontal* curve in the peak-stress region (Jaeger and Cook, 1979, p. 87; Paterson and Wong, 2005, pp. 212–217; Fig. 2.16b); it reflects a transition between localized and nonlocalized deformation due to the dominance of different deformation mechanisms within the rock.

An example of stress–strain curves for a porous sandstone, deformed under triaxial conditions, is shown in Fig. 2.17 (after Wong et al., 2004). Strain localization, expressed as faults, occurs in the sandstone for relatively small confining pressures (through 40 MPa, fine solid curves); shear-enhanced dilation that accompanies strain at low confining pressure (the 5 MPa curves) changes to shear-enhanced compaction at higher confining pressures (Fig. 2.17, lower panel). Nonlocalized distributed flow occurs in the sandstone for confining pressures exceeding 40 MPa (bold dashed curves), accompanied by progressively greater amounts of shear-enhanced compaction. The brittle–ductile transition occurs for this rock between 40 and 60 MPa (Wong et al., 2004).

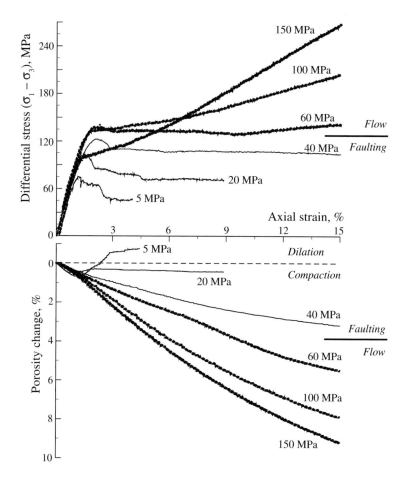

Fig. 2.17. Stress–strain curves (upper panel) and changes in sample porosity vs. axial strain (lower panel) resulting from triaxial testing of Adamswiller Sandstone (after Wong et al., 2004). The sandstone has an initial porosity of 23% and a constant pore-water pressure during testing of 10 MPa. Values of confining pressure ($\sigma_2 = \sigma_3$) shown by each curve.

The slope of the descending stress–strain curve in the post-peak region provides an additional pair of useful terms (Fig. 2.16c). A steep slope corresponds to a harder, *stiffer* rock conceptually analogous to the original definition of Young's modulus in region B. A shallower slope corresponds to a rock that is easier to deform, and is called *softer*. Because we are talking here about material response, and not the rock's ability to break or flow, these terms are used instead of "stronger" or "weaker."

There is an additional factor that is necessary to interpret a stress–strain curve: the **relative stiffnesses of the rock and its surroundings**. It is the *comparison* of these two properties that, in many cases, controls the development and expression of the brittle structures that you see and map in the field.

In the laboratory, you would compare the slope of the post-peak part of the stress-strain curve for a rock sample to the slope of the line representing the Young's modulus (actually, the "stiffness" from a force–displacement curve) of the testing frame (dashed line in Fig. 2.18; Brady and Brown, 1993, pp. 94–98; Paterson and Wong, 2005, pp. 60–62). Suppose the rock is soft and friable in comparison with the strength of the surroundings. Here the rock produces a *shallower* slope in the unloading, post-peak region than does

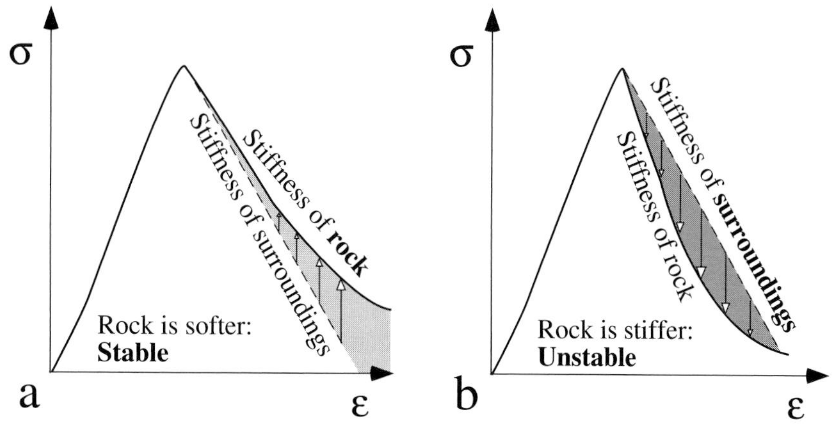

Fig. 2.18. Stress–strain curves relative to the unloading stiffness of the surroundings determine whether the rock will deform stably or unstably. Shaded areas show energy extracted from (b) or added to (a) the sample by unloading of its surroundings.

the stiff testing frame, which has a steeper slope (Fig. 2.16c). As the rock deforms inelastically in the post-peak region, it is restrained by the stiffer loading frame, so the deformation is *stable* and its rate is mostly controlled by the response of the surroundings. Rapid slip along faults and dynamic crack growth can be controlled in stiff testing machines, and you may recall that the folding of a sequence is also controlled by the stiff layer.

On the other hand, suppose that the sample is loaded in an oil-driven hydraulic press. If the rock is stiff (like an intact basalt) compared to the response of the oil in the testing frame, then the rock sample actually supports the load and eventually breaks. When it does break, there is no support from the surroundings to restrain its deformation, so in this case, the frame acts like a spring that adds energy to the sample, and the sample fails explosively (Fig. 2.15). You can see from Fig. 2.18b that *unstable* deformation like this arises when the rock is *stiffer* than the surroundings. In many geologically interesting exposures, the stiffer layer breaks (as in boudinage) or folds (as in buckling and multi-layer folding) when it is "more competent" than its surroundings (e.g., Price and Cosgrove, 1990, p. 425). Stiff inclusions in otherwise softer rock, such as zones of deformation bands in sandstone, may also act like this, leading to sudden faulting of the bands (Aydin and Johnson, 1983). An instability, or sudden change in deformation mechanism or rate, marks the point where the surroundings transfer strain energy to the stiffer rock, focusing and enhancing its deformation.

The concepts just discussed for rock response, such as strain hardening or softening, apply to particular small volumes of rock *in addition to* (or even in place of) the large-scale or macroscopic rock mass. For example, the elevated levels of stress generated near fracture tips can lead to a dense zone of micro-cracks, deformation bands, folds, or slip surfaces (e.g., Pollard, 1987; Rubin, 1993a; Mériaux et al., 1999) that absorb the stress and make it more difficult for the fracture to propagate into the surrounding rock mass. This is analogous to region C for the overall response (Fig. 2.12). The near-tip process zone (Chapter 9) is one of *local* strain hardening and only occurs around the fracture tip—the rest of the rock mass behaves normally. Faults can also be thought of as planar zones of strain softening since they can be weaker in shear (in many cases) than their surroundings; hence the classical term "zone of weakness." In fact, the Complete Stress–Strain Curve for Rock can be used to track the local processes, such as near a fracture tip (e.g., Palmer and Rice, 1973; Li and Liang, 1986) while a separate one, having different parameters appropriate to the larger scale, describes the overall, large-scale response of the entire rock mass.

2.6 Theoretical Strength of a Rock

Now that we have examined elements of rock deformation and elasticity, we turn to the topic of inhomogeneous stress, and stress concentration, in a rock mass. The theoretical strength of a rock is the amount of stress needed to break the bonds of an ideal crystal lattice in tension or shear (Wachtman, 1996, pp. 37–42). This has been known since at least the early twentieth century, and it was the subject of the famous works by A. A. Griffith in the early 1920s that led to the development of fracture mechanics as an instrument in engineering design and structural geology.

Using the crystal lattice shown in Fig. 2.19 (see Kanninen and Popelar, 1985, pp. 32–36, or Anderson, 1995, pp. 31–32 for details), the tensile stress, σ_{th}, needed to break an ideal rock is approximately

Size–strength relation
$$\sigma_{th} \approx \frac{E}{2\pi} \tag{2.33}$$

where E is the Young's (elastic) modulus. Using typical values for E of 10–100 GPa (1 Giga Pascal = 10^9 Pascals; values from Jaeger and Cook, 1979, p. 146), equation (2.33) suggests values of perhaps 2–20 GPa for σ_{th}. Because actual values of tensile strengths of rocks are commonly only 1–10 MPa, the theoretical strengths are about *1000 times too large*.

The solution to this paradox was provided by Griffith's experiments in the United Kingdom on the strength of thin glass fibers. Griffith (1921, 1924) showed that, oddly enough, the strength of a glass rod gets *smaller* as the rod gets *bigger*, which is contrary to what we would normally think of material behavior (where bigger is supposed to be better). The **inverse relationship between strength and size** (or rod diameter) is found in practically all materials, from the purest glass and metals, to concrete, to mineral crystals, to natural rocks and rock masses (e.g., Bieniawski, 1968; Bieniawski and Van Heerden, 1975; Paterson, 1978, pp. 33–35; Jaeger and Cook, 1979, pp. 196–197; Farmer, 1983, p. 189; Goodman, 1989, pp. 90–91; Scholz, 1990, p. 28–29; Schultz et al., 1994). The relationship between the uniaxial compressive strength ("UCS," in MPa) and sample size (in m) is given by (e.g., Farmer, 1983, p. 189; Scholz, 2002, pp. 35–37)

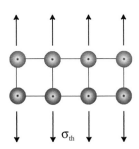

Fig. 2.19. Conceptual illustration of the theoretical strength of an ideal flawless crystal lattice.

*Size–strength
relationship*

$$\sigma_c = kL^{-\varsigma}$$

(2.34)

and plotted in Fig. 2.20. The exponent ς ranges from 0.17–0.32 for some coals, with values closer to 0.5 being more representative of the more common rock types that deform under brittle conditions (Jaeger and Cook, 1979; Scholz, 2002). Additionally, the exponent of 0.5 is consistent with the weakening effect of progressively longer joints (Farmer, 1983, p. 190; Scholz, 2002, p. 37) and with the dependence of strength on grain size, in which case the crack length scales with grain size (Scholz, 2002, p. 36; Mandl, 2005, pp. 17–18). The size–strength relationship thus underlies the field of fracture mechanics (Chapters 8 and 9) and explains why coarse-grained rocks can be weaker than fine-grained rocks under the same conditions (see Chapter 8).

This size–strength relationship explains why engineers use fiberglass, or small-diameter cables, in certain components: smaller sections are usually stronger than larger ones, and aggregates of small pieces glued together are much stronger than similarly sized chunks composed of single large pieces. It also explains why small, centimeter-sized laboratory samples of rock (the values most often quoted in textbooks) are much stronger than large outcrops of the same lithology.

The reason for the relationship between size and strength can be related to the existence of **flaws**. Griffith postulated that small flaws weakened the material in an amount proportional to their size, so that larger flaws had a more damaging effect than smaller ones. His insight was proven nearly 40 years later when high-magnification scanning electron microscope images of metals and other materials clearly revealed minute crack-like flaws, which were then named "Griffith flaws" in his honor. The discrepancy between theoretical and actual shear strengths of metals led to the identification during the 1930s of the *dislocation* as the type of flaw that reduces the plastic (shear) strength of metals, in parallel with *cracks* being responsible for reducing

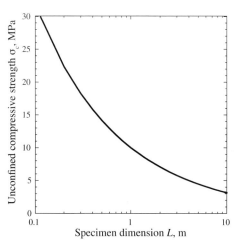

Fig. 2.20. Plot of the size–strength relation (equation (2.34)) using $k = 10$ kg m$^{3/2}$ s^{-2} and $\varsigma = 0.5$. Note semi-log scale.

the tensile strength of materials, during the period following World War II (Kanninen and Popelar, 1985, p. 33). An analogous discrepancy between the large strength of intact rocks and the much-reduced strength of rock masses led to the development of rock mass classification systems, and empirical strength criteria, in rock engineering during the 1970s (e.g., Hoek and Brown, 1980; Hoek, 1983; Bieniawski, 1989; Brady and Brown, 1993; Priest, 1993).

2.7 Flaws and the Strength of a Plate

In order to examine how flaws can weaken a rock, we first calculate the stress in a simple, rectangular plate of rock due to a tensile force acting on it. Next, we'll recalculate the stress around a circular hole in the plate, and we'll do that two ways: a simple, strength-of-materials approach and then a more sophisticated elasticity theory approach. Then we'll look into some of the implications for the strength and behavior of fractured rocks.

2.7.1 Simple Intact Plate

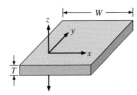

To begin we apply an in-plane tension to a plate as shown in Fig. 2.21. The x- and y-axes are shown on the surface of the plate, and the z-axis is perpendicular to those and extends in the (vertical) anti-plane direction. The plate is much larger in the x- and y-directions than in the z-direction. We also define **tension as positive** and compression negative, as is customary with elasticity theory, in this and the following exercises.

Fig. 2.21. Geometry of plate showing in-plane (*xy*) and out-of-plane (*z*) directions.

Suppose the plate (Fig. 2.22) is "thin," with a thickness T in the z-direction of 1.0 units. That way we can measure widths or distances on the plate, multiply by the thickness, and get an area ($W \times T = A$), which we'll need to calculate the stress in the plate. Let's make the plate 1 m wide.

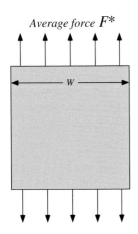

We'll pull on the plate in the y-direction with a force F^* of 10 MN. The average stress in the plate, σ^*, is the force divided by the area over which it acts. The area of the plate is the cross-sectional width (measured in the xz-plane) times the thickness, so $A = W \times T = (1.0\ \text{m}) \times (1.0\ \text{m}) = 1.0\ \text{m}^2$. So the average stress in the plate is

$$\sigma^* = \frac{F^*}{A} = \frac{10.0\ \text{MN}}{1.0\ \text{m}^2} = 10.0\ \text{MPa} \tag{2.35}$$

Fig. 2.22. Geometry of plate used to calculate the average stress within it.

The stress in the plate is an *average stress*: it has no position dependence with the same value, regardless of the xy location or position in the plate. As we'll see next, it is important and useful to identify whether the stress being considered is an average or if it varies in magnitude with position in the plate. Stresses that do not change in magnitude, like the average stress calculated above, are called *homogeneous* stresses—that is, they are constant on the scale of the material being analyzed. On the other hand, stresses that do vary with position are called *inhomogeneous*, or sometimes heterogeneous, stresses. These are considered next.

2.7.2 Circular Cavity ("Hole") in a Plate

Now let's place a circular hole in our plate (Fig. 2.23). The plate still has a width W of 1.0 m; the hole is half a meter in diameter, so that the diameter $D = W/2$. Let's focus on the area of the plate along the hole's diameter (dashed line in Fig. 2.23). The effect of the hole is to remove material from the plate that would have withstood the tensile force being applied to it. As a result, the full force of 10 MN is carried only by the two remaining pieces of the plate, which are called **ligaments**. There is one on either side of the hole, and each has a width d_1 of 0.25 m, or $d_1 = W/4$. The average stress carried by the remaining part of the plate (both ligaments) is F^*/A (where the new cross-sectional area is now $A = 2d_1$, $A = 2d_1/T$), so the stress is

$$\sigma^* = \frac{F^*}{A} = \frac{10.0\ \text{MN}}{2(0.25\,\text{m}^2)} = 20.0\ \text{MPa} \tag{2.36}$$

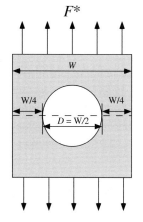

$F*$

Now the average stress is twice as high as it was before, when there was no hole. This makes sense because there is less area to carry the force, so the stress in the ligaments has to increase. The hole in the plate serves to *amplify*, or concentrate, the magnitude of stress—that's why cavities and holes are called **stress concentrators**. As we'll see, cracks, faults, and other types of structural discontinuities can also amplify the stress and serve as stress concentrators (Chapters 1 and 8). This is why they are so important in engineering design and structural geology.

In the previous example, the hole was chosen to have a diameter of one-half the width of the plate, leading to a stress of 20 MPa in the ligaments. If we change the size of the hole, the stress in the ligaments will change also. Table 2.2 shows how the value of stress for each will vary.

As you can see from Table 2.2 and Fig. 2.24, increasing the size of the hole reduces the size of the ligaments, and therefore the area over which the force acts; the average stress increases. Notice that as the hole becomes small

Fig. 2.23. Geometry of plate used to calculate the average stress within it with a circular hole.

Table 2.2 Stress changes due to cavity size

D, m	d, m	F^*/A	σ^*, MPa
0.5	0.25	(10)/(0.5)	20.00
0.75	0.13	(10)/(0.25)	40.00
0.1	0.45	(10)/(0.9)	11.10
0.001	0.4995	(10)/(0.999)	10.01

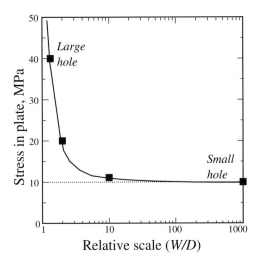

Fig. 2.24. Plot of the average stress in the ligaments due to a hole in the plate, along its diameter, illustrating the concepts of relative scale and stress amplification. Values from Table 2.2; ambient stress in the plate without the hole is 10 MPa.

relative to the size of the plate (i.e., W/D becomes large), its effect diminishes, so that for a very tiny hole (like $D = 0.001$ m above) the stress in the ligaments is almost indistinguishable from that in the case where there is no hole (our first example). This concept is called the **relative scale** of the hole to the plate.

2.7.3 Elasticity Theory Approach to the Cavity Problem

In the previous example, we saw that the average stress in the ligament near a circular hole changed with the width of the ligament, and that it was not dependent on the proximity to the hole. Now we perform a more detailed treatment of the problem that takes the *position dependence* of the stress into account. The result will be a specific, special state of stress for each point in the plate.

The approach used to obtain the state of stress around an elliptical cavity in a plate involves complex variables and partial derivatives (e.g., Westergaard, 1939; Kolosov, 1935; Paris and Sih, 1965; Rekach, 1979; Brady and Brown, 1993, pp. 174–176). Two problems will be illustrated here that show the major parts of the full solution. These are: (1) the maximum stress *at the edge* of an elliptical cavity; and (2) stress changes in the plate in relation to *proximity* to the hole.

2.7.3.1 Maximum Stress at a Cavity

In elasticity theory, each major solution in elasticity is named for the person who first solved it successfully. The stresses at the edge of an elliptical cavity were derived independently by Inglis in the United Kingdom in 1905, and his work laid the basis for engineering design in the West through the mid-twentieth century. The problem was also solved using a different

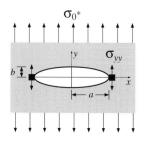

Fig. 2.25. Model setup for the Inglis/Kolosov problem.

theoretical framework in Russia by Kolosov in 1910, which was incorporated into the distinctive approach to elasticity adopted in the former Soviet Union. Since about 1960, the two approaches have joined, so that both Inglis and Kolosov share the credit for the work. The derivation of *Kolosov's problem* (1910) is well laid out in Rekach (1979), pp. 134–138 and in Timoshenko and Goodier (1970), pp. 191–194. The solution to this problem in stress analysis was later used by Griffith, Westergaard, Muskhelishvili, and Irwin among others during the mid-twentieth century to build the foundations of **Linear Elastic Fracture Mechanics**.

The problem is set up slightly differently than the previous examples. First, the cavity is represented as an *ellipse*, with the long (semi-major) axis a and semi-minor axis b. The length of the ellipse is $2a$ and its width is $2b$ (Fig. 2.25). An ellipse is used to represent the cavity because it is convenient mathematically and because it is versatile in capturing many natural situations. For example, when the semi-major and semi-minor axes are equal, the cavity becomes a circle, as in our previous example. Other ratios of a/b give us useful shapes as we'll see later on.

The elliptical cavity is placed in an *infinite* plate—a plate that is very large relative to the size of the cavity. The plate containing the cavity is loaded by a tensile stress (σ_0^*) this time, instead of a tensile force. The state of stress in the plate at the edge of the cavity may be written symbolically as

$$\sigma_{yy} = f(\sigma_0^*, a, b, x, y) \tag{2.37}$$

or perhaps more informatively,

$$\sigma_{yy} = f(\text{Remote stress, Cavity shape, Position in plate}) \tag{2.38}$$

As you can see, the state of stress at any point in the plate depends on five variables—more if we change the orientation of the uniaxial remote stress relative to the ellipse, or if we add a perpendicular component of stress (for a 2-D state of remote stress). However, a very simple and powerful result is obtained if we recast the full solution to find the *maximum* value of stress in the plate. This special value, σ_{yy}, corresponds to the points shown in Fig. 2.25, where $x = \pm a$ and $y = 0$. The result is

Inglis/Kolosov Relation

$$\sigma_{yy} = \sigma_0^* \left(1 + 2\frac{a}{b}\right) \tag{2.39}$$

This result is called the *Inglis/Kolosov Relation*. We can see that the stress at the edge of the elliptical cavity, along the semi-major axis, depends only on the value of the remote stress and the cavity shape. Because the local stress σ_{yy} has the same sign as the right-hand side of equation (2.39), if you apply a remote tensile stress, you get a tensile stress at the points shown in Fig. 2.25.

Table 2.3 Maximum stress changes at an ellipse: $a = 1.0$ m; $\sigma_0^* = 10.0$ MPa; $T_0 = 15.0$ MPa

b, m	a/b	σ_{yy}, MPa	σ_{yy}/σ_0^*	σ_{yy}/T_0
1.0	1.0	30.0	3.0	2.0
0.5	2.0	50.0	5.0	3.3
0.1	10.0	210.0	21.0	14.0
0.01	100.0	2,010.0	201.0	134.0
0.0001	10,000.0	200,010.0	20,001.0	13,334.0
0	∞	∞	∞	∞
2.0	0.5	20.0	2.0	1.33
5.0	0.2	14.0	1.4	0.93
10.0	0.1	12.0	1.2	0.80
1000.0	0.001	~10.0	~1.002	0.67

Let's insert some numerical values into equation (2.39) and see how the stresses are affected by the cavity. Table 2.3 lists the results using a value of remote stress of 10 MPa, the same value as used in the previous example of a hole drilled into a plate. The table also lists the ratio of local tension at the cavity edge to a representative value for the tensile strength of rock (last column) (see Fig. 2.26). When σ_{yy} exceeds T_0, local failure (cracking) should occur at that location. Notice that this last column, σ_{yy}/T_0, or stress normalized by strength, is a quantity known as **proximity to failure** P_f that plays a key role in determining how close the stress state is to producing failure of a rock (e.g., Segall and Pollard, 1980; Schultz and Zuber, 1994; see Section 8.4.3).

If we let $a = b$, the ellipse becomes a circle and $\sigma_{yy} = 3$ times the remote stress, or 30 MPa. As b gets smaller relative to a, the circle becomes a progressively

Fig. 2.26. Plot of the stress changes at the edge of an elliptical cavity; data from Table 2.3. Normalized maximum stress is σ_{yy}/σ_0^*; insets at top show ellipse orientations. The dashed horizontal line shows the level of stress change in the plate without a cavity present. Arrow shows value of stress concentration (3.0) for the circular hole ($a = b$). Shaded area shows where stress at cavity would exceed the tensile strength.

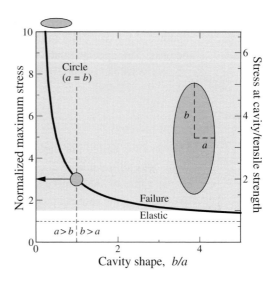

Flaw geometries:

♦ *Hole (a = b)*
♦ *Ellipse (a ≠ b)*
♦ *Crack (b = 0)*

slimmer ellipse, and the stress amplification at its semi-major axis becomes much larger. In the limit, as b approaches zero, the ellipse becomes a slit, or crack—this special case is called a *degenerate ellipse* (with $a = 1$ and $b = 0$). The stress amplification at the tip of this crack-like cavity is *infinity*. As we'll see in Chapter 8, the actual stress at a crack tip is not really infinite, but finite and directly related to the magnitudes of both the applied load and the crack length.

On the other hand, if $b > a$, then the cavity becomes an ellipse, but it is now oriented with its long axis parallel to the applied tensile stress direction. The stress at the same point ($x = \pm a$) is still larger than the applied stress, but by a much smaller amount. In the limit of a degenerate ellipse (last row of Table 2.3), corresponding to a flat crack oriented parallel to the direction of the remote tension, there is no significant stress change at the edge of the cavity at the location indicated. This means that cracks oriented parallel to the normal-stress direction have little or no effect on the stress state in the plate; only those oriented at other angles, and especially at 90° to the remote stress direction, can affect the state of stress in the plate.

2.7.3.2 Stress Changes Close to a Cavity

Let's look now at the more general problem—that of the state of stress surrounding a circular hole in an infinite elastic plate. This is *Kirsch's problem* (1898) and it is similar to, though more complex than, the Inglis/Kolosov problem. It will be useful for illustrating how the stresses near a hole or other cavity change as we get farther from the hole, and it is used extensively to evaluate the stability and failure of a borehole in petroleum (Aadnøy and Looyeh, 2010; Kidambi and Kumar, 2016) and other *in situ* stress applications (Engelder, 1993; Schmitt et al., 2012).

The problem is set up in Fig. 2.27. A remote stress is applied in the y-direction of σ_0^* (or σ_{yy}^* in previous figures) and the equations that follow also include a normal stress acting in the x-direction (Fig. 2.28) of $k\sigma_0^*$ (corresponding to σ_{xx}^*), where $k = (\sigma_{xx}^*/\sigma_{yy}^*)$ (Fig. 2.29). In order to compare these results with the Inglis/Kolosov Relation we've set $k = 0$ so that σ_{xx}^* also equals zero. The cavity has a radius of a and the position in the plate outside the hole is given by the polar coordinates (r, θ); the plate is specified with $r \geq a$.

The local stresses generated by the hole or cavity in the stressed plate are given by (Brady and Brown, 1993, p. 171–172):

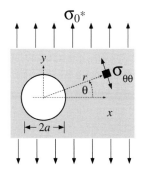

Fig. 2.27. Model setup for the Kirsch problem.

Kirsch solution

$$\sigma_{rr} = \frac{\sigma_0^*}{2}\left[(1+k)\left(1-\frac{a^2}{r^2}\right)-(1-k)\left(1-\frac{4a^2}{r^2}+\frac{3a^2}{r^4}\right)\cos 2\theta\right]$$

$$\sigma_{\theta\theta} = \frac{\sigma_0^*}{2}\left[(1+k)\left(1+\frac{a^2}{r^2}\right)-(1-k)\left(1+\frac{3a^4}{r^4}\right)\cos 2\theta\right]$$

$$\sigma_{r\theta} = \frac{\sigma_0^*}{2}\left[(1-k)\left(1+\frac{2a^2}{r^2}-\frac{3a^4}{r^4}\right)\sin 2\theta\right] \tag{2.40}$$

and the three stress components are as defined in Fig. 2.28.

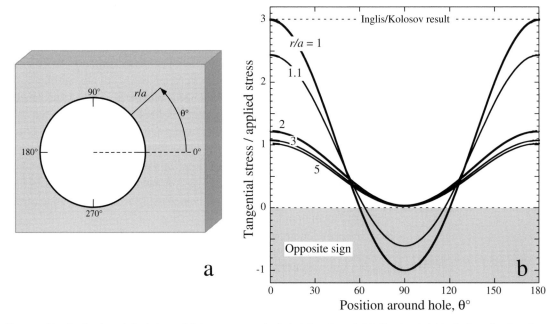

Fig. 2.28. Plot showing the variation in tangential stress $\sigma_{\theta\theta}$ near a circular hole in a stressed plate ($k = 0$; uniaxial tension) as a function of proximity to the hole. Stress changes are largest at the edge of the hole ($r/a = 1$) and decrease in magnitude with distance away ($r/a > 1$).

The state of stress in a rock mass depends on:

♦ *Applied stress*
♦ *Size of the fracture*
♦ *Proximity to the fracture*

The stress state in the plate near the hole, given by equations (2.40), depends on the ratio of the applied remote stresses (as well as their magnitudes and signs), the size of the hole, and the position in the plate—more precisely, the *proximity* to the hole. These variables are very much the same as those noted above in equation (2.38). In general, the state of stress in a rock mass containing flaws (joints and faults) depends on these same variables.

One of the three stress components from equations (2.40), the tangential normal stress $\sigma_{\theta\theta}$, is plotted in Fig. 2.28 for several normalized distances away from the hole. The tangential stress component near the hole (vertical axis) is shown divided by the constant value of remote far-field applied stress, $\sigma_0{}^*$ or $\sigma_{yy}{}^*$. This normalization process tells us how much greater (or less) the local stress component near the hole is relative to the stress in the plate in the absence of the hole (which is given by $\sigma_0{}^*$). The horizontal axis tracks the radial position around the circumference of the hole.

As can be seen from Fig. 2.28, the tangential stress at a particular constant radial distance from the hole, such as $r/a = 1.0$ (i.e., at the cavity's edge), varies depending on where you are along the circumference of the hole. For example, at $\theta = 0°$ the local stress is three times the remote stress. This point, illustrated in Fig. 2.25, corresponds to the particular solution of Kirsch's equations (2.40) given by the Inglis/Kolosov Relation (equation (2.39)). The local stress at the cavity edge here is three times whatever the applied stress may be.

As we travel around the circular hole's edge in a counterclockwise direction (corresponding to positive increases in the angle θ on Fig. 2.28a, the stress changes, decreasing in magnitude until it goes negative, or *compressive*, at $\theta = 45°$ (Fig. 2.28, right-hand panel). It attains its maximum value at the top of the cavity, at $\theta = 90°$, with a normalized value of –1: this means that the top of the cavity is being compressed in the x-direction by an amount equal to the load in the y-direction.

Just as in the case of the Inglis/Kolosov Relation, the sign of the applied stress is the same as that of the local stress—if the load is tensile, a magnified tension is produced at the cavity edge (at $\theta = 0°$) with a compression acting at the top of the cavity (shaded area in Fig. 2.28). Likewise, an applied compression acting in the y-direction on a circular cavity (perhaps generated by the weight of the overlying rock) produces an amplified compression at $\theta = 0°$ and an **induced tension** at the top, equal in magnitude to the vertical (y-direction) compressive stress.

This phenomenon is well known by the designers of tunnels and underground chambers. In this example, let's suppose that a tunnel having an approximately circular shape in cross-section is loaded from above by the weight of the rock above the tunnel. This generates an amplified compressive stress but, more importantly, it also induces a tensile stress acting tangentially to (i.e., in a horizontal direction) the top of the tunnel. This induced tension is generated even though the tunnel or chamber is far underground, and it may be great enough to cause the rock at the roof of the cavity to fail in tension.

Example Exercise 2.1

Using the elasticity (Kirsch's) solution for the state of stress around a circular cavity loaded by biaxial stress (x- and y-directions), let's calculate the induced tension at the roof of a tunnel and see if it is sufficient to promote tensile failure there.

Solution

We use the second of equations (2.38) and calculate the value of tangential stress at the roof of the tunnel (Fig. 2.29). We set the ratio of vertical to horizontal stress $k = 0.33$ (horizontal compression is taken to be one-third the vertical), a value typical of many underground situations in extensional tectonic (i.e., normal faulting) settings (e.g., Brady and Brown, 1993) having coefficient of friction of the surrounding rock of $\mu = 0.6$ (see Table 3.2). We'll let the tunnel radius be 1.0 unit. For the position around the tunnel corresponding to its roof, $r = 1.0$ and $\theta = 90°$, so that $\sigma_{\theta\theta}/\sigma_{yy}{}^* = -0.01$. If the vertical stress due to the overburden is given (as is customary in these situations) by its weight, then this vertical compressive stress $\sigma_{yy}{}^* = -\rho g z$, where ρ is the average rock-mass density and z is depth (assuming that compressive stresses are negative). Assuming hydrostatic pore-water conditions and using $\rho = 2800$ kg m^{-3}, $g = 9.8$ m s^{-2}, and $z = 300$ m, $\sigma_{yy}{}^* = -(2800 - 1000) \times (9.8) \times (300) = -5.29$ MPa (compressive). The tangential normal stress acting on the roof is then $(-0.01)(-5.29$ MPa$) = +0.05$ MPa (tension; shaded region in Fig. 2.29).

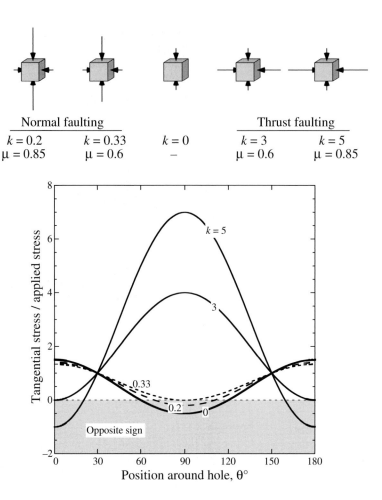

Fig. 2.29. Plot showing the variation in normalized tangential stress $\sigma_{\theta\theta}$ near a circular hole in a stressed plate ($r/a = 1.0$) as a function of stress ratio parameter k. Upper panel shows k for tectonic loading configurations.

This simple calculation shows that the roof of an underground tunnel loaded by the weight of the overlying rock (Fig. 2.30) may be subjected to a tensile stress (shaded area in Fig. 2.29) generated there as a result of the interaction between the remote stresses and the shape of the tunnel. Will the value we calculated be sufficient to fail the rock at the roof? This depends on the actual strength of the rock mass surrounding the tunnel. The tensile strength of rock masses is considerably smaller than that of the intact rock material—as much as 100 times smaller. Many rock engineers use a value of 0 MPa for the tensile strength of a rock mass, knowing that this underestimates the actual value (e.g., 0.1–5 MPa for typical rock masses) by only a small amount. As a result, we would suspect that the calculated value for the tangential stress acting along the roof of the tunnel, 0.08 MPa, may be sufficient to cause tensile failure (cracking) of the rock. In fact, it may be sufficient to cause the rock to fail dynamically and catastrophically in tension, producing an explosive failure known as a *rockburst*. A rock engineer would probably install steel netting and rockbolts into the roof to prevent its possible failure in tension (Fig. 2.30).

Fig. 2.30. Portal to underground tunnel in tuff at Hoover Dam, Nevada. Rockbolts (small dots above tunnel roof) reinforce the rock mass against failure.

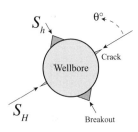

Fig. 2.31. Positions of failed wallrock surrounding a vertical circular wellbore. Remote horizontal stresses are compressive, so that $S_H > S_h > 0$.

The Kirsch solution is also widely used in the petroleum industry to place bounds on the state of stress in the subsurface (see Engelder, 1993; Jaeger et al., 2007; Zoback, 2007; Aadnøy and Looyeh, 2010; Schmitt et al., 2012). By identifying either tensile or compressive failure of rock surrounding a wellbore, the magnitude and orientation of the local *in situ* stress state in a particular stratigraphic interval can be obtained. Equations are given by Zoback (2007, p. 170) assuming now that compression is taken to be positive, and tension negative; for the case of a vertical wellbore (Fig. 2.31), the hoop stress (Fig. 2.27) at the edge of the wellbore is given by (Zoback, 2007, p. 174)

$$\sigma_{\theta\theta} = \left(S_H + S_h - 2P_p\right) \\ - 2\left(S_H - S_h\right)\cos 2\theta - P_p \tag{2.41}$$

in which S_H and S_h are the maximum and minimum compressive horizontal principal stresses in the surrounding rock, respectively; P_p is the internal fluid pressure, corresponding in this example to the mud weight at a given depth;

and θ is the angle measured from the azimuth of σ_H. Tensile failure of the rock can occur when the tangential stress component $\sigma_{\theta\theta}$ achieves its minimum (most tensile) value, whereas compressive failure (identified as borehole breakouts) occurs when $\sigma_{\theta\theta}$ attains its most compressive value (Zoback, 2007, pp. 167–205).

The type of failure can be identified by scrutinizing image logs from a well. The stress state and rock properties are thus bounded as follows (Fig. 2.31).

- **No rock failure**: orientation and magnitudes of local stress state are undetermined.
- **Drilling-induced tension cracks**: provide orientation and magnitude of the maximum compressive principal stress S_H where $\sigma_{\theta\theta}$ is a minimum (i.e., $\theta = 0°$ in Fig. 2.29); $\sigma_{\theta\theta} \geq$ the rock's **tensile strength** T_0 at this location. The location of these cracks indicates the direction of σ_H.
- **Borehole breakouts**: orientation and magnitude of the maximum compressive principal stress S_H where $\sigma_{\theta\theta}$ is a maximum (i.e., $\theta = 90°$ in Fig. 2.29); $\sigma_{\theta\theta} \geq$ the rock's **unconfined compressive strength** σ_c at this location. Borehole breakouts are oriented at about 90° to drilling-induced tension cracks and S_H depending on the degree to which the unconfined compressive strength of the rock adjacent to the wellbore was exceeded by the *in situ* stress state.

Calculation of stress state around a wellbore becomes complicated when the wellbore is deviated (i.e., inclined) from the vertical, and/or when the local stress state is also no longer aligned with the vertical axis, as can occur in areas of active tectonics or near faults or stratigraphic contacts in the subsurface. Differences in temperature between drilling fluid and host rock also contribute to the stress state near a wellbore. Standard treatises such as Engelder (1993), Zoback (2007), and Schmitt et al. (2012) provide detail on the methodology in such cases.

Example Exercise 2.2

Using the following values of pore pressure and remote stress state, determine whether drilling-induced tensile cracks or borehole breakouts would form adjacent to a vertical wellbore: $S_H = 78$ MPa, $S_h = 51.8$ MPa, and internal pore pressure (here taken to be equal to the mud weight) of 28 MPa. Assume values for rock tensile strength of $T_0 = 7.5$ MPa and an unconfined compressive strength $\sigma_c = 94$ MPa. What would the orientation of any resulting failure surface be?

Solution

Inserting the values for S_H, S_h, and internal fluid pressure into equation (2.41), we calculate the hoop stress as a function of angle θ. The maximum value of hoop stress occurs at $\theta = 90°$, in the direction of S_h, where $\sigma_{\theta\theta} = 98.2$ MPa (compression positive). Because this value exceeds the unconfined compressive strength of the rock (94 MPa), we would expect to see evidence of a borehole breakout in an image log of the wellbore wall. The minimum value of $\sigma_{\theta\theta}$ occurs at $\theta = 0°$, parallel to the direction of S_H, with a magnitude of –6.6 MPa. Since the rock has a stated tensile strength of –7.5 MPa, drilling-induced tension cracks should not be expected. However, if we do get these cracks in this orientation, then we would know that the value of tensile strength listed for the rock was in error, and would then more likely be less than |6.6|MPa for tensile failure in this area to have occurred.

The state of stress in the Earth's lithosphere has been determined by well-bore-based techniques during scientific drilling programs (e.g., Zoback and Healy, 1984; Zoback et al., 2003), resulting in the World Stress Map (Zoback, 1992; Heidbach et al., 2010; see www.world-stress-map.org for the current compilation) and definition of the critically stressed nature of the Earth's crust (e.g., see discussion by Zoback, 2007, pp. 127–137). *In situ* stress states determined in this way are also used routinely to design what are called completion strategies that create hydrofractures (Haimson and Fairhurst, 1967; Engelder, 1993, pp. 131–155) in particular reservoir intervals to aid in extraction of oil and gas (e.g., Cook et al., 2007; Kidambi and Kumar, 2016).

2.7.3.3 Maximum Stress and Distance from the Cavity

In the previous section we calculated the tangential stress component $\sigma_{\theta\theta}$ for a given radial distance from the hole ($r = 1$, actually right at the cavity wall) as a function of the angle θ. This allowed us to reproduce the Inglis/Kolosov result for one point on either side of the cavity and to see that the local stress actually changes sign, as well as magnitude, at the cavity roof. The sign change might be enough to permit tensile failure of the roof in an otherwise compressive stress field.

The final stage in examining the stress changes due to a cavity is to look at the radial variation of stress with distance from the hole. We'll again use the Kirsch solution (equations (2.40)), but this time plug in a constant value of angle into the expression for $\sigma_{\theta\theta}$. Using $\theta = 0°$, we solve for $\sigma_{\theta\theta}$ for a range of values of r. This gives us values of $\sigma_{\theta\theta}$ as we move out away from the hole along the x-axis (Fig. 2.27).

The results are shown in Fig. 2.32. At the cavity's edge, corresponding to a distance $r/a = 1.0$, and for $\theta = 0°$, we see that the value of the local stress, normalized by the value for the remote or far-field stress, is 3.0—the same result as we found for the Inglis/Kolosov Relation for the circular hole ($a = b$). This means that the local stress caused by the presence of the circular cavity, at its edge, is three times that of the applied stress. But as we get farther from the hole, the normalized stress decreases markedly. For a distance of one cavity radius away from the hole ($r/a = 2$), the local stress is only 20% greater than the applied stress. As the distance from the cavity increases further, the local stress decreases so that it is within 10% of the remote stress at $r/a = 2.5$ and 5% at $r/a = 3.5$. We can see that beyond $r/a = 5$ or so, the local stress is affected by the presence of the cavity by a negligibly small amount (2%). The

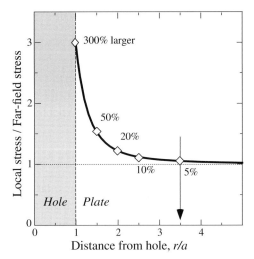

Fig. 2.32. Decrease in the local tangential stress due to a circular cavity with increasing distance from the cavity's edge. The arrow shows the approximate boundary between the local, inhomogeneous stress state (near the cavity) and the homogeneous stress state at larger distances.

cavity doesn't influence the stress beyond about 2–3 cavity diameters away, but within that zone the local stress is strongly affected by the cavity and the stresses are inhomogeneous there.

The results of the cavity analysis in this section have been stated as ratios of the distance from the cavity r and the cavity size, or radius, a. Let's plug in some values of cavity size and see how far away we have to go to escape the cavity's influence. For a pore in sandstone, the radius might be some fraction of the grain size, say 1 mm. This means that a stressed sandstone will have complex, inhomogeneous stresses within an area of about 3 mm surrounding the pore. In an outcrop, a train tunnel having a diameter of 10 m would influence its surroundings as much as 30 m away. A subterranean chamber containing a hydroelectric power station with a diameter of 200 m would alter the state of stress in a region of perhaps 500 m, or half a kilometer (0.3 miles), in all directions. We can see that the relative scale of a fractured rock mass really depends on the scale of the flaw that changes the stress state around it from average and well-behaved to inhomogeneous and variable.

2.8 Review of Important Concepts

1. Testing of rocks in the laboratory is the first step in determining the parameters under which the rock or rock mass had been deformed in the field. These parameters help to define several idealized rheological models that can approximate the behavior of the rock.
2. Several types of laboratory tests are widely employed, including uniaxial (or unconfined), triaxial, and polyaxial compression tests. Each of these is used to simulate particular sets of natural conditions in the laboratory.

3. A consistent suite of terms that describe aspects of rock deformation are commonly used in structural geology, tectonics, and rock engineering. Some of the most important terms include: elastic, linearly elastic, viscous, plastic (both the ideal material and the deformation mechanism), brittle, cataclastic, ductile, strong, weak, stiff, soft, strain hardening and strain softening, and various combinations of these.

4. Stresses and strains within a rock mass can be calculated by using the appropriate forms of Hooke's Law. These relations are especially useful if you impose a strain on a rock mass and want to know the stresses that are induced as a result of the applied deformation. Although the full 3-D expressions are not too difficult, the simpler 2-D set of expressions can be used if the physical problem can be represented as a 2-D one, invoking either plane stress or plane strain.

5. Many rocks can be represented by elastic materials for which the strain caused by an applied stress is fully recoverable. The elastic parameters E, G, S, and v work well in rock engineering and rock mechanics studies as long as rocks stay within their elastic range. As rocks become damaged progressively and crack or joint networks develop, the elastic parameters can still be used but their values will depend on the aggregate characteristics of the fracture networks.

6. The **Complete Stress–Strain Curve for Rock** is an elegant and compact way to represent and understand the response of a small rock sample, or field-scale outcrop, to a load. It can also be thought of as a tool for working out a progressive evolutionary sequence that may be recorded in a rock mass.

7. The theoretical strength of a perfect, flawless rock is some 1000 times greater than the actual strength that would be measured in the laboratory. The **theoretical strength is too large** because it ignores the effect of stress concentrating flaws that are typically present in rock.

8. The **size–strength relation** states that small rock samples are commonly stronger, not weaker, than larger samples. This means that fractured outcrops are generally not as strong as, and are also less stiff than, laboratory samples subjected to the same conditions of stress, pressure, temperature, and strain rate.

9. **Relative scale** is a fundamental concept that describes how closely the rock mass approximates a well-behaved, continuous, homogeneous material that has statistically uniform properties throughout. Thus *both* small intact rock samples and large fractured outcrops can *be considered as continuous materials* given the correct scale of observation.

10. The stress applied to a rock that contains a circular hole or cavity is amplified by three times at the cavity edge. The stress amplification

increases from three as the cavity flattens into an ellipse. The shape of the cavity in a rock has a profound, and predictable, influence on the level of inhomogeneous stress in the rock. This result is called the **Inglis/Kolosov Relation**. The stress amplification at the tip of a very thin ellipse, representing a crack, fault, or other type of structural discontinuity, becomes extremely large. This means that cracks and faults, for example, can serve as effective concentrators of stress in a fractured rock mass.

11. The values of the stress components *change with position* along the edge of a cavity in rock. In particular, the local, amplified stress can change from tension to compression along the cavity. Thus cavities or flaws in rock subjected to the normally compressive stress states in underground situations may generate a local tension that is sufficient to drive localized tensile cracking or failure, despite the overall compressive regime. Cavities and other types of flaws in rock are thus said to also *redistribute* the stresses in their vicinity.

12. The local, inhomogeneous stresses near a flaw in rock decay, or decrease in magnitude, with distance from the flaw. Thus **proximity to the flaw**, as well as the flaw's size and shape, are the important factors when it comes to characterizing the stress state in the rock mass. A limit of 3 cavity diameters is a reasonable value for the volume of the rock mass that is affected by the cavity's presence.

13. The state of stress in an outcrop can be thought of as having two parts: a *homogeneous* one, where the average stress does not vary with position or location, and an *inhomogeneous* one, where the magnitudes, orientations, and signs of the stress components do vary with location. **Mohr circles describe the homogeneous stress, but cannot represent the complicated inhomogeneous stresses** near a cavity, for example, except on a point-by-point basis.

14. Geologic structural discontinuities such as joints and faults interrupt the normal continuity of the rock mass; they can serve as **flaws**, or stress concentrators, that weaken the rock mass relative to an unfractured, intact rock sample. These "flaws" are not "imperfections" in the rock but are natural features; **geologic rock masses are inherently fractured**, making them a different type of material than steel, plastic, and other types of flawless, engineered materials.

2.9 Exercises

1. Define and contrast plane stress and plane strain, using both words and equations for the appropriate boundary conditions. Draw diagrams illustrating the differences in physical interpretation. Give an example showing how you would use these in engineering design. What differences occur in the equations for plane stress and plane strain?

2. Discuss the conditions under which a rock may respond elastically to stress. (Note: simply stating "high strain rates or cold temperatures" is not sufficient, although aspects of these may appear in a correct answer.)

3. Define Young's modulus and describe how values are obtained. For which geologic situations would you want to use this parameter?

4. (a) Suppose you apply a uniaxial compressive stress to a rock along the x-direction of 2.5 MPa. Using $E = 10$ GPa and $v = 0.25$, calculate the strains in the three principal directions and sketch the results. (b) Now add a confining stress in the y-direction. Recalculate and reinterpret the results.

5. Given the results from Exercise 4, calculate the volumetric strain for parts (a) and (b).

6. Using $E = 10$ GPa and $v = 0.25$, calculate the elastic shear modulus, stiffness, and bulk modulus of the rock.

7. Suppose you are given measurements of strains in the x-, y-, and z-directions from a geologist colleague's summer field season of 0.3%, 1.12%, and 15%, respectively. Using $E = 3$ GPa and $v = 0.2$, calculate the stresses induced in the rock mass in all three directions. Remember to use the actual numerical values for strain, not the convenient forms listed here.

8. Compare and contrast Young's modulus and deformation modulus. How would you determine when you should use either parameter?

9. Define peak strength and discuss how the stiffness of the testing machine or the surrounding rock mass can influence this value.

10. Describe how localization of strain within a deforming rock mass would be expressed in its overall response depending on the relative scale of the localization and the rock mass. Under what conditions would the deformation become unstable, both for the zone and for the rock mass?

11. Using the equations for stress variations near a cavity (the Kirsch solution), calculate the radial, tangential, and shear stresses on the periphery of an elliptical cavity in an elastic plate as a function of position angle θ (counterclockwise positive) for the entire circumference (θ ranging from 0° to 360°) for the two cases listed below. Discuss the significance of your results. Draw an accurate diagram of each case, showing all relevant parameters. A plot relating mean stress to position angle, showing all three cases, would be helpful.
   (a) $a = 1.0$ m, $r = a$.
   (b) $a = 1.0$ m, $r > a$ (just pick a value for r and go for it).
   (c) $a = 1.0$ m, $\theta = 0°$. Plot $\sigma_{\theta\theta} = \sigma_{yy}$ for $1 \leq r/a \leq 10$. Define the region of significant stress perturbation due to the hole as a function of the normalized distance away from it. Assume that stress changes of <5–10% from the remote, homogeneous, far-field values in the plate are negligible.

12. Using the the Inglis/Kolosov Relation, calculate and plot the stress concentration factor, σ_{yy}/σ_0^*, for an elliptical cavity having $0 \leq b/a \leq 1.0$. Produce a second plot for the opposite case (long axis of ellipse parallel to the tension direction) using $1.0 \leq b/a \leq 1000$. Draw an accurate diagram for each case, showing all relevant parameters. Discuss the significance of your results. What happens for $b = 0$ and what does this case approximate?

13. Summarize the effect of size on strength, based on the material given in this chapter. Why does this relationship exist?

14. Define "flaw" and discuss its significance for the related terms "discontinuity," "fault," "crack," and "fracture."

15. Using the equations in this chapter, define "regional stress" as it is used in tectonics and in relation to a crustal-scale fault of your chosen dimension. Discuss some of the implications for the *in situ* stress state and for geodetic measurements of crustal deformation in a faulted domain.

16. Suppose you're asked to design an underground tunnel in the extensional Basin and Range province of western North America. Would you orient the long axis of the elliptical tunnel opening vertically or horizontally, and why?

17. Using the same values of rock strength and *in situ* stresses as in Exercise 1, what would be the effect of increasing the internal pressure (corresponding to increasing the mudweight in the borehole)? How could you adjust the mudweight to prevent either drilling-induced tension cracks or borehole breakouts under these stress conditions?

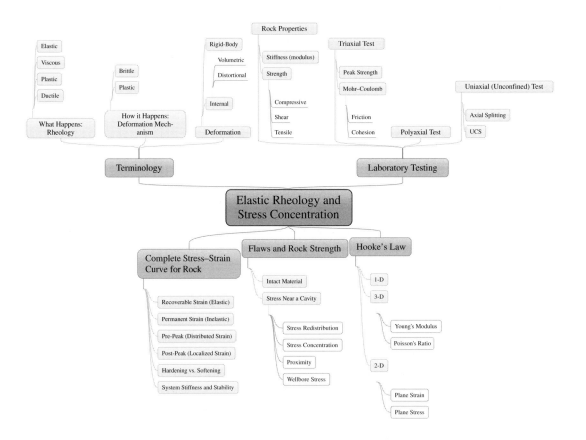

3 Stress, Mohr Circles, and Deformation at Peak Strength

3.1 Introduction

This chapter provides a synopsis of the use of 2-D (two-dimensional) stress in rock deformation. First, we'll look at how to apply these simple equations to problems in Coulomb frictional sliding along surfaces in rock. Then we'll introduce two other very useful, but somewhat more involved, failure criteria for rocks. By the end of the chapter you should be able to start with the stresses on a small piece of intact rock, know how to deal with either a simple Coulomb slip surface or a crack, and then apply this to understand the field-scale characteristics of large-scale fractured outcrops (Fig. 3.1).

In contrast to the **Coulomb criterion**, developed for frictional sliding along surfaces and valid only for compressive normal stresses, the **Griffith criterion** is based on a stability analysis of cracks loaded in tension. It provides a useful envelope for tensile normal stresses that dovetails into the Coulomb envelope. Many have used the Griffith criterion to infer the depth of jointing in the Earth's crust, for example. Following this, we'll investigate the **Hoek–Brown criterion**, which is used extensively to represent the shear strength of either intact rock or, more commonly, a large-scale, fractured rock mass. Simpler and more compact than the Griffith criterion, the Hoek–Brown criterion is perhaps the best empirical approach available that deals with the failure of rocks under either tensile or compressive stresses and for incorporating their fractured, weakened state that we encounter in outcrops. Hoek–Brown characterizes the strength and deformability of rock masses in the upper 1 km of the crust, below which a lithospheric strength envelope, based on the Coulomb criterion, becomes valid and appropriate. All three strength criteria (Coulomb, Griffith, and Hoek–Brown) require that one can calculate stresses in 2-D and compare them to an equation, and the skills needed to work with any of these are given in this chapter.

There are several good treatises available that provide a thorough introduction to 2-D stress analysis and/or Mohr circle problems (e.g., Johnson, 1970; Means, 1976; Fossen, 2010a,b). The structural geology text by Twiss and Moores (2007) gives a comprehensive exposé on the Mohr diagram, listing

Fig. 3.1. Numerous structural discontinuities, having a variety of orientations and spacings and caused by multiple episodes of faulting and jointing, control the rock-mechanical characteristics of this fractured rock mass.

the major properties and conventions. Clear discussions of the frictional properties of rock surfaces are given by Suppe (1985), Marone (1998b), and Scholz (1998, 2002). Derivations of the many key relationships for stress and rock friction are given in the works of Jaeger et al. (2007) and Price and Cosgrove (1990). Proficiency in stress and Mohr circle problems can only be gained by solving lots of problems in all permutations; any text on engineering strength of materials, such as Beer et al. (2004), can provide both worked examples and homework problems.

The material in this chapter draws from these and other sources, as well as from teaching experience with the material in upper division and graduate level university classes. The intent of this chapter is to present a streamlined coverage of the major operations provided to help either students or practicing geologists who deal with frictional slip problems to rapidly gain a basic working knowledge. The objective of this chapter is to help the reader build up a way to *reason out the correct answers physically*. It is useful to be able, with practice, to visualize what is being said with stresses and rotations in reality (called "physical space"), points and angles on the Mohr circle, and the equations themselves.

Elastic stresses and Coulomb sliding predict slip on potential planes. Once the planes slip, and displacements accumulate, the Mohr circle approach becomes invalid.

It is worth discussing some aspects of this approach to frictional sliding at this point. Elastic stresses are derived quantities, because they are calculated from other quantities (such as forces and areas or angles) that are real and measurable. The values of stress can be used either to understand past behavior or to predict future behavior. In either case, the stresses are guesses; although slip along a surface may be *predicted* from a Mohr–Coulomb analysis, the slip itself cannot be characterized by the simple relationships in this approach—only the predicted initiation of slip. Other ambiguities are built into the problem, including being able to specify only the orientations of *potential* slip planes, but not the actual planes themselves. As a result, the Coulomb criterion (or the related Griffith and Hoek–Brown criteria) cannot state which of a set of planes will slip, how much slip will occur and at what rate it will accumulate, or what happens in the rock after a particular surface slips. These questions can be addressed by using a separate set of tools, derived from fracture mechanics. However, even the simplest type of Coulomb slip problem—"will the fracture slip"—can be a very important and useful one in both structural geology and engineering design.

3.2 Sign Conventions

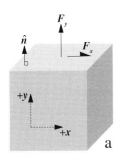

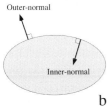

Fig. 3.2. (a) Use of force directions (force arrows F_y and F_y pointing in positive coordinate directions) and outer-normal directions (arrow n pointing in positive y-direction) to define stresses acting in the vertical (xy) plane. (b) Definition of outer-normal vectors for the tension-positive sign convention, and inner-normal vectors for the compression-positive convention.

In rock mechanics, soil mechanics, and structural geology, compressive stress is commonly taken to be positive, and tension negative—this can be called the **soil mechanics convention**. This convention is different than the tensor, or tension-positive convention, used in other types of stress analysis such as fracture mechanics. By keeping track of which type of problem you are solving, and its inherent sign conventions, you can switch back and forth as needed.

In the compression-positive sign convention used extensively in this chapter (and traditionally used for Coulomb friction problems), the greatest compressive stress is called σ_1, the intermediate stress σ_2, and the least compressive stress σ_3. If some or even all of the stresses are tensile, then σ_1 is the least tensile stress and σ_3 is the greatest tensile stress. On the other hand, in the tension-positive sign convention (called the **tensor convention**), the greatest compressive stress is called σ_3, the intermediate stress σ_2, and the least compressive (or most tensile) stress σ_1.

Using the tension-positive sign convention, a positive value of stress is defined by a two-fold criterion, that is, having the *force acting in the positive direction* and the *outer-normal vector* (Fig. 3.2a), which defines the orientation of the plane or surface on which the stress acts, *pointing in the positive coordinate direction* (Fig. 3.2a). Normal stresses that are positive according to this criterion are called tensile and the arrows point away from each other.

Shear stresses that are positive using this convention can be visualized using a simple graphical aid (Twiss and Moores, 2007). On the stress square, the top surface is the one to watch. If the *shear stress is positive*, the sense of shear will be *right-lateral*; for negative shear stress, it will look left-lateral (see Table 3.1 below). This visual aid works well for rotation angles of that plane of less than 45°; for more than that you can still use it, but with some care. For the compression-positive sign convention, one can define an inner-normal vector (Fig. 3.2b) to reverse the signs and produce positive

Table 3.1 Sign convention summary

Convention	σ_n	τ	θ or α	When used?
	C+	LL+	ccw+	
Soil mechanics (**compression** positive)				**Structural Geology, Rock Mechanics** Coulomb frictional sliding *In situ* stresses Fault-slip inversions Soil mechanics Initial brittle fracture type
	T+	RL+	ccw+	
Tensor mechanics (**tension** positive)				**Fracture Mechanics** Joint dilation and propagation Mixed-mode crack growth Computer simulations of rock-mass deformation

values for compressive normal stress and left-lateral shear stress, or else define this convention using an outer-normal vector which points in the *negative* coordinate direction (e.g., Crouch, 1976, p. 23).

Let's review how to name a stress component. Each stress, whether normal or shear, has a letter and two subscripts. Sigma (σ) is usually reserved for normal stress, and tau (τ) is used for shear stress; sometimes you'll see the normal stress written as σ_n, which is used to reinforce the fact that this is a normal stress. Stress is defined as force per unit area, or $\sigma = F/A$. Referring to Fig. 3.2a, we see a force F_y acting in the vertical or y-direction, and the plane's outer normal vector n also acting in the positive y-direction. Using the **on-in criterion** (the plane *on* which the force acts, *in* which direction), the stress is $+\sigma_{yy}$, a tensile stress (Fig. 3.3). Similarly, Fig. 3.2a defines a shear stress with an outer normal in the positive y-direction (the first subscript) and a force direction F_x in the positive x-direction; this gives us a shear stress called τ_{yx}. Again, its sign is positive (right-lateral shear) according to the tension-positive sign convention. In all cases and for both sign conventions, we will define **counterclockwise angles as positive** and clockwise angles as negative.

The last thing we need to do is talk about repeated subscripts. The normal stresses all have repeated subscripts, like these: σ_{xx}, σ_{yy}, σ_{zz} and for the principal stresses, σ_{11} and σ_{22}. These last two are pronounced sigma one-one (not "eleven") and sigma two-two (not "twenty-two"). Because it is understood that these subscripts are repeated, the normal stresses are typically written with only one subscript, like σ_y or σ_2, with the understanding that this normal stress really has two subscripts. The shear stress subscripts cannot be condensed like those for the normal stresses, so we retain both to show τ_{xy} or τ_{12};

Fig. 3.3. Use of the *on-in* criterion to name stresses in association with the planes on which they act.

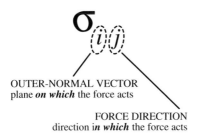

sometimes we choose to omit the subscripts for brevity and use the symbol τ by itself for the shear stress, understanding that the *on-in* subscripts are still included.

3.3 Equations for Stresses

The simplest way to calculate the stresses acting on a plane of interesting orientation is to start with the *principal stresses* as those that act on the material from far away. These are called remote, regional, or **far-field**, stresses to contrast with *local stresses* (discussed in Chapter 8) that may be associated with fault slip, deformation band shearing, or crack opening.

Remember that the **principal stresses** are *normal stresses that act on planes on which shear stress equals zero*; these are also known as the **principal planes**. The principal stresses represent the *maximum and minimum values* of normal stress that you can obtain for any problem, so that the values of normal stress that you may calculate must be *intermediate* between the values of the principal stresses. This provides a check on your answers, regardless of whether you used equations or a Mohr circle to obtain them.

3.3.1 Equations for 2-D Stress

We begin by writing down the equations that you'll need to calculate the two principal stresses from an arbitrary remote state of stress consisting of two normal stresses, σ_x and σ_y, oriented parallel to the *xy*-coordinate axes (Fig. 3.4), and one independent shear stress, $\tau_{xy} = \tau_{yx}$. The principal stresses, expressed here as the maximum and minimum algebraic normal stresses (assuming compression positive) are

$$\sigma_{max,\,min} = \left(\frac{\sigma_x + \sigma_y}{2}\right) \pm \sqrt{\left(\frac{\sigma_x - \sigma_y}{2}\right)^2 + (\tau_{xy})^2} \qquad (3.1)$$

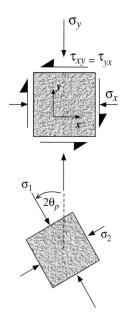

Fig. 3.4. Problem setup for calculating the maximum and minimum (principal) normal stresses.

and the orientations of the two principal planes are found by using

$$\tan 2\theta_p = \frac{2\tau_{xy}}{\left(\sigma_x - \sigma_y\right)} \qquad (3.2)$$

Example Exercise 3.1

Given values of $\sigma_x = 1.2$ MPa (compression positive), $\sigma_y = -0.85$ MPa, and $\tau_{xy} = 0.45$ MPa, find (a) the principal stresses, and (b) predict the orientations of the principal planes on which the principal stresses act.

Solution

(a) The problem is drawn in Fig. 3.4. Substitute the values of normal and shear stress into equation (3.1) to obtain:

$$\sigma_{max, min} = \left(\frac{1.2 + (-0.85)}{2}\right) \pm \sqrt{\left(\frac{1.2 - (-0.85)}{2}\right)^2 + (0.45)^2}$$

$$= 0.175 \pm \sqrt{1.051 + 0.203}$$

$$= 0.175 \pm 1.119$$

$$\therefore \sigma_{max, min} = 1.294 \text{ MPa and } -0.944 \text{ MPa}$$

(b) Now substitute the same initial values of stress into equation (3.2) to obtain

$$\tan 2\theta_p = \frac{2(0.45)}{\left(1.2 - (-0.85)\right)}$$

$$\tan 2\theta_p = 0.429$$

$$2\theta_p = 23.2°$$

$$\therefore \theta_p = 11.6° \text{ and } 101.6°$$

Remember that there are two sets of principal planes, so the answer to part (b) must include two angles: the calculated one (11.6°) plus another at that angle plus 90°. The maximum stress, 1.294 MPa, is σ_1 and the minimum stress, −0.944 MPa, is σ_2. Remember that the angle is positive for a counterclockwise sense of rotation (Fig. 3.4).

3.3.2 Resolving Stresses on a Plane: 2-D

Now we want to calculate, or *resolve*, the normal and shear stresses σ_n and τ on a particular orientation (that can represent a physical plane such as a joint, fault, or bedding plane) due to remote loading by a pair of principal stresses, σ_1 and σ_2. Recall that σ_1 here is the maximum compressive (or least tensile) stress, according to the compression–positive sign convention. The problem setup is shown in Fig. 3.5. Note the two angles that are used: α is the angle between the direction of maximum principal stress σ_1 and the *plane* of interest, while θ is the angle between σ_1 and the *normal* to the plane (Twiss and Moores, 2007, p. 176). These angles are defined to be more convenient in certain problems.

The stress transformation equations for 2-D, compression positive, are

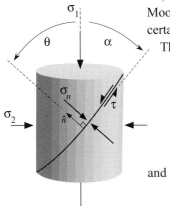

Fig. 3.5. Problem setup for calculating the normal and shear stresses on a given orientation from the remote principal normal stresses.

$$\sigma_n = \left(\frac{\sigma_1 + \sigma_2}{2}\right) + \left(\frac{\sigma_1 - \sigma_2}{2}\right)\cos 2\theta$$

$$\tau = \left(\frac{\sigma_1 - \sigma_2}{2}\right)\sin 2\theta$$

(3.3)

and

$$\sigma_n = \left(\frac{\sigma_1 + \sigma_2}{2}\right) - \left(\frac{\sigma_1 - \sigma_2}{2}\right)\cos 2\alpha$$

$$\tau = -\left(\frac{\sigma_1 - \sigma_2}{2}\right)\sin 2\alpha$$

(3.4)

Example Exercise 3.2

Let's try these equations out. We choose σ_1 = 5 MPa, σ_2 = 1 MPa, α = 30° (to the plane), and θ = 60° (to the plane's normal). Using equations (3.3) we get σ_n = 2 MPa (compressive) and τ = 1.7 MPa (positive, left-lateral). Using (3.4), noting that α = −30°, we also find that σ_n = 2 MPa (compressive) and τ = 1.7 MPa (left-lateral). We see that the answer we obtain is the same, regardless of how we define our angle.

Looking at the second term in the equations for normal stress, we can see why. Using θ, σ_n = 3 + 2 cos (120°) = 3 + 2 (−0.5). But by using α, σ_n = 3 − 2 cos (−60°) = 3 − 2 (0.5). The sign between the first and second terms compensates for the sign of the cosine term. It turns out that using (3.3) with the angle to the plane's *normal* vector is more convenient for the Mohr circle construction than (3.4).

3.3.3 Equations for 3-D Stress (pseudo 3-D)

The stress transformation equations for 3-D, using the same compression-positive sign convention, are obtained from equations (3.3) and (3.4) by substituting the least principal stress, σ_3, into them instead of σ_2. Then you can use the 3-D equations by noting that you're evaluating the largest stress difference—that between the greatest and least principal stresses. The intermediate principal stress σ_2 is implicit and doesn't enter into the calculations, although it can be very important to consider explicitly for some 3-D problems.

The equations now become

$$\sigma_n = \left(\frac{\sigma_1 + \sigma_3}{2}\right) + \left(\frac{\sigma_1 - \sigma_3}{2}\right)\cos 2\theta$$

$$\tau = \left(\frac{\sigma_1 - \sigma_3}{2}\right)\sin 2\theta$$

(3.5)

and we can also define a parallel pair of equations that uses the angle α to the plane.

3.3.4 Equations Using Tension Positive

For some problems you will want to calculate stresses using the other (tension-positive) sign convention. Many applications in civil or mechanical engineering (such as finite element analyses) and **fracture mechanics** (for example, crack propagation; Chapter 8; see also Pollard and Fletcher, 2005)) utilize this convention, so it is important to be able to cross over seamlessly between either one.

In the tension-positive sign convention, the subscripts of stress are positive when the force and outer-normal directions both point in the same, positive coordinate directions. As a result, tension and right-lateral shear stress are both positive in this convention.

The normal and shear stresses for the tension-positive convention in 2-D are calculated by using

$$\sigma_n = \left(\frac{\sigma_2 + \sigma_1}{2}\right) - \left(\frac{\sigma_2 - \sigma_1}{2}\right)\cos 2\alpha$$

$$\tau = -\left(\frac{\sigma_2 - \sigma_1}{2}\right)\sin 2\alpha$$

(3.6)

and the angle involved, α, is that from the maximum compressive principal stress direction (now called σ_2 instead of σ_1) to the plane. You can also write the expressions using θ or σ_3 for 3-D stress as done in the next section.

Example Exercise 3.3

Let's verify that these equations (3.6) give us the same answers as those from before. If we set the maximum compressive principal stress $\sigma_3 = -5$ MPa, the minimum compressive principal stress $\sigma_1 = -1$ MPa, and the angle between the plane and σ_3 of $\alpha = -30°$, we find that $\sigma_n = -2$ MPa (negative, compressive) and $\sigma = -1.7$ MPa (negative, left-lateral).

Notice in the above example exercise that, although the signs of the normal and shear stresses resolved onto the plane are opposite to those from the previous example, the absolute magnitudes of these stress components (2 MPa and 1.7 MPa) are the same. Note also that the directions of the stresses are the same too—compressive and left-lateral. So the physical problem doesn't change simply because you alter your sign convention. It also means that you can go from one sign convention to another just by changing the signs of all your input stresses before using the new set of equations. We'll do this when we take the results from a fracture mechanics exercise, in which the stresses are set up as tension-positive, and plot them on a Mohr diagram, which will use the compression-positive convention for consistency.

3.3.5 Summary of Stress Conventions

Let's summarize and compare these equations. We've seen equations for two angle conventions (α and θ), and any of the equations can be written using these. We've also written down the equations for 2-D and 3-D, and two sign conventions: compression-positive (let's call this C+ here) and tension-positive (T+; see Table 3.1). Let's see how it works.

$$\sigma_n = \left(\frac{\sigma_1 + \sigma_2}{2}\right) + \left(\frac{\sigma_1 - \sigma_2}{2}\right)\cos 2\theta \quad \text{2-D, C+,}\theta$$

$$\sigma_n = \left(\frac{\sigma_1 + \sigma_3}{2}\right) + \left(\frac{\sigma_1 - \sigma_3}{2}\right)\cos 2\theta \quad \text{3-D, C+,}\theta$$

$$\sigma_n = \left(\frac{\sigma_2 + \sigma_1}{2}\right) + \left(\frac{\sigma_2 - \sigma_1}{2}\right)\cos 2\theta \quad \text{2-D, T+,}\theta$$

$$\sigma_n = \left(\frac{\sigma_3 + \sigma_1}{2}\right) + \left(\frac{\sigma_3 - \sigma_1}{2}\right)\cos 2\theta \quad \text{3-D, T+,}\theta$$

First, all these equations are for the normal to the plane, angle θ. Second, the 2-D equations have σ_2, and the 3-D stress equations have σ_3. Also, look at the order of the stresses: for compression positive, σ_1 is always *first* (regardless of 2-D or 3-D), whereas for tension positive, σ_1 is always second. The same rules apply for the shear stress equations, so the equations haven't been listed again here.

Table 3.1 lists the sign conventions for the two normal stresses, the one (independent) shear stress, and the angle for both sign conventions that we'll use. A visual and geologic device, shown in Table 3.1, may be useful for remembering the sense of shear stress, on the horizontal faces of the stress square shown in Fig. 3.2, that is positive for each convention (e.g., Twiss and Moores, 2007, p. 158). These all work for either 2-D or 3-D stress.

As you can see, these sign conventions are easy to keep straight; if you forget them, you can always reconstruct what they should be by remembering how they are defined in Fig. 3.3, with the plane-force direction subscripts, when multiplied together, being positive for the tensor convention or negative for the soil mechanics convention.

3.3.6 The Mohr Circle for Stress

Developed by Otto Mohr (1835–1918), a German engineer, the Mohr circle technique is a simple and powerful graphical method for solving the stress transformation equations. It is equivalent to using a *stereonet* to find the intersection line between two planes oriented in space, instead of using 3-D mathematics. Elegant trigonometric identities and algebraic manipulations were used perhaps a century ago to simplify the solution of stress problems because calculators and computers were not yet available. Today, the Mohr circle is still very useful, if not indispensable, as a way to visualize complex stress states and for serving as a straightforward and rapid check on our calculations.

Let's look back over some of the equations for 2-D stress that we've covered. Equation (3.1) showed us how to take a set of normal and shear stresses and calculate the maximum and minimum normal stresses—the principal stresses—that act on specially oriented planes—the principal planes— which also have no shear stress acting on them. To do this we physically rotated the stress square a certain number of degrees—angle $2\theta_p$ (equation (3.2))—to simultaneously zero out the shear stresses, and maximize the normal stresses, on these special planes (see Fig. 3.4).

Let's look at equations (3.3). Here we used the principal stresses σ_1 and σ_2 (along with the compression-positive sign convention) to calculate, or *resolve*, these remote stresses onto a certain plane or orientation of interest. But see how systematic the terms of the equations are? In the expression for normal stress on a plane, the first term, $(\sigma_1+\sigma_2)/2$, is a simple average. This is called the **mean stress**. The next term, $(\sigma_1 - \sigma_2)/2$, is one-half the difference between the two principal stresses. This quantity $(\sigma_1 - \sigma_2)$ is called the **differential stress**, or stress difference. Both of these play an important role in understanding stresses and failure within the Earth. In particular, the mean stress describes the confining pressure experienced by rock at some depth, while the differential stress describes the greatest amount of stress change that a rock can withstand without breaking. Note that this quantity, differential stress, is quite different than the *deviatoric stress* (Engelder, 1994), which is the total value of stress with the mean stress subtracted

Don't associate differential stress with deviatoric stress.

Fig. 3.6. Summary of sign conventions used in the Mohr circle, as set up for frictional slip problems with compression positive.

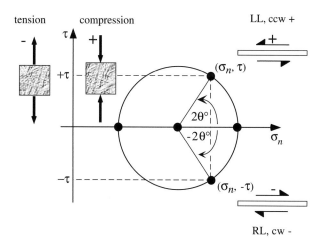

from the principal stresses. This can be obtained by centering the Mohr circle at the origin and examining the intercepts on the normal-stress axis, producing virtual values for the deviatoric tension or compression. Seismologists describe first motions of seismic slip events along faults by using deviatoric tension (T-axis) and compression (P (pressure)-axis). The angle function (θ in equation (3.3), α in equation (3.4)) serves to modulate the values of normal and shear stress, for given orientations or fault dips, around these two main quantities.

A summary of the sign conventions used in the Mohr circle is given in Fig. 3.6. Now we are ready to use the Mohr circle to calculate 2-D stress states. Just like using equations, you can use the Mohr circle to work a problem in either direction. Starting with the remote principal stress and the appropriate angle, you can find the normal and shear stresses resolved on that particular plane; these will plot as a point on the Mohr circle. Also, you can start with the normal and shear stresses on a plane of given orientation and use the Mohr circle to calculate the remote principal stresses. We'll do a simple example exercise for the first case.

Example Exercise 3.4

Suppose you have a general 2-D stress state like that shown in Fig. 3.4 (upper diagram), where $\sigma_{xx} = 0.8$ MPa, $\sigma_{yy} = 2.5$ MPa, and $\tau_{xy} = 1.2$ MPa (compression positive). Calculate the values for (a) the principal stresses and (b) the orientations of the principal planes on which they act.

Solution

The first step is to draw the problem (Fig. 3.7, upper diagram). We set up a stress element aligned with an xy-coordinate system and use the sign convention to get the stresses drawn correctly.

Fig. 3.7. Problem sketch (left-hand diagram) and four main steps in plotting stresses using a Mohr circle, keyed to example exercise 3.4.

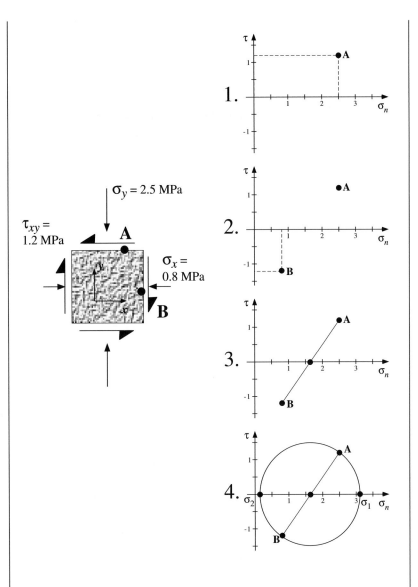

The stress states (normal and shear) acting on these planes are written down as ordered pairs of normal and shear stresses.

$$A: (2.5, 1.2) \text{ MPa}$$

$$B: (0.8, 1.2) \text{ MPa}$$

But before we can plot these, we need a point above and below the normal-stress axis, so we invoke the shear-stress sign convention given in Table 3.1 and assign the sign of the resulting shear stress to the ordered pairs listed above.

$$\text{A: } \left(+2.5, +1.2\right) \text{ MPa}$$

$$\text{B: } \left(+0.8, -1.2\right) \text{ MPa}$$

Notice that the signs of the (compressive) normal stresses have been added to reinforce where they should be plotted on the Mohr diagram.

Now we plot the correctly written ordered pairs on the diagram, as shown in Fig. 3.7, steps 1 and 2. Connecting the points with a line defines the diameter of the Mohr circle. Where the line (Fig. 3.7, step 3) crosses the normal-stress axis is the mean stress. Now take a compass, center it on the mean stress, and use the diameter to draw the Mohr circle (Fig. 3.7, step 4). By reading off values from your Mohr diagram, we can solve the example problem.

(a) The principal stresses are found by identifying the points where the Mohr circle crosses the normal-stress axis (and of course, where the shear stress is also equal to zero). These values are approximately

$$\sigma_1 \approx 3.2 \text{ MPa } \left(\text{compressive}\right)$$

$$\sigma_2 \approx 0.3 \text{ MPa } \left(\text{compressive}\right)$$

Using equations (3.1) to check our drawing abilities, we calculate values of 3.12 MPa and 0.18 MPa for the principal stresses.

(b) The principal planes are estimated by measuring the value for the angle formed between plane A and the σ_1-plane in Fig. 3.7, step 4. Because you move *clockwise* to get from the A-plane to the σ_1-plane, the angle is *negative*. We only plot double angles on the Mohr diagram, so $2\theta_p \approx -45°$, so $\theta_p \approx -22°$.

There are always two principal-plane orientations, because there are two orthogonal principal planes. We add 90° to θ_p (or 180° to $2\theta_p$) to obtain our final answer.

$$\theta_p \approx -22° \text{ and } 68°$$.

The first value, −22°, corresponds to the angle between the A-plane and the σ_1-plane (see Fig. 3.7, upper diagram). The second value, 68°, takes you from the A-plane all the way around, but in a counterclockwise fashion, to the σ_2-plane. Because the problem is symmetric, the answers apply just as well if you start with the stresses acting on the B-plane and then rotate to the principal planes. This finishes our exercise.

In this exercise, there are **two key things to remember**. *First*, because we retained the compression-positive sign convention from the start, we don't need any other conventions—this is a useful aspect of the soil-mechanics approach. Rotations stay *counterclockwise positive*, so the correspondence between rotations in physical space and in the Mohr circle is clear and direct (Fig. 3.6). *Next*, we rotate *plane to plane*; the planes (having their own outer-normal vectors) are plotted using their associated stress states as a proxy on the Mohr diagram. This is why we use the equations for 2-D stress that involve θ, rather than α, as the approach of choice when we need to calculate the stresses resolved on planes.

The final aspect is the concept of **effective stress** and its implementation on the Mohr diagram. The previous equations were written with normal and shear (or principal) stresses that implicitly incorporated information about the pore-water pressure within the cavities, porosity, or fractures within the rock mass. It is important to include the mechanical (stress) effect of pore pressure given its fundamental role in reducing the shear stresses needed for faulting (e.g., Sibson, 1985; Zoback et al., 1987; Jaeger et al., 2007), in opening and propagating dilatant cracks underground where all the remote stresses are compressive (e.g., Secor, 1965), in facilitating pressure solution (e.g., Fletcher and Pollard, 1981), and in the deformation of fold-and-thrust belts to form accretionary wedges (e.g., Davis et al., 1983), to note just a few applications. Chemical effects of the pore fluid, important in many problems such as slow, subcritical crack growth (e.g., Segall, 1984a,b; Atkinson and Meredith, 1987a,b; Olson, 1993; Renshaw and Pollard, 1994b; Chen et al., 2017), are considered in Chapter 9.

Effective stress can be calculated by using either the equations or the Mohr circle (Fig. 3.8). Physically, the pore-fluid pressure acts to *counterbalance the downward pressure of the overlying rocks*—think of it as a buoyant force. Operationally one can just subtract off the value of the pore pressure from each of the normal stresses (principal or resolved), leaving the shear stresses unaffected. The approach also assumes drained conditions, so that in the subsurface, the relative stress magnitudes can change if the rock behaves poroelastically (e.g., Zoback and Zinke, 2002; Zoback, 2007, pp. 381–384). For mud or other nonlinear fluids in the pores, the value of the shear stresses will also need some modification (e.g., Jaeger and Cook, 1979, p. 219–225; Suppe, 1985, p. 159–161; Brady and Brown, 1993, p. 89).

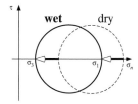

Fig. 3.8. The influence of pore-fluid pressure on reducing the effective stresses in a rock mass, shown on the Mohr diagram.

One of the most important uses of stress calculations, whether using the equations or the Mohr circle, is to predict whether the resolved stresses on a rock surface can produce frictional sliding along it. To do that, we use the results we have found in this section along with a rule for deciding on the stability of the surface. We'll apply these results now to frictional sliding using the Coulomb criterion as our rule.

3.4 The Coulomb Criterion for Frictional Slip

Coulomb's classic work in the late eighteenth century set the stage for what is probably the most widely used failure criterion for soils as well as for rocks. Although sometimes attributed to Navier and others, the Coulomb criterion was defined by him (Handin, 1969; Schofield, 2005) although his notation was different than what we use today. This simple **peak-strength relationship** gives us a useful starting point for understanding the frictional strength of planar geologic discontinuities such as joints, bedding surfaces, and faults. However, current understanding of the physical mechanisms of frictional sliding in rocks extends well beyond this idealized situation (e.g., Barton, 1976, Evans and Kohlstedt, 1995; Lockner, 1995; Marone, 1998a,b; Scholz, 1998, 2002; Zoback et al., 2012).

3.4.1 Relations for Local Fracture Stresses

In its simplest form, the Coulomb criterion defines the amount of shear stress needed to overcome the frictional resistance of the surface; the resistance is a combination of normal stress acting on the surface and some physical properties of the surface itself. The Coulomb criterion is expressed mathematically as

Coulomb criterion using local (normal and shear) stresses

$$|\tau| = C_0 + \mu\sigma_n'$$
(3.7a)

$$|\tau| = C_0 + \mu(\sigma_n - p_i)$$
(3.7b)

Here we recognize the normal and shear stresses, σ_n and τ. In equation (3.7b) the normal stress is explicitly reduced by the value of the pore-fluid pressure p_i, making the quantity in parentheses the *effective stress*; this is σ_n' in equation (3.7a). From now on we'll drop the superscripts and assume that all stresses are effective stresses unless otherwise specified.

On the right-hand side of equations (3.7a,b) are two material parameters that describe something about the resistance to slip of the interface being sheared. The **cohesion**, C_0, and **friction coefficient**, μ, are usually adequate to characterize the *strength of a small surface* under geotechnical (shallow depth, short loading duration, low temperature) conditions. Cohesion is a curve-fitting parameter that represents the *intercept* of the Coulomb criterion on the shear-stress axis—using a Mohr circle as in the following section. Cohesion describes the limiting or minimum amount of shear stress needed to get a surface slipping when the normal stress is very small (but still compressive). The friction coefficient is the *slope* of the Coulomb criterion on the Mohr diagram (Fig. 3.9). It is defined as τ/σ_n or the local slope using $C_0 = 0$ MPa.

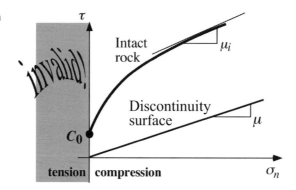

Fig. 3.9. The Coulomb criterion plotted on a Mohr diagram (Mohr circle omitted) showing definitions of cohesion C_0 and friction μ. The failure envelope for intact rock is commonly nonlinear, with the local slope giving a value for *internal friction μ_i* (see Savage et al., 1996, for clear discussion of intact-rock friction). Note that the Coulomb criterion is valid only for compressive normal stresses.

Equations (3.7a,b) are set up as an equality—meaning that when the two sides are equal, there can be incipient sliding. It is clearer to define the criterion as an inequality, as in the following equation pair:

$$|\tau| \geq C_0 + \mu\sigma_n \quad \text{:Frictional sliding} \tag{3.8a}$$

$$|\tau| < C_0 + \mu\sigma_n \quad \text{: No slip} \tag{3.8b}$$

This is very much a "yes-no" criterion: if the first equation (3.8a) is satisfied, you "get slip." In a calculation of resolved stresses on a particular plane, the Coulomb criterion may actually be exceeded, as in (3.8a); in this case, you would predict slip, with the amount related to the degree to which the Coulomb criterion was mathematically exceeded. If not, then the surface is stable and no frictional sliding occurs. Interestingly, if (3.8a) is satisfied, you have no way of knowing how much slip will occur, just which direction (given by the sense of resolved shear stress on the surface). We need some results from fracture mechanics to estimate the amount of slip, but even so, the Coulomb criterion is a reasonable first guess as to the frictional stability of surfaces in rock.

Notice that you have to specify the normal and shear stresses on the surface of interest in order to use the Coulomb criterion in this form. In order to do an actual problem, you'd first have to use the stress transformation equations from the previous section to calculate (or resolve) the remote stresses onto the surface, using the appropriate angle (α or θ). Then you'd plug those stresses into equations (3.8a,b) to predict if the surface should slip or not. You can also see from equations (3.7a,b) that it is easier to use a positive value of normal stress σ_n so that you don't have to mess with negative numbers on the right-hand side of the Coulomb criterion. This is the justification for the compression-positive sign convention.

Very often you see the Coulomb criterion written in its alternative form, using the equality here for brevity, as

$$|\tau| = C_0 + \sigma_n \tan\phi_f \tag{3.9}$$

where $\mu = \tan\phi_f$. Here ϕ_f is the **angle of friction**. The two forms are entirely interchangeable so the choice of equations (3.7) or (3.9) is totally up to you. This rock property (ϕ_f) differs from α in the stress transformation equations that we've just finished looking at ($\alpha = 45° - \phi_f/2$). The angles are related when we start looking at predicting frictional slip by combining the stress equations and the Coulomb criterion, although as we'll see they are not always identical. In more advanced applications, the friction angle ϕ_f is called the *basic friction angle* because it describes the contact strength of only small patches along a surface. Other terms can be added to include surface roughness and large-scale waviness or undulations in a geologic surface such as a bedding plane, joint, or fault (e.g., Barton, 1990).

Example Exercise 3.5

Given the values of remote principal normal stresses $\sigma_1 = 4.8$ MPa (compression positive), $\sigma_2 = 1.15$ MPa, and an angle to the normal to the plane from the σ_1 direction of 58°, with values for cohesion of 0.001 MPa and friction coefficient of 0.58, determine whether the fracture implied will slip (using the Coulomb criterion), and if so, what its sense of shear will be.

Solution

We use equations (3.3) to calculate the normal and shear stresses resolved on the fracture for the 2-D stress state given. We find that $\sigma_n = 2.175$ MPa (compressive) and $\tau = 1.640$ MPa (left-lateral). Plugging these numbers into the Coulomb criterion (3.7a,b), we find that

$$1.640 \text{ MPa} > 0.001 + (0.58)(2.175) \text{ MPa}$$

and 1.640 MPa > 1.363 MPa. So the left-hand side (the driving forces) is greater than the right-hand side (the resisting forces) and the surface will fail by frictional sliding, and in a left-lateral sense because the calculated shear stress was positive.

The previous example was an exercise in calculating the tendency for frictional sliding by using the Coulomb criterion in equation form. Let's run another example to look at how we handle slip direction and graphical implementation of the criterion.

Example Exercise 3.6

Given the values of remote principal normal stresses $\sigma_1 = 5.0$ MPa (compression positive), $\sigma_3 = 1.0$ MPa, and an angle to the normal to the plane from the σ_1 direction of 60°, with values for cohesion of 0.0 MPa and friction coefficient of 0.6, determine whether the fracture implied will slip (using the Coulomb criterion), and if so, what its sense of shear will be.

Solution

The stresses are the same as those used previously in Example Exercise 3.5. We know that the resolved normal and shear stresses, obtained from either equations (3.5) or the Mohr circle, are 2.0 MPa and 1.732 MPa, respectively. Let's do this problem two ways.

(a) Equations.

We plug the values of C_0 = 0.0 MPa, μ = 0.6, σ_n = 2.0 MPa, and τ = 1.732 MPa into equation (3.7a):

1.732 MPa > 0.0 + 0.6 (2.0) MPa
1.732 MPa > 1.2 MPa

Therefore yes, the plane will slip, and in a left-lateral sense, given a positive value of the resolved shear stress.

(b) Mohr diagram.

We can use the results from Example Exercise 3.4 again here. We just need to draw the Coulomb criterion on the diagram and see if it intersects the Mohr circle. Because the linear Coulomb criterion is within the Mohr circle, the stresses resolved on the surface are *more than sufficient* to cause slip; mathematically the criterion is not just met (equations (3.7a,b)), but *exceeded* (equation (3.8a)). Given the value for the remote normal stress resolved on the potential slip plane (σ_n = 2.0 MPa), the magnitude of slip will scale with the amount of *excess shear stress* acting on the surface (1.732 − 1.2 = 0.532 MPa). The sense of slip will be left-lateral, given that the shear stress has a *positive* value according to our compression-positive sign convention.

One more example will be sufficient to highlight the relevance of the Coulomb criterion. Let's see what happens with a different state of stress acting on a fracture.

Example Exercise 3.7

Now suppose the principal stresses acting on a bedding plane with angle α = −28° to σ_1 are σ_1 = 0.85 MPa and σ_2 = −1.6 MPa. Using values of cohesion of 0.0 MPa and a friction coefficient of 0.72, determine if the surface will slip, and if so, in what direction.

Solution

First we use equations (3.4) to calculate the normal and shear stresses acting on the surface. They are

$$\sigma_n = -0.20 \text{ MPa (tensile)}$$

$$\tau = -0.85 \text{ MPa (left-lateral)}$$

Because the resolved *normal stress is tensile*, and not compressive, the opposing sides of the bedding plane will not be in contact with each other, but instead will gape or move apart. As a result, frictional sliding will not be possible and we can't use the Coulomb criterion here. Instead we must use methods from fracture mechanics (see Chapter 8) to assess the opening across this *mixed-mode fracture*.

3.4.2 Relations for Remote Principal Stresses

Coulomb criterion using the remote principal stresses

What happens if you are working on a project and you don't yet know the orientation(s) of the potential slip surfaces? Then you need another formulation of the Coulomb criterion that doesn't require an explicit value for the angle between the remote stress state and the plane of interest. This is easily accomplished by recasting the criterion into its **principal-stress form**:

$$\sigma_1 > \sigma_c + q\sigma_3 \tag{3.10}$$

Here again it is traditional and useful to write the Coulomb criterion in its pseudo 3-D form, but you can substitute σ_2 for σ_3 to obtain the equivalent 2-D case. In equation (3.10) the two new parameters are the **unconfined compressive strength** of the rock mass, σ_c, and the **bulk friction parameter**, q.

$$\sigma_c = 2C_0\left(\sqrt{\mu^2 + 1} + \mu\right) \tag{3.11a}$$

$$q = \left(\sqrt{\mu^2 + 1} + \mu\right)^2 \tag{3.11b}$$

The unconfined compressive strength of the rock mass depends on the frictional resistance properties of the individual discontinuity surfaces, so we need to specify these parameters to obtain the rock-mass parameter. The orientations of the potential slip planes in this formulation are built into q as the *optimum slip plane orientations*—those surfaces that, when the Coulomb criterion (3.10) is exactly met, are just in a state of *incipient slip* (see equation (3.3)). Notice that (3.10) is written using the inequality for slip to occur.

The **orientation of the optimum frictional slip plane** θ_{opt} is related to the bulk friction parameter q by (Jaeger and Cook, 1979, p. 97)

$$q = \tan^2 \theta_{opt} \tag{3.12}$$

We designate θ_f as the angle between σ_1 and the *normal* to the optimum slip plane. We can also write the implied angle terms directly as

Orientation of the optimum slip plane using friction angle

$$\theta_{opt} = 45° + \frac{\phi_f}{2} \tag{3.13a}$$

and

Orientation of the optimum slip plane using friction coefficient

$$\theta_{\text{opt}} = 90° - \left[\frac{\tan^{-1}\left(\dfrac{1}{\mu} \right)}{2} \right] \qquad (3.13b)$$

using the *local* angle or coefficient of friction.

Let's run some typical numbers through these equations and see how the optimum orientation for slip depends on the ratio of the principal stresses and the frictional-resistance parameters. The results are shown in Table 3.2. The unconfined compressive strength of the rock mass, σ_c, is close to zero if the cohesion along the optimum slip planes is also zero, as is often the case for smooth surfaces. We know, however, that the unconfined compressive strength of a fractured outcrop is greater than zero, with the value depending in part on the frictional resistance of the fractures (e.g., Bieniawski, 1989). In such cases, Hoek-Brown may be used instead. The Coulomb criterion should be used primarily for assessing the tendency for frictional sliding only along a single smooth discontinuity surface, regardless of whether you use the local, orientation-based equations (3.8a,b) or the global, principal-stress-based (3.10) form.

One of the powerful discoveries in rock mechanics during the 1970s was that the frictional resistance of surfaces in rock is not very sensitive to lithology (Byerlee, 1978; see the wonderfully thorough discussion in Scholz, 2002). Measured values of friction coefficient μ range from 0.6 to 0.85 (Fig. 3.10), with some hydrous minerals such as clays having coefficients around 0.2. The virtual independence of friction on rock type is the basis for lithospheric strength envelopes (Goetze and Evans, 1979; Brace and Kohlstedt, 1980) and rock-mass strength (Hoek and Brown, 1980; Hoek, 1983) because the frictional properties of the discontinuities dominate the response to stress, rather than the intact-rock material (e.g., Bieniawski, 1989). An apparent frictional

Table 3.2 Calculated values for Coulomb friction parameters

C_0, MPa	μ	$\phi_f°$	σ_c, MPa	q	$\theta_{\text{opt}}°$
0.0	0.0	0.0°	0.0	1.00	45°
0.0	0.1	5.7°	0.0	1.22	47.9°
0.0	0.6	31.0°	0.0	3.12	60.5°
0.0	0.85	40.4°	0.0	4.68	65.2°
0.0	1.0	45.0°	0.0	5.83	67.5°
0.1	0.6	31.0°	0.35	3.12	60.5°
0.1	0.85	40.4°	0.94	4.68	65.2°
1.0	0.6	31.0°	3.53	3.12	60.5°

Fig. 3.10. The dependence of
optimum slip plane orientation
θ_{opt}, and bulk friction parameter
q, on frictional resistance ϕ_f of
the surfaces. Shaded box shows
typical "Byerlee" values of rock
friction.

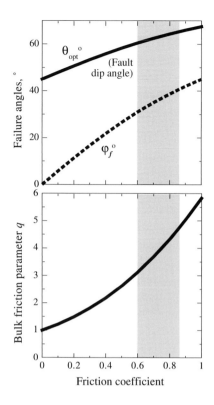

Fig. 3.10. The dependence of optimum slip plane orientation θ_{opt}, and bulk friction parameter q, on frictional resistance ϕ_f of the surfaces. Shaded box shows typical "Byerlee" values of rock friction.

resistance of nearly zero, likely due to high pore-fluid pressures within the fault zone (e.g., Rice, 1992), is common for many large plate-bounding faults such as the San Andreas system in California (e.g., Mount and Suppe, 1987; Zoback et al., 1987) and the Alpine fault of New Zealand (e.g., Scholz, 2002).

If $C_0 = 0$ (and therefore $\sigma_c = 0$), then the Coulomb criterion can be written very simply in either of two forms

Brittle strength envelope for the lithosphere

$$|\tau| < \mu\sigma_n \tag{3.14}$$

$$\sigma_1 < q\sigma_3 \tag{3.15}$$

The first of these is used for dry soils and for smooth fractures, for which $C_0 = 0$ is a reasonably good assumption. The second expression (3.15) is widely used as the **brittle, shear, or frictional strength of continental and oceanic crustal rocks** (Kohlstedt et al., 1995) applies for depths exceeding about 1 km (below the near-surface rock-mass zone) and when pore-fluid pressures are incorporated into the normal stresses (effective principal stresses). This **brittle strength envelope** (Goetze and Evans, 1979; Brace and Kohlstedt, 1980) works because the frictional strength of rock surfaces varies within a relatively small range, usually from 0.6 to 0.85 (Byerlee, 1978; Scholz, 1990) for low to moderate confining pressures, even for larger, crustal-scale faults (Sibson, 1994). Because the brittle part of the lithospheric strength envelope predicts a small compressive strength at the Earth's surface, it can be supplemented for use in the upper kilometer or so of the crust (Brace and Kohlstedt, 1980) by using a rock-mass strength criterion (Schultz, 1993, 1996) to account for the substantial, nonzero unconfined compressive strengths of near-surface rock masses (e.g., Bieniawski, 1989; Schultz, 1995a,b; Schultz et al., 2009).

Example Exercise 3.8

Given the principal stresses used in Example Exercise 3.4 and using values for cohesion and friction coefficient across the planar surface of 0.1 MPa and 0.6, respectively, determine if the rock mass will be stable or if any of its internal surfaces will slide frictionally.

Solution

We choose equation (3.10) and first calculate the unconfined compressive strength of the rock, σ_c, using equation (3.11a) and the bulk friction parameter q using equation (3.11b). These values are

$$\sigma_c = 0.353 \text{ MPa}$$

$$q = 3.12$$

Now we plug values of the principal stresses, along with these rock parameters, into the principal-stress form of the Coulomb criterion.

$$\sigma_1 > \sigma_c + q\sigma_3$$

$$5.0 \text{ MPa} > 0.353 + 3.12\left(1.0\right) \text{ MPa}$$

$$5.0 \text{ MPa} > 3.473 \text{ MPa}$$

Therefore the rock mass is expected to have internal planes that will slide frictionally. Notice that the greatest principal stress σ_1 (5.0 MPa) is much larger than the right-hand side (3.473 MPa). This implies that a greater range of plane orientations than just the optimum one will slip. The optimum orientation is given by equations (3.13a,b). We find:

$$\theta_{opt} = 90° - \left[\frac{\tan^{-1}\left(\dfrac{1}{0.6}\right)}{2}\right]$$

$$\theta_{opt} = 60.5°$$

Fig. 3.11. The conditions required for frictional sliding to occur based on the Coulomb criterion. The principal stress ratio is σ_1/σ_3.

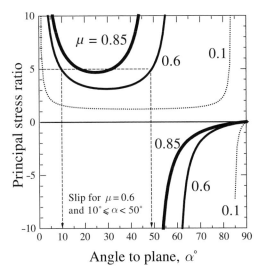

Equation (3.15) demonstrates that the frictional strength of a rock that contains potential slip surfaces depends on the ratio of the magnitudes of the principal stresses and on the frictional resistance of the surfaces (assuming that cohesion along them is negligible, as is commonly the case). We can write this explicitly to obtain a very useful form:

$$\frac{\sigma_1}{\sigma_3} = \frac{(1+\mu\cot\alpha)}{(1-\mu\tan\alpha)} \tag{3.16}$$

Notice that the friction coefficient μ is used here along with the angle between the σ_1-direction and the potential slip plane, so that there is no confusion over which parameter α represents. A plot of this slip criterion (e.g., Sibson, 1985, 1994; Schultz, 1992) clearly illustrates what levels of stress are needed to promote frictional sliding on surfaces having given values of friction coefficient. Most cases in the Earth's crust are represented in Fig. 3.11 by positive values of the stress ratio because both effective principal stresses are normally compressive. Only in unusual circumstances can the pore-fluid pressure exceed the least principal stress to make that stress component tensile, and therefore σ_3, negative (e.g., Sibson, 1985). This case plots in the lower half of the diagram.

In a more general case, if we retain a nonzero value of cohesion, then we can rewrite the principal-stress form of the Coulomb criterion using the basic parameters as (Brady and Brown, 1993, p. 107)

$$\sigma_1 = \frac{2C_0 + \sigma_3\left[\sin 2\theta + \mu(1-\cos 2\theta)\right]}{\sin 2\theta - \mu(1+\cos 2\theta)} \tag{3.17}$$

This relation reduces to the previous expression when cohesion equals zero. It turns out that the effect of cohesion on the stress levels required for slip is relatively small; you can verify this by setting a constant value of friction coefficient, cohesion, angle, and least-principal stress and calculating the relation for different values of cohesion.

The Coulomb criterion evaluates whether the stresses are sufficient to overcome the frictional resistance to slip along a surface. This is a *contact problem*—in order for two opposing surfaces, such as bedding planes, to slip frictionally, they must be in contact with each other. This requires that the normal stress resolved on the surface be compressive (positive here; McClintock and Walsh, 1962). If instead the resolved normal stress is tensile, the opposing surfaces will be pulled apart, creating a gap or crack with surfaces that are no longer in frictional contact. **The Coulomb criterion is not valid for cases in which the normal stress is tensile** (e.g., Brace, 1960; Paul, 1961; Jaeger and Cook, 1979; Farmer, 1983; Suppe, 1985; Goodman, 1989; Price and Cosgrove, 1990; Bayly, 1992; Brady and Brown, 1993; Engelder et al., 1993; see Fig. 3.9).

Coulomb criterion is valid for compressive normal stresses—not for tensile resolved stresses

3.4.3 The Coulomb Failure Function

From the material covered so far, it should be clear that there are two ways to make a Mohr circle become tangent to the Coulomb slip criterion. One can reduce the resolved normal stress (for example, by increasing pore-fluid pressure), and/or one can increase the value of shear stress (this also increases the diameter of the Mohr circle, and therefore the value of the principal stresses). The Coulomb failure function ("CFF") accommodates either or both of these.

Variations in both normal and shear stresses resolved on potential slip surfaces (or existing faults) can occur quite commonly in association with earthquakes and with slip along faults. Dilation across cracks and dikes, dissolution along an anticrack, and displacement along a deformation band will also change the normal and shear stresses in the rock surrounding them (e.g., Pollard and Segall, 1987). Here, we'll focus on faults, because this is where this technique has found its most widespread application; analogous results for dikes are presented by Rubin and Pollard (1988) and Rubin (1993a).

Displacements on structural discontinuities alter the normal and shear stresses around them

Because the pattern of stresses in the surroundings caused by slip on faults is inhomogeneous, the **actual values of normal and shear stress vary from place to place near the slipping fault**. Some representative examples of stress changes associated with faults are shown in the literature (e.g., Chinnery, 1963; Segall and Pollard, 1980; Pollard and Segall, 1987; King et al., 1994). Lobes of stress change can impinge on fractures having a range of orientations (in 3-D space), leading to the potential for some interpretation in evaluating the Coulomb criterion on these surfaces.

A compact formulation was developed in the late twentieth century to succinctly compare changes in the inhomogeneous stress state to the Coulomb criterion on a particular plane. Called variously the **Coulomb failure function** or Coulomb failure stress (Reasenberg and Simpson, 1992; Stein et al., 1992; Harris and Simpson, 1998; Stein, 1999), we start with the Coulomb criterion (3.7b) for frictional stability and set cohesion to zero to represent a fracture surface (e.g., equation (3.14)). We identify a particular plane in 3-D space (called the "receiver plane") and then explicitly calculate the *changes* in normal and shear stress on this plane from some background or reference state. The Coulomb failure function is written as (Bruhn and Schultz, 1996)

Coulomb failure function includes initial and perturbed stress states at each point in the rock mass

$$\Delta\sigma_{CF} = \left| \tau^* - \tau_0 \right| - \mu \left| \sigma_n^* - \sigma_n^0 \right| \qquad (3.18)$$

in which σ_n^* and τ^* are the new (or additional) normal and shear stresses (compression positive) resolved on the specified surface, σ_n^0 and τ_0 are the initial stresses resolved on the surface from the remote stress state, and μ is the coefficient of static friction of the surrounding rock mass. Normally any transient changes in pore-fluid pressure are ignored, along with any time dependence or decay in stress with distance from the source of altered stress. Here the calculation shows how much closer to (or farther away from) the Coulomb criterion the new stress increment moves the original state of stress on that specified plane.

Equation (3.18) can be written more succinctly as (Harris, 1998)

Coulomb failure function using effective friction coefficient

$$\Delta\sigma_{CF} = \Delta\left| \tau \right| - \mu' \Delta\sigma_n \qquad (3.19)$$

where $\mu' = \mu(1 - \beta')$ with β' being a factor for rock (or the fault zone) analogous to Skempton's coefficient β (defined for soils by Skempton, 1954; see discussions by Harris, 1998, and Beeler et al., 2000). The minus sign before μ in (3.18) and (3.19) would be changed to positive if the soil-mechanics convention for normal stress were used (i.e., compression negative gets the plus sign). Values of β for soils range from 0 (dry) to 1.0 (fully saturated), with values of β' for rock ranging from 0.5–1.0 (Harris, 1998). Using typical dry-rock values of static friction coefficient of 0.6–0.85, the effective friction coefficient μ' ranges from zero to 0.3–0.425, with the larger values associated with drier conditions (more nearly hydrostatic pore-fluid pressures) and the frictionless value (0.0) for fully saturated rock masses. The Coulomb failure function can also be written by using the remote principal stresses as (Crider and Pollard, 1998; Crider, 2001; Soliva et al., 2006, 2008)

Coulomb failure function using remote principal stresses

$$\Delta\sigma_{CF} = \left[\frac{(\sigma_1 - \sigma_3)}{2} \right] \sqrt{1 + \mu^2} - \mu \left[\frac{(\sigma_1 + \sigma_3)}{2} \right] \qquad (3.20)$$

Fig. 3.12. The distribution of Coulomb failure stress change shown in map view around: (a) a **normal fault** (line with tick, dipping at 60°), (b) a **thrust fault** (tooth on upper plate, dipping at 30°), and (c) a **strike-slip fault** (vertical plane). Values are calculated for planes having the same strike and dip as the main fault and are normalized to the background value. The areas of stress-change increase (values >1.0) would favor triggered earthquakes or fault slip on available planes parallel to the main fault, whereas decreases (values <1.0) would inhibit slip on those planes.

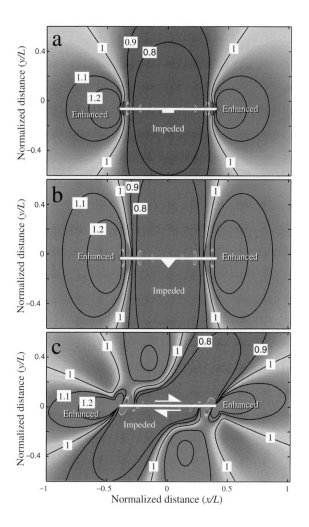

Using values for stress state far from a fault of $\sigma_1 = 17$ MPa, $\sigma_3 = 4.25$ MPa (compression positive), and $\mu = 0.6$ predicts a Coulomb stress on the order of 1.0 MPa, which corresponds to a background value to which local Coulomb stress changes near faults can be normalized (Soliva et al., 2006, 2008; Fig. 3.12).

Increased: trigger slip
Decreased: impede slip

Geophysical investigations demonstrate that **lobes of increased Coulomb failure stress** radiate outward through the surrounding rock mass in directions perpendicular to the plane of a normal (Hodgkinson et al., 1996) or thrust fault (Freed and Lin, 1998). These lobes can be correlated with the location and increased rates of occurrence of aftershocks following large earthquakes (e.g., Harris and Simpson, 1998; King et al., 1994; Freed, 2005). A stress change of only 0.01 MPa (1 bar) appears sufficient to **enhance** (or, for a reduction of –0.01 MPa, inhibit) **slip** on the specified planes in nature (e.g., Stein et al., 1992; King et al., 1994; Harris, 1998). Soliva et al. (2006) used this approach to investigate the spacing between normal faults in non-stratified and stratified sequences. An increased Coulomb failure stress may also be interpreted as potential earthquake ruptures being moved closer (in time) to failure, whereas decreases may be interpreted as surfaces being delayed.

Increased Coulomb failure stress promotes frictional sliding along favorably oriented surfaces, whereas decreased values inhibit slip along these surfaces. This formulation can be used to evaluate changes in stress state due to any type of nearby deformation beyond small-strain seismic events, including crack growth, dike injection, and formation of various tectonic patterns of large-offset faults, by providing a compact means for evaluating the associated stress changes (e.g., Cooke and Pollard, 1997; Schultz, 2000c; Wilkins and Schultz, 2003; Okubo and Schultz, 2004; Goudy et al., 2005; Olson and Cooke, 2005; Soliva et al., 2006, 2008; Fossen et al., 2010) including those associated with pre-peak inelastic yielding of porous rocks (Davis and Selvadurai, 2002, p. 95; Paterson and Wong, 2005, p. 252; Schultz, 2011a).

As can be seen in Fig. 3.12, the Coulomb stress changes are sufficiently reduced above the hanging wall of a normal or thrust fault to impede frictional sliding along potential slip planes or new faults there that would be parallel to the main fault plane. Lobes of reduced Coulomb stress change are generated normal to the fault strike for normal and thrust faults. Reduction of Coulomb stress change near strike-slip faults is associated with increases in mean stress in the compressional stress quadrants (see Chapters 5 and 8). On the other hand, Coulomb stress changes are increased near the tips of all three fault types, suggesting that **propagation** could occur by the coalescence and linkage of new parallel fault strands, or that seismicity could be triggered in these regions.

Coulomb failure function and fault propagation

3.5 More Complete Peak-Strength Criteria

Griffith criteria
Modified Griffith criterion
Hoek–Brown criterion

Extrapolation of a Coulomb-type failure criterion into the tensile regime requires explicit consideration of mode-I and mixed-mode crack growth (see Chapters 1 and 4), *rather than* frictional sliding and faulting. Two ways of extending a Coulomb-type Mohr envelope into the tensile-stress regime are currently used. The *Griffith envelope* (the "first way:" Griffith, 1921; Brace, 1960; Jaeger and Cook, 1979; Price and Cosgrove, 1990) is parabolic in form in the tensile-stress regime (where the resolved normal stress on the optimally oriented plane is tensile) and, in its *Modified Griffith form* (Jaeger and Cook, 1979, pp. 101–102, 280–282), either becomes linear (Brace, 1960) or joins the linear Coulomb sliding criterion (Secor, 1965) in the compressive regime (where the resolved normal stress is compressive). Although this formulation has been used extensively in rock mechanics, it has since been supplemented or **replaced by the Hoek–Brown criterion** and its several variants (the "second way:" Hoek and Brown, 1980; Brady and Brown, 1993, pp. 110, 132) due to the latter's greater flexibility in incorporating discontinuity sets found in most geologic outcrops and rock masses.

Both of these criteria are called **peak-strength criteria** (e.g., Farmer, 1983, pp. 81–85; Brady and Brown, 1993, p. 109; Paterson and Wong, 2005) that assume that "failure" (dilatant cracking or frictional sliding) is predicted when the differential stress reaches a limiting maximum value. Other peak-strength criteria include von Mises and Drucker–Prager (Jaeger et al., 2007). This approach is equivalent to representing rock-mass deformation by using a simple, linearly elastic rheology (see Chapter 2) until the peak stress is reached, where failure of the rock mass occurs. The **Complete Stress–Strain Curve for Rock** (Chapter 2) provides a comprehensive encapsulation of rock and rock-mass deformation because it is derived from stiff, well-controlled testing of rocks (e.g., Crouch, 1970; Wawersik and Fairhurst, 1970; Wawersik and Brace, 1971).

Many other criteria exist to model the strength of intact rock samples (e.g., Bieniawski, 1974; Carter et al., 1991; Wang and Kemeny, 1995; see reviews by Lockner, 1995; Colmenares and Zoback, 2002; Aadnøy and Looyeh, 2010; and Labuz et al., 2018). While these criteria in general are successful in representing the characteristically nonlinear shape of the envelopes in both the tensile and compressive regimes (e.g., see Lockner's (1998) discussion of Westerly granite), their utility to large-scale or fracture rock masses may be limited if they do not incorporate terms that account explicitly for scale (Barton, 1990) and for stress-concentrating fractures (e.g., Hoek, 1983; Germanovich and Cherepanov, 1995). Although created for intact materials, the Griffith criterion has been adapted, by using (the much smaller) rock-mass equivalents of the tensile strength, for field-scale conditions (Gudmundsson, 1992). However, the Hoek–Brown criterion and its variants remain the only ones in widespread use that explicitly consider fracture-related influences on rock-mass strength.

3.5.1 Griffith Criteria

The Griffith criteria are based on the hypothesis (Griffith, 1921) that minute crack-like flaws (later called "Griffith cracks") in any material concentrate the stress applied remotely to promote failure of the material at lower stress levels. Griffith sought to understand why the strength of a material such as glass decreased rapidly as its size was increased (we'll look at the reasons for this effect in Chapter 8; see also Kanninen and Popelar, 1985, pp. 30–37; Lawn, 1993, pp. 13–14; Anderson, 1995, pp. 36–38). Building on the prior work by Inglis (1913) that demonstrated the stress-amplifying effects of even tiny crack-like flaws, Griffith postulated that tensile failure occurred when the local stress at the most critically (or optimally) oriented flaw attained a value characteristic of the material. By noting that the propagation of a crack involves the creation of new surfaces in the material, which occurs by breaking atomic bonds in the crystalline lattice, Griffith formulated an expression for the energy balance between the applied loads and the resulting cracks that would grow in equilibrium within the material. He found that the *uniaxial tensile failure strength* (for plane-strain) is given by (Jaeger and Cook, 1979, pp. 337–340; Kanninen and Popelar, 1985, p. 35; Engelder et al., 1993, pp. 27–29; Anderson, 1995, p. 37)

Griffith criterion for 1-D uniaxial tension

$$T_0 = \sqrt{\frac{\pi E \gamma}{2(1-v^2)^a}} \tag{3.21}$$

in which E is Young's modulus, v is Poisson's ratio, γ is the energy required to create new crack-wall surfaces, and a is flaw half-length. Alternatively, if we define the critical value of energy release rate G that is associated with crack extension as

Critical strain energy release rate

$$G_c = \frac{\pi \sigma_{dl}^2 a(1-v^2)}{E} \tag{3.22}$$

then we obtain (after associating stress and surface energy terms; Broek, 1986, p. 23)

Griffith criterion for tensile strength

$$T_0 = \sqrt{\frac{E G_c}{\pi a}} \tag{3.23}$$

This pair of equations (3.21) and (3.23) is the well-known *Griffith criterion for tensile failure*. It is applicable strictly to a uniaxial tensile stress, oriented perpendicular to the optimum crack, applied to a cracked brittle solid. Notice that as the flaw size (length $L = 2a$) increases, the tensile strength decreases, as described in Chapter 2.

Although this relation (equation (3.21)) is a helpful result that leads us toward the field of fracture mechanics (Chapter 8), we now have an approach that is *complementary* to the Coulomb criterion—the Griffith criterion is used when the normal stress resolved on a discontinuity plane is **tensile**, and the Coulomb criterion for compressive resolved normal stress. As shown for example by Price (1966), pp. 29–35 and Jaeger and Cook (1979) pp. 340–341, Griffith's tensile strength can be recast by using two-dimensional stresses, as we did earlier in this chapter to resolve stresses onto a discontinuity surface in the Coulomb criterion, to obtain the widely used *Griffith criterion for rock strength* (e.g., Griffith, 1924; Jaeger and Cook, 1979, pp. 101–102, 277–280):

Coulomb criterion: Compressive normal stress
Griffith criterion: Tensile normal stress

Griffith criterion for 2-D stress

$$\begin{aligned} \left(\sigma_1 - \sigma_2\right)^2 &= 8T_0(\sigma_1 + \sigma_2) \quad \text{if} \left(\sigma_1 + 3\sigma_2\right) > 0 \\ \sigma_2 &= -T_0 \quad\quad\quad\quad \text{if} \left(\sigma_1 + 3\sigma_2\right) < 0 \end{aligned} \tag{3.24}$$

This criterion for *2-D plane conditions* (and effective stresses) is useful if you are working in a rock mass and don't know initially what the orientations of the discontinuities might be. It is not intuitive to tell easily if the resolved

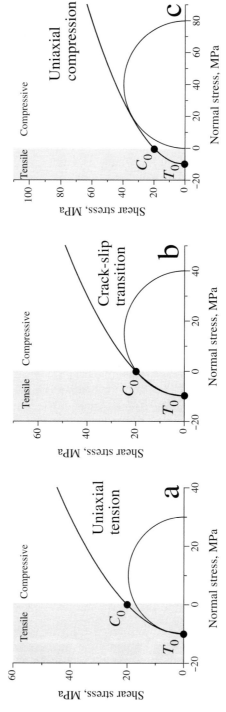

Fig. 3.13. The Griffith criterion and Mohr circles showing (a) uniaxial tensile failure, $\sigma_1 = 4.5T_0$; (b) uniaxial compressive failure, $\sigma_1 = 3T_0$ (at $\sigma_2 = T_0$); and (c) transition stress from crack growth to frictional sliding, $\sigma_1 = 8T_0$ (at $\sigma_2 = 0$).

stresses are tensile, as required, or are compressive instead. Equation (3.24) is the principal-stress form of the standard Griffith criterion, analogous to the principal-stress form of the Coulomb criterion (equation (3.10)).

Rewriting (3.24) using the (local) normal and shear stresses, the Griffith criterion becomes (Murrell, 1958; Jaeger and Cook, 1979, p. 102)

$$\tau^2 = 4T_0\,(\sigma_n + T_0) \tag{3.25}$$

Equation (3.25) is appropriate when the normal stresses on a potential flaw are *tensile* (Jaeger et al., 2007, p. 93). This relation would plot on the Mohr diagram (Fig. 3.13) as a parabola centered at the rock's tensile strength, T_0. By setting the normal stress $\sigma_n = 0$, we find that the value of cohesion C_0 is

$$C_0 = 2T_0 \tag{3.26}$$

A widely used application of the Griffith criterion is an estimate of the stress states associated with tensile normal stresses, producing jointing, and compressional normal stresses, resulting in frictional sliding. We see here that the cohesion is twice the (absolute) value of the tensile stress (equation (3.26)). If you work through the trigonometry (e.g., Secor, 1965; Suppe, 1985, pp. 190; Price and Cosgrove, 1990, pp. 28–34), then other limiting values of stress difference can be obtained. **Pure mode-I displacements** on a fracture (no resolved shear stress) are predicted where the Mohr circle touches the Griffith criterion at only one point—where the minimum (tensile) stress equals the tensile strength of the rock (T_0) along the shear-stress axis (for which shear stress equals zero, by definition). You must have a differential stress of $(\sigma_1 - \sigma_2) \leq 4\,T_0$ (along with setting $\sigma_2 = T_0$), giving $\sigma_1 = 3T_0$ for pure mode-I displacements (see Fig. 3.13a). Notice that the orientation of the plane containing T_0 is located at $2\theta = 180°$ to the principal plane containing the maximum compressive principal stress, so that the crack plane's outer-normal vector will point in the direction of the least principal (tensile) stress, and that the crack plane will parallel σ_1.

An approximate limit to the differential stresses associated with cracking is given by the geometry of the parabolic Griffith envelope. Following Suppe (1985), pp. 190–192, and Price and Cosgrove (1990), pp. 29–34, the differential stress that can be contained within the Griffith envelope, while retaining a tensile resolved normal stress (Fig. 3.13b), is $(\sigma_1 - \sigma_2) \leq 2\,T_0$, leading to a limiting value for jointing of σ_1 of 4–$5T_0$. Note that this value has been used previously to suggest various approaches to mixed-mode fracturing, but as we see below (Fig. 3.26), the earlier nomenclature serves to obscure the identification of mixed-mode cracks having bounded extents. These earlier approaches are no longer applied in the present context.

The limit to cracking can be recast to suggest a limit to the **depth of jointing** in the Earth's crust (e.g., Secor, 1965). If we define a fluid pressure ratio λ (e.g., Hubbert and Rubey, 1959; Suppe, 1985, p. 185; Price and Cosgrove, 1990, p. 168), using the vertical (lithostatic) stress $\sigma_v = \rho g z$ as

Fluid pressure ratio

$$P_f = \lambda \sigma_v \qquad (3.27)$$

then we find that the maximum depth of jointing (mode-I or mixed-mode cracking) is (Suppe, 1985, p. 192)

Maximum depth of jointing based on a Griffith envelope

$$z \cong \left| \frac{kT_0}{pg(1-\lambda)} \right| \qquad (3.28)$$

in which $k = 3$ for pure mode-I cracks or $(2+2\sqrt{2})$ for mixed-mode cracks. Setting the average rock density $\rho = 2800$ kg m^{-3}, $g = 9.8$ m s^{-2}, and $T_0 = -5$ MPa, we find values of 0.55–0.9 km for the depth of dry joints ($\lambda = 0$) and 0.85–1.4 km for joint networks growing in crust having *hydrostatic pore-fluid pressures* ($\lambda = 0.36$). A nearly *lithostatic* state of pore-fluid pressure can sometimes be found in overpressured rock strata (e.g., Engelder, 1993, p. 41) or in major fault zones (e.g., Zoback et al., 1987). Although (3.28) provides a reasonable first-order estimate for the depth of jointing, it should be used with care, because other processes can generate the stresses at depth needed to create joints (e.g., Lachenbruch, 1961; Pollard and Aydin, 1988; Engelder et al., 1993) and because the Griffith criterion normally assumes that the parameters for intact rock apply, following the original usage for intact, flawless glass.

Gudmundsson (1992) utilized the Griffith criterion in an innovative way to estimate the depth of faulting in Iceland. He re-interpreted the intrinsic flaw size of the basaltic lava flows that were faulted to be the scale of the columnar joint network, or about 0.5–2 m, rather than the μm characteristic of the intact rock material. The rock-mass tensile strength of the volcanic sequence measured *in situ* by Haimson and Rummel (1982), 1–6 MPa, was then used in equation (3.28) to infer depths of normal faulting (Opheim and Gudmundsson, 1989) around 0.5 km or greater (depending on the degree of crack closure at depth). Interestingly, the value measured for the rock-mass tensile strength is comparable to that predicted for basaltic lava flows by the Hoek–Brown criterion (Schultz, 1995a,b; Justo et al., 2010), which also provides a straightforward means of assessing the transition from jointing to faulting.

What happens if the cracks close under compressive stresses? We can recast the Griffith criterion and assess the frictional stability of a closed crack in the same way that we assessed the resistance to propagation under tensile stress. We find that, *for compressive resolved normal stresses*, the criterion (now called the compressive-stress side of the **Modified Griffith criterion**) becomes

Modified Griffith criterion for 2-D stress

$$\sigma_1\left[\sqrt{\mu^2+1}-\mu\right]-\sigma_2\left[\sqrt{\mu^2+1}+\mu\right]=4T_0 \qquad (3.29)$$

Having compressive normal stress resolved on the optimum slip surface (Fig. 3.13c) is ensured if the left-hand side of (3.29) is positive and nonzero. Following Brace (1960), McClintosh and Walsh (1962), and Jaeger and Cook (1979, pp. 95–97), recasting equation (3.29) into the normal and shear stresses results in

$$|\tau| = 2T_0 + \mu\sigma_n \qquad (3.30)$$

Equation (3.30) is the equation of a straight line—just like the Coulomb criterion. In fact, they are functionally equivalent, because the assumptions made to convert the Griffith criterion—designed originally for dilatant crack propagation—to a frictional one (McClintock and Walsh, 1962; Brace, 1960; Murrell, 1964; Hoek and Bieniawski, 1965) disable any attempt to interpret (3.30) as a propagation criterion (Jaeger and Cook, 1979, p. 341).

The *Modified Griffith criterion* is written for the Mohr diagram as

The Modified Griffith criterion on the Mohr diagram

$$\begin{aligned} \tau^2 &= 4T_0(\sigma_n + T_0) &&\text{if } \sigma_1 < 0 \text{ (tensile)} \\ |\tau| &= 2T_0 + \mu\sigma_n &&\text{if } \sigma_1 > 0 \text{ (compressive)} \end{aligned} \qquad (3.31)$$

or, equivalently but less transparently, by using the principal-stress forms (equations (3.24) and (3.29)). Plugging numerical values into these equations shows that the differential stress for unconfined compression of a rock sample, for which the least principal stress $\sigma_2 \equiv 0$ MPa (Fig. 3.13c), is given by (e.g., Brace, 1960; Jaeger and Cook, 1979, p. 102) $(\sigma_1 - \sigma_2) = \sigma_1 \leq 8T_0$. Larger differential stresses can only exist under the failure envelope when the Mohr circle for stress intersects the failure envelope under compressive resolved normal stresses, implying that compressive principal stresses lead to shear failure (properly interpreted in the present context as frictional slip along pre-existing surfaces) of the rock.

The Modified Griffith criterion is constructed (Fig. 3.14; Secor, 1965) to account for two very different aspects of fracture behavior—mode-I propagation and frictional sliding. As a result it is unsatisfying as a unifying criterion

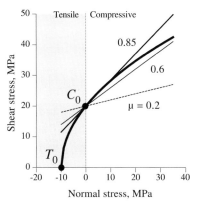

Fig. 3.14. The Griffith criterion (heavy parabolic curve) and Modified Griffith criterion (straight lines for various coefficients of friction) plotted in their respective regions of validity.

for brittle fracture. As noted by Jaeger (1971), Hoek (1983), Ramsey and Chester (2004), and others, the parabolic form of the Griffith criterion is a mathematical idealization of only one of several key processes, including crack nucleation, dilation (where the Mohr analysis applies), propagation, linkage, and arrest or termination (Hoek and Bieniawski, 1965; Horii and Nemat-Nasser, 1985; Pollard and Aydin, 1988).

3.5.2 Hoek–Brown Criterion

Probably the most widely used approach to representing the **response of fractured rock to stress** is that proposed by Hoek and Brown (1980). This empirical relationship, modeled after the Griffith criterion for tensile normal stress (Hoek, 1983; Figs. 3.15, 3.16), explicitly incorporates the degree of fracturing and rock alteration, among other factors, into a criterion that has been applied successfully to the design of geotechnical structures such as dams, roadcuts, and underground chambers in large-scale, fractured and faulted rock masses (see Brady and Brown, 1993, pp. 132–135; Franklin, 1993; Hoek et al., 1995; Hudson and Harrison, 1997, pp. 110–112). The criterion has been augmented and updated since its adoption, as reviewed by Hoek (2002), Hoek et al. (2002), and Langford and Diederichs (2015).

Fig. 3.15. Dr. Evert Hoek at the 47th U.S. Rock Mechanics/ Geomechanics Symposium, San Francisco, California, in June 2013; photo by Hill Montague.

Although rarely found in structural geology texts, the Hoek–Brown criterion has been applied to many geologic problems. Some representative examples include paleostress inversions from fault-slip data (Angelier, 1989; Moeck et al., 2009), brittle strength envelopes for shallow crust (Schultz,

Fig. 3.16. Mohr diagram showing common representations of intact and fractured rock strengths and the influence of relative scale on strength criteria. The Griffith and Modified Griffith envelope for intact rock, the linear Coulomb criterion for single slip surfaces, and Hoek–Brown envelope for an intact granite are illustrated.

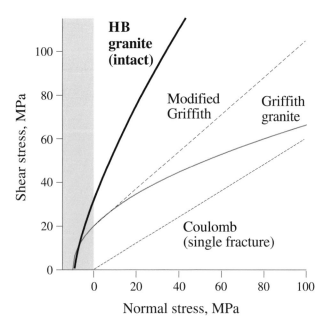

1993; Klimczak et al., 2015), predictions of faulting from elastic stress distributions (Schultz and Zuber, 1994; Andrews-Hanna et al., 2008), folding of basaltic lava flows (Schultz and Watters, 1995), nucleation of faulted joints above igneous dikes (Schultz, 1996), identification of subsurface igneous dikes from topography (Schultz et al., 2004), faulting of layered sequences (Ferrill and Morris, 2003), localization of forced folds above thrust faults (Okubo and Schultz, 2004), displacement–length scaling of faults (Schultz et al., 2006, 2008a), outcrop strength (Nahm and Schultz, 2007; Okubo, 2007), fractured petroleum reservoirs (Aadnøy and Looyeh, 2010), wellbore stability (Elyasi and Goshtasbi, 2016), compaction localization (Das et al., 2011), and rock slope stability (e.g., Schultz, 2002; Neuffer et al., 2006; Neuffer and Schultz, 2006; Li et al., 2008; Saade et al., 2016). This powerful and versatile strength criterion deserves wider use in structural and tectonic applications.

The **Hoek–Brown criterion** is given by

Hoek–Brown criterion

$$\sigma_1 = \sigma_3 + \sqrt{m\sigma_c\sigma_3 + s\sigma_c^2} \tag{3.32a}$$

or equivalently (Ucar, 1986),

$$\sigma_3 = \left(\sigma_1 + \frac{m}{2}\sigma_c\right) - \frac{1}{2}\sqrt{m^2\sigma_c^2 + 4m\sigma_1\sigma_c + 4s\sigma_c^2} \tag{3.32b}$$

in which σ_c is the unconfined (uniaxial) compressive strength of the *intact rock material*.

Two special parameters are also needed for the Hoek–Brown criterion: m and s. These depend on the initial values for the intact rock material, denoted m_i ($s_i = 1.0$, so it is not separately included). There are three cases to consider:

- **Intact rock**: use values for m_i and s_i (or s, either = 1.0).
- **"Undisturbed" rock mass** (i.e., natural conditions): use m and s calculated by using RMR (the Rock Mass Rating, described below) and equations (3.33a,b).
- **"Disturbed" rock mass** (i.e., one that had been damaged by blasting, excavation, seismic shaking, or impact cratering): use m and s calculated by using RMR and equations (3.34a,b).

Hoek–Brown parameters m and s

These relations for relating m and s to RMR are given by (Hoek and Brown, 1980; Brown and Hoek, 1988; Bieniawski, 1989)

$$\text{Undisturbed:} \quad m = m_i \, e^{\left(\frac{RMR-100}{28}\right)} \tag{3.33a}$$

$$\text{Undisturbed:} \quad s = e^{\left(\frac{RMR-100}{9}\right)} = \left(\frac{\sigma_c \text{ for rock mass}}{\sigma_c \text{ for intact rock}}\right)^2 \tag{3.33b}$$

$$\text{Disturbed:} \quad m = m_i \, e^{\left(\frac{RMR-100}{14}\right)} \tag{3.34a}$$

Fig. 3.17. Comparison between values of Hoek–Brown parameters m and s for undisturbed (equations (3.33)) and disturbed (equations (3.34)) rock masses.

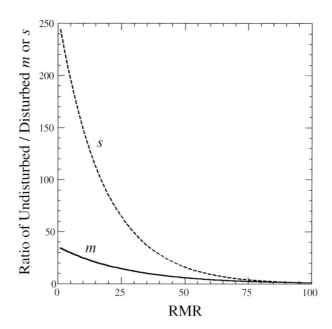

$$\text{Disturbed: } s = e^{\left(\frac{RMR-100}{6}\right)} = \left(\frac{\sigma_c \text{ for rock mass}}{\sigma_c \text{ for intact rock}}\right)^2 \quad (3.34b)$$

For any given value of RMR, the undisturbed values for m and s are consistently larger than the disturbed values (Fig. 3.17). Rock engineers consider the undisturbed values as "too conservative" in that they predict strength values greater than actually found in the field. The pair of estimates (3.33a,b) and (3.34a,b) typically bound the properties of rock masses at the field scale.

The Geological Strength Index (GSI) was introduced by Hoek et al. (1995) to replace RMR for weak rocks such as shale and flysch (e.g., Marinos and Hoek, 2001; Marinos et al., 2005). However, the Hoek–Brown criterion is emphasized in this section and RMR for three reasons. First, this approach is a consistent approach to use in the field, understand, and interpret. Second, GSI was intended to remain only semi-qualitative (e.g., Marinos and Hoek, 2001), which potentially limits its utility in structural geology, geomechanics, and rock engineering design. Third, GSI incorporates a replacement of the disturbed and undisturbed categories based on rock mass damage due (primarily) to tunneling (see Hoek et al., 2002) that must be determined or estimated for other situations. Hoek et al. (2013) have proposed methods to more systematically and quantitatively define GSI for a given rock mass.

At this point we must discuss how we obtain the parameters m_i, s, and RMR in equations (3.33a,b) or (3.34a,b). These values arise from an explicit, semi-quantitative way of *measuring the effect of fractures* on the strength and deformability of the rock mass. To do this we need a method that lets us see those properties of the rock mass—in the field—that are important

Table 3.3 Main categories of Rock Mass Rating system (after Bieniawski, 1989; Schultz, 1996)

Parameter	Description	Subtotal	%
Intact rock strength	>250 MPa	15	
	100–250 MPa	12	
	50–100 MPa	7	
	25–50 MPa	4	
	<25 MPa (weak rocks or strong soils)	0–2	0–15%
Linear fracture density	<6/m (90–100%)	20	
(Equivalent RQD[a])	6–10/m (75–90%)	17	
	10–16/m (50–75%)	13	
	16–26/m (25–50%)	8	
	>26/m (<25%)	3	3–20%
Fracture spacing	>2 m	20	
	0.6–2.0 m	15	
	0.2–0.6 m	10	
	0.06–0.2 m	8	
	< 0.06 m (60 mm)	5	5–20%
Fracture condition	Very rough fracture trace; discontinuous fractures; unweathered; fracture walls closed	30	
	Rough fracture trace; fracture opening <0.1 mm; slightly weathered rock adjacent to fractures	25	
	Slightly rough or irregular fracture trace; fracture opening <1 mm; highly weathered	20	
	Slickensides *or* gouge <5 mm thick *or* fracture opening 1–5 mm; continuous traces	10	
	Soft gouge >5 mm thick *or* fracture opening >5 mm; continuous traces; decomposed wall rock	0	0–30%

Table 3.3 (Cont.)

Parameter	Description	Subtotal	%
Groundwater	Completely dry	15	
conditions	Damp	10	
	Wet	7	
	Dripping	4	
	Flowing	0	0–15%
		RMR:	**0–100**

[a] Equivalent RQD ("Rock Quality Designation;" Deere, 1963) is obtained from the average number of fractures N per meter length using (Priest and Hudson, 1976; Brady and Brown, 1993, p. 54) RQD = 100 exp [–0.1 N (0.1 N + 1)]. A 10–20-m-long traverse length along a fractured outcrop is normally sufficient to capture a statistically valid sample of fractures for both categories of fracture spacing.

Fig. 3.18. Example of typical rock mass consisting of a sequence of basaltic lava flows with well-developed columnar joints and associated sub-horizontal fractures; photo taken south of Idaho Falls, Idaho, at the Snake River gorge. Individual flows are 1–2 m thick.

RMR: a rock mass classification system

mechanically, while simultaneously screening out the unimportant ones (like color and geologic age). These methods are called **rock mass classification systems** and are widely used and accepted by geological, mining, and civil engineers who must wrestle with the large-scale (i.e., outcrop) properties of rocks. The systems form a bridge between the properties of small, cm-scale rock samples, tested in the laboratory, and the properties of large-scale terranes of rock masses (Fig. 3.18; Table 3.3).

Several different varieties of rock mass classification systems have been developed, tested, and used in various applications such as slope design for

Fig. 3.19. Concept of **relative scale** in rock masses. Bedding-plane fractures and several less prominent joint sets at high angles to bedding in sub-horizontal Paleozoic rocks in the Kentland quarry, northwest Indiana. Smallest circle: intact rock, Griffith or Hoek–Brown criterion appropriate; medium circle, intact rock blocks separated by fractures, Coulomb or key-block criteria appropriate; large circle, rock mass, Hoek–Brown criterion appropriate.

Intact rock
(Griffith or
Hoek–Brown
criterion)

Key blocks
(Coulomb
criterion)

Rock mass
(Hoek-Brown
criterion)

Fig. 3.20. Comparison of RMR calculation for a basalt and a shale; categories are those listed in Table 3.3. Thin gray bars highlight fracture properties.

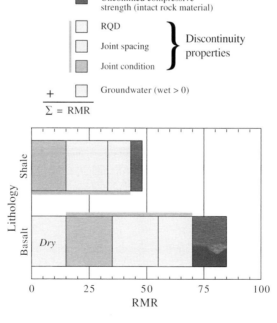

open-pit mines and highway roadcuts (Wyllie and Norrish, 1996) and tunneling of underground openings and chambers (Barton et al., 1974). Beginning with the simple criterion based on tunneling roof design support developed by Terzaghi (1946), classification systems have been enhanced and refined, culminating in the early 1970s with the two most widely used methods: the **Q-System** (Barton et al., 1974) and **Rock Mass Rating System**, or RMR (Bieniawski, 1974; Table 3.3). Informative treatments of the approaches used to assess the field-scale properties of fractured rock masses (Fig. 3.19) using these systems are given by Bieniawski (1989), Brady and Brown (1993), Priest (1993), and Schultz (1995a,b, 1996). Because of the direct connection between RMR and the Hoek–Brown criterion (e.g., Hoek, 1983), we have a workable empirical method for going to the field, measuring the fractures, and coming up with a **reliable Mohr envelope for fractured rock**. If you start with the Q-System, you can relate them by using RMR = 9 ln Q + 44 (Bieniawski, 1989, p. 89); as a result, we'll adopt the RMR system for convenience.

The end-result of applying RMR to an outcrop is a number from 0 to 100—think of this as a *percentage* of the properties of the rock mass relative to those of the (much smaller and more uniform) intact rock material. There

Table 3.4 Representative values of unconfined compressive strength for intact rock, in MPa

Material	Value or range
Sedimentary	
Berea Sandstone[a]	73.8
Navajo Sandstone[a]	214.0
Indiana Limestone[b]	48.9
Solenhofen Limestone[a]	245.0
Conglomerate	≈22
Metamorphic	
Baraboo Quartzite[a]	320.0
Igneous	
Nevada Test Site tuff[a]	11.3
Calico Hills tuff[c]	
Dry (nonwelded)[*]	26.34 ± 5.13
Wet (nonwelded)[*]	15.34 ± 0.70
Westerly Granite[b]	214–344
Palisades Diabase[a]	241.0
Hanford site basalt[d]	266 ± 98

[a] Data from Goodman (1989).
[b] Data from Hoek (1983).
[c] Data from Schultz and Li (1995).
[d] Data from Schultz (1995a,b).
[*] Samples tested normal to foliation.

are five main categories in RMR, as illustrated in Table 3.3. Once you arrive at a value for each category, summing them up gives the value of RMR for the rock mass, as shown graphically in Fig. 3.20.

As can be seen from Table 3.3, the unconfined (or uniaxial) compressive strength of the intact rock material (Table 3.4) has a contribution to RMR of only 0–15%. This means that the rock type—igneous, metamorphic, and sedimentary rocks—constitutes a minor constituent of the overall rock mass properties. The majority of the strength and deformability of field-scale outcrops of rock is controlled by factors *other than* what type of rock you have. Looking at Table 3.3, we see that the fracture population—the spacing between the fractures as well as their type and condition—contributes up to 70% of the total response of the rock mass. Groundwater conditions contribute the remaining 0–15%. Certainly, by following this tested framework you can appreciate why fractures are so important in rock-mass deformation.

You can probably anticipate that one could obtain the same value of RMR for a strong rock type (say, granite) with damp, altered, gouge-filled joints and for a weak rock type (say, sandstone) with fresh, rough, dry fractures. This indicates that these large-scale rock masses will have approximately comparable field-scale properties of strength and deformability. It also implies that RMR will vary with rainfall and climate, as well as the weathering or alteration state of the rock mass. Values (with ranges) of RMR for large-scale fractured exposures of various lithologies are shown in Fig. 3.21.

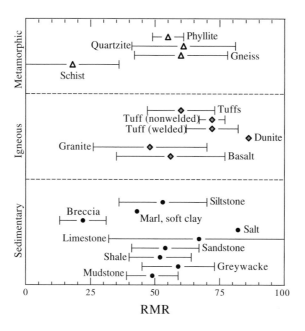

Fig. 3.21. Typical ranges (mean and standard deviation) of RMR for fractured large-scale rock masses from Bieniawski (1989), Lin et al. (1993), Schultz (1995a,b), and Sonmez et al. (1998).

Fig. 3.22. Histogram of RMRs compiled from engineering case histories by Bieniawski (1989). Note that very few unfractured or pervasively fractured outcrops were reported.

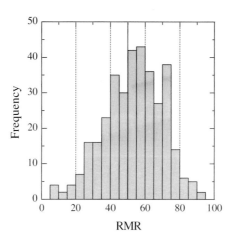

RMR was refined by using more than 350 case histories of rock-engineering projects (Bieniawski, 1989). A histogram of the RMR values obtained by *in situ* methods (Fig. 3.22) demonstrates that most rock masses considered have RMR values between 20 and 80, with most occurring around 45. The case studies show that **most outcrops** you will encounter will yield **RMR values between 35 and 75**.

The parameter m_i in equations (3.33a) or (3.34a) depends on rock type and decreases in value with the degree of fracturing or blockiness of the rock mass. For perfectly intact rock ($s = 0$, RMR = 100, and $m_i \gg 1$), m_i approximates the ratio between the unconfined compressive and tensile strengths (Hoek and Brown, 1980; see Richards and Read, 2013). This ratio is not a constant for rocks but ranges from about 10 for Nevada Test Site tuff to more than 167 for a shale, with typical values between 10 and 60 (Goodman, 1989, p. 61). The Griffith criterion predicts a constant value of 8 for this ratio (Jaeger and Cook, 1979, p. 102) while the Hoek–Brown criterion predicts a value dependent on the rock type, in better accord with laboratory test results.

Values of the parameter m_i, for intact rocks, are listed in Table 3.5; a compilation is given by Douglas (2002) and described by Richards and Read (2013). As illustrated on Fig. 3.23, wide ranges of m_i can be listed for a given rock type, such as sandstone, unless specific characteristics such as grain size, sorting, mineralogy, and diagenesis (e.g., presence and type of cement) are explicitly considered. A similar broad correlation between strength and deformability, stemming from a similar degree of imprecision in defining a rock type, was discussed in Section 2.5. These are the base values that must be modified by using RMR (specifically, fracture characteristics and groundwater conditions) to obtain the parameters for the rock mass. The intact-rock values of m_i range from 4–22 for sedimentary rocks, 6–33 for metamorphic rocks, and 15–33 for igneous rocks (Table 3.5). Typical values used for groups of rock types by engineers are $m_i = 7$ for intact limestone and carbonates, $m_i = 10$ for lithified argillaceous rocks (including shale), $m_i = 15$ for sandstones

Table 3.5 Representative values of Hoek–Brown parameter m_i for intact rock[a]

Material	Value or range
Sedimentary	
Claystone	2–6
Chalk	5–9
Siltstone	5–9
Shale	4–8
Limestone	8–15
Micritic limestone	7–10
Dolomite	6–12
Coal	8–21
Anhydrite	10–14
Gypsum	6–10
Sandstone	13–21
Breccia	≈20
Conglomerate	≈22
Metamorphic	
Slate[*]	3–10
Marble	6–12
Schist[*]	9–15
Phyllite[*]	4–10
Quartzite	17–23
Amphibolite	20–32
Gneiss[*]	23–33
Igneous	
Tuff	11–18
Rhyolite	20–30
Basalt[b]	20–30
Volcanic breccia	14–24
Andesite	20–30
Norite	15–25
Gabbro	24–30
Granodiorite	26–32
Granite	29–36

[a] Data from Hoek et al. (1995) and from Marinos and Hoek (2001).
[b] Data from Schultz (1995a,b).
[*] Samples tested normal to foliation.

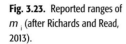

Fig. 3.23. Reported ranges of m_i (after Richards and Read, 2013).

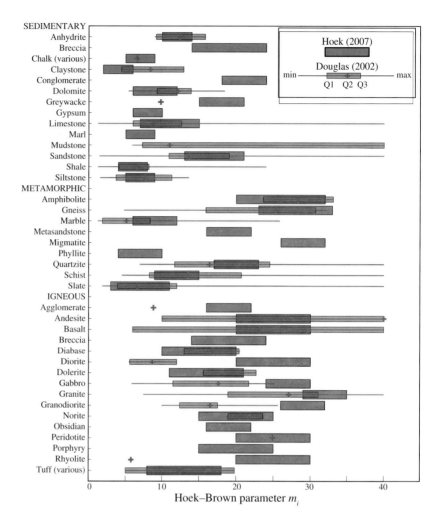

and quartzite, $m_i = 17$ for fine-grained igneous rocks, and $m_i \geq 25$ for coarse-grained igneous and metamorphic rocks (Hoek, 1983). The instantaneous friction angle (the local angle that depends on the value of normal stress resolved across a potential slip surface) increases with larger values of m_i (Hoek, 1983).

The other parameter, s, describes the effect of fracturing on the cohesion of the rock mass (Hoek, 1983). For $s = 1$ (the maximum value), the rock mass is completely intact; this value is used with values of m_i listed in Table 3.5 to predict the tensile strength, and Mohr envelope, for an intact rock sample. The Hoek–Brown criterion does a very good job of predicting the intact rock strength over a range of confining pressures, as well as accurately predicting the tensile strength (Hoek, 1983)—**a major advance over previous Coulomb and Griffith criteria**. A "pervasively fractured rock mass" (e.g., Jaeger and Cook, 1979, p. 98)—perhaps imagined as a mass of soft wet

Table 3.6 Representative values of rock-mass parameters[a]

	Rock mass		
	Poor	*Weak*	*Strong*
RMR:	25	45	75
m:	$0.07m_i$	$0.14m_i$	$0.41m_i$
s:	2.4×10^{-4}	2.22×10^{-3}	0.062
T_0^*, MPa[b]	−0.04	−0.2	−2.0
σ_c^*, MPa[b]	4.1	10	65
C_0^*, MPa[b]	0.25	0.94	6.6
$\mu^{*[b,c]}$	4.6 − 0.70	3.7 − 0.89	2.7 − 1.2
ϕ^*, °[b,c]	78– 35°	75– 42°	70– 51°
E^*, GPa	2.4	7.5	50

[a] Values calculated by using equations (3.33a,b). Compressive stress is positive.

[b] Values assume unconfined compressive strength of the intact rock material of $\sigma_c =$ 262 MPa, and $m_i = 22$, appropriate for basalt. **Other rock types, with lower values of compressive strength, will have smaller values for the rock-mass strength parameters than those listed here.**

[c] Instantaneous friction coefficients (μ) and friction angles (ϕ) calculated using equation (3.41) for normal stresses $0 \le \sigma_n \le 30$ MPa, typical for the upper km. Note that the Hoek–Brown envelope steepens toward the shear stress axis, leading to large values of friction at very low normal stresses.

material—is characterized by a limiting value of $s = 0$. Most rock masses have values closer to zero than to one.

The values of m_i and s for typical rock masses are shown in Table 3.6 along with representative values of the strength and deformability properties obtained from the Hoek–Brown equations given below. Values are also listed by Hoek (2007). The value of RMR = 75 is obtained from a relatively fresh, unweathered expanse of strong basaltic rock broken by cooling into columnar joints that are clean and rough (see Table 3.3; Schultz, 1995a,b). A weak rock mass (RMR = 45) is obtained from a basaltic rock mass that is hydrothermally altered, with gouge-filled fractures and wet conditions (Schultz, 1993). A "poor" rock mass—one that serves as a useful lower limit to ones you may encounter in the field—is approximated by RMR = 25. An altered, nonwelded pyroclastic tuff or a friable sandstone might fall into this category.

Look how rapidly the values of m_i and s decrease with RMR; this comes from the exponential dependence contained in equations (3.31). The rock mass parameter m is reduced to 41%, 14%, and 7% of the intact parameter m_i for RMR values of 75, 45, and 25 respectively. Similarly, the parameter s is reduced to 6%, 0.2%, and 0.02% of its intact-rock value for these RMR values. Because these rock-mass parameters are the basis of the Hoek–Brown criterion, and its associated Mohr envelope, you can anticipate that the values of tensile and shear strengths, for example, will be quite a bit different for the fractured rock mass than for an intact sample of the same rock material.

Deformation modulus for fractured rock masses using RMR

An additional benefit of RMR is that it allows one to calculate an estimate of the *field-scale deformability* of the rock mass. This large-scale counterpart to the Young's modulus (for small, intact, homogeneous, elastic rock samples; Chapter 2) is called the **deformation modulus**. It is calculated directly from the values of RMR you obtain in the field (Bieniawski, 1989, pp. 64–65; Brady and Brown, 1993, pp. 136–137; see also Hoek and Diederichs, 2006, for a GSI-based approach). The presence of cracks, faults, bedding planes, and other discontinuities in the field-scale rock mass not only weakens the rock mass, but increases its deformability (deformation modulus $\ll$ Young's modulus) relative to small intact rock samples (Priest, 1993, pp. 303–304; Schultz, 1996). Young's (elastic) modulus E is strictly appropriate for an intact rock sample, and not an effective continuum, because E varies inversely with fracture density (e.g., Walsh, 1965; Kulhawy, 1975; Segall, 1984b; Kachanov, 1992). The deformation modulus E^* of the rock mass is always less than the Young's modulus measured for the corresponding intact rock material.

The **deformation modulus** is given by (Bieniawski, 1989, p. 64)

$$E^* = (2\,\text{RMR}) - 100 \qquad \text{for RMR} > 55$$
$$E^* = 10^{\frac{(\text{RMR}-10)}{40}} \qquad \text{for } 10 < RMR < 55 \tag{3.35}$$

Young's modulus: Small, intact rocks Deformation modulus: Large, fractured rock masses

Comparison of equations (3.35) with case studies of deformation moduli measured *in situ* suggest uncertainties of about ± 20% in the predicted values (Bieniawski, 1989, p. 64; Fig. 3.24). Somewhat better correlations can be obtained by adding additional parameters such as Young's modulus and unconfined compressive strength (e.g., Gokceoglu et al., 2003; Hoek and Diederichs, 2006; Khabbazi et al., 2012). The more fractured a rock unit becomes, however, the less its modulus depends on rock type, so these

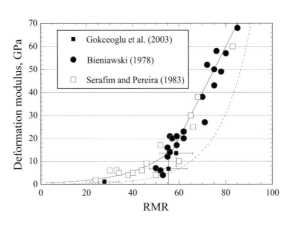

Fig. 3.24. Variation in deformation modulus E^* with RMR, as predicted by equations (3.35), solid curves and by Gokceoglu et al. (2003), dashed curve. The relationship has been verified by data from Serafim and Pereira (1983) and Bieniawski (1978) digitized from Bieniawski (1989).

relations become more reliable as fracture density increases (and as RMR decreases). Because Poisson's ratio does not appear to vary systematically with RMR, the shear modulus for the rock mass should also depend on RMR.

Bieniawski (1974, 1989) found that the deformation modulus was strongly correlated with RMR and insensitive to variations in rock type, as might be anticipated by the $\leq 15\%$ dependence of RMR on lithology. This insensitivity of deformation modulus on rock type parallels the findings of Byerlee (1978) and others that the Coulomb shear strength of individual fracture surfaces in rock also *varies little with rock type* (see Scholz, 2002, for an extended discussion). Thus RMR is quite useful in estimating the strength and deformability of near-surface rock masses where the Coulomb-based brittle strength envelope (Goetze and Evans, 1979; Brace and Kohlstedt, 1980; Kohlstedt et al., 1995) breaks down (Schultz, 1993, 1996; Schultz and Zuber, 1994).

We now have what we need to describe the strength and deformability of outcrops at the field scale. This lets us return to the **Hoek–Brown criterion** so we can calculate the various strengths of the rock mass. Just as we did for the Griffith criterion for 2-D stress, we can recast the Hoek–Brown criterion into useful forms. Setting $\sigma_3 = 0$ in equations (3.32), we obtain an estimate of the unconfined (uniaxial) compressive strength of the *rock mass*:

Compressive strength of the rock mass (unconfined)

$$\sigma_c^* = \sqrt{s\sigma_c^2} \qquad (3.36)$$

This very useful expression should not be confused with the parameter (σ_c, the unconfined compressive strength of the *intact* rock material) that is used in the calculation of both RMR (Table 3.3) and the Hoek–Brown criterion (equations (3.32)).

Similarly, the *rock-mass tensile strength* is found by setting $\sigma_1 = 0$ in equations (3.32) and solving for $\sigma_3 = T_0$:

Tensile strength of the rock mass

$$T_0^* = \frac{\sigma_c}{2}(m - \sqrt{m^2 + 4s}) \qquad (3.37)$$

This value is consistently smaller for rock masses than for the intact rock material (see Table 3.6).

Representative values of these two rock-mass strength parameters, for common ranges of RMR, are listed in Table 3.6. The unconfined compressive strength of the rock mass is reduced from 262 MPa (for the intact basaltic rock material used as a reference) to only 65, 10, or 4 MPa for the rock masses indicated. This is a **significant reduction from the laboratory scale to the field scale**. The tensile strength of the rock mass is reduced from about -10 to -15 MPa (for basalt) to -2, -0.2, and -0.4 MPa. You can see that the tensile strength of a typical rock mass is far less than you would expect—in fact, it is *so close to zero* that you could almost ignore it in a stress calculation, as many rock engineers do (e.g., Hoek et al., 1995).

Remember that 15%, or less, of the rock-mass behavior is dependent on the rock type—you can see immediately how important the fracturing is in reducing the strengths of the rock mass relative to those of the intact rock (see Table 3.6 for compressive strengths; intact-rock tensile strengths generally exceed several tens of MPa). A reduction of strength (tensile and compressive) by an order of magnitude (factor of ten) is not uncommon for rock masses (e.g., Hoek et al., 1995).

While these equations are useful when working in principal stresses, the Hoek–Brown criterion can also be seen more clearly by rewriting it as normal and shear stresses and plotting it on a Mohr diagram (e.g., Ucar, 1986; Hoek, 1990). The criterion is written in the familiar form as

Mohr envelope (2-D shear strength of the rock mass)

$$|\tau| = C_0^* + \sigma_n (\tan \phi_f^*) \tag{3.38}$$

in which the shear stresses are given by (Hoek, 1983, 1990)

Shear strength of the rock mass

$$\begin{aligned}
\tau &= (\cot \phi_f^* - \cos \phi_f^*) \frac{m \sigma_c}{8} \\
&= (\sigma_n - \sigma_2)\sqrt{1 + \frac{m \sigma_c}{2(\sigma_1 - \sigma_2)}}
\end{aligned} \tag{3.39}$$

in which the *instantaneous rock-mass cohesive strength* (shear-stress intercept of the local Hoek–Brown slope, on the Mohr diagram, as a function of normal stress or depth) is given by

*Cohesive strength of the rock mass (**instantaneous intercept** of nonlinear envelope's local slope at that normal stress)*

$$C_0^* = \tau - \sigma_n (\tan \phi_f^*) \tag{3.40}$$

and the *instantaneous friction angle* (as a function of normal stress or depth) is (Hoek, 1983)

*Friction angle of the rock mass (**local slope** of nonlinear envelope at that value of normal stress)*

$$\phi_f^* = \tan^{-1} \frac{1}{\sqrt{4h\cos^2 \left[30^\circ + \frac{1}{3}\sin^{-1}\left(h^{-\frac{3}{2}} \right) \right] - 1}} \tag{3.41}$$

$$h = 1 + \frac{16(m\sigma_n + s\sigma_c)}{3m^2 \sigma_c}$$

Of course, the value of the instantaneous friction coefficient, μ^*, is the tangent of (3.41). The value of normal stress used in these equations is specified, and fed into the equation for friction, through h (3.41) and directly into (3.40). Alternatively, one can calculate it (and plug into the above equations to get the Mohr envelope) by using (Hoek, 1990)

$$\sigma_n = \sigma_2 + \frac{(\sigma_1 - \sigma_2)^2}{2(\sigma_1 - \sigma_2) + \frac{1}{2}m\sigma_c} \tag{3.42}$$

Normal stress at failure

Fig. 3.25. Mohr envelopes showing effects of variations in the Hoek–Brown parameters.

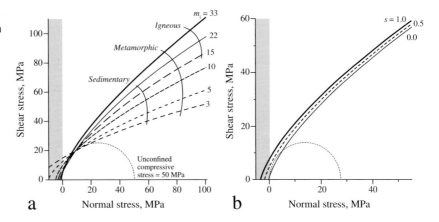

The **angle of the optimum failure plane that will be manufactured by the rock mass** in order to break along a single through-going shearing surface is (following Hoek, 1990)

Optimum shearing surface

$$\theta_{opt} = \sin^{-1}\left(\frac{2\tau}{(\sigma_1 - \sigma_2)}\right) \qquad (3.43)$$

Additionally, the *rock-mass cohesion* (shear-stress intercept for zero normal stress is given directly by equation (3.40) as

Cohesion of the rock mass

$$C_0^* = \tau\big|_{\sigma_n = 0} \qquad (3.44)$$

Equation (3.44) **is the value of "cohesion" for zero normal stress that is usually specified in other failure criteria** (e.g., the Coulomb or Griffith criteria). Notice that it is a special case of the instantaneous cohesion value (equation (3.40)) that is *projected* to the shear-stress axis from the specific value of normal stress at which the values of instantaneous friction and cohesion are calculated (using equations (3.40) and (3.39), respectively). This special case value is what you would calculate if someone asked you for values of "cohesion" for either an intact rock sample (calculated by using a nonlinear Mohr envelope such as Hoek–Brown) or for the rock mass.

Let's look briefly at how the Mohr envelopes obtained by using the Hoek–Brown equations (Hoek, 1983; Ucar, 1986) depend on changes in the parameters m and s. In Fig, 3.25a one can see that the Mohr envelope starts out relatively steep, for large values of m_i, with a small value of tensile strength. As m_i (or m, for the rock mass) decreases, the slope flattens out, predicting a larger value of tensile strength. All the curves shown assume that the unconfined compressive strength σ_c is held constant (here at 50 MPa). For other values one would scale the Mohr envelope by the unconfined strength; this

Fig. 3.26. Definition of
pressure-dependent,
"instantaneous" values of
cohesion (equation (3.40)) and
friction angle (equation (3.41),
or coefficient) for a nonlinear
Mohr envelope (equation
(3.32a)).

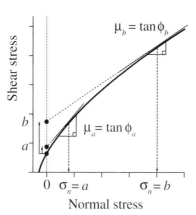

is why the Hoek–Brown criterion is often written in a normalized form (e.g., Brady and Brown, 1993, p. 112).

As m decreases, smaller values of the instantaneous friction angle are also implied. You can see how this works by examining Fig. 3.26. Here, one can see that, at low normal stresses (point a on the diagram), the Mohr envelope is steep, implying large values of friction angle (or coefficient) at that value of normal stress. At greater values of normal stress, though, the shallower slope yields smaller values of friction. At the same time, the instantaneous cohesion increases with normal stress, opposite to friction. So these two parameters trade off systematically as the normal stress or depth increase.

Returning to Fig. 3.25, the right-hand panel shows the effect on the Mohr envelope of changes in the parameter s. The effect is not nearly as pronounced as for m, but the effect is to lower the curve while preserving its slope. Reductions in s, for the rock mass relative to the intact rock material, contribute to a reduction in the cohesion while not greatly affecting the friction.

Now that we have these equations, how do we use them? First, we note that the Hoek–Brown criterion gives us a means to represent the strength of an intact rock sample—better, it turns out, than either the Coulomb or the various Griffith criteria (Hoek, 1983). For an intact rock, $s = 1.0$ and $m = m_i$, meaning that we can choose values of m_i from any table of these values—such as Table 3.5—in lieu of running a testing program on a set of rock samples. Along with this we also need to choose a value of the unconfined compressive strength of each rock type, such as those listed in Table 3.6. Then one can plot up the strength envelopes, and calculate relevant quantities like the intact rock's tensile strength, using the equations listed in this section.

Let's compare the strength envelopes for *granite* and *limestone*, as an example, by selecting the appropriate values and seeing what we get. First we'll look at the **intact rock material**. For the granite we choose values of $\sigma_c = 300$ MPa (Table 3.4), $m_i = 33$ (Table 3.5), and $s = 1.0$. For the limestone, $\sigma_c = 50$ MPa (Table 3.6), $m_i = 9$ (Table 3.4), and $s = 1.0$. It is acceptable to use rounded or approximate values because the RMR system and Hoek–Brown

Fig. 3.27. Mohr envelopes for intact samples of granite and limestone obtained by using the Hoek–Brown criterion. Stipple pattern indicates tensile stress regime.

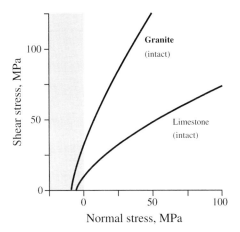

criterion are approximate by nature and take such uncertainties into account automatically.

The Mohr envelopes for the intact granite and limestone samples are shown in Fig. 3.27. Using equation (3.37) we also calculate the tensile strengths to be −9.1 MPa for the granite and −5.5 MPa for the limestone; these are the intercepts on the Mohr diagram just as in the previous cases of Coulomb and Griffith criteria discussed earlier in this chapter. Notice that the envelopes are *curved*—not straight lines. This is characteristic of intact rocks (see Scholz, 1990, p. 18–19; Lockner, 1995, 1998) and rock masses (Barton, 2013).

The slope of the envelope at each point is the instantaneous friction coefficient (or angle). See how the curve shallows out as normal stress (or confining pressure, relating the horizontal axis to depth) increases? The friction angles—called "internal friction" to make clear the distinction between intact rock and sliding surfaces (Handin, 1969; Savage et al., 1996)—also vary with normal stress.

The effect of characterizing rocks as intact or rock masses is clearly demonstrated in Fig. 3.28 for both rock types. In the left-hand column of Fig. 3.28 the Hoek–Brown envelopes for granite demonstrate the profound effect of discontinuities, groundwater, and (indirectly) the larger scale of an outcrop on the Mohr failure envelope. The envelopes for the limestone are shown to the right, at the same scale, for comparison. In the middle panels one can see how the instantaneous friction angle changes with normal stress, plotted—of course—for positive (compressive) values of normal stress. Friction cannot be calculated or measured for surfaces that are not in contact. The slopes become subparallel beyond modest values of normal stress (~20 MPa) suggesting little dependence of friction angle on confining pressure for depths exceeding about a kilometer.

Interestingly, rock mass characteristics reduce the value of friction for a given RMR. Frictional strength of surfaces in a rock mass in general depends on the scale of interest, as demonstrated by Barton and Choubey (1977),

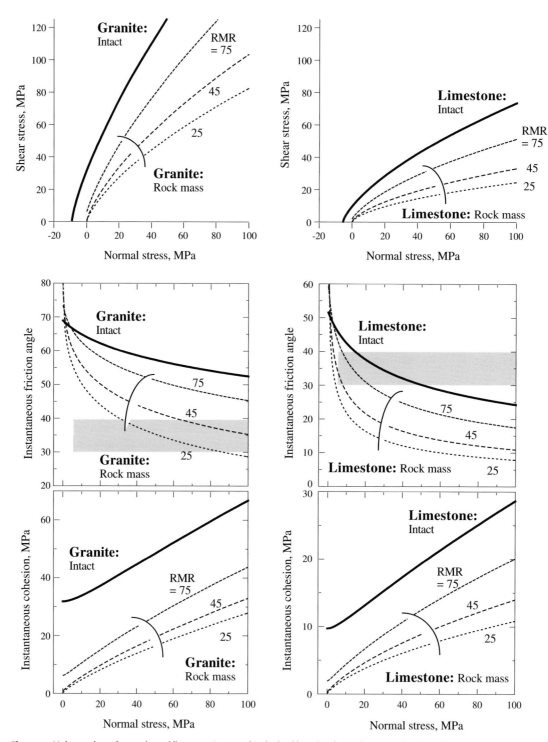

Fig. 3.28. Mohr envelopes for granite and limestone (top panels) obtained by using the Hoek–Brown criterion and common ranges of RMR. Middle panels show how the pressure-dependent (instantaneous) friction angle varies with RMR for both rock types. Horizontal shaded bands indicate typical ranges of Byerlee (basic) friction angle measured for intermediate pressures (5–200 MPa). Lower panels show the analogous dependence of instantaneous cohesion for these rock types.

Barton and Bandis (1990), and Barton (1976, 1990), rather than maintaining a constant value as sometimes suggested (horizontal bands in the figure). This subtle effect is not related to changes in the physics of friction along the rock surfaces, which is scale-invariant (Scholz, 1990, p. 44–47). Instead, it results from the addition of terms that include the discontinuous, sawtooth, or wavy geometry of the large-scale fracture surface and the pressure (normal stress) dependence of dilatancy during slip—created if the rock can ride up and down the bumps on the surface as is common at low pressures (e.g., Priest, 1993, p. 263–264; Brady and Brown, 1993, p. 122–124). The instantaneous cohesion also shows a normal-stress dependence for both rock types, with the value increasing with normal stress and depth; this can be understood by examining Fig. 3.28.

As normal stress or confining pressure on a rock mass increases, the nonlinear peak-strength envelope indicates that the local slope (instantaneous friction angle) decreases while the local, extrapolated intercept (instantaneous cohesion) increases. This trade-off in friction and cohesion is also apparent in the testing results presented by Byerlee (1978) for the intermediate and high pressure regimes, leading to his bilinear friction law (e.g., Kirby and Kronenberg, 1987) and the bilinear brittle strength envelope for planetary lithospheres (e.g., Brace and Kohlstedt, 1980; Kohlstedt et al., 1995; Kohlstedt and Mackwell, 2010). This trade-off is also the basis for an assessment of the stability against frictional sliding for rock slopes that bound highways, open pit gold mines, and landslide-prone slopes (e.g., Wyllie and Norrish, 1996).

The curves of instantaneous friction angle vs. instantaneous cohesion for the granite and the limestone, shown for the range of RMR values in Fig. 3.29, define two regimes: a combination of high friction and high cohesion is more stable against frictional sliding, and located in the upper right of the diagram, whereas low values would be associated with potential instability. The strength of the rock mass at a given value of RMR is indicated by the labeled curves in the figure. If one were to, say, pressurize an outcrop with snowmelt that percolates through the fracture network, one could invoke the concept of effective stress and reduce the frictional resistance of the rock mass; this would have the effect of shifting the strength values to the left in Fig. 3.28 and pushing them into an area in which the rock mass is too weak to support the new conditions. The design of rock slopes (e.g., Wyllie and Norrish, 1996) is accomplished routinely by first calculating the state of stress in the rock mass due to the rock density and slope geometry (angle and height)—analogous to calculating a Mohr circle—and then comparing that stress state to the strength of the rock mass—given by plots such as Fig. 3.29. Rock engineers invariably find that values of cohesion and friction angle for a large rock exposure under actual conditions—that may be as much as 1 km high—are significantly smaller than the intact-rock values and are commonly quite close to those predicted by a rock-mass strength criterion such as Hoek–Brown.

Fig. 3.29. Trade-space between inversely varying cohesion and friction angle for granite and the limestone. Note the differences between the intact and fractured (rock mass) values for the particular rock types and the overlap in values for these lithologies (for appropriate values of RMR).

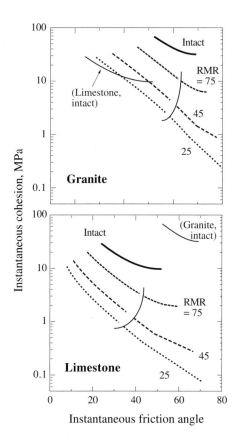

3.5.3 Time-Dependent Friction

Much more is known about friction in general, and for rocks in particular, than perhaps has come across in this chapter. In this section we'll briefly examine friction from a dynamic point of view. Excellent treatments are given by Scholz (1990, 1998, 2002), Lockner (1995), Savage et al. (1996), Harris (1998), Marone (1998b), Beeler et al. (2000), and Paterson and Wong (2005). What follows here is a summary of time-dependent rock friction to help us see how the coefficient and angle of friction used in the Coulomb and Hoek–Brown criteria vary as a function of time.

Friction is the shear strength, or resistance to sliding, of two rock surfaces in contact. Although it may seem odd, friction actually requires that the surfaces slide, so that the normal and shear stresses can be related across them. With two surfaces in contact, the frictional strength of the interface (the two opposing surfaces that are in contact) depends on two sets of factors—**slip rate** and the **physical state of the surface**. These factors exert a subtle effect on the frictional strength, but an important one nonetheless. Their main effects on friction are:

- **Frictional resistance depends on time**: both the *rate of sliding* (the slip velocity) and the *amount of time* that the surfaces were held in contact since the previous slip episode (see Dieterich, 1972, for an early and pioneering look). "Static" friction—when the frictional strength is achieved again after sliding—increases with the duration of stationary contact; "dynamic" friction decreases as the slip rate gets faster.
- The **stability of frictional sliding** depends on the interaction, or competition, between the interface and its surroundings. As we'll see in Chapter 8, following Scholz (2002), "strength" (whether frictional or otherwise) can be thought of as a comparison between the characteristics of the interface and those of its surroundings (the *loading system*). **Stable** frictional sliding is steady, creep-like motion of a "strong" interface, whereas **unstable** sliding is episodic, stick-slip motion of a "weak" interface (e.g., Brace and Byerlee, 1966; Byerlee, 1970; Dieterich, 1978; Linker and Dieterich, 1992).

What do static and dynamic friction mean?

This approach to strength and friction means that **"static" or "dynamic" friction are imprecisely defined for a particular rock** (Marone, 1998a,b), since the element of time is involved in each one. These two terms are now considered to be inaccurate, or at best incomplete (see Marone, 1998a,b), although they see continuing and widespread use. Let's examine what is meant by these common terms. To do this, we'll use what is known as a **rate-and-state variable friction law**. Lajtai (1991) paralleled much of this work in his paper on the time-dependent strength of *fractured rock masses*.

The frictional resistance of two rock surfaces in contact is given succinctly by (e.g., Marone, 1998b; Scholz, 1998)

Rate and state friction law

$$\tau = \left[\mu_0 + \left(a \ln \left(\frac{V}{V_0} \right) \right) + \left(b \ln \left(\frac{V_0 \theta}{D_c} \right) \right) \right] \sigma_n \tag{3.45}$$

$$\underbrace{\qquad\qquad}_{Rate} \qquad \underbrace{\qquad\qquad}_{State}$$

in which τ is shear stress acting on the surfaces, μ_0 is the basic (or intrinsic) friction angle given by Byerlee's rule (Scholz, 1998; typically 30°; Byerlee, 1978; Kohlstedt et al., 1995; Lockner, 1995) and used in this chapter to characterize the friction of discontinuity surfaces, V is the slip velocity, V_0 is a reference slip rate used in obtaining the value of μ_0, θ is a state variable (Ruina, 1983) that describes the healing rate and memory of the surface, D_c is a critical slip distance (e.g., Okubo and Dieterich, 1984, 1986) for slip velocity to change from V_0 to V, a and b are empirical constants obtained from rock friction experiments, and σ_n is the normal stress resolved across the surface (see Fig. 3.30a).

The term following a in equation (3.45) is an instantaneous slip rate sensitivity (Tse and Rice, 1986). The term following b represents the slip history, memory of faulting, average asperity (or contact) lifetime, and healing rate.

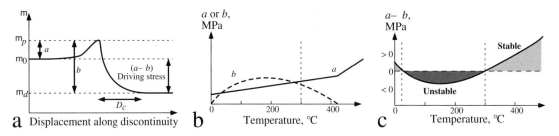

Fig. 3.30. Rate and state parameters a and b: (a) with displacement along a discontinuity surface (μ_p, peak friction, μ_0, basic friction, μ_d, dynamic friction, (b) as a function of temperature, and (c) showing their relationship to stable and unstable frictional sliding regimes (after Tse and Rice, 1986).

Thus, the first term (following a) describes the *rate effect* on friction, and the second term (following b) describes the *state effect* on friction.

Frictional stability in equation (3.45) comes from comparing the rate and state terms. Specifically, this involves calculating the quantity $(a - b)$, which, as a *long-term rate sensitivity term*, is **the key parameter in rate-and-state friction laws** (e.g., Tse and Rice, 1986; Marone, 1998b; Scholz, 1998). This term corresponds to the **driving stress** for discontinuity displacement. If $(a - b) \geq 0$, then slip is stable and aseismic; if $(a - b) < 0$, then slip is conditionally unstable, stick-slip, and seismogenic (depending on the normal stress and the magnitude of perturbations from the loading system). Earthquakes nucleate at the transition between high-temperature stable and adjacent unstable regimes (e.g., Scholz, 1998).

The key parameters (a and b) are shown in Fig. 3.30. Both a and b are roughly proportional to σ_n but depend on temperature, with a being positive for low temperatures ($T < 300$ °C for felsic rocks, 600 °C for mafic rocks: Abercrombie and Ekström, 2001; Scholz, 2002, pp. 320–322) and b being first negative below ~300 °C (or 600 °C) and then becoming positive (see Tse and Rice's treatment of the data of Stesky et al. (1975) and others). The temperature effects on $(a - b)$ are shown in Figs. 3.30b and 3.30c. The implication is that unstable, stick-slip behavior—and earthquake nucleation—is largely restricted to depths in the Earth and other planets (Schultz et al., 2010a–c) for which the approximate 300 °C (or 600 °C) isotherm is not exceeded (e.g., Tse and Rice, 1986; Marone, 1998b; Scholz, 1998; Abercrombie and Ekström, 2001).

The competing effects of *rate* (of sliding) and *state* (of the surface) are suggested by equation (3.45). The dependence of friction on sliding rate is well known to anyone who has stomped on the accelerator in their vehicle to get off an icy patch on the roadway—of course, the tires spun and there was no traction, thus defeating the purpose of this instinctive yet ultimately unproductive response. Getting going on a toboggan or other snow-worthy device is a comparable yet opposite example. In contrast, the regain of strength on a fault between earthquakes—called **fault healing**—is related to the recurrence interval between earthquakes and is a clear demonstration of the ability of rock interfaces to strengthen with stationary contact time (e.g., Cowie, 1998a,b).

What this means is that the "basic friction angle (or coefficient μ_0)" used in the Coulomb, Hoek–Brown, or other frictional strength criteria provides a good first-order measure of the peak frictional resistance of rock surfaces and faults (e.g., Barton, 1976, 1990; Byerlee, 1978; Sibson, 1994). Modulating this value are the competing, secondary effects of time, which either strengthen (through stationary contact time and static friction) or weaken (through slip velocity and dynamic friction) the interface slightly relative to the basic friction value.

Stress shadows: Coulomb vs. rate and state

Stress shadows around fractures related to changes in the proximity to failure (or factor of safety), assessed by using the Coulomb criterion (Section 3.5.3), also predict reasonably well where aftershocks will or will not occur after a large earthquake (e.g., Reasenberg and Simpson, 1992; Stein et al., 1992, 1994; King et al., 1994; Harris, 1998). Harris and Simpson (1998) showed how rate-and-state stress shadows can *advance or retard the time to failure* of faults sufficiently close to an earthquake. This example illustrates the importance and power of **time-dependent friction** in geophysical phenomena. Rate-and-state effects may also be important in interpreting microseismic activity caused by oil and gas production, where slow frictional sliding along faults may be induced (Zoback et al., 2012), and in seismicity induced by wastewater injection (Dieterich et al., 2015). Other important time-dependent fracture processes—known collectively as *subcritical fracture growth*—are discussed in Chapter 9.

How **stress triggering** of fault slip works is subtle, because stresses due to Earth or ocean tides, for example, are *of the same magnitude* (~0.01 MPa) as those due to earthquakes, but these do not generally trigger slip on continental faults (see Vidale et al., 1998; Lockner and Beeler, 1999; Beeler and Lockner, 2003; Brodsky et al., 2003; Scholz, 2003). The normal stress term in the Coulomb criterion (3.7a,b) need not represent static stress—in fact, earthquake physics is better understood for *dynamic* normal stress changes (e.g., Linker and Dieterich, 1992; Tworzydlo and Hamzeh, 1997; Freed, 2005; Hill and Prejean, 2007; Dieterich et al., 2015), such as **vibrations** radiated outward from an earthquake (e.g., Gomberg et al., 1997). It appears that abrupt changes in normal stress on a potentially triggered fault, of appropriate *amplitude and frequency* (Perfettini et al., 2003; Boettcher and Marone, 2004), with the magnitudes given by the static Coulomb or rate-and-state calculation (~0.01 MPa), can weaken faults sufficiently for an earthquake event to nucleate there.

In addition to natural earthquakes, time-dependent and rate-and-state friction laws can advance the understanding of seismicity induced by human activities (e.g., Dieterich et al., 2015). Extensive previous work has demonstrated that subsurface fluid withdrawal can induce seismicity and/or faulting within or near a reservoir (e.g., Yerkes and Castle, 1976; Segall, 1989; Grasso and Wittlinger, 1990; Teufel et al., 1991; Segall et al., 1994; Segall and Fitzgerald, 1998; McGarr et al., 2002; Zoback and Zinke, 2002). Conversely, a growing body of evidence indicates that induced seismicity can be caused by wastewater injection (Zoback et al., 2012; Zoback and Gorelick, 2012; Ellsworth, 2013; Kim, 2013; McGarr, 2014; Hornbach et al., 2015; Walsh and Zoback, 2015; Folger and Tiemann, 2016) and perhaps nucleate slip on nearby pre-existing faults if those faults are already close to failure from the pre-production *in situ* tectonic stress field and if injection volumes or rates are sufficiently large (Frohlich, 2012; Keranen et al., 2013; Sumy et al., 2014; Dieterich et al., 2015). Induced seismicity and fault slip can also occur in geo-thermal systems (Majer et al., 2007; Moeck et al., 2009). As noted by Walsh and Zoback (2015) and others, hydraulic fracturing during reservoir stimula-tion ("fracking") is not generally implicated in causing induced and felt seis-micity given the significantly smaller volumes of fluids involved as compared to wastewater injection and disposal, although work by several groups (Baig et al., 2012; Atkinson et al., 2016) suggests a relationship in western Canada.

The rate and state approach also applies to **gouge and granular aggre-gates** like sandstone (e.g., Karner, 2006). For example, Marone and Scholz (1988) showed that it is the fault gouge that often regulates whether faults generate earthquakes (i.e., slip unstably) or not. Other phenomena such as afterslip (Marone et al., 1991) and healing (e.g., Karner et al., 1997; Cowie, 1998a; Marone, 1998a) are also intimately related to the time-dependence of friction (see Tse and Rice, 1986, for a pioneering demonstration of this approach to the earthquake cycle, and Marone, 1995, for yield criteria for fault gouge). Mair et al. (2002) demonstrate how grain size distribution and shape control whether gouge deformation is stick-slip (for uniform grain size, rounded grains, and low confining pressure) or stable (for angular, interlocking, fracturing grains at higher confining pressures). In summary, any time there are particles in contact, rate-and-state friction—and its time dependencies—should be entertained in order to understand the kinematics and mechanics of the deformation.

Once the frictional strength is achieved (as assessed, for example, by the Coulomb or Hoek–Brown criteria), the *stability* of frictional sliding along the rock surfaces depends on the stiffness of the loading system (that is, the rocks adjacent to the frictional interface) and the competing effects of time, as portrayed in equation (3.45). The sliding can be either *stable or unstable* (stick-slip). Because a peak-strength criterion like Coulomb or Hoek–Brown can only assess the maximum values of stress that can be withstood before sliding starts, it says nothing about what happens afterwards. This is why we need the rate-and-state formulation—to assess the **stability of frictional sliding**. In addition, we need the tools of fracture mechanics (Chapter 8) to gauge **how much offset will occur** along a surface of given length (such as a joint or fault surface) once the frictional resistance has in fact been achieved.

3.5.4 Geologic Implications of Mohr Diagram Predictions

Joints are predicted, from a Mohr diagram (Fig. 3.31a), where the Mohr circle intersects the failure criterion at the normal-stress axis, on which shear stress equals zero. In this case, *pure opening displacements* are predicted across a fracture, bedding plane, or other suitably oriented surface (Fig. 1.14). It is recognized in the literature that the homogeneous elastic stress described by the Mohr circle is not a sufficient criterion for crack nucleation or its propagation (see Ingraffea, 1987; Pollard and Aydin, 1988, for additional criteria).

If, on the other hand, the far-field principal stresses cause the intersection of stress state and failure envelope to occur in the tensile regime, but with nonzero shear stresses, *oblique dilation of a crack* (or other discontinuity) will occur (e.g., Engelder, 1987, 1999), producing a **mixed-mode crack** (Fig. 3.31b; Ingraffea, 1987), with the sense of shearing mode (Fig. 1.15), I-II or I-III, depending on the 3-D state of stress (e.g., Pollard et al., 1982). Mixed-mode cracks (e.g., Pollard et al., 1982; Anderson, 1995) have been variously referred to as "transitional tensile joints" (Suppe, 1985, p. 154; Engelder,

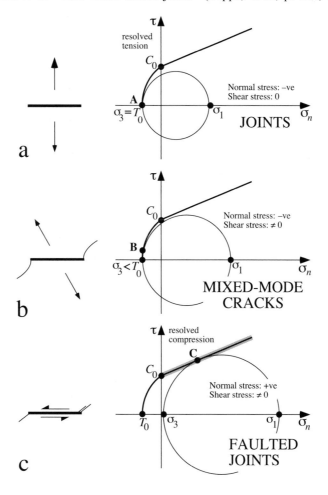

Fig. 3.31. Initial fracture modes associated with Mohr failure in relation to the parabolic Griffith-type envelope, in the tensile regime, and a linear Coulomb or Modified Griffith envelope under compressive conditions (after Schultz, 2000b). Initial displacements along fracture surface in the three main regimes are: (a) pure tensile opening (joints at point A); (b) oblique dilation (mixed-mode cracks at B); and (c) frictional sliding (faulted joints at C). Shaded area for C shows range of applicability of Coulomb and Modified Griffith criteria (compressive resolved normal stresses on existing surfaces).

1999) or "hybrid extension/shear fractures" (Hancock, 1985; Engelder, 1987; Price and Cosgrove, 1990, p. 44; Ramsey and Chester, 2004). According to the Mohr diagram prediction, such mixed-mode cracks, in an outcrop, would be oriented oblique to the remote principal paleostress directions because the cracks would not be associated with principal planes. Notice that the shear stress causes the crack walls to move *apart obliquely*—they cannot slip frictionally because the normal stress resolved on them is tensile and acts to pull them apart. Thus the sense of shear stress gives the direction of **oblique opening** but shear *stress* here does not imply shearing *displacements*, as on faults, because there is **no shearing of the crack walls** themselves (in the sense of frictional sliding and wear).

With continued extensional strain, mixed-mode fractures can either produce a fault zone oriented obliquely to the remote stress direction (for compressive normal stress acting across the incipient fault zone; see Lockner, 1995) or an array of cracks oriented perpendicular to the remote tensile stress trajectory. This sequence is shown in Fig. 3.32, which flows from the work of Hoek and Bieniawski (1965), Peng and Johnson (1972), Nemat-Nasser and Horii (1982), Horii and Nemat-Nasser (1985), and many others. Horii and Nemat-Nasser (1985) demonstrated that **the sign of the least principal remote stress strongly influences the subsequent evolution of an initially mixed-mode set of cracks**. They showed that a *tensile* least principal stress promoted *continued crack growth*, with wing cracks (Brace and Bombolakis, 1963; Erdogan and Sih, 1963; Hoek and Bieniawski, 1965; Pollard and Segall, 1987; Germanovich et al., 1994) forming in response to the oblique dilational displacements (modes I and II) along the parent cracks. The wing cracks nucleate at or near the tips of the mixed-mode cracks to accommodate the oblique kinematics there and propagate essentially normal to the remote least principal stress orientation. The resulting fracture array—that one would observe and map in the field—that has grown to progressively larger strains (e.g., Olson, 1993; Renshaw and Pollard, 1994b) would consist of a set of generally parallel dilatant cracks oriented, on average, normal to the tensile

Fig. 3.32. Schematic illustration showing the evolution of a tectonic joint set, from its initiation at a series of randomly oriented flaws, through growth of initially oblique mixed-mode cracks (second panel), as formally predicted by the Mohr diagram (Fig. 3.31b), through interaction, linkage, and continued growth as a pure mode-I joint set (right-hand panel).

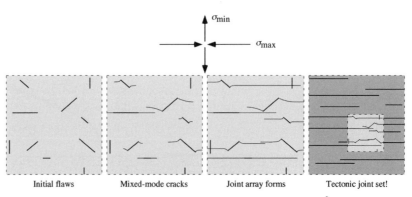

Initial flaws   Mixed-mode cracks   Joint array forms   Tectonic joint set!

→Increasing extension →

Fig. 3.33. Qualitative Mohr diagram for fracture prediction at large strains. Note that oblique (mixed-mode) cracks that are predicted initially, for infinitesimally small strains, grow into pure mode-I joint sets. Faults are predicted for the frictional sliding regime, where resolved normal stresses on potential surfaces are compressive.

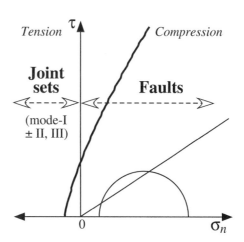

remote principal stress. A Mohr diagram for such larger-strain brittle structures (with these fracture growth processes implicit) is shown in Fig. 3.33. On the other hand, a *compressive* least principal stress promotes *linkage and coalescence of the crack segments into a fault zone* (Nemat-Nasser and Horii, 1982; Horii and Nemat-Nasser, 1985); this is the case on the right-hand side of the Mohr diagram.

Finally, as we've seen above, an intersection having a compressive resolved normal stress, and nonzero shear stress by construction, implies **frictional sliding along a pre-existing surface** (Fig. 3.31c)—this predicts *faulted joints* only, and not the *formation* of faults or fault zones (Martel and Pollard, 1989) having secondary fractures (Segall and Pollard, 1983b; Engelder, 1989), echelon geometries (Aydin, 1988), and gouge (Scholz, 1990; Marone, 1995). These other attributes are commonly associated with inhomogeneous stress states in the rock mass or with specific processes related to fault slip, neither of which is represented on the Mohr diagram.

An interesting case occurs for faults that form in mechanically heterogeneous stratigraphy (Ferrill and Morris, 2003; Fossen, 2010a,b). Here, frictional sliding occurs on segments that dip at 60–65°, whereas steeper fractures that form in surrounding strata—that failed initially to form dilatant cracks there—dip more steeply, 80–90°. As the system shears, the dilatant cracks open; they are also kinematically linked to the frictionally sliding fault segments and located in their trailing, dilational quadrants (see Fig. 6.16). Although the composite fault may exhibit a steep average dip angle which places it in the realm of mixed-mode cracks (Fig. 3.31b; Ferrill and Morris, 2003), the Mohr diagram in this case applies to the formation of the initial, **individual** fault segments, rather than the entire structure. On the other hand, an overall sense of the kinematics of the system—the opening and shearing displacements—*can* be appreciated from the Mohr diagram even though the mechanics of faulting would not be strictly correct in this case.

This well-known limitation of the continuum Mohr-circle approach to crack (Pollard and Aydin, 1988) and fault (Scholz, 1990; Martel, 1990) formation requires that other methods (e.g., Zhao and Johnson, 1991; Johnson, 1995) must be used to predict, for example, whether *conjugate sets* of joints or faults, having various dihedral angles implied by the optimum failure orientation θ_{opt} (e.g., Hoppin, 1961; Muehlberger, 1961; Hancock, 1985; Price and Cosgrove, 1990), will form in rock. Conjugate sets are inferred only when the assumptions built-in to Mohr circle analyses apply, or to certain classes of axisymmetric laboratory experiments that may predispose such sets to form (such as those conducted under plane-strain conditions: see Aydin and Reches, 1982). An investigation of the formation of conjugate fault sets in outcrops invites the use of more comprehensive methods (e.g., Price and Cosgrove, 1990, pp. 42–43; Davatzes et al., 2003) than are available from the Mohr-based approach. Further mechanical development of a discontinuity set, array, or population in rock (Hoek and Brown, 1980; Scholz, 1990, pp. 25–28; Schultz and Zuber, 1994; Ferrill and Morris, 2003) requires more complete treatments of the processes noted above, such as **fracture mechanics** (e.g., Pollard and Segall, 1987; Engelder et al., 1993; Anderson, 1995; Chapter 9).

3.6 Review of Important Concepts

1. The equations for 2-D stress are a powerful and easy-to-use way to solve problems in frictional sliding. Although several sign conventions are possible, if you retain the soil-mechanics, or compression-positive, convention throughout, you will minimize the number of conventions that you need to know. You can easily work with other sign conventions if you have a clear understanding of the approach itself.

2. The **Mohr circle** is perhaps best thought of as a *sketch*, not as the way to calculate numerically accurate answers. Once you understand and can construct Mohr circles, you can work out approximate answers, in real time, anywhere—including under harsh field conditions.

3. Stresses are named using the **"on-in" criterion**, with the subscripts referring to the plane and force direction, respectively. Normal stresses have repeated subscripts; shear stresses have mixed subscripts. Principal stresses are normal stresses acting on planes that have zero shear stress; these planes are the principal planes.

4. A state of stress, starting with normal and shear stresses or the principal stresses, can be **resolved** onto planes having other, specified orientations. This is accomplished using either equations or Mohr circles. This is needed, for example, to evaluate whether a surface, such as a bedding plane or pre-existing fracture, will *slip*

frictionally or crack open. Calculating stresses is a basic preliminary to many important and useful applications in structural geology and geomechanics.

5. Mohr circles are plotted on a diagram that shows the stresses, planes, and angles of interest. Once you define the two orthogonal sets of planes, and their stress states, you plot the two points, draw the diameter, and trace the circle. From this you can estimate rotation angles, resolved stresses, and other useful quantities.

6. The Coulomb criterion represents a **simple failure criterion for the initiation of frictional sliding** along a planar surface. It is superimposed onto the Mohr circle with the intersection point(s) defining the orientations of planes that are predicted to slip. These points can also be defined mathematically by using the Coulomb equations. This criterion forms the first step in understanding more subtle influences on frictional strength, such as rate, temperature, scale, and gouge properties, that are the subject of more sophisticated approaches.

7. Because the **Coulomb criterion** is only valid for cases in which the resolved *normal stresses on a surface are compressive*, other criteria are used for the tensile regime. The Griffith criterion is a parabolic limit on the amount of tensile stress that can be tolerated by flaws before they must propagate; pure mode-I and mixed-mode I-II crack growth is implicitly predicted. Continuation into the regime of compressive normal stresses requires consideration of frictional sliding of closed surfaces in contact with each other, resulting in the Modified Griffith criterion.

8. The **Hoek–Brown criterion** relates a remote two-dimensional stress state to the peak strength of a fractured rock mass. Modeled after the Griffith and Modified Griffith criteria, it has largely *replaced* these other approximations to rock-mass deformation, including for the intact rock material itself.

9. The **Rock Mass Rating system (RMR)** is a widely used method for systematically assessing the characteristics of large-scale rock exposures. It incorporates the strength of the intact rock material (its "lithology") along with the weakening influence of the joints, faults, and other discontinuities, plus groundwater conditions and pore-water pressure. Values of RMR can provide reasonable estimates of the tensile, compressive, and shear strengths, and the deformability ("deformation modulus"), of *large-scale rock exposures.*

10. **Deformation modulus** is a measure of the local slope of the loading part of the complete stress–strain curve for the rock mass. It includes the effects of fractures and other types of discontinuities, along with groundwater conditions and rock weathering or alteration, into the

value for the rock-mass deformability. These factors are not included in the corresponding value for the Young's modulus for the intact rock material. Deformation modulus brings rock mechanics into the realm of *field-scale applications* and projects by incorporating the scale effect on strength and deformability.

11. Peak-strength criteria including Coulomb or Hoek–Brown predict the values of elastic stress that are safe—in other words, the values of stress that do not cause frictional sliding (or jointing, for the tensile regime). Once the criterion for slip is met, then we need additional tools to evaluate what will happen next.

12. Once frictional sliding starts, it can be either stable or unstable. **Stable slip** requires energy input into the system to keep the surface slipping. In contrast, **unstable slip** (stick-slip behavior) can be seismogenic. Frictional stability depends on the *relative characteristics* (such as stiffness or velocity change) of the interface and its surroundings.

13. A rate-and-state friction law is a useful adjunct to more classical concepts of friction, such as the Coulomb criterion. **Rate-and-state friction describes how time influences and modulates the frictional resistance of surfaces in rock**. The widely used notions of static and dynamic friction are seen in this context to be particular values of the spectrum of frictional resistance in rock. The formulation works for any rock surfaces in contact, from planar ones (like bedding planes or faults) to irregular ones, as in gouge and granular (and porous) rocks such as limestones, tuffs, and sandstones.

14. The Mohr diagram and Coulomb criterion are most applicable to predictions of *initial* rock failure. Joint sets and faults that begin as smaller crack arrays grow into orientations consistent with the remote stress state as they accommodate progressively larger amounts of strain.

3.7 Exercises

1. Given each of the following remote principal stresses in parts (a) through (d), calculate: The resolved stresses (normal and shear) acting on the associated planes; the mean stress; and the maximum shear stress. Then draw a sketch showing the geometry of the problem. All stresses are given as compression positive. Include the sign conventions for each of your answers.

   (a) $\sigma_1 = 5.0$ MPa; $\sigma_2 = 1.0$ MPa; angle from σ_1 to outer-normal vector, $30°$.

   (b) $\sigma_1 = 2.5$ MPa; $\sigma_2 = 1.2$ MPa; angle from σ_1 to outer-normal vector, $-5°$.

   (c) $\sigma_1 = 1.0$ MPa; $\sigma_2 = -1.0$ MPa (simple shear); angle from σ_1 to outer-normal vector, $45°$.

   (d) $\sigma_3 = -3.5$ MPa; $\sigma_1 = -1.4$ MPa; angle from σ_1 to plane, $16°$.

2. For Exercise 1(a) above, calculate and plot the values of normal and shear stress, as a function of angle (from $0°$ to $90°$), using the 2-D stress equations. Then superimpose the results of your calculation from Exercise 1(a) (the resolved stresses) on your diagrams. Label all the important values on both diagrams.

3. Given the following remote stress components (compression-positive), calculate the values of the remote principal stresses and the angles to the principal planes. Next, draw a free-body diagram ("stress square") of the initial and final (principal stress) states.
   (a) $\sigma_{xx} = 2.5$ MPa; $\sigma_{yy} = 1.2$ MPa; $\sigma_{xy} = 0.83$ MPa.
   (b) $\sigma_{xx} = 1.3$ MPa; $\sigma_{yy} = -0.75$ MPa; $\sigma_{xy} = 1.1$ MPa.
   (c) $\sigma_{xx} = -1.6$ MPa; $\sigma_{yy} = 1.45$ MPa; $\sigma_{xy} = -2.13$ MPa.

4. Label the free body diagram for a 2-D state of stress, showing the two normal stresses and the two (equal and opposite) shear stresses. Hint: first choose your sign convention (compression positive or tension positive) and make all your stress components positive for clarity.

5. Draw a cube in an xyz coordinate system so that the edges of the cube are aligned with the coordinate axes. Now label the 3-D stress state on each of the cube's faces, including the three principal stresses (σ_{xx}, σ_{yy}, and σ_{zz}) and the six shear stresses ($\tau_{xy} = \tau_{yx}$, $\tau_{xz} = \tau_{zx}$, and $\tau_{yz} = \tau_{zy}$); first choose your sign convention and then make all your stress components act in the positive direction or sense.

6. You examine the state of stress near a fault and find values of $\sigma_x = 1.2$ MPa, $\sigma_y = 3.8$ MPa, and $\tau_{xy} = 0.15$ MPa.
   (a) Draw the Mohr circle corresponding to this stress state, labeling all the parts.
   (b) Using values of $E = 41$ GPa and $v = 0.38$, calculate the normal and shear strains in the rock mass, assuming
   (b1) plane stress and
   (b2) plane strain conditions. Draw a sketch and interpret your results for parts (b1) and (b2).

7. Using the following stress state and the Coulomb criterion written using normal and shear stresses, calculate whether or not the fracture will slip.

   $\sigma_1 = 3.5$ MPa

   $\sigma_2 = 0.52$ MPa

   $\theta = 28°$ (angle between the plane's outer normal and the σ_1 direction)

   $C_0 = 0.52$ MPa

   $\mu = 0.6$

8. Referring to the previous exercise (Exercise 7):
   (a) Redo Exercise 7 using the principal-stress form of the Coulomb criterion. Show values for all the parameters you need for this calculation.
   (b) Calculate the optimum orientation of the fault that would slip using the parameters given in Exercise 7. If σ_1 is vertical, what kind of fault would this be?

9. Redo Exercises 1 and 3 using only the graphical Mohr diagram method. Don't plot up the values you've already calculated. How do these diagrams compare with the values from those problems?

10. Plot up the Mohr circle and the Coulomb criterion from Exercises 7 and 8(a) on a Mohr diagram, and assess whether the surfaces would slip (i.e., re-examine your calculations above). Include all labels on your diagram and discuss any differences between the two cases.

11. Using the results of Exercise 7, recheck your optimum fault calculation from Exercise 8. You may need to either adjust the stress state as needed to ensure single-point intersection (tangency) with the failure criterion or, alternatively, state the range of orientations that can slip. An accurate drawing will improve your answers to this problem.

12. You're working on a fold and you suspect that the remote regional stresses are oriented approximately normal to the original bedding attitude (i.e., sub-horizontal). The situation is shown in Fig. 3.34. Using the Coulomb failure criterion as a guide, what type of deformation of the bedding planes might you expect to see? Estimate or calculate the normal and shear stresses resolved on the bedding plane and how they may vary from fold limb through hinge. Draw a sketch showing the structures you might see, including secondary structures.

13. Discuss the relative merits (strengths and challenges) in the Coulomb criterion, Griffith (and Modified Griffith) criteria, and Hoek–Brown criteria.

14. Describe the principal factors that need to be assessed or measured for RMR, including their significance. Which ones are the most important?

15. You're working on a project to design a new power plant that is sited in the Columbia Plateau sequence of layered basaltic lavas. Based on the following field description that your predecessor recorded in her field notebook:

> The rock is a fine-grained tholeiitic flood basalt, Grande Ronde flow. The intact rock material has a uniaxial compressive strength of 325 MPa (from prior test data available to the project), $m_i = 22$, cohesion of 66 MPa, and a tensile strength (from a Brazilian test) of 15 MPa. The rock mass consists of columnar joints forming prismatic blocks typically less than 2.4 m tall by about 0.5 m in diameter. RQD is 60% from prior core study and verified by the linear traverse measurements we did yesterday. There are several joint sets: vertical with variety of strikes (these bound the columns) and sub-horizontal that split many of the columns. The column faces are smooth

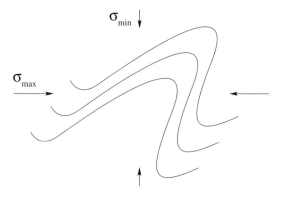

Fig. 3.34. Sketch map of folded bedding and the associated stress state now acting on the fold (for Exercise 12). The example is patterned after field examples studied by Ohlmacher and Aydin (1995) and by Cooke et al. (2000).

to rough (due to plumose structure developed better on some than on others), and several have coatings of secondary minerals, like epidote, but I haven't seen any serpentine on them. The columnar joints tend to gape open to ~1 mm, so only the roughest parts of the column faces are in contact. Also, column faces are not slickensided, so there was no faulting within the flow. The place is wet with groundwater seeping through most column joints.

(a) Calculate the RMR for the site, using a horizontal traverse. Use the average column width as the large-scale joint spacing parameter. Note any corrections to the field description that you find to become necessary to refine your estimate of RMR. Carry any ranges of RMR values over into parts (b) and (c).

(b) Calculate the associated deformation modulus E^* and strength parameters (m, s, unconfined compressive strength, tensile strength, and cohesive strength) for the basaltic rock mass.

(c) Construct a Mohr diagram showing the rock-mass strength envelope (using m and s), and the envelope for the intact (basaltic) rock material (using m_i and $s = s_i$). Then add a Mohr circle for tensile failure to each and discuss the differences in tensile strength of intact vs. fractured (i.e., outcrop-scale) basaltic rock. Label all important parameters and parts on your diagrams.

16. Now you are given *in situ* stress measurements for the basalt from exercise 15 near the cliff face. The vertical stress σ_1 is 0.55 MPa (compression positive) and the horizontal stress σ_3 is 28% of the vertical stress. Calculate the factor of safety (F_s) for the cliff face noting that F_s = (resisting forces)/(driving forces). In this case, the Hoek–Brown failure criterion represents the resisting forces. (Hint: If you know σ_3, compare the theoretical (calculated from the Hoek–Brown criterion) and actual (*in situ*) values of σ_1 to determine the factor of safety.)

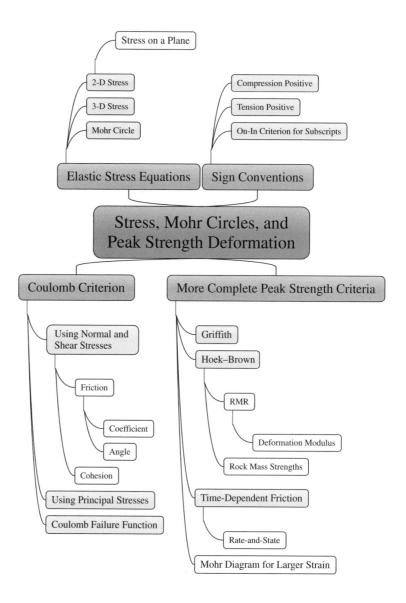

4 Cracks and Anticracks

4.1 Introduction

This chapter is devoted to geologic structural discontinuities that accommodate displacements perpendicular to their surfaces, including opening-mode fractures such as **cracks, joints, veins, and dikes** and closing-mode structures referred to as **anticracks** (Table 4.1). Opening-mode structures (mode-I, Fig. 1.16) are one of the most common types of geologic structural discontinuity. Cracks are defined as *sharp planar to curviplanar surfaces of opening-mode displacement discontinuity* (Table 1.1). During the process of crack growth, crack walls first were created, then were moved apart normal to the fracture trace to provide a slot-like opening in the rock. Frequently the crack is filled by mineral precipitates from hydrothermal solutions, such as quartz or calcite (producing veins), crystallized magma (producing igneous dikes and sills), or even petroleum or natural gas. Near the Earth's surface cracks or joints are often found gaping without any infilling solids (Fig. 4.1); these produce the fracture permeability necessary for efficient transport of groundwater, natural gas, and other fluids. Cracks form interesting and aesthetically pleasing patterns; these joint sets and echelon arrays contain information on the growth history of the cracks and, in turn, the brittle deformation of the host strata. Joint patterns can also provide clues to the geomorphologic and tectonic development of a region. In rock engineering, joints and other types of fractures divide an outcrop into an assemblage called a rock mass (e.g., Hoek and Brown, 1980; Chapter 3).

Anticracks have the opposite kinematic significance to cracks, and their stress and displacement fields can sometimes be represented by using a mode-I crack with a contractional sense of normal strain across them (e.g., Fletcher and Pollard, 1981; Tapp and Cook, 1988; Sternlof et al., 2005; Rudnicki, 2007; Schultz and Soliva, 2012), resulting in converging or interpenetrating discontinuity walls. Anticracks sometimes invoked as models to represent aspects of pure compaction bands or stylolites (see Chapter 1) can form sets and populations that in many ways are similar to those of opening-mode cracks (e.g., Rudnicki, 2007; Tembe et al., 2008; Schultz, 2009; Benedicto and Schultz, 2010). These closing-mode structures can provide an additional source of information on tectonic processes in the Earth's crust.

Fig. 4.1. An array of parallel joints exposed in an outcrop in the Finger Lakes Region of New York State (see Younes and Engelder, 1999).

Table 4.1 Common types of dilatant cracks

Joint	A sharp geologic structural discontinuity having field evidence for discontinuous and predominantly opening displacements between the opposing walls
Crack	A sharp structural discontinuity having field evidence for discontinuous and predominantly opening displacements between the opposing walls, usually whose faces are open and stress-free; same as "mode-I fracture"
Vein	A sharp geologic structural discontinuity having field evidence for discontinuous and predominantly opening displacements between the opposing walls and containing hydrothermal minerals such as calcite, silica, or epidote
Dike	A sharp geologic structural discontinuity having field evidence for discontinuous and predominantly opening displacements between the opposing walls, oriented sub-vertically and/or normal to bedding, and containing liquid or solid magmatic materials
Sill	A sharp geologic structural discontinuity having field evidence for discontinuous and predominantly opening displacements between the opposing walls, oriented sub-horizontally and/or parallel to bedding, and containing liquid or solid magmatic materials

Table 4.1 (Cont.)

Sheeting joint	A curved crack that forms close to, and subparallel to, a topographic surface beneath domes, ridges, and saddles in hard compact rock; thought to result from a combination of large compressive stress parallel to a curved topographic surface, rather than from uplift, reduction of overburden stress, weathering, pluton solidification, buckling, axial splitting, or residual stresses
Hydrofracture	A crack that opens in the subsurface due to internal fluid pressure
Anticrack	A sharp or tabular structural discontinuity having field evidence for predominantly closing displacements between the opposing walls

4.2 Why Study Cracks?

How does deformation occur in the Earth's crust? What happens when a fault slips or a fold begins to tighten? How does the crustal section thicken during continental convergence? What determines the geometry, dip angles, and map-view layout of grabens in igneous provinces like the Columbia River basalts? How is strain partitioned in different stratigraphic units? How does oil migrate from its source to the reservoir? What controls the transport, boiling, and depositional shape of gold-bearing fluids as ore deposits? How is mass transport in coordination with volume loss and stylolite growth in metamorphic terrains accomplished? How do porous rocks such as coarse-grained sandstone deform in thrust tectonic settings? These and other questions can be addressed given an understanding of opening- and closing-mode structural discontinuities.

The answers to these and many more questions can be obtained by studying cracks. Brittle deformation is one of the main ways that rocks and rock masses deform in both the shallow crust—where we live and work—and at deeper levels when the conditions are appropriate. By building up an understanding of the timing, spatial organization, and deformation mechanisms preserved in the geology, we can then construct and test those models and ideas to see if they are consistent with the evidence.

Let's look briefly at some examples where knowing about crack-related deformation can provide useful insight into deformation processes.

Paleostress Indicators. Cracks generally define principal stress planes and therefore can change their geometries under resolved shear stresses, including rotation, twisting, and breaking down into echelon arrays. *Cracks map out the trajectories* (or flowlines) of stress during deformation and provide reliable kinematic markers to help unravel progressive deformation under brittle conditions.

Coaxial Stress and Strain. Cracks provide a record of relatively small strains that can be modeled and understood by using the methods of infinitesimal strain theory (see Chapter 2). They are often (but not always) the initial phase of deformation that increases to much larger finite strains. Field information from cracks and their populations can document the *evolution of finite strains* during progressive deformation.

Deformation Mechanism. Cracks are probably the main manifestation of pressure-dependent (brittle) deformation of rocks. Many geologic processes such as cataclasis, brittle faulting, grain crushing, gouge formation, dike injection, hydrofracturing, and borehole breakouts all occur in association with the *same basic mechanism*: crack growth, underscoring the importance of crack-related deformation. Conversely, anticracks analogous to compaction bands and stylolites convey information about their respective host rocks as well as the environmental conditions (regional stress state, pressure, temperature) under which they formed.

Strain Indicators. Cracks track the local principal strain trajectories as they form since they generally support little, if any, shear stress across them. By measuring the opening displacements across cracks, and summing over the crack arrays, one can obtain a quantitative estimate of the magnitude and distribution of brittle strain in a rock mass along with related properties such as its deformability and the regional stress state that led to the deformation.

Fluid Flow in Crustal Rocks. Fracture networks composed of multiple cracks create ready pathways for fluid and natural gas flow through otherwise nearly impermeable lithologies. This *fracture porosity and permeability* is an important component of the brittle deformation.

Discontinuous Deformation. Why doesn't just one single crack form to take up all the brittle strain? Cracks develop into sets (Fig. 4.2), or populations,

Fig. 4.2. Jurassic Aztec Sandstone, exposed near Lake Mead, Nevada, showing two sets of outcrop-scale joints. The joints have grown into a left-stepping echelon configuration.

whose geometries and statistics carry key information about the physical, chemical, and environmental conditions under which they developed.

Rock Mass Deformation. Fractures change the stiffness of a stratigraphic layer or other geologic body, to first order in proportion to their number and size. Both fracture stratigraphy (Laubach et al., 2009) and mechanical stratigraphy (e.g., Gross et al., 1997) change with evolving fracture characteristics, thus profoundly influencing the resulting geologic structures.

Fracture Processes. As geologists we can *see and map out the causes* of deformation (in this case, cracks). Although it has been known for most of the past century that tiny crack-like flaws can significantly weaken materials such as steel, concrete, plastic, paper, wood, and also rock, these flaws are normally so minute in engineering materials that an electron microscope is often needed to see them. As a result, engineers have relied on *indirect methods* of design based on stress analysis, such as the Mohr circle and various energy methods. Geologic cracks can be large enough, however, for their dimensions, geometries, and related characteristics to be directly observed, leading to displacement-based, rather than stress-based, approaches to brittle tectonics.

4.3 Surface Textures and Their Meaning

Cracks have distinctive surface markings and textures that can enable a geologist or engineer to reconstruct the process of crack growth in the rock. Fracture surfaces can be beautiful and visually arresting studies in abstract art. Scientifically, however, the markings that decorate crack surfaces are systematic and the basic types are unambiguous. The study of crack surfaces is sometimes referred to as *fractography* (e.g., Bahat, 1991) and a vast literature on the interpretation of crack surface morphology exists.

Although the interpretation of crack-surface markings had been subject to considerable debate in the past and continues to motivate improved scientific interpretations (for example, see Chemenda et al., 2011), careful observation and laboratory experiments during the past several decades have motivated a consistent understanding of the morphology of crack surfaces. Excellent exposés on the interpretation of crack surfaces are provided by Kulander and Dean (1985, 1995), Engelder (1987, 1999), Pollard and Aydin (1988), Bahat (1991), McConaughy and Engelder (2001), and Engelder (2007). Pollard and Aydin (1988) present an insightful and comprehensive review of crack terminology, and they outline the evolution of thought during the nineteenth and twentieth centuries on how cracks form and what they mean.

As noted by Pollard and Aydin (1988) and by Engelder et al. (1993), among others, the surface morphology of mode-I cracks (joints) is unambiguously the result of *separation of the crack walls perpendicular to the fracture*. Previous notions that a high rate of crack propagation (Roberts, 1961) or shearing of the crack surface (e.g., Bucher, 1920; Parker, 1942; Hoppin, 1961) created plumose structure cannot be supported in the face of substantial observational and experimental evidence, as well as theoretical arguments, that convincingly demonstrate an opening-mode origin for geologic cracks (e.g., Kulander and Dean, 1985). In this regard, the pioneering work of Jean-Pierre Petit, Alexandre Chemenda, and coworkers (Chemenda et al., 2011) underscores the association of plumose structure with the opening mode of displacement, but explains this not as a result of abrupt separation of joint walls but as a more progressive and discontinuous debonding of the host rock, with dilation banding being an early stage of jointing in certain sedimentary rocks. In this chapter the term crack is used without distinction as to the details of its nucleation while retaining the principal opening-mode kinematics of interest to structural geology and rock engineering.

There are three main elements that are diagnostic of crack surfaces: *origins, hackle*, and *ribs*. Together these three morphologies are collectively referred to as **plumose structure** (Fig. 4.3). Although one may find several other ways to name the textures on crack walls in the geologic and engineering literatures, the basic framework set forth by Kulander and Dean (1985, 1995), DeGraff and Aydin (1987), Engelder (1987), and Pollard and Aydin (1988), and tested in the literature since then, is followed here. The resulting terminology for crack-surface fractography employed in this book represents a distillation of the key elements that are most useful in the interpretation of geologic cracks. Several common types of fracture *patterns*, and their terminology, are discussed in Chapter 5.

Plumose structure =
Origin + Hackle + Ribs

Fig. 4.3. Plumose structure is subtle yet clearly visible in this exposure of a siltstone in the Finger Lakes Region of New York State. Professor Terry Engelder (Pennsylvania State University) and cm ruler for scale (see Savalli and Engelder, 2005).

4.3.1 Origins

Origin = Location of crack nucleation

Cracks typically begin opening at small imperfections, or flaws, that modify the ambient stress state in the rock mass or stratum. If the flaw causes the local stress to become amplified or concentrated (see Chapter 2), then a crack can *nucleate* at that point (see Ingraffea, 1987, for a clear treatment of crack nucleation and growth). The record of crack nucleation can frequently be observed on a joint face as an "origin" (Kulander and Dean, 1985, 1995; DeGraff and Aydin, 1987; Pollard and Aydin, 1988; McConaughy and Engelder, 2001; see Fig. 4.4). You can make one of these yourself by hitting a slab of shale, parallel to the bedding, with a rock hammer. The split sample can display a clear origin (perhaps crushed at the point of impact) along with nicely developed hackle.

4.3.2 Hackle

Twist hackles increase in relief as they point in the direction of crack propagation.

Subtle curvilinear ridges that appear to radiate out and away from the crack origin are known as "hackle." Hackle has sometimes been referred to as "flame structure," "river patterns," "feather structure" (Woodworth, 1896), "plume" (Parker, 1942), "plumose" (Hodgson, 1961), and "fracture lances" (Sommer, 1969). The hackle can typically define the intersection surfaces that separate twisting parts of the crack wall. In this case, the *increase in twist*, or relief between crack wall segments, *points in the direction of the **local** crack propagation direction* (Figs. 4.3 to 4.5). Hackle can be associated with minor reorientations of the propagating crack front due to mixed-mode I-III stresses at the grain scale (Poncelet, 1958; Sommer, 1969; Pollard et al., 1982; Kulander and Dean, 1995). In the extreme, *echelon twist hackles*, also known as "fringe joints," can mark the lateral boundary of a crack where it meets another lithology or fracture surface. (Note that "hackle" can be used as either the singular or the plural form, like "fish.")

Fig. 4.4. Plumose structure cuts across bedding in the same locality as Fig. 4.3. An origin is visible in the middle-right part of the frame (circled) with hackle flowing to the left (mostly), top, bottom, and right.

Hackle fan out in the direction of crack propagation.

Hackle indicates the *direction of local crack propagation* (Michalske et al., 1981). The hackle pattern *diverges* in the direction that the crack surfaces peeled apart (the crack's *propagation* direction), and *converges* back to the initiation or nucleation point (at the origin). This pattern is shown in Figs. 4.5 and 4.6. One can often find surfaces in a jointed rock mass that are bounded by recognizable hackle, demonstrating the paleo-propagation direction.

Fig. 4.5. A symmetric plume is clearly preserved along a joint face in shale from West Virginia. Photo by the author (see discussion in Ayding, 2014).

Fig. 4.6. An elliptical joint tipline, or rib, is well exposed in this fine-grained sedimentary sequence from Provence, France, with subtleties in joint nucleation and growth being explained by Professor Jean-Pierre Petit.

Fig. 4.7. A series of conchoidal ribs outlines a large joint surface in Navajo Sandstone from Utah.

4.3.3 Ribs

Ribs are concave back to the crack origin and reveal the shape of the crack front.

Ribs (or "rib marks") are subtle to pronounced topographic changes on a fracture wall that are concave back to the crack origin (Preston, 1926; Murgatroyd, 1942; Kulander and Dean, 1985, 1995). Ribs are consistently perpendicular to the hackle, even when the opening of the crack is non-uniform (e.g., Bahat and Engelder, 1984; DeGraff and Aydin, 1987; Fig. 4.8). They trace out the transient *shape of the crack edge*, "front," "periphery," or "tipline" (borrowing terminology for the analogous 3-D fracture tip from faulting; Chapter 1) as it was preserved on the rock face. The shape of the transient crack front can be seen in Figs. 4.6 through 4.8; notice that it's curved, like a rib, or a wave moving radially outward in a pond. Although ribs likely aren't a wave phenomenon, in an analogous fashion the curved shape may reveal the relative propagation velocities of points along the crack tip. Identification of ribs and hackle on the same surface provides a robust determination of the local propagation direction of the joint (Fig. 4.8).

Ribs have also been called "arrest lines" (Kulander and Dean, 1985) or "hesitation lines" (Barton, 1983) in an attempt to relate the local crack propagation velocity to the morphology of the rib. There have been some studies in the laboratory that have attempted to relate changes in morphology of ribs (e.g., sharp or gently curved) to the local propagation velocity. Results for plastic blocks reported by Bai and Pollard (1997) suggested that rib marks having continuous hackle across them record variations in propagation velocity ("hesitation"), whereas those with hackle that stop at rib marks are thought to record a crack termination. Slow, quasi-static propagation velocities have been associated with both classes of plumose structure, rather than

Fig. 4.8. Example of a large joint surface in Aztec Sandstone, southern Nevada, that displays both ribs and hackle, giving a downward propagation direction. Rock hammer (circled) gives scale.

rapid, dynamic propagation (e.g., shattering of a window). It appears that variations in the local stress directions and magnitudes (stress gradients) are more important than velocity in producing the rib morphologies that we see, whereas the hackle–rib association may reveal relative propagation velocities.

4.3.4 Other Markings

A collection of very subtle markings can be found in amorphous or very fine-grained materials like glass, and sometimes rock, as a crack propagates away from its origin. These are the mirror zone, Wallner lines, and mist zone.

The *mirror zone* is a small quasi-planar surface—in amorphous materials like glass—that surrounds the crack origin (Kulander and Dean, 1995). It is formed as the incipient crack front accelerates as it grows, but the stresses at the crack front are low enough that atomic bonds oblique to the surface remain unbroken, inhibiting the formation of out-of-plane twisting and hackle (e.g., Poncelet, 1958; Engelder, 1987). The size of the mirror zone tends to decrease with increasing grain size, so it is rarely observed in typical rocks as well as it is in amorphous glass. Nevertheless, mirror radius scales inversely with the magnitude of the applied remote stress (in glass samples; Shand, 1959), barring other material inhomogeneities such as layering.

Wallner lines are subtle, rounded, ripple-like curves having very small relief that can form in the mirror zone of very fine-grained or amorphous materials (Wallner, 1939; Field, 1971). They are wave phenomena frozen in the mirror zone that result from sonic (sound) waves produced by the vibrations resulting from bond breakage and stress release during crack propagation (Frechette, 1972; Lawn, 1993, pp. 98–99; Kulander and Dean, 1995). Wallner lines are convex in the overall direction of crack propagation, pointing back toward the crack origin (Kulander and Dean, 1995). They have been observed in dynamically fractured ceramic insulators (see Kulander and Dean, 1995) and, if interpreted in this fashion, would indicate a *rapidly propagating crack*, with velocities on the order of half the speed of sound in the material. Such velocities may be rarely obtained for rock joints.

The *mist zone* records the transition from purely in-plane propagation to a kind of mixed-mode growth, where the crack begins to turn and deviate slightly from its original plane in the mirror zone (Bahat, 1979; Engelder, 1987). The outer margin of the mist zone is defined by an abrupt roughening of the fracture surface into a chaotic jumble of tiny cracks, as the end product of mist zone formation. Variously termed "velocity hackle" (Kulander and Dean, 1995) or simply "hackle" in the ceramics literature (Bahat, 1979; see Engelder, 1987), these structures record the forking, branching, or *bifurcation* of the crack itself as it achieves a (very fast) terminal velocity, perhaps half the speed of sound in the material (Yoffe, 1951; Lawn, 1993, p. 95; Kanninen and Popelar, 1985, pp. 205–207; Sharon et al., 1995). Branching is similar to kinking under mixed-mode I-II loading, rather than twisting and mixed-mode I-III breakdown into hackle. The crack surface splits into two kinks at a critical propagation velocity (Yoffe, 1951), eventually resulting in fragmentation of the material or cessation of crack growth (or both). Cracks tend to branch with angles (between both new branch cracks) of around 30° (Kalthoff, 1971) as a result of local stresses and dynamic fracture strengths at the branching crack tips (Ramulu and Kobayashi, 1983; Kanninen and Popelar, 1985, pp. 206–207; Lawn, 1993, pp. 95–99).

4.3.5 How Fast Do Joints Propagate in Rock?

Several of the fractographic textures discussed in this chapter can inform, to various extents, about joint propagation velocities. First, plumose structure (origin, hackle, and rib marks) by itself is not diagnostic of propagation rate; instead, it can form for a range of velocities, from very slow (subcritical) to perhaps dynamic (on the order of the rock's sonic velocity, referred to as postcritical by Savalli and Engelder, 2005 and Engelder, 2007). On the other hand, several elements have been attributed to fast crack propagation velocities in various geologic and non-geologic materials. These include a mirror zone ornamented with Wallner lines, a mist zone with velocity hackle and crack branching (Kulander and Dean, 1995), and crack branching itself. Wallner lines demonstrate that the speed of the crack front's advance through the material is comparable to, but somewhat less than, the sonic velocity. In other words, the crack tip must be propagating at roughly the same rate, within a factor of perhaps 2–3, as that of sound waves so that their interference can produce Wallner lines on the mirror zone (Kulander and Dean, 1995). This means that cracks would need to propagate at velocities exceeding hundreds of meters per second, given seismic velocities in typical rocks of >5000 m/s (Goodman, 1989, p. 41). This would be a comparatively rare event with tectonic rates of extension of millimeters to centimeters per year being more typical, and Wallner lines are not unambiguously identified on geologic crack surfaces.

Crack branching (also referred to as velocity hackle in engineering materials) is rarely identified in tectonic joint sets (Kulander and Dean, 1995; cf. Sagy et al., 2001), leading Segall and Pollard (1983a), Savalli and Engelder (2005), and others to conclude that the rate of tectonic joint propagation is characteristically slow, perhaps up to cm/s at the fastest in certain special cases of locally large driving stresses (see also DeGraff and Aydin, 1987, for slow rates of crack growth in cooling lavas to form columnar joint arrays). Crack branching in glass is found for crack tips propagating above 750 to 2155 m/s (Schardin, 1959).

The way in which cracks are loaded exerts a primary control on how they form. As developed further in Chapter 9, the two principal loading configurations, or outer boundary conditions, are referred to as **"fixed-grip"** and **"dead-weight" loading** (e.g., Lawn, 1993, pp. 20–23; Kanninen and Popelar, 1985, p. 185; Engelder and Fischer, 1996; Engelder, 2007; Gudmundsson, 2011, p. 210). A continuously increased remote stress corresponds to *stress-controlled*, or "dead-weight" loading, which is commonly used in the laboratory to study rock deformation and in the engineering analysis of cracked systems. It is also a manifestation of a **soft loading system** (Scholz, 2002, pp. 8, 122) that can lead to unstable and rapid crack growth. As emphasized by several workers including Engelder and Fischer (1996), Olson (2003), and Chemenda et al. (2011), **displacement-controlled loading** often represents the geologic framework of jointing, where joints can form in response to bending of layers during folding, for example. The distinction between loading types is important, as tension (or an effective tension; see Chapter 8) is required to open cracks under stress boundary conditions but not for displacement boundary conditions.

Geologic joints typically propagate very slowly, with velocities generally slower than millimeters per second (e.g., Segall, 1984a,b; Olson, 1993; Renshaw and Pollard, 1994b; Engelder and Fischer, 1996). The rates are so slow that chemical reactions at the joint tip can control, and therefore limit,

Remote loading conditions for joints

Subcritical crack growth

the propagation velocity of geologic joints. Chemically modulated joint propagation, called *subcritical crack growth*, is a well-known field of study in geologic fracture mechanics (Anderson and Grew, 1977; Segall, 1984a,b; Atkinson and Meredith, 1987a; Savalli and Engelder, 2005; Chapter 9); in the laboratory, joints grow *subcritically* at rates of less than $1-10^{-4}$ mm s^{-1}, depending on rock type and environmental factors that influence the chemical reaction rates such as humidity, temperature, and the ability of reactive species to gain access to the crack tip region.

Joints in clastic sedimentary rocks exhibit a range of diagenetic characteristics including cementation that may fill joints to varying degrees (e.g., Laubach et al., 2004, 2010). Joints that formed at depth during burial or uplift in a sedimentary basin can contain a record of pressure and temperature paths during their growth within their diagenetic cements that provides insight into joint propagation rates in these settings (e.g., Laubach et al., 2004). For example, detailed analysis of fluid inclusions contained within diagenetic quartz bridges that grew between joint walls enabled Becker et al. (2010) to demonstrate that joints in the East Texas basin formed primarily during burial, at temperatures between 130 °C and 154 °C, over an interval of ~48 million years. Comparable work on cemented joints in the Piceance Basin, Colorado, by Fall et al. (2012) corroborates these exceedingly long durations of jointing in sedimentary and hydrocarbon-bearing basins. Fall et al. (2012) demonstrated joint growth over a 35-million-year interval, at temperatures from 140 °C to 185 °C, and inferred that the joints grew as a result of hydrocarbon maturation during burial rather than in response to regional tectonism or compaction disequilibrium. This body of work illustrates the dependence of cementation growth rates on temperature, with the direct and useful result that joints that open at rates less than the growth rates of diagenetic cements will be filled, whereas joints that dilate faster than cement growth rates will retain some degree of fracture porosity and permeability. Partial infilling of joints by diagenetic cement has important consequences for joint propagation mechanics and can inform estimates of regional extension rates associated with the deformation (e.g., Olson et al., 2007).

Joint growth in sedimentary basins over tens of millions of years

Kulander and Dean (1995) suggest that the textures that are observed on geologic joints—origin, hackle, and rib marks—all occur within the mirror zone of a propagating crack, so that the high-velocity textures (Wallner lines, velocity hackle, crack branching) should consequently be rarely produced in rock. Savalli and Engelder (2005) emphasize the effects of grain size and near-tip strength on how cracks in rock and engineering materials such as glass differ in their surface morphologies. By implication, with driving stresses for joint growth in rock typically being relatively small (e.g., Rubin, 1993a,b; Engelder and Fischer, 1996), the mirror zone of joints should be quite large—perhaps spanning the entire crack surface. A synthesis of fracture surface textures and terms is given in Fig. 4.9.

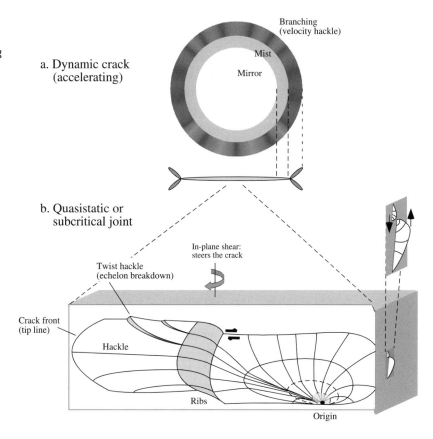

Fig. 4.9. Idealized fractography of (a) crack surface in a fine-grained, amorphous engineering material and (b) joint surface in rock. Inset shows relative propagation of echelon fringe joints locally toward each other (arrows).

a. Dynamic crack (accelerating)

Branching (velocity hackle)

Mist

Mirror

b. Quasistatic or subcritical joint

In-plane shear: steers the crack

Twist hackle (echelon breakdown)

Crack front (tip line)

Hackle

Ribs

Origin

4.4 Joint Geometries

The surface textures on joint surfaces can be readily used to infer the propagation directions of the joints, and from this, the sequence and pattern of joints can be worked out. In this way the smaller-scale details at the scales of a hand lens or outcrop can be connected to larger-scale tectonic processes such as folding or faulting.

In general, cracks and joints define principal stress surfaces, so that these structures will progressively reorient themselves as they grow in such a way as to reduce any shear stress resolved across them. Joints thus map out the stress trajectories in the rock mass at the time that they were growing. Depending on the sense of shear stress acting on a joint surface, the joint will generally either turn, forming wing cracks, or twist, forming echelon fringe joints.

4.4.1 Wing Cracks and Fringe Joints

If the sense of shearing on the joint's propagating edge is *in-plane shear* (Fig. 1.15a), then joints will reorient their actively propagating surfaces by **turning**. In this case, the joint is primarily a mode-I structure with a superimposed component of mode-II shearing displacement (see Fig. 1.16, center panel, lower diagram). A joint that is primarily opening, while also supporting a small component of in-plane shear stress, is then referred to as a **mixed-mode I-II** joint. Minor reorientations of a propagating mixed-mode I-II joint may be expressed as ribs or as wing cracks (discussed below). The fascinating and practical subject of mixed-mode cracks and joints is discussed extensively in the literature (e.g., Brace and Bombolakis, 1963; Pollard and Aydin, 1988; Engelder et al., 1993; Martel and Boger, 1998; Engelder, 1999; Schultz, 2000b; Scholz, 2002, p. 34).

The **rate of joint growth** relative to that of changes in the external stress state determines how smoothly the joint will be *steered* by the resolved in-plane shear stress. If the stress field changes slowly, or the joint is also propagating rapidly enough that its propagation rate is subequal to the rate of change of the local stress field, then the joint will follow the stress field and *curve smoothly* into a new orientation (Cooke and Pollard, 1996). On the other hand, if the stress field rotates suddenly or, more precisely, if the rate of joint growth is much less than the rate of change of the stress field, then the joint can make an abrupt turn to adjust its geometry to the new orientation; this produces a *kinked* crack geometry. The latter case is well documented in nature, where an earlier joint array is deformed later under a different tectonic regime that superimposes shear stresses onto the crack walls (e.g., Pollard and Aydin, 1988). If the joints are able to propagate under the new stress state, they will "kick out" as *wing cracks* into the new orientation (Cruikshank et al., 1991a; Cruikshank and Aydin, 1995). These near-tip structures are considered distinct from pinnate joints that propagate out into principal stress planes from the tips of faulted joints, whose surfaces are not gaping open but are in frictional contact with each other (e.g., Wilkins et al., 2001).

A second way that joints can accommodate small amounts of resolved shear stress is to break down into *echelon fringe joints* (e.g., Suppe, 1985, pp. 175–176; Pollard and Aydin, 1988; Fig. 4.5). These smaller joints are continuous with the hackle found on the parent joint surface closer to its origin (e.g., Kulander and Dean, 1985, 1995) and commonly are well developed where the propagating crack tipline encounters a markedly different stress field that resolves out-of-plane shear stress onto the tip. In the literature, such fringe joints, or **twist hackle**, have been interpreted as **mixed-mode I-III** (Fig. 1.16, lower right panel) breakdown of the parent crack as its propagating tip encounters an area of out-of-plane (Fig. 1.15) shear stress (e.g., Pollard et al., 1982; Lin et al., 2010; Leblond et al., 2011). More recent work by Chemenda et al. (2011) suggests, however, that hackle and fringe joints may preserve the details of the joint opening process in addition to the local stress state.

Twist hackles can be abrupt or gradual (e.g., Younes and Engelder, 1999). Gradual twist hackles have been associated with reorientation of the crack plane during its growth into a region of smoothly rotating stress state, leading to mixed-mode I-III loading resolved onto the propagating tipline. Abrupt twist hackles have been associated with a renewed propagation into a rotated stress state. Gradual echelon fringe joints are commonly found near bedding interfaces where a propagating crack senses the different lithology and deformation properties across the contact.

4.4.2 Common Joint Geometries in Rock

In this section we'll briefly examine several common geometries and patterns of joints that may be encountered in the field. Some common joint geometries are summarized in this section and in Fig. 4.10. Some of the terms come from engineering studies of cracks, but all can be used to infer important aspects of the deformation as presented in Section 4.2. The terms crack and joint are used interchangeably in this section as they often are in brittle tectonics.

Isolated cracks are those that are located sufficiently far from other structures that they function mechanically as a single fracture. Typically this means that the crack must be separated by at least 2–3 times its shortest dimension (height or length; Segall and Pollard, 1980; Olson, 1993; Gudmundsson, 2000) from anything else for it to be isolated. The reason for this is the decay in stress with distance from the crack, as developed in Chapter 2. Anything located

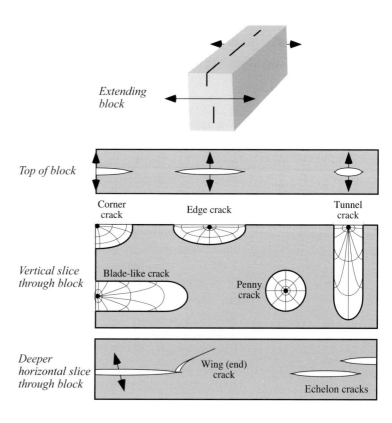

Fig. 4.10. Examples of mode-I crack and joint geometries. The upper two panels show 2-D renditions of joint geometries developed in the 3-D block (inset, above); the arrows show the direction of extension of the block. The lower panel shows a mixed-mode I-II obliquely opening joint exposed in a deep horizontal slice through the block along with its kinematically related wing crack in the trailing, dilational quadrant.

Extending block

Top of block

Corner crack   Edge crack   Tunnel crack

Vertical slice through block

Blade-like crack

Penny crack

Deeper horizontal slice through block

Wing (end) crack

Echelon cracks

beyond this approximate distance changes the stress state so little that the effects on the main crack are negligible. The scale depends on the size of each crack, so it can vary from millimeters to perhaps kilometers, depending on the scale of the system of interest.

Echelon cracks are sets of parallel but offset (non-collinear) fractures that interact mechanically by constructive or destructive interference of their near-tip stress fields (Pollard et al., 1982; Pollard and Aydin, 1988). The geometry of echelon structures is described more fully in Chapter 5.

Edge cracks are fractures that penetrate only partially into the rock mass from a free surface (or lithologic boundary). Their tip lines define partial ellipses (Fig. 4.10) and their opening displacements are greatest at the surface and decrease toward their peripheries. Often they nucleate at the surface and propagate into the rock mass, as demonstrated by the plumose structure on the crack planes.

Corner cracks are edge cracks that nucleate at corners and propagate into the rock mass. They show a quarter plume (Fig. 4.11) that may lead along the free surface, indicating relatively faster propagation there, and lag elsewhere, due to slower relative propagation velocities. These plume geometries can be used to map out changes in rheology and the overall direction of crack growth (e.g., DeGraff and Aydin, 1987; see Fig. 4.12).

Tunnel cracks are fractures having a height in the vertical direction that is much greater than the length in the horizontal direction (e.g., Olson, 1993). Fracture properties such as the magnitudes of opening displacement and the spacing between cracks both scale with the smaller, horizontal dimension, making tunnel cracks a common two-dimensional approximation (e.g., Lawn, 1993) for a joint that extends indefinitely far into a rock mass.

Blade-like cracks are fractures having a height in the vertical direction that is much smaller than the length in the horizontal direction (e.g., Olson, 1993). In this case, the magnitudes of opening displacement and the spacing

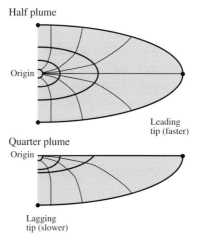

Half plume

Origin

Leading
tip (faster)

Quarter plume
Origin

Lagging
tip (slower)

Fig. 4.11. Schematic showing hackle and ribs for partial plumes. Quarter plumes lag in hotter or more ductile material.

Fig. 4.12. A beautifully preserved columnar joint surface (spanning the vertical width of the image frame) in lava flows from central Oregon. The columnar joint surface is a composite of many sub-horizontal stripes, each one a discrete cracking event (DeGraff and Aydin, 1987). As the flow cooled, the rock was able to fracture brittlely, to form a crack stripe. The overall cooling direction is vertical, with the local direction of crack propagation at right angles to this, as evident from the rib marks on several of the stripes. These quarter plumes show the local direction of crack propagation (to the left in this example), and also the origin point of the crack along the higher-temperature edge of the previous crack. By tracking the lagging and leading edges of the quarter plumes, the overall cooling direction was downward in this example.

between cracks both scale with the smaller, vertical dimension. This case is commonly observed as regularly spaced joints cutting the limb of a fold.

Penny cracks have circular tiplines, leading to the shape of a coin (US penny) when viewed along the axis of dilation. They indicate that the velocity of crack propagation was the same in the crack plane, implying consistent and uniform stress states and material properties in the rock.

Wing cracks are secondary fractures that nucleate near the tips of a parent fracture and propagate away from it at some angle (e.g., Anderson, 1995). These secondary fractures can form where remote tensile stresses applied obliquely to a parent mode-I crack cause oblique dilation and mixed-mode I-II propagation. They grow away from the parent crack along a curving trajectory: steep at first near the crack, then more gently until they become perpendicular to the direction of remote tension (and therefore defining the principal plane there). The point at which the wing crack stops curving and becomes normal to the remote tension marks the place where the amplified, inhomogeneous near-tip stresses that are altered by the parent crack decay to negligibly small values. Beyond this place, the wing crack tracks the remote stress trajectories. In general, wing cracks are one particular type of **end crack**.

Sheeting joints are curved cracks that form close to, and subparallel to, a topographic surface (Martel, 2017). They are prominent beneath domes, ridges, and saddles in hard compact rock, such as granite, gneiss, basalt, marble, and massive sandstone and decorate many exhumed plutons and batholiths (e.g., Twidale, 1973). Also referred to variously as exfoliation or unloading joints (e.g., Harland, 1957; Ramsay and Huber, 1987; Gudmundsson, 2011), sheeting joints are now thought to result from a combination of large compressive stress parallel to a curved topographic surface, rather than from uplift, reduction of overburden stress, weathering, pluton solidification, buckling, axial splitting, or residual stresses (Martel, 2017).

Various types of hackle, or plume, geometries may be identified on joint walls. Some of the major ones are shown schematically on Figs. 4.10 and 4.11. The plumes may show various degrees of symmetry, from a symmetric plume (Fig. 4.5) to an asymmetric plume (Fig. 4.12). Hackle may also define full plumes (as on a penny crack), half plumes, and quarter plumes. These plume shapes are revealing about the directions of propagation—both overall and local—as well as the local stress states associated with the jointing as it developed (Fig. 4.12).

Several patterns of joints and cracks are common. These are not defined on the basis of fractography but by their spatial relationship to related structures, such as faults, folds, or concentrated sources of stress. **Radial crack** geometries form when the stress state driving crack growth is described by polar coordinates, rather than Cartesian (*xy*) coordinates. Hydrofractures can be produced in this way, as the internal pressure in a drill hole locally exceeds the resistance of the adjacent wallrock to mode-I fracture (e.g., Aadnøy and Looyeh, 2010). Production of natural gas (methane) during hydrocarbon maturation has been suggested to drive joint growth in some shales (Lacazette and Engelder, 1992; Engelder et al., 2009). Enhanced cooling of lava flows can produce radial joint patterns at the base of columnar joints if rainwater can percolate down along the joint (DeGraff and Aydin, 1987). Explosions and blasting can produce radial crack patterns (Fig. 4.13), as can shock waves produced during formation of a hypervelocity meteorite impact crater (Melosh, 1989).

Other crack patterns are also common. For example, **cross joints** can form at high angles to an array of parallel joints, breaking the rock mass up into blocks (Gross, 1993; Bai and Gross, 1999). Detailed study of these patterns can reveal quite a bit about the nature of the stress state and, by extension, the tectonic environment, in which these joint sets developed. These will be examined in Chapter 5.

4.5 Stress States for Dilatant Cracking

The stress conditions required for jointing and related dilatant cracking in the Earth are well known and are treated extensively in the literature (e.g., Pollard and Aydin, 1988; Price and Cosgrove, 1990; Engelder, 1993; Olson, 1993, 2003; Mandl, 2005; Pollard and Fletcher, 2005; Zoback, 2007; Gudmundsson, 2011; see Chapter 3). In general, the opening of dilatant cracks at depth requires either a resolved tension or an effective tension produced by an internal fluid or gas pressure that is in excess of the least horizontal compressive

Fig. 4.13. Radial joints, created during road construction by blasting, are superimposed onto two sets of sub-vertical tectonic joints; location is near the crest of Going-to-the-Sun Highway, Glacier National Park, Montana.

stress resolved across the plane of potential opening (Secor, 1965), or else a relaxation in one or more of the far-field compressive stresses as can occur for example due to uplift or folding (e.g., Engelder and Fischer, 1996; Chemenda et al., 2011). Such conditions of relaxed compression can be generated through a range of mechanisms including folding, bending, faulting, and heterogeneities in material properties, such as stress concentrations at wellbores. Dilatant cracks are thus commonly developed in all tectonic settings and geologic environments.

Of particular importance is the **fluid-filled crack**, which is expressed as joints, veins, hydrofractures, and igneous dikes and sills. Under displacement loading conditions (see, e.g., Engelder and Fischer, 1996; Olson, 2003) common in the Earth's crust, an **internal pressure** associated with pore-fluids (including natural gas) within the rock can assist in the opening of fractures underground where the *in situ* stress state is compressive (Hubbert and Willis, 1957; Secor and Pollard, 1975; Pollard and Holzhausen, 1979; Pollard, 1987; Engelder and Lacazette, 1990; Rubin, 1993a; Cosgrove, 1995; Busetti et al., 2012). However, under certain remote stress conditions, incompletely lithified sedimentary rocks may develop joints at depth in the absence of internal fluid pressure (Chemenda et al., 2011) where the rocks are behaving more as a plastic than an elastic or brittle material. Damage zones produced near the tips of hydrofractures are sensitive not just to the internal pressure distribution within the hydrofracture but to its properties, such as viscosity, and those of the host rock and environmental conditions (pre-existing fracturing, pressure, temperature, remote stress state; Busetti et al., 2012).

Where does this fluid pressure come from? As is well known from deep drill cores and seismic investigations that involve shear-wave splitting, for example,

continental and oceanic crust are generally saturated with water and other fluids (e.g., Brace and Kohlstedt, 1980; Zoback et al., 1993; Kohlstedt and Mackwell, 2010). Metamorphic reactions leading to common assemblages, such as greenschist facies rocks, and growth of coupled vein-stylolite systems in carbonate rocks (Bons et al., 2012), require copious amounts of water flowing through crustal rocks (e.g., Nenna and Aydin, 2011). Tectonic joints are probably fluid-filled or perhaps even gas-filled (Engelder et al., 2009) when they form (e.g., Laubach et al., 2010), indicating that they are more akin to hydrofractures than to engineering (tensile) cracks if this internal pressure was the main driver for the joint dilation and propagation. Only above the water table, in the very shallow near-surface, are rocks found that are less than fully saturated. Here, blasting operations in mines and along highway roadcuts (Fig. 4.13) can provide an internal (gas) pressure that leads to dilatant crack growth.

Magma can also invade cracks or accumulate as pockets in a pluton, leading to a magma pressure gradient and flow out of the pluton into the surroundings (or country rock). Rubin (1995a) and Bunger (2008) eloquently describe this process. As the magma leaves the source area—the pluton or magma chamber—it opens a crack by the pressure of the magma against its surroundings, and if the propagation conditions are met (see Chapter 8), then this magma-filled crack can propagate outward as a dike or sill.

The characteristic displacement pattern of wallrock surrounding a large-scale igneous dike, or vertical hydrofracture, is shown in Fig. 4.14. This distribution of displacements corresponds to the "wedging" action noted by Anderson (1938, 1951) and it is reflected in the surface topography above the dike (Fig. 4.14, upper panel; see also Pollard et al., 1983; Rubin and Pollard, 1988; Schultz et al., 2004; Goudy and Schultz, 2005; Klimczak, 2014). The characteristic double peak is also represented in the distribution of stress near the tip of a fluid-filled crack, potentially leading to localized failure as a **process zone** of joints parallel to the main fracture plane (e.g., Delaney et al., 1986; Pollard, 1987; Rubin, 1993a). This stress distribution is calculated and shown in Fig. 8.18.

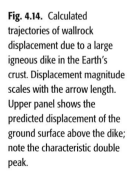

Fig. 4.14. Calculated trajectories of wallrock displacement due to a large igneous dike in the Earth's crust. Displacement magnitude scales with the arrow length. Upper panel shows the predicted displacement of the ground surface above the dike; note the characteristic double peak.

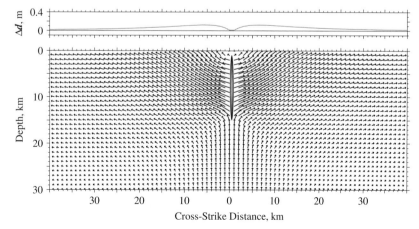

4.6 Anticracks

As discussed in Chapter 1, an anticrack is a planar or tabular structural discontinuity that accommodates primarily contractional normal strain (e.g., Fletcher and Pollard, 1981; Mollema and Antonellini, 1996; Pollard and Fletcher, 2005, pp. 16–19). The contractional normal strain can result from localized volume loss within the rock, such as porosity reduction associated with grain reorganization and crushing in sandstones (e.g., Mollema and Antonellini, 1996; Sternlof et al., 2005), mineralogical phase changes in deep mantle rocks (e.g., Green et al., 1990), or water-enhanced chemical dissolution that can produce disjunctive (spaced) pressure-solution cleavage and stylolites (e.g., Engelder and Marshak, 1985). Several geologic structures have been interpreted as anticracks including **pure compaction bands** (Sternlof et al., 2005; Eichhubl et al., 2010; Fig. 1.10) and **stylolites** (Fletcher and Pollard, 1981; Rispoli, 1981; Tapp and Cook, 1988; Zhou and Aydin, 2010; Nenna and Aydin, 2011; but see Toussaint et al., 2018; Fig. 4.15). Vein arrays oriented normal to many stylolites (e.g., Fletcher and Pollard, 1981; Tapp and Cook, 1988; Groshong, 1988), or at angles consistent with the local stress state (e.g., Soliva et al., 2010), demonstrate the corresponding volume increase and deposition of at least some of the soluble residue from the stylolite population (e.g., Geiser and Sansome, 1981; Rispoli, 1981; Fletcher, 1982; Pollard and Segall, 1987; Passchier and Trouw, 1996; Safaricz and Davison, 2005) in the vicinity of the stylolites.

Some of the mechanical attributes of anticracks in rock include:

- Anticracks are bounded in extent, forming discontinuities within the rock mass (Fletcher and Pollard, 1981; Tapp and Cook, 1988; Burnley et al., 1991; Mollema and Antonellini, 1996; Karcz and Scholz, 2003; Vajdova and Wong, 2003; Renard et al., 2004; Peacock and Azzam, 2006; Nenna and Aydin, 2011).

Fig. 4.15. Small fold in limestone from the Appalachian Mountains showing veins growing inward from the outer surface and stylolites (anticracks) growing inward from the inner surface (near tip of pencil).

- The magnitude of closing displacement (and contractional strain) is largest near the midpoint of an anticrack and decreases toward its tips (Fletcher and Pollard, 1981; Tapp and Cook, 1988; Mollema and Antonellini, 1996; Karcz and Scholz, 2003; Benedicto and Schultz, 2010; Toussaint et al., 2018).
- The maximum (closing) displacement across an anticrack, measured by the amplitude of "teeth" on a stylolite, is proportional to the dimensions of the surface (Fletcher and Pollard, 1981; Benedicto and Schultz, 2010).
- An "end zone" of enhanced compression-related activity is observed at the tips of many anticracks (e.g., Tapp and Cook, 1988; Raynaud and Carrio-Schaffhauser, 1992; Fueten and Robin, 1992).
- Anticracks propagate (i.e., lengthen) from their tips (e.g., Fueten and Robin, 1992; Raynaud and Carrio-Schaffhauser, 1992; Olsson, 2001; Toussaint et al., 2018).
- Mechanical interaction ("soft-linkage") between closely spaced echelon anticracks produces stepover geometries similar to those observed for cracks and other fracture types (e.g., Fletcher and Pollard, 1981; Vajdova and Wong, 2003; Zhou and Aydin, 2010; Fig. 1.10).

Field observations demonstrate that anticracks accommodate primarily closing-mode displacements, or contractional normal strains (e.g., Vajdova and Wong, 2003; Sternlof et al., 2005). This is consistent with the orientation of anticracks in experiments (Green et al., 1990; Riggs and Green, 2001; Green and Marone, 2002; Baud et al., 2004). It is also consistent with observations of anticracks in regions of large compressional stress in wellbore breakouts (e.g., Bessenger et al., 1997), folds (Fig. 4.15), and in the leading, compressional quadrants of faults (e.g., Lajtai, 1974; Rispoli, 1981; Pollard and Segall, 1987; Aydin, 1988; Bürgmann and Pollard, 1992, 1994; Ohlmacher and Aydin, 1995; Petit and Mattauer, 1995; Fig. 4.16).

Fig. 4.16. Classic faulted joints in limestone from the Languedoc region of southern France, with veins (V) and stylolites (S) organized at opposing quadrants of the faulted joint (see Rispoli, 1981; Petit and Mattauer, 1995). The faulted joints slip along kink folds, similar to those in granite (but without the stylolites) demonstrated by Davies and Pollard (1986).

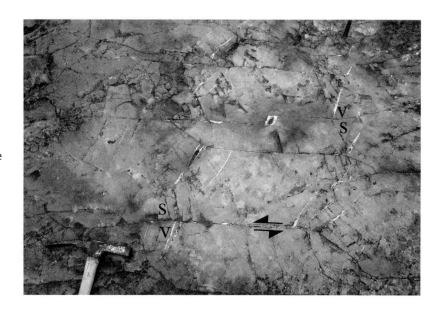

Fletcher and Pollard (1981) used a simple mechanical anticrack model to represent a stylolite in an elastic material and obtained a remarkably good correspondence with field observations. A similar approach that incorporated creep laws and stylolite surface roughness was used in numerical finite-element simulations of stylolite growth by Zhou and Aydin (2010, 2012). In this section we touch on some of the mechanical aspects of anticracks that draw from studies of both compaction bands and stylolites, paralleling approaches from the literature (e.g., Pollard and Fletcher, 2005, pp. 16–19).

Pure compaction bands are thin tabular chains of crushed grains perhaps 2–3 grain diameters, or perhaps 1 mm in width, having a wiggly or angular appearance (called crooked compaction bands by Mollema and Antonellini, 1996; see also Eichhubl et al., 2010; Fossen et al., 2011; Fig. 1.10) that sometimes can resemble chevron folds. The bands may form by collapse of **force chains** (Peters et al., 2005; Andrade et al., 2011) within a porous and well-sorted, coarse-grained host rock (e.g., Fortin et al., 2009; Stanchits et al., 2009). Pure compaction bands from the Buckskin Gulch site in Utah (Fossen et al., 2011) are smaller in wavelength than those from the Valley of Fire site in Nevada (Eichhubl et al., 2009). Shear enhanced and pure compaction bands from southern France (Fig. 4.17) are qualitatively similar to the previous two

Fig. 4.17. Pure compaction bands (PCB) formed in association with shear-enhanced compaction bands (SECB) in friable Upper Cretaceous sandstone from southern France (Soliva et al., 2013).

examples (Ballas et al., 2012; Schultz and Soliva, 2012; Soliva et al., 2013). Pure compaction bands appear to maintain the same width, and perhaps degree of compactional normal strain, regardless of their lengths, consistent with the predictions of Chemenda (2011) for deformation of a plastic material under loading conditions near general yield.

4.7 Review of Important Concepts

1. Dilatant cracks are one of the most common and important types of fracture in geologic materials. They create ready pathways for fluid migration through rock masses by increasing the fracture porosity and permeability, leading to many practical applications in groundwater aquifers, chemical waste storage, and petroleum extraction. Cracks also modify tectonic landscapes, providing windows into hidden geologic structures and processes.

2. Cracks are a key mechanism for accommodating **brittle deformation** in rocks. Stress concentrations and inhomogeneities in rock properties that are developed in the rock because of crack growth lead to reductions in the tensile, shear, and compressive strengths along with time-dependent deformation. Crack arrays increase the compliance (i.e., reduce the stiffness) of the rock, leading to smaller values of Young's modulus for fractured rocks. Cracks also provide substrates for the localization and nucleation of *slip patches* that lead to frictional sliding and faulting as a superimposed deformation mode.

3. Many important issues in structural geology and rock engineering can be addressed by understanding cracks. These include mapping out the paleostress directions (or trajectories) in a rock mass during its deformation, recording strain rotations, identifying paleostress magnitudes, and formation of faults and their damage zones. A working knowledge of cracks and the associated growth processes permits an improved understanding of the mechanisms of brittle deformation of geologic materials.

4. **Cracks are discrete surfaces of dilatant, opening displacement where the crack walls separate from each other normal to their planes.** They should not be thought of as either "surfaces of no displacement" or "fractures with small displacements."

5. Because a compressive normal stress resolved on a potential crack plane serves to squeeze the area shut, an effective tensile stress is commonly needed to dilate the crack. This **driving stress** can be supplied either by a true tension or, more commonly in geology, by the pressure of a fluid or gas phase within the crack that is large enough to overcome any resolved compression. It can also be supplied by a reduction in

confining pressure associated with processes such as uplift or bending. Cracks reflect extensional strain, but not necessarily a tensile stress.

6. The surface textures preserved on crack walls reveal the dominance of normal stress in their formation and growth. **Plumose structure** is the collective term for crack-wall (or crack-surface) textures composed of hackle, ribs, and origins. **Hackle** fan outward from the crack's **origin** point, showing the local direction of propagation of the tipline. As hackle get large they may be called "echelon twist hackle" or "fringe joints" especially where the parent joint interacts with stratigraphic contacts or other structures. **Ribs** point back to the origin and map out the shape of the tipline for various stages in the crack's growth.

7. Growth of cracks produces many attractive and diagnostic patterns. These include isolated cracks, planar cracks, echelon cracks, edge cracks, corner cracks, tunnel cracks, penny-shaped cracks, and end cracks. Each of these has its own distribution of opening displacement and its own characteristic stress field in its vicinity. Crack geometries contain information on the stress state and material properties, such as stratification, during their formation.

8. The rate of crack growth in the subsurface is generally slow to very slow. Joints may grow over periods of several millions of years or more, leading to exceedingly small values of opening and strain rates. Conversely, only in certain cases such as explosion or hypervelocity impact are joints inferred to propagate at velocities approaching an appreciable fraction of the sonic wave velocity and thereby exhibit branching.

9. **Anticracks** are structural discontinuities having a predominantly closing sense of displacement across them. Common types of anticracks in geology include pure compaction bands, although solution surfaces (stylolites) have also been interpreted in this kinematic sense. Given the contractional sense of normal strain, anticracks provide an additional source of information on kinematics, and their genesis (i.e., pressure solution or mechanical compaction) reveals insights into deformation mechanisms during the deformation.

4.8 Exercises

1. What are the main markings on a crack face, and what do they mean?
2. Describe the effect on the crack geometry of the relative rates of crack propagation and change in stress field. What phenomena can cause a gradual curving of a crack?
3. Define what is meant by twisting and breakdown of a crack. What can cause this common geometry?

4. Define "in-plane" and "out-of-plane" shear acting on a crack, tell what crack geometries they produce, and correlate them with the three engineering modes of fracture discussed in Chapter 1.

5. Describe what is necessary to open a crack underground. What complications must be considered when studying a field example of a crack that did not dilate due to a remote tensile stress?

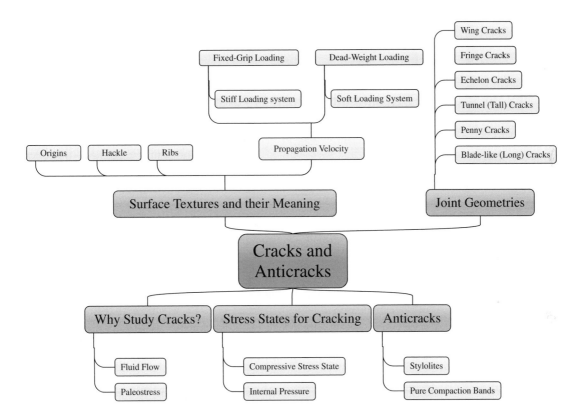

5

Discontinuity Patterns and Their Interpretation

5.1 Why Do Structural Discontinuities Occur in Patterns?

In This Chapter:

♦ *Discovery patterns: echelon discontinuities and end cracks*
♦ *Geometry of stepovers and relays*
♦ *Growth of fractures into echelon arrays, faults, and shear zones*
♦ *End cracks and near-tip stress*
♦ *Systematic and non-systematic joints*
♦ *Discontinuity dimensionality*
♦ *Riedel shear patterns*

Why doesn't one single, solitary structural discontinuity form and cut across a laboratory test specimen or an outcrop, rather than forming a network? Why isn't the San Andreas Fault just a single, continuous strand? Why are echelon arrays formed by the different structure types, such as joints, faults, or deformation bands?

Geologic structural discontinuities of all kinds characteristically form interesting and informative patterns. These patterns reflect their development as the structures grow from small flaws and interact mechanically with their neighbors as deformation progresses (Fig. 5.1). Discontinuities typically create a zone of influence around themselves, related to changes in stress and displacement around them, that scales in large measure with their dimensions (i.e., length, height, and displacement magnitude); as a result, discontinuities can either enhance or impede the growth of their neighbors. The patterns that form as a result reveal much about the kinematics and mechanics of these growth and interaction processes.

In this chapter we will explore some of the most common patterns that can be formed by geologic structural discontinuities. The use of patterns as a guide to the recognition or interpretation of these structures constitutes a vast literature (for example, see Pollard and Aydin, 1988, for a discussion of patterns interpreted in the context of joints, and Davison, 1994, for common patterns of faults). Here we examine several patterns that are informative about how structural discontinuities function in the context of geologic fracture mechanics, with an emphasis on their propagation to larger size with increasing strain.

Fig. 5.1. Breakdown of a parent joint into an echelon fringe array reveals the upward growth direction of the joint, in this case, Aztec Sandstone near Las Vegas, Nevada.

5.2 Typical Patterns

Although there are many discontinuity patterns that a geologist should be conversant in recognizing and interpreting physically, some of the more common or important ones are touched on in this chapter. Two of these patterns stand out as distinctive and historic: these are called "**discovery patterns**" in this chapter. The interpretation of these discontinuity patterns is aided by an appreciation of the role of structural discontinuities as stress concentrators. These two patterns are *echelon* discontinuities, and *end* (or wing) cracks found near joint and fault terminations.

Discovery Patterns:

♦ *Echelon discontinuities*
♦ *End cracks*

5.2.1 Echelon Discontinuities

Echelon arrays are sets of parallel, non-coplanar discontinuities having a consistent sense of step; the structures are soft-linked, with unbroken rock bridges located within the stepovers (see Fig. 5.8). As a rock mass initially deforms, growth begins at low discontinuity densities and, at progressively increasing amounts of remote strain (e.g., Segall, 1984b; Olson, 2003), begin to interact mechanically with neighboring discontinuities. The mechanical interactions lead to preferential propagation of some discontinuities and impeded propagation of others. This process has been well studied physically and mathematically (e.g., Kemeny and Cook, 1987; Ashby and Sammis, 1990; Kachanov, 1992; Lockner, 1995; Paterson and Wong, 2005). Echelon arrays are commonly observed in geology, both as sets of dilatant fractures (joints, dikes, or veins) or as precursory structures to faults in compact, low-porosity rock such as granite.

The term "echelon" has been used as an adjective to describe the geometry of closely spaced or spatially associated joints (see Pollard et al., 1982) and faults (Segall and Pollard, 1980). A set of closely spaced, parallel, non-collinear fractures (Fig. 5.2) or deformation bands may be referred to as echelon or, more commonly, en echelon (e.g., Twiss and Moores, 2007). The former term will be used for brevity in this chapter and to avoid an unnecessary tautology (e.g., fractures in an en echelon geometry).

In this section we will work through some studies from the literature that provide insight into the mechanics of echelon arrays. First, we will examine the propagation of an echelon array; i.e., why joint segments can grow with a consistent sense of step. Next, we will see how the geometry of an initial echelon crack array influences the orientation of faults that form as a result of such crack-related damage. Last, we will investigate the growth of echelon veins into sigmoidal geometries as a result of local stress interactions.

Fig. 5.2. Echelon joints from the Finger Lakes region of New York State. Note both right and left steps along the joint trace, so that they are parallel but not collinear or coplanar.

Structural discontinuities of all types and displacement modes tend to redistribute and concentrate stresses in their vicinities to some degree, leading to arrays of discontinuous structures that may have echelon geometries. Examples include **joints** (e.g., Pollard et al., 1982; Segall, 1984a; Acocella et al., 2000), **anticracks** (e.g., Fletcher and Pollard, 1981; Zhou and Aydin, 2010), **veins** (e.g., Olson and Pollard, 1991), **dikes** (e.g., Delaney and Pollard, 1981; Fink, 1985; Nicholson and Pollard, 1985), **fast-spreading oceanic ridges** (Pollard and Aydin, 1984), **strike-slip faults** (Christie-Blick and Biddle, 1985; Schultz, 1989; Aydin and Schultz, 1990; Peacock and Sanderson, 1994; de Joussineau and Aydin, 2009), **normal faults** (e.g., Peacock and Sanderson, 1994; Crider and Pollard, 1998; Long and Imber, 2011), **thrust faults** (e.g., Aydin, 1988; Nicol et al., 2002; Davis et al., 2005), and **deformation bands** (e.g., Fossen and Hesthammer, 1997; Davis, 1999).

Engineers have long recognized discontinuous cracks in their materials, calling them "fracture lances" or "cleavage steps" (Sommer, 1969; Swain et al., 1974; Fields and Ashby, 1976; Lin et al., 2010). Echelon cracks in geology have been interpreted in light of a remote shear stress (e.g., Hancock, 1972; Beach, 1975) or local stress changes resolved on these closely spaced opening-mode cracks (e.g., Roering, 1968; Pollard et al., 1982; Fink, 1985; Nicholson and Pollard, 1985; Du and Aydin, 1991; Olson and Pollard, 1991; Thomas and Pollard, 1993). Perhaps counter-intuitively, it is well established that if two cracks are opening along the same line (they are coplanar), they will not intersect tip-to-tip but will diverge and link up with a cross-fracture, leaving behind a bridged echelon pair of cracks (Swain et al., 1974; Melin, 1982; Du and Aydin, 1991). As a result, cracks or other dilatant structures, such as joints, veins, and dikes, in general are not perfectly planar structures, but have patterns that can reveal their sequence of development.

The geometry of echelon discontinuities is characterized by several geometric relationships. A viewing perspective taken perpendicular to the discontinuity trace, and its strike direction, makes the geometry of echelon structures clear.

A pair of echelon discontinuities is said to have a particular "**sense of step**" that depends on their position relative to one another. As shown in Fig. 5.3, discontinuities can be arranged in either *left-stepping* or *right-stepping* configurations. Discontinuities that are both parallel to each other and have no perpendicular separation between them (that is, no step) are said to be *collinear*. The perpendicular spacing between the discontinuities is called their **separation** (arrows in Fig. 5.3).

As is well known, echelon discontinuities grow toward each other because of **mechanical interaction** between their near-tip stress fields, very much like constructive interference between light waves. There are three configurations that can be produced, in sequence, as discontinuities grow toward each other and interact (e.g., Pollard and Aydin, 1984). These are called *underlapped*,

Fig. 5.3. The **sense of step** is defined between two echelon discontinuities. Imagine you're standing on the star and have to physically *step* to the left or right across the stepover to move onto the next one.

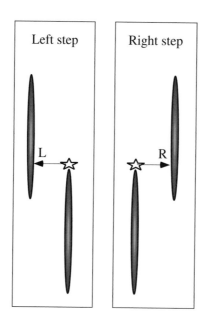

neutral, and *overlapped* configurations. As shown in Fig. 5.4, an underlapped configuration is produced by discontinuities whose tips have not yet interacted significantly with each other. As the discontinuities grow in length, their tips become aligned (the "neutral" configuration in Fig. 5.4; note different usage for relay strain by Watterson et al., 1998) and eventually may grow past each other into an **overlapped** configuration. When this happens, the discontinuities have great difficulty in propagating any further, so any increases in displacement accommodated by the structures leads to steeper displacement gradients, and perhaps linkage to the adjacent discontinuity, rather than continued in-plane propagation. This mechanical interaction between propagating echelon discontinuities provides the physical basis for interpreting phenomena such as **throw vs. length distributions** along fault systems (e.g., Peacock and Sanderson, 1994, 1995; Davis et al., 2005).

The area located between overlapping discontinuity tips is called the **stepover** (Fig. 5.4). This area of unbroken rock is also called a "rock bridge" or "ligament" in rock engineering and a transfer zone or **relay** between faults in structural geology (e.g., Walsh et al., 1999; Davison, 1994; Peacock et al., 2000; Nicol et al., 2002). The stepover is a highly stressed domain that deforms in order to transfer the displacement from one echelon discontinuity to the other. Echelon fault sets have been called "**soft-linked**" systems (e.g., Davison, 1994), which conveys the idea of stress (or displacement) transfer between the overlapping fault segments. "**Hard-linked**" systems are echelon faults that have generated secondary structures such as end cracks (see, for example, Section 5.2.2 and Fig. 5.16 later in this chapter), folds, pressure-solution cleavage, or faults that have propagated completely across the stepover. These connected discontinuities are called bridged or **linked echelon discontinuities** (Fig. 5.4).

A linked echelon pair of joints is shown in Fig. 5.5. Here, plumose structure preserved on the joint demonstrates that the structure was formed by the propagation of two closely spaced echelon joints toward each other; i.e., they both propagated into the stepover region. The shape of the stepover was controlled by the joint pair's pre-linkage echelon geometry, by the distribution of opening displacements along them, and by the material properties and stratification of the jointed unit.

Fig. 5.4. Sequential growth of two echelon discontinuities into underlapped, neutral, and overlapped configurations. The stepover is the region between the overlapped discontinuity tips. The final stage is linkage.

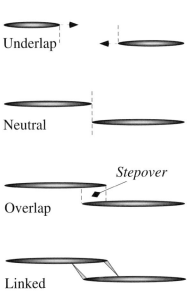

Underlap

Neutral

Stepover

Overlap

Linked

Fig. 5.5. A wonderful stepover between two echelon joints showing plumose structure and growth of both joints toward their shared stepover. The dark patches are water.

5.2.1.1 Formation of the Stepover

In this section we will examine how the stepover located between a pair of echelon discontinuities forms. Using the geometric relationships shown in Fig. 5.6 and the equations of Linear Elastic Fracture Mechanics (LEFM, Chapter 8), the influence of one discontinuity on another has been assessed in the literature by calculating the change in *crack or fracture propagation energy G* (also called "strain energy release rate" or "crack extension force" in the engineering literature; see Lawn, 1993, p. 22; Anderson, 1995, pp. 41–44) at the inner tips of the two echelon discontinuities. The approach involves setting up the problem geometry and then calculating the stress intensity factors at the inner tips of the echelon discontinuities; by incrementing the discontinuity length (L = 2a) while keeping the distance (2k) constant, inward growth of the echelon discontinuities into an overlapped configuration can be simulated.

This approach has been used, among others, by Pollard et al. (1982) to relate *echelon fringe joints* to the parent crack's mixed-mode I-III breakdown, by Pollard and Aydin (1984) to understand the overlapping and curving geometries of *fast-spreading mid-ocean ridges*, by Martel and Pollard (1989) to investigate the preferential growth of *splay cracks* between closely spaced faulted joints in the Sierra Nevada batholith, by Aydin and Schultz (1990) to understand the development of *pull-apart basins and push-up ranges* along echelon strike-slip faults, and by Willemse and Pollard (2000) to analyze the growth and *linkage of normal faults* in 3-D. Similar calculations provide the basis for determining the **effective lengths** of echelon discontinuity sets that are close enough to each other to be "soft linked" (e.g., Segall and Pollard, 1980; Schultz, 2000a; Gupta and Scholz, 2000a,b; Wilkins et al., 2002).

Representative results of this calculation are shown in Fig. 5.7. The calculations represented in this schematic diagram are based on those published for mode-I cracks (Pollard et al., 1982; Pollard and Aydin, 1984, 1988) and

Fig. 5.6. Idealized geometry of echelon discontinuities showing main parameters needed to characterize their geometry and mechanics. The discontinuities are shown in an underlapped configuration. Discontinuity length, 2a; perpendicular spacing, 2s; overlap of inner tips, 2o; distance between discontinuity midpoints, 2k; radial distance between inner tips, r; angle between inner tips, α. The "2"s come from the LEFM-based measure of length L as twice the semi-major axis a.

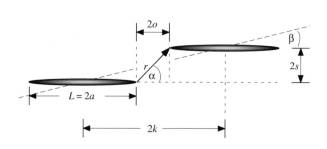

A stepover forms as echelon discontinuities grow progressively from an underlapped configuration into an over

for (mode-II) strike-slip faults (Aydin and Schultz, 1990), and they convey the principles that appear to be general for all types of geologic structural discontinuities. As the discontinuities grow (at a constant value of separation) one moves from left to right across the diagram. The vertical axis indicates the degree of mechanical interaction between the discontinuities that is developed at their inner tips—that is, in the *future stepover region*. The curves are shown for different values of discontinuity spacing, which is a parameter that is independent of the discontinuity overlap but crucial to their interaction. By normalizing the dimensions, overlap, and spacing of the discontinuities to the distance between their midpoints, the results can be applied to echelon discontinuities at any scale.

As a pair of echelon discontinuities grows laterally, their inner tips approach each other and begin to interact mechanically. Specifically, this occurs when their near-tip stress fields impinge on one another, leading to a new, composite stress field between them. This new stress field is spatially

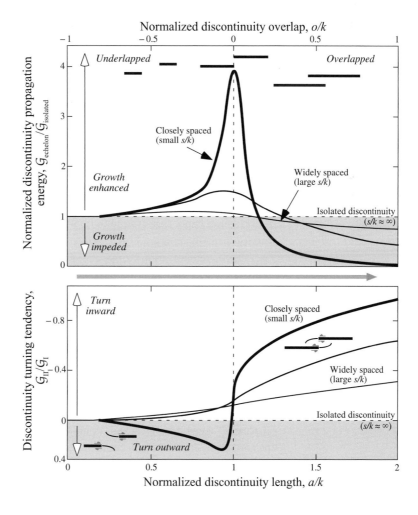

Fig. 5.7. Diagrams showing the roles of mechanical interaction between two echelon mode-I fractures on the lateral propagation of their inner tips. Fracture geometry parameters are given in Fig. 5.6. Upper panel shows fracture propagation energy of the echelon fracture tip normalized to that of the same but isolated fracture, stress state, and rock properties. Curves show results for different values of normalized fracture separation s/k. Lower panel shows the influence of in-plane (mode II) shear stresses that are locally generated and resolved on the opposing fracture's inner tip by the near-tip stress field. Oblique dilation of the inner tips leads to either convergent or divergent fracture paths depending on the degree of overlap o/k.

variable in all the components of stress (normal and shear) as well as in the derivative quantities like mean stress and the maximum shear stress. The trajectories of the local principal stresses are also different than those far from the discontinuities (which are given by just the remote stress orientations) or those in the near-tip region of the discontinuities. This means that the driving stresses at the tips will be different than those acting on a discontinuity that is widely spaced ($s/k > 10$) and effectively isolated from any neighboring ones. The associated discontinuity-wall displacements will also be different in magnitude (smaller or larger) from those along the same discontinuity but outside of the newly forming stepover. This means that slip vectors (for echelon faults) and fibrous minerals' c-axes (for echelon veins) may also differ in detail within the stepover as compared to those elsewhere along the discontinuity by some amount.

Echelon discontinuities **begin to form a stepover** when the discontinuities' inner tips achieve an *underlapped* configuration ($o/k > 0.2$ or so; Fig. 5.7) **and** when they are relatively *closely spaced* (approximately given by $s/k < 0.2$; Pollard and Aydin, 1984; Martel and Pollard, 1989; Aydin and Schultz, 1990; Schultz, 2000a). For any given separation, the peak energy occurs when the inner tips are aligned ("neutral" overlap, Fig. 5.7). Then the energy available to drive further inward growth decreases rapidly with larger values of overlap (Fig. 5.7). This means that the underlapped configuration tends to *attract the discontinuities into the stepover region*; the tendency to propagate increases until the tips pass each other, then decreases to effectively *freeze the overlapped discontinuity pattern* into the rock (e.g., Gupta and Scholz, 2000a,b).

Continued displacement accumulation on overlapped echelon discontinuities produces secondary fractures rather than larger overlaps

Formation of a stepover can be further understood by examining the stress state near discontinuity terminations. As noted in Chapter 3, the Coulomb failure stress provides a criterion for assessing the tendency for frictional sliding to be triggered on suitably oriented surfaces in the vicinity of a fault or deformation band. Correspondingly, the magnitude of tensile stress (or fracture propagation energy; Fig. 5.7) resolved onto a dilatant discontinuity, such as a crack, vein, or joint, by opening along a nearby structure determines the tendency for propagation into an overlapping stepover configuration.

As a fault, for example, propagates toward another fault (Fault 1 in Fig. 5.8) with a sufficiently small value of separation, the propagating fault (Fault 2 in Fig. 5.8) initially defines an underlapped configuration (Fig. 5.4; Region A in Fig. 5.8). As Fault 2 continues to lengthen (toward the right in the figure), it encounters a region of increasing Coulomb failure stress in the tip region of Fault 1 (Region B in Fig. 5.8), which corresponds to the area of increased mechanical interaction and fracture propagation energy calculated in explicit fault interaction studies for these areas (e.g., Region B in Fig. 5.7a). Increased shear driving stress along the inner tip of the propagating fault promotes larger shear displacements there, as noted by many including Aydin and Schultz (1990), Willemse (1997), Crider and Pollard (1998), and Gupta and Scholz (2000a,b). Fault 2 continues to propagate toward Fault 1, at some

Fig. 5.8. Plot of Coulomb stress change around a normal fault (Fault 1 in the figure) normalized by its background value far from the fault (see Fig. 3.12). Progressive stages of propagation of a second normal fault (Fault 2) into an overlapped configuration with Fault 1 are indicated by the labels A, B, and C.

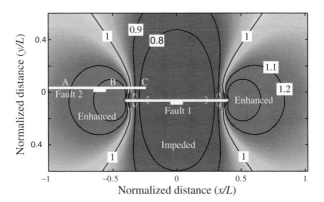

value of separation, into the particular overlapping configuration for which the propagation energy falls below the value required for continued in-plane propagation (Region C in Fig. 5.8). If displacement continues to accumulate on Fault 2, propagation will occur by creation of a splay crack or fault that grows obliquely toward Fault 1, eventually linking the stepover and facilitating kinematic coherence between the two formerly separate fault segments.

Many studies have quantified the **geometry of stepovers** between joints and faults. Several general findings have emerged from this body of work. First, the dimensions of the stepover (overlap and separation) scale proportionally with the maximum displacement accommodated along the joint (Pollard et al., 1982) or fault (Gupta and Scholz, 2000a,b; de Joussineau and Aydin, 2009; Long and Imber, 2011). Because displacement scales with length for joints, faults, and deformation bands (e.g., Cowie and Scholz, 1992a; Dawers et al., 1993; Clark and Cox, 1996; Schlische et al., 1996; Olson, 2003; Schultz et al., 2008a; Klimczak et al., 2010; Schultz et al., 2013), as well as other quantities such as strain or spacing (e.g., Marrett, 1996; Scholz, 1997; Cowie and Roberts, 2001; Soliva et al., 2006; Lamarche et al., 2018), the stepover dimensions also scale with length and down-dip height. Second, stepovers of any scale tend to form with a ratio of overlap to separation that falls within a relatively narrow range of values. The ranges for a particular dataset are typically smaller than those obtained by examining compilations of datasets from many lithologies, mechanical stratigraphies, and settings; this caveat parallels that noted for interpretation of displacement–length data (Clark and Cox, 1996).

Overlaps and spacings for strike-slip and normal faults from around the world, and some from Mars, are displayed in Fig. 5.9. In general, overlaps scale with fault separation, indicating growth due to fault interaction. For example, stepovers along joints exhibit **overlap/separation ratios** of about 3 (Pollard et al., 1982; Acocella et al., 2000); those along normal faults have ratios of about 4–5 (Soliva et al., 2006; Long and Imber, 2011; Fossen and Rotevatn, 2016); and those along strike-slip faults have ratios of about

Fig. 5.9. Plot of overlaps and separations measured along (a) strike-slip and (b) normal faults from the literature. Contractional stepovers (relays) along strike-slip faults are shown by filled symbols; extensional stepovers by open or gray symbols. Data for strike-slip faults compiled from Aydin and Nur (1982), Schultz (1989) [Mars], Aydin and Schultz (1990), Campagna and Aydin (1991), de Joussineau and Aydin (2009), Gürbüz (2014), and Micklethwaite et al. (2014). Data for normal faults from Soliva and Schultz (2008), Soliva et al. (2008) [Vallo di Diano fault, VdD], and Long and Imber (2011) [L&I; Table 1, T1, and Table 2, T2], and Awdal et al. (2014): Bartlett Wash and "near Goblin Valley" faults.

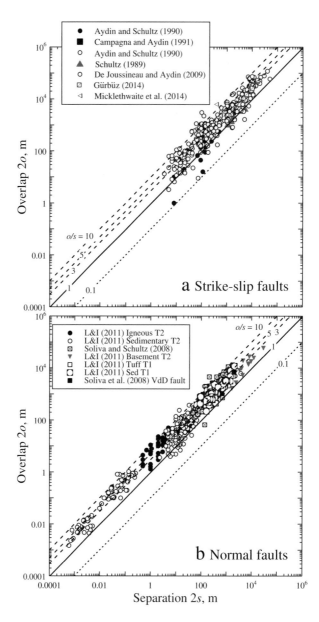

2.7 (Aydin and Nur, 1982; de Joussineau and Aydin, 2009; Gürbüz, 2014; Micklethwaite et al., 2014).

To first order, the ranges of overlaps and separations for stepovers or relays along normal and strike-slip faults are generally comparable when portrayed as a global dataset (Fig. 5.9). Along some fault systems, there may be a tendency for contractional stepovers (push-ups) along strike-slip faults to have smaller overlaps relative to extensional stepovers (pull-aparts) at the same value of fault separation (Aydin and Schultz, 1990). Both unlinked and linked (or breached) stepovers along normal faults are plotted in Fig. 5.9b using the

The Transition from Echelon to Parallel Discontinuities:
Echelon:
  Narrow spacing relative to length ($s/k < 1.0$) **and** interaction of near-tip stresses
Parallel:
  Wide spacing relative to length ($s/k > 1.0$) **and** no stress shadowing

same symbols, as the minimum fault dimension that controls step width is not considered (Soliva and Schultz, 2008). However, examination of a particular dataset that contains faults confined to a mechanical or rheological layer would reveal a maximum fault separation for linkage that scales with the faulted layer thickness (Fig. 5.10). **Parallel** fractures, faults, or deformation bands can form when their separations are larger than the maximum stepover width ($2s$ in Figs. 5.6 and 5.9) for a given set of structures.

Scatter in these relationships stems from several transient sources including fault interaction, stratigraphic or rheological restriction, and measurement technique. Overlap-separation ratios can also increase if faults exploit pre-existing anisotropies such as bedding, foliations, or other structures, or if the relays or stepovers merge into a single fault at depths less than the depth of faulting (Soliva et al., 2008). Because stepovers are 3-D structures, the 3-D geometry of the interacting faults, including their length, height, and tipline shape, are critical to identify if a more precise interpretation of stepover geometry is desired (e.g., Willemse, 1997; Crider and Pollard, 1998; Kattenhorn and Pollard, 1999; Willemse and Pollard, 2000; Soliva et al., 2008; Fossen et al., 2010; Long and Imber, 2011). The ranges of overlap/separation ratios obtained in these studies are consistent with processes of discontinuity propagation, interaction, and linkage that occur over a very broad range of scales, rates, and deformation conditions.

As displacements accumulate on the overlapped echelon discontinuities—whose continued propagation has been significantly impeded due to their mechanical interaction—the *near-tip displacement gradients increase*, leading eventually to sufficiently large near-tip stresses that the rock breaks. **Secondary**

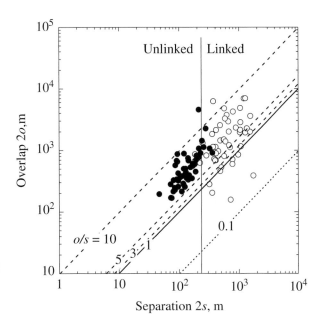

Fig. 5.10. Overlaps for stratigraphically restricted faults, after Soliva and Schultz (2008). Data for faults from Goba'ad rift segment, Afar depression, showing maximum fault spacing for linkage (or maximum separation for linked stepovers) there of ~220 m.

structures such as end cracks, stylolites, cross faults, or linking bands thus nucleate near the inner tips of the echelon discontinuities, propagating across the stepover to eventually link up the parent echelon structures (e.g., Segall and Pollard, 1980; Martel and Pollard, 1989; Aydin and Schultz, 1990; Davison, 1994; Okubo and Schultz, 2006; Fossen et al., 2010). The "soft-linked" echelon discontinuities then become "hard-linked," resulting in an effectively longer structure (Fig. 5.4, last panel) that can accommodate greater magnitudes of displacement (e.g., Tchalenko, 1970; Wesnousky, 1988; de Joussineau and Aydin, 2009). Called **growth by segment linkage** in the literature (e.g., Cartwright et al., 1995, 1996; Cladouhos and Marrett, 1996), this is a common way for discontinuity sets to grow in length and displacement as the far-field strains increase (e.g., Wesnousky, 1988; Dawers et al., 1993; Dawers and Anders, 1995; Fossen and Hesthammer, 1997; Acocella et al., 2000; Peacock, 2002; Scholz, 2002; Schultz and Balasko, 2003; Davis et al., 2005; Cowie et al., 2007; Malik et al., 2010; Schultz et al., 2010a,b,c).

The formation of hard-linked discontinuities described in this section informs the processes of discontinuity propagation. Discontinuities that propagate by increasing their lengths from their tip regions are said in the literature to undergo either **radial propagation** or **growth by segment linkage** (e.g., Watterson, 1986; Cartwright et al., 1996; Cladouhos and Marrett, 1996; Cowie, 1998b; Childs et al., 2017). It is evident from consideration of discontinuity interaction that radial (or tip) propagation of a discontinuity will occur unless it is impeded by a change in material properties or local stress state. In the former case, impeded propagation can be identified with stratigraphic or rheological restriction (e.g., Cowie et al., 1993a; Nicol et al., 1996; Gross et al., 1997; Wilkins and Gross, 2002; Soliva and Benedicto, 2005; Soliva and Schultz, 2008). In the latter case, faults that are relatively closely spaced will interact mechanically with each other, growing into an overlapped geometry followed by linkage across the stepover. As demonstrated explicitly for faulted joints by Martel and Pollard (1989), the mechanism that requires the **lesser amount of energy** will occur. It appears that the question of whether a structural discontinuity propagates from its tip or by segment linkage may be restated, more precisely, as whether the direction of propagation of the discontinuity is toward a nearby discontinuity or not. In both cases, the discontinuity propagates from its tip region, but the degree of interaction with other discontinuities (or material property changes) will determine the resulting pattern, such as linkage or parallel discontinuities.

The lower panel of Fig. 5.7 shows the local in-plane shear stresses that are generated and resolved onto the opposing discontinuity's inner tip by the discontinuity interaction. This diagram applies strictly to dilatant (mode-I) cracks although it can lend qualitative insight into the growth of other types of localized structures. In general, the normal and shear stresses being resolved on the opposing (mode-I) fracture will differ in magnitude (and often in sign) than those on the same fracture outside the stepover (and interaction) region.

Propagation occurs in the minimum energy direction

In Fig. 5.7, the echelon discontinuity geometry with non-neutral overlap (having $r \neq 90°$; see Fig. 5.6) leads to oblique opening of the inner tip's discontinuity walls and a curving propagation trajectory. Recalling from Chapter 4 that in-plane shear stress **steers a mode-I crack**, rather than leading to echelon twist and breakdown, this mechanical interaction and resulting curved growth trajectory well explains the hook-shaped geometry of many tensile cracks as well as the much larger **fast-spreading mid-ocean ridges** (Pollard and Aydin, 1984; Phipps Morgan and Parmentier, 1985; Sempere and Macdonald, 1986; Macdonald et al., 1988; see also Nicolas, 1995).

5.2.1.2 Growth of Faults or Shear Zones from Echelon Crack Arrays

Deformation of a laboratory sample (see Chapter 2) of compact rock illustrates the process of crack growth and coalescence into linked echelon arrays. Imagine a vertical slice through the cylinder onto which the deformation is projected (e.g., Hawkes and Mellor, 1970). The stress state in the rock cylinder is not uniform or homogeneous, but varies considerably with location within the test sample (Hawkes and Mellor, 1970; Brady, 1971a,b; Peng, 1971; Al-Chalabi and Huang, 1974; Olsson, 2000). As the sample is compressed, five areas of tensile stress concentration build up—one in the center and one near each of the four corners. Microcracks that are located in these areas grow preferentially to others located elsewhere in the cylinder, leading to a cloud of microcracks in the center and secondary concentrations near the corners (see Lockner, 1995, for a review). One or more **inclined zones of microcracks**—formed by linkage into a linear echelon array—typically develop in the center of the cylinder. As contractional strain builds

Why do some discontinuities grow into a linear array?

up and the sample shortens, the echelon cracks link up across their stepovers in order to accommodate larger values of displacement across them; this also serves to reduce their shear stiffness enough to promote array-parallel ("shearing") offsets. The linear echelon array selectively interacts with the corner zones of microcracks that maintain the consistent sense of step, and therefore the same planar orientation in the rock sample, leading to propagation of the linked echelon array—now a rough fault zone—diagonally across the sample and consequent failure of the cylinder. The central part of the test sample has the most homogeneous stress state and is considered to best represent the deformation of rock units in nature.

Using the analytical expressions for the near-tip stresses induced by dilation of an LEFM crack, Du and Aydin (1991) investigated what determines the sense of step for a set of underlapped echelon cracks, representing small, widely spaced microcracks in rock. The strength of mechanical interaction between fractures is shown to be a nonlinear function of the distance from a fracture tip, with interaction being strongest for closely spaced fractures and becoming weaker as the fractures are separated by larger distances. By examining the strength of mechanical interaction between the cracks and stepping through a range of overlaps and separations, Du and Aydin (1991) solved for the angle α (see Fig. 5.6) for which the interaction was maximized for a particular fracture or stepover configuration. They found that α_{max} varied systematically with the echelon geometry parameter r/a, suggesting a relatively narrow range of underlapped fracture geometries that could grow into most strongly interacting, overlapping echelon arrays.

Next, Du and Aydin (1991) demonstrated that an echelon crack pair will preferentially interact (and eventually link up) with a third crack that has the same, **consistent sense of step** (for which α_{max} and r/a are also satisfied). The interaction of an echelon pair with a third crack of the opposite sense of step is weaker, so if cracks are present in either location, the pair of interacting ("soft-linked") cracks will propagate preferentially to become a **linear zone of echelon cracks** (Figs. 5.11 and 5.12) rather than dissipate into a randomly stepping set of distributed cracks.

As echelon cracks interact with their neighbors to form a linear array having a consistent sense of step, one or more sets of linear arrays can form. The echelon cracks themselves were originally oriented parallel to the greatest compressive stress direction (σ_1). Given the small range of parameters for echelon fracture interaction and growth out of their initial underlapped configurations noted in the literature (see also Olson and Pollard, 1991, for explicit calculations), the orientation of the linear zone of echelon cracks to

Fig. 5.11. Photograph of echelon joints in shale from the Finger Lakes region of New York State. A three-dimensional view would be needed to interpret the origin of this echelon array.

Fig. 5.12. A pair of echelon cracks will grow preferentially into an array with a consistent sense of step.

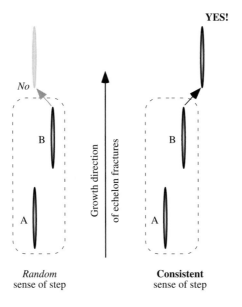

the remote stress state becomes a geometry problem. Du and Aydin (1991) give the relationship as

$$\beta = \frac{\sin\alpha}{\left(\cos\alpha + \dfrac{2a}{r}\right)} \tag{5.1}$$

with the parameters shown in Fig. 5.6. Recasting this relationship by using overlap $2o$ and separation $2s$ between two echelon fractures gives

$$\beta = \frac{s}{\left(o+a\right)} \tag{5.2}$$

A similar relationship was obtained by Reches and Lockner (1994) to relate the orientation of a fault, formed by linkage of echelon cracks, to the remote stress state. Defining $S = 2s$, their relationship between echelon fracture geometry and the angle to σ_1 is given by

$$\beta = \tan^{-1}\left[\frac{2\left(\dfrac{S}{L}\right)\sin\theta_m}{\sin\theta_m + 2\left(\dfrac{S}{L}\right)\cos\theta_m}\right] \tag{5.3}$$

in which θ_m is the angle β between the fracture tip and the direction of greatest tension, calculated in Cartesian coordinates (i.e., normal or parallel to the parent fracture). Reches and Lockner (1994) suggested that values of θ_m and S/L for which mechanical interaction between nearby echelon fractures was maximized were $\theta_m \sim 30°$ and $S/L \leq 2$. Values for the array angle β, shown in Fig. 5.13 calculated by using both approaches, and values of S/L from Du and Aydin (1991) for which interaction is maximized (i.e., $S/L = 0.5$–1.3), are 25–35°. Assuming Coulomb frictional sliding, this angle β would be called the "acute bisectrix" of a population of conjugate faults and is comparable to the **friction angle** of a planar surface.

The geometric relationship between the putative shear zones and the echelon fractures is interesting. As can be seen in Fig. 5.14, for the same North–South oriented maximum compression direction, the left-stepping cracks would define a zone having right-lateral shear resolved along it due to the far-field stresses, whereas the conjugate array of right-stepping cracks would define a zone of left-lateral shear. The sense of strain implied in the stepovers between the echelon cracks in either case is contractional (see Chapter 6) implying that the echelon cracks may remain soft-linked as they continue to grow.

As cracks grow into an overlapped configuration, they induce shear stresses onto the opposing crack's tip. This mixed-mode (I-II) loading steers the crack within this horizontal plane of observation (see Chapter 4) so that it tends to turn outward, then inward with increasing overlap (see Fig. 5.7, lower panel; Pollard and Aydin, 1984; Olson and Pollard, 1991), forming a "hook-shaped" geometry (Melin, 1982; Fender et al., 2010) of echelon fractures (Olson and Pollard, 1989; Thomas and Pollard, 1993). Linkage can thus occur when curved fracture tips intersect their neighbor in the array (Fig. 5.14). Curved fracture shapes resulting from local fracture interactions are to be

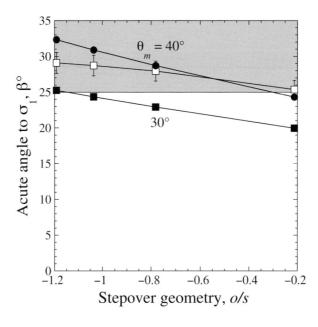

Fig. 5.13. Results of calculations by Du and Aydin (1991), open symbols, and by Reches and Lockner (1994), filled symbols, that relate the angle of the array β to the underlapped echelon configuration, o/s. Shaded area indicates typical values of β obtained from field measurements of faults.

Fig. 5.14. Illustration of crack growth into an oriented array, progressing from underlapped to overlapped echelon geometries. Linkage may occur once fracture tips have overlapped to a degree suggested by Fig. 5.7.

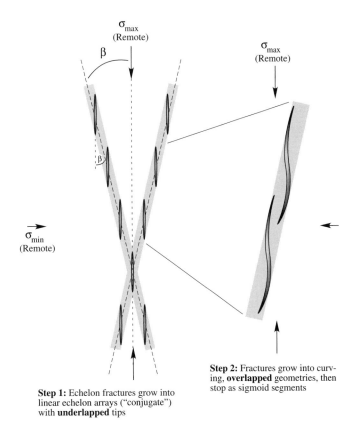

Step 1: Echelon fractures grow into linear echelon arrays ("conjugate") with **underlapped** tips

Step 2: Fractures grow into curving, **overlapped** geometries, then stop as sigmoid segments

distinguished from those produced by other mechanisms (Olson and Pollard, 1991), including remote shearing (e.g., Riedel, 1929; Lajtai, 1969; Price and Cosgrove, 1990, p. 45) and deformation of early-formed planar veins ("dragging") as a shear zone deforms internally (e.g., Roering, 1968; Lajtai, 1969; Hancock, 1972; Beach, 1975).

Hook-shaped fracture geometries are observed along fast-spreading mid-ocean ridges, in dikes and veins (e.g., Beach, 1975; Pollard et al., 1982; Nicholson and Pollard, 1985; Rogers and Bird, 1987), and in glass plates (Swain and Hagan, 1978). These curved fracture paths (Fig. 5.7, lower panel; Fig. 5.15) will be best expressed if the propagating fractures are able to turn in response to the local mixed-mode loadings. If the fracture walls are rough and the amount of dilation is less than sufficient to fully separate them, then the induced shear stress may be resisted by the strength of the rough fracture walls, reducing or even prohibiting any turning (Renshaw and Pollard, 1994a). Rough walls can also be created by diagenetic cements (e.g., Laubach et al., 2004, 2010), potentially leading to straighter geometries in siliceous rocks such as sandstone or carbonate, and more curved geometries, with smoother walls, in mudstones and shales. Thus, while curving crack patterns in a stepover may be reliable indicators of small values of the differential

Fig. 5.15. Crack propagation paths for in-plane loading. All cases assume a positive driving stress so that the cracks open against the remote perpendicular compressive stress. The magnitude of the crack-parallel stress component is zero (upper panel), equal and compressive (middle panel), and much larger and compressive (lower panel).

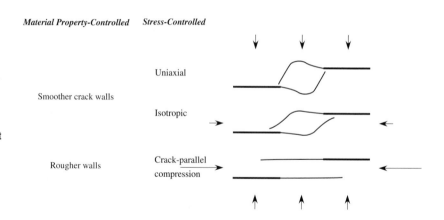

(crack-parallel minus crack-normal) stress acting on them, their *absence* may not be diagnostic of large differential stresses (Renshaw and Pollard, 1994a).

A single linear echelon array of cracks, veins, or dikes having sigmoidal shapes can also form by the propagation of a parent dilatant fracture into a region of different stress state (Pollard et al., 1982; Nicholson and Pollard, 1985; Rubin, 1993a). Here, the array also has a consistent sense of step (i.e., its sense of rotation, or "twist," relative to the plane of the parent fracture is constant along the array). Although the stress resolved onto the tipline of the parent fracture is mixed-mode (I-III) in this case, the fracture tip breaks down into an array of twist hackle (see Chapter 4) that, as they continue to propagate away from the parent fracture's inhomogeneous near-tip stress state, rotate to align themselves with the new stress state and minimize any resolved shear stress. For this mechanism, the *rate of twist* (Pollard et al., 1982) of the echelon fractures commonly increases with the distance from the parent fracture, whereas the orientation (or twist) of interacting fractures in a spatially distributed array would not show this type of variation.

Analogous stress rotations can occur in layered rock sequences or in mechanical stratigraphies that exhibit contrasts in stiffness or strength between adjacent layers. On a longer time scale, stress rotations commonly occur after a fault set is developed, for example above a basement fault that is subsequently reactivated (e.g., Frankowicz and McClay, 2010). In this case, reorientation of other types of structures such as faults can occur as they propagate upward or downward through layers having a different stress state. For example, **grabens** commonly also form echelon sets; echelon graben arrays having a consistent sense of step may be attributed to an oblique extension in rheologically stratified crustal rocks (e.g., Grimm and Phillips, 1991; Clifton and Schlische, 2001).

This situation is illustrated in Fig. 5.17 where upward or downward propagation of the normal fault (out of the plane of the page) into the rotated stress regime would impose a small component of left-lateral shear and promote breakup of the propagating parts of the fault into a series of echelon normal

Fig. 5.16. End cracks emerging from a (right-lateral) faulted joint, Sierra Nevada batholith.

Fig. 5.17. Schematic in map view showing effect of counterclockwise rotation of the far-field principal stresses about a vertical axis (represented here for simplicity by an initial horizontal strain direction in the East–West direction) on slip sense along a pre-existing normal fault.

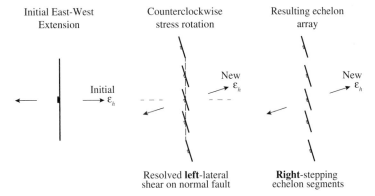

Initial East-West Extension   Counterclockwise stress rotation   Resulting echelon array

Initial ε_h New ε_h New ε_h

Resolved **left**-lateral shear on normal fault   **Right**-stepping echelon segments

fault segments that grow in orientations consistent with the new stress state (and new extension direction). In this case, the sense of stress rotation would lead to an array of consistently right-stepping segments. Conversely, a clockwise rotation of the remote stress state would resolve right-lateral shear onto the pre-existing fault and promote its breakup into left-stepping segments. Of course, the echelon segments see little right- or left-lateral shear, since their growth has oriented them to more efficiently accommodate extension in the rotated stress state; on the other hand, they can also be used

Shear Sense along
Inclined Echelon Array

♦ *Right-stepping, left-lateral*
♦ *Left-stepping, right-lateral*

to infer such a sense of lateral shear on a related pre-existing fault that is adjacent to them.

The geometric relationship developed for cases of fault reactivation under a rotated stress state is the same as that seen for echelon cracks that interact to form an inclined zone with a consistent sense of step, leading to a shear zone or ultimately, with hard-linkage, a fault (Fig. 5.14). A **rule of thumb** can be defined for inferring the sense of shear along an echelon crack or fault array based only on their sense of step (see Fig. 5.17, right-hand panel). To invoke this requires, however, that the origin of the echelon structures is known or can reasonably be inferred to be related to an inclined structure. As illustrated in Fig. 5.12 and the associated text, a pair of discontinuities can grow preferentially into an echelon array having a consistent sense of step in the absence of a far-field or resolved shear stress. Frequently, reliable information on the third dimension (above or below the plane of the echelon structures) can improve the reliability of a structural interpretation of echelon discontinuities.

5.2.2 End Cracks

End cracks are *secondary discontinuities* that nucleate at or near the ends of mode-I and/or mode-II discontinuities (Fig. 5.16). Also known in engineering fracture mechanics as *wing cracks* (Engelder et al., 1993, pp. 113–115; Lawn, 1993, pp. 48–49; Anderson, 1995, pp. 91–96), due to their curved geometries, end cracks that lack curvature are also sometimes referred to as *pinnate fractures* (Engelder, 1987; Wilkins et al., 2001), *splays*, or feather cracks in structural geology (see also Martel and Pollard, 1989; Flodin and Aydin, 2004a,b). They have their greatest degree of opening at the parent discontinuity, because this is where the greatest stress changes due to that discontinuity's displacement are localized. Their openings decrease away from their parent discontinuity, leading to wedge-shaped cracks in cross section. **End cracks demonstrate the importance—and the scale—of** *stress concentrations* **built up around discontinuity terminations**. End cracks typically do not develop if the parent discontinuity terminates against another one or if the displacement distribution along a discontinuity decays sufficiently gradually that large stress changes are not built up. Strongly curving wing cracks suggest tensile loading in strong rock, so that LEFM conditions apply, and imply large stress concentrations and gradients near the discontinuity tip (see Chapter 8). More planar end-crack geometries, and nucleation of secondary fractures along the discontinuity (not just at the tip), suggest a clearer association with the remote stress state and perhaps a smaller degree of stress change at the parent discontinuity's tip region (e.g., Wilkins et al., 2001) and smaller near-tip stress concentration, consistent with fracturing of weaker rock (see discussions of non-LEFM "cohesive" end zones by Cooke (1997), Martel (1997), Willemse and Pollard (1998), and in Chapter 9).

The existence and kinematic significance of end cracks has long been appreciated in the literature. Erdogan and Sih (1963) noted their growth near the ends of mixed-mode (I-II) cracks in engineering materials. Brace and Bombolakis (1963) and Hoek and Bieniawski (1965) produced them in the laboratory while deforming rocks under axial compression. Kemeny and Cook (1987) and Kemeny (1991) demonstrated the importance of end cracks in the compressive strength and time-dependent deformation of rocks. Several geologists including Rispoli (1981), Segall and Pollard (1983b), Martel et al. (1988), Engelder (1989), Schultz (1989), Cruikshank and Aydin (1994), Koenig and Aydin (1998), Willemse and Pollard (1998), Cooke et al. (2000), and Wilkins et al. (2001) have demonstrated the utility of end cracks and other secondary structures in reconstructing slip directions and paleostress orientations. The use of end cracks and, more generally, secondary structures as kinematic indicators will be exploited as a criterion for determining the sense of offset along faults in Chapter 6 when other criteria, such as pre-existing passive markers or bedding, are not available.

The formation of end cracks can be understood by calculating the maximum value of the tangential stress, $\sigma_{\theta\theta_{max}}$ as a function of position about the tip of a parent discontinuity (see Fig. 5.18; Pollard and Segall, 1987). For a pure mode-I crack, $\sigma_{\theta\theta_{max}}$ is located at $\theta = 0°$—directly ahead of the crack (see also Fig. 8.18, upper row). As a result, a mode-I crack (joint, vein, dike, sill, hydrofracture) will propagate in its own plane, without curving, branching, or twisting unless it is acted on by an additional source of stress perturbation (such as a nearby fracture or bedding plane).

What happens if a crack opens obliquely? The problem of mixed-mode (I-II) cracks is well studied (e.g., Erdogan and Sih, 1963; Anderson, 1995,

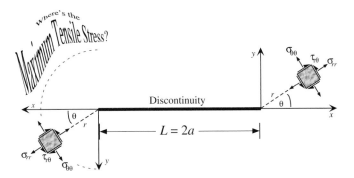

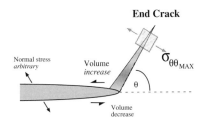

Fig. 5.18. Enhanced tensile stress is generated in the **trailing, dilational quadrants** of a discontinuity that slides (mode II) or dilates obliquely (mixed-mode I-II). End cracks nucleate at the parent discontinuity and propagate away at an angle

pp. 91–96; Engelder, 1999). The elastic near-tip stresses for several variants of mixed-mode loading are shown in Figs. 5.18 and 8.18 (lowest row). In this case, the angular position of $\sigma_{\theta\theta max}$ is no longer directly in front of the crack, but is located at an *acute angle* to the plane of the crack. For the particular case of zero normal stress and unit shear stress (e.g., Erdogan and Sih, 1963, and Pollard and Segall, 1987), the optimum angle for $\sigma_{\theta\theta max}$ is 70.5°, in either or both of the dilational quadrants; this case is shown in Fig. 8.18, center row. A simplified Mohr circle representation of these scenarios is shown in Fig. 5.19.

End cracks have important implications for the kinematics and the mechanics of rock deformation. First, an end crack can form at the tips of either mixed-mode (I-II) cracks or faults. In both cases, the fracture's displacement is stretching the rock near its tip into dilational quadrants, and it is there that the new, secondary fracture will nucleate. This means that secondary fractures (wing cracks, pinnate cracks, horsetail fractures, splays, even anticracks) can be used as a **kinematic indicator** of the sense of offset (or oblique opening direction) along a fracture.

Joints that later accommodated shearing offsets, such as those in Sierra Nevada granites studied by Segall and Pollard (1983b) and Martel and Pollard (1989), among others, present evidence for the existence and sense of shear along them in the form of cross joints (Fig. 5.20). In this case, irregular

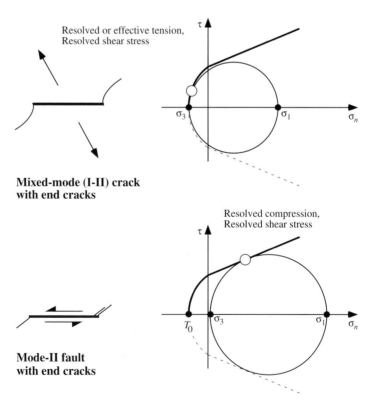

Fig. 5.19. Mohr circles showing stress states resolved on the *parent* fractures (open circles in the figure) that can lead to **end cracks** near their tips.

Resolved or effective tension, Resolved shear stress

Mixed-mode (I-II) crack with end cracks

Resolved compression, Resolved shear stress

Mode-II fault with end cracks

Fig. 5.20. A pair of closely spaced faulted joints having a right-lateral strike-slip shear sense in the Sierra Nevada batholith of central California; end cracks cut diagonally between the fault strands (separated by ~1 m) as well as outside of them with consistent angles relative to the faulted joints.

geometries along the joints produced stress concentrations during shearing that were sufficient to nucleate end cracks that propagated obliquely between closely spaced shearing joints. Observations of joint aperture and terminations demonstrate the end-crack origin of these cross joints, and, hence, establish their significance as kinematic indicators.

Second, the angle between an end crack and the parent fracture can be used as an approximate guide to the orientation of the local stress state near the parent fracture's tip. As noted above, the end crack may curve away from the parent fracture and become reoriented to the remote stress state. However, the angle may be smaller than the ideal 70.5° angle associated with pure mode-II loading and LEFM (small-scale yielding) conditions and the end crack may be planar away from the parent fracture's tip region. As developed in Chapter 9, the take-off angle contains information on the magnitudes and distributions of displacements along the parent fracture (i.e., particular combinations of opening or shear) and on the degree of stress change near the fracture tip (e.g., Cooke, 1997; Martel, 1997; Willemse and Pollard, 1998). These characteristics make end cracks into an extraordinarily informative attribute of rock deformation.

5.2.3 Discontinuity Dimensionality: 2-D vs. 3-D

Up to this point in the chapter, the geometries of structural discontinuities have been defined by using only one dimension, length. As noted in Chapter 1, length is taken to be the dimension as measured in the horizontal plane, such as in map view. This approach has roots in fracture mechanics in which discontinuities were idealized for convenience as cuts in a plane two-dimensional plate. Referring to Fig. 1.15, a discontinuity in a 2-D analysis is oriented perpendicular to the in-plane directions so that variations in stress or displacement in the z-direction are not introduced (e.g., Paris and Sih, 1965; Anderson, 1995, pp. 23, 36, 56). Looking down the thickness (z-) direction onto the horizontal surface of the plate, the discontinuity is represented as a line segment of length L within the xy coordinate system, as done in Fig. 5.6. Because the vertical (third, z-) dimension is not considered explicitly in a 2-D stress analysis, neither is the discontinuity's height (i.e., its dimension in the z-direction, H in Fig. 1.12), so that length and displacement are associated without directly considering the influence of discontinuity height. The calculations of echelon discontinuity growth and end cracks in the chapter, and those in Chapter 9, are largely based on this 2-D approach.

In contrast, a 3-D stress analysis of a fractured rock mass explicitly considers dimensions, stresses, and displacements in all three Cartesian (xyz) directions, including the discontinuity's height and shape. As noted in the literature, discontinuity heights can be smaller than, or comparable to, their lengths. For example, compilation of subsurface data by Nicol et al. (1996) indicates that many faults have length-to-height ratios of 2–3. Surface-breaking faults along plate margins or in plate interiors may have segments with $H > L$ but total lengths having $L > H$ (e.g., Segall and Pollard, 1980; Scholz, 1982; Willemse, 1997; Shaw et al., 2002; Soliva and Schultz, 2008). Calculating the strain from fault populations must therefore explicitly consider the 3-D fault dimensionality (e.g., Scholz, 1997; Watters et al., 1998; Twiss and Marrett, 2010a,b; Nahm and Schultz, 2011) as discussed in Chapter 9. Igneous dikes can exhibit similar three-dimensional geometries in length and height that are important to consider in analyses of dike-related deformation (e.g., Rubin, 1992).

The smaller of length or height controls offset, interaction, linkage, and spacing

Several studies have shown that the **shorter dimension** controls the displacement magnitude and associated discontinuity interaction distance (Segall and Pollard, 1980; Pollard and Segall, 1987; Olson, 1993; Pollard et al., 1993; Willemse, 1997; Cowie, 1998a,b; Crider and Pollard, 1998; Gudmundsson, 2000; Soliva et al., 2006). Although the maximum relative displacement (D_{max}) on a discontinuity depends explicitly on both length and height (e.g., Irwin, 1962; Kassir and Sih, 1966; Chell, 1977; Willemse et al., 1996; Willemse, 1997; Martel and Boger, 1998; Schultz and Fossen, 2002; Schultz et al., 2006), to a first approximation the dimension used in analyses of discontinuity interaction and growth corresponds to the shorter dimension of L or H. Key attributes of discontinuity populations such as **spacing**, as discussed below, thus depend on the shorter dimension, which may be different than the map-view length (e.g., Soliva et al., 2006).

The geometry of a 3-D discontinuity is illustrated in Fig. 5.21. "Tall" discontinuities have $H > L$, whereas "long" discontinuities have $L > H$. Discontinuities can also be described as having an **aspect ratio** given by L/H. Surface-breaking faults that are unrestricted at depth may be characterized by a **length ratio** that is twice the aspect ratio (Polit et al., 2009), since H is the half-height of the semi-elliptical fault. An indefinitely tall discontinuity, corresponding to one in a 2-D plane approximation, has $L/H < 0.2$ (see Chapter 9).

The subsurface faults measured by Nicol et al. (1996) as having an aspect ratio $L/H = 2$–3 correspond to long faults. Applied to crack geometries (e.g., Olson, 1993), tunnel cracks are tall, whereas blade-like cracks are long. In seismology and neotectonics, the down-dip fault dimension is commonly referred to as its width instead (e.g., Scholz, 1982, 1997; Wells and Coppersmith, 1994). Seismologically "small" faults can have a range of aspect ratios as long as the fault height is less than the thickness of the seismogenic crust, whereas "large" faults have heights limited to the seismogenic thickness. This statement implies a particular time scale relative to that of the seismic cycle.

5.2.4 Spacing of Discontinuities

Spacing distributions are closely related to the length distributions of the discontinuities, because spacing depends on the discontinuity's dimensions (length and height) and displacement, and because displacement and dimension are also related to each other (e.g., Chinnery, 1961; Lachenbruch, 1961; Savage and Hastie, 1966; Barnett et al., 1987; Pollard and Segall, 1987; Cowie and Scholz, 1992b; Dawers et al., 1993; Vermilye and Scholz, 1995; Fossen and Hesthammer, 1997). In general, the spacing scales with the smaller of the two discontinuity dimensions (height H or length L; Fig. 5.20; see, e.g., Olson, 1993; Willemse, 1997; Gudmundsson, 2000; Soliva et al., 2005, 2006). In the case of vertically tall discontinuities ($H > L$), the horizontal length L generally

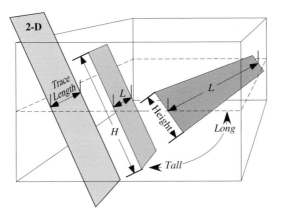

Fig. 5.21. Length and height of discontinuities which are shown for simplicity as having rectangular tiplines.

controls the spacing, whereas the down-dip height H controls the spacing when $L > H$. This relationship is shown in Fig. 5.22.

The physical dimensions of a particular layer can determine the spacing or, more generally, the degree of mechanical interaction between them (Fig. 5.23). Spacings less than some critical value of a relevant parameter, such as a contour of Coulomb stress change around faults (Fig. 5.8), lead to stress shadowing for particular echelon configurations. According to calculations of **stress shadowing** in the literature (Bai and Pollard, 2000a; Soliva et al., 2006) and neglecting the effects of fluids, significant mechanical interaction occurs between discontinuities having spacings less than ~1 discontinuity length or height, whichever is less. Discontinuity spacings less than this will be associated with mechanical interaction between them that can impede their in-plane growth, promoting the formation of stepovers and segment linkage.

A conceptual example of discontinuity growth within a layer having a given thickness T is shown in Fig. 5.24. The upper and lower contacts of the layer have properties of stiffness or strength that are sufficient to confine discontinuity growth to that layer, which then functions as a single mechanical unit. Contacts that slip frictionally, open to form bedding-plane joints, or are rich in clay minerals can restrict discontinuities to the layer (e.g., Helgeson and Aydin, 1991; Niño et al., 1998; Rijken and Cooke, 2001; Soliva et al., 2005; Schöpfer et al., 2011).

Numerical simulations of joint growth can reproduce many of the key aspects of jointing in layered stratigraphic sequences (e.g., Olson and Pollard, 1991; Olson, 1993, 2004). An example is shown in Fig. 5.25 in which the

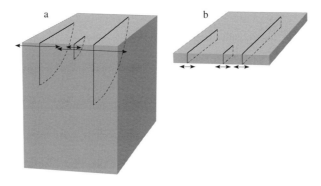

Fig. 5.22. Spacing of discontinuities is dependent on the controlling dimension. (a) Height H is comparable to or greater than horizontal length, L, and so is the controlling dimension. (b) Length exceeds height; these "long" discontinuities have spacings, s (arrows), that scale with their heights rather than their lengths.

Fig. 5.23. Composite photograph of Upper Cretaceous Frontier Formation sandstone, near the Oil Mountain anticline (Hennings et al., 2000), Wyoming, cut by systematic and non-systematic sets of joints. Photographs by Bob Krantz.

Fig. 5.24. Diagrams illustrating growth of discontinuities within a single mechanical layer or unit, after Soliva et al. (2006). (a) Early stage of growth with discontinuities having a range of lengths in relation to the layer thickness *T*. (b) Late stage in growth is characterized by constant spacing proportional to *T*.

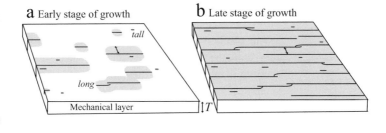

Fig. 5.25. Numerical simulations of joint growth demonstrating the effects of remote loading conditions and layer thickness on the pattern of joints (see Olson, 2007, for method and key parameters).

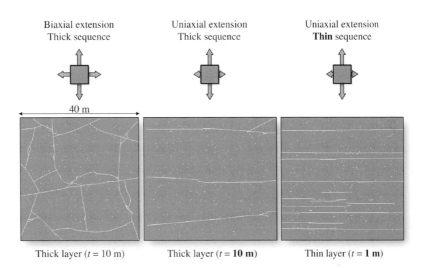

boundary element method is used to numerically grow a field-scale joint set from a series of started flaws (e.g., Olson, 1993, 2007). The model has horizontal dimensions of 40 m by 40 m, a layer thickness of 1 m or 10 m, and stress-state and joint-growth parameters that produce geologically realistic patterns (e.g., Olson, 1993, 2004). Two effects are illustrated in the figure. First, an isotropic or biaxial remote extension leads to randomly shaped joint-bounded polygons in a thick layer (Fig. 5.25, left-hand panel). However, a nearly uniaxial remote extension leads to parallel joints, which link when sufficiently closely spaced. The spacing between the parallel joints is wide for the thick layer (Fig. 5.25, center panel) but much narrower for the thin layer (Fig. 5.25, right-hand panel). These simulations show how joint sets can be understood by investigating the influence of particular key variables such as remote tectonic regime and layer properties.

In many cases, lengths and spacings are dependent on the degree of **vertical restriction** of the discontinuity population (Fig. 5.22). For the classical case of joints in a granitic pluton (e.g., Segall and Pollard, 1983a), the joints can grow unimpeded both in length and height (Fig. 5.22a); this implies that the joint aspect ratio (length/height) remains relatively constant as long as the properties of the granite remain homogeneous in all directions. Thus length can be used as a proxy for height, with just a constant of proportionality between them (e.g., Willemse et al., 1996; Gudmundsson, 2000). The opening displacements (kinematic apertures) along the joints scale with their lengths (e.g., Vermilye and Scholz, 1995; Olson, 2003; Schultz et al., 2008a; Olson and Schultz, 2011). As a result, both the length and spacing distributions of nonrestricted fractures tend to follow **power-law distributions** (e.g., Ackermann et al., 2001).

On the other hand, discontinuities in layered sedimentary sequences may become restricted at depth (Nicol et al., 1996), so that their aspect ratios increase as they grow progressively in length at a constant value of height (e.g., Segall and Pollard, 1980; Olson, 1993; Scholz, 1997; Gudmundsson, 2000; Schultz and Fossen, 2002). This promotes a stabilization of spacing because the interaction distance scales with the discontinuity's smaller (and constant) dimension (now the height; see, e.g., Olson, 1993; Willemse, 1997; Soliva et al., 2006). As a result, discontinuities that would interact and link in an unrestricted sequence would instead grow past each other to greater lengths in a restricted unit (e.g., Ackermann et al., 2001). This leads to an **exponential** length distribution and constant maximum spacing that scales with layer thickness (e.g., Soliva and Schultz, 2008). Power-law and exponential distributions are discussed further in Chapter 9.

Discontinuity spacings can vary systematically with position relative to a free surface or stratigraphic contact. For example, joints formed by dessication-related shrinkage or thermal contraction can grow away from the free surface with a spacing proportional to joint height (e.g., Lachenbruch, 1961; DeGraff and Aydin, 1987; Aydin and DeGraff, 1988; Goehring et al., 2009). A morphologically similar set of joints developed in tight gas shales in the Finger Lakes region of New York State has been interpreted to have propagated downward from a previously jointed layer (Engelder et al., 2009; Fig. 5.26). This pattern is referred to as **joint elimination**.

The mechanics of discontinuity growth and interaction can influence discontinuity patterns in a layer where other spatial scales, such as mechanical thickness or proximity to free or bedding surfaces, have not been established. For example, **joint clustering** has been associated with conditions of **subcritical crack growth** (Olson, 1993, 2004, 2007). In this case, the strength of rock near joint terminations is characterized by a Charles-type creep law (see Charles, 1958; Atkinson and Meredith, 1987a) and modulated by chemical and environmental conditions such as rate and temperature (e.g., Brantut et al., 2013). Both the velocity of joint growth and the degree of mechanical interaction between joints is modulated by parameters associated with the creep law, such as the stress corrosion index n. Because subcritical crack growth depends on the types and rates of chemical reactions near a crack tip (e.g., Chen et al., 2017), both fracture apertures (Olson et al., 2007), as well

Subcritical crack growth

Fig. 5.26. An example of joint elimination in sedimentary rocks. Dilatant cracks nucleated at a bedding contact and then propagated downward, increasing both in length and in spacing with distance from the contact (see also Younes and Engelder, 1999; Engelder et al., 2009).

as propagation rates and the associated fracture patterns, such as joint clustering, are sensitive to the degree of **diagenesis** (e.g., Laubach et al., 2010) and geochemistry of the deforming layers.

An additional factor in the spacing of faults is their slip rates (Cowie and Roberts, 2001). Because spacing depends on the magnitude of the displacement along the fractures, and because the displacement varies in magnitude along-strike, fracture spacings also depend on the displacement distributions along the fractures. For faults, the slip rate varies in proportion with the displacement magnitude (Cowie and Roberts, 2001), so for a large enough fault (having slip rates and earthquake recurrence intervals that are long compared to the time scale of observation), then the spacings between active faults should be greater normal to the fault midpoints and least near their ends. This relationship is borne out in several well-studied fault populations, such as the active extension in Greece and the Aegean Sea area of the eastern Mediterranean (e.g., Morewood and Roberts, 2002; Cowie et al., 2007).

5.2.5 Cross Joints

Perhaps one of the most intriguing and common joint geometries is cross joints. These are shorter, *non-systematic* (Pollard and Aydin, 1988) cracks that nucleate and grow between a pair of parallel, and much longer, *systematic* joints (e.g., Hodgson, 1961; Hancock, 1985; Dyer, 1988; Gross, 1993; Bai and Gross, 1999; Olson, 2007; Figs. 5.23, 5.27). There has been a lot of work done to understand what controls the range of patterns that are observed in the field, and the main results are summarized in this section. Joint sets are important in many areas including reservoir and basin scale fluid flow (e.g., Olson et al., 1998; Hennings et al., 2000, 2012; Lorenz et al., 2002).

The basic problem is sketched out in Fig. 5.27. First we cause a layer of given thickness to extend, leading to creation of a regularly spaced set of joints perpendicular to the layer (see next section for a discussion of what controls the joint spacing). These are the systematic joints because they are longer in the horizontal dimension (i.e., their length, L) than in the vertical one (their height, H). Many interbedded sedimentary sequences have a heterogeneous strength distribution, meaning that some beds joint more easily than others. This creates a set of systematic joints that are confined, or **restricted**, to that stratigraphic layer. In this case, the maximum joint height H equals the layer thickness, leading to constant joint heights H. In other cases, the systematic joints are not confined to a particular layer, but grow to a certain height and then stop, as in the case of a cooling basaltic lava flow (e.g., DeGraff and Aydin, 1987, 1993; Goehring et al., 2009). These systematic joints then define what is known as a *mechanical stratigraphy* (e.g., Gross, 1993; Gross et al., 1997; Laubach et al., 2009).

Many systematic joints are linked at regular intervals along their lengths by a different set of joints, called **cross joints**. These shorter cracks grow due to either the remote stress field alone—implying *two* directions of jointing and layer extension caused by a *single* direction of extension (e.g., Bai et al., 2002)—or a later, rotated stress state associated with a subsequent tectonic event (Bai et al., 2002). Cross joints nucleate in between the systematic

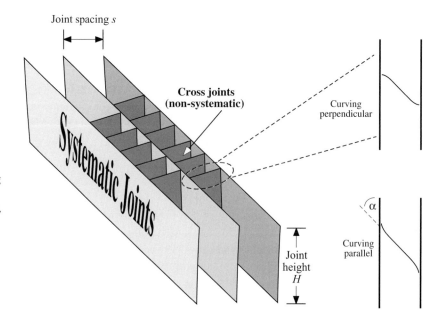

Fig. 5.27. Parallel long systematic joints have horizontal lengths L that are much greater than their vertical heights H, and a spacing between them of s. The cross joints may be either perpendicular or oblique to the systematic joints, with a take-off angle α. Cross joints will terminate either as curving perpendicular (inset, upper right) or curving parallel (inset, lower right).

joints at some high angle α to them, and then propagate toward and eventually may intersect them. Two common geometries of cross joints are **curving perpendicular** and **curving parallel** (Fig. 5.27), as defined by the intersection angle between the cross and systematic joints. In some cases, smaller joints that start out being parallel to the systematic joints will grow, filling in the space between the systematic joints and thus reducing the overall joint spacing (Bai and Pollard, 2000a,b).

Bai et al. (2002) showed how the nucleation of either cross joints or parallel, infilling joints depends on both the ratio of intermediate to most tensile remote stress (the "stress ratio" on Fig. 5.28) and the spacing/height ratio of the systematic joints (Fig. 5.27). Their analysis did not consider fluid flow from the host rock into the joints, which can also influence fracture spacings. Orthogonal (or very high angle) **cross joints are favored** when both remote stresses are compressive and when the systematic joints have spacings smaller than their heights (Fig. 5.28a). On the other hand, parallel (infilling) joints are favored to form when the spacing of systematic joints exceeds their vertical height (Fig. 5.28a, dark shading). For spacing/height ratios less than one, dilation of the systematic joints compresses the intervening rock, thus restricting the formation of cross joints to particularly favorable remote stress states such as an effective tension normal to the main joints.

The intersection angles of cross joints are influenced by the remote stress state and by whether the systematic joints are open (and thus functioning as principal planes) or closed (Bai and Gross, 1999). **High-angle intersections** (curving perpendicular on Fig. 5.27) are favored by a compressive remote stress state and open systematic joints for a wide range of initial take-off angles (Fig. 5.28b). **Low-angle intersections** (curving parallel) are favored by a tensile least principal remote stress and small take-off

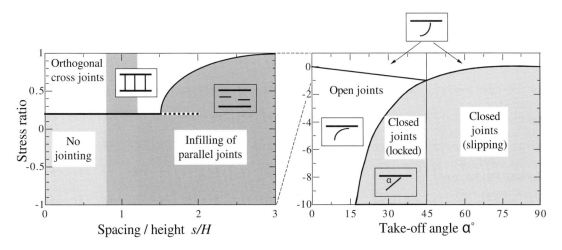

Fig. 5.28. (a) Growth fields of cross joints and infilling joints between systematic joints (after Bai et al., 2002). (b) Joint intersection map for cross joints (after Bai and Gross, 1999).

angles (Fig. 5.28b), so that the growing cross joints remain approximately parallel to the systematic joints. When the systematic joints are closed, and not frictionally sliding under the remote stress state, then the smaller cross joint is not significantly affected by stress perturbations due to them and grows unimpeded in a straight path (Fig. 5.28b). If the closed systematic joints slip, then the stress state near them becomes spatially heterogeneous, promoting more complicated and spatially variable growth directions of smaller joints near them (but see Rawnsley et al., 1992, for documentation and discussion of these).

How could one determine if the two orthogonal sets of joints were coeval, and related to the same remote extension, or if they formed as subsequent jointing events? Bai et al. (2002) found a simple and reliable tool that can provide insight into resolving this key question. Cross joints are favored under a unidirectional remote extension (that forms the systematic joints) for very **small spacing/height ratios** (see Fig. 5.28a). By implication, if cross joints are identified only between the closely spaced systematic joints in a set, and not between the wider systematic joints, then they are most likely synchronous (Fig. 5.29a). If for some reason the remote stress state rotates by perhaps 90° to create new joints orthogonal to the previous, relict set of systematic joints, then the cross joints that form between the widest pair demonstrate this second phase of extension and jointing (Fig. 5.29b).

Let's work through a simple example to illustrate these principles. Suppose you are perplexed by the cracked asphalt in the street outside your house. The asphalt layer that is cracked is 10 cm (4 inches) thick and a systematic fracture set has an average spacing of about 12 cm. The spacing/height ratio is then $s/H = 1.2$. Now you notice that a smaller set of open cracks is oriented nearly perpendicular to the first set. What can be said from these basic observations? First, the stresses acting on the asphalt are, most likely, both compressive and comparable in magnitude (Fig. 5.28a). Second, because the cross joints intersect the systematic ones at generally high angles, any small joint between the systematic ones will grow and reorient itself to become perpendicular to the

Fig. 5.29. A field-based diagram for determining whether the cross joints result from (a) the same remote stress state, and jointing event, as the systematic joints, or (b) a second, later event with a different direction of extension (after Bai et al., 2002).

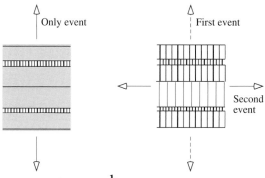

a One extension direction   b Two sequential extension directions

bounding joints. This example demonstrates how the basic principles outlined in this section can be used to begin to unravel the sequence and key controls of joint-related deformation, such as remote loading or stress conditions and relative scale between fracturing and layering.

5.2.6 Discontinuity Patterns from the Strain Ellipse

Shearing of homogeneous materials can, in many cases, produce a systematic array of structural discontinuities (e.g., Tchalenko, 1970). Propagation of a basement fault upward into an undeformed sedimentary sequence can produce patterns of faults, joints, or deformation bands that resemble Riedel (pronounced "ree-del") fracture sets (e.g., Riedel, 1929; Bishop, 1968; Wilcox et al., 1973; Harding and Lowell, 1979; Bartlett et al., 1981; Christie-Blick and Biddle, 1985; Fossen, 2010a, pp. 359–360; Stefanov et al., 2014; Stefanov and Bakeev, 2014; Chemenda et al., 2016), as can earthquakes that rupture sedimentary sequences (e.g., Tchalenko and Ambrasays, 1970; Quigley et al., 2012). Deformation of a circle into an ellipse during simple shear (Ragan, 2009, pp. 276–278), combined with information on the frictional strength of the material, leads to predictions of discontinuity orientations (Twiss and Moores, 2007, pp. 217–226) that may match those observed in the field and in experiments remarkably well (Harding, 1974; Christie-Blick and Biddle, 1985).

The Riedel shears are defined from a **strain ellipse** (Fig. 5.30a), implying a correspondence between the strain directions and initial discontinuity orientations. With ongoing strain, early-formed discontinuities can rotate out of their initial orientations to become overprinted by later ones (e.g., Wilcox et al., 1973). For simplicity in this section, only the orientations of the initial structures are presented.

The loading geometry, strain ellipses, and predicted discontinuities are illustrated in Fig. 5.30, upper panel; the relationship to strike-slip faults (lower panel) will be discussed below. Normally, the coordinate system is defined with respect to the direction of remote shearing (right-lateral in the example shown in the figure). This leads to a decomposition of the remote shear into principal normal stresses oriented at ±45° to the eventual through-going zone. This final orientation, parallel to the sense of remote shearing displacement, is commonly referred to as a Y-shear (Fig. 5.30b) or the principal-displacement shear.

The next step involves specifying the frictional resistance of the material. This can be given by the Coulomb or Hoek–Brown envelope (Chapter 3) and is typically invoked as the friction angle ϕ_f. The friction angle specifies the orientations of the initial planes of frictional sliding relative to the direction of maximum compression (σ_{max} in Fig. 5.30b), which are oriented at angles of $\pm\phi_f$ to σ_{max}. For typical angles of ϕ_f of 30°, the initial faults would form at angles of 15–45° to the remote shearing direction.

By using this geometric framework, a number of initial discontinuities have been proposed (Fig. 5.30b) including:

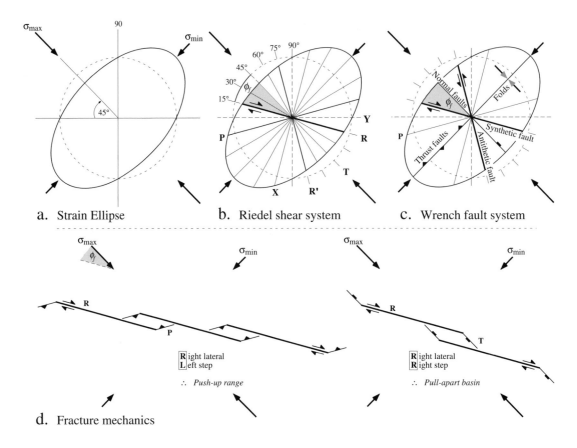

Fig. 5.30. Prediction of fault patterns using continuum (a–c) and discontinuum (d) mechanics. Sense of *remote shearing* is **right lateral** in all cases and parallel to the horizontal axis in (a–c). Shaded sector and ϕ_f show 30° angle of friction for typical rocks. (c) Geologic interpretation of the remote stress state, after Harding (1974). (d) Resulting fault pattern consistent with the remote stress state and its orientation, noting that certain orientations implied in (b) will form only in certain specific locations along the strike-slip fault array.

- *R-shears*. "Riedel shears" are the main shear surfaces that form by Coulomb frictional sliding in the rock or soil. They grow into echelon geometries at an angle to the remote principal stresses that is related to the frictional properties of the material. Given the oriented stress state shown in Fig. 5.30, the R-shears should form at initial angles of $\phi_f/2$, or $(45° - 30°) = 15°$ (for a friction angle of $\phi_f = 30°$) to the eventual through-going shear plane (labeled "Y" in the figure). Correspondingly, they form at $\phi_f = 30°$ to the direction of maximum compression (Fig. 5.30c) with a sense of slip that is the same (right-lateral) as that of the main or eventual through-going shear zone (due to the angle being acute), making them *synthetic* R-shears.
- *R'-shears*. "Conjugate Riedel shears" also form at angles of $\phi_f = 30°$ to the direction of maximum compression (Fig. 5.30c), but with a sense of slip that is opposite to that of the main shear zone (due to the angle being acute), making them *antithetic* (or conjugate) R-shears, and designated R'-shears. Their sense would be left-lateral in the example shown.

- *P-shears*. These surfaces form at angles of $\phi_f = 30°$ to the R-shears and thus $\phi_f/2 = 15°$ to the main zone. Given their orientations at high angles to the overall direction of shearing, they accommodate minor amounts of horizontal slip. Skempton (1966) named these Passive shears from the passive Rankine state in soil mechanics; they have also been called "thrust shears" (Tchalenko, 1970). Swanson (1984) identifies P-shears as strata-cutting *ramps* in ramp-and-flat thrust fault systems (with R-shears being the bedding-parallel *flats*).
- *T-shears*. "Tension shears" are now considered to be inappropriately named, given their mixed kinematic implication in this approach. The dilatant cracks that would form in this orientation—parallel to the direction of maximum horizontal compressive remote principal stress—would also be at 45° to the final orientation of the shear zone. At the field scale, normal faults can also form in this orientation because both cracks and normal faults strike parallel to σ_2.
- *Y-shears*. "Principal displacement shears" are the final surfaces along which rock moves along the through-going fault. These accommodate the largest amount of displacement and complete the decoupling between two blocks that were only partially or incompletely separated by the P- and R-shears. In some of the experimental literature on deformation of gouge (see Gu and Wong, 1994), these are called *B-shears* because they nucleate along the boundary between gouge and country rock (e.g., Marone, 1998b).
- *X-shears*. Swanson (1984) uses X-shears to relabel R- and R'-shears (if these are normal faults) and to relabel P- and P'-shears for thrust faults.

The strain-ellipse method predicts three main sets or classes of structural discontinuities resulting from the simple shear. A pair of **conjugate strike-slip faults** are predicted to form at angles of $\pm \phi_f°$ to the maximum horizontal compressive principal stress (Fig. 5.30c). These correspond to the R and R' structures. Parallel to the maximum compression direction, structures that accommodate extensional normal strain, and no shear strain, are predicted to occur: **joints and normal faults** would form at this orientation. Orthogonal to these structures, or perpendicular to the direction of maximum horizontal compression, **reverse faults or folds** would be predicted, accommodating contractional normal strains and no horizontal shear strain. This trio of structures defines what is sometimes referred to as the "wrench fault" tectonic style (e.g., Harding, 1974).

Although the orientations of the initial discontinuities can be well described by the strain-ellipse method, the timing and spatial arrangement of the structures is not obtainable by using this approach. However, the timing and fracture geometry are well understood in this context (e.g., Tchalenko, 1970; Aydin and Page, 1984; Christie-Blick and Biddle, 1985) although the Riedel sequence is not always followed by compact or strongly cohesive porous rock, or more complexly deformed rock sequences (Aydin and Page, 1984; Martel, 1990; Flodin and Aydin, 2004a,b). As a result, the strain-ellipse approach to predicting discontinuity localization can provide a useful first step that needs additional consideration in order to bring in spatial organization and relative timing.

The sequence of formation of Riedel shear structures is clear from experimental and theoretical work. The first discontinuities to form (in frictional materials like soil or porous rock) are the R-shears. These have been called the "Coulomb shears" because they lie in orientations of approximately $\pm 30°$ (corresponding to $\pm \phi_f°$) to the direction of maximum compressive principal stress (e.g., Tchalenko, 1970; Bartlett et al., 1981). These fractures (or sometimes deformation bands) correspond to **echelon fault segments** developed in strike-slip fault zones (e.g., Segall and Pollard, 1980; Aydin and Page, 1984). Continued shearing leads to formation of P-shears, but only in certain locations within the deforming shear zone. P-shears are short thrust faults (potentially with some synthetic right-lateral offset as well) that connect, or *link*, the R-shears, forming either "bull-nose" structures in the classical literature (Tchalenko, 1970) or **contractional stepovers** (e.g., Aydin and Nur, 1982; Davison, 1994; Fig. 3.6). A set of (right-lateral) R-shears can only grow into a left-stepping echelon array, suggesting contractional strain within their stepovers and P-shears that link them (see Fig. 5.30d, left-hand panel). Pull-aparts (or T-shears in the stepovers in place of P-shears) would form if the R-shears stepped to the right. T-shears—dilatant cracks or normal faults—form in the trailing, **dilational quadrants** of the R-shears at some point after the R-shears have begun to accumulate sufficient shearing displacements for the appropriate near-tip stress changes to have built up there. Ultimately, right-lateral strike-slip displacements build up along the **linked system** of R-shears (and connecting P-shears) to create a principal displacement zone or Y-shear.

5.3 Review of Important Concepts

1. Geologic structural discontinuities grow into a variety of distinctive patterns. These patterns are shared by the different types of discontinuities because the underlying principles that govern propagation, interaction, and coalescence of structural discontinuities are the same.
2. Two discontinuity patterns that require an understanding of stress concentration and mechanical interaction are called discovery patterns. These are wing cracks and echelon discontinuities.
3. End or wing cracks grow as a result of stress concentration generated near the tips of a larger discontinuity. The secondary discontinuities grow along orientations that maximize the stress component, such as tension, that controls their formation. Strongly curving wing cracks suggest tensile loading in strong rock and large near-tip stress concentration, whereas more planar end-crack geometries, and nucleation of secondary fractures along the discontinuity (not just at the tip), suggest a clearer association with the remote stress state and perhaps a smaller degree of stress change at the larger discontinuity's tip region. Secondary

structures provide a kinematic indicator of the sense of offset, or oblique opening direction, along a discontinuity.

4. Echelon discontinuities are defined by their sense of step and proximity to each other. The region between the overlapping (inner) discontinuity tips is called the stepover. These patterns form by mechanical interaction (or "soft-linkage") between the discontinuity pair as they both propagate mutually toward their stepover. Once a certain amount of overlap is achieved, further inward propagation is no longer possible, and the geometry freezes in. Continued straining of the rock mass around the overlapping echelon discontinuity pair causes the displacements along the discontinuities to increase, eventually promoting breakage of the rock in the stepover, at or near the trapped inner fault tips. This creates a secondary discontinuity that propagates obliquely into the stepover and eventually ("hard-") linking the echelon discontinuities.

5. Echelon joints can form an aligned array having the same sense of step, leading to a zone or through-going structure that accommodates a component of shear displacement. Because crack interaction and linkage are energetically favored for a relatively small range of stepover geometries, the resulting angle of faulting is also fairly constant for a wide range of rock types (i.e., about 30° to the maximum compression direction). The angle of friction for a fault composed of linked microcracks then depends on the interaction mechanics of the precursory cracks.

6. The relative dimensions of structural discontinuities profoundly influence the resulting patterns. Succinctly, the lesser of the two dimensions, height or length, controls the pattern. Long discontinuities form patterns governed by the horizontal dimensions (lengths) of the discontinuities, mimicking growth in a 2-D system. Height-restricted discontinuities create patterns governed by the third (height) dimension, resulting in mechanical interaction, spacing, and linkages that can scale with the height dimension.

7. In-plane shearing of homogeneous materials produces an evolving set of structural discontinuities whose initial orientations can be predicted by using a strain ellipse that is referenced to the remote principal stress orientations. The timing and locations of secondary discontinuities can be refined by considering the mechanical interactions between the echelon discontinuities.

5.4 Exercises

1. What discontinuity patterns might form for the following stepover parameters: (a) $r/a = 2.6$; (b) $r/a = 0.3$; (c) $o/s = -1.8$; (d) $o/s = 2.96$?

2. Describe the sequence of discontinuity growth into an echelon pattern. Be sure to note the controlling mechanical processes for each step.

Fig. 5.31. Pattern of normal faults exposed on a horizontal horizon, for Exercise 5.6.

3. What field observations might one need to collect in order to determine whether two orientations of discontinuities are conjugate?

4. Summarize how a wing crack might form at the end of a larger discontinuity, including its location, orientation, and curvature.

5. Describe the effect of mechanical stratigraphy on the patterns of discontinuities, giving specific numerical values.

6. Give an interpretation of the normal fault pattern shown in Fig. 5.31. Be sure to support your interpretation with drawings of stress states and geologic reasoning that incorporates fault mechanics, such as step sense, shear sense, and sequential development.

7. What are the fracture-mechanics names for the following Riedel shear structures: (a) R-shears; (b) R'-shears; (c) T-shears; (d) P-shears?

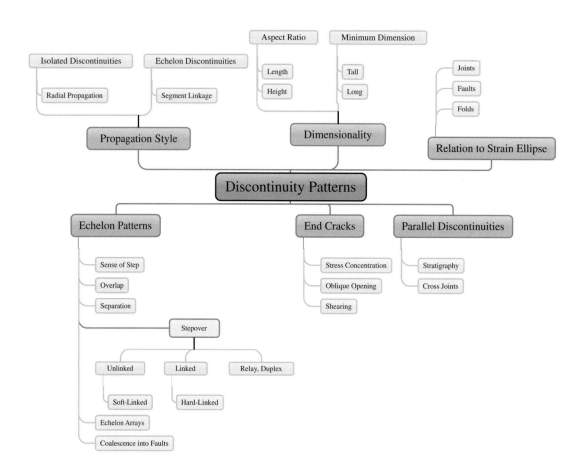

6 Faults

6.1 Introduction

Faults are an efficient mechanism for allowing large strains to accumulate in the upper lithosphere of the Earth and other planetary bodies. In general, faults are arrays of 3-D surfaces along which large shear offsets (fault-parallel displacements) have been accommodated by frictional sliding (see Chapter 1). Faults can redirect the flow of subsurface fluids (either channeling or restricting the flow), modify the transmission of seismic waves, and create spectacular surface topography—through steady or destructive slip events—that have attracted mankind over millennia for the associated resources and natural beauty.

The main characteristics of faults, such as their kinematics, fault surface textures, and relationships to larger-scale tectonic processes and terranes are well covered in standard textbooks on structural geology, and will therefore not be repeated here. In this chapter several aspects of faults related to geologic fracture mechanics will be stressed. Because faults are planes or zones of frictional sliding, fault patterns and other attributes depend in large part on what surfaces within the rock mass were available and used as **substrates** for the frictional sliding. Mechanical understanding of faults leads to a practical framework for understanding the **strains at fault tips** and for utilizing the kinematic distinctiveness of the **stepover regions** (Fig. 6.1) between closely spaced echelon faults. Patterns of deformation near fault tips and in the stepovers provide a reliable criterion for assessing the sense of slip along a fault when other criteria, such as surface textures (slickensides, etc.) or offset passive markers are not available.

Historically, faults have often been investigated by using a variety of approaches and tools. During the last decades of the twentieth century, a re-evaluation of faults as *fractures* having a displacement discontinuity (of either mode-II or mode-III type; see Chapter 1)—and understood using the principles of Linear Elastic Fracture Mechanics (e.g., Chinnery, 1965; Rudnicki, 1980;

Fig. 6.1. Field view of a world-class stepover (relay-ramp) between two graben-bounding normal faults, Devils Lane graben, Canyonlands National Park, Utah.

Pollard and Segall, 1987; Cowie and Scholz, 1992b; Cowie, 1998a,b; Cowie et al., 2007; Chapter 8)—has supplemented previous classifications that were more phenomenological and descriptive (see Reid et al., 1913; Hill, 1963; Billings, 1972). The discontinuity-based approach, adopted in this book, acknowledges that faults can be understood by applying the same techniques that have improved our understanding of joints and other opening-mode fractures over the past several decades (e.g., Lachenbruch, 1961; Rice, 1979; Rudnicki, 1980; Segall and Pollard, 1983a; Kulander and Dean, 1985; Engelder, 1987; Pollard and Aydin, 1988; Cowie and Scholz, 1992b; Engelder et al., 1993; Scholz, 1997; Aydin, 2000; Schultz and Fossen, 2008; Faulkner et al., 2010; Gudmundsson, 2011; Peacock et al., 2017).

Faults are discontinuities   Mechanically, faults are sharp structural discontinuities (Chapter 1) that interrupt the continuity of a rock mass and concentrate and redistribute the far-field stresses (e.g., Chinnery, 1961, 1963; Pollard and Segall, 1987). Although the surface markings of joints and faults differ due to their different displacement senses and the contact relations between the respective fracture walls, many of the principles of interaction, segmentation, and propagation that are well understood for cracks apply equally well to faults (e.g., Davison, 1994; Peacock, 2002; Scholz, 2002; Fossen, 2010a), as well as to deformation bands, as discussed in the literature and in Chapter 7. Similarly, the geometric configurations of joints and faults (such as spacing and echelon patterns) are related to the stress changes associated with the displacement discontinuities along their surfaces. Cracks and faults are close relatives in the world of rock deformation, and their common patterns were touched on in Chapter 5.

6.2 Fault Classification by Kinematics

As developed in standard texts on structural geology, fault types are defined by their sense of shear relative to the horizontal plane of the Earth's surface. Fault offsets, or displacements, are quantified by the rake (angle between the shearing vector in the plane of the fault relative to the horizontal plane) and fault dip angle (Fig. 6.2). The graphical portrayal of fault nomenclature in Fig. 6.2 follows the logical development enunciated by Professor Atilla Aydin in his undergraduate structural geology classes.

Faults with rakes less than 45° to the horizontal are strike-slip faults, since the shear offset is primarily along the strike direction of the fault. Faults with rakes steeper than 45° are dip-slip faults. Strictly, pure strike-slip faults have rakes of 0° and pure dip-slip faults have rakes of 90°, so that faults having rakes between these two extremes are referred to as oblique-slip faults, with both components needed to name them without ambiguity (e.g., right-oblique thrust fault). The two main types of dip-slip faults, normal and reverse, are separated based on the sense of relative motion of the hanging wall when viewed in cross section, with further division based on the fault dip angle.

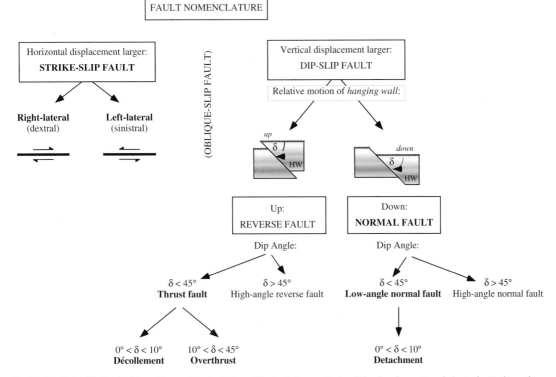

Fig. 6.2. Fault classification by kinematics, or the orientations of the fault plane and rake of the shearing vector relative to the Earth's surface. The fault names most frequently used by structural geologists are shown in **bold**.

Each of the main fault types also displays a characteristic suite of structural characteristics, both in map view and in cross section. For example, volumetric changes at the **tips**, and within the **stepovers** (see Chapter 5), of strike-slip and dip-slip faults are distinctly different, leading to diagnostic criteria for fault type identification by using this criterion (e.g., Rodgers, 1980; Segall and Pollard, 1980; Aydin and Nur, 1982; Davison, 1994; Peacock, 2002). Additionally, **topographic changes** of the Earth's surface provide an independent and diagnostic criterion for the sense of slip along a fault (e.g., Chinnery, 1961, 1965; Ma and Kusznir, 1993; Davies et al., 1997; Willemse, 1997; Davis et al., 2005). These additional characteristics, which are a product of the strain or displacement field associated with the fault, provide clues to processes of fault growth and linkage as well as to the fault kinematics that can be exploited in areas for which other criteria such as fault surface textures, rake, or offset of pre-existing passive markers such as bedding are not available (e.g., Schenk and McKinnon, 1989; Schultz, 1989; Flodin and Aydin, 2004b; Kattenhorn and Marshall, 2006; Schultz et al., 2010b). For example, surface topography is a well-known tool in **neotectonics** and aids in evaluating earthquake physics and seismic hazard (e.g., Wells and Coppersmith, 1994; Yeats et al., 1997; Cohen, 1999; Wesnousky, 2008).

Faults tips and stepovers as shear sense criteria
Structural topography indicates shear sense

6.3 Review of Fault Types

Faults are classified into three main types depending on their orientation relative to the Earth's surface (i.e., their strike and dip), the type of displacement accommodated across them (a discontinuous shearing offset), and the orientation of the shearing vector relative to the Earth's surface in the plane of the fault (see Fig. 6.2; Davison, 1994). In this section many of the principal characteristics relevant to their mechanics are summarized; more detail can be found in standard structural geology textbooks and in the general literature.

6.3.1 Normal Faults

Normal faults can be identified by surfaces that commonly dip 50° or more and have accommodated extensional strains normal to their strike directions. Normal faults occur either singly or paired with antithetic (Fig. 6.3 and 6.4) or synthetic (Fig. 6.5) normal faults; much of the rich literature on the resulting geometric patterns is illustrated and discussed, for example, by Davison (1994), Peacock et al. (2000), and Peacock (2002).

Individual normal faults can combine in three dimensions to create mechanically, aesthetically, and economically important geometries. Parallel normal faults can overlap partially, forming **echelon geometries** that are mechanically analogous to those documented for dilatant cracks (Pollard

Fig. 6.3. Rima Ariadaeus is an echelon graben pair on the Moon, shown clearly in this oblique Apollo 10 image. Notice the change in graben width (from ~3 to ~5 km) over the ridge by the impact crater, suggesting nonvertical fault dip angles, and the tilted ramps in the stepover between the two grabens.

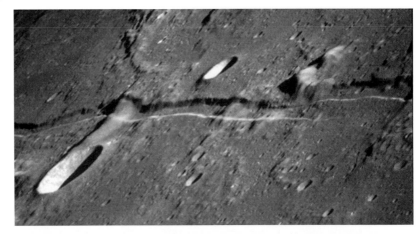

Fig. 6.4. Offset of distinctive stratigraphy (and confirmed by rake measurements) is perhaps the most reliable way to identify a normal fault. An aerial view of Devils Pocket graben in Canyonlands National Park, Utah, shows downward displacement of pre-existing Needles topography onto the graben floor (graben width ~100 m; photo by Matthew Soby).

Fig. 6.5. A pair of echelon normal faults from the Pyrenees Mountains near Berga, Spain, showing key geometric elements and definitions. FW, foot wall; HW, hanging wall.

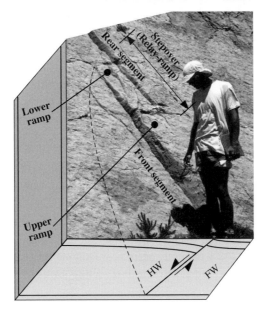

Fig. 6.6. View of the tip region of a single normal fault (northern Devils Lane graben, Canyonlands National Park, Utah) showing uplifted strata on the foot wall and subsiding strata (the same unit) in the hanging wall. Note the gentle, nearly linear increase of normal-fault offset from the tip near the skyline. Shrubs for scale; photo by Scott Wilkins.

and Aydin, 1988) and strike-slip faults (Aydin and Nur, 1985). Fault tips are sometimes observed, and demonstrate decreasing amounts of offset toward the tip (Figs. 6.1 and 6.6). Within the overlap, or **stepover** region (Figs. 6.3 and 6.5), large stress and strain gradients occur (Willemse et al., 1996; Crider and Pollard, 1998). Here, strata are flexed downward and faulted into patterns also commonly referred to as **relay-ramps** (Larsen, 1988; Ebinger, 1989; Bruhn et al., 1990; Morley et al., 1990; Gudmundsson and Bäckström, 1991; Peacock and Sanderson, 1994; Mack and Seager, 1995; Moustafa, 1997; Fig. 6.3). These areas define pathways for sediment transport (Anders and Schlische, 1994), and can form structural traps of importance in geothermal resources and in large oil and gas fields throughout the world (Morley et al., 1990; Jackson and Vendeville, 1994; Stewart et al., 1996).

Parallel, non-coplanar normal faults that completely overlap, with small separations relative to their lengths, define **grabens** that commonly exhibit a greater amount of vertical stratigraphic offset (throw) on one (master) fault than on the other (antithetic) fault (Fig. 6.7). Such *asymmetric grabens* (Gibbs, 1984; Rosendahl, 1987; Groshong, 1989; Jackson and White, 1989) can display a rich assemblage of topographic features (Davison, 1994) such as foot wall uplift (Weissel and Karner, 1989) and hanging wall subsidence (Gudmundsson and Bäckström, 1991). *Symmetric* grabens are also recognized (e.g., Fig. 6.3) and commonly occur in vertically stratified sections. These topographic elements increase in amplitude from zero at the graben terminations to maximum values near the middle regions of the fault, tracking the shape of the displacement distribution and location of maximum offset, D_{max} (e.g., Barnett et al., 1987; Pollard and Segall, 1987; Walsh and Watterson, 1987; Bürgmann et al., 1994).

Depocenters along faults and grabens are often closely associated with the position of maximum depth, or maximum displacement, along the faults (e.g., Gibbs, 1990; Roberts and Yielding, 1994; Gupta et al., 1998; Gawthorpe and Leeder, 2000; Gupta and Cowie, 2000; McLeod et al., 2000; Grosfils et al., 2003). These normally occur near fault midpoints and, with time, **shift toward the stepover of echelon or interacting faults as linkage begins** (e.g., Morley

Fig. 6.7. Two inward-dipping normal faults define a **graben** ~1 m wide in carbonate rocks from the Spanish Pyrenees. The asymmetry in displacement is clearly revealed here; the (left-hand) antithetic fault offsets a pre-existing dinosaur footprint.

et al., 1990; Morley, 1999). Each fault in a graben has its own maximum displacement, and depocenter, created prior to linkage and its incorporation into the observed graben. In addition, the depocenters will serve to collect any hydrothermal (or other) fluids that migrate into the graben, either along its faults or from surrounding rock units. The segmented geometry of normal faults provides clues to the best locations of depositional sinks within grabens.

Normal faults, as well as igneous dikes (Calais et al., 2008; Biggs et al., 2011), are fundamental components of **continental rift** systems (see McClay et al., 2005, for a review of fault relationships and patterns recognized in rift systems). Continental rifts mark the transition from continental crust to oceanic crust (Gibbs, 1984) and a complex variety of structures and rock types are found in these highly deformed, high-strain zones. Continental rifts are commonly defined by a series of subparallel rift basins that are asymmetric, can have significant foot wall uplift, and reflect the structural segmentation of the rift zone (e.g., Bosworth, 1985; Hayward and Ebinger, 1996).

The geometry of faulting in general depends critically on **how the growing fault interacts with the stratigraphy**, and on the characteristics of that stratigraphy. Normal faults grow by the progressive flexure and bending of stratigraphic sequences (or single lithologies, such as found in a granitic pluton or basaltic lava flow). The formation of joints, for example, in stiffer layers can be exploited as zones of weakness and perhaps reduced shear resistance, leading to nucleation of normal faults, having eventual dips steeper than 45°, on the joints in these layers (e.g., Peacock and Sanderson, 1992; McGrath and Davison, 1995; Gudmundsson, 1992; Childs et al., 1996; Ferrill and Morris, 2003; Schöpfer et al., 2006; Welch et al., 2009). On the other hand, thrust faults having dip angles less than 45° not only interact with the stratigraphy but are influenced by the **strengths of the stratigraphic contacts** and interfaces as well (i.e., with the mechanical stratigraphy). Strike-slip faults may be insensitive to the stratigraphy in their map-view expression but sensitive to it in their down-dip characteristics. As discussed below, the upward propagation of normal or reverse faults through a stratigraphic sequence, and the geometry and characteristics of folds and related structures that accommodate shearing sub-parallel to the layers, are strongly influenced by the *frictional resistance of the contacts* between the layers (Johnson, 1980; Erickson, 1996; Roering et al., 1997; Niño et al., 1998). As a result, the **strength anisotropy** of layered stratigraphic sequences exerts a fundamental control on the geometry, scaling, kinematic, mechanical, seismologic, and hydraulic characteristics of faults over nearly all scales.

Strength anisotropy can control fault development

6.3.2 Reverse and Thrust Faults

Reverse faults, and their more shallowly dipping counterparts, thrust faults, accommodate horizontal crustal or lithospheric shortening and are a key component in terrestrial contractional orogens such as convergent plate margins, as well as in terrestrial continental plate interiors (Fig. 6.8). They also are

Fig. 6.8. The trace of the surface-breaking thrust fault resulting from the 1969 Meckering, Australia, M_s 6.8 earthquake shows topography characteristic of many of these structures (image © Commonwealth of Australia (Geoscience Australia), used by permission).

found on other planets and satellites such as Mercury (Fig. 6.9), Venus, Mars, and the Moon (Golombek and Phillips, 2010; McGill et al., 2010; Nahm and Schultz, 2010, 2011 e.g., Schultz et al., 2010a; Watters and Johnson, 2010; Watters and Nimmo, 2010) where they accommodate contractional strain related to mechanisms such as planetary core formation, global cooling and contraction, and local to regional tectonism. Folding is an integral part of contractional fault-related deformation (Fig. 6.10).

The Meckering, Australia, thrust fault (Fig. 6.8) displays a rich assemblage of surface topographic features that can be used to probe the geometry,

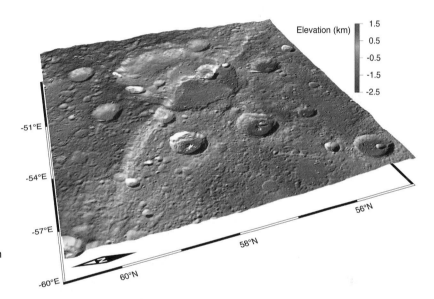

Fig. 6.9. Carnegie Rupes, a lobate scarp on Mercury, is interpreted to be a large surface-breaking thrust fault; scarp length is ~280 km and relief is ~2 km (digital elevation model produced by Christian Klimczak).

Fig. 6.10. Numerous small faults having reverse offsets are shown in this contractionally deformed section of sedimentary rocks from Oregon (photo by Marli Miller).

displacement distribution, and segmentation of subsurface thrust faults (e.g., Plescia and Golombek, 1986; Schultz, 2000c; Okubo and Schultz, 2003, 2004; Davis et al., 2005; see also Yeats et al., 1997). That fault resulted from intracontinental shortening (Gordon and Lewis, 1980; Clark et al., 2011), producing a 37-km-long arcuate surface trace with topographic relief as great as 4 m. Many of the morphologies of the Meckering scarp mirror those in other examples of terrestrial and planetary surface-breaking thrust faults (e.g., Plescia and Golombek, 1986; Schultz, 2000c; Okubo and Schultz, 2004; Tanaka et al., 2010; Fig. 6.9).

Fracture mechanics has improved the understanding of fold-and-thrust structures (e.g., King and Yielding, 1984; Aydin, 1988; Ohlmacher and Aydin, 1995; Cooke and Pollard, 1997; Roering et al., 1997; Cohen, 1999; Cooke et al., 2000; Cooke and Kameda, 2002; Okubo and Schultz, 2004), although other approaches to understanding the field relations, such as petrofabrics, geometric modeling, and strain analysis, may be more widespread (e.g., Boyer and Mitra, 1988; Wojtal and Mitra, 1988; Erslev, 1991). The key issue in using fracture mechanics in this context is identifying a part of the linked, large-strain structure that can be adequately treated by this approach. As documented in the literature, the **early stages of thrust faulting** provide fruitful avenues for mechanics that can supplement the more widely used kinematic models (including trishear) that find utility in geometrically more complex linked systems (e.g., Suppe, 1983; Geiser, 1988; Groshong and Usdansky, 1988; Suppe and Medwedeff, 1990; Erslev, 1991; Hardy and Ford, 1997; Allmendinger, 1998; Shaw et al., 1999; Welch et al., 2009; Hardy and Allmendinger, 2011).

Thrust faults are characteristically segmented and discontinuous in both the along-strike and down-dip directions (e.g., Aydin, 1988; Davison, 1994), giving rise to a diverse assemblage of map view and cross-sectional patterns (Figs. 6.8 and 6.10). Thrust faults may be associated with sets of smaller, echelon thrusts oriented either parallel or antithetic to the main thrust fault and that may also display small components of strike-slip displacement (e.g., King and Yielding, 1984). Systematic, localized rotations of the stress tensor about xyz axes are well known near the periphery of dip-slip faults in three dimensions (e.g., King and Yielding, 1984; Crider and Pollard, 1998), leading to echelon segmentation and a locally consistent sense of step of antithetic (backthrust) fault segments.

The development and architecture of **fold-and-thrust belts** (Nemčok et al., 2005) is closely related to the accommodation of dip-slip offsets along the faults. At a large scale, many fold-and-thrust belts assist an accretionary prism to change its cross-sectional shape (e.g., Mitra and Sussman, 1997); the study of **critical-taper wedge mechanics** (e.g., Davis et al., 1983; Dahlen et al., 1984; Davis and Engelder, 1985; Dahlen, 1990; Suppe, 2007) is closely related to the development of many fold-and-thrust belts and, more specifically, in the patterns and characteristics of individual reverse or thrust faults.

Although critical-taper orogenic wedges commonly occur on the Earth near subduction zones and therefore in association with plate tectonics, they represent a general process of crustal deformation that results in thrust nappes and overthrust terranes (e.g., Price and Cosgrove, 1990), since they also are known to occur on Earth in non-plate tectonic settings as well as on Venus (Suppe and Connors, 1992; Williams et al., 1994; McGill et al., 2010) and Mars (Nahm and Schultz, 2010, 2011). For example, fold-and-thrust belts are found at the base of sliding continental margins and deltas such as offshore Texas (Worrall and Snelson, 1989; Mount et al., 1990) and on the flanks of large volcanoes on Earth and Mars (Borgia et al., 1990; Morgan and McGovern, 2005). Reverse and thrust faults have been recognized on Mercury (e.g., Solomon et al., 2008), Venus, the Moon, and Mars (see Schultz et al., 2010a,c) but, with the exception of Venus, tend not to be organized into narrow or large-strain terranes as they more commonly are on the Earth. Thrust faults on Earth and Mars are commonly blind, but can be either blind (forming blind-thrust anticlines, referred to as **lobate scarps**; e.g., Schultz, 2000c; McGill et al., 2010; Tanaka et al., 2010) or surface-breaking on Mercury, Venus, and the Moon.

Field investigations and mechanical modeling of **blind thrust faults** associated with representative earthquake ruptures (e.g., 1980 El Asnam, Algeria; 1983 Coalinga, California; 1987 Whittier Narrows, southern California; 1994 Northridge, southern California) demonstrate that folding of near-surface strata, located between the blind thrust fault tip and the surface, is a fundamental and widespread process related to subsurface thrust faulting on the Earth (e.g., Philip and Meghraoui, 1983; King and Yielding, 1984; Stein and King, 1984; Lin and Stein, 1989; Taboada et al., 1993; Roering et al., 1997). Thrust faults in layered strata may not break through to the surface, but instead can be "trapped" at shallow depth as blind faults (Roering et al., 1997) by nucleating flexural slip along superjacent strata (Cooke and Pollard, 1997; Schultz, 2000c; McClay, 2011). Flexural-slip anticlines, located above the upper thrust fault tip, grow there as a result of displacement accumulation along the subjacent fault (Fig. 6.11). When combined with one or more small antithetic back thrust faults, the anticlines are referred to as **wrinkle ridges** on planetary surfaces.

Numerical modeling of thrust-related deformation reveals the importance of **layer characteristics** (mechanical stratigraphy, bedding-contact properties, and strain localization) and fault geometry in the resulting suite of near-surface structures. For example, calculations by Cooke and Pollard (1997), Roering et al. (1997), Schultz (2000c), and Okubo and Schultz (2004) demonstrate that frictional slip along a blind thrust fault can nucleate **slip patches** (e.g., Martel and Pollard, 1989) along portions of superjacent bedding interfaces, promoting localized bedding plane slip and the growth of flexural slip (forced) folds in these layers above the fault. Because stress perturbations associated with fault slip in surrounding rock are greatest near the fault and decrease with distance from it, fold amplitudes *decrease toward the surface*,

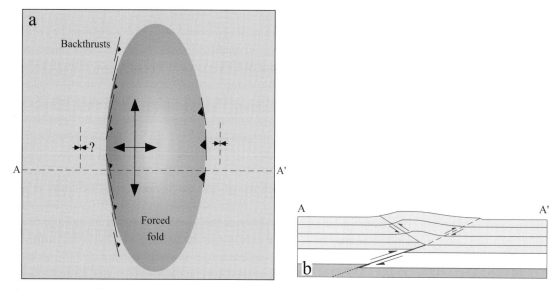

Fig. 6.11. Conceptual blind thrust model (after Schultz, 2000c); overall direction of thrusting is left to right. (a) Structural map showing major elements of blind thrust-forced fold morphology; along-strike length is greatly reduced, relative to fold width, in the figure. Section line A–A' shows position of cross section given in (b); surface breaks of thrust faults are shown with teeth on their particular upper plate; anticlines and potential synclines are shown by converging and diverging arrows, respectively. Upper plate of blind thrust-forced fold system includes the fold and continues to its left. (b) Cross section showing blind thrust fault (heavy line showing thrust displacement) beneath flexural slip anticline, backthrust faults, and up-dip prolongation of main fault surface (dashed).

leading to gentle folds nearer the surface and tighter folds at depth near the fault (Cooke and Pollard, 1997; Okubo and Schultz, 2004). Folding may not reach the surface if the fault is sufficiently deep. Although regional, far-field stresses associated with the thrust faulting may also promote layer-parallel shortening and folding of layers above, displacement along the fault facilitates folding by adding appropriate stresses and by localizing the flexure above it (e.g., Roering et al., 1997).

Whereas typical fold-and-thrust belts may be characterized by forward breaking thrust faults along ramp-and-flat systems appropriate to accretionary (Coulomb critical taper) wedges (e.g., Suppe and Connors, 1992; Mitra and Sussman, 1997; Shaw et al., 1999), the importance of **backthrusts** above blind thrust faults (Fig. 6.11) is well recognized (e.g., Rivero and Shaw, 2011), including in areas of active or single-phase deformation outside of accretionary prisms (e.g., Hill, 1984; Stein and King, 1984; Dunne and Ferrill, 1988; Okubo and Schultz, 2004). Niño et al. (1998) demonstrated that strain softening and localization in the overlying strata promote the nucleation and growth of backthrust faults. *Strain softening* is the time- and strain-dependent reduction from macroscopic peak strength to residual strength, related to the localization, growth, and linkage of joint and fault arrays in the layer. Formation of backthrust faults is prevented if strain localization is not allowed, as demonstrated in the physical experiments by Chester et al. (1991). These small, antithetic thrust faults initiate because of favorable stress states caused by the localized occurrences of bedding plane slip, and attendant flexural slip folding (Cooke and Pollard, 1997), above the blind fault, especially at bends in the fault trace (Ellis et al., 2004; Okubo and Schultz, 2004). The newly formed backthrust faults will nucleate with small displacements and lengths appropriate to fault-scaling relationships (e.g., Cowie and Scholz, 1992a; Clark and Cox, 1996; Davis et al., 2005) and increase in size and displacement as they propagate up toward the surface. The resulting fault that eventually breaks the surface may inherit a complex, echelon configuration from the inhomogeneous stress state developed above the blind thrust fault.

Many classes of kinematic models of thrust faults have been defined and used extensively in the literature (see McClay, 2011, for an overview). Two of the most common are called "fault-bend folds" (Fig. 6.12a) and "fault-propagation folds" (Fig. 6.12b; Suppe, 1983; Mitra, 1990; Suppe and Medwedeff, 1990; Chester et al., 1991; Suppe and Connors, 1992; Mercier et al., 1997; McClay, 2011; Bernal et al., 2018). Fault-bend folds are formed as a geometric response of originally horizontal layers to flexure up and around geometric irregularities of the underlying fault surface (such as ramps that change dip angle in the slip direction of a thrust fault, when viewed in cross section). Fault-propagation folds are geometric models of displacement transfer from an underlying fault to overlying beds, resulting in folding above the fault tip. The terms **forced folds** and **thrust anticlines** can also be used to label this geometry. The kinked fold geometries (Figs. 6.12, 6.13, and 6.14) implicitly would require relatively strong bedding strengths within the layered sequence (e.g., Johnson, 1980; Erickson, 1996).

Fig. 6.12. Basic kinematic models of thrust faulting (after Schultz, 2000c). (a) Fault-bend fold (Suppe, 1983) showing backlimb dip α and forelimb dip β. (b) Fault-propagation fold (Suppe and Medwedeff, 1990; Mercier et al., 1997); α and β as in (a).

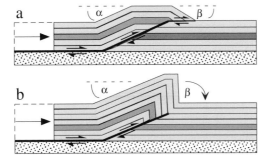

Fig. 6.13. Sequential development of a fold-and-thrust belt from the initial stages. (a) Upward and downward in-plane propagation of an early blind thrust fault. (b) Offset along the linked ramp-flat system produces a classic fault-bend fold, with continued shortening leading to nucleation of a new (incipient) blind thrust fault (dashed) in the foreland. The arrow on the left shows the amount of shortening associated with each stage.

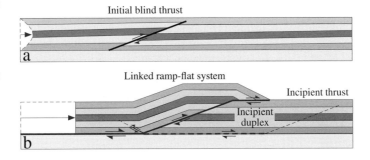

Fig. 6.14. Kink folds in
the Ordovician San Juan
Limestone, Precordillera thrust
belt, Argentina. These folds
occur near the outcrop of the
Talacasto thrust but may be
Paleozoic in age. Photography
by Richard W. Allmendinger ©
2006, used by permission.

The stacking, duplexing, and large-scale transport of stratigraphic sequences by linked thrust systems (or **imbricate sets**; e.g., Boyer and Elliott, 1982; McClay, 1992; Shaw et al., 1999) is well understood. Analog ("sand box") and numerical modeling of the growth of these structures is extensively applied (e.g., Ellis et al., 2004; Naylor et al., 2005; Buiter et al., 2006; Schreurs et al., 2006; Wu and McClay, 2011; Graveleau et al., 2012), providing a basis for the geometric or kinematic constructions (Boyer and Elliott, 1982; Suppe, 1983; Mitra, 1986, 1990; Suppe and Medwedeff, 1990; Erslev, 1991; Shaw et al., 1999) that have been applied to understand field relationships in fold-and-thrust belts from around the world.

A representative sequence is shown in Fig. 6.13, following Ellis et al. (2004) and others. First, a blind thrust fault nucleates and propagates up to the surface (Fig. 6.13a), and, perhaps, down to a slipping basal décollement (Tavani and Storti, 2011) within a horizon of weaker rock such as shale or salt. This sequence reflects the dependence of thrust-fault localization on the details of the stratigraphic section—whether it breaks first in stronger, stiffer units ("**strain partitioning**:" e.g., Eisenstadt and DePaor, 1987; Gross, 1995; Crider and Peacock, 2004) or if it propagates unidirectionally up from the slipping décollement (e.g., Niño et al., 1998). In either case, when the fault has grown large enough to cut through, and therefore decouple, the stratigraphic section in question, the system response shifts so that the sub-horizontal décollement (the "flat"), the dipping thrust fault (the "ramp"), and an upper décollement (the "flat") all accommodate contractional offsets (Fig. 6.13b). Shearing along the décollement is facilitated in many cases by transiently high *pore-fluid pressures* along them, producing elongate bedding-parallel pull-aparts (e.g., Ohlmacher and Aydin, 1997) that can resemble crack-seal veins. The intermediate stage of a **wedge thrust** (e.g., Banks and Warburton, 1986; McClay, 1992; Medwedeff, 1992; McClay, 2011; Rivero and Shaw, 2011) develops just before the upper décollement nucleates (not shown in Fig. 6.13). Backthrust faults can nucleate at upward bends in a fault trace (Ellis et al., 2004; Okubo and Schultz, 2004). As the sequence continues to shorten, the first linked fault system becomes inactive as slip along the basal décollement *propagates forward* (to the right on Fig. 6.13b), eventually cutting up-section to bound the future **duplex**, or horse, structure (Ellis et al., 2004). Continued stacking of these duplexes leads to imbricate sets developed along the Earth's convergent plate margins (e.g., Boyer and Elliott, 1982; Davis et al., 1983; Shaw et al., 1999; Nemčok et al., 2005).

6.3.3 Strike-Slip Faults

Faults that accommodate predominantly sub-horizontal shearing along sub-vertical surfaces are an important element in the accommodation of general or 3-D strains in the lithospheres of the Earth and other planetary bodies (Figs. 6.15 and 6.16). Strike-slip faults are an important component of continental transform plate margins (Woodcock and Schubert, 1994) although they occur in extensional and contractional plate margins as well. A rich literature exists on continental strike-slip faults (see Aydin and Page, 1984; Christie-Blick and Biddle, 1985; Woodcock and Fischer, 1986; Reches, 1987; Barka

Fig. 6.15. Trace of right-lateral Denali fault through the Canwell Glacier, central Alaska, formed during the 2002 earthquake; note consistent right steps along fault trace. U.S. Geological Survey/photo by Peter Haeussler.

Fig. 6.16. Sub-horizontal shearing of parallel joints along the Lonewolf fault, Valley of Fire State Park, southern Nevada (Flodin and Aydin, 2004b) creates linking cross-faults; strike-slip offset in excess of 60 m. Offset stratigraphy and fracture kinematics define offset sense; compare with Fig. 5.20, a smaller example in jointed Sierra Nevada granodiorite.

and Kadinsky-Cade, 1988; Sylvester, 1988; Wesnousky, 1988, 2005; Davison, 1994; and Stirling et al., 1996, for representative treatments). These structures play a key role in accommodating plate motions (i.e., horizontal translation and rotation of crustal terranes; e.g., Jarrard, 1986; McCaffrey, 1992; DeMets et al., 2010) as well as deformation of plate interiors (Storti et al., 2003) and margins. Earthquake rupture along these faults is modulated by the geometric irregularities such as bends and stepovers (e.g., King and Nábělek 1985; Scholz, 2002; Hill and Prejean, 2007). Strike-slip faults are also recognized on other planets and icy satellites including Mars and Europa (Schenk and McKinnon, 1989; Schultz, 1989, 1999; Pappalardo et al., 1998; Tufts et al., 1999; Kattenhorn and Marshall, 2006; Andrews-Hanna et al., 2008; Schultz et al., 2010a, c).

Similar to other types of discontinuities including dip-slip faults, strike-slip faults are characteristically segmented, especially in their earlier stages of development (e.g., Wesnousky, 1988; Davison, 1994). Slip along discontinuous surfaces with small rake angles leads to fractures having mode-II displacement distributions and near-tip stress states in map view (e.g., Chinnery, 1961, 1963, 1965; Rudnicki, 1980; Segall and Pollard, 1980). Near-tip stress states are antisymmetric, with areas of increased mean stress becoming concentrated in the leading quadrants (Fig. 6.17) and areas of decreased mean stress in the trailing quadrants (Fig. 6.17). Secondary deformation in the form of contractional-strain structures, such as anticracks, stylolites, reverse faults, and folds can localize in the leading quadrants of a strike-slip fault segment (Fig. 6.18), whereas extensional-strain structures, such as end cracks, joints, and normal faults can localize in the trailing quadrants. Characteristic structures also form within the stepovers depending on the sense of strain there, as discussed below.

Fig. 6.17. Kinematics of mode-II shearing along a strike-slip fault segment showing leading and trailing quadrants of enhanced contractional and extensional strain, respectively.

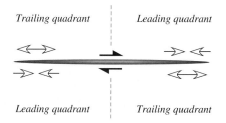

Fig. 6.18. Secondary and stepover structures formed along strike-slip faults.

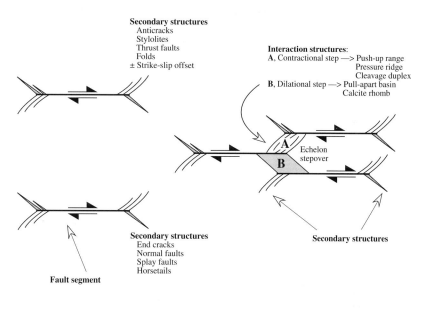

Fig. 6.19. Localized areas of oblique faulting along a strike-slip fault array define (a) *negative flower structures*, for strike-slip plus normal faulting, or (a) *positive flower structures*, for strike-slip plus reverse faulting. The dashed box shows the undeformed area.

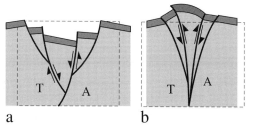

Field exposures and seismic sections show that strike-slip faults may not always dip vertically, as defined in the Anderson (1951) classification. The term **flower structure** has been used to describe non-plane fault geometries in cross-section (e.g., Harding and Lowell, 1979).

The sense of dip-slip fault offset defines the subclass of flower structure, as shown in Fig. 6.19. For example, a suite of **normal faults**, implying local crustal extension in addition to strike-slip offsets, defines a *negative flower structure*, in which the faults are concave upward (Fig. 6.19a). The opposite case of concave-downward **reverse faults**, indicating local crustal contraction along with strike-slip offsets, is called a *positive flower structure* (Fig. 6.19b). These geometries have also been called "tulip" or "palm tree" structure (e.g., Woodcock and Schubert, 1994).

Flower structures do not occur randomly along a strike-slip fault (cf. Christie-Blick and Biddle, 1985; Sylvester, 1988), but instead are localized near the fault terminations. Negative flower structures form as secondary structures in response to the oblique offsets along fault terminations (in their trailing or dilational quadrants; see Fig. 6.18; Woodcock and Fischer, 1986) and in their dilational stepovers (see Fig. 6.18). Similarly, positive flower structures form to accommodate oblique offsets along fault terminations (in their leading or contractional quadrants; see Fig. 6.18) and in their contractional stepovers. The stress changes near the tip of discontinuous strike-slip faults create the conditions for the coexistence of normal, reverse, and strike-slip faulting that over-rides the broader Anderson fault scheme in these areas (e.g., Segall and Pollard, 1980; Aydin and Page, 1984; Woodcock and Fischer, 1986; Davison, 1994).

Rotations of rocks, passive markers, and faults themselves are well documented along parallel faults. Perhaps the earliest example is that of rotation of normal faults during progressive, large-strain deformation that produced a series of crosscutting faults in the Yerington district of western Nevada (Proffett, 1977). In this case, as in others (e.g., Angelier et al., 1985; Angelier, 1994), the cross-sectional view of parallel normal faults is comparable to the map-view kinematics of parallel strike-slip faults (e.g., Aydin and Nur, 1985; Nur et al., 1986; Woodcock and Fischer, 1986; Davison, 1994). Block rotations along strike-slip fault systems are known from geologic, geodetic, and paleomagnetic evidence (e.g., Freund, 1970; Savage and Burford, 1973; Luyendyk et al., 1980; Ron et al., 1984, 1986, 1993; Christie-Blick and Biddle, 1985; Lyzenga et al., 1986; Nicholson et al., 1986; Nur et al., 1986; Sauber et al., 1986; Clark et al., 1987; Dokka and Travis, 1990) and are an important element of large-strain deformation of the crust.

Block rotations occur in association with shearing along the bounding faults (e.g., see the displacement trajectories calculated by Pollard and Segall, 1987; see also Martel, 1990; Katzman et al., 1995; ten Brink et al., 1996). The rotations accommodate shear strain in the rock adjacent to the fault (e.g., Mandl, 1987a,b; Price and Cosgrove, 1990, p. 9; Ramsay and Lisle, 2000, pp. 855–856). For a single strike-slip fault, left-lateral shear on the fault induces clockwise rotations about vertical axes in the fault's surroundings (Fig. 6.20a), whereas right-lateral offsets along the fault induce counterclockwise rotations in the surroundings (Fig. 6.20b).

As the surroundings rotate, planes within the rock can rotate into—or out of—the optimal orientations for frictional sliding (Sibson, 1974, 1985; Nur et al., 1986). As a result, cross faults can grow (i.e., slip) when planes have been rotated into the favorable orientations and will "lock up" when rotated beyond

Fig. 6.20. Dependence of cross-fault orientation and block rotations, relative to the bounding strike-slip faults, on the sense of shearing on the cross-faults. (a) Left-lateral fault, clockwise rotations; (b) right-lateral fault, counterclockwise rotations; (c) right-lateral fault and *forward-breaking cross-faults*, clockwise block rotations; (d) right-lateral fault and *backward-breaking cross-faults*, counterclockwise block rotations.

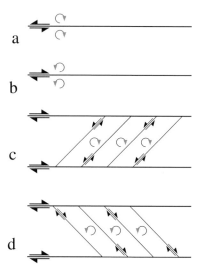

these orientations, leading to a regular and progressive sequence of rock rotation and fault sequence. The sense of block rotation between the cross faults depends on the sense of slip along the main border faults and on the relative orientation of the cross faults to the border faults. For right-lateral shear on the main border faults, and *forward-breaking* cross-fault orientations (Schultz and Balasko, 2003; Fig. 6.20c), the block rotations will be clockwise. Conversely, right-lateral shear on the main border faults, and *backward-breaking* cross-fault orientations (Fig. 6.20d), the block rotations will be counterclockwise. For left-lateral shear on the main bounding faults, the relations are reversed.

The **origin of the linking cross faults** is an important aspect of the block rotation process. In some cases, stress changes induced in the region surrounding the strike-slip faults may *trigger slip* along favorably oriented surfaces; continued offset along the bounding faults may drive block rotations in the faulted surroundings (e.g., Davies and Pollard, 1986; Flodin and Aydin, 2004a,b). In other cases when pre-existing surfaces are not available, secondary structures ("splay cracks") that *nucleate* near the ends of strike-slip fault segments (Martel and Pollard, 1989; Martel, 1990; Cooke, 1997) can define blocks between a pair of faults that may then begin to rotate as deformation proceeds. Block rotations have been inferred for sets of parallel strike-slip deformation bands by Katz et al. (2004); in this case, the linking bands may have nucleated as secondary structures as the bounding bands grew (Schultz and Balasko, 2003).

Transform faults accommodate local horizontal shearing of oceanic lithosphere as an important component of plate tectonics (e.g., Wilson, 1965; Sykes, 1967; Fox and Gallo, 1984; Nicolas, 1995; Gerya, 2012). As in continental tectonics, oceanic transform faults accommodate primarily horizontal shearing of the lithosphere, but with the opposite sense of offset since the spreading centers are not passive markers (e.g., Turcotte and Schubert, 1982, pp. 15–16; Nicolas, 1995, p. 56). However, as discussed by Gerya (2012) and others, oceanic transform faults are orthogonal to the local extension direction, given by the mid-ocean ridges, whereas continental strike-slip faults are steeply inclined to the regional extension direction. Orthogonality of oceanic transform faults has been related to factors such as thermal stress, fault weakness, progressive deformation, or energy dissipation that are more important in oceanic plates than in continental plates (e.g., see Oldenburg and Brune, 1975; Sandwell, 1986; Behn et al., 2002b; Chemenda et al., 2002; Taylor et al., 2009; Gerya, 2012). By implication, strike-slip faults in continental margins, such as California (e.g., Atwater, 1970; Aydin and Page, 1984), the Dead Sea (Garfunkel, 1981), and elsewhere **may not correspond** directly to an interpretation, formerly much in vogue, as continental transform faults (e.g., Davis and Burchfiel, 1973). The development of passive tectonic margins thus appears more complex and interesting than a pattern of normal and strike-slip faults inherited directly from the particular geometry of oceanic spreading centers and transform faults.

A schematic map-view representation of the principal structural patterns of the slow-spreading ridge setting is shown in Fig. 6.21, following Fox and Gallo (1984) and many others. There are several main elements to this structural system including a pair of **spreading centers** (the mid-ocean ridges), the transform domain between them that contains the array of linked **transform faults**, and the along-strike continuations of the transform faults, called **fracture zones** (see also Kastens, 1987). Areas of particular interest and importance (e.g., Karson and Dick, 1983; Tucholke and Lin, 1994), include the inside corners ("ICs") and outside corners ("OCs").

Several key principles have emerged that appear to govern **how oceanic transform faults work** (as part of the ridge–transform–ridge system):

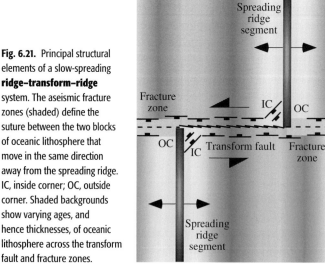

Fig. 6.21. Principal structural elements of a slow-spreading **ridge–transform–ridge** system. The aseismic fracture zones (shaded) define the suture between the two blocks of oceanic lithosphere that move in the same direction away from the spreading ridge. IC, inside corner; OC, outside corner. Shaded backgrounds show varying ages, and hence thicknesses, of oceanic lithosphere across the transform fault and fracture zones.

- Slow-spreading ridges are influenced more by faulting and less by volcanism, whereas fast-spreading ridges are influenced more by volcanism and less by faulting (e.g., Edwards et al., 1991; Carbotte and Macdonald, 1994; Behn et al., 2002a).
- The size of topography increases with the amount of offset along the transform fault (Fox and Gallo, 1984).
- Fault length, throw (or displacement) and topography, and spacing are all larger at slow-spreading ridges than at fast-spreading ridges (e.g., Searle and Laughton, 1981; Macdonald, 1982).
- Unstable frictional sliding occurs to greater depths, and generates much larger earthquakes, at slow-spreading ridges than at fast-spreading ridges (e.g., Solomon et al., 1988; Cowie et al., 1993a,b).

Earthquakes contribute only a *small fraction* ($\ll$20%) of the total fault-related deformation at oceanic transform faults (e.g., Cowie et al., 1993a,b; Okal and Langenhorst, 2000); this implies that the majority of brittle strain is accommodated **quietly**—for example, through aseismic creep, in apparent contrast to many strike-slip fault systems on the continents, such as the San Andreas and North Anatolian fault systems (e.g., Behn et al., 2002b).

6.4 Stress States for Faulting

Faults have dimensions along their generally planar surfaces, such as length and height, that far exceed the space between their shearing walls (e.g., Pollard and Segall, 1987). For example, a fault may have a horizontal length (Fig. 1.12) of 1 km, a down-dip height of 5 km, and a width of less than 1 m. They are 3-D structures (Nicol et al., 1996; Martel and Boger, 1998), yet the exposures of faults and fault arrays (Fig. 1.18) are often only 2-D. For example, the surface expression of a strike-slip fault, having an elliptical tipline, describes an irregular line or linear zone. The definition of the Earth's tectonic *plates* makes use of this 2-D simplification (e.g., Wilson, 1965; Morgan, 1968). In many geodynamic problems, the radius of curvature of strata or tectonic plates is sufficiently large that the shape can be regarded to a close approximation as a flat (planar) plate. In that case, a three-dimensional problem can then be formulated, using Cartesian (e.g., Fig. 1.15) rather than spherical coordinates for convenience.

6.4.1 Anderson's Classification

Anderson's (1951) classification scheme for faults associates the dominant sense of slip—normal, strike-slip, or thrust—with particular causative far-field (or "regional") stress states. With one principal stress vertical, the other two are necessarily horizontal (e.g., Brown and Hoek, 1978; McGarr and Gay,

Andersonian fault classification and stress state

1978; Angelier, 1994; Jaeger et al., 2007); with the fault strike defined to be parallel to σ_2 (the intermediate principal stress; Sibson, 1974), and making the assumption that only the extreme (maximum and minimum) principal stresses are important for driving frictional sliding in the Earth's crust (σ_1 and σ_3; see, e.g., Murrell, 1958, 1964; Brace, 1960; McClintock and Walsh, 1962; Hancock, 1985), **Anderson's classification scheme** (Fig. 6.22) is easily defined.

In this scheme, the Mohr diagrams (Chapter 3) for the regional, far-field tectonic stress state associated with each fault type (Fig. 6.22, left-hand column) is keyed to the stress and strain states for each type of fault. Depth is held constant in each case shown in the figure (so that the vertical stress, σ_v, is the same), illustrating the increase in stress difference or, equivalently, the diameter of the Mohr circle required for normal, strike-slip, and thrust faulting. The orientation of the fault to the maximum compressive principal stress σ_1, given by equation (3.13), depends only on the value of the host-rock friction angle.

The Coulomb criterion for frictional sliding on a pre-existing discontinuity surface is given in principal-stress form, following Zoback (2007), by

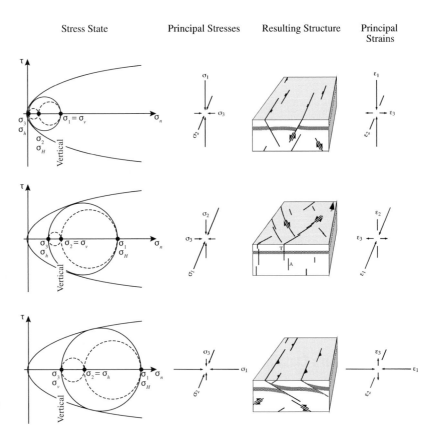

Fig. 6.22. The Anderson (1951) classification scheme for faults, based on the geometry of principal stresses relative to the Earth's surface, after Schultz et al. (2010a). The principal strains are also shown (right-hand column); note the *change in sign* of strain ε_1 for extension (normal faults) and ε_3 for strike-slip and contraction (thrust faults). This strain, with the *opposite sense* of the other two, is the **odd axis** of Krantz (1988). Its extensional sense is required when a rock mass deforms with constant volume. The sign of ε_2 is zero for conjugate faults and positive (extensional) for 3-D strain.

Stress State   Principal Stresses   Resulting Structure   Principal Strains

$$\frac{\sigma_1}{\sigma_3} = \left(\frac{S_1 - P_p}{S_3 - P_p}\right) \leq q = \left(\sqrt{\mu^2 + 1} + \mu\right)^2 \qquad (6.1)$$

in which σ_1 and σ_3 are the maximum and minimum effective principal stresses (for drained conditions, following Terzaghi; see Paterson and Wong, 2005, pp. 148–149), S_1 and S_3 are the maximum and minimum principal stresses (e.g., Zoback, 2007, p. 8), P_p is the pore-fluid pressure (Zoback, 2007, p. 132), and q is the bulk friction parameter that depends on the friction coefficient (equation (3.11b)). Effective stresses can also be written in the literature as primed symbols (σ_1'), with S being given in that case by σ (e.g., Jaeger et al., 2007, p. 98). In this book we continue to denote the effective principal stress as σ and specify S following Zoback (2007) for clarity when pore pressure P_p is considered separately, as in equation (6.1).

The three **Andersonian tectonic regimes** are rewritten explicitly from 6.1) to obtain (Zoback, 2007, p. 133)

$$\text{Normal faulting:} \quad \frac{\sigma_1}{\sigma_3} = \left(\frac{S_v - P_p}{S_h - P_p}\right) \leq q \qquad (6.2a)$$

$$\text{Strike-slip faulting:} \quad \frac{\sigma_1}{\sigma_3} = \left(\frac{S_H - P_p}{S_h - P_p}\right) \leq q \qquad (6.2b)$$

$$\text{Reverse faulting:} \quad \frac{\sigma_1}{\sigma_3} = \left(\frac{S_H - P_p}{S_v - P_p}\right) \leq q \qquad (6.2c)$$

in which the direction of the inequality indicates the maximum, limiting value of the *in situ* stress state before faulting can occur. In these equations, the greatest horizontal principal stress S_H corresponds to S_{Hmax}, and the minimum horizontal principal stress S_h to S_{hmin}, respectively, in Zoback (2007) and other works. Representative examples for the three Andersonian tectonic regimes are shown by Zoback (2007) and in Fig. 6.23.

Limiting values for the principal stresses for the three fault regimes (Zoback, 2007, p. 133) can be calculated from equations (6.2). The expressions are

$$\text{Normal faulting:} \quad S_h = \frac{\left(S_v - P_p\right)}{q} + P_p \qquad (6.3a)$$

$$\text{Strike-slip faulting:} \quad S_H = \left(S_h - P_p\right)q + P_p \qquad (6.3b)$$

$$\text{Reverse faulting:} \quad S_H = \left(S_v - P_p\right)q + P_p \qquad (6.3c)$$

Fig. 6.23. Stress magnitudes calculated as a function of depth for each of the three Andersonian tectonic regimes.

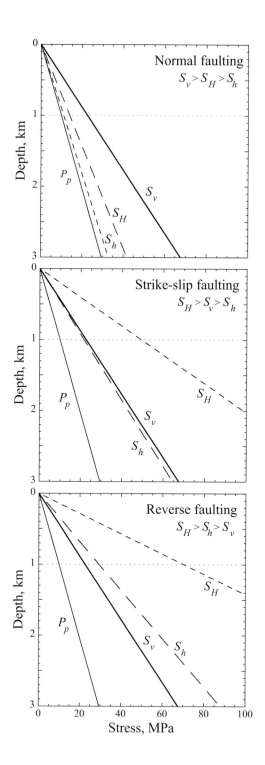

By setting $q = 3.1$ and assuming hydrostatic pore-fluid pressures with average dry-rock densities of 2300 kg m^{-3}, the limiting values for these parameters are $S_h/S_v = 0.62$ for normal faulting, $S_H/S_v = 2.2$ for strike-slip faulting, and $S_H/S_h = 2.2$ for reverse faulting. Using equations (6.3), the approximate limiting values of crustal stress can be calculated for given values of friction coefficient, pore-fluid pressure, and dry-rock density.

In the oil and gas industry and elsewhere, the minimum amount of pore pressure required to induce a hydraulic fracture is given approximately by $(S_h - P_p)$ (e.g., Hubbert and Willis, 1957; Breckels and van Eekelen, 1982; Desroches and Bratton, 2000). This is referred to as the **fracture gradient**, or "frac gradient" when stresses and pore pressures are recast as gradients (e.g., kPa m^{-1} depth). In many cases of fault-bounded or fractured hydrocarbon or geothermal reservoirs, however, the frictional resistance of the faults may govern the failure mode in the reservoir. For example, considering the frictional strength of faults by using the Coulomb criterion, the amount of pore pressure increase to trigger frictional sliding along optimally oriented faults may be given by rearranging equations (6.1) or (6.2) to obtain

$$P_p = \frac{\left(qS_3 - S_1\right)}{\left(q - 1\right)}$$

(6.4)

The numerical subscripts in equation (6.4) correspond to the Andersonian stress regimes written explicitly in equations (6.2).

The resulting relationship, which can be referred to as a **fault gradient**, is illustrated schematically in Fig. 6.24, where the five curves (or gradients) of

Fig. 6.24. Definition of fault gradient in relation to the *in situ* stresses and pore pressure condition. Shaded range of frictional strength associated with fault gradient (B) indicates a typical degree of uncertainty in the value of friction coefficient typical for a reservoir. The larger fracture gradient is indicated by arrow A starting from the initial pore pressure value.

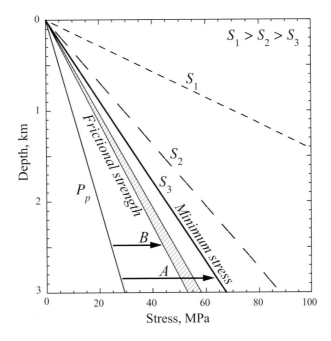

interest are the three principal (or normal) stresses S_1, S_2, and S_3 (corresponding to S_v, S_H, and S_h for a given tectonic regime), the pore pressure P_p, and the fault strength. A comparable fault-strength curve can be calculated for non-optimally oriented faults given knowledge of the *in situ* stress state and their attitudes (e.g., strike and dip). Finkbeiner et al. (1998, 2001) compared the fracture and fault gradients to assess hydrocarbon column heights in a field in the Gulf of Mexico.

A useful and informative variation on the standard Mohr diagram (e.g., Finkbeiner et al., 1998; Wiprut and Zoback, 2002) plots the total principal stresses (e.g., equations (6.3)) and the pore pressure separately (Figs. 6.23 and 6.25). The fault frictional strength is plotted so that the Coulomb frictional strength intercepts the normal-stress axis at the initial value of formation pore pressure P_{p0}, rather than at the origin as would be the case if effective stresses were plotted (e.g., Fig. 3.33). This convention clearly displays how pore pressure changes in relation to the stress state acting on a fault. For example, a reduction in pore pressure due to hydrocarbon production would shift the Coulomb failure line to the left, and farther from the stress conditions associated with faulting (whose magnitudes are unchanged on the diagram). The explicit display of principal stresses such as that due to the overburden, S_v, at given depth against pore pressure and frictional strength makes this version of Mohr diagram particularly informative in the

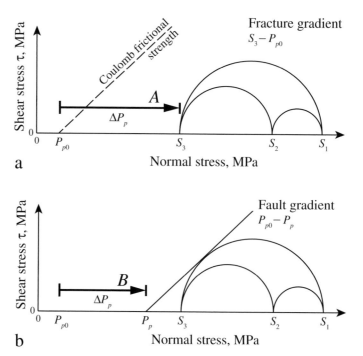

Fig. 6.25. Mohr diagrams showing (a) fracture gradient and (b) fault gradient.

oil and gas industry in which pore pressure is the main variable in fault stability that can be manipulated in subsurface operations (i.e., water injection or hydrocarbon production).

As can be seen from Figs. 6.24 and 6.25, the amount of pore pressure increase, ΔP_p, needed to trigger faulting (arrow labeled B on the figure) is less than that needed to induce hydraulic fracture (A on the figure). Because the Coulomb criterion always has a positive slope on the Mohr diagram, the fault gradient will always be less than the fracture gradient; this implies that the strength of faults in oil and gas fields and their overburdens need to be considered explicitly in discussions of brittle failure (tensile or shear) of a reservoir or geothermal field.

6.4.2 Stress Polygon

An elegant method for representing the bounding states of stress for each fault type (Moos and Zoback, 1990; Barton and Zoback, 1994; Zoback, 2007, pp. 137–139), referred to as a **stress polygon**, is shown in Fig. 6.26. The diagram is constructed by plotting values of the Coulomb criterion for frictional sliding along cohesionless surfaces, written using principal far-field total (not effective) stresses (equations (6.3); see Zoback, 2007, p. 133); i.e., $1.0 < S_1/S_3 \leq q$, where q is the bulk friction parameter given by equation (3.11b).

Stress polygon and faulting

The fields for normal, strike-slip, and thrust faulting are shown in Fig. 6.26 from equations (6.2) by assuming a depth of 1.5 km, average dry-rock density

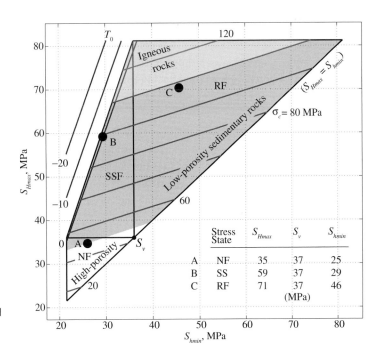

Fig. 6.26. Stress polygon showing fields for remote stress states for normal (NF), strike-slip (SS), and reverse or thrust faulting (RF), following Zoback (2007). Additional limits to S_H and S_h are provided by rock tensile strength (upper-left lines) and unconfined compressive strength (diagonal lines and shaded areas), both commonly assessed at the wall of a wellbore. Example stress states A, B, and C are shown and tabulated (stress magnitudes given in MPa).

Stress State		S_{Hmax}	S_v	S_{hmin}
A	NF	35	37	25
B	SS	59	37	29
C	RF	71	37	46
			(MPa)	

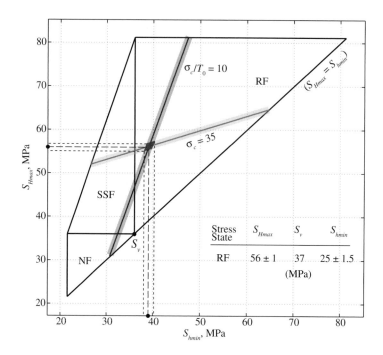

Fig. 6.27. Stress polygon characterizing a stress state for which tensile failure and unconfined compressive strengths have been used to bound the horizontal stress magnitudes. Uncertainties in tensile and unconfined compressive strength propagate into uncertainties for the stress magnitudes (dashed lines).

of 2250 kg/m³, hydrostatic pore-water conditions, and a friction coefficient of 0.6, corresponding to $q = 3.12$. Faulting will occur when the principal stresses in the rock mass achieve values given by the upper left-hand edges of the polygon in Fig. 6.26.

The strength of most rock types is additionally limited by their tensile or compressive strengths. The values of S_H and S_h are then further reduced from their frictional-equilibrium values in cases where either tensile failure (producing drilling-induced tensile fractures), or compressive failure (producing borehole breakouts) is interpreted from a wellbore. Curves of tensile strength T_0 (negative numbers, in MPa) and unconfined compressive strength σ_c (in MPa) are shown on Fig. 6.26, with approximate ranges of σ_c taken from Paterson and Wong (2005), p. 22. In practice, tensile strengths are considered by plotting the ratio of unconfined compressive strength to the absolute value of tensile strength; the intersection of this curve with that of σ_c, when both borehole breakouts and drilling-induced tensile fractures are identified in a wellbore, robustly bounds the *in situ* stress state at that depth and location (Fig. 6.27).

Successively weaker rocks (smaller values of σ_c) will permit successively smaller values of S_H; corresponding values of S_h may be determined by using *in situ* tests in the subsurface or by other means such as elasticity. Representative stress states are illustrated in Fig. 6.26 along with their tabulated values. This diagram succinctly portrays the 3-D regional stress states for faulting to occur in each of the Andersonian types. The diagram contracts self-similarly as pore-fluid pressure is increased (Zoback, 2007, p. 138) or if the frictional resistance along the faults (μ or q) is reduced. The stress polygon approach is widely used in the oil and gas and geothermal industries, along with wellbore data, to define the subsurface stress state (e.g., Rogers, 2003; Moeck and Backers, 2011; Hennings et al., 2012; Zakharova and Goldberg, 2014).

The stress polygon just described is created by choosing a particular depth, so that S_v is fixed on the diagram. The influence of pore pressure is also not included in the stress polygon plot, which can be important to consider in rock failure. Zakharova and Goldberg (2014) presented a normalized stress polygon in which the values of the three effective principal stresses were normalized by the magnitude of S_v'. This approach may facilitate the visualization and interpretation of stress paths associated with changes in depth, horizontal position, and time.

6.4.3   Strength Envelope

A second method for displaying the stress states for Andersonian faulting is a **brittle strength envelope**. Rather than assuming a particular depth, as in the stress polygon approach, the brittle strength envelope plots the difference between the maximum and minimum compressive principal effective stresses, ($\sigma_1 - \sigma_3$), as a function of depth, assuming particular values of rock-mass density and friction (e.g., Goetze and Evans, 1979; Brace and Kohlstedt, 1980; Kohlstedt et al., 1995; Kohlstedt and Mackwell, 2010). Brittle strength envelopes can be plotted either as departures from the lithostat (e.g., Suppe, 1985; Zoback, 2007; Fossen, 2010a,b; Fig. 6.28) or as stress differences ($\sigma_{max} - \sigma_{min}$)

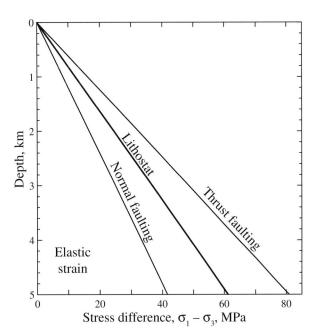

Fig. 6.28. Schematic strength envelope for continental (felsic rock) plotted as departure from lithostatic state of stress. Only the brittle, frictional strength is shown.

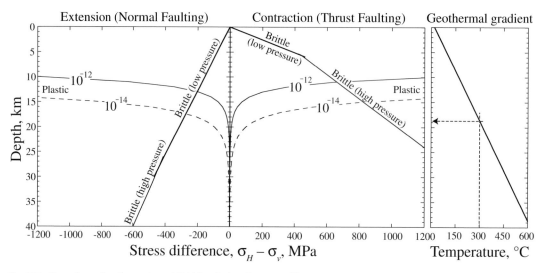

Fig. 6.29. Strength envelope for continental (felsic) rock plotted as stress differences ($\sigma_H - \sigma_v$), so that an extensional tectonic regime (a) is shown as negative values (since $\sigma_v > \sigma_H$; note that the stress difference remains compressive and positive, however) and a contractional regime (b) is positive (since $\sigma_H > \sigma_v$). A geothermal gradient is plotted in (c), so that the depth of seismogenic faulting or, more precisely, unstable frictional sliding, is given approximately by the depth to 300 °C.

keyed to tectonic environment (i.e., extension or contraction, Fig. 6.29). In Fig. 6.28, the lithostat is $\sigma_v = \sigma_3$, normal faulting is indicated by $\sigma_v - \sigma_h$, and thrust faulting by $\sigma_H - \sigma_v$, with $\sigma_1 = q\,\sigma_3$. These diagrams illustrate how *in situ* stress states within the Earth's crust compare to the curves for normal or thrust faulting in these tectonic regimes (e.g., Zoback and Healy, 1984; Engelder, 1993; Zoback et al., 1993; Zoback, 2007, pp. 273–279, 287–291).

Lithospheric strength envelope: Brittle and plastic parts

While the brittle strength of crustal rocks can be represented by its frictional resistance (Goetze and Evans, 1979; Brace and Kohlstedt, 1980), rock strength at greenschist-facies temperatures (~300 °C for felsic rocks) or greater, corresponding to mid-crustal depths of several kilometers, is controlled by temperature, rather than pressure (see Scholz, 2002, for review). Rock strength at such temperatures can be represented by a creep law (see Kohlstedt and Mackwell, 2010, for explanation and representative parameters) such as

Plastic creep law for lower crustal behavior

$$\left(\sigma_1 - \sigma_3\right) = \left(\frac{\dot{\varepsilon}}{A}\right)^{\frac{1}{n}} e^{\left(\frac{H}{nRT}\right)} \tag{6.5}$$

where $\dot{\varepsilon}$ is a regional tectonic strain rate, A is a material constant that depends on grain size, chemistry, and temperature, n is a stress exponent, H is the activation enthalpy for creep, R is the universal gas constant, and T is absolute temperature. For felsic rocks the plastic creep law plotted in Fig. 6.29 assumes $\dot{\varepsilon} = 10^{-12}$ (faster) or $10^{-14}\ \text{s}^{-1}$ (slower), $A = 2.0 \times 10^{-4}\ \text{MPa}^{-n}\ \text{s}^{-1}$, $n = 1.9$, $H = 137{,}000\ \text{J mol}^{-1}$, $R = 8.3145\ \text{J mol}^{-1}$, and $T = (273°\text{K} + 20°) + 15°$

km^{-1}. The lesser of the stress difference for brittle or plastic behavior defines which of these controls rock strength at any given depth. Analogous strength envelopes can be constructed for mafic oceanic lithosphere and can include complexities such as variations in rheology (e.g., crust vs. mantle rocks) and rock response (e.g., Goetze's criterion and other refinements that remove the sharp transition between brittle and plastic behavior). Numerical simulations verify the essential attributes of the strength envelope approach (e.g., Freed and Lin, 1998; Albert et al., 2002).

As evident on Fig. 6.29, rock is stronger in contraction than in extension (i.e., greater value of stress difference at any given depth). The plastic curves have smaller stress differences at depths greater than ~10 km (for the given values of creep parameters and geothermal gradient), indicating a non-brittle, temperature-controlled rock response at those depths. The 15 °C km^{-1} geothermal gradient plotted in Fig. 6.29c suggests that the threshold temperature for stable frictional sliding of ~300 °C for felsic rocks is attained at depths of about 19 km, thus defining the lower frictional stability transition (Marone and Scholz, 1988; Marone, 1998a,b; Chapter 3) and the approximate depths of earthquake nucleation and coseismic faulting.

It is worth emphasizing that the depth of coseismic faulting, represented on a strength envelope diagram, has little to do with a brittle–ductile transition (i.e., from localized faulting to distributed flow; see Scholz, 1998, for a clear discussion) and everything to do with the stability of frictional sliding. Because friction is temperature dependent, sliding at temperatures less than either 300 °C, for felsic rock compositions, or 600 °C, for mafic compositions, is unstable and therefore seismogenic, whereas it is stable and creep-like at higher temperatures and at greater depths (e.g., Tse and Rice, 1986; Marone, 1998b; Scholz, 1998; Abercrombie and Ekström, 2001). The geotherm, in concert with rate and stiffness effects, thus largely determines the depth of seismic faulting in a planetary lithosphere, with the strength envelope contributing to an understanding of the corresponding regional strain rates for the appropriate tectonic regime.

6.4.4 Critically Stressed Faults

Fractures such as faults and joints record the paleo-stress state associated with their formation and growth. If the structures are not neotectonic, then their paleo-stress state may differ from the contemporary *in situ* stress state. However, some of these structures may be suitably oriented for frictional sliding under the current stress conditions. Such faults whose combination of orientation (strike and dip) and resolved (normal and shear) stresses including the pore-pressure component are closest to frictional failure are referred to as **critically stressed faults** (Barton et al., 1995; Rogers, 2003; Zoback, 2007, pp. 340–349).

Critically stressed faults

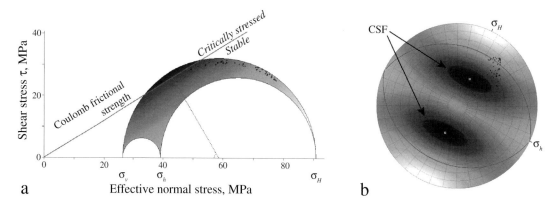

a

b

Fig. 6.30. Critically stressed faults can be identified on (a) the Mohr diagram, where the Coulomb frictional sliding criterion has been met or exceeded, or (b) on lower-hemisphere polar stereographic projections (Lisle and Leyshon, 2004, p. 42; Ragan, 2009, p. 104) of stress state and fault poles. The most critically stressed faults in (b) would plot near the poles of the optimum fault orientations (CSF, arrows) which would strike NW and dip NE and SW in the normal-faulting stress state illustrated.

Critically stressed faults can be identified on a Mohr diagram (Fig. 6.30a) or on a polar equal-area stereographic projection (Fig. 6.30b). In both cases, the frictional resistance ($\mu = 0.6$ in this case) and three-dimensional stress state are specified, and the magnitude of pore-pressure increase is calculated that would bring a certain part of the diagram, corresponding to the stress states or poles to a particular fault (points on Figs. 6.30a and b, respectively), into tangency with the Coulomb frictional sliding line. This approach is used routinely in studies of neotectonics, geothermal resources (e.g., Moeck et al., 2009; Moeck and Backers, 2011), and oil and gas exploration and production (e.g., Hennings et al., 2012).

The importance of critically stressed faults to oil and gas reservoirs lies in the empirical correlation, first published by Barton et al. (1995), between enhanced permeabilities measured along faults and their proximity to failure. Work by Barton et al. (1995) and by Townend and Zoback (2000) demonstrates that the majority of faults they studied having higher measured permeabilities plot on a Mohr diagram above the Coulomb frictional strength minimum (with the friction coefficient of $\mu = 0.6$), whereas the majority of faults that do not exhibit enhanced permeabilities plot below this threshold. The implication is that faults that slip under the contemporary *in situ* stress state generate enhanced values of fault-parallel fluid flow and that such fluid conduits can therefore be predicted from a fault-stress analysis. Examples of the connection between critically stressed faults and enhanced permeability abound throughout the world's petroleum basins and plate margins (e.g., Finkbeiner et al., 1998, 2001; Wiprut and Zoback, 2002; Johri et al., 2014). Work by Hennings et al. (2012) demonstrates that production was increased at the Suban, Indonesia, gas field in higher permeability zones that were consistent with critically slipping faults, and related fault damage zones, in the field. Measurement of ground motions at the surface above the field by satellite interferometry (InSAR) are consistent with the interpretation of active deformation at depth in association with critically stressed faults (Schultz et al., 2014).

As discussed by Zoback (2007), frictional sliding, or shearing, along a fault may lead to brecciation within the fault zone, or joint-dominated damage zones surrounding the fault core, in certain low-permeable lithologies, such as compact (crystalline) rocks. Other lithologies may not enable as strong a permeability increase, such as softer rocks (diatomites), more porous ones (sandstones), or clay-rich ones (such as shales) although an increase in slip-related permeability may occur in such rocks (e.g., Nygård et al., 2006). The reason why some critically stressed faults are not enhanced conduits, or conversely why some non-critically stressed faults are conduits, may relate to their geologic ages (see discussion by Wilkins and Naruk, 2007). According to Zoback (2007), common geologic processes including precipitation, cementation, and alteration of feldspars to clays can seal faults and related fractures over time. By implication, only critically stressed faults having the least amount of sealing, and therefore the smallest value of frictional strength, will slip under the contemporary stress state.

The critically stressed fault hypothesis may also be considered an expression of rate-and-state friction laws (see Chapter 3). As summarized for example by Marone (1998a,b), Scholz (1998), and Paterson and Wong (2005), the frictional resistance (or strength) of a fracture is time-dependent through the contributions of slip rate and the physical state or evolution of the slip surfaces. The maximum frictional resistance, or "static" friction, increases with the duration of stationary contact, or the amount of time between slip events. The minimum frictional resistance, or "dynamic" friction, decreases with increasing slip rate. In neotectonics, the time between successive earthquakes on a particular fault, the recurrence interval, scales with the magnitude of coseismic offset (e.g., Wells and Coppersmith, 1994). This interval is also identified with that required to heal faults (e.g., Cowie, 1998a,b) so that their static frictional strengths can return to the pre-slip values. Slip triggered on faults by high fluid pressures within low permeability rocks, such as shales, may also be more creep-like, rather than rapid and coseismic (Zoback et al., 2012). Applied to critically stressed faults, slip along a fault can generate enhanced permeability within it by fracturing the fault core and surrounding fault zone. If the slip magnitude is small enough, then permeabilities could be maintained within the fault zone since the recurrence interval would be less than that required to heal the fault. For large enough slip events, the fault can likely heal sufficiently for its strength to be marginally greater than the resolved stress state. Structural diagenesis (Laubach et al., 2010) and the contemporary rate of deformation thus compete to maintain enhanced fault-related permeabilities for critically stressed faults.

6.4.5 Orthorhombic Fault Patterns and 3-D Strain

The Andersonian classification for faults defined additionally by Coulomb friction (Section 6.4.1) is a 2-D one because it assumes that only the two extreme (i.e., the maximum and minimum) compressive stress magnitudes are sufficient and necessary for predicting faulting. Why, then, do we sometimes find four mutually crosscutting, synchronous, conjugate sets of faults in a single area?

As noted in the literature, a **conjugate fault pair represents a particular case of faulting** in which 2-D, *plane-strain* conditions prevail (e.g., Reches, 1983; Krantz, 1988, 1989). More general patterns of faults—**orthorhombic sets**—with three to four conjugate planes accommodate 3-D remote strain fields (Oertel, 1965; Mitra, 1979; Aydin and Reches, 1982; Reches and Dieterich, 1983; Crider, 2001; Healy et al., 2015). Interestingly, deformation bands can also form orthorhombic sets (e.g., Davis et al., 2000; Shipton and Cowie, 2001), implying that the analysis performed by Reches (1978, 1983) applies more generally to both pre-peak and post-peak conditions (i.e., it can explain deformation band orientation in addition to Coulomb slip plane orientation).

An orthorhombic fault pattern is a set of conjugate faults that no longer have parallel strikes (i.e., $\alpha \neq 0$; Fig. 6.31). For example, a graben or set of normal faults that are *not parallel* require that a nonzero magnitude of strain is being accommodated in the intermediate (σ_2 or ε_2) direction. If all the faults are mutually crosscutting, and therefore are synchronous, then this demonstrates a case of 3-D strain. The relationship between 3-D strain and number of fault sets has been demonstrated from theory (Reches, 1978, 1983), experiment (Aydin and Reches, 1982; Reches and Dieterich, 1983), and field observation (e.g., Krantz, 1988, 1989; Davis et al., 2000; Shipton and Cowie, 2001). The **departure of the faults from parallelism** is given by the angle α as defined in equations (6.6) and shown in Fig. 6.31. Following Krantz (1988), the relations are

Degree of three-dimensional (3-D) strain represented in discontinuity sets

$$\left(\frac{\varepsilon_2}{\varepsilon_3}\right) = \tan^2\alpha = k$$

$$\left(\frac{\varepsilon_2}{\varepsilon_1}\right) = -\sin^2\alpha$$

$$\left(\frac{\varepsilon_3}{\varepsilon_1}\right) = -\cos^2\alpha \tag{6.6}$$

Fig. 6.31. Orthorhombic faults that accommodate three-dimensional strain states. Two-dimensional (2-D) strain states are indicated when $\alpha = 0°$, and three-dimensional (3-D) ones are demonstrated for $\alpha > 0°$.

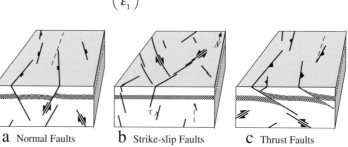

a Normal Faults   b Strike-slip Faults   c Thrust Faults

Fig. 6.32. Plot of the first of equations (6.6) showing relationship between strain ratio k and angle α. Two-dimensional (2-D) strain states are indicated when $\alpha = k = 0°$, and three-dimensional (3-D) ones are demonstrated for α or $k > 0°$.

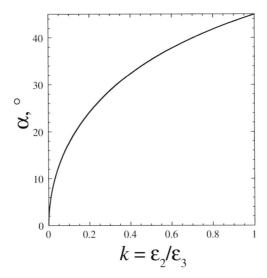

$$k = \varepsilon_2/\varepsilon_3$$

where k ranges from -0.5 to 1.0 (Reches, 1983; Krantz, 1989) and assuming that the ratio of slip to fault separation is small. In this equation set, the odd strain axis is ε_1, the intermediate axis is ε_2, and the minimum (or similar) axis is ε_3. This parameter α (Fig. 6.32) is the key measurement for the orthorhombic faults, and Krantz's (1988, 1989) odd-axis model has been employed in determining fault-related strain ellipsoids for fault sets as well as a precursory deformation band array (e.g., Shipton and Cowie, 2001). Several fault sets that have been suggested to have formed in a three-dimensional (i.e., not plane-strain) strain field, based on their mutual cross-cutting relations and set intersection angles α include normal faults and strike-slip faults (e.g., Krantz, 1989).

In this model, the faults rotate about the odd axis through an angle α; for normal faults, the axis of rotation is the vertical axis. Similarly, if thrust fault strikes are not parallel, then a component of strain (extension or contraction, depending on the sign of α) is implied in the direction parallel to the trend of the orogen (i.e., along a horizontal axis oriented normal to thrusting). For strike-slip faults, a conjugate pair in map view has vertical (parallel) dips for 2-D (plane) strain in the vertical direction; non-vertical dips (in both directions; Fig. 6.31b) then imply non-plane-strain in the vertical direction (i.e., thickening or thinning) along with the horizontal shear strain.

Oblique slip and 3-D strain

The determination of 3-D strain fields sheds light on the origin of faults having **oblique slip** (e.g., Williams, 1958; Bott, 1959); that is, having rakes between 0° (horizontal, corresponding to pure strike-slip faults) and 90° (parallel to the dip direction, corresponding to pure normal or reverse faults). Although such departures from the pure (2-D) Andersonian fault types have been suggested on grounds including stressing of rocks having pre-existing structures (e.g., Simpson, 1997), fault reactivation or growth into a new stress regime (e.g., Higgins and Harris, 1997; Clifton and Schlische, 2001; Crider,

2001; Frankowicz and McClay, 2010; Giba et al., 2012), or fault interaction (Maerten et al., 1999), the odd-axis model directly predicts that faults having oblique slip vectors will occur whenever $\alpha \neq 0°$. Thus for non-plane-strain conditions, formation of fault sets having oblique offsets (i.e., $90° <$ rake $< 0°$) can occur in the absence of stress-state rotations at various temporal or spatial scales.

The slip direction along a fault is a reliable indicator of the stress geometry in its local vicinity, but it also contains information about the overall, *regional* (or far-field) stress state (see Aydin and Nur, 1982, for clear examples and rationale). In western California, for example, all three Andersonian types of faults (i.e., normal, strike-slip, and thrust) are found in this transform plate margin (e.g., Aydin and Page, 1984; Christie-Blick and Biddle, 1985; Woodcock and Schubert, 1994). This can also occur in contractional or extensional orogens. For example, normal and strike-slip faults are common in Tibet, a contractional orogen formed by subduction of the Indian plate beneath the Asian plate (e.g., Molnar and Tapponnier, 1975). Faults and other types of discontinuities characteristically change the magnitudes and distribution of stress in their vicinity. A classic example is the formation of end cracks, or other types of secondary structures, at the tips of strike-slip faults (e.g., Segall and Pollard, 1980; Rispoli, 1981). Rispoli (1981) and Petit and Mattauer (1995) demonstrated the coexistence of strike-slip faults, veins, and stylolites in an outcrop in southern France (shown in Fig. 4.16); the secondary structures were created locally at the tips of the sliding (mode-II) fault surfaces (see also Pollard and Fletcher, 2005, pp. 16–19). As a result, any combination of the three basic Andersonian fault types can coexist in a region or in a particular fault zone if the *local* stresses—at the scale of a particular fault or other structural discontinuity—satisfy the stress geometry and magnitudes defined by rock physics (e.g., King and Yielding, 1984; Aydin and Page, 1984; Davison, 1994) or, more generally, by Reches (1983) and Krantz (1989).

6.5 Faults as Sequential Structures

In keeping with the understanding in the literature of faults being formed and evolving as sequential structures, faults can be defined as "a sharp structural discontinuity, defined by the fault planes (surfaces of discontinuous displacement) and related structures including fault core and damage zones (e.g., cracks, deformation bands, slip surfaces, and other structural discontinuities) that formed at any stage in the evolution of the structure. Commonly associated ductile deformation structures, recognized as drag or faulted fault-propagation folds, are associated elements not included in the term fault, although clay smearing or other early forms of strain localization may be included."

Faulting thus involves a suite of progressively evolving processes, but, perhaps most importantly, frictional sliding along a planar element. The **plane of displacement discontinuity** (of negligible thickness) having offset parallel to the plane (e.g., Pollard and Segall, 1987; Aydin, 2000; see Chapter 1) is called a **slip surface**. This process accounts for the ability of a fault to accommodate large displacements, and therefore shear strain, along it.

In order for frictional sliding to occur, the rock needs to first have a plane of weakness, or some planar elements which exhibit a suitable contrast in physical properties, such as stiffness or strength, on which the sliding can occur. This requires that the surface is created first, with sliding occurring on it later. This situation has been referred to as a **pre-existing surface** by Crider and Peacock (2004). On the other hand, the surface can be created shortly before sliding occurs; i.e., during the same deformation event (referred to as a **precursory surface**). Johnson (1995) referred to pre-fault deformation bands as a premonitory shear zone, noting how they influenced the later development of a through-going fault.

Substrates for faulting

The timing and nature of what may be referred to as the slip-surface substrate is important to the nucleation and evolution of the fault structure. The **substrate** can be defined as the planar or tabular zone having different mechanical properties than the surrounding host rock that serves as the nucleation site for frictional sliding. The two classes of substrates of interest to faulting are:

(a) *Pre-existing substrate* – the pre-faulting zone of stiffness or strength contrast that formed prior to, and under a different far-field stress state from, the later slip surface; and

(b) *Precursory substrate* – the pre-faulting zone of stiffness or strength contrast that formed during the same deformation event, and under the same far-field stress state, as the later slip surface.

The **nature of the substrate**—specifically, its strength or properties relative to its surroundings—exerts an important control on the characteristics of the eventual fault. For example, a common scenario occurs when the substrate is relatively weaker than the surroundings; i.e., a "plane of weakness." Some examples of these include joints, bedding planes, geologic contacts having smaller coefficients of friction than the surrounding rock, or thin clay layers. Because these substrates are relatively weaker than their surroundings, they serve as the nucleation sites for subsequent shearing, leading to **strain localization**. Once the deformation of the substrate becomes incompatible with that of its (relatively stiffer) surroundings, then the system becomes **unstable** and a slip surface nucleates, usually on the edge of the substrate at its contact with its surroundings. This scenario generates structures known as **faulted joints**. On the other hand, a relatively stiffer substrate, like a cataclastic deformation band, strains at a slower rate than its (softer or weaker) surroundings, leading eventually to a mechanical instability and nucleation of a slip surface. This path leads to **faulted deformation bands**.

6.5.1 Pre-existing Substrates

Pre-existing Substrates: Formed before faulting (slip surface nucleation) under a different remote stress state

Joints and other geologic cracks can be frictionally weaker than the surrounding rock mass by nature of their discontinuous surfaces, but also when filled by hydrothermal minerals (e.g., Segall and Pollard, 1983b; Segall and Simpson, 1986). Much work has been done on this most common of fault substrates (e.g., Segall and Pollard, 1983b,c; Martel et al., 1988; Martel and Pollard, 1989; Martel, 1990; Martel and Boger, 1998; Flodin and Aydin, 2004a; see also McGill and Stromquist, 1979; Trudgill and Cartwright, 1994; Cartwright et al., 1995, 1996; Moore and Schultz, 1999; Wilkins et al., 2001; Crider and Peacock, 2004; Schöpfer et al., 2006). Under an appropriate remote stress state that resolves sufficient shear stress to overcome the joint's frictional resistance, the joints can shear. The resulting structures are called **faulted joints**.

Precursory joints have a range of spacings, lengths, and apertures which all relate to each other physically (e.g., Olson, 1993; Renshaw and Pollard, 1994b; Renshaw, 1997; Gudmundsson, 2000). The **longer joints tend to be weaker in shear** because the joint thickness (aperture) increases with length (Vermilye and Scholz, 1995; Schultz et al., 2008a, 2013), making the joint walls progressively less likely to frictionally interlock (e.g., Bieniawski, 1989; Martel, 1990).

Faulting may be said to occur once **slip patches nucleate on the pre-existing joints**. This occurs if the remote stress state has rotated (with appropriate stress magnitudes) to resolve enough shear stress onto the joint planes to overcome any compressive normal stress acting perpendicular to the joint plane, or if bending of layers, for example, contributes to shearing along them. In the latter case normal faulting in a layered sequence commonly begins when the stiffer (and perhaps stronger) layers form joints, which then shear as the deformation proceeds (Fig. 6.33). Slip patches nucleate on these joints (if

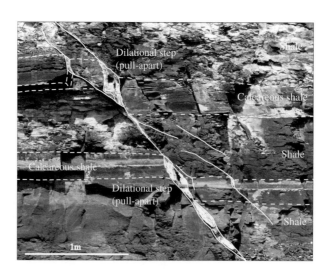

Fig. 6.33. Normal faults growing within a heterogeneous layered mudrock sequence near Kimmeridge Bay, Dorset, UK, nucleate initially on joints formed in the more brittle calcareous shale layers and propagate into the shale layers, eventually linking up to form a fault (after Schöpfer et al., 2006).

they are in frictional contact; otherwise, they gape and open obliquely; Ferrill and Morris, 2003), leading to fault segments that then propagate up or down, out of these layers into more ductile or deformable layers (e.g., Peacock and Sanderson, 1992; McGrath and Davison, 1995; Childs et al., 1996; Schöpfer et al., 2006; Welch et al., 2009).

As slip spreads along the former joints, **secondary structures** (end cracks) nucleate at several possible locations, including: joint segment tips, bends, and between closely spaced joints due to mechanical interaction and impedence of echelon slip patches (Segall and Pollard, 1983b; Martel and Pollard, 1989). These end cracks (and, for contractional stepovers, cleavage duplexes (Bürgmann and Pollard, 1992)), serve to **link** the (soft-linked) echelon fault segments, allowing larger offsets to accumulate along the now-longer fault segments (Figs. 6.33, 6.34, and 6.35). This process can scale up to produce wider zones of shearing along strike-slip faults ("fault zones" and "compound fault zones;" Segall and Pollard, 1983b; Martel, 1990; Flodin and Aydin,

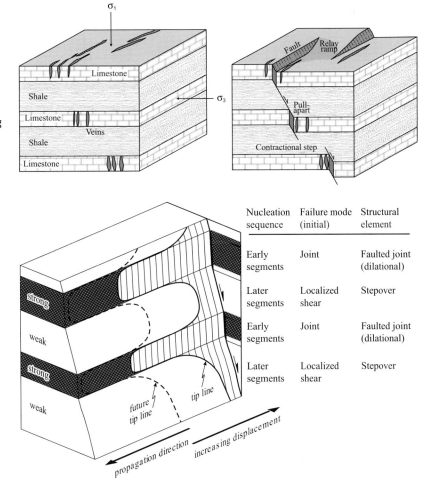

Fig. 6.34. Growth of normal faults within a heterogeneous layered sedimentary sequence illustrating (a) initial tensile failure of the more brittle layers and (b) dilation and/or shearing of dilatant structures and propagation into shaley layers (after Crider and Peacock, 2004).

Fig. 6.35. An evolutionary sequence leading to normal faults within a heterogeneous layered sedimentary sequence (figure after Schöpfer et al., 2006).

Nucleation sequence	Failure mode (initial)	Structural element
Early segments	Joint	Faulted joint (dilational)
Later segments	Localized shear	Stepover
Early segments	Joint	Faulted joint (dilational)
Later segments	Localized shear	Stepover

2004b). In normal fault systems, linkage zones are likely to occur in the more ductile layers, such as shales. Linkage zones and dilatant cracks (e.g., Petrie et al., 2014; Aydin, 2014; Raduha et al., 2016) can therefore define seal-by-pass structures (Cartright et al., 2007) that can compromise the integrity of caprock/top-seal sequences in hydrocarbon, geothermal, carbon dioxide, and natural gas storage zones and reservoirs.

A comparable process and sequence have been documented for **bedding-plane slip** by Cooke et al. (2000). They show how slip can nucleate along a bedding plane subjected to an appropriate remote stress state, creating secondary end cracks that mark the ends of the slipping sections of the bedding interface. A similar sequence was demonstrated by Graham et al. (2003), but with the normal faults nucleating on pre-existing stylolites instead of joints.

Another common substrate geometry is that of twist hackles, or echelon fringe joints (see Chapter 4). Here, nucleation of strike-slip patches along the main pre-existing crack can produce **shearing along the fringe surfaces** (Flodin and Aydin, 2004a) that, with sufficiently large strike-slip displacement, can produce a nested set of faults that closely resembles the fault zones, with attendant block rotations, documented in the Sierra Nevada (Segall and Pollard, 1983b; Martel, 1990).

Faulting of poorly lithified clay-rich sediments (and weakly lithified silty sedimentary rocks) occurs by a different mechanism than in the previous examples (e.g., Maltman, 1988). Here, the bedding contacts between the foliations, or layers, are considerably *stronger*, frictionally (even under wet conditions), than in the previous case of joints (which corresponded to relatively weak surfaces). Under appropriate remote stress states (e.g., Johnson and Fletcher, 1994, pp. 273–277), the finely laminated sequence forms a set of narrow **kink bands** because of the strong contacts that, as deformation proceeds, can form the substrate for slip patches and attendant faulting. The kinks probably nucleate at small imperfections in the material. This class of faulting is common in "soft-sediment deformation" (e.g., Maltman, 1984, 1994) although slip surfaces may not always be present. The "faults" in some of these materials have sometimes been referred to as "neutral porosity deformation bands" (e.g., Fossen, 2010b).

6.5.2 Precursory Substrates

How does a fault cut its way through a hard, low-porosity, crystalline rock, like a granite or a dolomite, if there are no pre-existing joints to nucleate upon? First, when a granite (for example here) is subjected to a compressive remote stress state (either in the laboratory or in the field), dilatant cracks grow and align themselves with the direction of maximum compression (e.g., Brace et al., 1966; Scholz, 1968a,b; Peng and Johnson, 1972; Gowd and Rummel, 1980). As the cracks lengthen, an echelon pair forms (see Chapter 5), thus defining the **acute angle** of the linked crack array to the maximum

*Precursory Substrates:
Formed during faulting
(slip surface nucleation)
under a related remote
stress state*

compressive stress direction (Du and Aydin, 1991; Reches and Lockner, 1994). Because this linked crack array is softer, and weaker frictionally than the surrounding granite, it can begin to accommodate shearing displacements along it; continued shearing further connects the fractures and brecciates the zone (e.g., Mollema and Antonellini, 1999). When the strain rate within the weaker shear zone becomes sufficiently greater than that in the host granite (e.g., Aydin and Johnson, 1983), *slip surfaces* nucleate and a fault in granite is born.

Interestingly, the shear zone or fault can lengthen, or propagate, in-plane by the nucleation and coalescence of microcracks in a **damage zone** at its tips (e.g., Lockner et al., 1991; Reches and Lockner, 1994). Fault propagation in this case does not involve the growth of a "shear crack" but instead, the fault is a *hybrid* of shearing overprinting an array of coalesced (i.e., linked) micro-cracks. This may suggest an explanation as to why efforts to measure a **critical shear fracture toughness for mode-II** (K_{II}, analogous to the mode-I fracture toughness; see Chapter 8) have met with mixed results (e.g., Cox and Scholz, 1988a,b). Under these conditions, the fault propagates further into the granite by a *combination* of tensile and shearing mechanisms. On the other hand, faults may also propagate in their own planes in materials that are weaker in shear than in tension; examples include laboratory-scale faults in wet plaster (Fossen and Gabrielsen, 1996) and field-scale faults in porous sedimentary rocks (Du and Aydin, 1993, 1995).

Many limestones contain beautiful sets of pressure solution surfaces, or **stylolites**. These planar structures have a *closing or interpenetrating sense of displacement across them*, due to dissolution and volume loss under large compressive normal stress (Stockdale, 1922; Durney, 1974; Passchier and Trouw, 1996, p. 26; Benedicto and Schultz, 2010; Zhou and Aydin, 2010; Nenna and Aydin, 2011; Toussaint et al., 2018). Stylolites can form parallel to layering, due to overburden pressure, and at high angles to layering if formed by tectonic stress states.

In this example, a set of tectonic stylolites and associated veins serves as the substrate for fault nucleation. First, echelon arrays of veins form as purely dilatant structures in an orientation parallel to the remote maximum compression direction. In their stepover regions, bending of the rock bridge fosters the nucleation and growth of short stylolites (Nicholson and Pollard, 1985; Willemse et al., 1997) that *link* them near their tips. These stylolites can apparently be weak enough relative to the surrounding limestone, and compatibly oriented to the remote stress state, for *slip surfaces* to nucleate upon them. The kinematic combination of echelon slip surfaces (in the veins' stepovers) and adjacent veins leads to their conversion into **pull-apart** structures with an attendant change in the shearing direction to vein-parallel. Arrays of sheared vein/stylolite sets are linked by new stylolites (later sheared) within contractional stepovers. Finally, continued displacement accumulation along the composite structure leads to a more continuous fault that slices through the limestone.

The deformation sequence described above for fault development in limestone contrasts with that in which the fault slips first, with veins and stylolites forming as secondary structures (Fletcher and Pollard, 1981) in the dilational and contractional quadrants of the fault, respectively (e.g., Rispoli, 1981; Pollard and Segall, 1987; Ohlmacher and Aydin, 1995; Petit and Mattauer, 1995; Willemse and Pollard, 1998). Detailed and careful field work is required in order to reconstruct the sequence of deformation in limestone as well as, more generally, in any faulted lithology. Recognition of pre-existing or precursory structures is critically important for understanding the timing and locations of structural elements that comprise the eventual fault.

Faults in **porous rocks** such as sandstone, many limestones, and non-welded pyroclastic tuffs commonly nucleate on a precursory substrate formed earlier in the deformation sequence defined by a zone of **deformation bands** (Aydin and Johnson, 1978, 1983; Shipton and Cowie, 2001, 2003; Schultz and Balasko, 2003; Davatzes et al., 2005; Schultz and Siddharthan, 2005; Aydin et al., 2006; Fossen et al., 2007; Faulkner et al., 2010). In this case, accompanying or following initial *strain softening* and attendant accumulation of shearing offsets within the bands (e.g., Schultz and Balasko, 2003; Katsman et al., 2004; Okubo and Schultz, 2006), a zone of (cataclastic) deformation bands can become *stiffer than its surroundings* due to grain-size reduction due to crushing and/or tighter grain packing within the band (Aydin, 1978; Kaproth et al., 2010; Wong and Baud, 2012). The final deformation bands become stiff inclusions within the host rock that can serve as precursory nucleation sites for later brittle structures, such as joints and faults.

When the strain rate within the stiffer zone of deformation bands becomes sufficiently incompatible relative to that in the adjacent host rock (e.g., Aydin and Johnson, 1983), *slip surfaces* nucleate at a band–host rock interface (Shipton and Cowie, 2001), link up, and, eventually, a fault in porous rock is born (Schultz and Siddharthan, 2005). Experimental work reported by Ikari et al. (2011) demonstrates that weak gouges that initially deform in a stable, velocity-strengthening rate-and-state manner (Chapter 3) can become velocity-weakening, and therefore may begin to slip unstably, as shear strain and gouge strengthening occur. By implication, deformation bands may also transition from stable, velocity-strengthening shearing at lower shear strains to unstable, velocity-weakening frictional sliding at sufficiently large values of shear strain as grain fragmentation, grain angularity, and band strength increases toward some critical value.

6.6 The Stepover Region: Key to Interpretation

Segmented faults function mechanically as discontinuity arrays, just as joints, veins, and dikes were seen to be dilatant mode-I cracks in Chapter 4. When discontinuous fault segments approach each other, they interact mechanically in several interesting ways, depending on the sense of slip, the sense of step, and the three-dimensional geometry (e.g., Segall and Pollard, 1980). Where the appropriate fault-tip stresses at one fault segment are *enhanced* by slip along the neighboring fault, fault propagation and growth of secondary fractures is *encouraged*. On the other hand, the converse inhibits propagation. Continued slip accumulation along faults whose terminations are restricted by the near-field stress state, and thereby prevented from propagating, produces steeper displacement gradients near the fault-segment tips (e.g., Segall and Pollard, 1980; Peacock and Sanderson, 1995), displacement maxima shifted away from the fault midpoints (Bürgmann et al., 1994 ; Peacock and Sanderson, 1995), and spectacular large-strain deformation. Nowhere is this more pronounced than in the *stepover regions* between the echelon fault segments (Fig. 3.13), where pull-apart basins, push-up ranges, and duplexes nucleate and grow along strike-slip faults (e.g., Kingma, 1958; Lensen, 1958; Quennel, 1958; Crowell, 1962; Burchfiel and Stewart, 1966; Clayton, 1966; Dibblee, 1977; Aydin and Nur, 1982, 1985; Mann et al., 1983; Woodcock and Fischer, 1986) and relay-ramps nucleate and grow along dip-slip faults (e.g., Peacock and Sanderson, 1991; Davison, 1994; Davies et al., 1997; Crider and Pollard, 1998; Crider, 2001; Peacock, 2002; Davis et al., 2005).

The near-tip stress field surrounding mode-I fractures (joints, cracks, dikes, and veins) is *symmetric* about the fracture plane when the displacement is normal to the fracture. We will see in Chapter 8 that oblique opening of a mixed-mode crack leads to stress fields that are not symmetric, and thereby promotes crack propagation directions that are not parallel to the strike of the parent crack. Stylolites and anticracks having negligible shear components can also have symmetric near-tip stress fields (e.g., Fletcher and Pollard, 1981; Zhou and Aydin, 2010; Nenna and Aydin, 2011). Resulting stepover kinematics, mechanics, and fracture set geometries were discussed in Chapter 5.

Conservative and Nonconservative Stepovers

Segmented faults link differently depending on the orientation of the slip vector relative to the plane. Two kinematic classes of stepovers can be defined depending on the degree of volumetric deformation that occurs there. Following King and Yielding (1984), stepovers in which area or volume are not conserved are called **nonconservative or discordant stepovers**, whereas those in which area or volume is conserved are called **conservative or concordant stepovers**. For example, relays between synthetically dipping normal or reverse faults are sites of predominantly distortional strain; these would correspond to conservative stepovers. Pull-apart basins and push-up ranges that occur along strike-slip faults are associated with significant volumetric deformation; these would correspond to nonconservative stepovers. In general, mode-II shearing leads to nonconservative stepovers (i.e., along-strike stepovers between strike-slip faults in map view and down-dip stepovers between normal or reverse faults), whereas mode-III shearing promotes conservative stepovers (i.e., along-strike stepovers (relays) between normal or reverse faults and down-dip stepovers between strike-slip faults). The sense of shearing relative to the stepover geometry determines the stepover kinematics (Aydin, 1988).

6.6.1 Dip-Slip Faults

Stresses surrounding strike-slip or (dipping) dip-slip fault tips are characteristically not symmetric (e.g., Chinnery, 1963; Segall and Pollard, 1980; Crider et al., 1996; Crider and Pollard, 1998). The tearing sense of shearing along dip-slip faults leads to distortional strain predominating within the stepover, as opposed to predominantly volumetric strain within strike-slip stepovers.

Numerical models by Crider and Pollard (1998) and others reveal that a cross fault between echelon normal fault tips (Fig. 6.36) can nucleate at depth where the mechanically interacting faults' elliptical tiplines are in closest proximity; in map view, this location is near the middle of the relay-ramp (Fig. 6.37). As offsets along the faults increase, the cross fault grows in 3-D, eventually both linking the faults and breaking the surface. The relay-ramp is now considered **breached** (Fig. 6.36b) and a zig-zag fault geometry is created. This pattern causes no significant impediment to further dip-slip offsets, however (in contrast to strike-slip faults), so the linked fault segments (called **kinematically coherent** by Willemse et al., 1996) can continue to accumulate dip-slip offsets. As part of this process, the former fault tips located at greater values of overlap than the linking fault are no longer favorably situated to accommodate further slip and become **inactive tips**, kinematically analogous to the "horns" on linked joint and dike arrays (e.g., Pollard et al., 1982; Nicholson and Pollard, 1985).

Many normal fault systems have faults with rakes <90°, corresponding to an oblique sense of dip-slip offset; i.e., the normal faults have a small

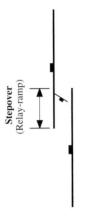

a Incipient linkage
across echelon step

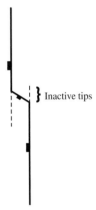

b Kinematically coherent
(linked) fault

Fig. 6.36. Evolution of a conservative stepover between synthetically dipping normal faults (relay-ramp) illustrated as a function of increasing remote strain and increasing offsets along the normal faults (after Crider and Pollard, 1998).

Fig. 6.37. An unlinked relay-ramp in a conservative stepover developed between two synthetically dipping normal faults in limestone near Kilve, United Kingdom (photo by Martin Schöpfer, used by permission); length of stepover (overlap) is 89 cm.

Fig. 6.38. Evolution of a normal-fault stepover that has a small component of oblique (strike-slip) offset along the faults (after Crider, 2001).

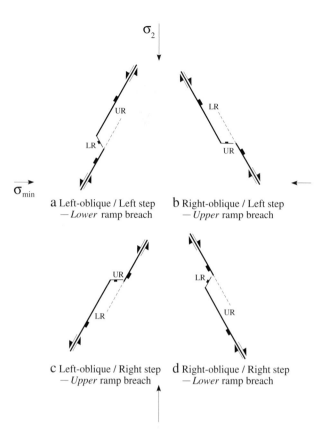

a Left-oblique / Left step
 — *Lower* ramp breach

b Right-oblique / Left step
 — *Upper* ramp breach

c Left-oblique / Right step
 — *Upper* ramp breach

d Right-oblique / Right step
 — *Lower* ramp breach

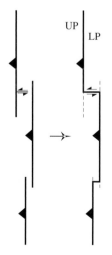

Fig. 6.39. Evolution of an echelon stepover between synthetically dipping reverse or thrust faults. UP, upper plate; LP, lower plate.

strike-slip component of offset along them. Investigation of this common fault geometry by Crider (2001) reveals important systematics within the stepover (Fig. 6.38). A **lower ramp breach** (see Fig. 6.5) occurs for left-stepping, left-oblique (or right-stepping, right-oblique) faults. An **upper ramp breach** occurs for left-stepping, right-oblique (or right-stepping, left-oblique) faults. Thus, the relay-ramp links at its lower part for a dilational sense of strain in the stepover (Crider, 2001), whereas a contractional sense of strain in the stepover fosters a cross fault in the upper part of the relay-ramp.

Stepovers between **echelon thrust faults** (called "frontal ramps;" Butler, 1982; McClay, 1992; see also Aydin, 1988; Fig. 6.39) appear to share the same first-order kinematics as the normal-fault relay-ramp examples discussed above and shown in Figs. 6.37 and 6.38. Research on actively deforming thrust fault systems such as that by Shaw et al. (1999, 2002) and Davis et al. (2005), especially in 3-D, has clarified the role of fault linkages in contractional fault systems given its importance to seismic hazard analysis and prediction. However, because the offsets along thrust faults can be so much larger than those along normal faults, the inactive tips either can be carried along (if located on the upper plate of the thrust sheet) or are difficult to observe (if located on the lower plate; Fig. 6.38).

Fig. 6.40. Relationship
between the sense of slip for a
pair of echelon (a) normal or
(b) reverse faults (map view;
strike-slip faults in cross section)
and the horizontal fluid flow
paths in their stepover.

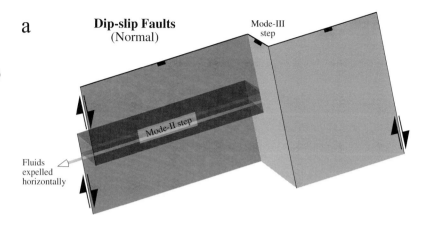

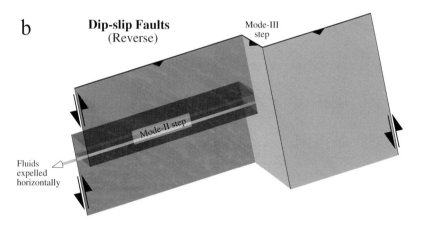

Slip vectors that are parallel to linked faults do not induce significant volume changes in the surrounding rock; these conservative stepovers are important in fluid flow along the fault. Fluid flow in conservative stepovers is directed horizontally along these systems (Fig. 6.40). As a result, one could expect substantial fluid flow to occur along the strike direction of normal-fault and thrust-fault systems.

6.6.2 Strike-Slip Faults

A Simple Mnemonic—
Left lateral
Left step
... Dilational strain in step
Left lateral
Right step
... Contractional strain

The relationship between step and slip sense for strike-slip faults is systematic: **dilational stepovers** have either *left*-lateral faults with *left* stepovers, or *right*-lateral faults with *right* stepovers. **Contractional stepovers** have either *left*-lateral faults with *right* stepovers (Fig. 6.18), or *right*-lateral faults with *left* stepovers. This mnemonic contributes to the location of (upper or lower) ramp breaches along oblique normal fault sets as illustrated in the previous section.

For a pair of closely spaced echelon mode-II discontinuities, the relationship between them is given by the following:

$$\text{Sense of strain } = f(\text{sense of step, sense of slip}) \qquad (6.7a)$$

$$\text{Sense of step} = f(\text{sense of slip, sense of strain}) \qquad (6.7b)$$

$$\text{Sense of slip} = f(\text{sense of step, sense of strain}) \qquad (6.7c)$$

Knowledge of *any two* of these allows one to easily solve for the third. This relationship was exploited by Schultz (1989) to identify strike-slip faults on Mars based only on the sense of step and sense of strain (i.e., topography) within the stepover. In that case, the absence of offset passive markers along the echelon faults precluded their identification by traditional means. This approach was also used to identify strike-slip faults on the icy satellites of Jupiter (Schenk and McKinnon, 1989; Pappalardo et al., 1998; see also Andrews-Hanna et al., 2008; Schultz et al., 2010a). In another example, the right-stepping echelon fault set illustrated in Fig. 6.40 should separate jogs or areas of dilational strain if the sense of slip on the overall fault set is right-lateral, as it is elsewhere along the Denali fault. The combination of these three stepover characteristics provides a powerful tool for determining the sense of slip, for example, along a fault that cut rocks in which passive markers, or unambiguous fault-surface textures that might otherwise indicate the slip sense, were unavailable.

Stepovers along strike-slip faults lead to volumetric increases or reductions there; these stepovers are commonly recognized as pull-apart basins (Aydin and Nur, 1982, 1985; Mann et al., 1983, 1985) and push-up ranges (Fig. 6.41) along strike-slip systems (Fig. 6.18) and duplexes or rhombohedral voids in the down-dip direction of dip-slip faults (Fig. 6.42). These structures illustrate how the sense of strain in a fault stepover, along with assessment of the sense

Fig. 6.41. A well-exposed example of a push-up range is the Echo Hills (Campagna and Aydin, 1991) located in southern Nevada between a right-stepping pair of echelon left-lateral strike-slip faults. Curved line shows approximate location of the southern left-oblique fault strand.

Fig. 6.42. Small composite **pull-apart in carbonate rocks** from the Pyrenees fold and thrust belt, northeast Spain. This nonconservative stepover is developed between echelon normal faults exposed in cross section (i.e., rocks are offset top down to the right); the final rhomb-shaped void was built up from a series of smaller ones having comparable geometries.

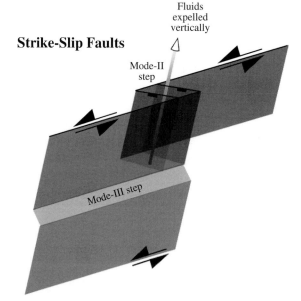

Strike-Slip Faults

Fluids expelled vertically

Mode-II step

Mode-III step

Fig. 6.43. Relationship between the sense of slip for a pair of echelon strike-slip faults (in map view, dip-slip in cross section) and the **vertical fluid flow paths** in their stepover.

of step, can predict the sense of slip along a fault system of any orientation or scale.

Slip vectors oriented normal to the linking fault (Fig. 6.43) lead to volumetric increases or reductions, corresponding to nonconservative stepovers. Sibson's (1986a,b,c; 1989) work in particular fostered significant progress in the understanding of how fluid flow can be modulated by, and related to, fault slip. Martel and Boger (1998) have identified these classes of stepovers in their studies of the 3-D geometry and mechanics of strike-slip faults. Connolly and Cosgrove (1999) showed experimentally how fluid flow can be enhanced at nonconservative stepovers, with corroborating evidence from large-scale field examples. Fault slip will produce areas of locally higher mean stress near the fault, from which groundwater will flow, and areas of locally reduced mean stress and volume increase, into which groundwater flow will be promoted. Nonconservative dilational stepovers are common and important conduits for fluid flow (Fig. 6.43), leading to a predominantly vertical flow directionality for nonconservative (strike-slip in map view, dip-slip in cross section) stepovers. As a result, one could anticipate subvertical fluid breakouts occurring in areas of strike-slip fault segmentation (e.g., Royden, 1985).

6.6.3 Plane vs. 3-D Deformation

Previous studies have recognized the relationship between the thickness of a deforming section and the fault pattern at the surface (e.g., Tchalenko, 1970; Bartlett et al., 1981; Naylor et al., 1986; Mandl, 1988; Katzman et al., 1995; ten Brink et al., 1996; Soliva et al., 2008). This effect can be an important one to consider in fault-related tectonics. The development and expression of secondary structures and stepovers along faults depends on the relative scale of the faulted domain. Specifically, the **degree to which the vertical cross-section of a faulted layer is able to deform vertically** (or not, for sufficiently thick sections) helps to determine the contributions of vertical (out-of-plane) strains and horizontal (in-plane) strains in a faulting environment. In this context, the faulted section is considered to be "thick" mechanically if its thickness is *sufficiently greater than* the horizontal length of the fault array that no out-of-plane strain is significantly occurring (Fig. 6.44b; see also Schultz and

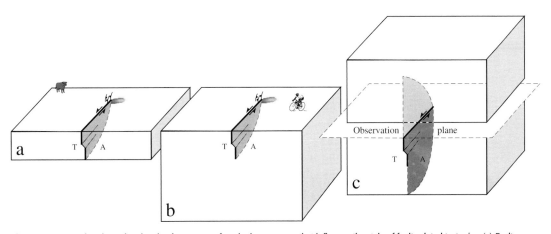

Fig. 6.44. Perspective views showing the three types of geologic exposures that influence the style of fault-related tectonics. (a) Faults exposed at the Earth's surface and cutting a relatively thin section. (b) Faults exposed at the Earth's surface and cutting a relatively thick section (plane-strain, half-space). (c) Blind faults now visible, for example at the Earth's surface or in seismic section, and cutting a (formerly) relatively thick section.

Fossen, 2002). This case corresponds to one of engineering plane strain (see Chapter 2). On the other hand, a "thin" section (thickness comparable to, or *smaller than*, the fault array length) allows out-of-plane (vertical) strain to readily develop (i.e., the plane-stress approximation in engineering, or else fully 3-D strain; Fig. 6.44a).

The proximity to the Earth's surface, or to other boundaries or free surfaces, is well known to influence certain aspects of faulting. At the Earth's surface, the stress changes due to faulting are amplified because of the free-surface effect (e.g., Chinnery, 1963; Pollard and Holzhausen, 1979; Pollard and Aydin, 1988). Displacements are similarly increased along surface-breaking or near-surface structures for the same reason (e.g., Chinnery, 1961; Bilham and King, 1989). Secondary structures are also influenced by the reduction of system stiffness at the Earth's surface. Oblique fault arrays having components of both "out-of-plane" dip-slip (normal or thrust) and "in-plane" strike-slip offsets are commonly developed near bends, irregularities, and tips of fault segments. Folds can grow more efficiently in contractional (leading) quadrants and near bends or faults in thin plates and/or in proximity to the Earth's surface. Interestingly, using an approximate guide based on fracture mechanics, faults that are shallower than their down-dip heights are strongly influenced by the proximity to the Earth's free surface. This implies that fold-and-thrust belts and critical-taper accretionary wedges are thin-skinned phenomena when viewed in cross section.

If the faults are **blind** and they are sampled at some arbitrary horizontal plane of observation (Fig. 6.44c), as along an outcrop surface, then the kinematics remain, but their vertical (out-of-plane) displacements—perpendicular to the shearing direction and parallel to the fault plane—are reduced. Resulting structures include pull-apart cavities accommodating areas of volume increase and commonly filled by calcite or other hydrous minerals (Figs. 6.42 and 6.45; e.g., Aydin and Nur, 1985; Ohlmacher and Aydin, 1997), petroleum (Dholakia et al., 1998), or other precipitates from the fluid-charged faulted section (e.g., Aydin, 2000; Eichhubl and Boles, 2000). Similarly, instead of push-up ranges, cleavage duplexes can be formed in contractional echelon stepovers in thick sections (e.g., Aydin, 1988) with anticracks or stylolites generated in the contractional quadrants of the faults (e.g., Rispoli, 1981; Willemse et al., 1997; Willemse and Pollard, 1998; Fig. 6.45, upper panels).

Some startling and useful rules of thumb for assessing or interpreting the relative plate thickness have been revealed by 3-D mechanical models. For example, Katzman et al. (1995) showed that *reducing the faulted-section thickness* so it is smaller than the horizontal length of the fault array produces a full graben within the pull-apart (Fig. 6.46b), instead of a warping of the basin floor into *two depocenters* (Fig. 6.46a) for subsidence between two tall echelon strike-slip faults. You can see this warping in larger field examples, such as the Ridge Basin along the San Andreas fault system in

Fig. 6.45. Schematic diagrams showing difference in expression of structures in nonconservative (extensional or contractional) stepovers, and at fault-segment terminations, due to the relative thickness of the faulted section.

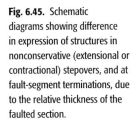

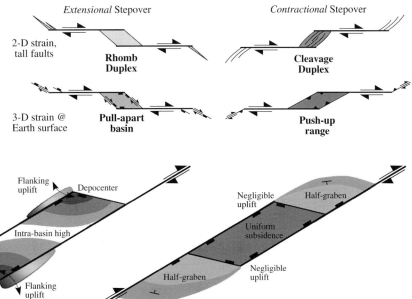

Fig. 6.46. Pull-apart basins develop differently depending on whether the faulted section is (a) thick, as in Fig. 6.44b, or (b) thin, as in Fig. 6.44a, relative to the horizontal length of the fault array.

California (see Christie-Blick and Biddle, 1985). The pull-apart is confined to the stepover for tall (2-D) faults, and an uplifted (folded) flank is also commonly observed in the bounding echelon faults' contractional quadrants (Fig. 6.46a). In addition, basin subsidence in a thin faulted section extends beyond the confines of the stepover, producing a gentle asymmetric graben, dipping toward the bounding strike-slip fault, on either end of the stepover (Katzman et al., 1995; ten Brink et al., 1996; Fig. 6.46b). Their mechanical model also suggests that push-up ranges also change with faulted-section thickness, so that contractional deformation in a stepover between strike-slip faults cutting a relatively thin sequence might be more symmetric than those in thicker sections.

Normal-fault stepovers are also influenced by the geometry of the relay in three dimensions. Soliva et al. (2008) showed by using a combination of field and numerical approaches that synthetically dipping normal faults that bound a stepover (or relay) and terminate against each other at depths shallower than the depth of faulting (i.e., branched faults) created stepovers having larger overlaps for a given separation, and more asymmetric stress changes within them, than would relays between parallel normal faults that are unlinked at depth. It appears that fault branching tends to reduce the amount of fault interaction in the relay, implying a reduced tendency to form breached relays.

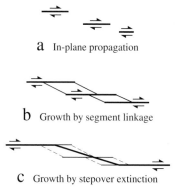

Fig. 6.47. The three stages in the growth of faults. (a) Growth by in-plane, "radial" propagation. (b) Growth of closely spaced echelon faults by segment linkage. (c) Smoothing of linked echelon (composite) fault traces by new faulting through the stepovers.

a In-plane propagation

b Growth by segment linkage

c Growth by stepover extinction

6.6.4 Stepover Extinction

Many of the classic studies in the literature on faults deal with so-called meso-scale structures—those that span an outcrop. This scale enables study of the nucleation and growth of faults from perhaps millimeters to perhaps tens of meters. Using *displacement–length scaling relations* (e.g., Scholz and Cowie, 1990; Cowie and Scholz, 1992a; Clark and Cox, 1996; Schultz et al., 2008a, 2013; see Chapter 9), fault offsets can be on the order of 1/10 of the longest fault length, or perhaps a few meters.

Faults tend to **smooth their surface traces** as they grow larger (Wesnousky, 1988) although the details of earthquake rupture length, segmentation, and fault growth are complex in detail (e.g., Wesnousky, 2008). They do this by slicing diagonally through their stepovers (Fig. 6.47)—cutting through pull-apart basins to make new, through-going faults (Zhang et al., 1989), or by faulting through a push-up range (e.g., see geologic map of the Echo Hills range by Campagna and Aydin, 1991). This is equivalent to creating the "principal displacement zone" from a Riedel shear zone that slices through the linked network of earlier-formed faults (e.g., Tchalenko, 1970; Bartlett et al., 1981). This **three-stage sequence of fault growth** is shown schematically in Fig. 6.47. Whereas the first two stages have been emphasized in many fault studies (e.g., Crider and Peacock, 2004) a comprehensive understanding of fault growth at all scales can be further informed from study across the scale range (e.g., Graham et al., 2003).

6.7 Review of Important Concepts

1. Faults are a class of structural discontinuity whose essentially planar surfaces interrupt the continuity of a rock mass, thereby concentrating and redistributing stresses. Faults are characterized by having displacements (offset of passive markers on either side of the fault plane) chiefly parallel to the plane, in contrast to cracks and joints that have offsets normal to their planes.

2. Faults are classified according to their kinematics relative to the Earth's surface into strike-slip and dip-slip faults. The latter are distinguished as normal and reverse faults depending on the sense of displacement relative to the Earth's surface. The former includes oceanic transform faults although the orientations of transform and strike-slip faults to the direction of horizontal extension are distinctly different.

3. The kinematic fault types are keyed to particular Andersonian **stress states**, so that the three principal stress orientations, and sometimes their relative magnitudes, can be inferred from fault type; 2-D and 3-D strain states associated with fault types can also be inferred.

4. Stress states for faulting can be quantified by using the mechanics of **frictional sliding** on a pre-existing plane. The resulting stress magnitudes can be associated with the Andersonian stress orientations and portrayed on a stress polygon diagram if all three principal stress components are known. Limiting values of stress magnitudes by friction in a rock mass leads to well-known concepts including **lithospheric strength envelopes** and **critically stressed faults**.

5. **Faults are segmented**, just like other types of discontinuities including joints; the region between the fault segments is called a **stepover**. The *type of deformation* generated within the stepover depends on the sense of displacement along the fault.

6. The **sense of displacement** along a segmented, echelon fault can be inferred in the absence of offset markers by utilizing the kinematics of the fault-tip and stepover regions. Three inter-related quantities are the **sense of slip** along the faults, their **sense of step**, and the **sense of strain** in the stepover. Knowing any two quantities reveals the third.

7. Fault slip requires a surface on which to slide frictionally. Such "substrates" for faults can be recognized as defining two general classes. **Pre-existing substrates** formed before, and commonly under a different far-field stress state, than the slip surfaces that later nucleated on them. A tectonic joint set is a very common pre-existing substrate that did not form in shear and thus may be unrelated to the later shearing deformation that triggered slip surface growth on the earlier-formed joints. **Precursory substrates** form in an early stage of the same deformation event that eventually produces slip surfaces. A faulted zone of deformation bands is a common example of slip surfaces that nucleated on a precursory substrate.

8. The **anisotropy** in strength and/or stiffness of a layered stratigraphic section exerts a profound control on the types and sequence of failure. Stiffer layers that fail first provide pre-existing substrates (e.g., joints) on which fault slip can nucleate. Continued fault growth may involve

propagation through intervening layers and linkage of segments into a through-going fault. The relative orientation of the remote stress state and the layer-induced mechanical anisotropy strongly influences the style of the resulting faults.

9. Near the tips of strike-slip faults, volume is not conserved, leading to dilational and contractional strains in their **trailing and leading quadrants**, respectively. **Extensional stepovers** house calcite rhombs, vein arrays, breccias, joint networks, and normal faults at smaller scales and *pull-apart basins* at the Earth's surface. At **contractional stepovers** cleavage duplexes, stylolites, pressure ridges, thrust faults, folds, and *push-up ranges* can occur. Although volume is generally conserved near the tips of dip-slip faults, bending of strata in the stepovers leads to relay-ramps and related structures there.

10. **Fluid flow in linked systems of echelon faults follows their stepovers**. For example, strike-slip faults can have vertical conduits, whereas dip-slip (normal and thrust) faults can have horizontal conduits. *Ore mineralization* can also be enhanced in these stepover configurations, motivating exploration models based on fault mechanics for these distinct tectonic regimes.

11. **Block rotations** about vertical axes are quite common between a pair of parallel strike-slip faults. Rotation-related stresses may trigger slip along suitably oriented pre-existing or precursory planes, producing cross faults that facilitate the block rotations. Continued shearing along the bounding faults can rotate the cross faults out of optimal orientations, locking the faults and, later, requiring the cutting of new cross faults. Block rotations become more complicated near fault terminations and stepovers, with rotation axes varying from horizontal to vertical depending on position.

12. The **thickness of the faulted section**, relative to the map length of the fault array, exerts an important control on strike-slip tectonics. *Thick sections* (with relatively short fault arrays) inhibit vertical strains near faults, so a more 2-D deformation state is developed. Stepovers in thick sections have two displacement maxima set diagonally across, leading to two depocenters in pull-apart basins and two uplift peaks in push-up ranges. In contrast, relatively thin sections (with relatively long fault arrays) can deform readily in the vertical direction, promoting full grabens in pull-aparts and more constant uplifts in push-up ranges. Thus, the degree and expression of uplift and subsidence can change dramatically as a fault array grows in length across a region.

6.8 Exercises

1. Summarize the key similarities and differences between faults and joints.
2. What are the types of stepovers called for (a) normal faults, (b) reverse or thrust faults, (c) strike-slip faults? Sketches of each of these would improve your answer.
3. Describe the criteria that one might use to discern the sense of slip, shearing, or offset along a fault.
4. Provide several lines of evidence that would indicate that fault segments were mechanically interacting with other fault segments.
5. What is a fault segment, and describe what criteria one might invoke to decide whether it was a segment or an isolated fault.
6. Describe the interaction between faults and layering for (a) shallow fault dips, (b) steep fault dips, and (c) vertical fault dips.
7. What properties control whether folds are curved or kinked?
8. Define (a) fracture gradient and (b) fault gradient and describe what factors or processes control each.
9. Describe a strength envelope for the Earth's lithosphere including variation with depth, tectonic regime, pressure, and temperature. How might this relate to the depth of seismogenic faulting?
10. Compare conjugate and orthorhombic fault patterns and tell what one might measure to discriminate between them.

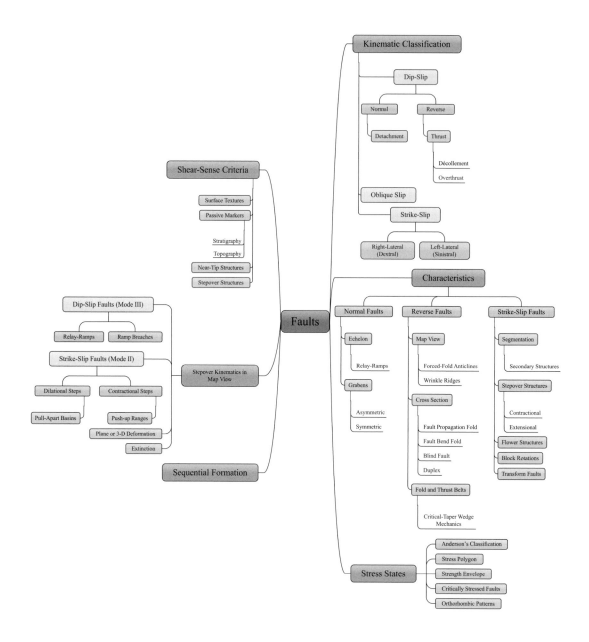

Kinematic Classification

Dip-Slip

Normal   Reverse

Detachment   Thrust

Décollement

Overthrust

Oblique Slip

Strike-Slip

Right-Lateral   Left-Lateral
(Dextral)   (Sinistral)

Shear-Sense Criteria

Surface Textures

Passive Markers

Stratigraphy

Topography

Near-Tip Structures

Stepover Structures

Dip-Slip Faults (Mode III)

Relay-Ramps   Ramp Breaches

Strike-Slip Faults (Mode II)

Dilational Steps   Contractional Steps

Pull-Apart Basins   Push-up Ranges

Plane or 3-D Deformation

Extinction

Stepover Kinematics in
Map View

Sequential Formation

Faults

Characteristics

Normal Faults   Reverse Faults   Strike-Slip Faults

Echelon   Map View   Segmentation

Relay-Ramps   Forced-Fold Anticlines   Secondary Structures

Grabens   Wrinkle Ridges   Stepover Structures

Asymmetric   Cross Section   Contractional

Symmetric   Fault Propagation Fold   Extensional

Fault Bend Fold   Flower Structures

Blind Fault   Block Rotations

Duplex   Transform Faults

Fold and Thrust Belts

Critical-Taper Wedge
Mechanics

Stress States

Anderson's Classification

Stress Polygon

Strength Envelope

Critically Stressed Faults

Orthorhombic Patterns

7

Deformation Bands

7.1 Introduction

In This Chapter:

♦ *Types of deformation bands*

♦ *Deformation of porous granular rocks*

♦ *From deformation bands to faults*

Deformation bands are a common and important type of tabular geologic structural discontinuity that results from strain localization in porous granular rocks (e.g., Aydin et al., 2006; Fossen et al., 2007). First recognized in sandstone (e.g., Aydin, 1978; Aydin and Johnson, 1978; Hill, 1989), they were subsequently identified in other porous rock types including carbonate grainstones (Tondi et al., 2006), nonwelded tuffs (Wilson et al., 2003; Evans and Bradbury, 2004), chalk (Wennberg et al., 2013), and even sedimentary sequences on Mars (Okubo et al., 2009). Similar structures, sometimes called *Lüders' bands* (e.g., Friedman and Logan, 1973; Olsson, 2000), have been noted and investigated in engineering materials such as polystyrene plastic (Argon et al., 1968; Bowden and Raha, 1970; Kramer, 1974) and mild steel (Nadai, 1950, p. 279) long before deformation bands were recognized as such in rocks (Aydin, 1978).

Deformation bands provide precursory substrates for faulting (e.g., Aydin and Johnson, 1978, 1983; Shipton and Cowie, 2001, 2003; Fossen et al., 2007, 2017; Chapter 6) and thereby represent an important deformation mechanism that leads to faulting in sedimentary basins. Pioneering work by Davis and colleagues demonstrated the utility of deformation bands in regional structural geology and tectonics (e.g., Davis, 1999) and helped move the study and application of these structures from somewhat esoteric to mainstream (e.g., Davis and Reynolds, 1996; Fossen, 2010a,b). Their utility encompasses geohydrology given that their petrophysical properties can baffle or redirect subsurface fluid flow important in groundwater aquifers and hydrocarbon reservoirs (e.g., Ogilvie and Glover, 2001; Sternlof et al., 2006; Fossen and Bale, 2007; Torabi and Fossen, 2009; Bense et al., 2013; Deng et al., 2015).

7.2 Deformation Band Classification

Formerly part of a seemingly arcane branch of structural geology and rock mechanics, deformation bands have become recognized as a widespread and important class of geologic structure (e.g., Davis and Reynolds, 1996; Davis, 1997, 1998, 1999; Aydin et al., 2006; Fossen et al., 2007; Wong and Baud, 2012; Fossen et al., 2017). Of the many nomenclatures and classification schemes available in the literature, those by deformation mechanism and by deformation-band kinematics are perhaps the most applicable to field-based studies of deformation bands. These classification schemes may facilitate establishing correspondences between field observations, strain-localization theory (e.g., Papamichos and Vardoulakis, 1995; Bésuelle, 2001a,b; Johnson, 2001; Issen, 2002; Borja and Aydin, 2004) and experiments (e.g., Friedman and Logan, 1973; Olsson, 1999, 2000; Saada et al., 1999; Mair et al., 2000, 2002; Olsson and Holcomb, 2000; Chemenda, 2009, 2011; Chemenda et al., 2014; Soliva et al., 2016).

7.2.1 Classification by Deformation Mechanism

Two primary approaches have been adopted to classify deformation bands: mechanistic and kinematic (e.g., Fossen et al., 2007, 2017; Fossen, 2010a,b; Wong and Baud, 2012). Each of these has proponents in the structural-geology literature and each has substantial intrinsic merit. **Mechanistic** approaches rely on the identification of fabrics and similar characteristics to distinguish types of deformation bands in relation to their petrophysical properties (such as grain size, shape, and mineralogy, sorting, degree of diagenesis, and magnitude of strain accommodated by the bands; see Fig. 7.1). Following Fossen et al. (2007), four main types can be recognized:

- **Disaggregation bands** (formed by grain rotation and grain-boundary sliding; also called "invisible bands").
- **Phyllosilicate bands** (formed by reorganization and shearing within thin zones of platy clay minerals).
- **Cataclastic bands** (formed by grain fragmentation within a narrow tabular zone).
- **Solution and cementation bands** (formed by localized dissolution or cementation within narrow zones).

Disaggregation bands are commonly recognized in deformation of soft sediments (e.g., Maltman, 1988, 1994; Fossen, 2010b; Fossen et al., 2011; Brandes and Tanner, 2012) but have also been reported in deltaic gravels (Exner and Grassemann, 2010). Sometimes referred to as microfaults or, simply, shear bands in the geotechnical engineering literature (e.g., Davis and Selvadurai, 2002), they accommodate slippage of soft, water-saturated or otherwise unconsolidated sediments and similar materials along surfaces of

Fig. 7.1. Deformation bands of various types crop out in this classic exposure of Navajo Sandstone at Buckskin Gulch, Utah (Mollema and Antonellini, 1996; Schultz, 2009, 2011a; Fossen et al., 2011).

shearing that display relatively little volumetric strain. These thin bands tend to be porosity-neutral and therefore are transparent to fluid flow in rocks that contain these structures.

Phyllosilicate bands are common in mudrocks and shales. They can be recognized as thin tabular zones filled by material resembling fault gouge; accordingly the bands may impede fluid flow and contribute to an anisotropy in the host-rock permeability. These bands commonly form in mudrocks that are grain supported (i.e., clay contents less than ~35–40%); deformation of mudrocks having larger clay fractions by matrix-supported shearing (e.g., Sone and Zoback, 2014) may produce **clay-smear zones** that may be associated with reduced frictional strength and stable, creep-like frictional stability (Chapter 3).

This type of deformation band is distinctive and important in several respects. First, phyllosilicate bands lack the grain crushing and size reduction so typical of cataclastic deformation bands (Antonellini and Aydin, 1994; Gibson, 1998). Phyllosilicate bands can be identified in the field as closely spaced arrays, or colorful bundles, of subparallel strands, perhaps 10–50 cm thick, along which clays from the host rock are *smeared*. A high clay content (perhaps 15%) in the host rock is considered necessary for these bands to form (Antonellini et al., 1994), so faults that nucleate in *silty sandstones and mudstones* may fall into this category. Second, clay-assisted bands that occur near faults can be much narrower than cataclastic deformation bands in clay-poor sandstones. As shear displacement along phyllosilicate bands increases (beyond some tens of meters in the Entrada Sandstone (Antonellini et al., 1994)), a through-going slip surface—a true fault—tends to localize and propagate along the array. Faults that have impermeable clays or a cataclastic inner fault core can form effective seals against the flow of subsurface fluids such as petroleum and groundwater (Antonellini and Aydin, 1994; Aydin, 2000); these faults are referred to as *sealing faults* in the petroleum industry (e.g., Downey, 1984; Knott, 1993; Gibson, 1998).

Cataclastic bands are analogous to the cataclastic compactional shear bands discussed below. Cataclastic textures within a deformation band illustrate the reliance of deformation mechanism (i.e., grain crushing during band development) and factors such as displacement magnitude within this mechanistic classification scheme. The processes of grain-size reduction, shape change (i.e., increase in grain angularity; Fig. 7.2), and interlocking can lead to **strain hardening** of cataclastic compactional shear bands (e.g., Mair et al., 2002; Katsman et al., 2004; Kaproth et al., 2010; Wong and Baud, 2012) even at shallow depths (i.e., «1 km), leading eventually to the cessation of displacement accumulation along them. Cataclastic deformation bands are thus important to recognize as they can represent evidence of a change in deformation mechanism, from strain-hardening deformation banding to strain-softening faulting in sedimentary sequences.

Solution and cementation bands record the interplay between grain properties, fluid composition, diagenesis, and pre-existing deformation bands. These *secondary structures* have been recognized and discussed by several workers including Antonellini et al. (1994), Antonellini and Aydin (1994), Gibson (1998), Parnell et al. (2004), Sample et al. (2006), Fossen et al. (2007), Onasch et al. (2010), Parnell (2010), and Exner et al. (2013). Grain dissolution and precipitation or aggregation of minerals, such as pyrite, calcite, silica, and various clays, can either strengthen or reduce the permeability of deformation bands due

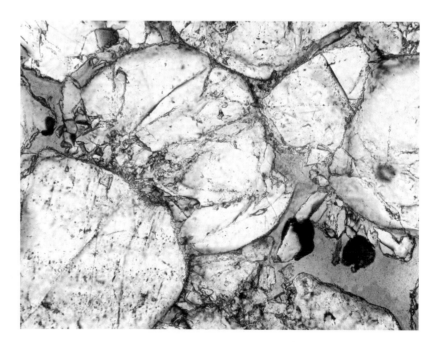

Fig. 7.2. Grain cracking, crushing, and increased angularity are common hallmarks of cataclasis in deformation bands. Photo by Haakon Fossen.

to this secondary diagenetic alteration (e.g., Exner et al., 2013). The spectacular exposures of deformation-band arrays at Devils Wall, in the Harz Mountains of Germany, may also be considered as an example of cementation bands formed during or after cataclastic deformation bands (Klimczak and Schultz, 2013b).

7.2.2 Classification by Band Kinematics

The kinematic classification scheme (Table 1.2) involves identifying the sense of offset of passive markers across a deformation band. There are two major classes (e.g., Aydin et al., 2006; Schultz et al., 2008a): bands that accommodate mostly volumetric strain (i.e., opening or closing senses—**dilation or compaction bands**) and mostly shearing strain across them (**shear bands**). Although the terminology may vary, five main kinematic varieties of deformation band are recognized; the latter two varieties are further divided into two sub-types as listed here.

- **Dilation bands** (predominantly opening displacements).
- **Dilational shear bands** (opening and shearing displacements in various proportions).
- **Shear bands** (predominantly shearing displacements).
- **Compactional shear bands** (predominantly shearing with closing displacements in various proportions)
  - (a) *Cataclastic* compactional shear bands
  - (b) *Non-cataclastic* compactional shear bands.
- **Compaction bands** (predominantly closing displacements)
  - (a) *Shear-enhanced* compaction bands
  - (b) *Pure* compaction bands.

Dilation bands were first identified in poorly consolidated marine sands in coastal California by Du Bernard et al. (2002a,b). The bands in this example may represent dilatant secondary structures formed in the extensional quadrants of cataclastic deformation bands, perhaps in relation to seismic activity (Cashman and Cashman, 2000). Exner et al. (2013) reported another example from the Matzen oil and gas field, Vienna Basin, Austria, in which diagenesis transformed the bands into cementation bands. Dilation bands may be colorful in outcrop, due to the presence of pore spaces filled with authigenic clay minerals and/or iron oxide minerals precipitated there by subsurface fluid flow (Antonellini et al., 1994). They exhibit an increase in porosity of several percent above that of the host rock.

  Dilational shear bands are predicted by a wide range of theoretical models for pre-peak yielding with positive values of dilation angle and friction angle (e.g., Johnson, 1995; Bésuelle, 2001a; Bésuelle and Rudnicki, 2004; Borja and Aydin, 2004; Aydin et al., 2006; Schöpfer and Childs, 2013). Because such structures would be less stiff and weaker in shear than their surroundings due to their localized porosity increases and reduced grain coordination number, they may be transient structures formed early (pre-peak) in the deformation of a porous granular solid (Davis and Selvadurai, 2002). When dilational shear bands provide precursory substrates for faults, they could be recognized as expressions of the earliest stages of shearing deformation of porous granular geomaterials.

Continuing along the kinematic classification scheme, **shear bands** accommodate simple shear with no volume change across them (i.e., isochoric shear; Borja and Aydin, 2004; Aydin et al., 2006). They then would correspond to disaggregation bands formed in weak sediments. From basic plasticity theory, shear bands may correspond to flow of the material along characteristic lines, or *slip lines*, that correspond extremely well with the yielding of soils and other materials (Davis and Selvadurai, 2002, pp. 152–184). Interestingly, the propagation of isochoric shear bands according to slip-line theory would smoothly follow the planes of the bands, rather than propagating abruptly out-of-plane as is common for brittle rocks (i.e., wing-cracks; Segall and Pollard, 1980; Petit and Barquins, 1988). Such in-plane propagation of shearing deformation bands (not just isochoric ones) may be inferred from, for example, the wet plaster experiments reported by Fossen and Gabrielsen (1996); laboratory experiments on sand by Schreurs and Colletta (1998), Saada et al. (1999), McClay et al. (2005), and many others; the development of landslide slip surfaces in soils (Palmer and Rice, 1973; Muller and Martel, 2000); and by field observations of deformation bands in sandstone (Schultz and Balasko, 2003; Okubo and Schultz, 2006). Laboratory experiments by Riedel and Labuz (2007) demonstrate in-plane propagation of a cataclastic deformation band in Berea Sandstone.

Shear bands are very common in many geologic settings, being recognized as linear zones of grain rolling and realignment into the direction of shearing displacement (Antonellini et al., 1994). They have been identified in many localities (Fossen et al., 2007) including the top (uppermost 15 m) of the Navajo Sandstone near Utah's Bryce Canyon National Park (Davis, 1999), in Utah's Arches National Park (Zhao and Johnson, 1991; Antonellini et al., 1994), in sedimentary rocks from coastal California (Cashman and Cashman, 2000), and in Triassic/Jurassic sandstones from the Gullfaks Field in the North Sea (Fossen and Hesthammer, 2000). In all cases, shallow burial of the unit (less than a few hundred meters; e.g., Fossen and Hesthammer, 2000), and by implication low confining pressures (Antonellini et al., 1994) and probably small values of differential stresses, are implicated in the formation and growth of these structures.

By far the majority of deformation bands described in the geological literature are **compactional shear bands** (e.g., Fossen et al., 2007). Of the two main varieties, **cataclastic compactional shear bands** are probably the most widely recognized type of deformation band, and they were among the first to be studied in detail (Engelder, 1974; Aydin, 1978; Aydin and Johnson, 1978; Jamison, 1989). These structures are easy to recognize in the field as resistant tabular surfaces (Fig. 7.3), fins (Fig. 7.4; Davis, 1998), or spatially distributed arrays (Fig. 7.5) projecting from the softer country rock.

Fig. 7.3. Deformation bands crop out as erosionally resistant tabular structures from eolian sandstone near (a) Goblin Valley, Utah, and (b) Buckskin Gulch, Utah. Note modern analog in (a) to paleodunes in foreground.

Fig. 7.4. Fault-fin landscape (Davis, 1998) formed from thick zones of deformation bands in southwest Utah. The thick zones that weather out in positive relief may also display slip surfaces.

Fig. 7.5. Deformation bands from the Buckskin Gulch site in southern Utah define an organized array of bounding and linking bands that have accommodated shear deformation. Photo by Haakon Fossen.

Cataclastic deformation bands form as tabular zones of grain reorganization and crushing along points of grain contact (Aydin, 1978; Zhang et al., 1990). The porous rock is supported by its framework of grains (not the open pore space or matrix minerals), which fracture into progressively smaller fragments as shear strain is accommodated. The pore space collects the fragmented debris (quartz, feldspar, and clays; e.g., Exner and Tschegg, 2012) and allows the process to continue until the pore space is completely filled. Re-sorting of the fine-grained debris further increases its packing, leading to an increasing degree of strain hardening within the band (Rudnicki and Rice, 1975; Aydin and Johnson, 1983; Bésuelle, 2001a,b). Both porosity and permeability are reduced within the deformation bands (e.g., Gibson, 1998; Mair et al., 2000; David et al., 2001), leading to reduced permeability across the band (e.g., Crawford, 1998; Aydin, 2000; Shipton et al., 2005; Fossen and Bale, 2007; Torabi and Fossen, 2009; Ballas et al., 2015; Philit et al., 2017).

This common band type tends to form most readily in well-sorted coarse-grained rock with high porosity, such as paleo-sand dune deposits (i.e., Fig. 7.3), and least well in rocks with very strong grains, poor sorting (Cheung et al., 2012), and/or low porosity (e.g., Dunn et al., 1973; Aydin, 1978; Smith, 1983; Antonellini et al., 1994). Cataclastic deformation bands have even been identified in meteorite impact structures (e.g., Grieve, 1987; Hergarten and Kenkmann, 2015) such as Kentland (Indiana: A. Aydin, pers. comm., 1983) and Upheaval Dome (Utah: Kenkmann, 2003; Okubo and Schultz, 2007; Key and Schultz, 2011).

Non-cataclastic compactional shear deformation bands lack the extensive degree of grain crushing and accumulation of fine debris found in cataclastic deformation bands (Antonellini et al., 1994). Their identification hinges on the recognition of either porosity differences or subtle alignments of grains that produce a preferred orientation, or fabric, within the band. These bands are characterized by reorganization of grains within a loosely packed granular rock to form tabular surfaces having a tighter grain packing, minor grain cracking, and a porosity reduced by several percent from that of the host rock (Antonellini et al., 1994). Lateral offsets along these bands may be small (millimeter to centimeter) and the offsets scale in proportion to the average grain size of the rock (Antonellini et al., 1994), in accord with theory and observation for loosely consolidated granular materials (e.g., Mühlhaus and Vardoulakis, 1988). Confining pressures are thought to be smaller for these bands to form in rocks than for cataclastic bands, again implying relatively shallow depths of burial during their formation.

Shearing deformation bands, including dilational and, especially, compactional shear bands and their cataclastic subtype, provide arguably the major mechanism for fault formation in porous sedimentary rocks that have gained much attention in the study of petroleum reservoirs and groundwater aquifers (e.g., Aydin et al., 2006; Fossen et al., 2007; Holcomb et al., 2007). As a sedimentary sequence becomes strained (i.e., folded, flexed, faulted, extended, or shortened), shear deformation bands can form, increase in size and displacement, and eventually "break loose" to form a true fault, with slip surfaces and associated structures, that can attain regional extent and tectonic importance (Aydin and Johnson, 1978, 1983; Davis, 1999; Shipton and Cowie, 2001, 2003; Soliva et al., 2016; Fossen et al., 2017; Fig. 7.6).

Formation of faults in porous sedimentary rocks

Shear deformation bands represent an **important mechanism of fault formation** in porous granular rocks such as sandstones, carbonates, and nonwelded tuffs (e.g., Wilson et al., 2003; Evans and Bradbury, 2004), as well as some chalks (Wennberg et al., 2013) and siliceous mudstones (Ishii, 2012), but not in compact rocks such as granite, welded tuff, many limestones, and basalt. Whereas faults can channel water, waste, and petroleum through their dilatant damage zones and related structures, fluid flow is often *restricted or baffled* by deformation bands. It is common to mis-identify deformation bands in field exposures as joints or veins, but close inspection with a hand lens or in thin section will reveal the distinctive characteristics that set deformation bands apart from those opening-mode dilatant cracks.

Compaction bands are a type of deformation band characterized by compactional (closing) strains that form in high-porosity sandstone (e.g., Mollema and Antonellini, 1996; Aydin et al., 2006; Fossen et al., 2007; Schultz and Fossen, 2008; Tembe et al., 2008). These thin tabular structures define a geologically interesting class of strain localization in rock that can impede subsurface fluid flow and can be important to ground water and petroleum production (e.g., Holcomb et al., 2007). Compactional strain within compaction bands is typically accommodated by grain reorganization with variable degrees of fragmentation and plastic grain deformation (Mollema and

Fig. 7.6. Imbricate sets of deformation band arrays accommodate contractional strain in the footwall of the East Kaibab monocline, southern Utah; view is toward the southwest from the Buckskin Gulch site (Schultz, 2011a).

Antonellini, 1996; Sternlof et al., 2005; Eichhubl et al., 2010; Schultz et al., 2010b; Fossen et al., 2011).

Compaction bands have been produced in a range of physical experiments on highly porous sandstones (e.g., Olsson, 1999; Baud et al., 2004; DiGiovanni et al., 2007; Townend et al., 2008) and limestones (Baxevanis et al., 2006), and can occur as borehole breakout structures (Haimson, 2001). Laboratory experiments and theoretical work suggest that compaction bands may form under a particular set of evolving host-rock properties, stress states, and loading paths (e.g., Olsson, 1999, 2000; Issen and Rudnicki, 2000; Bésuelle and Rudnicki, 2004; Chemenda, 2009, 2011; Nguyen et al., 2016), and field relations demonstrate that compaction bands can be restricted to certain layers within the same stratigraphic unit (e.g., Mollema and Antonellini, 1996; Eichhubl et al., 2010; Schultz et al., 2010b; Fossen et al., 2011; Soliva et al., 2013).

Two main types of compaction bands are recognized from field observations (Fig. 7.7). **Shear-enhanced compaction bands** were defined by Eichhubl et al. (2010) as deformation bands with approximately equal amounts of shear and compaction accommodated across them. Called *thick compaction bands* by Mollema and Antonellini (1996) to distinguish them from pure compaction bands, shear-enhanced compaction bands are characteristically planar to gently curving, tabular zones of variable thickness (ranging in thickness from millimeters to several centimeters) and length (ranging from centimeters to several meters; Sternlof et al., 2005; Schultz, 2009). Grain crushing is commonly observed in these bands (Fossen et al., 2007; Holcomb et al., 2007). Shear-enhanced compaction bands commonly form two oppositely dipping sets that mutually crosscut each other and therefore define conjugate sets (e.g., Solum et al., 2010; Fossen et al., 2011; Soliva et al., 2013). The angle between conjugate sets is characteristically larger than that for sharp structural discontinuities that have accommodated frictional sliding in the same rock type (i.e., faults and slip surfaces; e.g.,

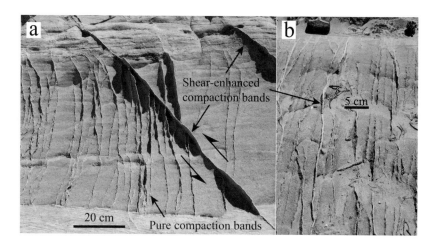

Fig. 7.7. Outcrop expression of both recognized types of compaction bands in (a) cross-section and (b) map view, both from outcrops at the Buckskin Gulch site, southern Utah. Note distinct morphology, attitude, and scaling of shear-enhanced and pure compaction bands.

Friedman and Logan, 1973; Olsson, 2000; Eichhubl et al., 2010; Fossen et al., 2011), as discussed below, implying a difference in formation mechanism between shear-enhanced compaction bands and faults (e.g., Baud et al., 2006; Eichhubl et al., 2010). Shear-enhanced compaction bands are sometimes referred to simply as "compaction bands" in the literature (e.g., Sternlof et al., 2005; Aydin and Ahmadov, 2009; Meng and Pollard, 2014; Lyakhovsky et al., 2015).

Pure compaction bands, called crooked compaction bands by Mollema and Antonellini (1996), are recognized from several field localities from around the world, including Valley of Fire (Nevada), Buckskin Gulch (Utah; Fig. 7.7), and multiple sites in southern France. Typically forming only in well-sorted layers having large porosities (e.g., >20%), this band type has a characteristic wavy or crooked geometry (Figs. 1.10, 7.7), which may imply formation by collapse of **force chains** (Peters et al., 2005; Andrade et al., 2011) and coalescence of the pore collapse clusters (Fortin et al., 2009; Stanchits et al., 2009). Pure compaction bands show no detectable shear offset of sedimentary lamination or structures that they cross, and their irregular wiggly or angular geometries are kinematically incompatible with significant shear displacement at scales much larger than that of the individual grains (Mollema and Antonellini, 1996; Fossen et al., 2011). In thin section, pure compaction bands show a marked, although variable, reduction in porosity along them that involves some grain fracture and dissolution at grain contact points. Most pure compaction bands range from a few centimeters to a few decimeters in length and are stratigraphically restricted to the most porous, and presumably most compliant, layers (e.g., Schultz et al., 2010b; Fossen et al., 2011; Fig. 7.7a).

Compaction bands of both types are documented from an increasing number of sites from around the world. In addition to what is probably the best-known site, in Valley of Fire State Park, southern Nevada (Hill, 1989; Sternlof et al., 2005; Eichhubl et al., 2010), these structures are also known from Buckskin Gulch, southern Utah (Mollema and Antonellini, 1996; Schultz, 2009, 2011a,b; Schultz et al., 2010b; Solum et al., 2010; Fossen et al., 2011), and at several sites in southern France including Boncavaï, Montmount, and Tresques (Ballas et al., 2013; Soliva et al., 2013). Compaction bands that are subparallel to bedding have been recognized and documented from the Valley of Fire by Aydin and Ahmadov (2009), Deng and Aydin (2012, 2015), and Deng et al. (2015).

The thicknesses of shear-enhanced compaction bands in porous sandstones scale with band length (Rudnicki, 2007; Schultz et al., 2008a; Tembe et al., 2008; Schultz, 2009). In contrast, the thicknesses and compactional displacements of pure compaction bands appear to be relatively constant for different lengths (e.g. Fossen et al. 2011), implying that the latter structures do not follow displacement–length scaling relations as they grow (Schultz et al., 2013).

7.3 Deformation Bands as Structural Discontinuities in Rock

Deformation bands share many characteristics with other types of geologic structural discontinuities such as joints and faults. These commonalities underscore the theme that fractures and deformation bands both represent types of structural discontinuities in rocks (Chapter 1), with their differences pinpointing key aspects of the deformation process.

- **Shape.** Deformation bands are narrow (tabular) zones of localized volumetric and/or shearing deformation in the rock mass. The strain is localized into discrete structures that may grow in size with increasing strain, rather than becoming distributed throughout the deforming unit. As a result, deformation bands function as "flaws" or discontinuities in the mechanical and hydraulic properties of the unit.

- **Displacement sense.** Passive markers in the host rock, such as bedding, minerals, or pre-existing structural elements, demonstrate the sense of displacement across deformation bands, just as they can with geologic fractures (e.g., joints and faults). Deformation bands accommodate some combination of opening (more precisely, dilation), shearing, or closing modes of displacement. Cataclastic compactional shear bands, the most common type, accommodate mostly shear offsets with smaller amounts of closing offsets, whereas shear-enhanced compaction bands accommodate subequal amounts of shearing and closing offsets.

- **Displacement distribution.** The amount of offset (dilation, closing, and/or shear) accommodated along a deformation band varies systematically with position along it, with minimum values (nearly zero) at the band tips and maximum values near their midpoints (Antonellini et al., 1994; Fossen and Hesthammer, 1997; Schultz, 2009). Displacement maxima are commonly observed to be shifted toward stepovers between closely spaced echelon bands, just like we see along dilatant cracks and faults, implying stress and displacement transfer and mechanical interaction between them.

- **Linkage and mechanical interaction.** The field expression of deformation bands is replete with clear examples of discontinuous bands (Figs. 1.10, 8.1; e.g., Fossen and Hesthammer, 1997; Davis, 1999; Davis et al., 2000; Nicol et al., 2013). Where these bands impinge on one another, they adopt the many characteristic forms of echelon discontinuity geometry seen in other discontinuity types, such as stepovers, linkage, and curvature (e.g., Cruikshank et al., 1991b; Antonellini et al., 1994; Fossen and Hesthammer, 1997; Okubo and Schultz, 2006; Schultz, 2009). Linked bands can accommodate greater displacement magnitudes than smaller bands, reflecting a typical "discontinuity-like style" of growth into larger-strain sets.

- **Spacing.** A wide variety of spacings can be documented for deformation bands, although the controls on spacing are not well understood. Many bands tend to localize adjacent to one another, in contrast to the patterns of dilatant cracks and faults, forming clusters, zones, and lenses (or lozenges) of deformation bands (Aydin and Johnson, 1978; Shipton and Cowie, 2001, 2003; Fossen et al., 2007; Awdal et al., 2014; Philit et al., 2018; Fig. 7.14). These zones accommodate larger offsets and strains than the individual bands. Larger numbers of bands can accommodate larger strains, and spacings have been related to non-uniform bending, proximity to faults, and other producers of strain gradients (Jamison and Stearns, 1982; Jamison, 1989; Antonellini et al., 1994).

- **Displacement–length scaling.** Except for pure compaction bands, deformation bands demonstrate a proportionality between the maximum displacement and band length. For example, cataclastic compactional shear bands, the first band type investigated, show displacement–length scaling with a slope of $n = 0.5$ (Fossen and Hesthammer, 1997; Schultz et al., 2008a, 2013). Shear-enhanced compaction bands also show the same scaling, with $n = 0.5$, as do opening-mode fractures such as joints, veins, and igneous dikes. On the other hand, shear bands scale with $n = 1.0$, as do faults (e.g., Cowie and Scholz, 1992a; Scholz, 2002, p. 117). The scaling exponent has been associated with the mechanics of fracture and deformation band propagation (e.g., Cowie and Scholz, 1992b; Fossen and Hesthammer, 1997; Olson, 2003; Schultz et al., 2008a, 2013; Schultz and Soliva, 2012) and remains an active area of current study.

- **Growth into larger-strain arrays.** Deformation bands are a progressive deformation—their morphology and properties change systematically with ongoing deformation and strain of the host rock. Throughout their development, deformation bands remain a type of *strain localization* which collects and focuses it along discrete tabular zones, leaving the porous host rock remarkably untouched. The accumulation of offsets, and attendant focusing of strain, along deformation bands may be attributed to a strain-softening behavior during their growth (e.g., reduction of stiffness during grain fragmentation and reorganization). Strain hardening within deformation bands (related to grain lock-up and cessation of displacement accumulation within a band; Aydin, 1978) and within the band network (Schultz and Balasko, 2003) has been inferred to lead to a spreading of the deformation outward into the host rock.

7.3.1 Orientations of Deformation Bands

The orientations of geologic structural discontinuities are organized by stress state and material properties (such as frictional resistance for faults). As noted in any structural geology or rock mechanics text, joints open perpendicular to the least principal stress and strike parallel to the greatest principal stress (e.g., σ_1). Faults are commonly oriented at an angle to σ_1 that is related to the frictional resistance to sliding along the surface or zone. For example, the optimum angle for frictional sliding is given by

$$\theta_{\text{opt}} = 45° + \frac{\phi_{\text{f}}}{2} \tag{7.1}$$

where ϕ_{f} is the angle of friction of the sliding surface or zone (see Chapter 3, equation (3.13a)). For friction coefficients of $\mu = 0.6$, $\phi_{\text{f}} \sim 30°$ and a fault is oriented at about 30° to σ_1. For conjugate sets of shearing surfaces, this angle is one-half of the dihedral angle θ_{d} measured between the opposing sets of surfaces. Anticracks are oriented approximately perpendicular to σ_1, which for these structures represents a macroscopic average orientation of the smaller, grain-scale irregularities.

The orientations of deformation bands are similarly related to the local stress state and material properties, but for these tabular structures the simple Coulomb criterion does not adequately predict their orientations. The orientations of the non-shearing bands, namely dilation bands and pure compaction bands, appear to be adequately predicted in field examples by the orientation of σ_1, similar to joints and stylolites. Compactional shear bands and shear-enhanced compaction bands, however, form at angles that are not adequately predicted by the Coulomb criterion. This discrepancy arises because the Coulomb criterion does not consider volumetric changes (dilation or contraction) across a shearing surface (e.g., Davis and Selvadurai, 2002; Zhao and Cai, 2010).

Dilation angle

The volumetric changes across a shearing zone can be included by using a parameter called **dilation angle** ψ (e.g., Hansen, 1958; Roscoe, 1970; Johnson, 1995). The dilation angle is the angle of motion of the sliding blocks relative to the plane of failure (Arthur et al., 1977; Bolton, 1986) and represents the ratio of plastic volume change to plastic shear strain (Vermeer and De Borst, 1984). It relates shear band formation to volumetric changes within the deformed material, where a positive ψ represents volumetric increases and a negative ψ represents a volume loss (see also Davis and Selvadurai, 2002; Paterson and Wong, 2005). Positive dilation angles occur for rocks such as granite, other compact rocks, and dense soils that expand volumetrically during shear deformation; negative values of dilation angle are found for rocks such as sandstone and loose soils that contract during shear deformation. For dense granular materials, which tend to dilate when they are sheared (Reynolds, 1885), the Roscoe relationship is found to yield good results in coarse-grained rock (Vermeer, 1990) and for shear-enhanced compaction bands (Klimczak et al., 2011). This parameter has found wide application in soil mechanics and geotechnical engineering, where accurate representation of soils as they dilate or compact in response to shear deformation is central to the design, stability, and failure prediction of foundations, earthen dams, bearing capacity, landslides, and related structures (e.g., Davis and Selvadurai, 2002; Schofield, 2005).

Vermeer and de Borst (1984) developed a useful criterion to estimate the orientation of a shear band in soil that takes into account both the friction and

Orientation of the optimum deformation band using friction and dilation angles

dilation angles. The predicted shear band would be oriented to σ_1 by (see also Klimczak and Schultz, 2013a)

$$\theta_{opt} = 45° + \frac{\phi_f}{2} \qquad (7.2)$$

which reduces to

Orientation of the optimum deformation band for associated plasticity

$$\theta_{opt} = 45° - \frac{\phi_f + \psi}{4} \qquad (7.3)$$

for comparable values of friction and dilation angles. If the dilation angle is equal to zero, then there is no appreciable volumetric change across the shearing surface or zone (i.e., no significant dilation or compaction) and (7.2) reduces to the optimum angle for frictional sliding in a perfectly plastic material as given by the standard Coulomb criterion (7.1). The case of **associated** plasticity flow laws, used in soil mechanics and geotechnical engineering and some geomechanical applications such as deformation of chalk (e.g., Gutierrez and Homand, 1998; Hickman and Gutierrez, 2007), assumes that $\phi_f = \psi$. But as emphasized by Vermeer and de Borst (1984) and many others, **non-associated** plasticity does a better job of describing deformation of porous granular aggregates including the localization and growth of deformation bands (e.g., Borja, 2002; Xie and Shao, 2006).

When comparing both parameters, the difference between ψ and ϕ_f, not their measured magnitudes taken separately, has been used to determine whether a fault forming in sedimentary rock would be dilating or compacting (Zhao and Cai, 2010; Klimczak and Schultz, 2013a). For example, a material having $\phi_f = 30°$ and $\psi = 35°$ is dilational, whereas a material having $\phi_f = 30°$ and $\psi = -35°$ is compactional. This relationship reduces to the well-known expression for optimum dip angle when both angles are equal (i.e., $\psi/\phi_f = 1.0$, or the difference $\phi_f - |\psi| = 0$; equation (7.1)). Several authors including Johnson (1995), Zhao and Cai (2010), and Shin et al. (2010) discuss the role and implications of dilation angle for faults.

Measurements of the orientations of several examples of deformation bands as quantified by the dihedral angle between conjugate band sets θ_d, from laboratory experiments and from the field, reported and tabulated by Klimczak and Schultz (2013a) and Fossen et al. (2017), are plotted in Fig. 7.8. As is well known (e.g., Friedman and Logan, 1973; Olsson, 2000; Bésuelle, 2001a; Aydin et al., 2006; Fossen et al., 2007, 2017; Eichhubl et al., 2010; Chemenda et al., 2014), the orientations of deformation bands relative to the orientation of the maximum principal compressive stress σ_1 change systematically with the sense and degree of volumetric strain accommodated across them. Dilation bands are oriented at small angles to σ_1, with shear bands, compactional shear-enhanced compaction bands, and finally pure compaction bands being oriented at successively larger angles to σ_1. An example of conjugate shear-enhanced compaction bands is shown in Fig. 7.9. As shown by Klimczak and Schultz (2013a) and others, this variation in orientation may be related to the dilation angle, with band dilation during shear contributing to a smaller angle to σ_1 and compaction during shearing in the band leading to larger angles to σ_1. In detail, the band angles relative to the stress state are governed by a considerably more complex interplay between stress state and host-rock and band properties (e.g., Rudnicki and Rice, 1975; Bésuelle and Rudnicki, 2004; Borja and Aydin, 2004; Chemenda, 2009) although the use of friction and dilation angles continues to be widely and successfully used in practical applications of soil mechanics and geotechnical engineering.

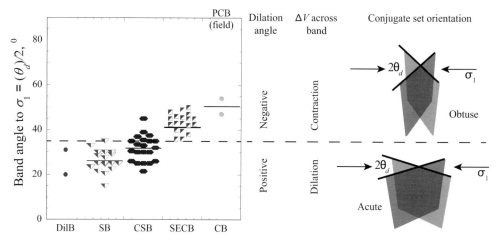

Fig. 7.8. Plot of relative angle between direction of maximum compressive principal stress and deformation bands; data from Klimczak and Schultz (2013a) and Fossen et al. (2017). Average values for band orientations of a given type shown by solid horizontal lines; approximate change related to dilation angle from dilatant to compactive shear (Klimczak and Schultz, 2013a) shown by dashed line. DilB, dilation band; SB, shear band; CSB, compactional shear band; SECB, shear-enhanced compaction band; CB, compaction band; PCB, pure compaction band.

Fig. 7.9. Conjugate sets of shear-enhanced compaction bands, Buckskin Gulch site, Utah, shown in both foreground and at distance; backpack gives scale.

The angular relationships between conjugate sets of shearing deformation bands, depicted in Fig. 7.8, show that the dihedral angles between the conjugate sets increase as the degree of compaction across the bands increases. The angles between each set and the maximum compressive principal stress σ_1 (plotted in Fig. 7.8, and one-half the dihedral angle between the conjugate sets) range from approximately 20–40° for shear bands and compactional (and cataclastic) shear bands, and at least 80–100° for shear-enhanced compaction bands. Displaying these angular ranges on a Mohr diagram (Fig. 7.10), shear bands and compactional (shear) deformation bands define an *acute angle* between their conjugate sets and the direction of σ_1, implying, for a given value of friction coefficient or angle, that they are oriented *favorably* to slip frictionally and to accumulate larger shear displacements. Conversely, shear-enhanced compaction bands display *obtuse angles* between their conjugate sets, implying *unfavorable* orientations for frictional sliding. These relationships are in accord with results from the field and the laboratory in which shearing deformation bands may be accompanied by, or serve as the substrates for, faults of comparable orientations, whereas shear-enhanced compaction bands generally are not reactivated to form faults (e.g., Soliva et al., 2013).

Acute angles: Favorable for later slip
Obtuse angles: Frictionally locked

Deformation bands can range in orientation to σ_1 (Fig. 7.11) from subparallel (dilation bands), to acute angles that are favorably oriented for shearing (shear bands, compactional/cataclastic shear bands), through obtuse angles (shear-enhanced compaction bands) to sub-perpendicular (pure compaction bands). Deformation bands thus tend to display a wider angular range within the causative stress field than do fractures (joints and faults) and stylolites.

7.3.2 Displacements Accommodated by Deformation Bands

Deformation bands are recognized in the field (and at thin-section scales and smaller) by interruptions of passive markers on either side of a band. Passive markers such as individual grains, beds, and pre-existing deformation bands are offset continuously across a band so that, in many cases, the marker can

Fig. 7.10. Schematic Mohr diagram showing approximate ranges of dihedral angle between conjugate sets of shearing deformation bands. Structures such as compactional/cataclastic shear deformation bands, and also faults, can exhibit dihedral angles between 40° and 80° (140° and 110° on the diagram). Shear-enhanced compaction bands can exhibit dihedral angles between 80° and 100° (110° and 80° on the diagram).

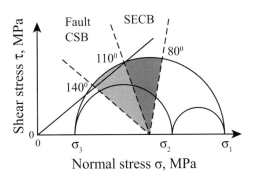

Fig. 7.11. Schematic illustration showing typical angles of deformation bands to the direction of maximum compressive principal stress, after Klimczak and Schultz (2013a). In the figure, DilB, dilation bands; CSB, compactional shear bands; SECB, shear-enhanced compaction bands; PCB, pure compaction band.

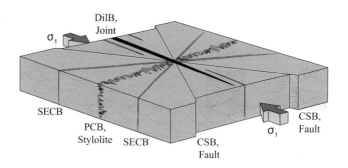

be traced through the cross-cutting deformation band. This characteristic distinguishes deformation bands from fractures which are characterized by discontinuous displacement of pre-existing markers across them (Chapter 1).

Both normal and shear displacements can be observed across deformation bands. By far the most common sense of normal displacement across deformation bands is **compaction** (e.g., Aydin et al., 2006; Fossen et al., 2007, 2017). The normal displacement due to compaction across a deformation band can be estimated by measuring the mechanical reduction in pore volume between the host rock and the band (e.g., Sternlof et al., 2005; Schultz and Soliva, 2012; Soliva et al., 2013). In cases involving diagenesis of host rock and/or bands, the degree of syn- or post-kinematic cementation must also be taken into account to assess the degree of chemical porosity reduction due to compactional rearrangement of the grains (e.g., Elliott et al., 2014). Neglecting diagenetic effects, the compactional displacement increases with the difference in porosity between band and host rock and with the band thickness.

Shear displacements across deformation bands are widely observed and must be distinguished from apparent shear offsets caused by normal (compactional) offsets of markers across a band (Soliva et al., 2013). The magnitude of the latter can be obtained from trigonometry. Shearing within deformation bands can progressively fragment the initial grains, changing the size-frequency and sorting relations within the band and producing various degrees of crushing and cataclasis.

Normal (compactional) and shearing displacements (corrected for apparent offsets) have been measured for deformation bands of several types and from a variety of locations, host rocks, and tectonic settings from around the world. Compilations of these data by Soliva et al. (2013) and Fossen et al. (2017) motivate some useful and intriguing generalizations. A log–log plot of compactional vs. shear displacements by Soliva et al. (2013) shows that ratios of shear to compaction are largest for shear bands, then decrease for compactional shear bands, and are smallest for shear-enhanced compaction bands. Eichhubl et al. (2010) first noted sub-equal magnitudes of compaction and shearing for this band type with larger ratios of shear to compaction being characteristic of compactional shear bands. Increasing ratios of shear/compaction that correlate with band type were suggested to be consistent with increasing differential stress, decreasing mean stress, smaller dihedral angles between conjugate band sets, increased grain fragmentation, and perhaps a change in **tectonic regime** from contractional to extensional (Soliva et al., 2013). The importance of tectonic regime (essentially as a proxy for the subsurface stress state during band growth) on deformation band types and patterns was also drawn by Solum et al. (2010), Klimczak et al. (2011), Fossen et al. (2011), and Ballas et al. (2014).

The compilation by Fossen et al. (2017) shows how shearing and compactional displacement magnitudes are related as a function of band type and tectonic regime (Fig. 7.12). Cataclastic compactional shear bands display ratios of shear to compactional displacements that generally exceed about 100; shear-enhanced compaction bands show sub-equal ratios. Compactional shear bands and band clusters show intermediate ratios that exceed at least four. In general, compactional shear bands accommodate considerably more shearing than compaction, likely related to lesser values of compliance along the bands (in the shearing direction) than across them during displacement accumulation.

7.3.3 Spatial Organization of Deformation Bands

A rich literature exists that documents and analyzes the spatial organization of many types of deformation bands. Individual compactional shear bands having a range of lengths and spacings (e.g., Davis, 1999) are observed

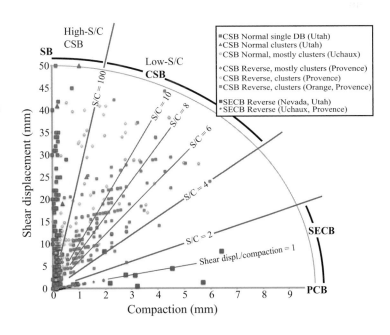

Fig. 7.12. Diagram comparing magnitudes of normal displacement (compaction) and shear displacement for a variety of deformation band types and settings (after Fossen et al., 2017). Shear-enhanced compaction bands (SECB) plot in the lower part of the diagram, close to the $S/C = 1$ line where the contribution of shear and compaction are of comparable magnitude. Cataclastic compactional shear bands (CSB) plot in the upper part. SB, shear bands; CSB, compactional shear bands; SECB, shear-enhanced compaction bands; PCB, pure compaction bands.

to interact with their neighbors, forming geometries typical of many other geologic discontinuities including echelon (soft-linked) and linked (e.g., Fossen and Hesthammer, 1997) stepovers. Geometries of closely spaced deformation bands have been described as *inosculating* (Aydin, 1978) when viewed down the shearing direction; an example of this geometry is shown in Fig. 7.13. Map-view exposures of deformation bands having either normal or reverse senses of shear displacement commonly display these relationships.

The term "inosculating bands" describes adjacent structures that are collinear and coplanar. As discussed below, this viewing geometry was used by Aydin and Johnson (1978, 1983) to formulate their model of deformation band growth, in which strain hardening within one band leads to a cessation of shear displacement accumulation along it and a jump in new band formation at some distance into the adjacent undeformed host rock. Continuation of

Fig. 7.13. Lenses (or lozenges) form as stepovers between deformation bands as viewed looking down the shearing direction (indicated by pencil). In this view the bands have a mode-III sense of shearing with a normal sense of offset. (a) Partially formed lenses between individual bands. (b) Lenses forming between zones of bands. The lens segments are surrounded by and enclose less-deformed host rock.

Fig. 7.14. Zones (or clusters) of compactional shear deformation bands exposed in three dimensions near Goblin Valley, Utah, looking across the normal-sense shearing (and slip) direction. Note closely spaced inosculating bands and the lozenges between them. Photo by Sven Philit.

Zones and clusters

this process would produce a thick, closely spaced set of bands, called a **zone** (Aydin and Johnson, 1978; Davis et al., 2000; Figs. 7.14 and 7.4) or **cluster** (Soliva et al., 2016; Philit et al., 2017, 2018), that accommodates an amount of shear offset subequal to its thickness. Lastly a slip surface nucleates at an edge of the zone, leading to fault localization there. Slip surfaces located on or within deformation-band zones (or deformation band damage zones around faults) have been identified and analyzed by many including Davis (1998), Shipton and Cowie (2001, 2003), Wibberley et al. (2007), Johansen and Fossen (2008), and Awdal et al. (2014).

Lozenges and lenses

Awdal et al. (2014) adopted the term "lozenge" for lenses of rock bounded by deformation bands, reserving lenses for those bounded by faults and slip surfaces (following Candela and Renard, 2012). Zones can be observed to be composed of numerous closely spaced lozenges (Aydin and Johnson, 1978; Davis et al., 2000; Fig. 7.14). Several lozenges in various stages of development are visible in Figs. 7.13 and 7.14. The terms lozenges and lenses will be used interchangeably in this book (despite some differences in detail, such as overlap-separation ratios; Awdahl et al., 2014) in order to emphasize their commonalities in formation mechanism (i.e., deformation band or fault interaction).

Deformation bands can occur in a variety of geometrical arrangements. In many areas, individual deformation bands occur in sub-parallel arrangements (Fig. 7.3) separated by a range of spacings (e.g., Aydin and Johnson, 1978; Fossen and Hesthammer, 1997; Davis, 1999; Davis et al., 2000). Systematic relationships can sometimes be identified that suggest growth of deformation band arrays in dimension, displacement magnitude, spacing, scaling, and relative timing analogous to joint and fault arrays (e.g., Aydin and Johnson, 1978; Fossen and Hesthammer, 1997; Schultz and Fossen, 2002; Fossen et al., 2007; Schultz et al., 2008a, 2013).

Deformation bands are known to define zones resembling Riedel or overstepping, duplex-like arrangements (e.g., Fig. 1.5). These will be discussed in more detail below.

Conjugate sets of shear-enhanced compaction bands in porous sandstone, exposed in three dimensions, were shown in Fig. 7.9. A wonderful large-scale example of deformation bands defining conjugate sets displayed in map view is shown in Fig. 7.15. These structures formed in soil in association with a large M_w 7.1 earthquake in New Zealand in 2010 (Barrell et al., 2011; Quigley et al., 2012). Two conjugate sets of deformation bands can be seen in Fig. 7.15a, bisected by the inferred direction of the maximum horizontal compressive principal stress, giving a right-lateral sense of shear to the resulting zone (Fig. 7.15a) and a contractional sense of strain within the stepovers between the right-lateral, left-stepping echelon segments (Fig. 7.15b). Barrell et al. (2011) discuss the humorous and telling graffiti produced by a local farmer in Fig. 7.15b.

Fig. 7.15. Riedel and echelon patterns of shear deformation bands formed during an earthquake rupture in New Zealand, after Barrell et al. (2011).

Clusters

Clustering of deformation bands is commonly observed, especially in extensional tectonic environments. In general, spacings that are not uniform imply some degree of clustering even in statistically random systems, so formal assessments of departures from randomness are necessary to interpret the physical significance of clusters (e.g., Kamb, 1959; Ragan, 2009, pp. 478–480; Marrett et al., 2018). Statistically significant clusters of deformation bands may not be clearly associated with larger-displacement faults (e.g., Johnson, 1995, 2001; Fossen and Hesthammer, 1997; Davis, 1999) and might then be related to other processes, such as layer property contrasts of an extending sedimentary stratigraphic section (e.g., Chemenda et al., 2014), similar conceptually to boudinage but at a larger scale.

More commonly, (statistically significant) clusters of deformation bands are associated with faults in a clear deformation sequence (Aydin and Johnson, 1978, 1983; Philit et al., 2018). These precursory **deformation-band damage zones** contain a record of strain accumulation and progressively larger degrees of host-rock deformation within a tabular volume that eventually encloses a fault (e.g., Shipton and Cowie, 2001, 2003; Wibberley et al., 2007; Johansen and Fossen, 2008; Schueller et al., 2013). Indeed, the classic evolutionary model of individual deformation bands to zones to slip surfaces to faults developed by Aydin and Johnson (1978), and discussed below, outlines the sequential formation of a precursory deformation-band damage zone around a normal fault.

Damage zones

As noted previously, certain types of deformation bands can be associated with a particular set of environmental conditions (beyond lithology and host-rock properties), such as stress state and confining pressure. Interestingly, the patterns and spacings of deformation bands may also be related in a general or macroscopic way to extensional or contractional tectonic regimes. For example, to date shear-enhanced compaction bands have primarily been reported from contractional tectonic regimes (e.g., Eichhubl et al., 2010; Schultz et al., 2010a,b,c; Solum et al., 2010; Fossen et al., 2011, 2015, 2017; Klimczak et al., 2011; Schultz, 2011a,b; Ballas et al., 2013, 2014; Saillet and Wibberley, 2013; Soliva et al., 2013). Deformation bands in extensional settings tend to be concentrated into zones or clusters in the vicinity of faults, separated by regions of less deformed host rock (i.e., with fewer deformation bands per length, area, or volume). In contrast, deformation bands in contractional settings tend not to be strongly clustered but define spatially distributed arrays or networks (Fig. 7.16; see also Figs. 1.11, 7.1, and 7.7). It appears that stress state influences not only the types of deformation bands, as is widely appreciated (e.g., Aydin et al., 2006), but also to a large degree their interactions and larger-scale patterns

Relationship to tectonic regime

Fig. 7.16. Examples of distributed shear-enhanced compaction bands from contractional tectonic settings. Note oblique angles between bands and the horizontal plane. Exposure from Orange quarry, southern France.

Orthorhombic Sets

and distribution. Some of the basic observations and ideas that suggest a tectonic control on band type and pattern are collected in Fig. 7.17.

Superimposed on these basic arrangements (clustered or distributed) are patterns related to larger and nearby structures, such as faults, monoclines, and various types of folds as well as the dimensionality or symmetry of the imposed deformation state (Davis, 1999; Fossen et al., 2017). Similar to the description in Chapter 6 for faults, deformation bands can form in conjugate or **orthorhombic sets** (e.g., Aydin and Reches, 1982; Davis et al., 2000; Shipton and Cowie, 2001; Fig. 7.18). Contrasts in layer properties during deformation of porous rocks can also affect band patterns and spacings (Chemenda et al., 2014). Deformation bands can thus provide detailed records of the evolution of the imposed stress state within structures such as fault relay zones (e.g., Johansen and Fossen, 2008) and in stratigraphic sequences at a variety of spatial scales.

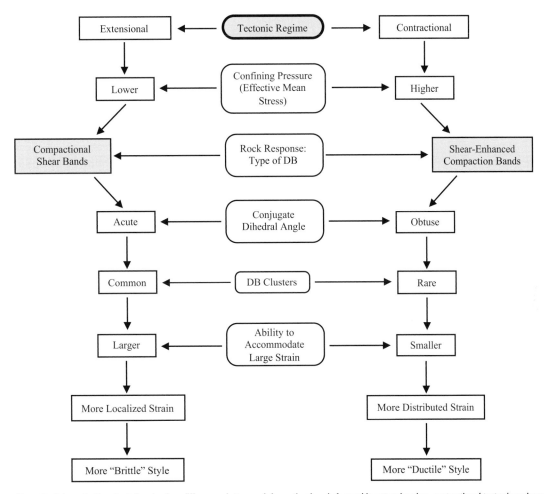

Fig. 7.17. Schematic flowchart showing key differences between deformation bands formed in extensional vs. contractional tectonic regimes.

Fig. 7.18. Shear-enhanced compaction bands at the Buckskin Gulch, Utah, site displaying an approximate three-dimensional or orthorhombic outcrop pattern.

7.4 Deformation of a Porous Granular Rock

Deformation bands form differently than faults growing within crystalline (or compact) rocks due to the petrophysical characteristics of the typical host rocks (e.g., porosity, grain size, sorting). There is a rich diversity of evolutionary paths from deformation band to fault, depending on the initial type(s) of band and how it develops as the stress state and host-rock properties evolve during continued deformation. Accumulation of shear offsets along many types of deformation bands may be followed by strain hardening and frictional lock-up, leading to a spreading of deformation into the host rock. Some bands may later fail and then provide precursory substrates for faults, which serve as "planes of weakness" that slip frictionally according to a Coulomb-type friction law. The types and evolution of deformation bands hinge on how the porous, granular host rock deforms. As we'll see, this behavior has much in common with **soil mechanics**, elements of which that are relevant to deformation bands will be touched on in this section.

7.4.1 Shearing and Volume Change of Loose and Dense Granular Rocks

A porous granular rock is a weakly cohesive open framework that has grains in contact with each other surrounded by open spaces (the porosity). Cementation if present is relatively minor, so that the porosity appropriate to deformation band growth commonly exceeds 5–10%. A weakly cohesive rock has grain-to-grain contacts that permit grain slip and rotation under small applied loads, which then can undergo volumetric changes (dilation and compaction), given sufficient pore volume, during shearing deformation (e.g., Paterson and Wong, 2005, pp. 134–136). This response contrasts with those of a strongly cohesive porous rock (having stronger grain-to-grain contacts and/or significant diagenetic cementation within the former pore volume), or a compact rock (having little to no porosity) in which the nucleation and propagation of cracks and flaws dominates the deformation style.

A: "Loose" granular rock
Higher porosity
Coarse grained
Well sorted
Distributed DBs

B: "Dense" granular rock
Lower porosity
Fine grained
Poorly sorted
Localized DBs

The deformation of the latter two rock types is treated exhaustively in classical rock mechanics (e.g., see overview in Chapter 2). For brevity, weakly cohesive porous granular rocks will be referred to in the rest of this chapter as porous granular rocks.

Drawing from the literature of soil mechanics and geotechnical engineering, porous granular rocks can be divided into two basic types (Fig. 7.19, top): "loose" or "dense" (Desai and Siriwardane, 1984, p. 283; see also Johnson, 2001).

Fig. 7.19. Deformation of porous granular rocks. (a) Stress-strain curves. (b) Volumetric strain curves.

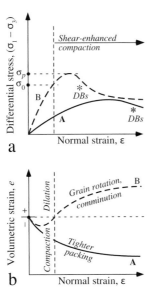

a

b

- A **"loose"** granular rock is coarse grained, well sorted (i.e., uniform grain size), high porosity (>15% or so), with grains that do not interlock but can move freely past each other during deformation. This is referred to as a **"normally consolidated"** soil.
- A **"dense"** granular rock is poorly sorted (having a wide range of grain sizes), of moderate porosity (> 5% or so), with tighter packing that causes grains to move over or through adjacent ones during deformation. This is called an **"over-consolidated"** soil.

These granular-rock types are important because they largely determine the types and major characteristics of deformation bands that can form within them as the rocks are strained. They are also useful to help understand the circumstances under which other porous granular rock lithologies—such as limestone, chalk, or tuff—can form deformation bands.

Now let's examine what happens when a granular rock is deformed under compression and shear. A *loose granular rock* (solid curves "A" in Fig. 7.19) will exhibit a gently rising, strain-hardening curve on a stress-strain diagram (Fig. 7.19a). Deformation is associated with a reduction in pore volume due to closer packing of the grains (Fig. 7.19b), called **compaction** in the soil mechanics literature (Das, 1983, p. 36). The volumetric strain increases with axial strain to larger values of compaction and volume reduction (Figs. 7.19b and 7.20). As the loose granular material strain-hardens, it deforms along distributed surfaces of grain rotation, instead of localization of just a few main crack-like surfaces (e.g., Ménendez et al., 1996). Although a peak strength may not be revealed in a typical test (e.g., Desai and Siriwardane, 1984,

p. 283), it can become apparent at larger strains (>20%, Antonellini et al., 1994). Once the peak strength is attained, however, localization of strain onto fewer discrete surfaces can occur (Antonellini et al., 1994).

The characteristics exhibited by a *dense granular rock* as it deforms are somewhat different. As stress is increased, the dense granular rock (dashed curves "B" in Fig. 7.19) also begins to strain-harden and attain a peak strength, but the peak strength is larger than that for a loose granular rock (e.g., Dunn et al., 1973; Antonellini and Pollard, 1995) and it occurs at much smaller values of axial strain—typically < 10 percent (Dunn et al., 1973; Mair et al., 2000). The changes in volumetric strain are interesting for the dense granular rock (Fig. 7.19b). As stress is applied and the rock deforms, there is an initial volume decrease associated with tighter packing of the grains—just as for a loose granular rock. Once strain hardening sets in, though (dashed line at σ_0 in Figs. 7.19 and 7.20), the dense granular rock undergoes a volume increase and overall *dilatancy* of the rock. This dilatancy continues past the peak strength and continues as strain is localized in the post-peak region (asterisk in Fig. 7.19a). The cause of this dilatancy is attributed to the movement of closely packed grains (Rowe, 1962). At low differential stress levels (to the left of the dashed line in Fig. 7.19), the sample compacts to smaller values of porosity by pore space reduction and grain rolling (Desai and Siriwardane, 1984, pp. 283–285; Antonellini et al., 1994; Fig. 7.20). At higher differential stresses (to the right of the dashed line in Fig. 7.18 and to the right in Fig. 7.19), grains can move past each other only by climbing up over adjacent grains, leading to: (a) **localized increases in porosity** and (b) **increased contact stresses on opposing grains** (Antonellini et al., 1994).

The role of **confining pressure**—or equivalently, depth—is two-fold. First, confining pressure squeezes the rock in all directions, making it more difficult for dilatant behavior (i.e., due to grain motion) to occur. This will tend to stiffen a dense granular rock as it strain hardens into the peak strength region. Second, a larger confining pressure will promote grain cracking by increasing the ambient stress level within the grains. If the confining pressure is increased sufficiently, the free movement of the rolling grains is inhibited, leading to amplified compressive stresses at grain-to-grain contacts and, potentially, cracking of the grains (Gallagher et al., 1974; Aydin, 1978; Zhang et al., 1990; Fig. 7.20). Grain cracking will become pervasive throughout the rock, leading to grain-size reduction and compactional strain. Both effects tend to inhibit shear localization and promote distributed cataclastic flow and macroscopically "ductile" deformation of the rock (Ménendez et al., 1996). Larger values of initial rock porosity have the same effect as increasing the confining pressure on a particular rock—to inhibit strain localization (e.g., Handin et al., 1963; Antonellini and Pollard, 1995; Ménendez et al., 1996).

Hall–Petch relation   As **grain size** decreases within a band, the finer-grained material increases in frictional resistance, modulus, and yield strength compared to the coarser-grained surroundings. This strain-hardening behavior, due to both grain-size

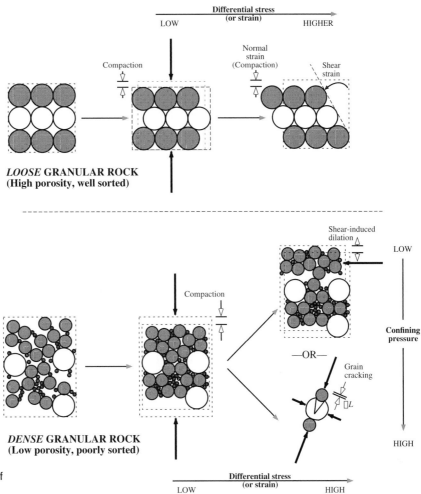

Fig. 7.20. Deformation styles of a porous granular rock.

reduction and changes in grain sorting or packing (Felbeck and Atkins, 1984, p. 184–185; Gu and Wong, 1994) is well known from materials science as the *Hall–Petch relation* (Callister, 2000, p. 167; see also Felbeck and Atkins, 1984, p. 145, for the "Petch" equation). A simplified form of this type of relationship is

Hall–Petch relation

$$\sigma_y = \frac{k}{\sqrt{D}} \tag{7.4}$$

in which σ_y is yield strength, D is the average grain diameter (or porosity), and k is a material constant. You'll recognize that this equation is quite similar to the *size–strength relationship* that we discussed in Chapter 2, in which smaller rock samples, such as cores, tend to be stronger than larger ones, such as outcrops (see also Scholz, 1990, pp. 28–29). In the context of deformation bands, the *yield strength of the band is increased* according to (7.4) through two mechanisms:

- *Grain-size reduction* (Felbeck and Atkins, 1984, pp. 145, 189; Callister, 2000, p. 167) combined with
- *Grain sorting* (uniform to variable grain size) or *packing* (e.g., cubic to hexagonal) (Felbeck and Atkins, 1984, pp. 184–185; Callister, 2000, p. 167).

In soil mechanics, the process of **compaction** (also called consolidation) strengthens a granular material by reducing its pore volume through compression, shaking, and/or shearing (e.g., Das, 1983, p. 36). Strengthening can further occur by adding finer particles as a matrix around the larger grains (Sammis et al., 1987; Antonellini and Pollard, 1995), leading to *poorer sorting* of the material (Biegel et al., 1989). Either or both of these mechanisms can lead to **strain hardening of cataclastic or clay-assisted deformation bands**, as inferred by Aydin (1978), Aydin and Johnson (1983), and others. Although the Hall–Petch relation was developed for polycrystalline metals and alloys that have minimal porosity (Felbeck and Atkins, 1984, p. 145; Callister, 2000), its simplified form shown as equation (7.4) suggests conceptually how porous materials such as sandstone and gouge can strain harden as their average grain size decreases within the band (e.g., Sammis et al., 1987; Antonellini et al., 1994).

Let's see how the simplified Hall–Petch relation (7.4) might apply to deformation bands. If the yield strength σ_{y0} of a coarse-grained sandstone, like the Navajo, is, say, 100 MPa (for $D = 1.0$ mm and $k = 100$ MPa m$^{1/2}$), then the yield strength within a deformation band σ_{yB} that has experienced a grain-size reduction (with an associated and implicit increase in packing) of a factor of 10 (to $D = 0.1$ mm) would be 316 MPa (using equation (7.4) with a constant value for k). An order-of-magnitude reduction in grain size within the band leads, in this approach, to a three-fold increase in its yield strength. The band then might be considered to become *more brittle* (see Fig. 2.16) than the surrounding rock because it could reach its strength maximum (the peak or yield strength) at a smaller value of strain than would the host rock. The band could then fail before the host rock under the same remote load because of its reduced grain size.

From this simple example, it appears that fracturing and crushing of grains within a deformation band could lead to significant strengthening (via the yield strength) of the band relative to its surroundings. But grain size changes are not the whole picture—it is also necessary to consider porosity reduction and grain fracturing within the bands, since these also contribute to increased strength and strain hardening (e.g., Zhang et al., 1990; Wong and Baud, 2012). Specifically, the increased angularity of fractured grains within a cataclastic deformation band leads to a substantial strengthening relative to bands having more rounded grains with comparable grain size and porosity (Mair et al., 2002; Cheung et al., 2012). Thus, cataclastic deformation bands are expected to strain harden to a greater degree than non-cataclastic deformation bands that form from reduced porosity alone.

7.4.2 Plastic Yield Envelopes and the *q–p* Diagram

Yielding vs. failure

Yielding and failure are related but separate properties of engineering and geologic materials. In general, **yielding** refers to the change in behavior from elastic to inelastic deformation (e.g., Schofield and Wroth, 1968; Rudnicki, 1977; Muir Wood, 1990; Davis and Selvadurai, 2002, p. 52). The point that marks the onset of inelastic, permanent deformation is called the *yield strength* and can be indicated on a stress–strain curve as a departure in linearity (see Fig. 2.12). Various mechanisms including microcrack growth and interaction; intra-grain plasticity (i.e., twinning and dislocation growth); grain comminution, rotation, and translation; and growth of localized shear zones can "loosen" the rock fabric (Paterson and Wong, 2005, p. 221) and permit larger deformation at given values of applied stress, leading to **strain hardening** with a reduction in the slope of the stress–strain curve (Fig. 2.16). Yielding is commonly discussed in connection with plasticity, ductility, creep, and flow (e.g., Jaeger et al., 2007, p. 252) in which large-scale or catastrophic reductions in strength (or load-carrying capacity) do not occur. Such material breakage occurs at the material's **peak** or **ultimate strength**; this is referred to as **failure** of the material. Failure is identified with the maximum load on a stress–strain curve, which separates the pre-peak (loading, distributed deformation) and post-peak (unloading, localized deformation) regions on this diagram (Fig. 2.12).

In a perfectly brittle material, which fails at a peak stress following a linear loading path on the stress–strain diagram, no yield strength is defined so there is no ambiguity. More generally, however, as noted by many including Christensen (2013), yielding and failure are often used somewhat interchangeably in describing rock deformation, leading to some imprecision in the literature. For example, common yield criteria include Tresca, von Mises, and Coulomb criteria (plus many others; Davis and Selvadurai, 2002, pp. 52–82) whereas failure criteria may include Griffith, Mohr, and Coulomb criteria (Gudmundsson, 2011, pp. 19–20). Frictional sliding along discontinuities in rock may then be associated with either yielding or failure. One way to address the difference in usage is to note the *scale dependence* that is implied in the terms (e.g., Rutter, 1986; Evans et al., 1990; Khan et al., 1991; Karato, 2008, p. 115). During yielding a material maintains its macroscopic continuity while changing its shape (i.e., necking or flow), which can occur as a result of any number of physical mechanisms including brittle (crack growth, grain fracturing) and ductile (grain rotation, translation, or distortion). The stress–strain curve for a yielding material is continuous and smoothly varying. In contrast, failure involves a separation of the material at the scale of observation. In this sense, microcracks and slip surfaces may be related to rock yielding whereas macrocracks and faults may be related to rock failure.

Yield envelopes are mathematical representations of the stress states over a range of conditions (including confining pressures, loading and deformation rates, water content, mineralogy, temperature, and many others) that separate elastic from inelastic deformation. A plot of peak strength of granites and limestones, as a function of confining pressure, would be an example of yield envelopes for these rocks. One can easily anticipate that these yield envelopes may also be constructed to include the effects of other important factors, such as temperature or rate, producing a set of yield envelopes that characterize a rock's response to stress under the specified environmental conditions. Yield envelopes for weakly cohesive porous granular rocks may similarly depend on additional factors including porosity; grain size, shape, mineralogy, and sorting; and diagenesis (e.g., Borja and Aydin, 2004). A Mohr envelope is an example of a yield envelope that separates elastic from inelastic (cracking or frictional sliding) behaviors. The shape of this yield surface depends on the physical characteristics, and the deformation mechanisms, of the porous rock being deformed (e.g., Wong et al., 1992, 2004; Borja and Aydin, 2004; Nova, 2005).

7.4.2.1 The q–p Diagram

Yield envelopes appropriate to porous granular rocks and soils come in a variety of types and degrees of complexity; the choice of which one to use in a given situation depends on the intended use and the availability of input parameters obtained from laboratory tests. Excellent treatises and readable entries into this approach are given by Antonellini et al. (1994), Davis and Selvadurai (2002), Borja and Aydin (2004), and Wong and Baud (2012). We'll make use of an approach from soil mechanics called a q–p **diagram** (Muir Wood, 1990, pp. 112–118; Nova and Lagioia, 2000; Davis and Selvadurai, 2002, pp. 68–71; Fig. 7.21). It's similar in many respects to a Mohr diagram except that the horizontal axis is mean effective stress, rather than the normal stress resolved onto a particular plane, which is given by the effective mean stress $p = ((S_1 + S_2 + S_3)/3) - P_p$, with P_p being the pore pressure (e.g. Bésuelle, 2001a,b; Rudnicki, 2004; Wong and Baud, 2012). The vertical axis is (for 2-D loading) the differential (deviatoric) stress, $q = (S_1 - S_3)$, rather than the resolved shear stress on a plane. The value of q equals the diameter of the Mohr circle (or twice the maximum shear stress) and provides a measure of the shear stress supported in the rock mass; it is also recognized as the differential stress on the y-axis of a stress–strain curve. The q–p diagram is used instead of the Mohr diagram for comparing 2-D or 3-D loading (stress) paths to yielding or failure in a rock, especially on the high-pressure "yield cap" side (Fig. 7.21; Risnes, 2001); q–p diagrams are used especially often in the deformation-band literature (e.g., Antonellini et al., 1994; Wong and Baud, 1999; Cuss et al., 2003; Baud

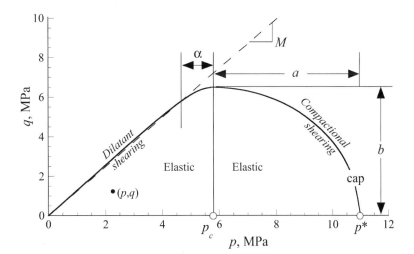

Fig. 7.21. Schematic of a two-yield surface model for a porous granular rock; parameters defined in text.

et al., 2004; Borja and Aydin, 2004; Wong et al., 2004; Okubo and Schultz, 2005; Schultz and Siddharthan, 2005; Aydin et al., 2006; Wibberley et al., 2007; Soliva et al., 2013; Philit, 2017). Analogous to a Mohr envelope, the porous-rock yield envelope defines stress states (p,q) that are elastic, below the yield envelope, and those that are inelastic, having stress states that intersect the envelope.

A representative yield envelope for a porous granular rock is shown schematically in Fig. 7.21. It is separated into two inelastic regimes. At lower confining pressures (smaller values of p), the yield surface has a positive slope and is associated with permanent dilatant shear within the host rock. At higher confining pressures (i.e., at greater depths; $p > p_c$ on Fig. 7.21), the yield surface has a negative slope and permanent compactional shear occurs in this region. The latter region, called a **cap**, distinguishes yield envelopes for porous granular rocks from those for compact rocks such as granite or completely cemented sandstones that do not have negative slopes or close at high confining pressures.

The yield surface need not intersect at the origin (i.e., for rocks having sufficient cohesive strength; Wong et al., 1997; Borja, 2004; Borja and Aydin, 2004). At the other extreme, the yield surface intersects the mean-stress axis p (having zero shear stress there) at a point called the **critical pressure** p^* (e.g., Wong et al., 1992, 1997). This is the pressure at which compaction, volume loss, and/or grain crushing occur in the absence of shearing (i.e., under isotropic or "hydrostatic" loading). This important value scales in a general way with the product of average grain size R and porosity n (Zhang et al., 1990; Wong et al., 1997; Wong and Baud, 2012; see Vajdova et al., 2004a, for results from carbonates). This Hertzian fracture criterion is given by

Critical pressure

$$p^* = (nR)^{-1.5} \tag{7.5}$$

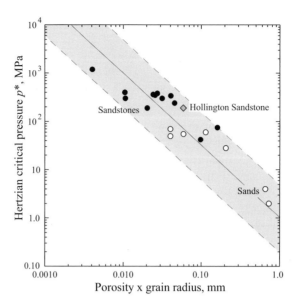

Fig. 7.22. Dependence of the critical pressure on the product of grain size and porosity (data from Wong et al. (1997), Vajdova et al. (2004a), and Rutter and Glover (2012)). Solid symbols, consolidated rocks; open symbols, unconsolidated sands; solid line, eq. (7.5) from Wong et al. (1997). Schematic uncertainties of a factor of five are shown (dashed lines).

This relation, shown plotted against data on quartz sands and sandstones in Fig. 7.22, was derived from the mechanics of grain crushing within a band, for spherical, monolithologic grains, but is considered good to first order for reasonably well-sorted grain-size distributions (e.g., Wong and Baud, 1999, 2012; Wong et al., 2004; Cheung et al., 2012). Rutter and Glover (2012) note that critical pressures measured for their sandstone data are also consistent with this equation within the stated uncertainties.

The elliptical yield surface shown in Fig. 7.21 is one of a class that includes Modified **Cam Clay** (Roscoe et al., 1958, 1963; Roscoe and Poorooshasb, 1963; Roscoe and Burland, 1968), named after the Cam River at Cambridge, England, where the original work on plastic yielding of clay-rich and granular soils was done (Muir Wood, 1990, p. 113; Davis and Selvadurai, 2002, p. 70). Nice treatments of this technique are given by Schofield and Wroth (1968), Muir Wood (1990), pp. 112–138, and Davis and Selvadurai (2002), pp. 190–210.

Critical state line

In soil mechanics, the **critical state line** (e.g., Schofield and Wroth, 1968; Farmer, 1983; Muir Wood, 1990, pp. 139–213) reflects large shear deformation of the soil with no volume change. The critical state line separates the two fields of the yield surface discussed above—volume increase (i.e., dilational shearing) to the left and volume reduction (compactional shearing) to the right. The slope M of this line on the q–p diagram is related to the friction angle in soils and rocks by (Muir Wood, 1990, p. 178)

$$M = \frac{6\sin\phi}{3-\sin\phi} \tag{7.6}$$

with the friction coefficient $\mu = \tan \phi_f$. For typical friction coefficients of $0.6 < \mu < 0.85$ (light shading in Fig. 7.23), $31° < \phi_f < 40°$ and $1.25 < M < 1.65$ (Fig. 7.23, dark shading).

The dilational side of a yield surface (Fig. 7.21) is called the **Hvorslev surface** in soil mechanics (e.g., Farmer, 1983, pp. 90–94). It is analogous to a set of frictional sliding curves that track the material's water content (Schofield and Wroth, 1968, pp. 207–215) and physical state as it yields inelastically down toward the critical state line. The compactional side of a yield surface, called the **yield cap**, is referred to as the **Roscoe surface** in soil mechanics (Farmer, 1983, pp. 90–94). As noted by Professor Teng-fong Wong (personal correspondence to the author, 2004), a one-to-one correspondence is not established between the homogeneous deformation implied by critical state soil mechanics and yield caps constructed for porous granular rocks (see also Rutter and Glover, 2012), so the parts of the yield surfaces discussed in this chapter are not referred to by the names used in soil mechanics (i.e., Hvorslev, Roscoe surfaces). Nevertheless, many authors have exploited the yield-envelope approach to explain the growth of shear (or deformation) bands in soils and porous rocks (e.g., Antonellini et al., 1994; Papamichos and Vardoulakis, 1995; Saada et al., 1999; Nova and Lagioia, 2000; Wolf et al., 2003; Wong et al., 2004; Wibberley et al., 2007; Aydin et al., 2006).

A series of yield surfaces can be drawn for rocks of different porosities and grain sizes (Fig. 7.24). As grain size increases, porosity increases or the critical pressure decreases (moves to the left on Fig. 7.21) and so does the size of the yield surface (Fig. 7.24a). On the other hand, as porosity goes to very small values for a compact (crystalline) rock like a granite, basalt, or quartzite, the critical pressure theoretically goes to infinity, meaning that it is

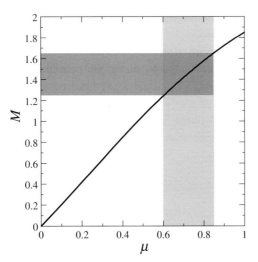

Fig. 7.23. Relationship between the friction coefficient μ on a Mohr diagram and the slope M on the q–p diagram (eq. 7.6). Typical ranges are shaded (see text).

Fig. 7.24. Basic elements of a yield cap diagram applied to porous rocks. (a) The rock's yield cap increases in size as the host rock's porosity, average grain size, and/or water content decrease. Arrow drawn through points of zero volume change (local horizontal slope of yield surface) implicitly defines the critical state. (b) The yield surface for the dilating regime (left of the critical-state line) decreases in size as the host rock strain softens, whereas the cap for the compacting regime increases in size as the host rock strain hardens toward the critical-state line.

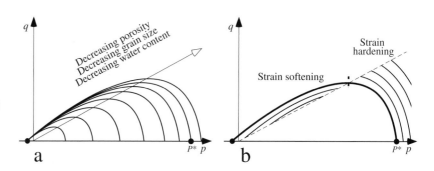

no longer a relevant property of a low-porosity rock. Mineralogy also affects the critical pressure, as does grain shape, since angular grains have a higher frictional resistance than do rounded ones (e.g., Mair et al., 2002). The **water content** influences the yield surface of a porous rock (Fig. 7.24a). The yield surface and cap both *contract in size*, while retaining their shapes, as the pore water content of the host rock increases (e.g., Wong and Baud, 1999; Wong et al., 2004).

Yield envelopes also expand or contract as inelastic shearing occurs within a soil or porous rocks (e.g., Davis and Selvadurai, 2002, pp. 198–200; Nguyen et al., 2016). For example, dilatant shearing locally increases the porosity, leading to a reduction in stiffness. As a result, the amount of differential stress q that is needed for inelastic strain to accumulate at a given value of effective confining pressure p decreases, leading to a shrinking of the yield envelope toward the critical state line. This behavior is called **strain softening**. Conversely, compactional shearing locally decreases the porosity, leading to an increase in stiffness and in the amount of differential stress q needed for inelastic yielding to continue. This behavior is referred to as **strain hardening**. Some types of deformation bands, such as cataclastic bands, accumulate progressively larger shear offsets in association with strain softening, whereas shear-enhanced compaction bands undergo strain hardening, eventually becoming too stiff for further offsets to accumulate along them.

7.4.2.2 Building Yield Envelopes for Porous Granular Rocks

A variety of physical mechanisms may operate within a rock during various stages of its deformation (e.g., Paterson and Wong, 2005). Yield surfaces may be defined for many of these deformation mechanisms (e.g., Borja and Aydin, 2004). Although single-yield surface envelopes (e.g., Papamichos et al.,

1997) can be simpler to define and implement, they generally do not match experimental data as well as two-surface envelopes defined by (a) a linear or nonlinear envelope for dilatant shearing (i.e. frictional sliding), and (b) an elliptical envelope for compactional flow (referred to as a cap; see Issen, 2002; Nova, 2005; Schultz and Siddharthan, 2005; Aydin et al., 2006; Xie and Shao, 2006; and Wong and Baud, 2012, for examples and discussion). Gutierrez and Homand (1998) and Hickman (2004) note that the extensive laboratory data available for chalks are fit better by a two-yield-surface model than a single-yield-surface model. This requires that the shear failure criterion is functionally independent of deformation on the cap, paralleling results for porous sandstone (e.g., Wong and Baud, 2012). Plastic yielding in a porous rock (i.e., formation of deformation bands) is taken to occur when the stress path intersects the lesser of the two envelopes.

Both theory and laboratory experiments on a variety of porous granular rock types including sandstone, limestone, and chalk support an elliptical shape in q–p space for the cap (e.g., Wong and Baud, 1999, 2012; Risnes, 2001; Cuss et al., 2003; Rudnicki, 2004; Grueschow and Rudnicki, 2005; Karner et al., 2005; Xie and Shao, 2006; Omdal et al., 2010). An elliptical cap shape has been used to interpret the kinematics and the geometry of deformation bands (e.g., Fossen et al., 2007, 2017) that were developed in porous sandstones and carbonates and observed in the field (e.g., Antonellini et al., 1994; Schultz and Siddharthan, 2005; Aydin et al., 2006; Eichhubl et al., 2010; Schultz et al., 2010a,b,c; Fossen et al., 2011) and accordingly is described here.

The yield cap can expand during deformation due to hardening of the plastically yielded rock (e.g., Xie and Shao, 2006; Wong and Baud, 2012), in association with growth of deformation band networks (e.g., Schultz and Siddharthan, 2005). Time-dependent plastic deformation such as creep can sometimes also occur within the host rock before or after yielding (e.g., Karner et al., 2003; Hickman and Gutierrez, 2007; Brzesowsky et al., 2014).

A yield envelope for a porous granular rock can be constructed by using any of several formulations (e.g., Davis and Selvadurai, 2002; Borja and Aydin, 2004; Schultz and Siddharthan, 2005; Wong and Baud, 2012; Nguyen et al., 2016). In this section we follow the approach of Gutierrez and Homand (1998) and Hickman (2004) who utilized a two-yield-surface model for porous-rock deformation. Their equations for yielding in the shear and cap regions are given in this section. Although a tensile strength could be included in the yield criterion for porous rocks such as chalk (e.g., Gutierrez and Homand, 1998; Hickman, 2004) and sandstone, available yield envelopes for porous sandstones obtained by Rutter and Glover (2012) and Wong and Baud (2012) suggest that yield envelopes intersect close enough to the origin that tensile strength can probably be neglected.

Yielding in shear can be expressed by a linear Mohr–Coulomb criterion by using $q = Mp + q_0$, where $M = q/p$ is the slope on the q–p diagram (equation (7.6)) (Gutierrez and Homand, 1998) and thereby to a host-rock friction angle

ϕ_f or coefficient μ (Rutter and Glover, 2012), and q_0 is the host-rock cohesion in MPa. The cohesion parameter q_0 is related to the laboratory cohesion C_0 by $q_0 = (6\,C_0 \sin \phi_f)/(3 - \sin \phi_f)$. One could alternatively choose a nonlinear, hyperbolic shear yield criterion in order to define a continuous and smoothly varying yield envelope from the host-rock tensile strength through the critical pressure. Following Gutierrez and Homand (1998) a hyperbolic shear yield envelope can be given by using

Nonlinear shear failure envelope

$$q = Y_0 - \sqrt{\alpha^2 + \left[k_f M \left(p - p_c \right) \right]^2},$$

$$Y_0 = q_0 + \sqrt{\alpha^2 + \left[k_f M p_c \right]^2}$$

$$(7.7)$$

in which the dimensionless parameter α describes the degree of nonlinearity of the shear yield envelope, with $\alpha = 0$ corresponding to a linear envelope and $\alpha/p^* \leq 0.1$; k_f is a dimensionless parameter related to the host-rock friction, where $k_f < M$; p_c is the location of the cap vertex; and q_0 is the cohesion, both of the host rock. Values of host-rock friction coefficients of $\mu = 0.6$ and 0.5 correspond to $M = 1.242$ and $k_f = 1.051$, respectively. The fitting parameters $\Gamma = p_c/p^* = 0.523$ and $\alpha/p^* = 0.1$ (Gutierrez and Homand, 1998) in this approach ensure a continuous slope along the yield surface (Fig. 7.21).

A critical pressure p^* as defined by host-rock grain size and porosity is commonly used to define an elliptical yield surface (cap) for the host rock that smoothly joins the shear yield envelope at $p = p_c$ (i.e., both envelopes are defined to have the same slope at that point). Following Gutierrez and Homand (1998) the compactional (cap) yield envelope is given by

Compactional yield cap

$$q^2 = b^2 \left[1 - \frac{\left(p - p_c \right)^2}{a^2} \right]$$

$$(7.8)$$

in which a is the horizontal semi-axis of the cap. Aspect ratios a/b for caps in porous sandstones from the literature range approximately from 0.75 to 1.0 (Wong and Baud, 2012; Rutter and Glover, 2012). The vertical semi-axis of the cap b is given by $b = Y_0 - a$ (Gutierrez and Homand, 1998) so that the horizontal semi-axis a can be calculated given values of b and a/b. The value of critical pressure used in this example and shown in Fig. 7.21 is 11 MPa.

An approach for quantifying the yield envelope for a porous rock such as that just outlined provides a useful framework for analyzing the occurrence, development, and patterns of deformation bands. Once the yield envelope has been defined, stress states that might be associated with porous rock strata at particular depths, or with deformational processes such as folding or faulting, can be plotted on the q–p diagram (e.g., Wibberley et al., 2007; Soliva et al., 2013). **Stress paths** describe the evolution of stress states with processes such as burial, uplift, changes in formation pore pressures (e.g., due to hydrocarbon production or injection of carbon dioxide or wastewater), and tectonic forces. The comparison between stress states and yield envelopes defined for each rock layer can be used to understand the origin, timing, and spatial relationships between deformation bands formed in the same layer (e.g., cross-cutting or superposed sets) and in adjacent layers (e.g., Schultz et al., 2010a,b,c; Fossen et al., 2011).

Just as a layer's mechanical properties at the time of deformation may promote faults in one layer and joints in an adjacent one (e.g., Gross, 1995), both fractures (joints) and deformation bands can form in adjacent layers depending on their mechanical properties (e.g., Fossen et al., 2007, 2017; Raduha et al., 2016); q–p diagrams can facilitate exploration of the structural development of such layered stratigraphic sequences.

7.4.2.3 Interpretation of Yield Envelopes for Deformation Bands

Let's probe the evolution of yield envelopes farther by explicitly considering what a stress–strain curve might look like for each of these two regions. Suppose we take a porous rock such as a sandstone or a limestone and differential stresses q are applied to it at a shallow depth (corresponding to a small value of p). As the rock compresses, the stress–strain curve departs from linearity, marking the onset of inelastic deformation, shear-induced dilation, and the rock's yield strength; a population of dilational shear bands might begin to grow within the sample (point A in Figs. 7.25a and 7.25c). The inelastic part of the stress–strain curve continues to increase toward the peak strength (point B) and then decreases, perhaps unstably, toward the ultimate failure line (Muir Wood, 1990, p. 124), which corresponds to a smaller (frictional) strength at failure. A **faulted deformation band** forms at point C, which corresponds to the case of deformation bands that serve as precursory substrates to slip surfaces (Aydin and Johnson, 1978; Shipton and Cowie, 2001, 2003) that link up to form larger-offset faults (e.g., Rudnicki and Rice, 1975; Aydin and Johnson, 1978, 1983).

If instead a porous rock is loaded starting at a higher confining pressure p (for the same rock porosity and grain packing geometry), then it may follow the path D–E–F on Fig. 7.25a. As the differential stress increases and the rock begins to compress and shear, the grain-to-grain contacts support a much larger compressive stress, leading eventually to grain rotation, translation, fracturing, and crushing with an associated decrease in pore volume within a tabular volume within the host rock (Aydin, 1978; Zhang et al., 1990). This has the triple result of: (a) reducing the average grain size within the growing deformation band, (b) producing a tighter packing geometry, and (c) making the grains more angular and consequently less able to roll under shear stress (Mair et al., 2002). These three factors all make it more difficult to accommodate shearing displacements within a deformation band formed of such material; as a result, the band strain hardens. This corresponds to point D on Fig. 7.25a and 7.25d,

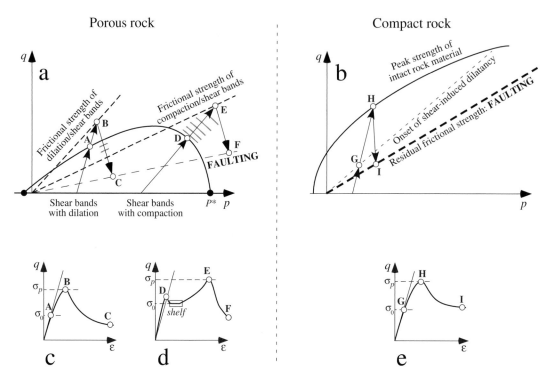

Fig. 7.25. Association of yield envelopes and stress-strain curves for (a) porous granular rock and (b) compact rock (after Schultz and Siddharthan, 2005). Lettered stress paths in (a) correspond to the stress–strain diagrams in (c), for dilatant shear bands, and (d), for compactional shear bands. The stress path in (b) corresponds to the stress–strain diagram for compact rock in (e). Dashed gray lines in (a) depict moving yield surfaces in the direction of the critical state line.

corresponding to the onset of inelastic yielding and volume reduction in the rock. However, because grain crushing depends on several factors including grain size, packing geometry, grain composition and shape (Wong et al., 1997; Wong and Baud, 1999), the same stress state on the yield envelope may produce compaction *without* cataclasis in one rock type and compaction *with* cataclasis in another.

There is a region of the stress-strain curve for porous soils and rocks subjected to a sufficiently large confining pressure called a **shelf** (Fig. 7.25d). Here, strain accumulates under constant stress, indicating a plastic strain that corresponds to compaction within the rock (Wong et al., 1992; Olsson, 1999, 2001; Nova and Lagioia, 2000; Olsson and Holcomb, 2000). The width of the shelf (i.e., the amount of strain accommodated) is proportional to the amount of *porosity reduction* in the rock, so higher porosity rocks have a longer shelf than do lower porosity rocks. In the limit of negligible porosity, the stress-strain curve for a compact rock will not show a shelf since there is no pore volume to collapse.

Once a population of compactional shear bands has formed and collapsed the available pore volume within them, the bands have become stronger and stiffer than the surrounding host rock. The stress–strain curve must then increase with increasing strain, as has been demonstrated in many experiments on rock (e.g., Wong and Baud, 1999; Cuss et al., 2003; Vajdova et al., 2004a). Strain hardening moves the rock off the initial yield cap (point D) and upward (since it is strain *hardening*) until the stress state associated with frictional sliding is achieved. At this point (point E), the frictional strength of the compactional shear deformation band would be reached and the band would fail unstably. However, the magnitudes of q and p at point E in nature appear to be so large that the compactional shear bands formed on the cap rarely fail to become through-going faults (e.g., Soliva et al., 2013). The bands formed on the cap in this regime are also misaligned with optimum orientations for frictional sliding, as discussed above, further reducing the likelihood of failure.

Interestingly, the loading of a compact, low-porosity rock is akin to the loading of a dense granular rock. Both rock types undergo dilational shearing (point G on Figs. 7.25b and 7.25e) at yield, strain hardening up to peak strength (point H), and strain softening in the post-peak region until a through-going frictional sliding surface (fault) is formed (point I).

7.4.2.4 Implications for Deformation Bands

Based on the foregoing discussion, some general conclusions may be drawn. A loose granular rock tends to undergo compactional shearing, whereas a dense granular rock undergoes dilational shearing, all other factors being equal. On the basis of this characteristic alone, the yield envelope for a dense granular rock should have a positive slope (given appropriately small values of p), with implications of strain softening behavior and perhaps formation of dilational shear bands. Cataclasis may occur if the rock has sufficiently few grain-to-grain contacts for a given stress state, so coarser-grained, well-sorted rock units would favor the development of cataclasis (e.g., Fossen et al., 2017). Conversely, the yield envelope for a loose granular rock should have a negative slope (again for sufficient values of p), implying strain hardening behavior and perhaps formation of shear-enhanced and/or pure compaction bands.

Loose vs. dense granular rock

Dilation angle

The dilation angle also influences the type of deformation band that forms within a given porous granular rock. All factors being equal, a dense granular rock should tend to have positive dilation angles, promoting dilational shear bands and cataclastic deformation bands. A loose granular rock, with negative dilation angles, should favor compaction bands. As noted above, the difference between the friction angle of host-rock grains and their dilation angle exerts a control on the type of deformation band that develops, including cataclastic bands, so that factors such as grain angularity, roughness, rugosity, cementation, grain sorting and packing, mineralogy (including clay content), hardness, and the like will influence both angles and, thereby, the type of deformation band.

Stress state and stress path

Stress state and loading conditions exert a primary influence on porous rock deformation and hence on the types of deformation band that can form. For the same porous granular rock (i.e., loose or dense), the yield envelope changes slope, from positive to negative, with increasing values of the effective confining pressure. This implies that dilational shearing may be expected at shallower depths, with a change to contractional shearing occurring at greater depths (e.g., Exner and Tschegg, 2012). By implication, the types and dihedral angles of deformation bands should change systematically with depth for a particular far-field stress state. Particular depths for this transition could be calculated from the yield envelope (i.e., point p_c on Fig. 7.21), which tends to be about one-half the value of the critical pressure for many porous rock types examined in the literature (e.g., Gutierrez and Homand, 1998).

Pore pressure

Pore pressure changes the effective confining pressure p without changing the differential stress q. By implication, overpressured rocks deform as through they were shallower, perhaps promoting non-cataclastic deformation to greater depths than in normally pressured sequences (Fossen et al., 2017).

Extensional vs. contractional tectonic regime

Related to stress state is the tectonic regime. Stress magnitudes at any given depth tend to be smaller under normal-faulting regimes than under thrust-faulting regimes (Wibberley et al., 2007), which has been shown to influence the types and spatial organization of deformation bands (Solum et al., 2010; Soliva et al., 2013, 2016). In contractional (thrust-faulting) tectonic regimes therefore, cataclastic deformation bands and compaction bands may occur to quite shallow depths as compared to those forming in extensional (normal-faulting) tectonic regimes (Ballas et al., 2014).

Diagenesis

A final factor to consider is the degree of diagenesis or, specifically, of syn-kinematic cementation within a deformation band. Precipitation or secondary-mineral growth within host rock and/or deformation bands leads to a reduction in their porosity and increases their stiffness and grain-to-grain contacts. Diagenesis then essentially transforms a loose granular rock into a dense(r) one, promoting yielding under less negative or even positive slopes, inhibiting cataclasis while favoring dilational shearing or dilatant cracking rather than deformation banding. The diagenetic overprint produces cementation bands with different physical and fluid properties than deformation bands which can be important in hydrocarbon fields (Exner et al., 2013), geothermal fields, or other subsurface environments in which chemical reactions between rock and aqueous fluids are active.

7.4.3 Fracturing and Yielding of Mudrocks

Mudrocks are a broad class of fine-grained, variably indurated rocks with large clay fractions and generally low permeability (e.g., Aplin and Macquaker, 2011). Shales are mudrocks that have developed bedding-parallel foliations that cause the rocks to cleave readily into sheets. Siliceous mudrocks are economically important rock types as both unconventional (self-sourced) hydrocarbon reservoirs (Fig. 7.26) as well as the caprock and top-seal sequences above conventional and heavy-oil reservoirs (e.g., Ingram and Urai, 1999; Petrie et al., 2014; Prost and Newsome, 2015; Loizzo et al., 2017). Similar sequences are also important to long-term sequestration of radioactive waste (Hansen et al., 2010) and CO_2 (Hawkes et al., 2005; Rutqvist, 2012; Zoback and Gorelick, 2012; Altman et al., 2014; Pawar et al., 2015) in the subsurface.

The yielding and failure of mudrocks depends on several characteristics, including: (a) composition, such as percentage of carbonate; and (b) diagenetic maturity (i.e., illitization, pore-filling cement precipitation), which can influence stiffness and strength of the sequence (e.g., Katsube and Williamson, 1998; Ingram and Urai, 1999; Nygård et al., 2006; Hansen et al., 2010; Aplin and Macquaker, 2011). As noted above, siliceous mudrocks can yield initially to form compactional shear deformation bands with some cataclasis (e.g., Ishii, 2012) that can serve as precursory nucleation substrates for faulting.

Under low to moderate confining pressures and temperatures, mudrocks and shales can also fracture and transmit fluids through existing or newly formed joint and fault networks (e.g., Petrie et al., 2014; Aydin, 2014; Raduha et al., 2016). Fracture sets and fault damage zones that have not been sealed with diagenetic cements (e.g., Laubach et al., 2010) are commonly recognized in the caprocks and top-seal sequences above hydrocarbon reservoirs (e.g., Nygård et al., 2006; Ferrill et al., 2014; Petrie et al., 2014). Anisotropy in the physical and hydraulic properties of bedded or heterogeneous mudrock sequences (e.g., Fidan et al., 2012) can cause fracture networks to increase in complexity within them down to the finest scales of the bedding laminations (Petrie et al., 2014).

The capacity of shales and mudrocks to seal against fluid and gas migration in the subsurface is of critical importance in oil and gas fields, storage of natural gas in underground chambers, and long-term storage of carbon dioxide

Fig. 7.26. Outcrop expression of Eagle Ford siliceous mudrocks, south Texas, showing finely laminated layers and low-amplitude folding (photograph by Peter Hennings).

(e.g., Altmann et al., 2014). Several approaches to estimating the seal capacity or degree of brittleness of caprock/top-seal sequences have been developed and applied in this rapidly evolving field, including over-consolidation ratio (OCR) and brittleness index.

The **over-consolidation ratio** has been exploited in geotechnical engineering (e.g., Terzaghi, 1943; Terzaghi et al., 1996; Strokova, 2013) and hydrocarbon exploration and risking (e.g., Ishii et al., 2011) for many years. As clay-rich soils and rocks are loaded to increasing burial depths, the void ratio (or porosity) decreases as the particles deform elastically. Continued burial leads to inelastic deformation of the rock by particle rearrangement and fracturing or deformation banding. The transition point (Casagrande, 1936) between the linear (elastic) and nonlinear (plastic) regimes is the *yield stress*, referred to in this context as the *preconsolidation stress, $\sigma_{eff\,Max}$*, which can be determined experimentally from odeometer or field vane tests (Chang, 1991; Davis and Selvadurai, 2002, pp. 44–46). It may also correspond to the *critical pressure p** (Skurtveit et al., 2013) discussed above as used in studies of deformation bands. The idea is that stronger, overconsolidated ("brittle") rocks will dilate during deformation (such as shearing) to a greater degree than weaker ones (that are either normally or under-consolidated). This ratio is taken to be an estimate of the degree of uplift of the potential seal sequence from its original depth of burial.

The over-consolidation ratio (OCR) can be defined as

Over-consolidation ratio

$$OCR = \frac{\sigma_{eff\,Max}}{\sigma_{eff\,Actual}} \qquad (7.9)$$

in which $\sigma_{eff\,Max}$ is the maximum past effective pressure (equivalent to the effective vertical stress ($S_v - P$) and $\sigma_{eff\,Actual}$ is the present-day effective pressure (Ingram et al., 1997). The ratio has been interpreted to imply that values of OCR much greater than one might become dilatant during deformation, due to an increase in void ratio (or porosity relative to the earlier consolidation state), and therefore may define poor seal sequences. By implication, poorly cemented, normally compacted shales and mudrocks might be anticipated to be ductile over most depths of interest, with dilatancy and embrittlement becoming important only for significant uplift magnitudes (Ingram et al., 1997). This approach continues to find application today.

Brittleness

More recent explorations of **brittleness** center on the use of rock-mechanical properties to quantify, and potentially predict, the degree of brittleness or ductility of a mudrock sequence. These approaches have been rejuvenated with the success of unconventional ("self-sourced") hydrocarbon reservoirs in shale basins throughout the world, which relies to a great extent on reservoir stimulation by hydraulic fracturing (e.g., Valkó and Economides, 1995) from horizontal wells. The drillability and stability of wellbores also can be influenced by brittleness (Denkhaus, 2003; Holt et al., 2015). However, determination of brittleness is not straightforward, in part because a unique and uniformly agreed-upon definition of brittleness remains elusive despite considerable effort.

The use of well logs that record values for compressional- and shear-wave velocities and density can be used to calculate values for Young's modulus and Poisson's ratio for a given stratigraphic layer (e.g., Sheriff and Geldart, 1995; Mullen et al., 2007; Kidambi and Kumar, 2016). These dynamic values can then be converted into static values, analogous to those obtained in laboratory uniaxial or triaxial tests, if suitable conversion relations are available (e.g., Barree et al., 2009). This pair of elastic properties was interpreted as a simple means for estimating brittleness by Rickman et al. (2008, 2009), who associated Young's modulus with the ability of a rock to fracture and Poisson's ratio as its ability to fail under stress. A plot of Young's modulus vs. Poisson's ratio is interpreted following this approach as indicating increasing degrees of brittleness for larger values of Young's modulus combined with smaller values of Poisson's ratio for a given rock (Fig. 7.27). The approach that they pioneered and helped to popularize has found widespread use in the oil and gas industry (e.g., Gray et al., 2012), given its ready availability from typical subsurface logging tools, despite lacking a clear physical basis for equating elastic properties with brittle rock failure.

More complete and robust investigations of brittleness have centered on parameters extracted from a rock's stress–strain curve, such as strain to failure, peak strength, post-peak stress drop, residual strength, compressive vs. tensile strengths, friction angle, and other approaches to characterizing rock strength such as over-consolidation ratio (e.g., Ingram and Urai, 1999; Holt et al., 2011, 2015; Yang et al., 2013; Rybacki et al., 2016; Zhang et al., 2016). Because the stress-strain curve and related properties depend on the petrophysics of the rock, such as its chemistry, mineralogy, porosity, and texture,

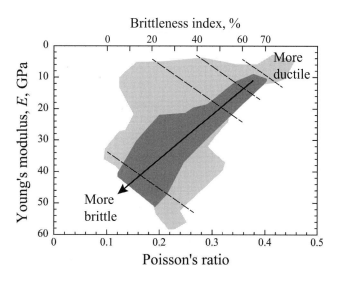

Fig. 7.27. A brittleness index as used in the oil and gas industry results from plotting Poisson's ratio vs. Young's modulus for various rock types. Note reversed scale for Young's modulus following Rickman et al. (2008) and scale in GPa following Rybacki et al. (2016). Shaded areas show approximate ranges of points plotted in Rickman et al. (2008): high point density, dark shading; low point density, light shading. Dashed lines show approximate contours of brittleness index.

brittleness can also be investigated as a function of these properties (Rickman et al., 2008; Rybacki et al., 2016; Zhang et al., 2016).

In general, the response of a rock to applied forces, and brittleness in particular, is **not a rock property**, and thus is difficult to characterize and quantify from a small set of parameters. One common approach used in materials science and engineering identifies a brittle material as one that fails at a well-defined peak load at a small strain with dilatancy, strain-softening behavior, and strain localization. In rocks and soils, brittleness depends on several sets of factors including: (1) scale, (2) rock properties (composition, mineralogy, porosity, sorting, texture, degree of cementation and diagenesis), and (3) environmental conditions (e.g., loading and strain rates, temperature, confining pressure, differential stress, rate-and-state and time-dependent friction, and system stiffness) (Hucka and Das, 1974; Evans and Kohlstedt, 1995; Scholz, 2002; Wong and Baud, 2012; Sone and Zoback, 2014; Rybacki et al., 2016).

7.5 The Evolutionary Sequence to Faults

The study of deformation bands is enjoying a resurgence in recent years, thanks to work by several research groups on aspects as diverse as kinematics, laboratory simulation, and petroleum engineering. In this section we'll briefly examine how shear deformation bands are thought to grow in scale and complexity, from the earliest stages to big, honking, regional-scale fault sets.

7.5.1 The Classic Aydin–Johnson Model

Aydin (1978) and Aydin and Johnson (1978) proposed a developmental sequence associated with the accommodation of progressively larger strains (and band-parallel shearing offsets) by cataclastic (compaction with shear) deformation bands (Fig. 7.28). Initially, cataclastic deformation bands localize shearing displacements on the order of the width of the band, as viewed parallel to the direction of shearing displacement (Aydin, 1978; Fig. 7.28a). This observation implies an initially "loose" and porous granular rock (e.g., Fig. 7.20). The individual bands tend to weather out in positive relief and clearly demonstrate band-parallel offsets in finely laminated rocks such as cross-bedded sandstones (Fig. 7.3). Because the bands are laterally discontinuous—just like other types of discontinuities—they create distinctive stepover geometries where adjacent tiplines approach each other. These geometries are quite distinct for the respective displacement modes along the bands.

Let's examine Fig. 7.28a. As originally defined from the field observations, cataclastic deformation bands observed in a direction *parallel* to the shearing direction (down the displacement direction) have wavy traces and inosculating geometries (Fig. 7.13). The stepovers between collinear bands form cup-shaped **lenses** separating less deformed rock (Antonellini et al.,

1994). However, the other view of the band shown in Fig. 7.28a is obtained by looking *normal* to the shearing direction. Now the individual bands appear as straighter segments instead of wavy ones (e.g., Aydin and Johnson, 1978) with

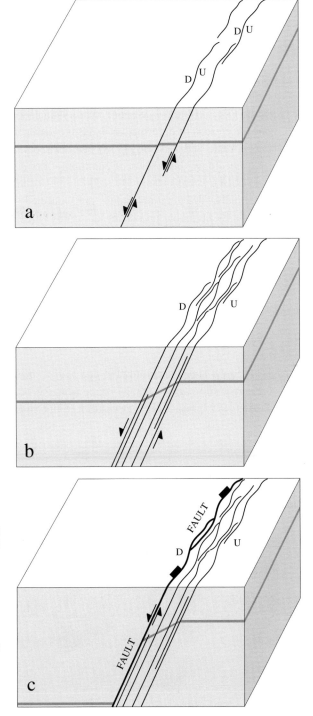

Fig. 7.28. The sequential development of faults from undeformed porous sandstone according to the Aydin–Johnson model. (a) **Single** compactional shear deformation band. U, footwall up; D, hanging wall down. (b) **Zones or clusters** of closely spaced deformation bands. (c) Localization of a Coulomb slip surface (heavy lines) and **frictional sliding** on the zone of deformation bands. Fault displacements, D (down), U (up).

stepovers more like strike-slip **duplexes** than lenses (Fig. 6.15). Cataclastic deformation bands viewed in this orientation function as in a manner analogous kinematically to mode-II fractures (e.g., Mair et al., 2000).

Porosity decrease within the individual cataclastic deformation band (Antonellini and Aydin, 1994) during compaction and grain crushing (Zhang et al., 1990) contributes to strain-hardening of the band (Aydin and Johnson, 1983). Formation of a new band close to the previous one (Aydin and Johnson, 1978) occurs because the strain-hardened band "locks up" and cannot accommodate further displacements as easily as the surrounding rock can (Aydin and Johnson, 1983); this eventually produces a *zone of deformation bands* (Figs. 7.14 and 7.28b) having very small spacing/length ratios. Here, a zone (or cluster) of deformation bands is a series of closely spaced cataclastic bands (Aydin and Johnson, 1978) that are separated by *lens-shaped* regions and exceedingly thin sheets of commonly less deformed rock (e.g., Awdahl et al., 2014).

Normally, deformation bands are confined to particular stratigraphic intervals by favorable properties like porosity (e.g., Aydin, 1978; Antonellini et al., 1994; Fossen and Hesthammer, 1997; Davis, 1999; Fossen et al., 2007). A beautiful set of bands, confined to a set of crossbeds in the eolian Navajo Sandstone near Sheets Gulch, Utah, is shown in Fig. 7.29. The bands exhibit duplex-style linkages in map view. These bands (and others like them) can continue to build up strains until the strength properties of the bounding stratigraphic interfaces (e.g., Cooke and Underwood, 2001) are exceeded. Once this occurs, the band can cut up and down section as a fault (e.g., Soliva et al., 2005).

Continued strain accumulation can lead to instability and nucleation of a corrugated *slip surface* (fault) on or near the edge of the zone (Aydin and Johnson, 1978; Fig. 7.28c) or within it (Shipton and Cowie, 2001; Fig. 7.30).

Fig. 7.29. Deformation bands having nearly strike-slip offsets confined to a steeply dipping cross-bedded sequence near Sheets Gulch, Utah.

Fig. 7.30. A fault (normal sense, down to the left) nucleated within a zone of deformation bands in Entrada Sandstone, east-central Utah. Offset is ~6 m (after Schultz and Fossen, 2002).

The fault is a surface of discontinuous displacement, on which rocks on either side of the surface have been displaced by several meters or more. Notice in Fig. 7.28c that the continuous displacement within the zone of cataclastic deformation bands is "frozen in" when the fault develops, so the total displacement on the fault is the sum of the continuous and discontinuous offsets. Nucleation of a fault represents a change in the style of deformation, from one involving strain hardening of bands and zones of bands to one involving strain softening and formation of a weaker slip plane (Aydin and Johnson, 1983).

As cataclastic deformation bands or other types of discontinuities accumulate displacements, they also grow in size (length and height) in order to maintain a ratio of displacement/length within certain ranges (Fossen and Hesthammer, 1997; Schultz et al., 2008a, 2013). As an elliptical (or, in three dimensions, an ellipsoidal) deformation band (Sternlof et al., 2005; Meng and Pollard, 2014) strain hardens relative to its surroundings (forming what's known as an Echelby inclusion; see, e.g., Rice, 1979; Meng and Pollard, 2014), the shearing displacement it can accommodate becomes increasingly incompatible with—much less than—that of the enclosing rock. Eventually the interface between the band and the rock *breaks unstably* to form a Coulomb surface of frictional sliding (Aydin and Johnson, 1983)—a **fault**.

This is the now-classical model for the growth of cataclastic deformation bands into faults. It works generally well if the viewing direction is along (i.e., parallel to) the shearing (or offset) direction (Fig. 7.13). However, the geometry of the zones is very different when they are observed at right angles (perpendicular) to the shearing direction, so that the band is viewed in cross-section (Figs. 7.4 and 7.9). We'll discuss this *viewing perspective* in Section 7.5.3. Zones of cataclastic deformation bands are often clustered, with approximately planar zones several cm thick separated by less-deformed sections of sandstone, forming the commonly observed "radiator rock" configuration documented by Davis (1999) and re-interpreted as stepovers by Schultz and Balasko (2003) and Okubo and Schultz (2006).

7.5.2   Deformation Bands with Slip Surfaces (Faults)

Although cataclastic deformation bands accommodate localized shearing displacements, they are not faults in the strict sense. This is because faults are surfaces of displacement discontinuity, with essentially zero thickness, leading to a theoretical value of shear strain (displacement/thickness) of infinity. However, faults can and do nucleate on zones of deformation bands, as noted by Aydin (1978), Aydin and Johnson (1978), Johnson (1995), Fossen and Hesthammer (1997), Shipton and Cowie (2003), and many others. This distinction becomes important if the deformation mechanisms as a function of strain or displacement magnitude (or time) are of interest.

First, the down-dip surfaces of cataclastic deformation band zones are characteristically *corrugated*, with the plunge of the topography paralleling the offset direction. An example is shown in Fig. 7.31, and from correlations of offset bedding within the host sandstone, the displacement across the band is continuous, not sharp. Thus, **corrugations are not in themselves sufficient evidence for faulting**. The early growth and development of corrugations along deformation bands in the geometry of mode-III lenses or some lozenges (see also Awdal et al., 2014)—formed apparently before the development of pre-fault slip surfaces—can be observed at many localities (e.g., Figs. 7.4 and 7.14).

Fault surfaces within zones of cataclastic deformation bands are easily distinguished from non-slipped bands and identified in the field (e.g., Aydin and Johnson, 1978; Shipton and Cowie, 2003; Okubo and Schultz, 2006) as polished surfaces overprinting the pre-slip corrugations. Thus, slickensides can

Fig. 7.31. Linked echelon bands in Entrada Sandstone showing the development of down-dip corrugations that are later exploited by frictional sliding to produce polished, slickensided fault surfaces.

follow the previously formed corrugations as long as the straining direction within the deforming network remains the same.

7.5.3 Deformation Bands as Riedel Shear Structures?

In seeming counterpoint to the work of Aydin and colleagues, Davis (1999) and Davis et al. (2000) documented echelon arrays of cataclastic deformation bands in southwestern Utah that resemble conjugate Riedel sets of partially linked bands (Fig. 1.8). Open networks of parallel bands are bounded by zones of deformation bands, forming what Davis (1999) calls "radiator rock" (Figs. 7.5, 7.9, and 7.32). This geometry, as it turns out, is entirely consistent with the classic Aydin–Johnson model but reveals how important the viewing geometry is in the interpretation of deformation band and fault zone architecture. These patterns were observed previously by Aydin and Johnson (1978, 1983), Antonellini and Aydin (1994), and produced experimentally by Mair et al. (2000), but they were not elaborated upon until later (Davis, 1997, 1998; Davis et al., 2000; Schultz and Balasko, 2003; Katz et al., 2004; Okubo and Schultz, 2006). To see them most clearly, one needs to look *perpendicular to the shearing direction* of a network of bands (Fig. 7.32).

Davis (1999) and Davis et al. (2000) emphasized that the cataclastic deformation bands form in overlapping echelon configurations, rather than the inosculating and parallel arrangements focused upon by Aydin (1978), Antonellini et al. (1994), and others. Field observations demonstrate that the linking bands form preferentially within overlapping (echelon) deformation bands (Davis et al., 2000; Ahlgren, 2001; Fig. 7.33b), suggesting that increased stress magnitudes between overlapping bands (McGarr et al., 1979; Schultz and Balasko, 2003) facilitated localization of the linking bands. Ahlgren's (2001) field

Fig. 7.32. The striking "radiator rock" geometry (Davis, 1999; Davis et al., 2000) formed between zones, or **stepovers** (Schultz and Balasko, 2003; Okubo and Schultz, 2006), of cataclastic deformation bands (Antonellini and Aydin, 1994) are clearly revealed in outcrop when viewed normal to the shearing direction.

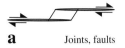

a Joints, faults

b Deformation bands

Fig. 7.33. (a) Forward-breaking orientation of linking fractures formed under peak-stress conditions. (b) Backward-breaking orientation of linking fractures found in cataclastic deformation band arrays.

observations also imply a low overall strain magnitude and a symmetry of the associated stress and strain states (e.g., Tchalenko and Ambrasays, 1970).

The echelon band geometry (Fig. 7.33b) is reminiscent of (strike-slip) fault strands (Fig. 7.33a) that link and transfer displacement by nucleating cross fractures that bridge them at non-90° angles (e.g., Segall and Pollard, 1980; Martel, 1990; Peacock and Sanderson, 1995; Roznovsky and Aydin, 2001), forming *duplexes* (e.g., Woodcock and Fischer, 1986; Aydin, 1988). Similar geometries are observed in gouge (e.g., Bartlett et al., 1981; Gu and Wong, 1994; Marone, 1998b) although the orientation of the linking bands differs for these structures.

Deformation bands having the same orientation as the linking bands occur in isolation and far from stepovers (the "**conjugate**" bands of Davis, 1999, and Ahlgren, 2001; Fig. 7.32). This suggests that the orientations of both linking and conjugate bands (as opposed to their particular location (Davis, 1999)) in these examples are closely related to the far-field stress state (Schultz and Balasko, 2003). In these examples, as illustrated in Fig. 7.34, the linking bands form only between overlapping or parallel ("bounding") bands as long as the spacing between the bounding bands is sufficiently small (e.g., Ahlgren, 2001, his Fig. 4a); overlapping bands having wide spacings are not generally associated with linking bands. Conjugate bands may crosscut a bounding band (to which they are not related) whereas linking bands are contained within a pair of bounding bands.

These and other deformation band arrays have been described as Riedel shear zones by many including Davis (1999), Davis et al. (2000), Ahlgren (2001), Katz et al. (2004), Barrell et al. (2011), Quigley et al. (2012), and Chemenda et al. (2016). Under this interpretation, the bounding bands would correspond to R-shears (e.g., Tchalenko, 1970; Bartlett et al., 1981) and the linking bands to (conjugate) R′-shears. This interpretation is somewhat different than what we see as classic Riedel shearing in rocks (Fig. 5.30), in which R-shears are characteristically connected by P-shears (e.g., Tchalenko,

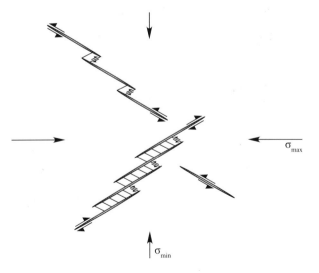

Fig. 7.34. Conjugate arrays of deformation bands can form under an appropriate far-field stress state. Note that the orientations of linking bands of one set are the same as those of the conjugate bounding bands.

1970; Bartlett et al., 1981). The growth of particular types of Riedel shears (e.g., R′- or P-shears) depends on the physical properties of the deforming material (e.g., Bartlett et al., 1981). The patterns discussed here and seen in nature suggest a response of the host sandstone to shearing that favors growth of DBs in conjugate R-shear orientations, most likely as a result of particular (pre-peak) localization mechanisms of bounding and linking bands (Okubo and Schultz, 2005, 2006).

What controls the *sense of step* of the putative R-shears (i.e., the bounding bands)? In Fig. 7.35a, the right-lateral bounding bands are left-stepping, producing a contractional sense of strain within the stepover (see Figs. 3.5 and 7.4). This configuration also leads to strain localization (i.e., a fault zone) in a plane. In the absence of a planar shear zone—to which the Riedels would be oriented—the configuration in Fig. 7.35b would seem to be equally plausible, with right steps between the bounding bands, ideally producing a dilational stepover (Fig. 7.35b). Once a contractional stepover between echelon deformation bands is formed, however, it will continue to grow parallel to its overall strike with the same sense of step (Du and Aydin, 1991); this underlying mechanism of near-field mechanical interaction is how a Riedel discontinuity geometry is formed, as illustrated schematically in Fig. 7.36.

What appears to be required is a mechanism to form the bounding bands (heavy lines in Fig. 7.35a) first, then the linking bands between them. Schultz and Balasko (2003) inferred that the sense of step and timing relations are related to the mechanical interaction between the echelon bounding bands (e.g., Fig. 7.36). Using the distortional strain energy density criterion, they, and Okubo and Schultz (2006), showed that in-plane growth of bounding bands leads to a contractional echelon stepover that, with increasing values of overlap (see Fig. 5.4), leads naturally to the ladder geometries described by Davis (1999) and Davis et al. (2000). The sequence is shown in Fig. 7.36 where one can also imagine how Riedel arrays could be produced from the successive addition of ladders. One might envision such a sequence creating characteristic patterns of deformation bands such as that shown in Figs. 7.9 and 7.32.

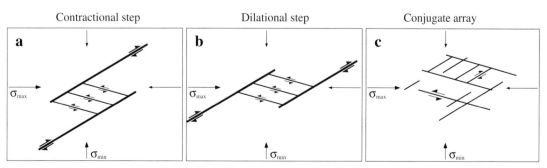

Fig. 7.35. Possible geometries of spaced deformation bands when viewed normal to the shearing direction. (a) Contractional step–observed; (b) dilational step–not observed; (c) synchronous and cross-cutting bands in both orientations–observed for widely spaced bands (e.g., Ahlgren, 2001).

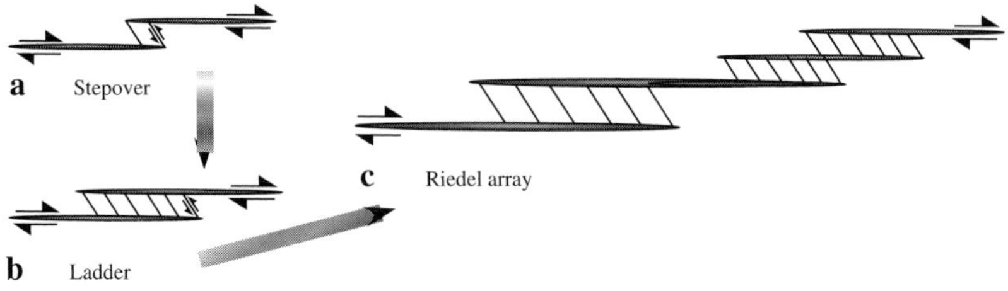

Fig. 7.36. Sequential growth of a contractional echelon stepover into "ladder" geometry and, in turn, into a "Riedel" array of deformation bands.

7.5.4   Damage Models, or Which Came First?

The concept of **damage to brittle materials** as they deform progressively is an icon of engineering mechanics (e.g., Bazant, 1976; Mazars and Pijaudier-Cabot, 1996; Lemaitre and Desmorat, 2005). Suppose that the local stress state or near-tip yield properties are such that a single dominant fracture cannot propagate within the material. In this case, the material may instead develop closely spaced arrays of microcracks, which cause the material to respond nonlinearly to stress, with reduced values of stiffness and modulus, so that it can deform to larger values of strain (i.e., become "more ductile") (Lawn, 1993, p. 361). In a somewhat different usage, damage models have been used to predict the response of a stressed material to dilatant fractures, whether single ones (the largest cracks related to *damage tolerance*) or populations (e.g., Lawn, 1993, pp. 335–362). One can easily calculate the largest opening-mode fracture that can be stable (i.e., won't propagate) under a given load (e.g., Chapter 8, exercise 6) as an example of damage tolerance (see also Anderson, 1995, pp. 552–554). These are standard approaches in engineering.

The idea that rocks can accumulate damage as they deform under stress is not new: it is, in fact, a basic tenant of the **Complete Stress–Strain Curve for Rock** as defined by laboratory experiments for a wide range of conditions (see Chapter 3). Distributed microcracking generally begins to become recognized or important above approximately 50% of the peak stress value (e.g., Hawkes and Mellor, 1970) and continues into the post-peak region, which is characterized by the coalescence of this damage into larger macrocracks that then divide the deforming sample into partially connected blocks (e.g., Wawersik and Fairhurst, 1970). At outcrop scales and larger, these macrocracks may correspond in a general way to **tectonic joints**, and their role in weakening the rock mass, and in making it less stiff, is well known (e.g., Bieniawski, 1989; Kachanov, 1992; Schultz, 1996; Hudson and Harrison, 1997; Hoek, 2002, 2007).

Faults can also localize within a zone of microcracks in otherwise unfaulted or unjointed rock, such as granite (Lockner et al., 1991; Lockner, 1995, 1998). Here, as in many engineering materials, the microcrack arrays define a "damage zone" of reduced stiffness at certain locations within the rock (e.g., Caine et al., 1996; Kim et al., 2004; Faulkner et al., 2010; Choi et al., 2016). Because this zone is softer than its surroundings, it will localize strain and lead eventually to an instability that produces a slip surface within the zone, which we identify as a fault (e.g., Goodman, 1989, pp. 69–71). Laboratory testing on Berea Sandstone by Riedel and Labuz (2007) showed that microcracking and local transient dilatancy led to the nucleation and (in-plane) propagation of a compactional shear deformation band.

Once a fault or other macrocrack forms, however, the host rock may sustain additional damage. One may postulate **two classes of strain localization** occurring within the rock:

Two components of damage zones:

♦ *Precursory damage*
♦ *Resultant damage*

1. **Precursory damage** of the host rock that leads eventually to a new macrocrack (or, more generally, a new macrofracture); and
2. **Resultant damage** of either the host rock, the precursory damage zone, or (most likely) both, by displacement and strain accumulation along the macrofracture.

In engineering materials with a multitude of tiny cracks, it may not be feasible to distinguish the products of these classes. In geology, these classes may also be combined (e.g., Kim et al., 2004). Even so, the distinction between precursory damage and the more localized damage associated with the resulting discontinuities is considered important (e.g., Kanninen and Popelar, 1985; McGrath and Davison, 1995). In common geologic situations, discontinuities in rock can be large enough compared to the scales of observation and sampling methods (e.g., mapping, boreholes, geophysical methods, fluid flow) that a distinction between these two classes can be made.

Let's start with precursory damage zones made up of fractures that principally undergo **strain-softening** during strain accumulation (i.e., cracks, joints, and faults). In addition to the mechanical effect of the damage zone (i.e., strain softening), the hydrologic effect is to provide increased fracture porosity and *fracture permeability* (if the microcracks interconnect, as can often be observed or inferred). As a result, a damage zone created from microcracks can provide a conduit that facilitates fluid flow within the rock. Fluid flow within the fracture network can be reduced to varying degrees by diagenetic infilling (cementation) that can restrict flow through fracture apertures that are sufficiently small (e.g., Olson et al., 2007).

Now let's consider precursory damage zones made up of structures that **strain-harden** during strain accumulation (i.e., certain types of deformation bands). Here, both the mechanical and hydrologic properties are opposite to those in the previous case, since the damage zone strain-hardens (Schultz and Balasko, 2003) and also can develop reduced hydraulic conductivity (e.g., Antonellini and Aydin, 1994; Crawford, 1998; Matthäi et al., 1998; Shipton et al., 2002; Fossen and Bale, 2007; Rotevatn et al., 2009) that can be important in modeling production in oil and gas reservoirs (Qu and Tveranger, 2016).

Building on the classic sequence (Aydin, 1978; Aydin and Johnson, 1978), cataclastic deformation bands form first as individual strain-softening structures that accommodate progressively larger amounts of strain, strain-harden, and lock up. Further deformation of the host rock is then taken up by formation of a new band adjacent to the previous one (Aydin and Johnson, 1983); in this way, a wider zone of deformation bands is built up by accretion of new bands and consequent thickening of the zone with increasing strain accumulation (Aydin and Johnson, 1978). This sequence is shown in Figs. 7.28 and 7.37. At some point, the zones "break loose" from the rest of the deforming rock to serve as the nucleation site for a fault having discontinuous displacement across its surface (Aydin and Johnson, 1983) (not shown in Fig. 7.37). Although slip surfaces may indeed be observed to have formed adjacent to zones of deformation bands, the resulting fault, composed of hard-linked slip surfaces, commonly cuts through the middle of the precursory damage zone (e.g., Shipton and Cowie, 2001, 2003). This sequence requires that *individual bands form first*, as the earliest structures in a rock—**relicts of the initial strain**. Measurements of this precursory zone of deformation bands, from several porous sandstones, suggest that widths exceeding ~5–8 m may be associated with faults, while smaller widths lack faults (Fossen and Hesthammer, 2000; Shipton and Cowie, 2001; Schueller et al., 2013).

As a fault in otherwise porous rock is approached, the frequency of deformation bands (i.e., number per meter along a linear traverse) characteristically *increases* (Antonellini and Aydin, 1994; Fossen and Hesthammer, 2000; Shipton and Cowie, 2001, 2003; Schueller et al., 2013). The fault thus occurs within a swarm of deformation bands whose frequency decreases away from it. Most importantly, however, the *bands came first*, and the *fault came later*. Hesthammer et al. (2000) and Rotevatn et al. (2009) interpreted arrays like those described here as (at least partly) pre-fault damage zones.

An alternative view is based on the increased level of stress that is generated near discontinuity tiplines, or the smaller slip patches nucleating along them, as they accommodate their localized displacements. Here, faults (or the smaller slip patches along them) and the other discontinuity types act as stress concentrators

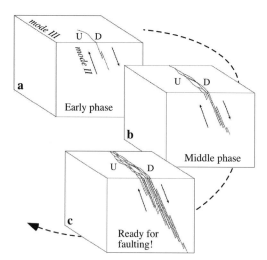

Fig. 7.37. Growth of strain-hardening "precursory damage zone" of deformation bands in porous rock in 3-D (after Schultz and Balasko, 2003).

that increase the stress magnitudes in some places and soak up the stress in others (creating a "zone of influence" or shadow zone around the structure). Elevated stress magnitudes near the discontinuity tip, if sufficient to fulfill a particular strength criterion, can produce a "process zone" of localized deformation, akin to the *plastic zone* ahead of a fracture in a perfectly elastic material (Delaney et al., 1986; Pollard and Aydin, 1988; Engelder et al., 1993). Propagation of the fault (or more generally, any discontinuity type) to greater lengths would essentially bisect the process zone (along with generating a new one adjacent to the new tip), leading to a **"process zone wake"** (Pollard and Aydin, 1988; Lawn, 1993, p. 57; Scholz et al., 1993; Broberg, 1999, pp. 27–29; Scholz, 2002, pp. 118–121; Gudmundsson, 2011, pp. 299–302) and, consequently, a relict zone of deformation, or breccia, surrounding the discontinuity. Zones that meet this description are sometimes observed or inferred around faults (e.g., Scholz et al., 1993; McGrath and Davison, 1995; Caine et al., 1996; Vermilye and Scholz, 1998; Shipton and Cowie, 2001, 2003; Du Bernard et al., 2002a,b) and igneous dikes (Mériaux et al., 1999), leading to the concept of a **damage zone** *being produced by the fault or fracture itself.*

How can a geologist discern which came first? Several avenues have been suggested in the literature that can be used; these include:

1. **Presence of a large discontinuty (fault, slip surface, dike, etc.).** If zones of deformation bands are identified in a host rock and there are no significant slip surfaces or large faults associated with them, then it would be difficult to infer that the deformation bands had formed due to a process zone wake. In this case, they would be identified as *precursory* damage zones. This may be most common in low-strain settings or in strata located up- or down-section from a pre-existing fault.

2. **Consistency of local strain fields with the regional one.** Early damage-related discontinuities (microcracks or deformation bands) grow largely in isolation from others and predate the through-going large-offset structure; these small structures should thus track the regional stress and strain state and therefore be consistent in orientation throughout the host rock. In contrast, damage localized near the tips of large-offset discontinuities is spatially heterogeneous, with distinct orientations from place to place that depend on the position relative to the dominant structure.

3. **Correlation of damage width with throw or offset.** Several workers have demonstrated a positive correlation between damage zone width and the displacement magnitude along the large-offset structure within it (e.g., Shipton and Cowie, 2001; Scheuller et al., 2013). This may be considered compelling evidence for a *resultant* damage zone.

Two trajectories for strain localization:

♦ *Initial softening, then localized softening*
♦ *Initial hardening, then localized softening*

As can be seen, there are several lines of evidence that can help a geologist to infer whether the structures arrayed about the principal one formed generally *first* (as precursors to strain localization) or *second* (as a consequence of strain localization).

Shipton and Cowie (2003) identified an important set of mechanical consequences for damage zones in rock, which have since been supported by numerous subsequent studies by other investigators. In general, there are *two main trajectories for strain localization of a host rock* into damage zones and, in turn, into larger-offset structures:

- **Strain softening followed by strain softening.** This is the traditional one that has been associated with "damage" in engineering materials and also in compact, low-porosity rocks (like granite). First, microcracks form in the rock prior to peak strength, which make this dilatant damage zone *less stiff* (i.e., softer or more compliant) than its surroundings. As a dominant fracture grows within this zone, it takes over and continues as a "zone of weakness" by further softening the rock mass. Faults and joints are good examples of structures that tend to follow this trajectory.

- **Strain hardening followed by strain softening.** This path is followed in high-porosity rocks like sandstones and others that localize strain initially as cataclastic deformation bands, noting that the bands while actively accumulating displacements must be temporarily softer than their surroundings. The resulting damage zone of strain-hardened and frictionally locked deformation bands becomes stiffer, on average, than the host rock, so it acts as a more rigid inclusion (e.g., Aydin and Berryman, 2010; Meng and Pollard, 2014) within its less deformed surroundings. After sufficient strain the inclusion breaks, forming a few large-offset fractures such as faults. The change in deformation mechanism from deformation bands to frictional sliding marks a fundamental change in the dominant deformation behavior from initial strain hardening to later strain softening.

The *hydrologic implications* of each path are different too. In the first case, the fracture permeability increases in the damage zone and again in the through-going fracture. But in the second case, the fluid flow becomes increasingly reduced as the less-permeable damage zone grows, leading to compartmentalization and local barriers to flow (e.g., Shipton et al., 2002). Once a large-offset fracture—like a fault—slices through the hard damage zone, however, it opens up a high-permeability conduit for fluid flow that may be even better than the host rock itself. Much interesting and useful work has demonstrated the evolution of fault zones along these lines (e.g., Davatzes and Aydin, 2003). For example, building on the mapping results of Myers and Aydin (2004), Flodin and Aydin (2004a,b) showed how a damage zone due to fault slip can exploit echelon twist hackles ("fringe joints") formed during an earlier period of jointing.

In both cases, a **fault core** may eventually develop at large values of shear strain; this core and its accumulated pulverized debris, slip surfaces, and gouge can substantially reduce the permeability of the core (e.g., Chester et al., 1993; Caine et al., 1996; Billi et al., 2003; Shipton and Cowie, 2003) for either path described above. However, the core may still be a strain softening feature depending on its physical conditions and pore-fluid state during slip.

7.5.5 Paths to Faulting for the Different Types of Deformation Bands

In the first part of this section, the results of the classic Aydin–Johnson model and the Davis model are correlated and integrated into a coherent package. Far from being different and distinct characterizations, geometries, or sequences, these models are in fact complementary and simply describe different aspects of the same problem. The architecture and sequence outlined here also best apply to high-porosity rocks, such as well-sorted sandstones, with little diagenetic overprint, and to cataclastic deformation bands, implying sufficient confining pressure or differential stress for grain crushing to occur. However, it gives clues as to the development of deformation band characteristics, patterns, and networks in other rock types. It is also becoming increasingly recognized to apply most directly to deformation bands and faults forming in *extensional* tectonic settings (Soliva et al., 2013, 2016). The relationship between type of deformation band, faulting, and tectonic regime in a given rock type is emphasized in the latter part of this section.

The 3-D geometry of the simplest cataclastic deformation band array is best described as a pair of deformation bands that interact in a way that depends on the viewing direction (Fig. 7.38) but which share the same kinematics (e.g., overall shearing sense). As a pair of echelon cataclastic deformation bands grow in-plane into an overlapped configuration (with a contractional stepover), the mechanical interaction between them becomes sufficient to promote the growth of linking bands into the stepover but insufficient to shut down propagation of the bounding bands (Schultz and Balasko, 2003; Okubo and Schultz, 2006). This sequence parallels that for other echelon arrays such as dilatant cracks and faults as described in Chapter 5. As a result, echelon cataclastic deformation bands that have appropriate values of perpendicular spacing relative to their amount of overlap can grow into ladders. There seems to be a natural progression in many field examples from contractional echelon stepovers (Fig. 7.36) to ladders and Riedel arrays by successive in-plane addition of DB sets in the down-dip (shearing) direction (Fig. 7.39).

Strain hardening of individual cataclastic deformation bands leads to nucleation of new bands adjacent to previous ones and consequent widening of the resulting zone (e.g., Aydin and Johnson, 1978; Antonellini and Aydin, 1994). This scenario was originally identified for a cataclastic deformation band array when viewed parallel to shearing (the mode-III sense; Fig. 7.38). Schultz and Balasko (2003) inferred that widening of the zone in mode III also requires the addition of new contractional echelon stepovers in mode

Fig. 7.38. The 3-D geometry of the interacting pair of echelon cataclastic deformation bands showing the importance of viewing geometry on kinematic interpretations of these arrays (after Schultz and Balasko, 2003).

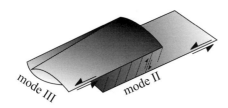

Fig. 7.39. A synoptic model for fault growth in porous rocks (after Schultz and Siddharthan, 2005). (a) **Single** cataclastic deformation bands with incipient linkage of closely spaced bands. U, footwall up; D, hanging wall down. (b) **Zones or clusters** of closely spaced cataclastic deformation bands: linking bands in contractional echelon mode-II stepovers (*duplex* formation) and inosculating bands in mode III (eyes or *lenses*). (c) Localization of a Coulomb slip surface (heavy lines) and **frictional sliding** on a **fault** that slices through the zone of cataclastic deformation bands.

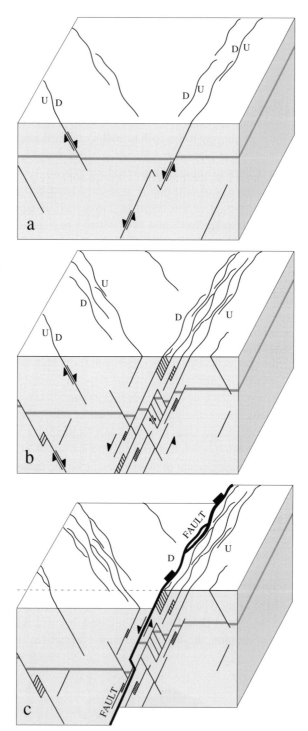

II (i.e., ladders; see Fig. 7.37). Because the "backward-breaking" stepover geometry of DBs (Fig. 7.33b) is thought to be geometrically less efficient in accommodating large shearing displacements than the "forward-breaking" geometry that is characteristic of strain-softening structures (Fig. 7.33a), the mode-II arrays may themselves lock up and, in conjunction with strain hardening of the deformation bands, further promote widening of the cataclastic deformation band zone. By implication, cross-strike widening of the array as strain hardening and kinematic lock-up proceed may lead eventually to the spatially distributed "radiator rock" geometry described by Davis (1999) and identified as the pre-fault "damage zone" (e.g., Shipton and Cowie, 2003; Schultz and Siddharthan, 2005; Fig. 7.39b). As a result, formation of fluid compartments and flow barriers, along with the damage zone and overall fault-zone architecture, may follow a systematic and progressive sequence (Fig. 7.39a–c) that hinges on the mechanism of fracture or deformation band localization.

Unraveling the architecture and sequence of deformation-band networks requires that a distinction be made between *formation* of a deformation band and its later *faulting*. These processes are referred to in the literature as **yielding** (for band nucleation and growth) and **failure** (for faulting of the now-formed bands), respectively. Yielding is defined in this context as the change in deformation behavior of the rock from elastic to permanent, inelastic (i.e., plastic) strain. Because cataclastic deformation bands commonly have reduced porosity, different grain packing, and crushed, angular, and more poorly grains, they are permanent features of the host rock and therefore are consistent with inelastic yielding of the rock.

A representative and schematic yield surface for a porous rock was shown in Figs. 7.21, 7.24, and 7.25. Dilation bands with shear are thought to form at smaller values of confining pressure and/or depth to the left on the diagram, whereas compaction bands with shear (the most common type of deformation band) would form farther to the right, on or near the cap, at higher values of confining pressure and/or depth (Schultz and Siddharthan, 2005; Philit, 2017; Robert et al., 2018; Fig. 7.40). Compaction bands with shear that also have cataclasis additionally require a specific rock type that includes large grains, minimal cement or matrix, and an open (loose) packing configuration (e.g., Zhang et al., 1990; Antonellini and Pollard, 1995)—all of which are consistent with eolian sandstones. Pure dilation bands (Du Bernard et al., 2002a,b) apparently form at very low pressures near the left side of the diagram, whereas pure compaction bands (Mollema and Antonellini, 1996) form quite far down along the cap (Grueschow and Rudnicki, 2005), perhaps with confining pressures at some moderate to large fraction of the critical grain crushing pressure of the host rock (Zhang et al., 1990; Wong et al., 1997, 2004; Wong and Baud, 2012). As discussed previously, this value depends on the host rock's porosity, grain size (for a given mineralogy and grain shape),

Fig. 7.40. Schematic yield envelope for porous granular rocks showing approximately where the main types of deformation bands are thought to localize (DiB, dilation band; SB, isochoric shear band; CSB, compactional shear band; SECB, shear-enhanced compaction band; PCB, pure compaction band), and kinematic regimes (NF, normal faulting; SSF, strike-slip faulting; TF, thrust faulting). Locations and extents are schematic. Schematic consolidation curve and tectonic loading paths to extensional (NF) and contractional (TF) tectonic regimes shown by heavy dashed lines.

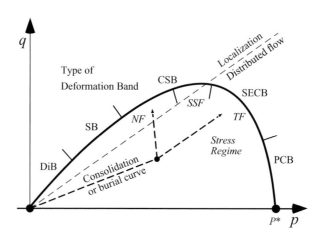

and sorting (Cheung et al., 2012) and may be related to its maximum depth of burial (Skurtveit et al., 2013).

The synoptic architecture developed here works best for arrays of cataclastic deformation bands that form in high-porosity rocks. As porosity and average grain size decrease within (for example) a sandstone, deformation bands become less planar and the arrays less geometrically regular, so that in a porous silty sandstone, the arrays look like filamentary networks of intersecting nonparallel fine deformation bands. Because the **host-rock properties and the stress state** exert a fundamental pair of controls over the mechanisms—and therefore the field expression—of strain localization in rocks (e.g., Olsson, 2000; Issen, 2002; Wong et al., 2004; Soliva et al., 2016), it may be anticipated that arrays of deformation bands should be developed and expressed differently in different lithologies, depths, and tectonic settings.

So how do faulted deformation bands occur? Here the types of deformation bands found in nature that are commonly faulted (or not), and their angular relations, provide us with some clues. Cataclastic shear deformation bands, for example, form at acute angles to the inferred maximum compressive principal stress direction, which places them in favorable orientations for frictional sliding to occur, given suitably larger stress differences (Figs. 7.8 and 7.10). Such favorably oriented bands, whether cataclastic or not, may become faulted without much difficulty. One possibility is a nearly spontaneous nucleation of frictional sliding (Sheldon et al., 2006) during the strain-softening stage that should accompany the early stages of displacement accumulation, associated with grain mobilization and early fragmentation, along the deformation bands.

Similarly, a precursory damage zone of **cataclastic deformation bands** growing within an extensional tectonic setting may require only somewhat greater levels of differential stress and confining pressure (q and p, respectively) for faulting to occur, given strain hardening of the bands and perhaps the damage zone band network as well (Schultz and Siddharthan, 2005). This inference is based on results from soil mechanics that suggest that strain hardening of individual shear bands in a soil can be associated with an expansion of the yield surface to the upper right on a q–p diagram (Davis and Selvadurai, 2002, p. 200). Physically this would require a greater amount of differential stress to fail a strain-hardened shear band in the soil; a corresponding situation may be envisioned for the failure (faulting) of strain-hardened networks of cataclastic deformation bands in porous rock. Faulting of a precursory cataclastic deformation band zone might then be predicted to occur when the expanded yield envelope intersects the frictional sliding line on the q–p diagram (diagonal dashed line to the right on Fig. 7.25a), which would represent the peak frictional strength of the precursory array at that point in its evolution. Faulting could then occur given appropriate stress paths that could lead to an intersection with the failure criterion.

The scenario just outlined implies that rather small increases in differential stress (i.e., tectonic loading) and/or confining pressure (i.e., depth) may be sufficient to nucleate faulting along deformation band arrays that initially could be located at stress values corresponding to the upper part of the yield envelope for the host rock (Fig. 7.25a). This may correspond to the most common examples found in nature—faulted arrays of strain-hardened cataclastic deformation bands.

Faulting of dilatant and compactional shear bands

In contrast, **shear-enhanced compaction bands** are observed to form with more oblique dihedral angles and seemingly less favorable orientations relative to the maximum compressive principal stress direction for frictional sliding (Figs. 7.8 and 7.10). They also tend to form in contractional tectonic settings that would promote larger values of confining pressure at relatively shallow depths (Soliva et al., 2013, 2016; Robert et al., 2018). As these bands accumulate contractional and shear strains, it could be inferred that the volume of host rock that contains them stiffens and strain hardens, leading to an expansion of its yield envelope to the right on the q–p diagram. One could envision that faulting in such contractional cases could occur only for very special, steep loading paths that have sufficiently large q/p ratios, so that in general, faulting may not be expected to be common in such situations (Soliva et al., 2013). Instead, the yield surface might expand more horizontally to the right (Fig. 7.24b), with loading stress paths that potentially never intersect the frictional-sliding line; these bands would grow into a nonlocalized, distributed network of shear-enhanced compaction bands instead (e.g., Wong et al., 2001) corresponding to **macroscopic flow** of the rock. In this scenario, compaction bands that have the largest ratios of compaction/shear displacements, such as pure compaction bands, might have the hardest time providing well oriented precursory substrates for faulting.

Faulting of shear-enhanced compaction bands?

7.6 Review of Important Concepts

1. Deformation bands are an important class of localized geologic structural discontinuity that occur widely in **porous rock types** such as sandstones, limestones, chalks, poorly consolidated sediments, and pyroclastic tuffs. These structures were previously considered to be rare or arcane but are now widely recognized to be common and important structures.

2. Deformation bands are **tabular zones of grain rearrangements**. The strain localization and grain rearrangements can occur by several mechanisms including grain reorientation into the shearing and/or compaction directions, reduction of pore space by tighter packing of grains, and grain fracturing and crushing (also leading to reduced pore space by filling the voids with debris). Deformation bands become less stiff as they accommodate shear and/or normal offsets relative to the surrounding host rock; continued deformation may lead to stiffening within the band as grains interlock.

3. Both mechanistic and kinematic approaches have been used to classify and describe deformation bands. The former relies on fabric and petrophysical characterization whereas the latter requires identification of the sense and magnitude of offsets along the band.

4. The five main kinematic classes are: pure dilation bands, dilational shear bands, shear bands, compactional shear bands (with or without cataclasis), and compaction bands (either shear-enhanced compaction bands or pure compaction bands). Each of these carries both mechanistic and kinematic implications and systematic relationships between their displacement distributions, orientations, host-rock properties, and far-field stress states can be surmised.

5. Many interesting patterns can help to uniquely identify deformation bands relative to other kinds of structure, including Riedels, ladders, radiator rock, positive-relief fins, clusters and zones of deformation bands, and backward-breaking echelon linkages. These patterns and modes of expression can be used to infer the presence of deformation bands and, by extension, porous granular rock types.

6. Mechanically, deformation bands function as a type of weak discontinuity, announced by continuous displacement gradients across the bands. Deformation bands concentrate and redistribute stress much as cracks and faults do, but may lack singular or highly concentrated stress states near their tips. The effect is to reduce the magnitude and spatial extent of mechanical interaction between the bands. Nevertheless, many of the same characteristics of fractures, such as bounded lengths and heights, displacement variation (reducing to minima toward the tips), mechanical interaction and linkage, spacing,

and displacement–length scaling are demonstrated for deformation bands.

7. The orientations of deformation bands are related to their mode of occurrence and hint at the remote stress state. Orientations may depend on rock properties such as friction and dilation angles, and on the host rock's stress–strain relations, such as associated or non-associated plastic flow laws. Orthorhombic patterns of deformation bands are common and may indicate 3-D far-field strains if the bands are mutually crosscutting (i.e., synchronous). Because these structures typically preserve the displacement sense and magnitude across them, unlike faults that can record multiple and different slip events, deformation bands provide a graphic record of rock deformation at various stages in its sequence.

8. Deformation bands grow into distinct collections of structural elements with ongoing deformation, from individual bands at small strains through band networks, zones of bands, clusters, and sometimes to through-going faults at larger strains. Each stage carries implications for fluid flow depending on the types of structures produced, their connectivity, and their spatial extent.

9. The deformation of soft or porous granular rock shares many commonalities with that of soils and unconsolidated materials, motivating a transfer of concepts from soil mechanics into structural geology. Description of porous granular rocks as loose or dense, normally or over-consolidated, and attention paid to porosity and microstructure and their evolution with deformation, for example, have become part of the structural geologist's lexicon when working with deformation bands and, more generally, with the deformation of many types of sedimentary materials.

10. Yielding refers to the change in behavior from elastic to inelastic deformation, whereas failure refers to breakage at the rock's peak strength. Yielding in rocks always occurs before failure. The q–p diagram has become a common means for relating stress state, material properties, yielding, and failure for soils and sedimentary rocks. Concepts such as critical state lines, yield envelopes, yield caps, critical pressure, localized dilatant yielding, and distributed compactional flow all can inform the interpretation of deformation bands.

11. Yielding and failure of mudrocks can be related to their composition, degree of consolidation, and diagenetic maturity. Both fractures and deformation bands can form in these rock types, providing clues about practical issues such as sealing capacity. Brittleness (a failure property) of mudrocks has been related to pre-failure properties such as those mentioned as well as stiffness, fostering predictions in the oil

and gas industry, for example, of the ability of hydraulic fractures to propagate, link, and thereby stimulate recovery from unconventional reservoirs.

12. Deformation bands and faults can be intimately related parts of a progressive deformation sequence in sedimentary rocks. Assemblages of deformation bands associated with faults can form as either precursory or resultant damage depending on the relative timing of fault slip within the zone.

7.7 Exercises

1. Describe the main characteristics of deformation bands following both classification systems, including specific criteria that one could use to distinguish them in the field.

2. What factors contribute to strain hardening of deformation bands, and why? How might one tell if a particular band had strain hardened relative to the surrounding host rock?

3. What factors contribute to strain softening of deformation bands, and why? How might one tell if a particular band had strain softened relative to the surrounding host rock?

4. Compare and contrast yielding and failure in rock deformation.

5. Describe the implications for fluid flow of deformation bands in a porous host rock, including each of the different types of bands that have been recognized. Be sure to include an explanation of the causes of the change in flow characteristics in or along the band.

6. Summarize the evolutionary sequence from deformation bands to faults, assuming deformation bands with cataclasis. How would this sequence differ if the bands did not develop cataclasis? How would it differ if the bands were dilational with shear, instead of compactional?

7. Define "damage zone" and describe how one is made. Sketch out the anatomy of a damage zone, labeling all the relevant parts. Why is it important to know when damage zones are composed of strain-softening discontinuities as opposed to strain-hardening ones? Give an example of each of these along with their large-strain consequences.

8. Summarize the q–p diagram as used in deformation band studies including the effect of factors such as rock friction, water content, critical state, and their evolution with increasing deformation. Describe how each of the main kinematic types of deformation bands can form using this diagram as the context. Include orientations of bands, conjugate sets if appropriate, degree of strain hardening, types and relative amounts of offsets accommodated, tectonic setting, and inferred depth of burial during formation in your answer.

9. What is the difference between strong and weak discontinuities? List several ways that these might be identified and distinguished at the field, hand sample, or thin section scales.

10. Describe the relationships between deformation bands and faulting including spatial and temporal factors.
11. Describe the main characteristics of porous and compact rocks and what the implications are for their deformation style.
12. Outline the characteristics of porous rock and how they influence its deformation, including loose, dense, degree of compaction, consolidation, and brittleness.
13. Define brittleness and evaluate the physical basis for each approach.

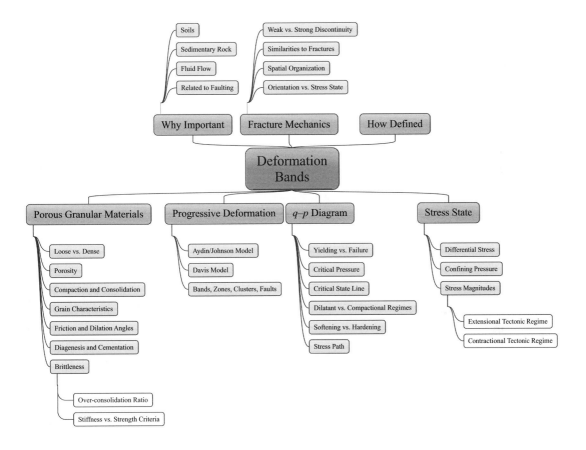

8 Fracture Mechanics: A Tour of Basic Principles

8.1 Introduction

This chapter lays out a *simplified roadmap* to the main aspects of fracture mechanics that are important for geologic structural discontinuities. Fracture mechanics is a branch of engineering that nevertheless provides tools, concepts, and a mechanical context for interpreting the major characteristics and patterns of geologic discontinuities that are seen in the field (e.g., Rudnicki, 1980; Pollard and Segall, 1987; Engelder et al., 1993; Pollard and Fletcher, 2005; Segall, 2010; Gudmundsson, 2011).

Chapter 1 outlined a basic overview of what discontinuities are, how to describe and name them, and several important aspects of their geometry and kinematic significance. Chapters 4 and 6 sketched out many of the important field relationships for opening-mode (cracks, joints, veins, and dikes), closing-mode (stylolites and compaction bands), and shearing-mode (fault) structures (Fig. 8.1), all of which can be treated and understood by using fracture mechanics. The kinematics, patterns, and propagation of deformation bands (Chapter 7) into complex networks can also be understood by using fracture mechanics. This chapter provides many of the fundamentals of this exciting approach.

As a discipline, engineering fracture mechanics began in the early 1900s (e.g., Inglis, 1913; Griffith, 1921; Kolosov, 1935; Westergaard, 1939) with rapid development occurring in the latter half of that century (e.g., Irwin, 1957, 1960, 1962; Sih et al., 1962; Erdogan and Sih, 1963; Muskhelishvili, 1963; Paris and Erdogan, 1963; Paris and Sih, 1965). Several works including Rice (1968), Kanninen and Popelar (1985), Broek (1986), Lawn (1993), Anderson (1995), Landes (2000), and Newman (2000) elegantly describe the development of engineering fracture mechanics and its critical importance to the design and reliability of buildings, welded plates, aircraft, spacecraft, nuclear reactors, and ships. These works all highlight the central importance in typical engineering materials such as steel and ceramics of the near-tip stress state that leads to the use of the **stress intensity factor** and **dynamic fracture toughness** in assessing or predicting fracture stability and propagation (e.g.,

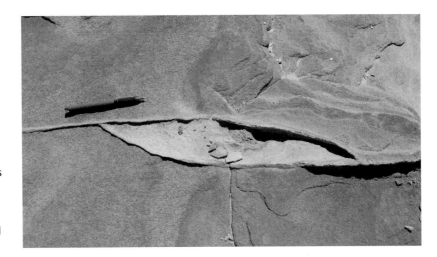

Fig. 8.1. Stepover between two echelon compaction bands in eolian sandstone in Valley of Fire State Park, Nevada, implying mutual propagation of the bands into a hard-linked stepover geometry.

Sih et al., 1962; Paris and Sih, 1965; Tada et al., 2000; Ritchie, 2011). As a result, the starting point of engineering fracture mechanics invariably begins with stress analysis of cracks with an emphasis on the near-tip stress state and continuing with propagation criteria such as stress intensity factor, strain-energy release rate, or the J-integral (Rice, 1968; Kanninen and Popelar, 1985; Anderson, 1995).

A distinguishing attribute of engineering fracture mechanics is discontinuity size. In general, fractures in engineering materials are quite small, typically visible only under substantial levels of magnification. In contrast, many geologic discontinuities are large enough for their dimensions and physical properties to be identified and measured. The sense of displacement across the geologic structure, such as crack or fault and, indeed, their relative orientations to the Earth's surface are required to correctly identify and name the structure (e.g., joint, sill, or dike; normal, strike-slip, or thrust fault). Accordingly, the **displacement magnitudes and distribution** on a geologic fracture or deformation band are of primary importance in the study of geologic structural discontinuities, with the stress, displacement, or strain field in the surrounding rock mass being a closely related attribute (e.g., Chinnery, 1961; Lachenbruch, 1961; Pollard and Segall, 1987; Cohen, 1999; Segall, 2010; Gudmundsson, 2011).

In this chapter we focus on the mechanics of individual geologic fractures, noting that fracture mechanics applies equally well to all types of structural discontinuities. Starting with the concept of driving stress, we then move on to the simplest method of fracture analysis, namely Linear Elastic Fracture Mechanics (LEFM), which has been used extensively and successfully in the analysis of geologic structural discontinuities (e.g., see Chinnery, 1961, 1963, 1965; Lachenbruch, 1961; Secor, 1965; Palmer and Rice, 1973; Rice, 1979, 1980; Engelder and Geiser, 1980; Rudnicki, 1980; Segall and Pollard, 1980, 1983a,b; Fletcher and Pollard, 1981; Aydin and Nur, 1982; Atkinson, 1984; Aydin and Page, 1984; Kemeny and Cook, 1986, 1987; Engelder, 1987,

1993, 1999; Ingraffea, 1987; Pollard, 1987; Lin and Parmentier, 1988; Petit and Barquins, 1988; Pollard and Aydin, 1988; Kemeny, 1991, 2003; Engelder et al., 1993; Pollard and Fletcher, 2005; Segall, 2010; Gudmundsson, 2011).

Following the introduction of LEFM, we examine the displacement distribution along ideal LEFM fractures which then exert a profound influence on the stress and displacement fields surrounding them; these inhomogeneous fields are well described by analytical solutions (Pollard and Segall, 1987) that are discussed in this chapter. We then examine the rather special conditions predicted at a fracture tip under conditions of LEFM and explore how the infinite theoretical stress concentration at fracture tips leads to the concepts of stress intensity factor and dynamic fracture toughness that are critical to evaluations of fracture propagation.

8.2 What Makes a Fracture Open, Close, or Shear?

Few rocks or rock masses deform as completely continuous materials. Under the pressure, temperature, and strain-rate conditions prevalent in the upper crust (e.g., Kohlstedt et al., 1995; Lockner, 1995; Kohlstedt and Mackwell, 2010), deformation tends to *localize*, or *focus*, into discrete narrow zones that accommodate a large fraction of the total strain, with a smaller amount held in the matrix or host rock between these zones (e.g., Segall, 1984a). While the processes of strain localization into fractures and deformation bands are many and complex (e.g., Rudnicki and Rice, 1975; Lockner, 1995; Evans and Kohlstedt, 1995; Bésuelle and Rudnicki, 2004; Borja and Aydin, 2004; Paterson and Wong, 2005; Aydin et al., 2006), we'll focus here on **fractures**—cracks and faults—because they are some of the most important structures that can be understood and modeled by using fracture mechanics. However, other structures such as deformation bands and various types of anticracks can also by analyzed by using these concepts, as illustrated elsewhere in this book.

8.2.1 Driving Stress

Fracture walls displace because the tractions or stresses acting on them exceed the resistance to displacement built into the fractured material. The amount that the value of resistance is exceeded by is called the "**driving stress.**" Let's look at two main classes of driving stress—one for **cracks** (σ_{dI}, opening-mode fractures, based *on the normal-stress components*) and one for **faults** (σ_{dII}, shearing- or tearing-mode fractures, based *only on the shear-stress components*). In both cases the driving stress must be positive for displacements to build up along the fracture. Physically this means that the driving forces—such as shear stress—must exceed the resisting forces—such as frictional resistance—for displacements to occur along the fracture.

8.2.1.1 Cracks I–Pure Opening

The driving stress on a crack wall can be defined as the *net force that acts to pull the crack walls apart*. This net force, given as a stress, can be a tension applied at some distance from the crack, a compression applied to the internal part of the crack wall (like a fluid-pressurized vein or dike) that exceeds any remotely applied compressive stress, or some combination of the two. To cause the crack walls to separate, producing an opening displacement, the amount of driving stress is calculated by comparing the normal stress resolved on the plane of the crack (see Chapter 3) with the amount of stress applied internal to the crack (if any). For pure opening (mode-I) cracks, there is no shear stress resolved across the crack faces, so the **shear driving stress is zero**. However, the normal driving stress must be greater than zero for the crack walls to separate the crack to open (Pollard and Aydin, 1988).

In calculating the driving stress acting on a crack, a sign reversal is necessary due to the opposite directions of outer-normal vectors for remotely applied and internally applied stresses, as shown in Fig. 8.2. What physically

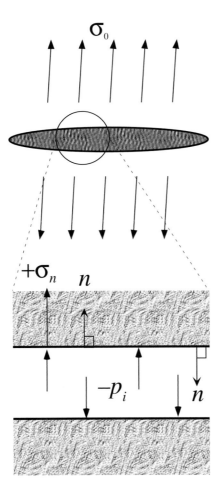

Fig. 8.2. Detail of force and outer-normal directions for the calculation of normal driving stress for a mode-I fracture.

happens is that the internal pressure, being compressive, serves either to augment (add to) the value of remotely applied tensile stress resolved on the crack plane or to partially reduce (subtract from) the amount of compression resolved on the crack plane.

The driving stress σ_{dI} is defined as the difference between the remote resolved stress acting on the fracture and the internal pressure p_i. Symbolically, it's written as

$$\sigma_{dI} = \sigma_n^\infty - p_i \tag{8.1a}$$

for compressive or tensile normal stress, and

$$\sigma_{dII} = 0 \tag{8.1b}$$

for a **pure mode-I crack with no resolved shear stress** across the crack tip.

In contrast to the compression-positive sign convention used in structural geology and frictional sliding along faults, we now switch to the **tension-positive sign convention**. This is because engineering fracture mechanics traditionally uses the tensor convention rather than the soil mechanics convention that we relied on for the Coulomb criterion and Mohr circle constructions (see Chapter 3 and Table 3.1). The following example illustrates the similarity between the two sign conventions by calculating the driving stress acting on a crack.

Example Exercise 8.1

Given a water-filled hydrofracture oriented normal to the maximum (tensile) remote stress, of magnitude 1.1 MPa (tension-positive with $\theta = 90°$), a least principal stress of 0 MPa, and having an internal fluid pressure of −0.15 MPa (compressive), calculate the driving stress acting to dilate the hydrofracture.

Solution

First we use the 2-D stress transformation equations (3.5) to calculate the normal stress acting on the crack wall due to just the remote stresses (see Fig. 8.3). Because the stresses are given as tension positive, we have to first switch the signs to match what we need for compression positive. By multiplying the stresses by −1, we get:

Tension positive	Compression positive
$\sigma_1 = +1.1$ MPa	$\sigma_3 = -1.1$ MPa
$\sigma_3 = 0$ MPa	$\sigma_1 = 0$ MPa
$\tau_{xy}^\infty = 0$ MPa	$\tau_{xy}^\infty = 0$ MPa
$\theta = 90°$ (from σ_1)	$\theta = 0°$ (from σ_1)
$p_i = -0.15$ MPa	$p_i = +0.15$ MPa

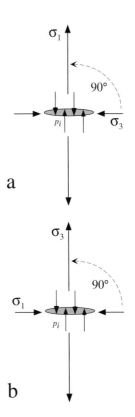

Fig. 8.3. Stress geometry for Example Exercise 8.1. (a) Tension-positive sign convention; (b) compression-positive convention.

Next, we plug the stresses and angle into equations (3.5)—using those listed in the compression-positive column—and find the value of the normal stress resolved on the crack wall. Assuming the compression-positive sign convention used in those equations, we get:

$$\sigma_n = -1.1 \text{ MPa } \left(\text{tensile}\right)$$

$$\tau_{xy}^{\infty} = 0.0 \text{ MPa}$$

Now we calculate the driving stress:

Tension positive	Compression positive
$\sigma_{dI} = (1.1 - (-0.15))$ MPa	$\sigma_{dI} = (-1.1 - 0.15)$ MPa
$\sigma_{dI} = 1.25$ MPa	$\sigma_{dI} = -1.25$ MPa

and $\sigma_{dII} = 0.0$ MPa. So the normal driving stress that causes dilation of the crack walls is 1.25 MPa of effective tension, as listed in the tension-positive (left-hand) column above.

8.2.1.2 Cracks II—Oblique Opening

In the previous section, we were interested only in whether or not the crack walls will separate, so only the normal stress was needed for that pure mode-I case. On the other hand, if the crack walls separate but the in-plane shear stress (Fig. 1.15) resolved on the crack plane is nonzero, then the shear stress serves to open the crack *obliquely*. This corresponds to what is known as a **mixed-mode I-II crack** (Pollard et al., 1982; Pollard and Aydin, 1988; Engelder, 1999) and which is occasionally referred to in the literature as "transitional tensile joints," "shear joints," or "hybrid extension/shear joints" (see Section 3.5.4). On the other hand, if the driving stress σ_{dI} on a crack were zero or negative, and the crack walls would not open, the shear stress would then need to overcome the frictional resistance of the walls for the fracture to slide frictionally and become a fault (and there, $\sigma_{dII} > 0$). This case is considered in section 8.2.1.3.

Example Exercise 8.2

Given a water-filled hydrofracture obliquely to the maximum (tensile) remote stress, again of magnitude 1.1 MPa (tension positive), but now at 5° to its outer-normal vector ($\theta = -5°$), with a least principal stress of -0.5 MPa (compressive), and having an internal fluid pressure of -0.15 MPa (compressive), calculate the driving stress acting to dilate the mixed-mode I-II hydrofracture.

Solution

Again we use the 2-D stress transformation equations (3.5) to calculate the normal stress acting on the crack wall due to the remote stresses (see Fig. 8.4). Because the stresses are given as tension positive, we could use the stress transformation equations that were derived by assuming this sign convention (equations (3.6)). However, in keeping with the previous example, we'll use the same compression-positive equations as before.

First we switch the signs to match what we need for compression positive. Multiplying the stresses by -1, we get:

Tension positive	Compression positive
$\sigma_1 = +1.1$ MPa	$\sigma_3 = -1.1$ MPa
$\sigma_3 = -0.5$ MPa	$\sigma_1 = +0.5$ MPa
$\tau_{xy}^{\infty} = 0$ MPa	$\tau_{xy}^{\infty} = 0$ MPa
$\theta = -5°$ (from σ_1)	$\theta = 85°$ (from σ_1)
$p_i = -0.15$ MPa	$p_i = +0.15$ MPa

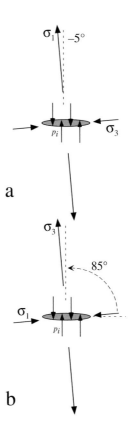

Fig. 8.4. Stress geometry for Example Exercise 8.2. (a) Tension-positive sign convention; (b) compression-positive convention. Note the changes in stress labels and angles between the two parts of the figure that flow from the choice of sign convention.

We then plug the stresses and angle into equations (3.5)—using compression positive—and find the value of the normal stress resolved on the crack wall. We get:

$$\sigma_n = -1.09 \text{ MPa (tensile)}$$

$$\tau_{xy}^\infty = 0.14 \text{ MPa (left-lateral)}$$

Notice that if you didn't check the angle using the new sign convention and used 5° in the equations (3.5), instead of 85°, you would have calculated a compression-positive normal stress of +0.49 MPa, which would be compressive and therefore incorrect, as evident by inspection from the geometry of the problem (Fig. 8.4). Now we calculate the driving stresses.

Tension positive	Compression positive
$\sigma_{dI} = (1.09 - (-0.15))$ MPa	$\sigma_{dI} = (-1.09 - 0.15)$ MPa
$\sigma_{dI} = 1.24$ MPa	$\sigma_{dI} = -1.24$ MPa
$\sigma_{dII} = -0.14$ MPa	$\sigma_{dII} = 0.14$ MPa

So the driving stress that causes dilation of the mixed-mode hydrofracture walls is 1.24 MPa of tension—slightly less than what we found in the previous problem for a pure mode-I crack because of the small stress rotation. There is also a shear driving stress of –0.14 MPa (left-lateral shear sense for this tension-positive sign convention; see Table 3.1). The combination of tensile and shear driving stresses produces an obliquely opening mixed-mode crack. Because the crack walls are pulled apart by the tensile driving stress, they are not in contact with each other any longer, so the shear driving stress does not produce frictional sliding; because of this, a mixed-mode crack is not a fault.

In our final example of driving stresses that lead to crack opening displacements, we'll use compressive stresses that would *close the crack* in the absence of any internal fluid pressure (see Secor, 1965, for a clear illustration of this important problem).

Example Exercise 8.3

Given another water-filled mixed-mode hydrofracture oriented with its outer-normal vector at $\theta = 5°$ to the least remote stress (tension-positive) $\sigma_1 = -0.15$ MPa (compressive), greatest principal stress σ_3 of -1.3 MPa (compressive), and having an internal fluid pressure of -0.35 MPa (compressive), calculate the driving stress acting to dilate the hydrofracture.

Solution

First we use the 2-D stress transformation equations (3.5) to calculate the normal stress acting on the crack wall due to just the remote stresses (Fig. 8.5). Because the stresses are given as tension positive, we again choose to switch the signs for this example exercise to match what we would use for compression positive. We get:

Tension positive	Compression positive
$\sigma_3 = -1.3$ MPa	$\sigma_3 = +0.15$ MPa
$\sigma_1 = -0.15$ MPa	$\sigma_1 = +1.3$ MPa
$\tau_{xy}^\infty = 0$ MPa	$\tau_{xy}^\infty = 0$ MPa
$\theta = -5°$ (from σ_1)	$\theta = 85°$ (from σ_1)
$p_i = -0.35$ MPa	$p_i = +0.35$ MPa

Next, we plug the stresses and angle into equations (3.5)—using compression positive and the correct angle—and find the value of the normal stress resolved on the crack wall. We get:

$$\sigma_n = +0.16 \text{ MPa (compressive)}$$

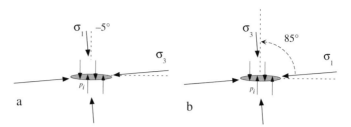

Fig. 8.5. Stress geometry for Example Exercise 8.3. (a) Tension-positive sign convention; (b) compression-positive convention.

$$\tau_{xy}^{\infty} = 0.1 \text{ MPa (left-lateral)}$$

The values imply that the crack walls are closed by the compressive normal stress. Because there is also a nonzero shear stress, the closed crack may potentially slide as a fault, but only if the frictional resistance is small enough to permit sliding. But because there is also an internal fluid pressure, we need to incorporate this into the calculation of net normal stress acting on the crack. Now let's calculate the driving stresses.

Tension positive	Compression positive
$\sigma_{dI} = (-0.16 - (-0.35))$ MPa	$\sigma_{dI} = (0.16 - 0.35)$ MPa
$\sigma_{dI} = +\textbf{0.19 MPa}$	$\sigma_{dI} = -0.19$ MPa
$\sigma_{dII} = -0.1$ MPa	$\sigma_{dII} = 0.1$ MPa

The driving stress that causes dilation of the crack walls is 0.19 MPa (tensile), even though the crack is trying to open in an overall *compressive* stress field. In essence, the internal pore-fluid pressure pushes the crack walls apart harder than the remote compression is squeezing them shut; it also means that the crack walls are not in frictional contact so a further assessment of frictional stability is not warranted. We therefore call this an **effective tension** because it conveys the value of the *net* force or *net* normal stress acting across the crack. This relationship defines the important role of crustal water, and other fluids, in dilating joints in the subsurface (Secor, 1965; Pollard and Aydin, 1988) and in maintaining hydrostatic to nearly lithostatic pore-fluid pressures in crustal rocks (e.g., Brace and Kohlstedt, 1980), as discovered in deep drill cores down to 6 km depths (Zoback et al., 1993).

8.2.1.3 Faults I—Individual Slip Events

Now let's examine the calculation of driving stress as it applies to the inception of frictional sliding along a pre-existing planar surface. Here we are concerned with both the normal and shear stresses resolved across a potential

Fig. 8.6. Detail of remote resolved shear stress and Coulomb frictional resistance for the calculation of driving stresses acting on a mode-II fracture. Notice that pore fluid pressure p_i within the fault zone motivates the use of an effective normal stress in equation (8.2). The net shear stress $\Delta\tau$ is that above the frictional resistance.

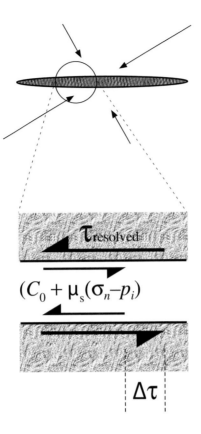

slip surface, along with the frictional strength, or *resistance*, of the surface. The relevant quantities are shown in Fig. 8.6. The frictional resistance of the fracture serves to reduce the amount of shear stress acting on the fracture, but it doesn't affect the level of (effective) normal stress σ'_n. The initiation of sliding occurs when the resolved shear stress is equal in magnitude to the **static**, or maximum (Byerlee, 1978; Okubo and Dieterich, 1984), **frictional resistance** of the surface:

$$\text{Pre-slip strength} = \left(c_0 + \mu_s \sigma'_n\right) \qquad (8.2)$$

Here μ_s is the maximum (static) friction coefficient (see Chapter 3). Next we need to specify how much the frictional strength is reduced by sliding along the surface—this value is the fracture's **residual frictional strength** (the "dynamic" or "kinetic" strength in seismology):

$$\text{Post-slip strength} = \left(c_0 + \mu_r \sigma'_n\right) \qquad (8.3)$$

The driving stress for frictional sliding is calculated by subtracting the residual frictional strength from the static frictional strength:

$$\sigma_{dII} = \left[\left(c_0 + \mu_s \sigma_n' \right) - \left(c_0 + \mu_r \sigma_n' \right) \right]$$
$$= \sigma_n' \left(\mu_s - \mu_r \right)$$
$$= \sigma_n' \left(\Delta \mu \right) \tag{8.4}$$

The **shear driving stress** (8.4) represents the difference between the maximum and residual frictional strengths (before and after sliding), respectively (e.g., Palmer and Rice, 1973; Rudnicki, 1980; Li, 1987; Martel and Pollard, 1989; Bürgmann et al., 1994; Cooke, 1997; Fig. 8.7). It is also referred to in seismotectonics as the **stress drop** across the fault (Pollard and Segall, 1987; Scholz, 2002, pp. 202–206; Segall, 2010) as a measure of the reduction in shear stresses associated with the frictional sliding during an earthquake (Rice, 1980; Cooke, 1997). There are **two levels of stress drop** that are commonly inferred:

Partial stress drop: $\tau_{dII} = \left[\sigma_n' \left(\mu_s - \mu_r \right) \right]$

Total stress drop: $\tau_{dII}^* = \sigma_n' \left(\mu_s - 0 \right) = \left[\sigma_n' \mu_s \right]$ $\qquad$ (8.5)

Martel and Pollard (1989) inferred that frictional sliding along faulted joints in Sierra Nevada granite was driven by a *partial stress drop*, given the non-90° intersection angles between the slipping joints and the abutting end cracks. Total stress drops may be approximated for slip along very weak faults such as those reported along bedding planes in the Appalachian fold and thrust belt by Ohlmacher and Aydin (1997).

Now let's calculate the driving stress for a potential fault.

Fig. 8.7. Correspondence between the strengths displayed on the Mohr diagram (left panel) and those important in considering frictional sliding along a fracture surface (right panel). Partial stress drop, τ_d; total stress drop, τ_d^* (equations (8.5)).

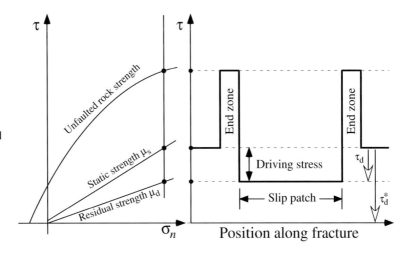

Example Exercise 8.4

Given a joint oriented at $\alpha = 30°$ to the maximum remote compressive principal stress of 5.0 MPa (compression positive), a least principal stress of 1.0 MPa (compressive), and having values of cohesion of 0.02 MPa, static friction coefficient of 0.6, and residual friction coefficient of 0.5, calculate the driving stress acting on the joint and determine if the stress state is sufficient to initiate frictional sliding. The principal stresses are effective stresses.

Solution

First we calculate the normal and shear stresses resolved on the joint using the compression-positive sign convention and equations (3.4), noting the relevant angle. We find

$$\sigma'_n = 2.0 \text{ MPa } \left(\text{compressive}\right)$$

$$\tau = 1.732 \left(\text{left-lateral}\right)$$

We previously saw that, for opening-mode cracks, an effective tension leads to a positive normal driving stress. For faults, however, there is commonly little or no significant component of opening or closure across a fault (especially when compared to the magnitude of the fault-parallel component), so the **normal driving stress (σ_{dI}) for faults is approximately zero** (but see Aydin et al., 1990, for an exception along strike-slip faults). There are two reasons or this. First, a resolved compressive normal stress is required for frictional sliding (see Chapter 3), so the fault walls would not open or dilate under the resolved compression (in this example, 2 MPa). Second, the stiffness of rock precludes any interpenetration or significant closing of the opposing fault walls.

Using the relation for Coulomb frictional resistance (equation (3.8)), and noting that σ'_n is compressive along the joint (and therefore that the joint is closed), we can see that

$$1.732 > 0.02 + 0.6\left(2.0\right)$$
$$1.732 > 1.22$$

and so the fracture should slip. The driving stress is the *difference* between the initial and final frictional strengths; using equation (8.4) gives

$$\sigma_{dII} = \tau_{dII}\left[\left(c_0 + \mu_s \sigma'_n\right) - \left(c_0 + \mu_r \sigma'_n\right)\right]$$
$$= \sigma'_n\left(\mu_s - \mu_r\right)$$
$$= 2.0(0.1)$$
$$= \boxed{0.2 \text{ MPa}}$$

and $\sigma_{dI} = 0$

> The shear driving stress, or stress drop, is 0.2 MPa. This positive number implies that the joint will indeed slip to become a fault, since the resolved shear stress exceeds the joint's frictional resistance. The sense of sliding (normal, strike-slip, or thrust), however, depends on the relative orientation of the maximum shear stress to the Earth's surface as resolved along the fault plane, rather than on the sign of the driving stress itself. If the driving stress is negative, the maximum static frictional resistance of the closed joint is not exceeded, the Coulomb criterion is not met, and the surface will be stable against frictional sliding.

What happens if the sign of the resolved shear stress is negative (right-lateral)? We'd still subtract off the frictional resistance, again as a positive number. This is because it always acts in the direction *opposite* to the resolved shear stress (e.g., Cooke, 1997). Conceptually,

$$\tau_{dII=}\left[\left(\text{left-lateral}\right)-\left(\text{right-lateral}\right)\right]$$
$$\tau_{dII=}\left[\left(\text{right-lateral}\right)-\left(\text{left-lateral}\right)\right]$$

This sign change is analogous to the action of internal fluid pressure in a crack that opposes a resolved compressive stress, as we saw in the previous example. In a friction problem, however, the frictional resistance *always* opposes the action of the resolved shear stress.

8.2.1.4 Faults II—Cumulative Geologic Offsets

Faults that we see and map in the field are the result of numerous slip events, some of which are seismic and some of which are aseismic (that is, stable or creep). The accumulation of many small shear displacements—each of which is calculated by starting with the methods outlined in the previous section—leads to the total, cumulative displacement (or geologic offset) that we measure along faults. As a result, we need to specify relationships between shear driving stress and total fault offset.

Two issues arise when dealing with this subject: *how to sum the displacements* and *how the fault grows in length* as its displacement magnitude increases. Excellent treatments of these subjects are given by Cowie and Scholz (1992c), Scholz (1997), Cowie (1998a,b), and Cowie et al. (2007) among others. We will deal with the first issue here, and the second one in Chapter 9.

Individual slip events along a fault—referred to as *earthquakes* if they involve unstable, stick-slip frictional sliding (e.g., Bridgman, 1936; Brace and Byerlee, 1966; Byerlee and Brace, 1968; Byerlee, 1970; Dieterich, 1972; Rice, 1979; Li, 1987; see Lockner, 1995, Marone, 1998a,b, and Scholz, 1998, for informative reviews)—are associated with individual driving stresses and individual stress drops for each slip event (equations (8.5)). The small driving stresses, on the order of perhaps 0.1–10 MPa, lead to small magnitudes of shear offset, per slip event, along the growing fault (e.g., Martel and Pollard, 1989). The effect of the fault on its surroundings (the changes in stress and displacement) is also small, due to the small value of induced offset per slip event.

On the other hand, faults having larger values of maximum offset (i.e., centimeters to hundreds of meters) appear to have required larger values of driving stress unless the stiffness (i.e., Young's or shear modulus) of the surrounding rock decreases systematically with ongoing slip accumulation (see discussion by Gudmundsson, 2004). Although faulting can reduce the effective stiffness of the surrounding crust on both short-term and long-term time scales, repeated earthquakes, and therefore a repeated, or cumulative driving stress, appear necessary (Cowie et al., 1992c; Cowie and Roberts, 2001; Cowie et al., 2007).

Cumulative driving stress for geologic fault offsets

A useful approach to modeling fault growth is to explicitly relate the total, cumulative fault offsets to a total, **cumulative driving stress** (e.g., Cowie and Scholz, 1992b; Scholz, 1997; Gupta and Scholz, 2000a; Schultz, 2003b). The ratio between cumulative driving stress and individual (i.e., seismic) stress drop can be calculated by dividing the geologic offset by a seismic offset (e.g., Cowie and Scholz, 1992c); typical values of this ratio are about 100. The cumulative driving stress for geologic fault offset calculations can then be found from using the relation for displacement magnitude along a fault (see equation (8.12)) and solving for the cumulative driving stress σ_d.

8.2.1.5 Summary–Driving Stress for Cracks and Faults

The procedure for calculating the normal or shear driving stress acting to promote dilation (jointing) or shearing (faulting) is straightforward.

- First, evaluate the sign of the *normal stress* resolved on the fracture. If it is tensile, then joint-like dilation may occur and the structure becomes a **crack**. Calculate the normal driving stress σ_{dI} from equation (8.1). Now evaluate the value of shear stress resolved on the crack plane. If it is zero, then so is the shear driving stress and the crack is **pure mode-I**. If it is nonzero (positive or negative), then the shear driving stress for this open, **mixed-mode crack** is the magnitude of the resolved shear stress.

- If the sign of the *normal stress* is compressive, then calculate the effective normal stress by subtracting off any internal pore-fluid pressure. A crack loaded by compressive tectonic stress requires $(p_i > \sigma_n)$ to open. If the resulting effective normal stress is *still compressive*, then you could have a **fault** instead of a joint, with zero normal driving stress. The shear driving stress σ_{dII} is then due to friction (equation (8.4)). Compactly, the first condition for a fault is given by $(p_i < \sigma_n)$. Now evaluate the value of shear stress resolved on the potential fault plane. If it is less than what is required for frictional sliding by the Coulomb criterion (equations (3.7)), then the shear driving stress is less than zero and the fracture remains **frictionally locked**. If on the other hand the frictional strength is met or exceeded, then the fracture will slide to become a **fault**.

Table 8.1 summarizes the fracture driving stress. Calculating the driving stress along either a joint or a potential fault is the first step in determining how much displacement would be predicted to occur along it; the relationships for this are given later in this chapter. Frictional resistance is a very interesting and subtle collection of properties (Rudnicki, 1980; Li, 1987; Marone, 1998b). In detail, friction depends not just on the resolved stresses, as we saw in Chapter 3 and embodied in the Coulomb criterion, but on scale (Barton, 1990), time-dependent elements such as stationary-contact duration (Rabinowicz, 1958), sliding velocity (Dieterich, 1978), lithology, displacement magnitude, gouge properties (Marone, 1995), and temperature (Rice, 1979, 1983; Scholz, 1990; Lockner, 1995; Marone, 1998a,b). In addition, because geologic displacements are typically built up from a series of numerous smaller slip events (e.g., Sibson, 1989; Cowie and Scholz, 1992b; Scholz, 1997) the differences between dynamic and static friction coefficients (e.g., Harris and Day, 1993; King et al., 1994; Marone, 1998b; Scholz, 1998), as well as maximum (Byerlee, 1978) and residual static friction (Okubo and Dieterich, 1984, 1986), must be included in expressions for the driving stress (e.g., Cowie and Scholz, 1992c; Bürgmann et al., 1994) and the eventual relationships for stress and displacement near the fault (Martel, 1997).

The driving stress can also be calculated for other types of geologic structural discontinuities. The driving stress for an anticrack, such as a **stylolite**, can be related to the selective dissolution or removal of material (including pore volume) across the stylolite (Zhou and Aydin, 2010). This involves both a compressive stress component (Fletcher and Pollard, 1981) along with terms that are sensitive to stress, temperature, fluid chemistry, and other factors related to this process (Fueten and Robin, 1992). The driving stress for **deformation and compaction bands** also involves not just the remote stress state but also the details of compaction, grain crushing, and displacements within the band (e.g., Cleary, 1976; Aydin and Johnson, 1983; Menéndez et al., 1996; Rudnicki, 2007; Tembe et al., 2008; Marketos and Bolton, 2009;

Table 8.1 Driving stress for joints and faults

Fracture	Normal driving stress	Shear driving stress
Joint (pure mode-I)	$\sigma_{dI} = \sigma_n^\infty - p_i$	$\sigma_{dII} = 0$
Joint (mixed-mode)	$\sigma_{dI} = \sigma_n^\infty - p_i$	$\sigma_{dII} = \tau_{xy}$
Fault	$\sigma_{dI} = 0$	$\sigma_{dII} = \sigma_n'(\Delta\mu)$

Schultz, 2009). As a result, the driving stress for deformation bands and anti-cracks can be related to the applied load and the appropriate set of internal boundary conditions.

8.2.2 Fractures as Stress Concentrators

In Chapter 2 we worked through a very useful relationship that was derived independently in Russia and England during the early part of the twentiethth century—the *Inglis/Kolosov Relation*. Expressed in equation (2.39), this result elegantly encapsulates the crucial way in which the size and shape of a fracture determine the level of stress change at the special location of the fracture termination (or "tip"). If we rewrite this equation using the present terminology for driving stress needed to make a fracture's walls displace either apart (mode I) or in shear (mode II or III), we have:

$$\sigma_{tip} = \sigma_d \left(1 + 2\frac{a}{b} \right)$$

(8.6)

and substituting the expressions for driving stress (mode-I for cracks, mode-II for faults having in-plane shear), we obtain

$$\text{Cracks: } \sigma_{tip} = [\sigma_{dI} + \sigma_{dII}] \left(1 + 2\frac{a}{b} \right)$$

$$= [(\sigma_n^\infty - p_i) + \tau_{xy}^\infty] \left(1 + 2\frac{a}{b} \right)$$

(8.7)

Driving stress causes displacements that produce high stress levels near a crack or fault tip.

$$\text{Faults: } \sigma_{tip} = \sigma_{dII} \left(1 + 2\frac{a}{b} \right)$$

$$= [\sigma_n'(\mu_s - \mu_r)] \left(1 + 2\frac{a}{b} \right)$$

(8.8)

The left-hand side of these equations (8.6)–(8.8) represents the elevated, or *concentrated*, level of stress produced at the fracture tip as a result of displacement along the fracture and, in turn, by redistribution of stress in the surrounding rock. The driving stress causes the fracture walls to displace relative to each other, producing an opening or shearing displacement that we can measure in the field, and it is this **displacement along the fracture that produces the high stress levels near the fracture tip**. Recall that a is the half-length of the fracture (so it has length $L = 2a$) and that b in these expressions is the cross-sectional width or thickness of the crack or fault (taken to be very small relative to a and different from the variable b used to characterize the vertical or down-dip fracture height, $b = H/2$, in Chapter 9).

Remember the problem that occurred in applying the Inglis/Kolosov Relation to fractures? As b gets very small, the stresses calculated for the fracture tip get unrealistically large, regardless of the fracture's length. In the limit of negligibly thin fracture width, $b = 0$ and the tip stresses would become infinitely large. This theoretical situation does not occur in nature. The predicted mathematical *singularity*, or infinite stress, is mitigated by what is known as the **Irwin plastic-zone correction** (e.g., Irwin et al., 1958; Kanninen and Popelar, 1985, p. 62; Anderson, 1995, pp. 72–75), that describes a **process zone** ahead of a fracture tip. We'll investigate this concept later in this chapter.

8.3 Fractures in Perfect Rock: Linear Elastic Fracture Mechanics

The simplest, and still the most generally useful, way to model fractures in rock is to use the equations derived for Linear Elastic Fracture Mechanics (LEFM). This approach is a subset of the more general field of Elastic–Plastic Fracture Mechanics (EPFM) that can include, as extensions to LEFM, rapid or dynamic (explosive) fracture propagation, such as sometimes encountered in underground mines (known as rockbursts; Goodman, 1989, p. 102; Brady and Brown, 1993, pp. 399–400), and inelastic yielding of rock surrounding the fracture, especially near the tips (called process zones), as explored in Chapter 9. Except for earthquake ruptures, where it is very important, dynamic fracture propagation is relatively uncommon in geologic examples of joint and fault sets (e.g., Segall and Pollard, 1983a; Segall, 1984a,b; Lawn, 1993, pp. 86–105). Inelastic processes near crack and fault tips can be fundamental observables that can tell us quite a lot about the growth and development of geologic fractures and fracture sets. We will examine several basic LEFM approaches to fractures and deformation bands starting in the next section.

8.3.1 Basic Assumptions and Rock Properties

LEFM works well when applied to fractures in a homogeneous elastic material that is indefinitely strong. An ideal elastic material is one in which the strain is recoverable and no fracturing or additional damage occurs during the application of stresses. An elastic material can be compressed, pulled, or twisted without limit because there is no built-in provision for failure. Its deformability is given by Young's modulus, Poisson's ratio, and the shear modulus, but its strength is not specified. A separate criterion for failure of the material must be chosen and superimposed for the "strength" of an elastic material, such as a rock, to have any meaning.

As discussed in Chapter 2, the behavior of rock can be adequately represented, or idealized, as an elastic material under a wide range of conditions that we encounter in the Earth. Elastic behavior of rocks is promoted by temperatures low enough that thermally activated processes, such as twinning, melting, creep, and flow, are slow enough to be unimportant on typical time scales of deformation. Stress levels well below those required to begin fracturing or otherwise damaging the rock surrounding a parent fracture also are consistent with those associated with elastic deformation of the rock (Lockner, 1995). Low stress levels applied to rocks typically also lead to small strains that are also recoverable, another hallmark of elastic behavior. Although rapid loading is sometimes invoked as a requirement for elastic behavior, loading rate is only one of several factors noted here that can promote material elasticity. Indeed, rapid loading has been shown to produce folding and related ductile flow during the formation of a meteorite impact crater (e.g., Laney and Van Schmus, 1978; Kenkmann, 2002, 2003); in these cases folding is promoted by the addition of high confining pressures (in excess of tens of GPa; e.g., Melosh, 1989) and stratification (Johnson, 1980; Erickson, 1996) in sedimentary sequences.

For a geologic fracture or deformation band to be idealized by using LEFM, the discontinuity (see Chapter 1) must additionally be larger than the scale of any inhomogeneities in the rock mass, such as grains (e.g., Ingraffea, 1987); otherwise, these can interact with the discontinuity and cause the rock to deform non-uniformly. An isotropic (CHILE) material (having the same properties in any direction) is widely used for LEFM problems, although many solutions exist for fractures in layered and otherwise anisotropic (DIANE) materials (e.g., Li, 1987). The use of anisotropic elastic equations, and the applicable material parameters, in LEFM would require substantial modifications to the stress equations presented in this chapter; the interested reader is referred to the research literature for the derivations and applications of this approach.

LEFM Assumptions:

♦ *Homogeneous*
♦ *Isotropic*
♦ *Isothermal*
♦ *Large relative scale*
♦ *Plane strain*
♦ *Fracture modes are additive*

The basic set of assumptions required for LEFM analysis of geologic fractures includes (e.g., Anderson, 1995, p. 101): homogeneous, isotropic, and isothermal (constant temperature) conditions, so that the simplest elastic equations apply. Additionally, it is often convenient to idealize a fracture or deformation band further as a two-dimensional (2-D) structure, instead of using the full three dimensions (see Pollard and Segall, 1987; Cohen, 1999; and Segall, 2010, for representative examples). This idealization commonly captures the major properties of the process, permitting the use of the much simpler plane-strain or plane-stress forms of the elastic equations (see Section 8.6).

Plane-stress fracture problems occur as thin sheets or panels with cracks that are large compared to their thickness. In this case the equations for fracture-tip stress and energy must be modified to account for the mechanical interaction between the fracture and the surface(s) or edges, which can become an intricate problem (e.g., see Broek, 1986, pp. 116–120). Most geologic problems that are 2-D are plane-strain (see also Chapter 1), so the subset of plane-stress LEFM problems is available to explore from other sources, such as materials engineering (e.g., see Kanninen and Popelar, 1985, pp. 172–188; Broek, 1986, pp. 110–113; Lawn, 1993; Anderson, 1995, pp. 82–84). We'll concentrate on

plane-strain LEFM problems here and note some of the changes needed in certain relationships to apply them to plane-stress and three-dimensional (3-D) situations as needed in this chapter.

One advantage of the LEFM approach is that the stresses or displacements in the rock, determined by each displacement mode, are *additive*. This means that one can separately calculate the stress component, say, normal to the fracture, for its opening displacement (mode I) and then calculate the same stress component due to the fracture's oblique, in-plane displacement (mode II), then add them to obtain the total values for the stress normal to the fracture. This approach works because the equations assume small (infinitesimal) strains and linear elasticity (Hooke's Law). As we'll see, these apparent restrictions do not prevent us from obtaining quite useful and powerful results, for typical large-strain geologic problems, from these ideal equations. In particular, the displacements and strains predicted by fracture mechanics can lead to large-strain, geologically important results, for example in areas of continental tectonics (e.g., Segall and Pollard, 1980; Aydin and Nur, 1982; Aydin and Page, 1984; Pollard and Segall, 1987; Schultz, 1989; Schultz and Aydin, 1990; Willemse, 1997; Crider, 2001; Cowie et al., 2007).

8.3.2 Displacements for the Three Modes

In geology we normally observe the end results of deformation, so it is typical to measure (or calculate) strains and infer the causative stresses. Similarly, we can identify and measure the displacements associated with structures such as faults, deformation bands, joints, veins, and igneous dikes. As a result, it is more natural for us as geologists to **base our fracture-mechanics methods on observable quantities—the displacements recorded along geologic discontinuities**. In contrast, engineers routinely deal with stress (and energy) methods because the flaws in their materials are so small that direct observation is rarely practical. Let's look first at the deformation associated with displacements along fractures. Then we'll work through the stress equations. As we'll see, both of these approaches (displacements and stresses) are complementary and useful for understanding discontinuity patterns and occurrences in rock.

8.3.2.1 Displacements on the Fracture

This calculation is one of the easiest to do for geologic fracture mechanics and one of the most powerful. The geometry of the problem is shown in Fig. 8.8. It is common when deriving displacement and stress equations from the formal elasticity theory to utilize the elastic shear modulus rather than Young's modulus. As long as the material maintains its elastic behavior, however, the parameters are interchangeable. The equations that follow are written *using Young's modulus*, however, to retain the connection between the LEFM results

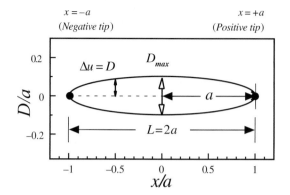

Fig. 8.8. Normalized displacement profiles for the opposing crack walls showing the three critical points (two tips, crack midpoint), calculated using $E = 10\ \sigma_{dI}$.

and the more familiar processes of Coulomb slip and experimental rock deformation. The shear modulus form is obtained by replacing $2(1-v^2)/E$ by $(1-v)/G$ in the stiffness term (e.g., Bürgmann et al., 1994).

The displacement of fracture walls relative to each other (opening or slip) is given, for each crack wall, by (Pollard and Segall, 1987; see also Broek, 1986, p. 96)

$$\Delta u_{\mathrm{I}} = \sigma_{dI} \frac{2(1-v^2)}{E} \sqrt{a^2 - x^2} \tag{8.9a}$$

Relative displacement of fracture walls

$$\Delta u_{\mathrm{II}} = \sigma_{dII} \frac{2(1-v^2)}{E} \sqrt{a^2 - x^2} \tag{8.9b}$$

$$\Delta u_{\mathrm{III}} = \sigma_{dIII} \frac{2(1-v^2)}{E} \sqrt{a^2 - x^2} \tag{8.9c}$$

In these equations, the fracture displacements Δu (denoted D in Fig. 8.8) are listed with their fracture-mode suffixes (see Chapter 1), along with the appropriate driving stress terms for that fracture mode. Notice how similar these three expressions are? The displacements depend on the elastic properties, the driving stress, and the position of interest along the fracture (the final, square-root term). The stiffer the surrounding host rock (larger E), the less able the fracture surfaces will be to displace, so the same length and driving stress will promote smaller displacements in stiffer materials, like granite, than in softer materials, like tuff. The larger the driving stress and or fracture length, the larger the displacements will be.

As one can infer from equations (8.9) and see from Fig. 8.8, the fracture surface displacements vary as a function of position along the fracture. The elastic displacement profile, or displacement distribution, embodied by those equations describes an **ellipse**, with minimum values of opening displacement located at either crack tip ($x = -1$ or $x = +1$) and the maximum value located at the crack's midpoint ($x = 0$). The along-strike locations are typically recalculated for fracture mechanics applications by normalizing them to the crack half-length a, leading to crack tips at $x/a = \pm 1$. You'll see this approach frequently in this chapter and the literature—it is a convenient way to apply fracture mechanics ideas to geologic discontinuities regardless of their actual physical scale, because the mechanics can be very much the same regardless of scale. The displacements are also normalized by the crack half-length (D/a in Fig. 8.8) and scale with the driving stress and the Young's modulus.

Setting $x = 0$ in equations (8.9) gives an expression for the displacement at the center of the fracture:

Maximum relative displacement on a 2-D fracture within an elastic material

$$D_{max} = \sigma_d \frac{4(1-v^2)}{E} a \qquad (8.10)$$

Equation (8.10) shows us the **maximum value of displacement D_{max}**, that would be produced along a two-dimensional fracture, given appropriate values for the other parameters. This can be the first step in assessing maximum-displacement vs. length data for a fracture population (see Clark and Cox, 1996; Chapter 9). This expression (and equations (8.9)) works well for many types of cracks but is only approximate for geologic discontinuities such as faults and deformation bands, where a different, non-elliptical displacement distribution appears, primarily due to differences in their internal boundary conditions relative to cracks (e.g., Cowie and Scholz, 1992b; Bürgmann et al., 1994; Fossen and Hesthammer, 1997; Rudnicki, 2007). Similarly, joints, veins, and dikes that have grown within a population of closely spaced structures can have non-elliptical displacement distributions that are important to recognize when examining outcrops containing these fractures (e.g., Rubin, 1993a, 1995b).

At the fracture tips, $x = \pm a$ and the displacement magnitude is then equal to zero. So we see that the **displacement distribution is not constant** along a bounded fracture, but varies from a maximum value at or near the fracture's center (or midpoint) and *decreases toward zero* at its terminations (e.g., Barnett et al., 1987; Segall and Pollard, 1980; Pollard and Segall, 1987; Segall, 2010; Gudmundsson, 2011). Notice how this result contrasts with the basic question in structural geology, "how much slip is there on the fault?" The answer *has to be*, "the value depends on where you are along the fault" with the wrong answer being a simple constant number. An example of displacements that vary in magnitude along a fault is shown in Fig. 8.9. Here you can correlate the beds across the normal fault and see that the beds in the middle of the fault are offset more than those near its upper tip.

How much offset is there on that fault?

Let's calculate the maximum displacements for two examples: a mode-I crack (a joint) and a mode-II fracture (a strike-slip fault).

Fig. 8.9. Photograph of a bounded normal fault near Kodachrome State Park in Utah showing variations in the amount of displacement accommodated along the fault.

Example Exercise 8.5

Estimate the amount of maximum displacement expected at the center of a fracture for the following cases: (a) *a dry joint* loaded by a remote tension of magnitude 0.5 MPa oriented perpendicular to the joint having length of 1.8 m, and (b) *a fault* of length 4.7 m oriented at $\alpha = 30°$ to the maximum remote compressive stress = 0.5 MPa (compression positive), least principal stress of 0.1 MPa (compressive), and having values of cohesion of 0.02 MPa, static friction coefficient of 0.6, and residual friction of 0.5. The rock has a Young's modulus of 12 GPa and a Poisson's ratio of 0.25.

Solution

We first calculate the driving stresses on the fractures, then plug the numbers into the relation for maximum displacement (8.10).

(a) The driving stress on the **joint** is easy to find, being just the magnitude of remote normal stress (tensile) on the joint, since there is no internal pore pressure. Given zero resolved shear stress on the joint, the shear driving stress is zero. Remembering that the half-length $a = L/2$ is used in the equation and keeping our units straight (m and MPa), we have

$$D_{max} = \left(0.5\,\text{MPa}\right)\frac{4(1-(0.25)^2)}{12,000\ \text{MPa}}(0.9\ \text{m})$$

$$\therefore D_{max} = 1.41 \cdot 10^{-4} \, m = 0.14 \, mm$$

The joint will open a bit more than one-tenth of a millimeter under these conditions, forming a hairline crack in the rock.

(b) The resolved normal and shear stresses on the **fault** are found using equations (3.4) to be

$$\sigma_n = 0.2 \text{ MPa (compressive)}$$

$$\tau = 0.1732 \left(\text{left-lateral} \right)$$

Now that we see that the normal stress resolved on the fracture is indeed compressive, as required, the shear driving stress for frictional slip is (from Example Exercise 8.4)

$$\sigma_{dII} = \tau_{dII} = \sigma_n (\mu_s - \mu_r) = 0.02 \text{ MPa}$$

Using equation (8.10) we have

$$D_{max} = (0.02 \text{ MPa}) \frac{4(1 - (0.25)^2)}{12,000 \text{ MPa}} (2.35 \text{ m})$$

$$\therefore D_{max} = 1.47 \cdot 10^{-5} \, m = 0.02 \text{ mm (left-lateral)}$$

The fault will have a maximum left-lateral offset of about two-hundredths of a millimeter near its center for this single episode of loading and slip. Additional slip events will build up a larger value of the cumulative offset on the fault (Cowie and Scholz, 1992c).

Now let's revisit the concept of cumulative driving stress for the total, cumulative (i.e., measured) geologic offset along a fault (i.e., Fig. 1.9).

Example Exercise 8.6

Given a fault that is 1.2 km long, a cumulative (maximum) offset measured in the field of 10.3 m, and the rock-mass properties from the previous example exercise, estimate: (a) the cumulative driving stress and (b) the number of slip events that the fault would have experienced.

Solution

(a). First we rearrange equation (8.10) to solve for (cumulative) driving stress to obtain

$$\sigma_{dII} = \frac{D_{max}}{a}\frac{E}{4(1-v^2)}$$

Substituting the appropriate values, we find

$$\sigma_{dII} = \frac{10.3 \text{ m}}{600 \text{ m}}\frac{12{,}000 \text{ MPa}}{4(1-(0.25)^2)}$$

$$\therefore \sigma_{dII} = 54.9 \text{ MPa}$$

Using equation (8.5) and the appropriate parameters for a *partial stress drop*, we find

$$\# slip\ events\ = \frac{[54.9-(0.02+0.5\cdot2.0)]}{[2.0(0.6-0.5)]}$$

$$= \frac{54.8}{0.2} \approx 274 \text{ events}$$

The fault should slip about 274 times in order to accumulate 10.3 m of offset, assuming no changes between (or due to) each slip event. We won't deal with the size (e.g., moment or magnitude) of individual slip events here. To do that we would need additional information on fault lengths for each stage of slip accumulation and an assessment of the stability of frictional sliding (seismic or aseismic slip events). Standard works on **seismotectonics** (e.g., Cowie and Scholz, 1992c; Scholz, 2002) will get you started with this important and interesting subject.

The foregoing analysis was based on the solution to a plane-strain 2-D fracture that corresponds essentially to a vertically dipping fracture in map view whose vertical dimension (i.e., its height) is much larger than its horizontal dimension (its length). In many situations, however, it is important to consider a fracture whose horizontal dimensions are both bounded and known, so that its tipline has a regular shape, such as rectangular or elliptical. Assuming an elliptical tipline for convenience (and because it approximates the geometries of many cracks and faults in outcrop), the fracture becomes an ellipsoid (Fig. 8.10) having axes of length ($L = 2a$), height ($H = 2b$), and displacement (normal and/or shear, depending on its displacement mode, $D = 2u$).

The stresses and displacements associated with an elliptical fracture in an infinite homogeneous elastic material are well known from LEFM (e.g., Green and Sneddon, 1950; Irwin, 1962; Kassir and Sih, 1966; Xue and Qu,

1999; Zhu et al., 2001). The maximum (relative) displacement D_{max} on the elliptical fracture surface is given by

$$D_{max} = \frac{4(1-v^2)}{E}\sigma_d \frac{a}{E(a,b)} \tag{8.11}$$

in which a and b are the semi-major and semi-minor axial dimensions of the fracture (Fig. 8.10), respectively.

$E(a,b)$ in (8.11) is the complete elliptic integral of the second kind (e.g., Irwin, 1962; Kanninen and Popelar, 1985, p. 153; Lawn, 1993, p. 33), which is given by

$$E(a,b) = \int_0^{\frac{\pi}{2}} \sqrt{1 - \left(\frac{a^2 - b^2}{a^2}\right)^2 \sin^2 \phi} \, d\phi \tag{8.12}$$

where ϕ is the amplitude of the elliptic integral. Particular values of (8.12) can be obtained by specifying a and b and using standard mathematical tables to evaluate the integral.

Equation (8.12) can be approximated for fracture mechanics problems by a **flaw shape parameter** (Anderson, 1995, pp. 115–116), Ω

$$E(a,b) = \Omega \cong \sqrt{1 + 1.464\left(\frac{a}{b}\right)^{1.65}} \tag{8.13}$$

(Fig. 8.11). This approximation is within 5% of that from equation (8.12) (Schultz and Fossen, 2002). Ω can be substituted into equation (8.11) for

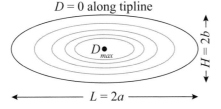

Fig. 8.10. Major and minor axes, in plan view, for an elliptical, 3-D fracture.

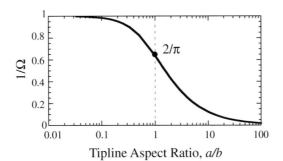

Fig. 8.11. Inverse of flaw shape parameter plotted as a function of log fracture aspect ratio a/b.

$E(a,b)$, permitting calculation of D_{max} for arbitrary fracture lengths and aspect ratios. Using (8.13) or example, for a circular, "penny-shaped" fracture ($a = b$), $E(a,b) = \pi/2$; a tall fracture ($b \gg a$, or $a/b < 0.2$) has $E(a,b) = 1$; and a long fracture ($a \gg b$, or $a/b = 10$) yields $E(a,b) = 8$ (see also Gudmundsson, 2000, 2011). This factor reduces both D_{max} and the associated stress intensity factor relative to those for a tall 2-D fracture (e.g., Lawn, 1993, pp. 31–33; Savalli and Engelder, 2005). The maximum displacement on an elliptical discontinuity (eq. 8.11) thus depends explicitly on both a and b (Irwin, 1962; Willemse et al., 1996; Gudmundsson, 2000).

Replacing a by $L = 2a$, the resulting expression for the maximum displacement at the center of a fracture having an elliptically shaped tipline is

Maximum displacement on a fracture with elliptical tipline

$$D_{max} = \frac{2\left(1 - v^2\right)}{E} \sigma_d \frac{L}{\Omega} \tag{8.14}$$

This expression forms the basis for 3-D displacement–length scaling for geologic cracks, deformation bands, and faults (Schultz and Fossen, 2002; Soliva et al., 2005; Schultz et al., 2006, 2008a). For example, the maximum shear displacement on a fault having a typical L/H ratio of 3 (Nicol et al., 1996) would have about one-third ($\Omega = 3.12$, as plotted in Fig. 8.11) of the displacement that would be calculated from a purely 2-D plane-strain approach (i.e., using equation (8.10)). Stratigraphically restricted faults would accommodate still smaller values of offset, producing nonlinear growth paths on a displacement–length diagram (see Schultz and Fossen, 2002; Soliva et al., 2005; Polit et al., 2009; Chapter 9).

8.3.2.2 Displacements Away From the Fracture

Many analytical solutions exist to describe the variation in particle displacement and stress in the vicinity of a plane (2-D) fracture surface (e.g., Westergaard, 1939; Sneddon, 1946; Irwin, 1957; Williams, 1957; Erdogan and Sih, 1963; Eftis and Liebowitz, 1972). Historically most of these start with the region closest to one of the fracture tips—the "near-tip solution" (see the Inglis/Kolosov results, Chapter 2). Here we'll use the equations, derived by Pollard and Segall (1987), for the **displacement field near a 2-D fracture** that explicitly account for both fracture tips, rather than just one tip as in many previous formulations (see Engelder et al., 1993, for a clear discussion). Then once the nature of the displacement field near a fracture is established, we'll examine the **state of stress near a fracture**. Last, we'll zoom in to the **tip region** where much of the interest as far as fracture interaction and propagation are concerned occurs.

The problem is set up in Fig. 8.12 by using a standard tri-polar coordinate system that tracks the distances to a material point in the rock from both fracture tips and the fracture midpoint, giving us three radii and three angles to keep straight. These are consolidated by defining two spatial parameters

$$R = \sqrt{r_1 r_2} \tag{8.15a}$$

$$\Theta = \frac{\theta_1 + \theta_2}{2} \tag{8.15b}$$

The *displacement field* in the rock near a fracture loaded by uniform driving stress, and for all three modes, is given by (Pollard and Segall, 1987, p. 294):

Displacement fields near a tall 2-D fracture

$$u_y = \sigma_{dI} \frac{(1+v)}{E}$$

$$\left\{ 2(1-v)(R\sin\Theta - r\sin\theta) - r\sin\theta \left[\frac{r}{R}\cos(\theta-\Theta) - 1 \right] \right\}$$

$$-\sigma_{dII} \frac{(1+v)}{E} \tag{8.16a}$$

$$\left\{ (1-2v)(R\sin\Theta - r\sin\theta) + r\sin\theta \left[\frac{r}{R}\cos(\theta-\Theta) - 1 \right] \right\}$$

$$u_x = \sigma_{dII} \frac{(1+v)}{E}$$

$$\left\{ 2(1-v)(R\sin\Theta - r\sin\theta) + r\sin\theta \left[\frac{r}{R}\cos(\theta-\Theta) - 1 \right] \right\}$$

$$+\sigma_{dI} \frac{(1+v)}{E} \tag{8.16b}$$

$$\left\{ (1-2v)(R\sin\Theta - r\sin\theta) - r\sin\theta \left[\frac{r}{R}\cos(\theta-\Theta) - 1 \right] \right\}$$

$$u_z = \sigma_{dIII} \frac{2(1+v)}{E}(R\sin\Theta - r\sin\theta) \tag{8.16c}$$

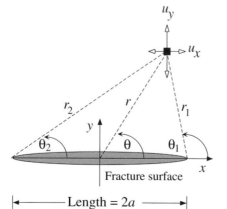

Fig. 8.12. Geometry and parameters for calculating the displacement field in the vicinity of a tall, 2-D, plane-strain fracture.

The first equation (8.16a) is for the component of the displacement field that moves rock perpendicular to the fracture (toward or away), and it depends on both the mode-I and mode-II displacements, through their positive driving stresses, of the fracture. For pure opening mode (no mode II), the second term in equation (8.16a) is zeroed out. Likewise, the fracture-parallel displacement component (8.16b) depends on mode-I and mode-II displacements; for pure strike-slip (mode II) displacements, the mode-I terms are zeroed out. These equations define the magnitude and direction of the displacement field and can be represented as vectors or contoured magnitudes.

The third expression (8.16c) describes the displacement field surrounding mode-III fractures. Anti-plane (vertical) slip on the mode-III fracture only induces anti-plane (vertical) displacements in the surrounding rock; because neither of the first two equations (8.16a, 8.16b) contain the mode-III driving stress, no horizontal, in-plane displacements occur for the vertical mode-III crack. In geologic situations, however, normal and thrust faults can be idealized as dipping mode-III cracks; because these mode-III cracks are nonvertical, shearing along dip-slip faults does induce a component of horizontal, in-plane displacements in the surrounding rock (e.g., Savage and Hastie, 1966; Mansinha and Smylie, 1971; Cohen, 1999).

The displacement field surrounding a fracture can be represented by a series of line segments, or tick marks, that record the local direction and magnitude of displacements (e.g., Pollard and Segall, 1987). These line segments trace out the **trajectories** that map out how markers in the rock mass will be moved by displacement along the fracture, and by how much. Another way to show this is to make an *originally square mesh deform around the fracture*; three of these deformed meshes are shown in Fig. 8.13. As one can see in the figure, the displacement directions can be traced along the gridlines; the displacement magnitudes are found by seeing how much the square grids have been stretched, squeezed, or sheared by the fracture.

As is apparent from Fig. 8.13, the **displacement field surrounding a fracture is not homogeneous** and uniform. The amount of deformation is greatest near the fracture and dies out with distance away from it (compare with kinematic models that do not include gradients in strain with distance from a fault (e.g., Woodcock and Fischer, 1986)). Notice that the displacement field surrounding an internally pressurized fracture—a hydrofracture or igneous dike—is *the same* as that surrounding a crack opening in response to a remotely applied tensile stress. This differs from the state of stress surrounding each, which as we'll see in Fig. 8.14, is different. One can also see the *oblique dilation* of the mixed-mode crack (middle panel in Fig. 8.13) and the sense of *rotation* of the rock mass on either side of a mode-II fracture (strike-slip fault in this view; note the "reverse drag" of the passive markers oriented at a high angle to the fault; Grasemann et al., 2005). The most important observation from the deformation field here is that **the strain in the rock is maximized near a fracture and dies out to zero far from the fracture**.

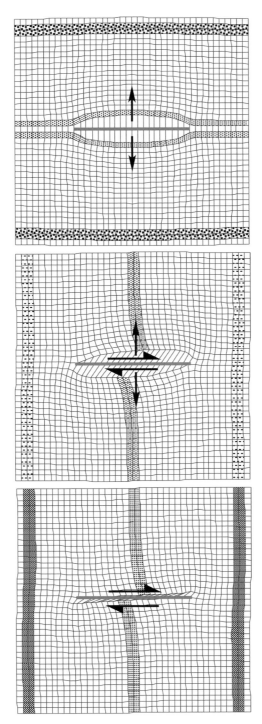

Fig. 8.13. Displacement fields in the vicinity of a fracture (equations (8.16)) portrayed as deformed grids. Fracture shown by solid bar; arrows on the fracture represent displacements, not resolved stresses.

8.3.3 Stress Changes Near Fractures

Using the same formulation as in the previous section, the stress components near a tall, 2-D fracture can be obtained (Pollard and Segall, 1987). For the general case of a fracture having an arbitrary combination of displacement modes (I, II, and/or III), there are six components of stress that each can have nonzero values, although not all are always applicable to the fracture of interest. For example, in-plane mixed-mode displacements (I-II) contribute only to the normal stresses σ_{xx}, σ_{yy} and a shear stress σ_{xy}; in plane-strain cases a vertical normal stress σ_{zz} is also generated by these mixed-mode displacements. No anti-plane stresses are generated by mixed-mode (I-II) displacements. On the other hand, mode-III slip on a vertical fracture surface only generates the vertical shear stress components σ_{xz} and σ_{yz}, with no changes for the in-plane stresses.

Using the spatial parameters given in equations (8.15), the equations for stress components, in 3-D, due to mixed-mode fracture displacements (I, II, and III) are (Pollard and Segall, 1987, pp. 305–306):

$$\sigma_{yy} = \sigma_{yy}^{\infty} + \sigma_{dI}\left[\frac{r}{R}\cos(\theta - \Theta) - 1 + \frac{a^2 r}{R^3}\sin\theta\cos 3\Theta\right] + \sigma_{dII}\left(\frac{a^2 r}{R^3}\sin\theta\sin 3\Theta\right)$$

(8.17a)

Stress state near a tall, 2-D fracture

$$\sigma_{xx} = \sigma_{xx}^{\infty} + \sigma_{dI}\left[\frac{r}{R}\cos(\theta - \Theta) - 1 - \frac{a^2 r}{R^3}\sin\theta\cos 3\Theta\right]$$
$$+ \sigma_{dII}\left(\frac{2r}{R}\sin\theta - \Theta\right) - \frac{a^2 r}{R^3}\sin\theta\sin 3\Theta$$

(8.17b)

$$\sigma_{xy} = \sigma_{xy}^{\infty} + \sigma_{dII}\left[\frac{r}{R}\cos(\theta - \Theta) - 1 - \frac{a^2 r}{R^3}\sin\theta\cos 3\Theta\right] + \sigma_{dI}\left(\frac{a^2 r}{R^3}\sin\theta\sin 3\Theta\right)$$

(8.17c)

$$\sigma_{xz} = \sigma_{xz}^{\infty} + \sigma_{dIII}\left[\frac{r}{R}\cos(\theta - \Theta) - 1\right]$$

(8.17d)

$$\sigma_{xz} = \sigma_{xz}^{\infty} + \sigma_{dIII}\left[\frac{r}{R}\sin(\theta - \Theta)\right]$$

(8.17e)

$$\sigma_{zz} = v\left(\sigma_{xx} + \sigma_{yy}\right)$$

(8.17f)

In these equations (8.17a–f) the local state of stress surrounding a fracture, given by the left-hand side of each expression, depends on the far-field, or *regional*, stress state—usually this far-field stress is the objective to be

Stress near a fracture depends on:

♦ *Driving stress*
♦ *Remote stress*
♦ *Length, and*
♦ *Position relative to fracture*

determined from a paleostress inversion of fault-slip data (e.g., Angelier, 1994) or borehole breakouts and hydraulic fracturing tests (e.g., Rummel, 1987; Zoback et al., 1993). But what happens if you're relatively close to the fracture? The stress components are a combination of local and regional stress states (e.g., Mount and Suppe, 1987; Zoback et al., 1987; Pollard et al., 1993). Notice, too, that if the driving stress for fracture displacement is zero for all three modes, then the local state of stress near a fracture is the same as the regional stress. This baseline result indicates that **a joint must dilate, or a fault slip, to have any effect of the fracture on the regional stress or strain field.**

The equations are plotted and contoured for several fracture types in Fig. 8.14. The diagrams all assume unit (1.0 MPa) driving stress across each fracture, and the **tick marks record the local orientations (or trajectories) of the principal stresses as perturbed by the displacements across the fracture**. A mode-I crack opens against the surrounding rock mass, leading to increased values of **mean stress** (see Pollard and Segall, 1987; Chapter 3); this is equivalent to saying that the confining pressure is locally increased near the crack as it tries to push open the rock. This is even more true for an internally pressurized crack.

The mode-II crack—analogous to a right-lateral strike-slip fault—compresses the rock in the first and third quadrants—the "leading" quadrants—while dilating it in the "trailing" quadrants (Chapter 6). Dilational zones in these quadrants are likely places for secondary cracks—**end cracks**—to nucleate in the surrounding rock (Rispoli, 1981; Segall and Pollard, 1980; Willemse and Pollard, 1998). The mixed-mode I-II crack (lowest panels) represents a crack having both opening and shearing components (see Olson and Pollard, 1991; Fig. 8.14).

The right-hand column in Fig. 8.14 shows contours of the **maximum shear stress as redistributed about the fracture**. Lighter-toned areas represent the smallest values, and darker areas are elevated values. Both mode-I cases have zero shear stress at their tips, meaning that they will **propagate** straight ahead as planar fractures in their own principal planes. The other two cases that involve shearing are more complicated, and their propagation directions will not be simply in-plane, given the non-symmetric mean stress distributions.

Why are values of maximum shear stress important for cracks? For stationary cracks, this *in-plane shear stress* causes any other cracks in the proximity of the main one to open obliquely, with the direction given by the sense of shear stress resolved upon it. For cracks propagating into the region of the main crack, the *shear stress steers the crack*, leading to curved or kinked crack paths and joint traces. We therefore need both the normal stress (shown here as the mean stress) and the shear stress to understand, or predict, what can happen to fractures when the local stress environment changes due to opening (or shearing) displacements along the main crack (e.g., Erdogan and

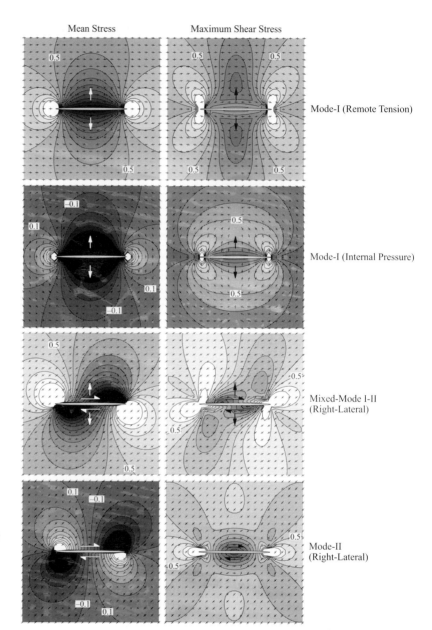

Fig. 8.14. Stress fields in the vicinity of a fracture (equations (8.17)). Fracture shown by white bar. Contour interval, 0.1 σ_d, with maxima at 1.0 and minima at –0.5. Tick marks show orientations of local principal stresses; maximum stress given by longer tick.

Sih, 1963; Olson and Pollard, 1989; Cruikshank et al., 1991a,b; Renshaw and Pollard, 1994a,b; Cooke and Pollard, 1996; Engelder, 1999).

As can be seen in Fig. 8.14, **the stresses are redistributed within the surrounding rock by the fracture**. The effect is analogous to that demonstrated in Chapter 2, where a circular cavity in a stressed plate both focuses and redistributes the stress in its vicinity. But the effect of a sharp, planar fracture is more pronounced around its terminations, where the elastic stress is theoretically infinite (b approaches zero in equations (8.6)–(8.8)). We'll concentrate on this region in a later section, on the *stress intensity factor*, when we look at its critical role in crack propagation (e.g., Lawn, 1993; Anderson, 1995; Pollard and Aydin, 1988) and the breakdown into echelon fracture sets (e.g., Pollard et al., 1982; Cooke and Pollard, 1996).

8.3.4 Summary of Fracture Displacement and Stress Fields

The first main result centers on the key observation that we are dealing with **bounded fractures**—they have clear terminations within the rock mass (e.g., Segall and Pollard, 1980; Mandl, 1987a; Pollard and Aydin, 1988; Davison, 1994; Crider et al., 1996). Unbounded fractures are like dominos, wooden block models of faults, or a card deck, where the surfaces can open or slip an unlimited amount and there are no volume problems at their edges (Mandl, 1987b). Faults and joints that reached the Earth's surface when they were active, like the San Andreas fault, are unbounded there. However, the structures are probably bounded at depth and along strike. The assumptions built into LEFM assume that fractures are bounded along at least part of their periphery and that the surroundings can deform continuously.

Dominos and card decks vs. discontinuous joints and faults

The second result is that displacements along a fracture **modify** the stress or displacements (i.e., strains) in a rock mass. The driving stresses resolved on a fracture surface cause the fracture surfaces to displace relative to one another; this, in turn, generates new stresses, and displacements, in the surrounding rock mass.

The third result emphasizes the variation in stress or displacement with **distance from the fracture**. For example, stress changes due to the excavation of a cavity in a stressed plate (Chapter 2) were largest closest to the cavity, and decreased as we moved farther away. Beyond about three cavity radii, the influence of the cavity is not much more than a minor fluctuation in the ambient stress state there. This led to the concept of **proximity**—that is, the closer we get to an imperfection, or *flaw*, the greater the stress or displacement changes we can expect to see. This is called "St. Venant's Principle" in solid mechanics (Timoshenko and Goodier, 1970, pp. 39–40; Davis and Selvadurai, 1996, pp. 180–187), where the point of application of a load is the region of greatest effect, and that the effect decreases with decreasing proximity. This is why a nail hammered into lumber is so effective, and why paleostress methods, such as fault-slip inversion or borehole breakouts, must consider the position or *context* of the measurements in relation to the scale of the faulted domain (e.g., Pollard et al., 1993).

Fractures change the stress state in their vicinity

The influence of a fracture decays with distance away from it

The three main results summarized here collectively provide the physical context for treating geologic structural discontinuities as crack-like flaws and for invoking the methods of Linear Elastic Fracture Mechanics.

8.3.5 Stress Concentration at Fracture Tips

Now that we've gained some familiarity with the overall effects of a fracture on the stress and displacement fields in its vicinity, let's focus in on the special region near the fracture tip. In this section we'll continue to examine LEFM approaches that assume a sharp tip and therefore predict infinitely large near-tip stresses.

8.3.5.1 Elastic Near-Tip Stress States

The expressions given previously were derived for the entire crack for arbitrary mixed-mode displacements (I, II, and III). In order to predict fracture propagation, a simplified subset of the whole-crack stress equations is typically derived that retains only the terms most important at the fracture tip. This approximation to the full stress field is called the **Near-Tip Solution** for the stresses and displacements within a tiny region at the tip.

Using the problem geometry shown in Fig. 8.15, the near-tip stresses in Cartesian (xy) and polar ($r\theta$) coordinates for the K-dominant region are given following Pollard and Segall (1987), p. 324, and Lawn (1993), p. 25. First, the *Cartesian form*:

$$\sigma_{yy} = \sqrt{\frac{a}{2r}}\left\{\sigma_{dI}\left[\cos\left(\frac{\theta}{2}\right)+\frac{1}{2}\sin\theta\sin\left(\frac{3\theta}{2}\right)\right]+\sigma_{dII}\left[\frac{1}{2}\sin\theta\cos\left(\frac{3\theta}{2}\right)\right]\right\} \quad (8.18a)$$

$$\sigma_{yy} = \sqrt{\frac{a}{2r}}\left\{\sigma_{dI}\left[\cos\left(\frac{\theta}{2}\right)+\frac{1}{2}\sin\theta\sin\left(\frac{3\theta}{2}\right)\right]+\sigma_{dII}\left[\frac{1}{2}\sin\theta\cos\left(\frac{3\theta}{2}\right)\right]\right\} \quad (8.18b)$$

Cartesian form
$$\sigma_{xy} = \sqrt{\frac{a}{2r}}\left\{\sigma_{dI}\left[\frac{1}{2}\sin\theta\sin\left(\frac{3\theta}{2}\right)\right]+\sigma_{dII}\left[\cos\left(\frac{\theta}{2}\right)\frac{1}{2}\sin\theta\cos\left(\frac{3\theta}{2}\right)\right]\right\} \quad (8.18c)$$

$$\sigma_{yz} = \sqrt{\frac{a}{2r}}\,\sigma_{dIII}\cos\left(\frac{\theta}{2}\right) \quad (8.18d)$$

$$\sigma_{xz} = \sqrt{\frac{a}{2r}}\,\sigma_{dIII}\left[-\sin\left(\frac{\theta}{2}\right)\right] \quad (8.18e)$$

$$\sigma_{zz} = \sqrt{\frac{a}{2r}}\left\{\sigma_{dI}\left[2v\cos\left(\frac{\theta}{2}\right)\right]+\sigma_{dII}\left[-2v\sin\left(\frac{\theta}{2}\right)\right]\right\} \quad (8.18f)$$

Now the *polar form* (which is written in cylindrical $r\theta\zeta$ coordinates):

$$\sigma_{\theta\theta} = \sqrt{\frac{a}{2r}}\left\{\sigma_{dI}\left[\cos^3\left(\frac{\theta}{2}\right)\right]+\sigma_{dII}\left[-3\sin\left(\frac{\theta}{2}\right)\right]\cos^2\left(\frac{\theta}{2}\right)\right\} \quad (8.19a)$$

$$\sigma_{r\theta} = \sqrt{\frac{a}{2r}} \left\{ \sigma_{dI} \left[\sin\left(\frac{\theta}{2}\right) \cos^2\left(\frac{\theta}{2}\right) \right] + \sigma_{dII} \left[\cos\left(\frac{\theta}{2}\right) \left\langle 1 - 3\sin^2\left(\frac{\theta}{2}\right) \right\rangle \right] \right\} \quad (8.19b)$$

Polar form

$$\sigma_{rr} = \sqrt{\frac{a}{2r}} \left\{ \sigma_{dI} \left[\cos\left(\frac{\theta}{2}\right) \left\langle 1 + \sin^2\left(\frac{\theta}{2}\right) \right\rangle \right] + \sigma_{dII} \left[\sin\left(\frac{\theta}{2}\right) \left\langle 1 - 3\sin^2\left(\frac{\theta}{2}\right) \right\rangle \right] \right\}$$

$$(8.19c)$$

$$\sigma_{r\zeta} = \sqrt{\frac{a}{2r}} \sigma_{dIII} \sin\left(\frac{\theta}{2}\right) \quad (8.19d)$$

$$\sigma_{r\zeta} = \sqrt{\frac{a}{2r}} \sigma_{dIII} \cos\left(\frac{\theta}{2}\right) \quad (8.19e)$$

$$\sigma_{\zeta\zeta} = \sqrt{\frac{a}{2r}} \left\{ \sigma_{dI} \left[2v \cos\left(\frac{\theta}{2}\right) \right] + \sigma_{dII} \left[-2v \sin\left(\frac{\theta}{2}\right) \right] \right\} \quad (8.19f)$$

Polar stresses: Propagation

Both polar and Cartesian forms of these near-tip equations are listed because each serves a specific application. The **polar form is most relevant to fracture propagation studies**, because it is centered on the fracture tip and tells us what the fracture "sees" just ahead of it. The **Cartesian form is useful**

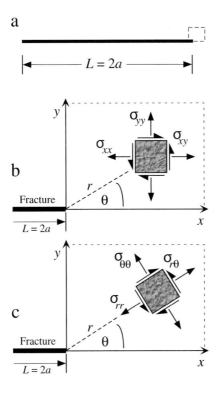

Fig. 8.15. Geometry and parameters for calculating the near-tip stress field at the tip of a fracture. (a) Fracture and approximate near-tip region; (b) enlargement showing near-tip stress for Cartesian local coordinates; (c) near-tip stress using local polar coordinates.

for resolving the influence of the near-tip stresses on objects or boundaries nearby. As in the previous whole-crack equations for stress and displacement, these are written without the arbitrary $\sqrt{\pi}$ term in the denominator (e.g., Lawn, 1993).

Figure 8.16 shows how the *proximity to the fracture tip* influences the near-tip stresses. The level of circumferential stress ($\sigma_{\theta\theta}$ in the figure) close to the tip is 7–10 times that of the driving stress. This means that a small internal fluid pressure or resolved tension is magnified by about an order of magnitude, in certain locations, by the presence of the crack. If the rock was just on the verge of failing in tension, then the stress concentration at the tip would have caused the rock to fracture, at the crack tip first, long before the rest of the surrounding rock was even close to failing. If we were to look even closer to the crack tip, the level of stress would become so large that the magnitude would not have any physical meaning. A limiting value of stress at the tip region is given by the *stress intensity factor K* discussed below. Although *K* is more convenient in many applications because its use can vastly simplify the near-tip stress equations, either method (near-tip stress state or stress intensity factor) lends insight into the tendency and direction of crack propagation.

Geologic fractures concentrate and redistribute stresses around them as do circular and elliptical cavities in rock (Chapter 2), but the rate of change of stress magnitudes away from them is different. In particular, the stress components calculated from the near-tip solution decrease away from the tip following a characteristic $1/\sqrt{r}$ decay, as shown in Fig. 8.17. Stresses decay much more rapidly with distance from a fracture than from a cavity. The *strength of the stress concentration* ($1/\sqrt{r}$ in this case for a sharp-tipped fracture) governs the lateral extent of mechanical interaction between a fracture and neighboring structures—that is, the length and spacing of fractures and the patterns that they generate are controlled by the rate of stress change near the fracture tip.

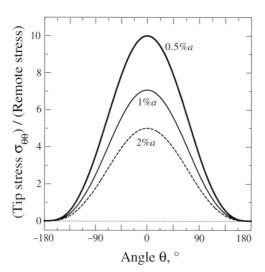

Fig. 8.16. Variation of the magnitude of the near-tip stress field ($\sigma_{\theta\theta}$) at the tip of a mode-I fracture, shown for r/a = 0.05, 0.1, and 0.2 (equation (8.19a)).

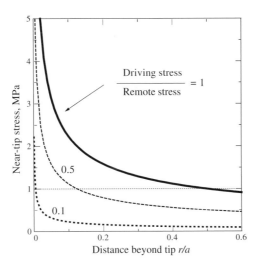

Fig. 8.17. The near-tip stress field (NTS) displayed as a function of distance from the tip of a fracture. The σ_{yy} component is shown for three values of driving stress; the level of remote stress (1.0 MPa) is the horizontal dotted line.

The near-tip stresses (calculated for $r = 0.01a$) for three cases of cracks are plotted for both coordinate conventions in Fig. 8.18. In order to examine the potential propagation directions of the cracks, we'll focus on the polar coordinate plots. This corresponds to what the crack tip "sees" just ahead; propagation can occur in the orientation that has the maximum value of the circumferential stress, $\sigma_{\theta\theta}$ (Erdogan and Sih, 1963; Ingraffea, 1987, pp. 90–94).

The relationships developed in the engineering literature for mixed-mode (usually I-II) crack growth, and presented here, assume a 2-D, plane-stress or plane-strain crack. This approach can provide a useful illustration of basic tendencies for the simplest incipient cases (see Lawn, 1993, pp. 44–50), such as the formation of twist hackle (Pollard et al., 1982; mixed-mode I-III) or end-cracks (e.g., Cotterell and Rice, 1980; Anderson, 1995, pp. 91–96; mixed-mode I-II). However, the propagation of a crack sustaining anti-plane shear stresses at the tip (I-III) cannot be simply described or predicted fully by using the 2-D LEFM equations because this process requires an explicit consideration of the 3-D stress state near the fracture front (e.g., Cooke and Pollard, 1996). This means that the results presented here and in the literature provide a reasonable guide to in-plane fracture propagation geometries.

There are other methods available in the engineering literature for predicting the initial increment of crack growth, including **maximum strain energy release rate** (or fracture propagation energy: Erdogan and Sih, 1963; Anderson, 1995, p. 93; Pollard and Aydin, 1988; Aydin and Schultz, 1990), and **minimum strain energy density** (Sih, 1973, 1991; Ingraffea, 1987, pp. 94–96) that can, in certain cases, provide a somewhat closer match between the predicted angle and that observed for cracks propagating in perfectly brittle materials (see discussion by Engelder et al., 1993). However, each of these methods provides nearly the *same level of accuracy*, but with some discrepancies between the predicted and measured values (Ingraffea, 1987, pp. 96–98; Engelder et al., 1993, p. 83; Cooke and Pollard, 1996). These will be discussed in Chapter 9.

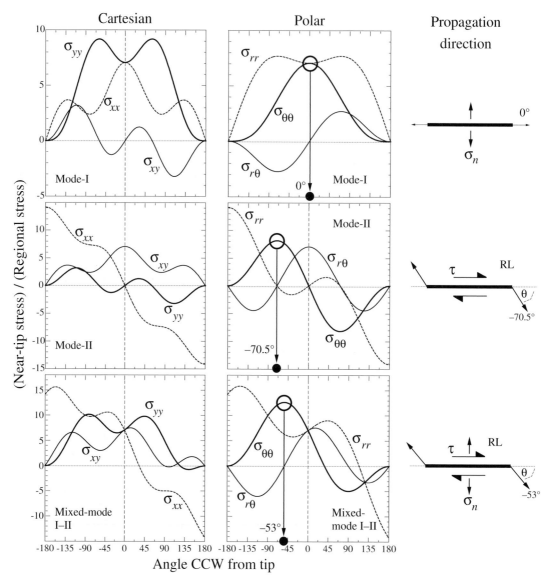

Fig. 8.18. Near-tip stress components in Cartesian and polar coordinates at the tip of a mode-I, mode-II, and mixed-mode I-II fracture. The circumferential stress component $\sigma_{\theta\theta}$ is the most useful for assessing the direction in which the crack will propagate (second column, arrows) due to the resolved stresses acting on it. Note that in-plane (i.e., along-strike) growth is predicted only for a pure mode-I crack; any amount of resolved shear stress will cause the crack walls to dilate obliquely and the crack to propagate at an angle to its original orientation.

Although the equations (8.18) and (8.19) for elastic near-tip stresses establish values for the angles at which a crack may initially propagate, more sophisticated formulations can be used to account for continued growth of secondary fractures near fracture tips (Cotterell and Rice, 1980; Horii and Nemat-Nasser, 1985). Similarly, the propagation of faults having frictional boundary conditions (Lin and Parmentier, 1988; Du and Aydin, 1993, 1995), and the inelastic near-tip stresses inferred for geologic fractures such as dikes (Rubin, 1993a, 1995b) and faults (Cowie and Scholz, 1992b; Bürgmann et al., 1994; Willemse et al., 1996; Cooke, 1997; Martel, 1997; Willemse and Pollard, 1998) may require more sophisticated treatments than that summarized in this chapter. These topics will also be explored in Chapter 9.

8.3.5.2 The Stress Intensity Factor

The stress intensity factor, K, was developed to recover the dependence of crack propagation on the applied stress and crack length (e.g., Paris and Sih, 1965; Broek, 1986; Lawn, 1993, pp. 1–17; Landes, 2000; Newman, 2000). Because the Inglis/Kolosov Relation for a degenerate ellipse predicts infinite stress at a fracture tip regardless of its length, applied (remote) stress, or loading configuration (displacement mode), some mathematical manipulation is needed to get around this problem. The important results for geologic fractures will be presented in this section.

Analytical derivation of the stress intensity factor traditionally involves conformal mapping of stress functions written using complex variables (e.g., Westergaard, 1939; Paris and Sih, 1965). This approach starts with a solution to a particular problem, like an elliptical hole in an elastic plate loaded by remote principal stresses (e.g., the Inglis/Kolosov problem from Chapter 2); this solution is an equation referred to as the *stress function* (e.g., Westergaard, 1939; Muskhelisvili, 1963; Rekash, 1979; Broek, 1986, pp. 75–80). Complex variables (e.g., Churchill and Brown, 1984) and conformal mapping (e.g., Nehari, 1952) are typically used to condense the math, reducing expressions and derivatives of two spatial variables (x and y, for a Cartesian coordinate system) to one ($z = x + iy$, with real and imaginary parts). Following standard treatments such as those given by Paris and Sih (1965) and Kanninen and Popelar (1985), the mixed-mode (I-II) stress intensity factor is defined as the limit of the near-tip stress state as the fracture tip is reached. The essence of this approach is illustrated in equation (8.20) using the notation shown in Fig. 8.19.

$$K = \left(K_{\mathrm{I}} + iK_{\mathrm{II}}\right) = 2\sqrt{2\pi}\lim_{z \to a}\left[\sqrt{(z-a)}\phi'(z)\right] \tag{8.20}$$

Fig. 8.19. Main parameters for defining the stress intensity factor.

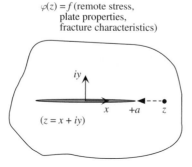

$\varphi(z) = f$ (remote stress, plate properties, fracture characteristics)

In the above expression the stress function (given by its derivative) is evaluated at the crack tip(s) for which $z = \pm a$ (z_1 in the derivation).

After setting up the problem and executing the math, we obtain remarkably compact expressions for the dependence of the local stresses very close to the fracture tip on the remote stress and fracture length. The general form can be written as (Lawn, 1993, p. 31)

$$K_{\mathrm{I}} = Y\sigma_{\mathrm{dI}}\sqrt{\pi a} \tag{8.21a}$$

Stress intensity factors
$$K_{\mathrm{II}} = Y\sigma_{\mathrm{dII}}\sqrt{\pi a} \tag{8.21b}$$

$$K_{\mathrm{III}} = Y\sigma_{\mathrm{dIII}}\sqrt{\pi a} \tag{8.21c}$$

The value of K for a given mode depends on the magnitude and sign of the driving stress and on the fracture half-length a. By taking the limit of the stress changes near the fracture tip, we were able to eliminate the theoretical singularity (infinitely large value) in the values of the stress components as r approached 0 in equations (8.18) and (8.19). Thus K recovers the dependence of stress concentration at a fracture tip on the fracture length and resolved (driving) stress, analogous to the Inglis/Kolosov Relation for $b \neq 0$.

$\phi(z) = f$ (remote stress, plate properties, fracture shape)

Most commonly the stress intensity factors are written assuming plane-strain conditions; this holds for dipping geologic fractures or deformation bands as well when viewed in cross-section. The arbitrary $\sqrt{\pi}$ term (see historical discussion by Landes, 2000) is included in these expressions, following standard practice (e.g., Broek, 1986, p. 82; Pollard and Segall, 1987; Lawn, 1993). The assumption of linear elasticity suggests that the stress intensity factors are additive (Lawn, 1993, p. 28). The units of stress intensity factor are MPa $m^{1/2}$, which are informally defined in this book as the "Irwin," abbreviated "Ir."

Unit for stress intensity factor:
1 Irwin = MPa m$^{1/2}$

The term Y in the above expressions for stress intensity factor is a convenient factor that accounts for the influence of boundaries or other nearby stress-changing structures on the value of K for the fracture tip of interest. Particular values of Y are listed in Table 8.2; these parameters or polynomial expressions (e.g., Broek, 1986, p. 85) are generally available for relatively simple problem configurations. As can be seen, K_{I} is increased for an edge crack by 12% above that for the same crack (half-length) and same value of driving stress. A penny-shaped crack has K_{I} of only 64% that for an isolated, 2-D "tunnel crack" having the same half-length and driving stress (see equation (8.13) and Fig. 8.11). The Y factor is used extensively in engineering design of practical objects to compensate for specimens or components that are either small relative to the crack(s) contained within them (finite size correction) or that have multiple collinear cracks, such as bathroom tissue. Approximate values for Y can also be calculated either analytically (e.g., Paris and Sih, 1965; Broek, 1986, pp. 79–94; Lawn, 1993, pp. 31–33; Anderson, 1995, pp. 58–64) or numerically for more complex problems such as fracture

Table 8.2 Typical Values for Configuration Parameter Y

Problem	Y	Reference
Single crack in infinite plate	1.0	Westergaard (1939)
Edge crack	1.12	Paris and Sih (1965)
Circular (penny) crack	$2/\pi$	Sneddon (1946)

interaction (e.g., Pollard et al., 1982; Pollard and Aydin, 1988; Aydin and Schultz, 1990; Fig. 5.7) or end cracks (e.g., Bürgmann et al., 1994). Although used less frequently in engineering problems, a comparable factor that accounts for 3-D fracture shape can also be used to modify the displacement equations (e.g., Willemse et al., 1996; Gudmundsson, 2000; Schultz and Fossen, 2002; Savalli and Engelder, 2005), as shown by equation (8.14) and the third entry in Table 8.2.

Example Exercise 8.7

Calculate the stress intensity factors for the following cases:

(a) A dry **joint** in an infinite rock mass loaded by a remote tension of magnitude 0.25 MPa oriented perpendicular to the joint having length of 1.8 m.

(b) An isolated **fault** of length 16 m oriented at $\alpha = 30°$ to the maximum remote compressive stress of 0.5 MPa (compression positive), least principal stress of 0.1 MPa (compressive), and having values of cohesion of 0 MPa, and friction coefficients of 0.2 (static) and 0.1 (residual).

Solution

We first calculate the driving stresses on the fractures, note the configuration factor Y for each fracture, then insert these values into equations (8.21).

(a) **The dry joint**: The geometry is ambiguous as written. In this case we assume that the field conditions didn't permit a more precise evaluation of the parameters we need for a fracture mechanics approach. We'll work this problem twice, first using a rectangular (tunnel) crack shape and then using a circular (penny) crack shape to address the ambiguity in crack shape. Y is 1.0 and $2/\pi$, respectively. Using $\sigma_{dI} = 0.25$ MPa, $\sigma_{dII} = 0$ MPa (no mode-II component), and $a = L/2 = 0.9$ m, we have

$$K_I = (1.0)(0.25\,\text{MPa})\sqrt{(0.9\,\text{m})}$$

$$\therefore K_I = 0.42\,\text{MPa m}^{0.5} = 0.42\,\text{Ir (2-D tunnel crack)}$$

and

$$K_I = (0.637)(0.25\,\text{MPa})\sqrt{\pi(0.9\,\text{m})}$$

$$\therefore K_I = 0.27\,\text{MPa m}^{0.5} = 0.27\,\text{Ir (3-D penny crack)}$$

(b) **The fault**: Here we have the same stress conditions (outer boundary conditions) as in Example Exercise 8.5, giving $\sigma_n = 0.2$ MPa (compressive) and $\tau = 0.1732$ MPa (left-lateral). The mode-II driving stress (assuming a partial stress drop) $\sigma_{dII} = 0.02$ MPa; there is no mode-I or mode-III component from the information recorded. Noting the same values of Y as in part (a) above, we let $Y = 1.0$ and $2/\pi$, $\sigma_{dII} = 0.02$ MPa, and $a = 8$ m, to obtain

$$K_{II} = (1.0)(0.05\,\text{MPa})\sqrt{\pi(8.0\,\text{m})}$$

$$\therefore K_{II} = 0.1\,\text{MPa m}^{0.5} = 0.1\,\text{Ir (left-lateral, 2-D fault)}$$

and

$$K_{II} = (0.637)(0.02\,\text{MPa})\sqrt{\pi(8.0\,\text{m})}$$

$$\therefore K_{II} = 0.064\,\text{MPa m}^{0.5} = 0.064\,\text{Ir (left-lateral, 3-D fault)}$$

The faults are both left-lateral as was the case in Example Exercise 8.5. Note that $K_{II} = K_{III} = 0$ for the joint cases and $K_I = K_{III} = 0$ for the fault cases.

A related quantity is the *strain energy release rate, G*. Also called the **fracture propagation energy** (e.g., Pollard and Aydin, 1988; Aydin and Schultz, 1990) or, for compaction bands, the **compaction energy** (Rudnicki and Sternlof, 2005; Tembe et al., 2006, 2008; Rudnicki, 2007; Schultz, 2009; Stanchits et al., 2009; Schultz and Soliva, 2012), this quantity represents the energy available at a fracture or deformation band tip to drive propagation or interaction. G is an energy with units of J m^{-2} (joules per meter squared) and can be related to the stress intensity factor (when LEFM applies) by using (Lawn, 1993, p. 29)

Strain energy release rate

$$G_i = \frac{K_i^2(1-v^2)}{E} \text{ (plane strain)} \tag{8.22}$$

Here, the subscript $i =$ I, II, or III for the appropriate fracture displacement mode. The "rate" in this expression is per unit of new fracture length, not time, so it's the energy available along the fracture tip. Alternatively, if the driving stresses (instead of K) are known, one can calculate the propagation energy by using (e.g., Engelder et al., 1993, p. 100)

$$\mathcal{G}_i = \left(\sigma_d^2\right)_i \pi a \frac{(1-v^2)}{E} \qquad (8.23)$$

where the subscript i again indicates the appropriate displacement mode and the flaw shape parameter Ω has been omitted for this plane-strain solution (but could be included for a 3-D problem). The *total strain energy release rate* for general, mixed-mode loading is the sum of the individual propagation energies (Lawn, 1993, p. 29)

$$\mathcal{G}_{total} = \frac{K_I^2(1-v^2)}{E} + \frac{K_{II}^2(1-v^2)}{E} + \frac{K_{III}^2(1-v^2)}{E} \qquad (8.24)$$

written for plane-strain conditions. These expressions are valid for *initial* loading of planar fractures, and not their later geometries.

Example Exercise 8.8

Calculate the fracture propagation energy for the joint and fault that you used in Example Exercise 8.6, assuming 2-D geometry, a thick plate, and a Young's modulus of 15 GPa, and a Poisson's ratio of 0.3, for the rock mass.

Solution

Noting that the plane-strain equations are to be used as an approximation to the thick-plate geometry and that all units must be in MPa, equations (8.18) give us

$$\mathcal{G}_I = \frac{(0.42\,\mathrm{MPa\,m}^{0.5})^2\left(1-(0.3)^2\right)}{15{,}000\,\mathrm{MPa}}$$

$$\therefore \mathcal{G}_I = 1.07 \cdot 10^{-5}\,\mathrm{J\,m}^{-2} \text{ for the 2-D joint}$$

and

$$\mathcal{G}_{II} = \frac{(0.064\,\mathrm{MPa\,m}^{0.5})^2\left(1-(0.3)^2\right)}{15{,}000\,\mathrm{MPa}}$$

$$\therefore \mathcal{G}_{II} = 2.49 \cdot 10^{-7}\,\mathrm{J\,m}^{-2} \text{ for the 3-D fault}$$

The strain energy release rate was the cornerstone of Griffith's (1921) thermodynamic energy-balance formulation for crack problems (Rice, 1978). Originally designed as a global parameter to encapsulate the total strain energy of the strained material, such as a rock mass, without regard for the details of crack-tip processes or material rheologies (e.g., linearly elastic), you will most often see it calculated and used as a fracture-tip parameter. While this is legitimate as long as LEFM applies, the practice may not be valid if the plastic zone is too large for LEFM, as it may be for many geologic fractures (e.g., Cowie and Scholz, 1992b; Rubin, 1993a; Willemse et al., 1996), as demonstrated below. Alternative methods such as the *J*-integral (Rice, 1968; Kanninen and Popelar, 1985, pp. 282–291; Lawn, 1993, pp. 66–70; Rudnicki and Sternlof, 2005) or *R*-curve (crack resistance) graphs (e.g., Engelder et al., 1993, pp. 99–103; Broek, 1986, pp. 130–136) may be used for cases in which the assumption of small-scale yielding is not appropriate. We'll look at some of these approaches in Chapter 9.

8.3.5.3 Stress Intensity Factor from Fracture Displacements

In geology, we observe and measure the offsets (opening and shear) across fractures much more often than we can infer stress states. This places a limit on the utility of the stress-based equations for stress intensity factor just discussed. We can rewrite these equations by replacing the driving stress in equations (8.21) with the value of **maximum displacement** from equation (8.10). Solving for driving stress in (8.10) and substituting this into (8.21), we obtain

K from fracture displacement

$$K_i = \left[\frac{E}{4(1-v^2)} \right] D_{\max} \sqrt{\frac{\pi}{a}} \qquad (8.25)$$

where i = I, II, or III for the engineering mode of fracture displacement and the geometry factor $Y = 1.0$. This expression provides a very useful alternative to the usual equations for K that come to us from engineering (8.21), where stresses are employed instead of fracture-surface displacements.

8.3.5.4 Take-Off Angle for Secondary Fractures

The angle between the parent fracture and secondary fractures (joints, stylolites, and folds) provides a useful parameter for assessing the kinematics of fracture growth in rocks. We can use the stress intensity factors (or alternatively, the values of normal or shear displacements) to relate the take-off angle, or the angle of propagation of the fracture (see also Fig. 5.18, where this angle for that case was denoted θ), to the amount of opening and shearing displacements that the main fracture has accommodated. Several workers have evaluated or exploited this approach in their studies of sheared joints (e.g., Cruikshank et al., 1991a,b; Wilkins et al., 2001; Flodin and Aydin, 2004a,b; Kattenhorn and Marshall, 2006) and faults (e.g., Martel, 1997).

Following Ingraffea (1987), pp. 90–93, and Schultz and Zuber (1994), the take-off angle of initial crack growth Δ is given by

Take-off angle using K

$$\Delta = 2 \tan^{-1} \left[\frac{-K_{\mathrm{I}} + \sqrt{K_{\mathrm{I}}^2 + 8 K_{\mathrm{II}}^2}}{4 K_{\mathrm{II}}} \right] \qquad (8.26)$$

in which the normal and shear stress intensity factors are $K_{\mathrm{I}} = \sigma_{\mathrm{dI}} \sqrt{\pi a}$ and $K_{\mathrm{II}} = \sigma_{\mathrm{IId}} \sqrt{\pi a}$ (setting $Y = 1.0$) and a is the half-length of the parent fracture. A new crack will nucleate and propagate into the angle Δ if the total stress intensity (mode-I plus mode-II) equals or exceeds the dynamic fracture toughness of the rock, K_{Ic}. The problem setup is shown in Fig. 8.20 and the results of the calculation are shown in Fig. 8.21. One could also calculate the stress intensity factors **directly from fracture-surface (maximum) displacements** using (8.25) and then apply (8.26) to predict take-off angles. The expression is

Take-off angle using D_{max}

$$\Delta = 2\tan^{-1}\left[\frac{-D_I + \sqrt{D_I^2 + 8D_{II}^2}}{4D_{II}}\right] \qquad (8.27)$$

in which D_I and D_{II} are the maximum mode-I and mode-II fracture-surface displacements, respectively. Interestingly, the rock stiffness and driving stress terms from (8.10) and (8.21) cancel out in (8.27), indicating that the take-off angle is a function only of the (maximum) fracture-surface displacements.

As can be seen from Fig. 8.21, there are *three sets of values* for the take-off angle:

- **0°**: this is **in-plane growth** of a pure mode-I fracture, such as a crack, vein, hydrofracture, or dike loaded with only a normal stress; it has no shear stress at its tip.
- **70.5°**: this corresponds to **pure-mode-II loading** conditions of no normal stress and any amount of shear stress. Faults are often considered to propagate under this condition (e.g., Segall and Pollard, 1980; Pollard and Segall, 1987, and many others).

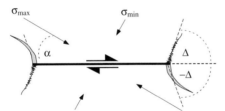

Fig. 8.20. Geometry of remote loading and take-off angle Δ.

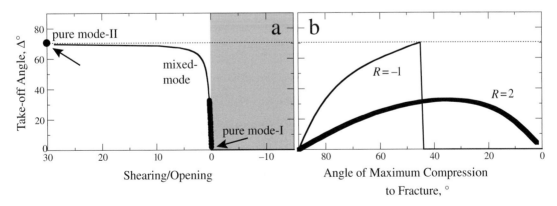

Fig. 8.21. Relationship between the take-off angle of a secondary fracture at the tip of a parent fracture on: (a) the ratio of shearing to opening displacements on the parent fracture; shaded area, closed surfaces; and (b) the angle of the remote maximum compressive stress to the parent fracture plane. Two cases are shown: greatest-to-least compressive remote principal stress ratio $R = -1$ (thin curves in (a) and (b)) and 2 (bold curves). Opening displacements are taken as positive.

- **Anything in between**: this may indicate **either a mixed-mode crack** with both normal and shear displacements along it (Cruikshank et al., 1991a,b; Kattenhorn and Marshall, 2006), **or a fault** (Martel, 1997) with a non-LEFM end zone (see Chapter 9; Martel, 1997; Willemse and Pollard, 1998; Koenig and Aydin, 1998).

Additionally, one can see that the relationship between take-off angle and the far-field, remote stress state is not straightforward (see Anderson, 1995, pp. 91–96). As shown in the right-hand panel of Fig. 8.21, particular combinations of normal and shear stress can be resolved onto a fracture from different remote stress states by varying (a) the ratio of the remote principal stresses; (b) the angle of the principal stresses to the fracture plane; or (c) both of these. This is why the ratio of displacements (normal and shear) on the fracture surface can be used as the main criterion for interpreting take-off angle. It would take some additional work to relate the displacements to the remote stress state, although this can sometimes be done.

Some representative values of take-off angle measured from the field (Fig. 8.22) include:

- ~50° along slipped bedding planes in sandstone (Cooke et al., 2000)
- 4–44° in layered clastic rocks (Wilkins et al., 2001)
- 35–50° in massive sandstone (Cruikshank et al., 1991a,b)
- 20–70° in layered clastic rocks (Kim et al., 2004)
- <50° in faulted granite (Segall and Pollard, 1983b)

The values less than 70° may be indicative of a combination of opening and shearing displacements along the deforming joint (Fig. 8.22). For the case discussed by Segall and Pollard (1983b) and Martel (1997), however, no opening of the sheared joint can be demonstrated, suggesting that the low take-off angle (<70°) may result instead from non-LEFM conditions at the tip of the shearing joint. This point is discussed at length in Chapter 9.

Fig. 8.22. Calculated take-off angles near the ends of a parent fracture plotted as a function of the ratio of the fracture-surface maximum displacements (opening displacements are taken as positive). Take-off angles less than 70° (horizontal dashed line) imply that shearing displacements may be comparable in magnitude (e.g., vertical dashed arrow) to opening displacements along the fracture—a prediction readily tested in the field.

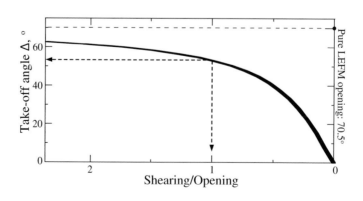

8.3.6 The Near-Tip Process Zone and Small-Scale Yielding

The near-tip solution (equations (8.18) and (8.19)) is valid only for the **K-dominant region** (Kanninen and Popelar, 1985, pp. 146–147; Rubin, 1993a), which is the area over which the amplified fracture-tip stresses exceed the level of applied remote stress, resulting in a region of locally modified stress around the fracture. This is the fracture's **"zone of influence."** A simple estimate of the extent of the K-dominant region, of radius $r = r_K$, in front of a fracture is obtained by rearranging a near-tip component of stress to solve for r. For example, setting $\theta = 0°$ and $\sigma_{dII} = 0$ (to evaluate r in a line directly ahead of a pure mode-I dilatant crack) in (8.18a) gives

K-dominant region

$$r_K = \frac{a}{2}\left(\frac{\sigma_d}{\sigma_{yy}}\right)^2 \tag{8.28}$$

where a is the fracture half-length. Similarly, the size of the K-dominant region for a fault can be estimated by setting $\theta = 0°$ and $\sigma_{dI} = 0$ in (8.14b) and solving for r; the same result (equation (8.28)) is thus obtained regardless of fracture mode (except that the remote stress for a fault would be σ_{xy} instead of σ_{yy} for joints in (8.28)). This procedure is illustrated in Fig. 8.23, where it becomes apparent that the size of r_K scales with the magnitude of the driving stress, with fracture length being a second-order factor. Notice that $\sigma_{dI} = \sigma_{yy}$ for

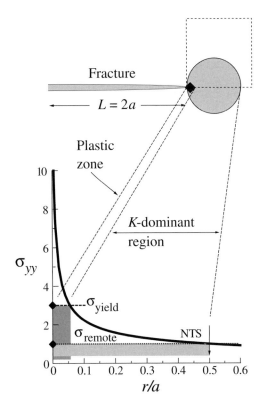

Fig. 8.23. Relationship between the near-tip stress field (NTS) near the tip of a fracture and the plastic zone (dark shading, shown for yield strength of 3 MPa) and the enclosing K-dominant region (light shading, shown for NTS ≥ remote stress). LEFM is valid as long as the plastic zone is smaller than the K-dominant region and small relative to the fracture dimensions.

tensile loading of mode-I cracks, whereas $\sigma_{dI} \neq \sigma_{yy}$ for geologic cracks loaded by compressive stress (that therefore require internal pore-fluid pressure to dilate).

An estimate of the size of the **plastic zone** can be obtained in similar fashion by comparing the level of near-tip stress to a specific value for the yield strength of the rock ahead of the fracture tip. For **mode-I cracks**, the yield strength σ_{yield} is equivalent to the *tensile strength* T_0, with values typically around 10–50 MPa (Jaeger and Cook, 1979). For **faults** (mode-II and/or mode-III) the yield strength is the *frictional resistance of rock at the fault tip*. Using a constant value of yield strength (e.g., tensile or shear), a first-order estimate of plastic zone radius ahead of a fracture (the "Irwin plastic zone correction") is given by (Broek, 1986, pp. 13–14, 99–102; Anderson, 1995, pp. 72–75)

Size of the near-tip plastic zone

$$r_p = \frac{a}{2}\left(\frac{\sigma_{di}}{\sigma_{yield}}\right)^2 \tag{8.29}$$

where the subscript "i" in the driving stress term denotes the fracture displacement mode.

As can be seen by comparing equations (8.28) and (8.29), the only difference is the specification of the level of remote stress (for the size of the K-dominant region) or the yield strength (for the plastic zone). For the fracture to be described by LEFM, the entire region surrounding the fracture must be idealized as an isotropic, homogeneous, linearly elastic material—that is, any embedded areas of plasticity must be so small relative to the fracture's dimensions that they can be neglected. Although different estimates of plastic zone size (e.g., Schmidt, 1980; Anderson, 1995, p. 73) and shape (Engelder et al., 1993, pp. 89–97) can be made, these are merely variations on the theme presented here, rather than significant differences.

SSY: $r_p < r_K$ and $r_p \ll a$

LEFM is valid when SSY applies

Small-scale yielding ("SSY"), a requirement of LEFM, is satisfied as long as the plastic zone is: (a) smaller than the K-dominant region, and (b) small relative to fracture dimensions (Kanninen and Popelar, 1985, p. 146), so that the elastic stresses describe the state of stress near the fracture tip.

We can compare the sizes of the plastic zone (equation (8.29)) and that of the K-dominant region (equation (8.28)) by dividing them—this gives us useful estimates for the limiting values of stress in the rock mass that would be consistent with small-scale yielding, and, therefore, LEFM. We find that

$$\frac{r_p}{r_K} = \left(\frac{\sigma_{yy}}{\sigma_{yield}}\right)^2 = \left(\frac{\text{Remote stress level}}{\text{Rock strength}}\right)^2 \tag{8.30}$$

We can see that, for r_p to be much less than r_K, and so for LEFM to apply to fractures in rock, the remote stresses (resolved on a fracture surface) must also be considerably less than the yield strength (tensile or shear, for mode I or II/III). This is illustrated in Fig. 8.24. For example, using a criterion that

Fig. 8.24. Dependence of the assumption of small-scale yielding (SSY) on the relative magnitudes of the resolved remote stress and the rock's yield strength. Dashed and dotted boxes show approximate range for small-scale yielding and brittle behavior, respectively.

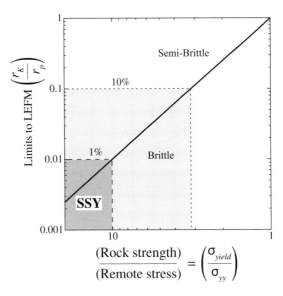

the plastic zone must be no larger than 1% of the K-dominant region (dashed line in Fig. 8.24), small-scale yielding is achieved for rock yield strengths that exceed 25 times the value of the remote stress. For r_p/r_K of 10% (dotted line in Fig. 8.24), corresponding perhaps to a reasonable upper limit, SSY is met for yield strengths exceeding 2–3 times the remote stress level.

The first requirement for small-scale yielding (given by equation (8.30)) is met automatically when $\sigma_{yield} > \sigma_{yy}$ (in which case $r_p \leq 0.5a$). As the level of remote stress increases relative to the value of rock yield strength—moving to the right on the diagram—the plastic zone takes up a larger share of the K-dominant region. As a rock therefore gets farther away from having small-scale yielding valid, it is also moving away from having elastic descriptions of fracture behavior—like the displacement profile and near-tip displacements—adequately representing the observations.

For the second requirement, the plastic zone must be so small relative to the fracture length (and the K-dominant region) that it can be ignored in the elastic equations for stress (recall that an isotropic, *homogeneous* material on the scale of the fracture is assumed). For many geologic fractures, the maximum size of this region can be approximated by being less than 10% of the fracture half-length (dotted line in Fig. 8.24, or defining $C = 0.1a$). Setting $r_p < C$ in (8.29) gives

Condition for small-scale yielding

$$\sigma_{yield} > \frac{\sigma_{dl}}{\sqrt{2C}} \qquad (8.31)$$

Equation (8.31) indicates that the yield strength must exceed approximately 2.2 times the driving stress on a fracture for the size of the plastic zone to be 10% or less than the fracture half-length, corresponding approximately to brittle fracture. It appears that driving stress magnitudes that are comparable

Fig. 8.25. Examples of the crack-normal stress component, calculated using a pure mode-I crack and σ_{yy}, for several configurations. NTS, σ_{yy} calculated using the near-tip approximations (equation (8.18a)) to the whole-crack equations for σ_{yy} (equation (8.17a)). The K-dominant region (stippled, upper panels) is defined where the amplified stresses caused by crack opening displacement exceed the level of the resolved remote stress, σy_y (set to 1.0 MPa in the figure). The plastic zone r_p (shaded, lower panels) is suggested where the level of near-tip stress exceeds that of the rock's tensile strength (10 MPa).

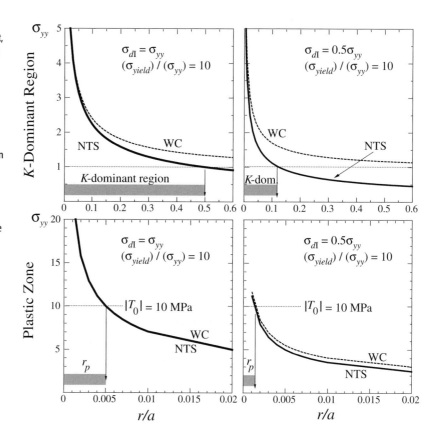

to the magnitudes of the remote stresses (i.e., within a factor of two) would produce plastic zones that are too large relative to fracture size for small-scale yielding, and therefore LEFM, to apply in more than an approximate way (i.e., semi-brittle, "post-yield" fracture mechanics; Chapter 9).

Specific values for r_p depend on the magnitudes of the applied remote stress and rock strength, as demonstrated in Fig. 8.25. There, one can see how the level of remote stress must remain small relative to the magnitude of the rock's yield strength, or else distributed (nonlocalized) fracturing and yielding can occur and the assumptions of LEFM would not apply.

Now let's work through a representative example exercise for the opening and shearing modes.

Example Exercise 8.9

Calculate the size of the plastic zone at the tips of the joint and fault from Example Exercise 8.7, assuming 2-D geometry and an appropriate yield strength for the rock surrounding the fracture.

Solution

We first check to see that small-scale yielding is met, and therefore that LEFM is approximated. Recalling that the joint was loaded by a joint-normal tensile stress of 0.25 MPa and with the rock having a tensile strength $T_0 = 10$ MPa for the **joint**, and comparing the magnitudes of the remote stress and the (tensile) yield strength using equation (8.30), we find that

$$\frac{r_p}{r_K} = \left(\frac{\sigma_{yy}}{\sigma_{yield}}\right)^2 = \left(\frac{0.25\,\text{MPa}}{10\,\text{MPa}}\right)^2 = 0.00063 = 0.06\%$$

so SSY is met and we can proceed to use LEFM to infer plastic zone size. Using equation (8.29) and $a = L/2 = 0.9$ m, we find that

$$r_p = \frac{0.9\,\text{m}}{2}\left(\frac{0.25\,\text{MPa}}{10\,\text{MPa}}\right)^2$$

$$\therefore r_p = 2.8 \cdot 10^{-4}\,\text{m} = 0.3\,\text{mm for the 2-D joint}$$

By normalizing the result, $r_p/a = (2.8 \times 10^{-4}\,\text{m})/(0.9\,\text{m}) = 3.125 \times 10^{-4} = $ **0.03% a**, we see that the plastic zone is indeed negligibly small relative to the crack, as required.

For the **fault**, the yield strength is now the *static (maximum) frictional resistance* of the fault. Using equation (3.7b) and noting that the coefficient of maximum static friction along the fault is $\mu_s = 0.6$, the frictional yield strength is

$$\sigma_{yield} = C_0 + \mu_s(\sigma_n - p_i)$$

$$0\,\text{MPa} + 0.6(0.2 - 0.0\,\text{MPa}) = 0.12\,\text{MPa}$$

First we check for validity, just as we did for the joint example. Comparing the values for driving stress and yield strength, we obtain

$$\frac{r_p}{r_K} = \left(\frac{\sigma_{yy}}{\sigma_{yield}}\right)^2 = \left(\frac{0.02\,\text{MPa}}{0.12\,\text{MPa}}\right)^2 = 0.028 = 2.8\%$$

The predicted plastic zone size for the fault is nearly 3% of the size of the K-dominant region: larger than for the joint but still reasonably small. Using equation (8.29) and $a = 8.0$ m, along with the mode-II driving stress (or partial stress drop) of $\sigma_{dII} = 0.02$ MPa, we see that

$$r_p = \frac{8.0\,\text{m}}{2}\left(\frac{0.02\,\text{MPa}}{0.12\,\text{MPa}}\right)^2$$

$$\therefore r_p = 0.11\,\text{m for the 2-D fault}$$

Again, normalizing the results gives us $r_p/a = (0.11\,\text{m})/(8.0\,\text{m}) = 0.125 = $ **1.4% of a**, corresponding to a reasonably small plastic zone for the fault.

The result for the fault is valid for only one single coseismic slip event. For a fault having geologically reasonable values of displacement (see Fig. 1.9), the cumulative driving stress may be 100 times that of the coseismic stress drop for an individual slip event. In that case, using a *cumulative* driving stress of $(100)(0.02) = 2.0$ MPa, we would obtain

$$\frac{r_p}{r_K} = \left(\frac{\sigma_{yy}}{\sigma_{yield}} \right)^2 = \left(\frac{2.0\,\text{MPa}}{0.12\,\text{MPa}} \right)^2 = 277.8 > 1.0$$

The predicted plastic zone size for the fault, calculated by using a *cumulative* driving stress, far exceeds the size of the K-dominant region. Because the ratio of r_p/r_K exceeds one, the fault appears not to obey LEFM any longer, since small-scale yielding is clearly violated. In fact, the need to model faults, as opposed to earthquakes, as non-LEFM fractures was one of the primary conclusions of the classic paper by Cowie and Scholz (1992b). In addition, differences between LEFM fractures (such as joints, veins, and igneous dikes) and non–LEFM fractures (such as faults) are easily identified by their displacement–length scaling relations (e.g., Scholz, 2002, p. 116; Olson, 2003; Schultz et al., 2008a) as we will explore in Chapter 9.

The values of cohesion, friction coefficient, and pore-fluid pressure that contribute to the fault's frictional resistance affect the elastic conditions near the fault tip. Increasing the pore-fluid pressure, for example, can decrease the effective normal stress and thereby increase the driving stress on the fault, leading to a larger plastic zone. Fig. 8.26 shows that pore-fluid pressure and friction coefficient each act in opposite directions, implying that the plastic zone size represents a balance between these two parameters, modulated by the fault length.

Numerous studies have demonstrated that the levels of *in situ* stress in the Earth's crust are generally close to that required for frictional sliding along optimally oriented faults to occur (Sibson, 1974; McGarr and Gay, 1978; Zoback and Healy, 1984; Engelder, 1993; Townend and Zoback, 2000; Zoback et al., 2003; Zoback, 2007, pp. 127–137). For example, Martel (1997) notes that the stress drops across faults during individual slip events are within ~10% of the ambient level of shear stress acting on the fault, again implying a large plastic zone and that small-scale yielding might no longer apply (see supporting conclusions by Cooke, 1997, and by Willemse and Pollard, 1998). Using an Andersonian tectonic stress state for normal faulting (see Chapter 6) of 90% of the shear yield strength in the above example exercises would imply

Fig. 8.26. Dependence of plastic zone size on the competing effects of (upper panel) pore-fluid pressure and (lower panel) friction coefficient along a fault. The horizontal line represents an approximate limit of LEFM at $r_p/a \leq 0.1$ (LEFM is valid below the line). Shaded regions show typical ranges of pore-fluid pressure (upper panel) and friction coefficient (lower panel); dry curve is bold.

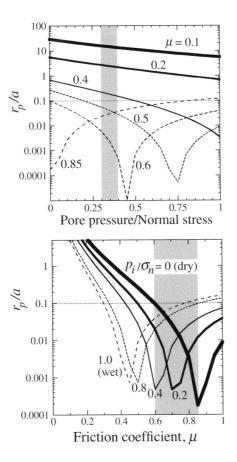

$$\frac{r_p}{r_K} = \left(\frac{\sigma_{yy}}{\sigma_{yield}}\right)^2 = \left(\frac{0.108\,\text{MPa}}{0.12\,\text{MPa}}\right)^2 = 0.81 \approx 1.0$$

The critically stressed crust

In other words, the *crust is nearly in general yield* since the values of remote stress and shear yield strength are comparable in magnitude. This set of calculations implies that, in general, the ratios of remote stress and driving stress to yield strength for faults in the crust are relatively large, and that the **conditions of small-scale yielding and LEFM may not strictly apply to faults**. In general, the stronger the rock ahead of the fracture, the better an LEFM-based approach will fit the observations.

8.4 Rules for Discontinuity Propagation

It is probably a truism that all geologic structural discontinuities—cracks, joints, faults, deformation bands, anticracks, stylolites—start out smaller than their final dimensions and have grown in length, or *propagated*, into the final configurations that we see. Discontinuity growth is an integral part of progressive brittle deformation. Tectonic strains continue to evolve, as do plate boundaries, as long as the conditions remain favorable for this continued progressive development. Because discontinuities are an effective means for accommodating—if not controlling—large strains in upper crustal rocks, an understanding of the controls on their propagation is of first-order importance.

In this section we'll examine two sets of rules for the propagation of fractures (cracks and faults) in crustal rocks. Although both of these rely on the assumptions and concepts of LEFM, they remain useful as a first-cut approach to studying discontinuity growth. Both can also be modified to account for inelasticity near discontinuity tips that is an important aspect of geologic fracture mechanics.

8.4.1 Basic LEFM Fracture Growth

Engineering mechanics contains a thorough account of the conditions required for the propagation of fractures that obey LEFM. The application of LEFM to geologic fracture propagation has resulted in a considerable improvement in our understanding of brittle fracture (e.g., see Pollard and Aydin, 1988; Engelder et al., 1993; and Gudmundsson, 2011, for many examples). While many of the concepts presented here have already been explored in the geologic literature, many others from materials science, ceramics, and fatigue, remain to be utilized.

Engineering fracture mechanics was developed mainly in the post-World War II period and before engineers could actually observe the terminations and shapes of flaws that they hypothesized weakened their materials (e.g., Kanninen and Popelar, 1985, pp. 36–39; Anderson, 1995, pp. 3–18). As a result, LEFM-based methods for crack propagation are dominated by the use of the *stress intensity factor*, and its derivative quantities. Observation and measurement of flaws was generally not required (but see Wells, 1963, for an exception), and indirect methods involving stress analysis were relied on instead.

8.4.2 Fracture Toughness: The Critical Dynamic Stress Intensity

The basis for fracture propagation studies based on LEFM is a comparison between the amplified stresses at a fracture tip and the inherent resistance to fracture propagation maintained by the rock. As we've seen, as long as small-scale yielding is satisfied then the stress intensity factors K_i can be used to describe the stress state at the fracture tip (just beyond the small plastic zone). One can also convert these into G_i, the strain energy release rate, which describes the amount of energy that is made available at the fracture tip to drive propagation. By themselves, these quantities are very useful in mapping out the degree of *mechanical interaction* (akin to constructive or destructive interference of the near-tip stress fields) between fractures and other structural elements.

However, just like any other type of strength calculation, the calculated elastic (or plastic, or viscous) stresses are not all that useful unless they are referenced to how much stress the rock can withstand before it breaks or shears. For large samples, the applied stress might be a tensile stress, and the limiting value for the average rock would be its average tensile strength. For a fracture, however, the appropriate strength is not the large-scale tensile strength for the rock mass but some other, analogous quantity that measures the tensile or shear strength of the rock *at the scale of the fracture tip*. This strength is called the rock's **fracture toughness**, or critical stress intensity, and denoted K_c (note that K_{Ic} is strictly the dynamic mode-I fracture toughness; K_{IIc} and K_{IIIc} may be defined but this is rarely done in engineering). As described by Kaninen and Popelar (1985), Olson (2003), Savalli and Engelder (2005), and others (see also Chapter 9), K_c represents the threshold for *dynamic* fracture propagation, so fracture toughness should be equated with the critical stress intensity factors at a fracture tip that is sufficient to initiate dynamic fracture propagation.

For room temperature and low-stress conditions appropriate to civil and mining engineering applications (upper km or less; e.g., Rubin, 1993a), the dynamic fracture toughness of rock is a *material constant*, independent of crack size, shape, and the state of stress (or displacement) acting on the rock mass. K_c must be determined experimentally (e.g., Atkinson and Meredith, 1987b; Anderson, 1995, pp. 423–458). In general, however, K_c depends on temperature and the ambient (confining) stress state (Kanninen and Popelar, 1985, p. 15; Rubin, 1993a; Khazan and Fialko, 1995; Bunger, 2008) so is not a material constant, but rather a **material parameter**.

Fractures such as joints within the Earth's crust propagate at values of stress intensity factor that are less than the rock's dynamic fracture toughness (e.g., Anderson and Grew, 1977; Atkinson, 1984; Segall, 1984a,b; Swanson, 1984; Olson, 2003). Such **subcritical crack growth** has been recognized for many decades (see discussions by Atkinson, 1984; Atkinson and Meredith, 1987a; Savalli and Engelder, 2005; and Engelder, 2007), with threshold values being perhaps on the order of 0.1 times the dynamic fracture toughness (e.g., Segall, 1984a,b; Swanson, 1984). The term **critical stress intensity factor** is sometimes used instead of fracture toughness with the value in rock understood to be generally less than the dynamic fracture toughness (e.g., Engelder, 2007; Olson and Schultz, 2011). Growth of geologic cracks then occurs when the mode-I stress intensity factor at the crack tip reaches the critical value of stress intensity factor that is defined by the lithologic and environmental conditions in the host rock, such as fluid-rock chemistry, temperature, and strain rate (e.g., Atkinson and Meredith, 1987a; Holder et al., 2001). The velocity of crack growth also depends on this critical value, as does the ultimate fracture pattern (e.g., Olson, 1993, 2004; Renshaw and Pollard, 1994b; Renshaw,

Unit for dynamic fracture toughness:
1 Irwin = MPa m$^{1/2}$

Critical stress intensity factor

1996, 1997). When considering the propagation and displacement–length scaling of geologic cracks (see equation (8.33) below), the value of critical stress intensity factor or dynamic fracture toughness quantify the resistance to crack growth in the rock.

The basic principles behind fracture propagation are quite simple, and in many ways analogous to what we've already seen for frictional sliding using the Coulomb criterion (Chapter 3). There, we calculated elastic (normal and shear) stresses and compared them to a criterion that incorporated material properties (cohesion and friction coefficient). Here, we see that fracture propagation is governed by

To propagate, or not to propagate

$$K_i < K_c : \text{Fracture is stable}$$
$$K_i \geq K_c : \text{Fracture will propagate}$$

(8.32)

Thorough treatments of (dynamic) fracture propagation using equation (8.32) can be found in most textbooks on engineering fracture mechanics (e.g., Kanninen and Popelar, 1985; Lawn, 1993; Anderson, 1995; see also Ingraffea, 1987; Sih, 1991; Engelder et al., 1993; Gudmundsson, 2011). Equation (8.32) gives a useful baseline for assessing fracture stability by invoking only the simplest possible set of tools that might also apply reasonably well for many situations. For example, geotechnical projects in civil and rock engineering commonly make use of LEFM-based crack propagation rules in their successful engineering designs, demonstrating the power and utility of this approach.

Determining a value for the dynamic fracture toughness is important for geologic problems involving, for example, propagation of a hydrofracture or an igneous dike. Atkinson and Meredith (1987b) have compiled dynamic fracture toughness data (called "stress intensity resistance" by them) for many rocks and minerals. These values provide a starting point for understanding the resistance of a rock mass to fracture propagation. Representative values, for room temperature and pressure, are listed in Table 8.3. There is a range in values for a particular rock type, due to several factors including the type of test apparatus used and the natural anisotropy of the material (such as coal). But sedimentary rocks, such as coal or salt, tend to have lower values of dynamic fracture toughness ($K_c < 1.5$ Ir) than do igneous rocks, such as gabbro ($K_c > 1.5$ Ir). Larger values tend to be associated with higher modulus minerals along with a more complex crystalline structure, similar to the roles played by bond strength and lattice arrangements in the *cleavage* of single crystals and their polycrystalline aggregates (Schultz et al., 1994). Most rocks have dynamic fracture toughnesses K_c between 0.2 and 3 Ir.

The LEFM propagation criterion can also be written using fracture-surface displacements instead of driving stresses (equation (8.25)) for stability under LEFM conditions. Measured values of rock properties (modulus and Poisson's ratio) and fracture-surface displacements (instead of calculated values of stresses) can be used more directly in this relationship to assess the tendency of a fracture in rock to propagate.

Table 8.3 Dynamic fracture toughness values for rocks[a]

Material	Value or range, Ir
Arkose	0.62
Berea Sandstone	0.28
Hohensyburg Sandstone	1.17–1.33
Mojave Quartzite	2.10
Chalk	0.17
Balmholtz Limestone	1.77
Klinthagen Limestone	1.03–1.41
Falerans Micrite	1.01
Carrara Marble	0.64–1.26
Tennessee Marble	0.62–0.67
Black Granite	2.80
Chelmsford Granite	0.59–0.64
Sierra White Granite	0.79
Stirpa Granite	1.83–2.36
Westerly Granite	0.90–2.50
Icelandic Tholeiite	0.87–0.99
Jinan Gabbro	3.75
Preshal More Basalt	2.50–2.58
Shale (various)	0.37–1.34
Pittsburg Coal	0.047–0.063
Corris Clate	1.08–2.25
Welsh Slate	1.14–2.32
Salt	0.23–0.57
Nevada Tuff	0.35–0.50
Woodford Shale[b]	0.76 ± 0.07

[a] Data from Atkinson and Meredith (1987b).
[b] Chen et al. (2017), tested at ambient air humidity.
1 Irwin (Ir) = MPa m$^{1/2}$.

Equation (8.25) also provides a revealing test of LEFM's applicability to geologic discontinuity. First we assume that the criterion for propagation under LEFM conditions is met (i.e., $K = K_c$). Then, solving for the maximum displacement across the discontinuity of length $L = 2a$ and using Ω in (8.14), we have (see also Olson, 2003; Savalli and Engelder, 2005; and Schultz et al., 2008a)

$$D_{max} = \frac{K_{Ic}(1-v^2)}{E}\sqrt{\frac{8}{\pi}}\frac{\sqrt{L}}{\Omega} \qquad (8.33)$$

This equation predicts that D_{max} scales with $\sqrt{L}$ for LEFM conditions (see Scholz, 2002, p. 116). Interestingly, if one examines actual data on displacement and length for faults and shear deformation bands (e.g., Clark and Cox, 1996; Schultz et al., 2008a; Schultz et al., 2013; Chapter 9), one finds that D_{max} is linearly proportional to L, implying non-LEFM conditions for these types of geologic discontinuities. On the other hand, Olson (2003) showed that opening-mode fractures such as veins and igneous dikes scale according to (8.33), or as $\sqrt{L}$; additional datasets for joints, veins, and dikes acquired since his study support his conclusion (Schultz et al., 2008a,b; Klimczak et al., 2010; Schultz et al., 2013). Further, certain types of deformation bands—those that involve grain crushing—also scale as $\sqrt{L}$, as do shear-enhanced compaction bands (Rudnicki, 2007; Tembe et al., 2008; Schultz, 2009), whereas deformation bands in which shearing greatly predominates over volumetric changes scale linearly with length, just like faults (Schultz et al., 2008a, 2013). This simple yet powerful test provides a powerful motivation to explore the applicability to LEFM for geologic structural discontinuities—a topic that is considered in Chapter 9.

Tensile strength of a rock

Another useful application of the dynamic fracture toughness is its role in determining the value of **tensile strength** (see Chapter 3). As laid out for example by Mandl (2005), pp. 17–18, the tensile failure strength of a rock can be understood by equating the initial flaw size with the average size of grains (or equivalently, the grain-boundary cracks) and solving for the remote stress that would lead to dilatant crack growth and, therefore, macroscopic tensile failure of the rock. Using (8.19a) with (8.30) and equating the remote (mode-I driving) stress to the tensile strength T_0 gives

$$K_{Ic} = K_I = \sigma_{dI}\sqrt{\pi a}$$
$$K_{Ic} = T_0\sqrt{\pi a}$$
$$\therefore T_0 = \frac{K_{Ic}}{\sqrt{\pi a}}$$

$$(8.34)$$

One can see that the tensile strength is predicted to increase as the average grain size (given by a) decreases. This is shown in Fig. 8.27 for several plausible values of dynamic fracture toughness (see Table 8.3) with the values of tensile strength generally ranging from 10 to 100 MPa, in accord with published values (e.g., Lockner, 1995; Paterson and Wong, 2005, p. 33).

The expression for tensile strength in (8.34) has the form which can be recognized as the size–strength relation discussed in Chapter 2 (in particular, see equation (2.33) and Fig. 2.20; Brace, 1961; Scholz, 2002, p. 36). Although the size–strength relation (2.33) relates strength (commonly the unconfined compressive strength) to the size of the rock *specimen*, and not to the rock's inherent grain size as done here for tensile strength, the principle holds that larger specimens can accommodate larger cracks and other discontinuities. By implication, the strength of successively larger rock specimens, from the laboratory to the field scale, and ultimately to rock masses (Chapter 3),

Fig. 8.27. Tensile strength of rocks predicted from grain size and dynamic fracture toughness. Some common grain sizes are shaded.

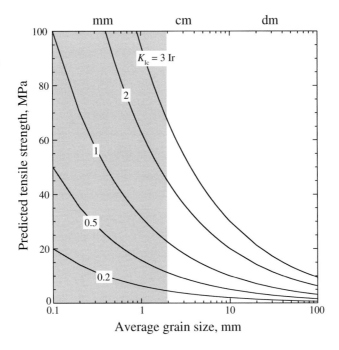

Rock as a low-toughness, low-strength material

is controlled under brittle conditions by the stability against propagation of the rock's flaw population. As discussed by Paterson and Wong (2005), pp. 30–32, the dependence of strength on specimen size is augmented when stress gradients and/or stored elastic strain energy are produced in the sample, as can be the case for certain types of laboratory tests or material types. In sufficiently fractured rock masses the strength is controlled by the frictional resistance of the discontinuities (i.e., their shear strength) rather than by their ability to propagate as mode-I cracks. This leads to a scale-independent shear strength (Scholz, 2002, p. 37) that depends on the characteristics of the discontinuity population including orientation and friction (e.g., Hoek, 1983; Brady and Brown, 1993, p. 132).

Another way to describe rock is to compare dynamic fracture toughness to yield strength (e.g., Ritchie, 2011). Engineered materials that are both strong and tough possess large values of both of these parameters, leading to damage-tolerant behavior. Property pairs such as these, as well as Young's modulus vs. unconfined compressive strength (Fig. 2.13), can be plotted to compare the performance of different materials (e.g., Ashby, 2016). Given typical values for (dynamic) fracture toughnesses ranging between perhaps 0.5 and 3 MPa m$^{1/2}$ (e.g., Meredith and Atkinson, 1983; Li, 1987; Table 8.3) and yield strengths of perhaps 1–20 MPa (tension, absolute value) and up to perhaps 200 MPa (in compression, with shear (frictional) yield strengths also being in this range; Lockner, 1995; Gudmundsson, 2011), rock would be considered a low-toughness, low-strength material (Fig. 8.28).

Fig. 8.28. Ashby material property plot showing ranges of yield strengths and dynamic fracture toughnesses for engineering materials, after Ritchie (2011), used by permission. Box shows approximate ranges of values for rocks.

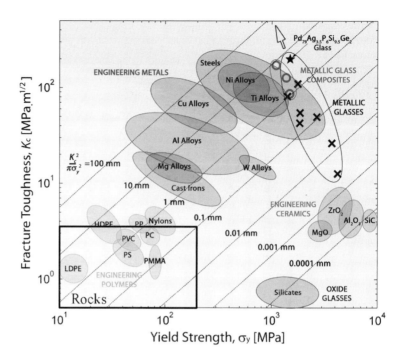

8.4.3 Fracture Stability Estimation

In Chapter 3 we saw how the stability of a particular stress state was defined by how close it was to the strength of the rock. This was called either the **proximity to failure** P_f (stress/strength) or the **factor of safety** F_s (strength/stress). Let's use the same two approaches here to estimate the stability of fractures in rock.

Using the concept of dynamic fracture toughness or critical stress intensity factor K_c, we can define

Proximity to failure

$$P_f = \frac{K}{K_c} \qquad (8.35)$$

and

Factor of safety

$$F_s = \frac{K_c}{K} \qquad (8.36)$$

(e.g., Whittaker et al., 1992, p. 499). In these expressions, K is calculated by using equation (8.21) or (8.25) since this approach implicitly assesses whether the discontinuities would propagate or not.

Discontinuities are stable against propagation (i.e., they will not grow in length) when $F_s > 1.0$ or $P_f < 1.0$. For example, by choosing values of $K_I = 1.35$ Ir and $K_{Ic} = 2.2$ Ir, we have a proximity to failure $P_f = 0.62$, and a factor of safety $F_s = 1.6$, both indicating that *the joint will dilate but not propagate* (since $K_I < K_{Ic}$). In the former case, the mode-I stress intensity factor is only 62% of what it would need to be for propagation of the joint to occur. This approach was used by Tharp and Coffin (1985) to investigate the stability of rock slopes that contain joints and other geologic discontinuities, whereas Singh and Sun (1990) applied it to surface and underground mining.

8.4.4 Closing Thoughts on Fracture Tips in LEFM

In Chapter 2 we saw that the Inglis/Kolosov Relation for the stresses at the tip of an elliptically shaped flaw predicted infinitely large stresses for $b = 0$. This ideal, mathematically convenient shape, called a degenerate ellipse, or *slit*, is used here as the basis for calculating crack-tip stresses. In fact, this sharp-slit approach is the basis for Linear Elastic Fracture Mechanics (e.g., Lawn, 1993, pp. 24–26) and underlies the concepts of the stress intensity factor and plastic zone. But it may also have become apparent that the shape of the fracture tip is a parabola (displacement varies with $\sqrt{r}$) as calculated by using the equations for near-tip displacement (e.g., Paris and Sih, 1965; Pollard and Segall, 1987, pp. 322–325; Anderson, 1995, p. 113), or an ellipse using equations (8.16) for the whole fracture (Anderson, 1995, p. 113), to represent the opening or sliding displacements along the fracture walls. Although neither a parabola nor an ellipse provide very good representations of a sharp crack, this geometry is analytically related to the stresses associated with a degenerate ellipse (Inglis, 1913; Lawn, 1993, p. 26).

The assumptions of constant driving stress across a discontinuity (that can arise for uniform remote stress acting on a discontinuity having constant stress boundary conditions along it, such as a crack under uniform remote tension), causes an unrealistically large stress and strain gradient, sometimes known as an instability or *singularity*, to be produced at the discontinuity tip. Physically, the material at the discontinuity tip is trying to rip itself apart but remains elastic and therefore intact because it lacks an explicit strength or failure criterion. Mathematically, the resulting infinite (singular) stresses are associated with the discontinuous slope of the sharp slit's elliptical profile at the tip, where the derivative of the stress function is not defined (e.g., see discussions by Nehari, 1952, p. 5; Jaeger and Cook, 1979, pp. 264–273; and Pollard and Segall, 1987, p. 293). The singularity in stress implies that the rheologic model—linear elasticity in this case—was no longer strictly appropriate in our analysis.

The solution is to add a more realistic rheologic model that can accommodate the large near-tip stresses predicted to be generated near a fracture tip. For LEFM a small *plastic* zone can be used, as described in this chapter. If small-scale yielding is not met, however, we need a more comprehensive way to describe how the rock can adjust to the large near-tip stresses generated by displacements of the fracture walls. The choice of approach depends on the rheologic and strength properties of the material—specifically, on how weak it is relative to the level of stress that is applied. Because many engineering materials such as steel are stressed at levels well below their failure or yield points, LEFM can readily apply. But when geologic materials are stressed at levels comparable to their yield strengths (e.g., Rubin, 1993a), then these *low-toughness, low-strength materials*—similar to concrete (Anderson, 1995, pp. 313–361)—may be better characterized as something other than purely elastic materials, using more general approaches such as **Elastic-Plastic Fracture Mechanics** (e.g., Heald et al., 1972; Kanninen and Popelar, 1985), as defined for engineering materials, but increasingly augmented for the special situations encountered with rocks (e.g., Haddad and Sepehrnoori, 2015).

8.5 Review of Important Concepts

1. Fracture mechanics provides the physical basis for understanding brittle deformation in rocks, outcrops, and the Earth's crust. The tools of fracture mechanics come to us from engineering applications such as the design of bridges, ships, and aircraft. Research and applications over several decades have demonstrated, however, that these approaches and equations provide a relevant and powerful tool for investigating and understanding how geologic structural discontinuities work.

2. Geologic structural discontinuities respond to the stresses and displacements that are applied remotely, far from them, by a two-fold process. First, *they themselves deform*, leading, for example, to crack opening and fault offsets. Second, they cause the stress and displacement state in their local vicinity to become *inhomogeneous*, so that the magnitudes and directions are locally changed—sometimes radically—from the remote or regional stress or displacement state.

3. Boundary conditions are sets of basic physical constraints that are specified mathematically to clarify what factors are acting on a discontinuity. There are two kinds: *inner* and *outer*. Inner boundary conditions apply to the discontinuity itself; they include, for example, internal fluid pressure, grain crushing or rolling resistance, and frictional resistance. Outer boundary conditions describe what is happening far from the discontinuity, including remote or regional stresses and displacements. A pre-faulting state of stress in the crust (e.g., lithostatic) is an example of an outer boundary condition.

4. *Driving stress* is the net stress state acting on a structural discontinuity. It is related to both the outer and inner boundary conditions and must be of positive sign for the discontinuity to displace. Driving stress is the difference between the remote stress resolved on the discontinuity plane and the internal pore-fluid pressure (for mode-I) or change in frictional resistance of discontinuity walls in contact (for modes-II and -III). Most cracks in the Earth's crust are squeezed shut by the compressive state of stress underground (the driving stress here is negative); internal pore-fluid pressure is a very common mechanism for producing a positive driving stress that leads to crack and joint opening and propagation in rocks. Similarly, the resolved stress acting on a potential fault plane

must exceed that plane's frictional resistance (producing again a positive driving stress) in order for the plane to slip and become a fault. The value of driving stress in this case is often referred to as the stress drop in seismotectonic applications. Cumulative driving stress can be related to the cumulative geologic offset along a mappable fault.

5. Structural discontinuities of all types *concentrate and redistribute stresses in their vicinity.* The spatial changes in stress and displacement around discontinuities may be quite subtle away from the discontinuity, with the changes being most pronounced and important near the discontinuity tips. **Propagation and linkage** are direct results of the discontinuity tip's influence on its surroundings.

6. **LEFM provides the basic equations that describe the amount of displacement that a structural discontinuity might accommodate.** This explains and quantifies why the displacement along a discontinuity that is bounded (contained within the rock mass) is maximum near its center and minimum (i.e., zero) at its ends, assuming constant inner and outer boundary conditions. LEFM thereby relates the displacements accommodated along a particular discontinuity to the much smaller displacements of the surrounding rock. Estimates of quantities such as magma pressure or remote tectonic stress state can also be obtained by modeling discontinuity displacements using LEFM.

7. The region close to a discontinuity tip has traditionally been the focus of engineering fracture mechanics and LEFM. Because the stress magnitudes in this region are so much larger than those in the surroundings, the stresses may produce local deformation around the tip, leading to large numbers of closely spaced microcracks (or other types of structures) that are localized within a *process zone* ahead of a fracture or deformation band. LEFM is valid when several criteria are met, including negligibly small plastic zone and small-scale yielding, both associated with sufficiently large ratios of yield strength to driving stress.

8. The *stress intensity factor, K*, is a parameter that describes the amplified stresses, near the tip, due to opening, closing, or slip along a discontinuity. It recovers the information on discontinuity length and level of driving stress on which the near-tip stress state depends from the mathematical artifact of infinite stress calculated from LEFM. There is a different value of K for each displacement mode, and the total tendency for a discontinuity to propagate is the sum of all Ks applicable to the loading configuration of interest. Propagation of an LEFM discontinuity occurs when its stress intensity factor reaches a threshold value, called the *dynamic fracture toughness, K_c*, or critical stress intensity, of the rock. The stability of fractures provides an important design tool for rock engineering, underlies the size–strength relationship for rocks, and explains the dependence of tensile strength on the grain size of a rock.

8.6 Exercises

1. Calculate the mode-I and mode-II stress intensity factors for discontinuities having lengths 0.86 m, 2.0 m, and 3.7 m subject only to a normal stress (equivalent here to the remote normal stress σ_{yy} as resolved onto the discontinuity plane) of 1.2 MPa tension. Assume a tall tunnel crack (2-D) so that $Y = 1.0$.

2. For the discontinuities and resolved remote stresses given in Exercise 1 and assuming a tensile strength of 15 MPa for the intact rock material, calculate the sizes of the plastic zone and K-dominant region associated with each discontinuity.

3. Now the 2-m long crack from problem 1 is subjected to a normal stress of 1.2 MPa tension and crack-parallel compression of –5.5 MPa. Calculate the mode-I and mode-II stress intensity factors for this case and discuss the role of crack-parallel compression in the near-tip stress field. $E = 10$ GPa and $v = 0.25$ in case you need them to calculate more precisely the normal stress generated onto the discontinuity by the biaxial remote stress state.

4. A discontinuity 4.5 cm long is loaded by 1.3 MPa (tensile) normal stress and 4.0 MPa shear stress (in-plane).
    (a) Calculate K_I and K_{II}. For this mixed-mode (I-II) crack, the shear driving stress will equal the resolved shear stress.
    (b) Knowing that the crack will tend to propagate into a principal plane, calculate the "take-off angle" for the next increment of crack growth if this crack were to propagate (a sketch may help here).

5. Calculate K_I and K_{II} using $\mu_s = 0.6$ and $\mu_r = 0.5$ for the discontinuity in exercise 4 using, instead, a *compressive* normal stress of –1.3 MPa. In this case, be careful of the sign of K_I, making the correct physical interpretation of your calculated value. For K_{II}, be sure to choose the correct sign convention for the shear driving stress, given that this is now a frictional sliding problem.

6. Suppose that a rock has a dynamic fracture toughness of 1.25 Ir. What is the largest stress that can be applied to a discontinuity having length 2.0 m so that mode-I propagation of the discontinuity does not occur? What is the largest crack that can be tolerated in a rock mass for an applied tensile stress of 0.85 MPa? You should show all your work and discuss the meaning of your results. In both cases, assume a tall tunnel crack (2-D) so that $Y = 1.0$.

7. Calculate the shear stress close to the tip of a mode-I crack assuming a crack length of 0.6 m, unit remote stress (biaxial tension), $\theta = 0°$, and a distance from the crack tip of 10^{-5} m. Discuss the meaning of your results.

8. What is the maximum dilation of a 2-D tunnel crack with length 10 m in a linearly elastic rock mass having a Young's (or deformation) modulus of 10^9 Pa, a Poisson's ratio of 0.25, and a remote normal stress of 0.5 MPa? How would this value change if, instead, the crack is propagating under LEFM conditions, assuming a dynamic fracture toughness of 1.2 Ir?

9. Calculate and sketch the propagation direction (take-off angle) for a discontinuity 4.5 cm long, loaded by 1.3 MPa (tensile) normal stress and 4.0 MPa shear stress, in a linearly elastic rock mass having a Young's (or deformation) modulus of 10^9 Pa, a Poisson's ratio of 0.25, using both stress intensity factors and maximum discontinuity displacements; be sure to show all your work.

10. Calculate the critical crack opening displacement, associated with a crack for which the crack-tip propagation criterion is just satisfied, for the crack given in exercise 8 and assuming the same value for dynamic fracture toughness. Compare this critical value to that calculated above and discuss your results.

11. Using the near-tip solutions, calculate the stress state close to a crack tip. The crack has length $2a$ and the plane-strain plate is infinite in the xy plane. Use the equations for the two normal stresses and one shear stress using polar coordinates with an origin at the positive crack tip. Plot the three stress components near the crack tip as a function of r and θ. Assume unit remote tensile stress and unit half-length. Use intervals of $0°$, $30°$, $45°$, $60°$, and $90°$ and $r = 0.1a$, $0.5a$, and $1a$. Discuss the variations in stress with radius and position about the crack tip for each of your three plots.

12. Suppose you have identified two faulted joints in your field area. They are in two different rock types and the ages of deformation are different, judging from the cross-cutting relations.

    (a) In the sandstone, you can identify and measure the separations of a passive marker across your joint, and you determine that there was 1.2 mm of opening and ~2 mm of strike-slip offset across the structure. You verify that nearby joints in the sandstone that didn't shear only show openings that are so small that you can't measure them (i.e., < 0.1 mm). You also measure the take-off angles of end cracks along the faulted joints to be 55–63°. What can you conclude quantitatively about the deformation of the faulted joint?

    (b) In the granite pluton exposed within the western part of your field area, you also find some faults that grew as faulted joints. Here, the take-off angles are 30–40° and you can only find evidence for shearing offsets of 4–5 mm. What might you infer about the deformation along these faults?

13. Using the whole-crack equations, calculate and plot all three Cartesian stress components surrounding a pure mode-I and mode-II fracture. First establish your spatial coordinates and then choose values for driving stress (for example, 1.0 MPa, and then reference the calculated values to this), fracture length (let $L = 2a = 2.0$ m), and rock properties (assume $E = 10$ GPa and $v = 0.25$). Next, set your values for r and θ so that you cover the area surrounding the fracture (let r range from 0 to $2a$ and θ from 0° to 360°). Now plot and contour your matrix of values using Cartesian (xy) coordinates and interpret the results. How far from the discontinuity does its influence extend (defined as stress changes decreasing to less than 10% above the 0.0 MPa background value)?

14. A series of active, magma-filled dikes are located near your winter cottage in Hawaii. The dikes occur in a homogeneous, isotropic basaltic rock mass far from other geologic structures and are widely separated from each other. Dike lengths are 0.5 m, 2.7 m, and 12.0 m. Suppose $K_{Ic} = 1.0$ Ir.

    (a) Calculate the internal magma pressure P_i necessary to propagate each dike. Assume that temperature plays no role in the problem, that $K_I = \sigma \sqrt{\pi a} = -P_i \sqrt{\pi a}$, and tension is positive (compression, or pressure, negative). Hint: for propagation, $K_I = K_{Ic}$. Briefly discuss your results.

    (b) The remote stress field has changed since part a due to upward flexure of the ground surface. Now a uniaxial remote tensile stress $\sigma°$ also acts on the pressurized dikes. Solve part a again using $\sigma° = 0.05$ MPa. Note that

$K_I = (\sigma_{total}) \sqrt{\pi a}$, $(\sigma_{total}) = \sigma^\circ - P_i$. Briefly discuss your results. Will your cottage survive?

15. Using the three cracks and their mode-I stress intensity factors from exercise 1, along with a dynamic fracture toughness of 1.5 Ir, calculate the proximity to failure and the factor of safety for these fractures. Which ones would propagate using these criteria?

16. Define "driving stress" for an episode of fault slip (include both normal and shear displacement modes). How does it compare to "stress drop" for seismic (earthquake) slip? How does it relate to the driving stress for mode-I cracks? How does it relate to the cumulative driving stress inferred for large-offset faults?

17. Given values of $\sigma_3 = -2.5$ MPa (tension positive), $\sigma_3/\sigma_1 = 3.1$, $\phi = 30°$ (between σ_3 and the fault plane), $C_0 = 0$ MPa (along the fault), $L = 14.5$ m, $Y = 1.0$, $\mu_s = 0.6$, $\mu_r = 0.59$ (both for the fault), $E = 10$ GPa, and $v = 0.25$, calculate:
    (a) The shear driving stress for fault slip.
    (b) The maximum offset (shear displacement) along the fault.
    (c) K_{II} at the fault tips.
    (d) The strain energy release rate (fault propagation energy).
    (e) The size of the plastic zones at the fault tips. Is LEFM valid?

18. Suppose you map a tall (tunnel shaped) fault that has $L = 25$ m, $D_{max} = 18$ cm, $E = 5$ GPa, and $v = 0.3$. Calculate the cumulative driving stress that caused this fault to accumulate displacement to the given value of D_{max}. Calculate the displacement–length ratio and discuss how these values compare to those plotted in Fig. 1.9. How would your result change if the fault had an aspect ratio of $L/H = 2$?

19. Define the specific conditions that are necessary for LEFM to apply to a fracture or deformation band, using both words and equations. Plug some representative numbers into your equations and demonstrate the range or extremes of values that would make LEFM apply. Are these values realistic for fractures in rock? If so, under what conditions or situations could you then apply LEFM to fractures in rock? If not, how might you modify the equations to make them better represent geologic discontinuities?

20. Calculate the tensile strength of a porous sandstone having a dynamic fracture toughness of 0.28 Ir and an average grain size of 1.1 mm. How does your result compare with values in the literature?

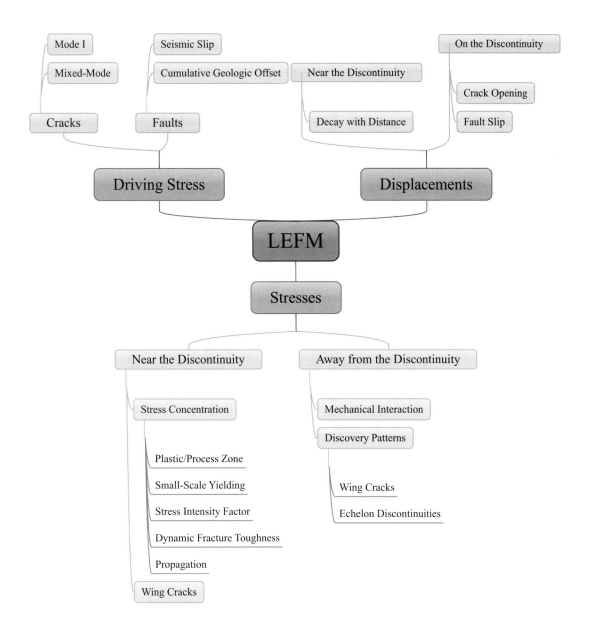

Beyond Linear Elastic Fracture Mechanics

9.1 Introduction

Geologic fracture mechanics (GFM) can be thought of as an interdisciplinary field combining approaches from engineering, materials science, and geology. It includes Linear Elastic Fracture Mechanics (LEFM) but relaxes some of the assumptions that are required for LEFM to apply to geologic structural discontinuities (i.e., fractures and deformation bands). LEFM is widely regarded as the most simple and restrictive special case of fracture mechanics (see discussions by Latzko, 1979; Kanninen and Popelar, 1985, p. 13; and Anderson, 1995, p. 117). Upon close examination, it may be seen that many of the predictions of LEFM do not match geologic observations as well as might be desired, suggesting the need for a more general approach that includes material from chemistry (to better consider diagenesis (Fig. 9.1) and subcritical fracture propagation) and plasticity (to better represent near-tip processes). Elements of some of these approaches are described in this chapter.

It is useful at this point to collect and list several common predictions of geologic discontinuities based on LEFM. Some of the more telling ones include:

- The displacement distribution along a geologic structural discontinuity (assuming uniform loading, for example) is predicted by LEFM to be elliptical, whereas it can be observed to be more nearly linear or ogive (i.e., decreasing from a central maximum to a gentle gradient at the tips).
- The near-tip opening displacement along many geologic discontinuities is not parabolic or elliptical, as implied by LEFM.
- The relationship, or scaling, between the maximum displacement magnitude and the length of many types of geologic discontinuities, such as faults, tends to follow a steeper slope (around unity) than that predicted by LEFM (i.e., a slope of 0.5 for geologic discontinuities whose propagation is modulated by stress intensity factor, such as joints).

Fig. 9.1. Partially cemented fracture in sandstone core (Travis Peak Formation, East Texas Basin, after Lander and Laubach, 2015) showing quartz cement bridges extending across, and filling, the tip region but incompletely sealing the fracture where its aperture is wider. Image courtesy of Steve Laubach; scale bar in inches.

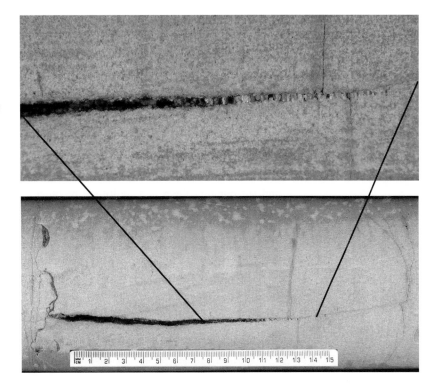

- The spacings between geologic discontinuities in many different rock types are much less than what might be predicted by LEFM and layer-thickness models, implying additional mechanisms for discontinuity spacing.
- The rates of opening-mode fracture propagation are typically many orders of magnitude slower than those implied by dynamic wave speeds and models of cracks loaded in tension, implying some combination of far-field loading (e.g., displacement-controlled) or near-tip processes that promote stable, quasi-static propagation.
- Observations of joints and their bridging cements imply rates of opening and propagation that can be exceedingly slow and likely are modulated by the rates of chemical processes in the deformed rocks.

In this chapter we'll look at several topics that apply fracture mechanics to the range of geologic phenomena that are observed in association with geologic structural discontinuities. First we'll examine the roles that crustal fluids can play in regulating the slow growth of geologic structural discontinuities over geologic time. This process is called *subcritical fracture growth* and we'll demonstrate how the stress intensity factor can be tied into this class of concepts. Then we'll examine a more general way to represent discontinuity terminations—that of the "post-critical" *end zone* concept. Next, the amount of opening, closure, or slip along a structural discontinuity represents a balance between displacement accumulation and the propagation, or increase in physical size, of the structural discontinuity's surface, leading to an examination of *displacement–length scaling* and its implications for the many types of structural discontinuities. Last, we'll explore some basic concepts of *discontinuity population statistics* and how they can quantify strains in the crust that are associated with large numbers of geologic discontinuities.

A brief outline of the historical development of engineering fracture mechanics and its diffusion into geology is listed in Table 9.1. This table provides the merest glimpse into the names and topics of some of the giants who have worked to advance the field of fracture mechanics. Accompanying and surrounding them is a multitude of other fine scientists and engineers in academia and industry—past and present—from around the world who have also worked to better understand why geologic structural discontinuities operate the way they do. This body of work clearly indicates the importance of crustal fluids, chemistry, strain rates, rock properties, and other factors that can interact with the mechanics to modulate fracture-related processes.

Table 9.1 Selected milestones in geologic fracture mechanics

Period	Worker/group	General area	Topical advance
1905–1913	Inglis/Kolosov	Stress concentration	Engineering mechanics
1920	Griffith	Crack thermodynamics, flaw size vs. strength	Griffith flaws
1930s	Muskhelishvili	Theory of elasticity, stress functions	Green's functions for cracks
	Westergaard	Crack-tip stresses and displacements	Crack-tip stresses
1948–1950s	Irwin	Stress intensity factor K, LEFM	**LEFM for brittle materials**
	Orowan	Near-tip plastic zone correction	
	Mott	Dynamic fracture	
	Yoffe	Dynamic K, crack-tip branching	
1960s	Dugdale	Cohesive-zone model: mode I	LEFM for yielding materials
	Barenblatt	Plastic end-zone model: mode I	
	Paris	Fatigue, subcritical crack growth	
	Paris and Sih	Compendium of Ks	
	Rice	J-integral	

Table 9.1 (Cont.)

Period	Worker/group	General area	Topical advance
	Hodgson	Natural geologic mode-I cracks	LEFM for geologic cracks
	Lachenbruch	Crack stresses and shielding	
	Secor	Fluid pressures and jointing	
1970s	Johnson	Linkage of field observation and theory for fracture processes	**Development of GFM**
	Pollard	Mechanics of igneous intrusions	
1980s	Pollard	Geologic fracture theory	Rapid growth in development of GFM and acceptance as a tool in modern structural geology
	Segall	Joint and fault studies	
	Rice	Friction, stick-slip, and faulting	
	Atkinson	Subcritical crack growth	
	Engelder	Mechanically based joint studies	
	Aydin	Faults and deformation bands	
1990s	Cowie	Cohesive-zone model for faults; displacement-length scaling	Inelastic GFM; numerical modeling of complex fracture systems; fusion between lab, theory, seismology, and faulting
	Rubin	Cohesive-zone model for dikes	
	Scholz	Seismotectonics, fault growth	
2000s	Marone	Laboratory studies of friction	Stability of frictional sliding
	Zoback	Reservoir geomechanics	Fault mechanics in oil and gas fields; critically stressed faults and permeability
	Laubach	Fractures and diagenesis	Coupling of fluids, chemistry, and deformation

9.2 The Realm of the Geologic Fracture

Succinctly stated, the material in Chapter 8 described how stresses or displacements applied remotely (far) from a fracture, for example, could cause the fracture's walls to displace, leading to a redistribution of stress and displacement in the vicinity of the fracture. In general this applies to other types of geologic structural discontinuities as well. We then examined how the rock adjoining the discontinuity at its tipline responds to the increased magnitude of stress there. That is, the rock breaks or yields in the infinitesimally small areas where the stresses locally exceed the strength of the rock. In LEFM, large finite deformations near the fracture tipline were not admitted by the assumption of an ideal elastic rheology for the rock. This in turn prevents any significant strain from being generated in a volume near the fracture tip while motivating two mathematical implications: infinitely large stress magnitudes at the tip and unphysically steep (or parabolic) near-tip displacement gradients. The propagation of an LEFM fracture is then predicted by taking the mathematical limit of the infinite stress (to recover a dependence on fracture length and driving stress magnitude) and comparing that to a measurement of the rock strength in that near-tip environment, the dynamic fracture toughness (K_c).

As may be anticipated, relaxing some of the assumptions and restrictions noted here in the LEFM-based approach has provided a better correspondence between the theory and many geologic observations, such as those tallied above and as demonstrated in the literature. Accordingly, the growth of geologic structural discontinuities may be considered to be modulated by three primary groups of factors:

- The external boundary conditions, such as remote stress or displacements that can drive discontinuity growth;
- The characteristics of the geomaterials that are being deformed; and
- The environmental conditions that might include pressure, temperature, and chemical interactions between geologic fluids and the solid geomaterials.

Various combinations of these factors have been applied by engineers to other materials such as soils and concrete with great success.

The effect of including these groups of factors in the geological sciences is profound. Their inclusion permits the description and modeling of subcritical fracture growth, which also carries powerful implications for discontinuity spacing and clustering during such low-velocity propagation. Removal of the mathematical stress singularity from the structural discontinuity's tip enables more realistic displacement distributions and stress magnitudes in the deformed region. Consideration of temperature and fluid–rock chemistry are important for understanding the coupling and feedback processes between deformation (i.e., crack growth) and diagenesis (i.e., formation of cement that bridges the crack walls), which may also be manifested in discontinuity patterns.

Stated another way, structural discontinuities grow in a geologic environment (Fig. 9.2) having some subset of the following general characteristics:

- **Hot.** Temperatures may range from below the freezing point of water (in unusual cases near the surface) through several hundred degrees Celsius at depth. Elevated temperatures in the geologic subsurface are generally the norm, and there may also be large thermal gradients in the rock mass. Temperature exerts a fundamental control on both rheology and chemical kinetics in the subsurface.

- **Wet.** Water and dissolved species like acids, gases, and chemicals potentially available to precipitate out of solution saturate the pore spaces of the rock mass. Deformation of a rock mass may be considered to occur under "drained" or "undrained" conditions if the deformation rate is sufficiently slow or rapid, respectively, relative to the hydrologic characteristics of the rock mass (e.g., its permeability). Fluids of various saturations may react with host rock constituents at rates influenced by temperature, giving rise to solution, precipitation, and creep within the rock mass.

- **Confining.** The rock mass is under pressure on all sides, causing stresses to be built up in various places. The large tensile stresses that can cause accelerating, "running" cracks to zip along at large fractions of the limiting elastic-wave velocity (under "dead-weight loading") in engineered materials are rarely generated in the geologic environment. Instead, displacement-controlled, compressive loading may be more common. Increasing ratios between stress magnitude and rock strength foster a change from more brittle, localized deformation to more plastic, distributed deformation.

- **Sluggishly deforming.** The rock mass is slowly deforming through mechanical response to stress and through chemical reactions with the fluids within it.

Fig. 9.2. Schematic illustrating some of the factors affecting the development of structural discontinuities in geologic environments.

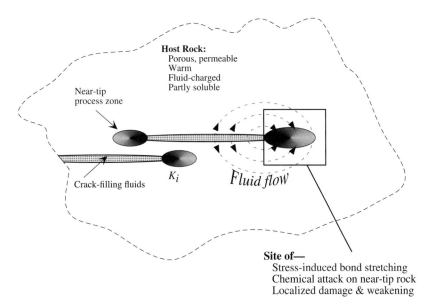

From this brief account it should be apparent that the **ambient environment exerts a significant control on the rate** of growth of geologic structural discontinuities (e.g., Unger, 1995, pp. 207–259). In particular, subcritical fracture growth, as we'll see, is an important process in the slow, time-dependent deformation of fractured crustal rocks (e.g., Anderson and Grew, 1977; Atkinson, 1984; Segall, 1984b; Swanson, 1984; Atkinson and Meredith, 1987a; Kemeny, 1991, 2003; Hatton et al., 1993; Lockner, 1995; Main, 2000; Brantut et al., 2013). Subcritical fracture growth has also been implicated in time-dependent degradation of compressive rock strength (Haimson and Kim, 1971; Attewell and Farmer, 1973; Costin, 1987; Lajtai, 1991; Masuda, 2001; Brantut et al., 2012), joint network spacing and development (Segall, 1984b; Hatton et al., 1993; Olson, 1993, 2004; Renshaw and Pollard, 1994b; Renshaw, 1997) and the prediction of volcanic eruptions (Kilburn and Voight, 1998) and earthquakes (Main and Meredith, 1991; Sammonds et al., 1992). In addition to subcritical fracture growth, other processes such as pressure solution (e.g., Rutter, 1983; Kohlstedt and Mackwell, 2010; Brantut et al., 2012) can also contribute to the time-dependent deformation of rocks in a planetary lithosphere.

9.2.1 LEFM and Critical Fracture Growth

The work of Griffith (1921, 1924), and especially Irwin (1957), laid the cornerstone of fracture mechanics by relating values of strain energy release rate, G, and stress intensity factor K, respectively, calculated within a fractured solid (see also Section 8.4). Under far-field tensile constant-stress loading conditions, LEFM would predict that a crack having $K = K_c$ (or equivalently, $G = G_c$) would propagate dynamically, eventually reaching a terminal velocity approaching the Rayleigh wave velocity V_r (e.g., see general discussions by Kaninnen and Popelar, 1985, pp. 54–56; Freund, 1990; Marder and Fineberg, 1996; Gudmundsson, 2011) of perhaps several kilometers per second in rock. The condition of $K = K_c$ constitutes an *instability*, and cracks propagating under these conditions are referred to as *unstable fractures*, or fractures that propagate unstably. The value of K_c is thus understood to represent the material's **dynamic** fracture toughness.

Such dynamically propagating, "running" cracks have been studied extensively in the engineering and geologic literatures. Once a running crack accelerates to a large fraction (~20–30%) of V_r, it is predicted to bifurcate or branch, leading eventually to fragmentation (e.g., Yoffe, 1951; Kanninen and Popelar, 1985, pp. 205–207; Grady and Kipp, 1987; Sharon et al., 1995; Sagy et al., 2001). This is called a "running" crack and its analysis requires the inclusion of inertial terms to handle acceleration and stress wave effects in the near-tip stresses (see Kanninen and Popelar, 1985, pp. 192–207, for a clear discussion of dynamic crack problems). However, it is fair to say that dynamically propagating, running cracks in the Earth's crust would be considered to be controversial or rare at best (e.g., Segall, 1984a,b; Engelder, 2007). Instead, geologic discontinuities are inferred, or sometimes observed, to propagate *slowly*, or in what is known as a "quasi-static" or "stable" manner (in which inertial terms do not enter into the propagation equations). Several explanations that have been offered for stable fracture propagation in rocks include:

- Chemically assisted stress corrosion of atomic bond strength near fracture tips, leading to propagation at values of stress intensity factor less than the (dynamic) fracture toughness;
- Fixed-grip, displacement-controlled far-field loading conditions (e.g., progressive bending into a fold), rather than "dead-weight," constant remote stress conditions; and
- Fatigue.

These factors will be considered in the following sections to explore their implications for the propagation of geologic discontinuities. Additionally, fracture propagation can be modulated by hydraulic diffusivity and fluid pressures, as discussed by many including Olson (2003) and Lecampion et al. (2018).

9.2.2 Subcritical Fracture Growth

Many disasters have occurred through the twentieth century because small cracks in engineered structures were less stable than predicted by LEFM, leading to the destruction of ships, aircraft, missiles, trains, and high-pressure steam boilers, often with substantial loss of life (e.g., Latzko, 1979; Kanninen and Popelar, 1985, p. 47; Broek, 1986, pp. 3–5). Under LEFM conditions, the value of stress intensity factor K achieved at a fracture tip when the propagation criterion is met ($K \geq K_c$) is called the *critical* stress intensity factor. However, the cracks noted in these studies were apparently able to propagate when subjected to stresses that were much smaller than the dynamic fracture toughness of the material, K_c.

In a sense, this problem paralleled Westergaard's (1939) investigation of the failure of ships under loads that should, under a strength-of-materials design approach, have been safe. Why the cracks in the former set of structures might have propagated at stress intensity factors less than the critical value K_c—i.e., *subcritically*—was the focus of intense research in the United States and elsewhere following World War II.

A convenient way to discuss subcritical fracture growth is to refer to Fig. 9.3, which displays several important and well-known relationships. The diagram plots the velocity of fracture propagation (or the log of this value) against the stress intensity factor at a fracture tip. This shows immediately that, in the subcritical realm, where $K < K_c$ (the dynamic fracture toughness), the velocity of fracture propagation is *variable*, and depends on the value of stress intensity factor at a fracture tip. Such propagation velocities can be exceedingly slow, on the order of 10^{-1} to 10^{-10} m s^{-1} (e.g., Atkinson, 1984;

Swanson, 1984; Atkinson and Meredith, 1987a; Brantut et al., 2013; Bergsaker et al., 2016). For example, it would take 10^5 seconds, or about 28 hours at a rate of 10^{-5} m s^{-1}, for a crack to lengthen by 1 m. With geologic strain rates being even slower—less than 10^{-17} s^{-1} or so for continental interiors (e.g., Anderson, 2017; Zoback et al., 2002) and somewhat faster rates for plate boundaries (e.g., Kreemer et al., 2014)—it could take as long as 31 Ma (in the craton) for the same crack to lengthen by 1 m. The latter rate is in broad accord with propagation rates inferred from fluid inclusions in crack-sealing diagenetic cements for some non-tectonic joints (in gas-producing hydrocarbon basins, such as the Piceance and East Texas basins in the US mid-continent region) (Becker et al., 2010; Fall et al., 2012). This is the realm of **stable, quasi-static** fracture growth.

Most commonly, subcritical fracture growth in the Earth's crust has been associated with **stress corrosion**, which describes how atomic bonds that are strained by stress amplification near fracture tips can degrade in strength over time by the action of chemically reactive pore fluids (e.g., Anderson and Grew, 1977; Atkinson, 1984; Atkinson and Meredith, 1987a; Costin, 1987; Meredith et al., 1990). Other, less common mechanisms for stress corrosion include atomic diffusion, dissolution, ion exchange, and microplasticity (Atkinson and Meredith, 1987b). Deformation mechanism maps show that stress corrosion tends to occur at larger values of differential stress, for any given temperature,

Stress corrosion depends on differential stress, temperature, confining pressure, and relative chemistries of pore fluids and host rock

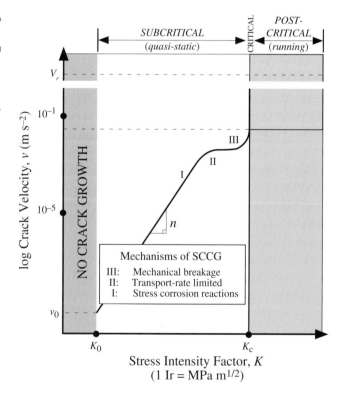

Fig. 9.3. Schematic relationship between stress intensity factor K (in log scale) and propagation velocity of the crack or fracture tip (after Anderson and Grew, 1977; Atkinson and Meredith, 1987a; and Savalli and Engelder, 2005). Mechanisms controlling crack growth in the subcritical regime ($K_0 \leq K \leq K_c$) are listed (I, II, III; note that these are not the fracture-displacement modes; SCCG, subcritical crack growth). For $K \geq K_c$, LEFM implies dynamic propagation (for dead-weight, stress-controlled remote loading).

than pressure solution (e.g., Renard et al., 2000), suggesting that stress corrosion is favored (for the porous sandstones investigated) in higher (differential) stress, lower temperature environments than pressure solution (Brantut et al., 2012).

Clearly, large differential stresses tend to be generated around fracture tips, making them prime sites for enhanced rock deformation, regardless of mechanism. Fluid-rock interactions (e.g., Steefel et al., 2005; Laubach et al., 2010) are thus seen as necessary contributors to fracture propagation in the heated, stressed, fluid-rich Earth's crust. As a result, considerable work has been done to characterize and quantify the chemical processes and kinetics involved. Generally speaking, most work on stress corrosion cracking in rock involves chemical reactions as a (or the) rate-limiting process in fracture propagation (e.g., Atkinson and Meredith, 1987a; Olson et al., 2007; Brantut et al., 2013).

Subcritical fracture growth depends on the local stress state, rock type, the relative chemistries of fluid (usually water and multiple dissolved species) and rock mass, a reaction rate at the fracture tip, and the far-field strain rate. At sufficiently low rates of tectonic strain, the rate of propagation of the fracture tip through the rock mass is limited (and governed) by the chemical reaction rates, leading to subcritical crack growth (Fig. 9.3, region I; see Lawn, 1993, pp. 119–128; Evans and Wiederhorn, 1974; Savalli and Engelder, 2005). As the far-field strain rate exceeds the reaction rate-limited values, however (Fig. 9.3, region II), the chemical processes at the fracture tip cannot keep pace with the applied strain, leading to somewhat more rapid (but still quasi-static) crack growth (Engelder et al., 1993; Anderson, 1995, pp. 55–58; Fig. 9.3, region III). Models based on the Charles Law and stress corrosion lose validity in this region (Brantut et al., 2012). The remote-load boundary conditions (fixed-grip or dead-weight loading), and probably the ambient hydraulic conditions within the host rock, may determine whether the LEFM crack continues to propagate stably and quasi-statically or accelerates into a rapid "running" crack (Segall, 1984b; Kemeny, 1991; Anderson, 1995, p. 48–51; Engelder et al., 1993).

Several crack growth simulations, performed numerically, have adopted the Charles (1958) static-fatigue law as an approximate model for subcritical crack growth in rock (e.g., Meredith et al., 1990; Kemeny, 1991; Olson, 1993, 2003, 2004, 2007; Renshaw and Pollard, 1994b; Olson et al., 2007; Brantut et al., 2013). The mechanism of stress corrosion used in this approach assumes that fracture propagation is facilitated by chemically assisted weakening of atomic bonds in the near-tip region (Atkinson and Meredith, 1987a).

The **Charles Law** can be given by (Atkinson and Meredith, 1987a)

Subcritical fracture propagation rates in rock tend to be limited by the rates of bond-weakening chemical reactions near the fracture tip

Charles Law for time-dependent dilatant deformation

$$v = v_0 \exp\left(\frac{-H}{RT}\right) K^{n_0} \tag{9.1}$$

in which v is the velocity of fracture propagation (usually for mode I), H is the activation enthalpy (e.g., Atkinson and Meredith, 1987b), R is the universal gas constant (Turcotte and Schubert, 1982, p. 301), T is absolute temperature, K is the stress intensity factor, v_0 is the *threshold velocity* for subcritical crack growth, and n_0 is the **stress corrosion index**.

Available data for n_0 under ambient conditions have been compiled and plotted in Fig. 9.4. Typical values for n_0 in rock obtained by a number of researchers (e.g., Atkinson and Meredith, 1987b; Holder et al., 2001; Nara et al., 2013; Gale et al., 2014; Chen et al., 2017) are approximately $25 < n_0 < 80$ (exclusive of chalk). Important influences on this exponent include lithology, anisotropy, temperature, confining pressure, and chemical environment (i.e., fluid composition relative to host rock composition), although these factors are not readily apparent in the first-order grouping of data as depicted in Fig. 9.4.

Holder et al. (2001) showed how n_0 decreased from 35 to 25, and 95–124 to 20–42, when water, rather than air, was the fluid that permeated Scioto Sandstone and Austin Chalk, respectively (corresponding to reductions of ~30% and ~75%). Similar results were reported by Nara et al. (2013) for igneous rocks (Kumamoto andesite, Oshima and Inada granites) and by Chen et al. (2017) for Woodford Shale. Nara et al. (2013) noted that n_0 also decreased with temperature. The value of n_0 increases with the complexity of rock-mass structure, indicating a greater resistance of polylithologic rock types to subcritical crack growth relative to monomineralic rocks under the same environmental conditions (Atkinson and Meredith, 1987a).

A simple expression for the predicted rate of subcritical fracture propagation follows from equation (9.1). It is thought that the maximum propagation velocity v_{max} will likely be attained when $K = K_c$ (Olson, 1993); v_{max}

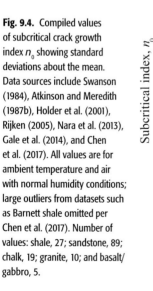

Fig. 9.4. Compiled values of subcritical crack growth index n_0 showing standard deviations about the mean. Data sources include Swanson (1984), Atkinson and Meredith (1987b), Holder et al. (2001), Rijken (2005), Nara et al. (2013), Gale et al. (2014), and Chen et al. (2017). All values are for ambient temperature and air with normal humidity conditions; large outliers from datasets such as Barnett shale omitted per Chen et al. (2017). Number of values: shale, 27; sandstone, 89; chalk, 19; granite, 10; and basalt/gabbro, 5.

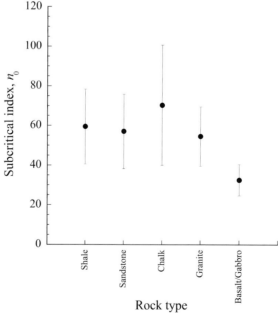

is typically in the range of one-tenth to one-half the rocks' Rayleigh wave velocity V_r. In this case, the terms in the Charles Law (equation (9.1)) for temperature, threshold velocity, and rock composition (via the universal gas constant) simplify to $v_{max}/K_c^{n_0}$ (Olson, 1993). We then have the simple criterion (Main and Meredith, 1991; Olson, 1993) for the predicted velocity of subcritical fracture propagation

Velocity of propagation for subcritical fractures

$$v = v_{max} \left(\frac{K}{K_c} \right)^{n_0} \tag{9.2}$$

Using a representative value for $v_{max} = 1$ km s^{-1}, the time to failure depends on the value of n_0 and on how close in magnitude K is to K_c. For small values of n_0 that are commonly taken to be appropriate to rapid loading and a benign chemical environment ($n_0 = 1$–3), a fracture network may develop in perhaps seconds to minutes. For more chemically active environments ($n_0 = 6$–12), the fracture set may take days to perhaps years to grow (Olson, 1993). Numerical simulations of subcritical crack growth (e.g., Olson, 1993, 2004, 2007; Olson et al., 2007) suggest a qualitative correspondence between the value ranges noted here and those that provide results in the simulations that mimic natural fracture patterns. However, the experimental results show important systematic decreases in the value of the subcritical index, for a given rock type, when water saturation and temperature are considered.

9.2.3 Remote Loading Conditions

Fixed-grip vs. dead-weight loading

Two basic types of loading configurations generally considered, called "**fixed-grip**" (or displacement controlled) and "**dead-weight**" (load controlled) **loading** in the literature (Lawn, 1993, pp. 20–23; Kanninen and Popelar, 1985, p. 185; Engelder and Fischer, 1996; Gudmundsson, 2011, pp. 207–211). The paper by Engelder and Fischer (1996) provides probably the most comprehensive exploration of loading states, including fluid-related, for geologic cracks. These configurations can be thought of as useful end-members of loading conditions in the Earth's crust.

A continuously increased remote stress, applied to drive crack growth, is referred to as *stress-controlled* or "dead-weight" loading. It is also, in a very interesting and useful departure from previous usages, a manifestation of a **soft loading system**, such as a Griffith crack loaded by a remote tensile stress (Scholz, 2002, p. 7–8, 122) that can lead to unstable (i.e., rapid) crack growth.

For example, suppose one loads a rock by gradually increasing the amount of stress applied to it. This can be done, for example, by folding, uplift, or other processes that deform progressively; in the lab, this can be done by increasing the tensile, compressive, or shear load on a sample. The key here is that the rate of increase of the load must be less than, essentially, the ability of the crack(s) or other geologic discontinuities to propagate fast enough to absorb the increasing load (Segall, 1984a,b; Scholz, 2002, p. 8). If the crack is unable to keep up with the load, then it accelerates under dead-weight loading to approach high subsonic speeds. If, on the other hand, the load is applied slowly, it may take some time (or several small increments of bending, for example) for the stress to get large enough for the crack(s) to grow. This is *displacement-controlled*, or "fixed-grip" loading and it can lead to *stable crack growth* (Segall, 1984b; Lawn, 1993, p. 23; Olson, 2003). This case is analogous to a **stiff loading system** (Scholz, 2002, p. 8, 122) that can also lead to stable (i.e., slow and controlled) crack growth.

Tectonic loadings may be more nearly displacement-controlled and cracks grow stably, at low velocities (Brace and Bombolakis, 1963; Hoek and Bieniawski, 1965; Segall, 1984a,b; Olson, 1993, 2003; Renshaw and Pollard, 1994b; Renshaw, 1996). The trend has been to switch over from stress-controlled to displacement-controlled formulations to investigate the propagation of geologic fractures. This precisely parallels the recognition that dead-weight compressive loading of unconfined rock cores in the laboratory tends to fail them explosively, leading to a stress–strain curve that is truncated slightly before peak stress is attained due to dynamic, unstable crack growth and, consequently, the need for stiff testing machines and displacement-loading conditions (e.g., Crouch, 1970; Wawersik and Fairhurst, 1970; Wawersik and Brace, 1971; Hudson et al., 1972; Brady and Brown, 1993, pp. 94–98). As a result, cracks that propagate when $K < K_c$ are best referred to as quasi-static cracks—if they lack evidence in the field for branching.

Scholz's insight that the stability of crack propagation (or of frictional sliding; see also Dieterich and Linker, 1992) is controlled by the response of the loading system provides a fresh and useful perspective on the deformation of geologic materials in general. From this point of view, the stability of a stiff structural member, such as a beam, plate, cylindrical rock core, or geological layer, depends to a large extent on the stiffness of the loading system. Let's take two common examples: buckling and boudinage. Both of these are promoted when the layer of interest is contained by layers having smaller relative stiffnesses (or viscosities, for rate-dependent systems; Johnson and Fletcher, 1994). As the stiffness contrast decreases (i.e., a stiffer system), buckling or boudinage is suppressed. Stability of deformation in the post-peak region was discussed in Chapter 2 as being dependent upon the stiffness contrast between the deforming rock and its surroundings, whether defined by a mechanical testing frame (e.g., Brady and Brown, 1993) or by the geology (e.g., Aydin and Johnson, 1983; Paterson and Wong, 2005). One may imagine this concept extending to the peak compressive strength of rock increasing from its unconfined value with increasing temperature or confining pressure (i.e., increasing stiffness of a test sample's surroundings).

9.2.4 Fatigue Fracture Growth

Fatigue is one of the mechanisms that can lead to slow, time-dependent, subcritical fracture growth (e.g., Haimson and Kim, 1971; Attewell and Farmer, 1973; Kanninen and Popelar, 1985, pp. 47–50; Costin, 1987). In this scenario, instead of applying a constant stress or displacement to the rock mass, a cyclic, repetitive load is applied. This cyclic load can be described as a waveform, with frequency, amplitude, and wavelength (Anderson, 1995, pp. 513–514; see Dowling and Thangjitham, 2000, and Dowling, 1999, for clear and in-depth treatments).

As the atomic bonds near the fracture tip are alternately stretched and relaxed due to the cyclic loading, they weaken and eventually break, leading to fracture growth. You can perform an easy exercise in fatigue fracturing by taking a metal bar, or better yet, a mica book, and flexing it up and down in your hands. You'll notice after a short time that the metal bar heats up as the bonds reject the bending stress (as heat). More interestingly, the mica book begins to crackle (the audible reports are called acoustic emissions: e.g., Cox and Meredith, 1993; Lockner, 1993), weaken, and finally it breaks. You've perhaps used fatigue fracture growth to break large pieces of wood for a campfire. Fatigue fracturing may be an important mechanism of time-dependent weakening in rocks (Haimson and Kim, 1971; Attewell and Farmer, 1973; Costin, 1987) beneath highways or railroad trestles. Common processes such as freeze-thaw (e.g., Lachenbruch, 1961), the earthquake cycle, or fracture growth on Europa (an icy satellite of Jupiter; Hoppa et al., 1999) provide additional examples of fatigue cracking.

A basic relationship between cyclic loading and the stress intensity factor is shown in Fig. 9.5. Here one can see how the amount of fracture growth per cycle, da/dN, depends on the value of the stress intensity factor ΔK (this value being the variation in K with the amplitude of the load; Anderson, 1995, p. 514). Here, K_c is the dynamic fracture toughness and crack growth to that point along the curve is slow, quasi-static, subcritical growth. This curve is quite similar to creep curves, in which time-dependent creep strain is plotted against time (Hobbs et al., 1976; Main, 2000; Brantut et al., 2013) and in fact there may be a correspondence between the two for materials like concrete

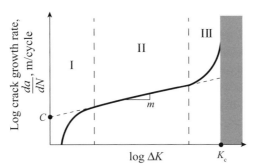

Fig. 9.5. Typical fatigue fracture growth due to cyclic or periodic loadings.

or rock (Bazant and Hubler, 2014). On such a diagram (and keying to the analogous regions on Fig. 9.5), region I corresponds to transient (primary) creep, region II to steady-state (secondary) creep, and region III to accelerating (tertiary) creep.

The linear portion in Fig. 9.5 (region II) can be fit by the empirical expression (Paris et al., 1961; Paris and Erdogan, 1963; Anderson, 1995, p. 517; Pugno et al., 2006)

Paris Law for fatigue fracture propagation

$$\frac{da}{dN} = C \, \Delta K^m \qquad (9.3)$$

in which C is the extrapolated intercept and m is the slope of the curve in region II (Fig. 9.5). Equation (9.3) has generally survived the test of time (and many experiments) and is now traditionally referred to as the **Paris Law** for fatigue fracture growth. Nguyen et al. (2001) showed that such fatigue cracking is generally much better predicted by an elastic-plastic (cohesive end zone) fracture model than by a simple LEFM model.

9.3 Better Terminations: Yield Strength and the End-Zone Concept

Terms for elastic–plastic (non-LEFM) process zone models:

♦ *End-zone model*
♦ *Cohesive end zone*
♦ *Strip yield model*
♦ *Slip-weakening model*
♦ *Fictitious crack model*
♦ *Tension-softening model*
♦ *Dugdale–Barenblatt model*
♦ *Post-yield fracture*

While LEFM does a remarkably good job at capturing the major characteristics of geologic structural discontinuities, it needs some help when the near-tip region is concerned. This help is provided by the concept of the **end zone** (e.g., Goodier, 1968; Palmer and Rice, 1973; Kanninen and Popelar, 1985, pp. 282–283; Lawn, 1993, pp. 59–65; Anderson, 1995, pp. 75–77; Scholz, 2002, pp. 15–17; Gudmundsson, 2011, pp. 299–302). A "cohesive" end zone (Dugdale, 1960; Park and Paulino, 2011) is one type, developed for mode-I cracks, that has wide applicability to both concrete (e.g., Kaplan, 1961; Li and Liang, 1986; Elfgren, 1989) and geologic fractures of various modes (e.g., Cowie and Scholz, 1992b; Hashida et al., 1993; Rubin, 1993a, 1995a,b; Yao, 2012; Haddad and Sepehrnoori, 2014, 2015; Wang et al., 2016). Although more comprehensive (and realistic) approaches to modeling the inelastic processes near fracture tips exist (e.g., Scholz, 2002, pp. 16–17), the cohesive end-zone model presents us with a fairly simple yet elegant way to deal with geologic fractures under non-LEFM conditions.

The strength of the rock ahead of a structural discontinuity, and the levels of remote and driving stress, jointly determine the character of near-tip deformation. For example, driving stress leading to fracture-surface displacement within a linearly elastic material produces **singular (infinite) stress magnitudes at the fracture tip, and steep near-tip displacement gradients**, because the material is defined implicitly as indefinitely strong (Lawn, 1993, pp. 52–53). The small area over which the amplified near-tip stresses are predicted to exceed the rock's yield strength defines the size of the Irwin plastic zone (Broek, 1986, pp. 13–14, 99–102; Anderson, 1995, pp. 72–75); within this small area the predicted level of elastic stress would become so large that its magnitude would not have any physical meaning. By setting a separate criterion for yield strength, the predicted stresses within this area are reduced and limited to the yield strength value appropriate to the near-tip stress state (e.g., tension, shear, or compaction). Using an elastic material rheology sets up physically unrealistic conditions at the fracture tip that drive the near-tip stresses to infinitely large values and the near-tip displacement profile to similarly elliptical or parabolic shapes.

Here are two geologic examples that demonstrate (or at least imply) that LEFM analyses might not be correct and must therefore be supplemented by a more comprehensive approach.

- Many geologic discontinuities **do not reflect conditions of small-scale yielding** and LEFM (Cowie and Scholz, 1992b; Rubin, 1993a; Scholz, 2002, p. 116; Yao, 2012) due to a combination of factors such as finite (small) rock strength at the tipline and relatively large ratios of driving stress to yield strength (e.g., Rubin, 1993a). Measurements of the near-tip displacement profiles for isolated joints and dikes (Khazan and Fialko, 1995; Rubin, 1995b), faults (Dawers et al., 1993; Cowie and Shipton, 1998; Cartwright and Mansfield, 1998; Moore and Schultz, 1999; Manighetti et al., 2001, 2004), and deformation bands (Fossen and Hesthammer, 1997) reveal large departures from the ideal steep near-tip displacement gradients required by LEFM, such as tapered or linear displacement distributions. Sufficiently large end zones serve to limit the magnitude of tipline stresses to that of the surrounding rock's yield strength, leading to finite, non-singular near-tip stress magnitudes, and shallow or nearly linear near-tip displacement gradients (e.g., Scholz, 2002).
- End zones are one type of model that provides a **physical basis for the linear displacement–length scaling** of geologic discontinuities (Cowie and Scholz, 1992b; Scholz, 2002, p. 116; Fig. 9.5). A linear strength increase along a fault (Bürgmann et al., 1994) can also achieve the same result. By replacing dynamic fracture toughness (having units of MPa $m^{1/2}$) by yield strength (having units of MPa), a linear (proportional) relationship between the maximum displacement and fault length (e.g., Clark and Cox, 1996) is achieved, with a slope of $n = 1.0$. D_{max} would scale with $\sqrt{L}$, instead of with L^1, for an LEFM fracture propagating under conditions limited by the rock's (dynamic) fracture toughness (see Fig. 9.16, below). As shown on Fig. 9.6, a shallow slope of $n = 0.5$ (dashed line in the figure) is inferred for fractures that scale with the LEFM dynamic fracture toughness, rather than the yield strength as in many cohesive end-zone models (solid line in the figure, $n = 1.0$; e.g., Schultz et al., 2008a, 2013). Displacement–length scaling of geologic structural discontinuities is discussed later in this chapter.

Fig. 9.6. Predicted scaling exponents between maximum relative displacement D_{max} on a discontinuity and its length L, using two approaches (after Schultz et al., 2008a). For propagation under constant driving stress, akin to non-LEFM conditions with non-singular near-tip stress states, the slope $n = 1.0$; for propagation under LEFM conditions that assume constant dynamic fracture toughness, the slope $n = 0.5$. The constant of proportionality is $D_{max}/L = 0.001$ in both cases.

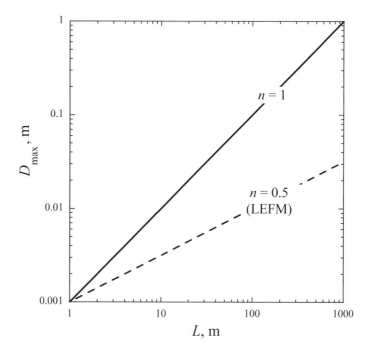

9.3.1 Length of the End Zone

In order to balance the singular near-tip elastic stresses generated near an ideal LEFM discontinuity, opposing stresses act across a narrow zone to cancel the near-tip singularity, leading to modified discontinuity displacement profiles, finite near-tip stress magnitudes, and end zones that may attain appreciable lengths relative to the discontinuity (e.g., >10%). Following Rudnicki (1980), Martel and Pollard (1989), and Rubin (1993a), end zone length is given approximately for a constant yield strength within the zone and under plane strain (Fig. 9.7) by

$$s = \frac{\pi^2}{8}\left(\frac{\sigma_d}{\sigma_y}\right)^2 c \qquad (9.4)$$

in which σ_d is the driving stress, σ_y is the yield strength of the surrounding rock ahead of the discontinuity tip, and c is the discontinuity half-length (Fig. 9.7). End zone length scales with driving stress and (as a second-order factor) discontinuity length $2c$, and inversely with yield strength.

How long is the end zone?　　The relationship between rock strength at the tip (σ_y) and end-zone length s is shown in Fig. 9.8. One can see that, for large values of σ_y/σ_d (i.e., >10), the end-zone length decreases asymptotically toward zero (horizontal dashed line in Fig. 9.8). This is the condition of *small-scale yielding* that is required for LEFM to apply (see Section 8.3.6). Because the strength of the end zone must be larger than that of the actual discontinuity, the ratio must exceed one

Fig. 9.7. Length of end zones (s) near discontinuity, relative to the discontinuity length (L = 2c or 2a; inset) as a function of the strength of rock at the discontinuity tip σ_y/σ_d), after Schultz and Fossen (2002). Equation from Heald et al. (1972), plotted, is comparable to those of Rudnicki (1980) and equation (9.4). Horizontal dashed line is the case for the LEFM discontinuity and infinitely strong rock.

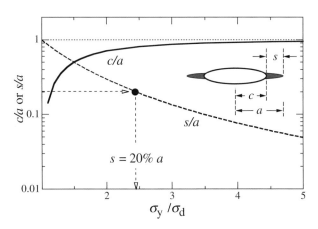

Fig. 9.8. Relationship between stresses, strengths, and end zone bounding a discontinuity tip (after Schultz and Fossen, 2002). Amplified stress due to discontinuity-wall displacements (dashed curve) decays in magnitude with distance away from the discontinuity tip to the peak strength σ_p in the surrounding unfractured rock, defining length s of the end zone for a constant value of yield (peak) strength of the rock in the end zone. Driving stress σ_d is the difference between resolved remote stress σ_r (or static frictional strength) and the internal boundary value σ_i (e.g., pore-fluid pressure or residual frictional strength). Yield strength σ_y is the difference between σ_p and the internal boundary value σ_i. The initial near-tip elastic stress (dashed line) is redistributed by explicit plastic yielding in the end zone to the stippled area.

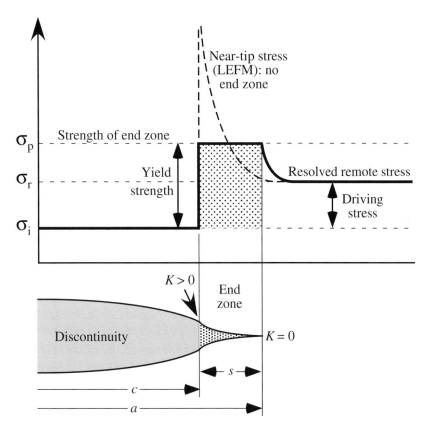

for realistic cases. For an end zone that is 20% of the discontinuity length ($s/a = 0.2$ (Cowie and Scholz, 1992b)), $\sigma_y/\sigma_d = 2.5$ (dot in figure). For deformation bands in porous sandstone, σ_y/σ_d can be perhaps 1.5–2.5 (Schultz and Fossen, 2002).

One can also write an expression for end zone length using a critical value of near-tip displacement δ_c (opening for mode I, shear for modes II and III, closing displacements), as illustrated by Palmer and Rice (1973), Rice (1979), and Rudnicki (1980):

$$s = \frac{\pi}{4} \frac{G}{(1-v)} \left(\frac{\delta_c}{\delta_y}\right) \qquad (9.5a)$$

following Rubin (1993a) and Cooke (1997) for modes I and II, respectively. Equivalently, using Young's modulus, the expression is

$$s = \frac{\pi}{2} \frac{E}{(1-v^2)} \left(\frac{\delta_c}{\delta_y}\right) \qquad (9.5b)$$

Here the driving stress has been recast as its associated near-tip displacement δ_c, using the stiffness term (eq. (2.17)) and modulus relationships as intermediaries. These expressions are most useful either when the critical near-tip displacement value δ_c has physical meaning (as from laboratory friction experiments; see Rudnicki, 1980; Marone, 1998b) or can be observed (as fracture tip opening displacement, and "CTOD" or "FTOD" in engineering; Rubin, 1993a; Khazan and Fialko, 1995; Park and Paulino, 2011; Yao, 2012; or for faults, see Cowie and Scholz, 1992b; Bürgmann et al., 1994; Martel, 1997; Cooke, 1997). In both versions, however (equations (9.4) and (9.5a,b)), one can see that it takes a longer or stronger end zone to counteract the stress singularity generated (ideally) at the discontinuity tip.

Example Exercise 9.1

Calculate the length s of the inelastic end zone (or process zone) that would be generated near (a) the joint, and (b) the fault, from Example Exercise 8.5.

Solution

We choose equation (8.4) and we know all the values for it. In LEFM $a = c$, but here we must be careful to specify the fracture half-length exclusive of the end zone, and this is value, now called c, is plugged into equation (8.4). The yield strengths must also be specified (see Example Exercise 8.9). For the dry joint we use the rock's tensile (yield) strength $T_0 = 10$ MPa. For the fault, the fracture's frictional (yield) strength ($c_0 + \mu_s\sigma_n$) was calculated to be 0.04 MPa. Using the remaining parameters for driving stress and fracture half-length from Example Exercise 8.9, we find:

(a)
$$s = \frac{\pi^2}{8} \left(\frac{0.25}{10.0}\right)^2 (0.9)$$

$\therefore \sigma = \mathbf{0.7\,mm}$, or

$s/a = 0.08\%$ for the joint

and

(b)

$$s = \frac{\pi^2}{8}\left(\frac{0.02}{0.04}\right)^2 (8.0)$$

$\therefore \sigma = \mathbf{2.5\,m}$, or

$s/a = 31\%$ for the fault.

The lengths of these end zones are 2–3 times as large those predicted by the simpler LEFM approach. These values should be a better measure of the inelastic zone near the fracture tip because they incorporate less restrictive assumptions about the material behavior there.

Because the magnitude of fracture surface displacement, and its maximum value D_{max}, are proportional to the level of driving stress acting on the fracture (e.g., Pollard and Segall, 1987; equations (8.10) and (8.16)), and for a given fracture length and yield strength, end zone length increases with D_{max}. Fractures of comparable length but with smaller displacements require shorter end zones (e.g., Li and Liang, 1986). A diagram that illustrates the difference in shape between LEFM fractures and the more commonly found "post-yield" discontinuities is shown in Fig. 9.9. End-zone length is known to depend additionally on the **fracture aspect ratio**, a/b (Chell, 1977), so that tall, 2-D fractures can develop end zones perhaps five times longer than circular, 3-D penny fractures having identical values of driving stress, yield strength, and fracture length (Schultz and Fossen, 2002). Some suggested geometries of end-zone sizes surrounding discontinuities are shown in Fig. 9.10.

The **displacement distribution** along a discontinuity, for simple cases of fracture geometry and loading configurations, can be described mathematically. We did this in Chapter 8 for the LEFM fracture (see Section 8.3.2, equations (8.11)). If the rock at the fracture tips is not perfectly elastic and therefore not infinitely strong, as is typical in rocks and rock masses, then the displacement distribution along the post-yield (or elastic–plastic) fracture will be different from the ideal LEFM case. In particular, the weaker rock ahead of the fracture tip yields (breaks, shears, flows, compacts) and thereby prevents steep displacement gradients from developing at the fracture tip. We can see this from Fig. 9.11, where the elegant equations of Bürgmann et al. (1994) are plotted for useful values of σ_y/σ_d.

The ideal elastic (LEFM) displacement profile is shown as the bold curve in Fig. 9.11. See how it defines an ellipse from the fracture's center (midpoint) to its tip, and notice how steep the displacement gradient is near the tip. This case is best associated with fractures in *very strong rock*, with negligibly small end zones ($s/a = 5 \times 10^{-4}$). More typically, fractures may be represented by $\sigma_y/\sigma_d = 2$–3, with end zones from 13% to 30% of the fracture half-length. *Very weak rock* at the fracture tip—such as a pre-existing joint or bedding

Fig. 9.9. Schematic illustrating more displacement across an LEFM fracture (top panel, with plastic zones) than a "post-yield" fracture having a non-negligibly small end zone (lower panel); remote stress state is the same for both cases. Displacements and end-zone sizes are exaggerated for clarity.

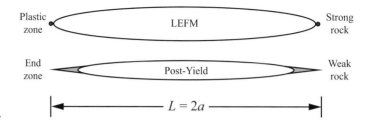

Fig. 9.10. End zone length depends on the displacement gradient along a fracture. This schematic cartoon (based on results from Chell, 1977) shows relative lengths of end zones for each crack geometry. The end zones are shown here as lighter gray fringes surrounding the darker fracture with shades hinting at the displacement distribution.

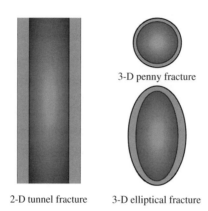

3-D penny fracture

2-D tunnel fracture   3-D elliptical fracture

Fig. 9.11. Displacement distributions along fractures as a function of the strength of rock at the fracture tip (σ_y/σ_d); $G = 10$ GPa, $v = 0.25$. Equations from Bürgmann et al. (1994); LEFM case is the bold curve. Shaded areas near fracture tips are the end zones; note how they restrict the buildup of displacement along the fracture, leading to a smaller D_{max}.

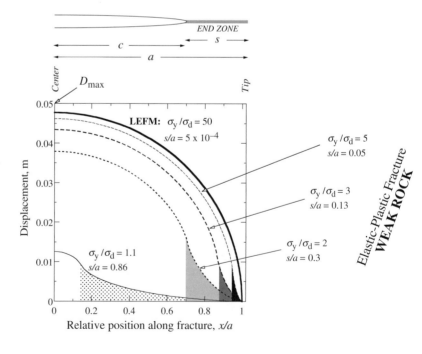

plane—may produce relatively flat displacement profiles ($\sigma_y/\sigma_d = 1.1$ in Fig. 9.11). Such a case is shown in relay ramps like that of Devils Lane graben, Utah (Schultz and Moore, 1996; Moore and Schultz, 1999), pictured on the cover of this book. Notice, too, that the amount of displacement that can be accommodated by a fracture decreases as the rock gets weaker. In this case, the end zone lengthens to blend the fracture displacement into the much tinier magnitudes in the adjacent rock.

A word of clarification is on order regarding the equations for end-zone length and the associated displacement distribution. These equations are derived analytically by adding the contributions to stress intensity factor from the fully yielded fracture and from the end zone so that the net $K = 0$ at the tip (e.g., Palmer and Rice, 1973; Rudnicki, 1980; Cowie and Scholz, 1992b; Rubin, 1993a; Cooke, 1997; Martel, 1997; Willemse and Pollard, 1998). However, for certain combinations of parameters, the displacement can actually go negative near the tip—implying either interpenetration or a reversal of slip sense. Cooke (1997) has shown from numerical solutions, however, that when frictional resistance is included (for example), the reversal of slip sense disappears. Similarly, the negative values for mode-I cases disappear when interpenetration is prevented. Because problems in mechanics involving objects in contact and with frictional resistance must be solved by iteration (e.g., Crouch and Starfield, 1983, pp. 232–249; Lin and Parmentier, 1988; Aydin and Schultz, 1990; Cooke, 1997; Tada et al., 2000, pp. 31–38), simple addition of terms to produce $K = 0$ is not appropriate for these cases.

These end-zone equations also assume that the near-tip damage (or process) zone is largely confined to a narrow zone coplanar with the main fracture (e.g., Rubin, 1993a). Instead, what is more often found (e.g., Vermilye and Scholz, 1998) is a zone of damage that is distributed throughout a volume surrounding the fracture tip (e.g., Gudmundsson, 2011, pp. 299–302); this observation has motivated the use of "small-scale yielding" models that explicitly incorporate spatially distributed damage (e.g., Kanninen and Popelar, 1985; Scholz, 2002, pp. 13–17; Manighetti et al., 2004). These models may also predict linear displacement distributions along fractures, rather than elliptical or bell-shaped ones (Scholz, 2002).

9.3.2 What is the End Zone?

The end zone is a region around and adjacent to the discontinuity tip (or tipline) that has partially broken (yielded) due to the elevated stress magnitudes generated away from the discontinuity tip by displacement sustained along the discontinuity. You can think of it as an *incipient fracture* or the part of the mature, existing discontinuity that is starting to *break out into the surrounding rock* as part of the process of discontinuity propagation. End zones have been described for many types of geologic structural discontinuities.

- **Cracks:** Large (cohesive) end zones have been documented adjacent to cracks in concrete and other engineering materials for a long time (e.g., Dugdale, 1960; Barenblatt, 1962; Hillerborg et al., 1976; Li and Liang, 1986; Ingraffea, 1987; Hillerborg, 1991; Park and Paulino, 2011). Nguyen et al. (2001) showed how cracks modeled with cohesive end zones can better explain fatigue-related deformation (including the Paris Law) than the standard LEFM-based approaches that rely on K.
- **Joints:** Tectonic joints can be understood by analogy with their cousins from engineering materials (e.g., Ingraffea, 1987; Pollard and Aydin, 1988). The development of process zones (microcrack swarms; Swanson, 1987) and process zone wakes that the propagating joint splits (Costin, 1987; Broberg, 1999; Gudmundsson, 2011, p. 301) is thought to be an important aspect of joint development, including the related toughening of partially molten rock with elevated temperatures (DeGraff and Aydin, 1993).
- **Hydrofractures:** The use of classical LEFM equations (e.g., Rummel, 1987; Engelder, 1993, pp. 131–170; Rutqvist et al., 2000) can underestimate the injection fluid pressures required to propagate a hydrofracture laterally through subterranean strata (e.g., Cleary et al., 1991; Rubin, 1993a; Romijn and Groenenboom, 1997; Yao, 2012; Haddad and Sepehrnoori, 2015). This discrepancy can be attributed, in part, to the toughening effect of newly generated microcracks within the growing hydrofracture's process zone (e.g., Warpinski, 1985; Johnson and Cleary, 1991).
- **Dikes:** Pollard (1987) demonstrated how a sizable end zone can better explain the observations of dike segments at Ship Rock, New Mexico, than can the LEFM model. The tip cavity (beyond the magma-filled central portion of the dike; open area in Fig. 9.6, lower panel) may contain heated groundwater or steam (Delaney, 1982), hydrothermal fluids exsolved from the magma (Currie and Ferguson, 1970; Rubin, 1993a), or brecciated host rock (Johnson and Pollard, 1973). End zones that are larger than the LEFM plastic zones appear required to adequately explain dike propagation and growth (e.g., Rubin, 1992, 1993a,b, 1995a,b, 1998; Khazan and Fialko, 1995; Fialko and Rubin, 1998; Mériaux et al., 1999).
- **Faults:** The utility of an end zone in "smearing out" the shear stress concentration near fault tips was first pointed out by Chinnery and Petrak (1967), while Palmer and Rice (1973), Elliott (1976), Rudnicki (1980), King and Yielding (1984), Okubo and Dieterich (1984), Li (1987), Cox and Scholz (1988a,b), Martel and Pollard (1989), Cowie and Scholz (1992b), Cooke (1997), Martel (1997), Vermilye and Scholz (1998), Willemse and Pollard (1998), and Manighetti et al. (2004) later emphasized the key role of end zones in fault propagation. The end zone adjacent to fault tips consists in many compact rock types of incompletely linked crack arrays that accommodate progressively smaller amounts of shear displacement away from the tip (Cowie and Scholz, 1992b). However, a linear displacement distribution along the fault can also be achieved *without an*

end zone (Bürgmann et al., 1994) by allowing the frictional resistance to vary continuously along it. This approach provides a plausible alternative to end zones where a bell-shaped (ogive) profile along a fault is not observed (and considered common by Manighetti et al., 2001, and Scholz, 2002).

- **Stylolites:** Raynoud and Carrio-Schaffhauser (1992) identified a zone of locally enhanced porosity ahead of stylolites in micritic limestone. The process zone there is a region of diffusive mass transfer, where soluble grains are beginning to dissolve at the high-stress contact points between them. As insoluble residue (e.g., clay minerals) accumulates within the end zone, compactional strain is impeded and the stylolite tip migrates laterally (propagates) as a new end zone is formed. The numerical simulations reported by Fueten and Robin (1992) provide a basis for these observations, and displacement gradients along some stylolites are consistent with propagation processes (Benedicto and Schultz, 2010).

- **Deformation bands:** Aydin (1978) and Antonellini et al. (1994) documented a zone of increased porosity ahead of cataclastic deformation bands in porous sandstone. Here, compactional and shearing deformation within the band promotes reshuffling of grains to larger porosities, which in turn leads to greater grain contact stresses, crushing, and comminution. This process of band propagation was simulated numerically by Antonellini and Pollard (1995), Rudnicki and Sternlof (2005), Okubo and Schultz (2006), and Meng and Pollard (2014).

The evidence and arguments suggest that a non-negligibly small end zone may be an integral part of fracture or deformation band architecture. The end zone connects the relatively larger displacement along the discontinuity to the smaller displacement in the surrounding rock mass.

9.3.3 Nucleation of Secondary Structures

In Chapter 5 we examined two "discovery patterns" of fractures (echelon fractures and end cracks) that can best be understood by using fracture mechanics. Here we'll modify the prediction of end cracks slightly from that given previously by invoking a post-yield, elastic–plastic end zone approach instead of LEFM.

Rubin (1993a), Cooke (1997), Martel (1997), and Willemse and Pollard (1998) all demonstrate how the presence of a **cohesive end zone at a fracture tip changes the near-tip stress state** and the near-tip displacement distribution (see also Bürgmann et al., 1994; Fig. 9.10). The end zone has several key functions here, including:

- Removal of the stress singularity
- Relocation of maximum tangential stress ($\sigma_{\theta\theta}$ in Fig. 5.16) from the tip to the junction between end zone and yielded fracture.

Additionally, several researchers explicitly included terms for the remote stress into expressions for take-off angles for secondary fractures from pure mode-I (Cotterell, 1966) and mixed-mode I-II (Maiti and Smith, 1983, 1984; Lee and Olson, 2017) cracks. These terms have the effect of modifying the predicted take-off angles to bring the theoretical predictions into better accord with experimental observations (Smith et al., 2001), especially when a finite distance away from the fracture tip is considered (e.g., Williams and Ewing, 1972).

Secondary fractures still nucleate near the tips of elastic–plastic fractures (with end zones) because the stress changes (and magnitudes) are significantly larger there than elsewhere in the rock mass. However, the end zone introduces some new twists in secondary fracturing that can be used in the field to infer what's happening along the fracture.

In LEFM, the near-tip stress field is inhomogeneous, varying greatly in magnitude with position (r and θ) around the fracture tip. Near-tip stress magnitudes are so large that all the action (secondary fracturing, tip propagation) must occur right at the tip. This implies that end cracks (Segall and Pollard, 1983b), stylolites (Fletcher and Pollard, 1981), and perhaps deformation bands (Rudnicki and Sternlof, 2005; Okubo and Schultz, 2006; Schultz and Soliva, 2012; Meng and Pollard, 2014), folds, cleavages, thrust faults, and normal faults (Aydin and Page, 1984; Davison, 1994) all will begin at the tip and propagate away along paths that first are controlled by the near-tip stress field, then perhaps curve gently into orientations more closely aligned with the regional, far-field remote stress state (see Fig. 5.19, upper panel).

In the case of an elastic–plastic (post-yield) discontinuity with a non-negligibly small end zone, however, the near-tip stress field is homogeneous (Rubin, 1993a; Willemse and Pollard, 1998) with stress components that can be comparable in magnitude to the remote stress state. The stress magnitudes are still larger there than elsewhere in the rock mass, as required for discontinuity propagation from their tips, but the configuration of the near-tip stress field is different than for the simpler LEFM one. In particular, the location of $\sigma_{\theta\theta}$ has moved from the fracture tip back—inward—along the elastic–plastic fracture to the area where the fully yielded fracture grades into its end zone (Cooke, 1997; Martel, 1997). Secondary structures such as end cracks and stylolites will nucleate there instead of at the fracture tip (Cooke, 1997). Depending on the details of what is physically happening within the end zone, swarms of secondary fractures can localize within the end zone region, forming sets sometimes called "horsetail fans" or "splays" in the literature. These fractures will interact and interfere with each other as they grow outward from the parent fracture, resulting in fewer but longer fractures at larger distance from the main fracture (Cooke, 1997; Koenig and Aydin, 1998). In the case of a fault whose frictional resistance decreases from tip toward center (Bürgmann et al., 1994), secondary fractures may nucleate close to the fault tip (as would fault propagation itself), with specific locations depending on the existence of geometric or mechanical irregularities in the fault-tip region.

The incremental distance that a fracture propagates has been related to the size of the plastic or process zone r_p at its tip (e.g., Williams and Ewing, 1972; Aliha et al., 2010; Lee and Olson, 2017). Under conditions of small-scale yielding, r_p is smaller than perhaps 1% of the fracture half-length a, so that propagating fractures with sufficiently small process zones can appear to propagate smoothly and continuously. Including the process zone changes the predicted take-off angles for mixed-mode I-II cracks, as shown in Figure 9.12.

Willemse and Pollard (1998) demonstrate how important the remote stress state is in secondary fracturing near the tips of post-yield fractures with still larger end zones. In the case of elastic-plastic (post-yield) fractures (having non-negligibly small end zones) there are two factors that come into play. First, because the near-tip stress magnitudes are comparable to those of the remote stress, no longer can we just use the LEFM near-tip stresses to predict the angles of these secondary fractures (see also Lee and Olson, 2017). This leads to some ambiguity in relating the angles between end cracks and parent fault to stress state or fracture displacement mode (Willemse and Pollard, 1998). Second, but perhaps more important, we now have a **key to determining how large** (and therefore how significant) the end zone actually is for any given fracture and, implicitly, the stress state within it). In the explicit post-yield formulation by Willemse and Pollard (1998), the angles between

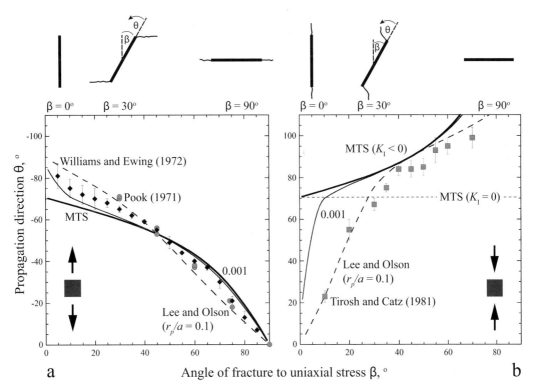

Fig. 9.12. Predicted crack propagation angles for a crack having a range of orientations to remote uniaxial stress (after Lee and Olson, 2017): (a) uniaxial extension; (b) uniaxial compression; mixed-mode cracks have $0 < \beta < 90°$. MTS, maximum tensile stress criterion (Erdogan and Sih, 1963); curves show predicted propagation (take-off) angles calculated for cracks having end zones of $r_p/a = 0.1$ and 0.001; the open-crack cases in (b) consider negative K_I due to fracture closure under remote compression, not considered in LEFM case (dashed line for MTS). The experimental data of Tirosh and Catz (1981) show how crack-tip curvature (blunting, with curvature ratio 0.031) tend to yield smaller take-off angles, especially at small to moderate values of β, than would be predicted for the sharp-tip cases shown by the curves.

dilatant (e.g., crack) and contractional (e.g., stylolite) secondary structures, in the trailing and leading quadrants of a fault, respectively, are consistently 90°. In contrast, the inter-fracture angle in LEFM is 141° (70.5° + 70.5°) for pure mode-II fractures. Considering the take-off angles calculated for cases involving crack closure by Lee and Olson (2017), the inter-fracture angle appears to lie between 90° and 141° (H. Lee, personal communication, 2017) depending on the resolved stresses on the parent fracture (i.e., its orientation in the remote stress state) and the size of its end zone. One implication of this is that one could **measure this angle in the field** (assuming that both sets of secondary structures are present) and, having some information about the remote stress state, infer the relative and importance size of the **end zones**, with their size increasing from LEFM-like values as the inter-fracture take-off angles increase from 90° toward 141°. Correspondingly, the increase in angle suggests a change in host-rock response from brittle (with rapid failure near peak stress, and LEFM-like conditions) to semi-brittle (having post-yield fracture propagation).

Key to identifying end zones:
< 141° angle between dilatant and contractional secondary structures

The basic or end-member patterns of secondary fractures predicted for mode-II fractures are shown in Fig. 9.13. Here the LEFM case (Fig. 9.13a) is based on standard parameters for a "strong" fault, having typical values for friction of $\mu_s = 0.6$, so that the maximum remote compressive principal stress (σ_{max}) makes an angle of 30° to the fault. Because only the shear stresses at the tip are important in $\sigma_{\theta\theta}$ (because opening or closure along the fault is minimal or zero compared to the fault-parallel displacement), the predicted angle of end cracks (or normal faults) is 70.5° in the fault's trailing quadrants. Contractional secondary structures (stylolites, cleavages, folds, thrust faults) will be oriented at 70.5° in the fault's leading quadrants. Both sets of secondary structures must nucleate very close to the tip of the parent fracture. The structures also **curve** out of the near-tip stress zone (from angles of 70.5°) to become parallel to the remote principal stress orientations (dashed lines in Fig. 9.13a).

The predicted orientations of secondary dilatant fractures (cracks or normal faults) are different for post-yield faults with non-negligible end zones (following Willemse and Pollard, 1998). For a "strong" fault having friction $\mu_s = 0.6$ along it and an end zone 20% of its half-length, end cracks are oriented at 15° to the fault, while stylolites (or other contractional structures) are oriented at 90° − 15° = 75° to the fault in the leading quadrant (Fig. 9.13b). It would seem easy to interpret the end cracks as **Riedel shears** (see Sections 3.5.2 and 7.5.3) because the 15° angle of the smaller fractures to the main fault plane is the same for both scenarios. The *timing*, of course, would tell you definitively, however, which case applies (the fault is last for Riedel shearing, but first here).

Secondary structures growing due to slip along a "weak" fault (Fig. 9.13c) also share the 90° inter-fracture angle under this formulation. Here, though, the dilatant structures should grow at 35° angles to the fault, with the contractional ones oriented at 65° to the fault in the leading quadrant. One can see

Fig. 9.13. Prediction of end cracks and other secondary structures due to mode-II slip along faults. (a) LEFM case with negligibly small end zones (shaded lines near fault tips). The maximum remote principal (compressive) stress is oriented at 30° to the fault, given $\mu_s =$ 0.6 ($\sigma_{max}/\sigma_{min} = 3.0$; $\sigma_{max} = 3.0$ MPa). (b) Strong post-yield (PYFM) fault with end zone $s = 20\%$ of a. Same parameters as in LEFM case; note shallow (acute) orientation of end crack (15°) relative to parent fault. (c) Weak post-yield fault with end zone $s = 20\%$ of a. Same parameters as in (b) except that $\sigma_s = 0.1$ ($\sigma_{max}/\sigma_{min} = 1.22$; $\sigma_{max} = 1.22$ MPa) along the fault, so that the principal stresses are reduced in magnitude and rotated from (b).

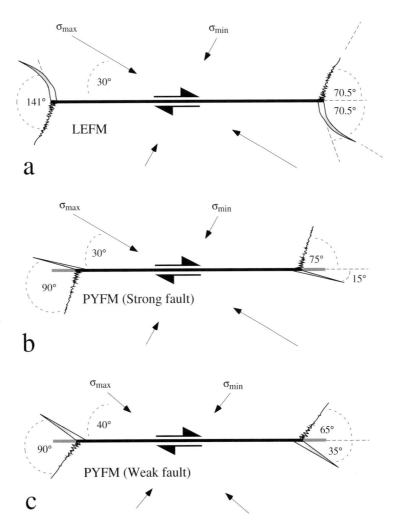

that, for both cases of faults with end zones (Figs. 9.13b and c), there is still a mismatch between the orientations of secondary structures and the remote principal stresses, but the discrepancy is less ($30° - 15° = 15°$ for strong faults, $40° - 35° = 5°$ for weak faults) than that for the LEFM fault ($70° - 30° = 40°$). The secondary structures at post-yield fault tips may also be fairly straight, whereas those at the tips of an ideal LEFM fault should curve more strongly as they emerge from the K-dominant region.

9.3.4 Stress Concentration at Post-Yield Fracture Tips

Infinite levels of stress are predicted theoretically to occur at the tips of LEFM fractures (see Chapter 8). In order to accommodate the predicted infinite elastic stress magnitudes there, the stress intensity factor K was defined as the mathematical limit of the asymptotically increasing near-tip stress state. In the end-zone approach, an end zone was prescribed with given length and strength so that its contribution to the near-tip stress exactly cancels that of the main (fully yielded) fracture; this produces $K = 0$ at the tip (e.g., Goodier, 1968; Heald et al., 1972; Palmer and Rice, 1973; Chell, 1977; Rudnicki, 1980; Cowie and Scholz, 1992b; Rubin, 1993a; Bürgmann et al., 1994; Cooke, 1997; Martel, 1997; Willemse and Pollard, 1998). We now don't have a stress intensity factor at the tip of this elastic–plastic, post-yield fracture, but we still have a stress *concentration* there.

How large is the stress concentration at the tip of the post-yield fracture? Explicit calculations of the stress state near a post-yield fracture by Willemse and Pollard (1998) demonstrate that the strength of stress concentration is comparable to the ratio of yield strength (of the end zone) to driving stress. We define the **stress concentration factor for post-yield fractures** with a cohesive end zone as

Stress concentration at post-yield fracture tips

$$S_{PY} = \frac{\sigma_y}{\sigma_d} \qquad (9.6)$$

and show in Fig. 9.14 how it relates to end-zone length. **For typical end zones and yield strengths, S_{PY} = 2–5 times the values of ambient stress in the rock mass.** This is much smaller than the infinitely large, singular LEFM value because, for example, both quantities have magnitudes of perhaps a few tens to hundreds of MPa. Because the magnitude of stress concentration is now on the same order as that of the ambient stress state in the rock mass, post-yield fractures (and, by implication, discontinuities) have a milder effect on stress perturbations than did their hypothetical LEFM predecessors.

As a final point about applying end-zone models to post-yield fractures. The Dugdale–Barenblatt end-zone model implies that the entire fracture displaces in one episode or event; this end-member case produces the classic bell-shaped (ogive) profile (Fig. 9.11) found, for example, in some igneous dikes (e.g., Rubin, 1993a) or for certain faults (e.g., Schlische et al., 1996).

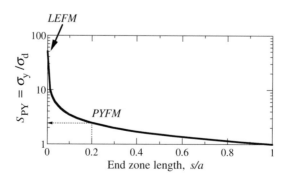

Fig. 9.14. Comparison between the end-zone length, s/a, and the degree of stress concentration S_{PY} at a post-yield fracture tip.

At the other extreme, however, successive addition of small slip events on a fault can lead to more *linear* displacement profiles that are considered to be more commonly recognized than bell-shaped ones (Cowie and Shipton, 1998; Manighetti et al., 2001). Here, a different model is needed so that the fault remains governed by the rock's yield strength, not its dynamic fracture toughness. A linear increase in frictional resistance along the fault, from a weaker central portion toward the stronger tips, can also produce a non-singular near-tip stress concentration (Bürgmann et al., 1994), and therefore, a linear displacement-length scaling relationship. As a result, *D–L* scaling for faults can be understood using a model that explicitly **removes the near-tip singularity** (Cowie and Scholz, 1992b), as done by both the linear strength increase model and the end-zone model.

9.3.5 Summary of LEFM vs. PYFM Approaches

To summarize the cohesive end-zone, post-yield fracture mechanics (PYFM) approach to geologic structural discontinuities, the main attributes and points are laid out in Table 9.2. One can see that relaxing the assumption of small-scale yielding (which follows from the assumption of an indefinitely strong elastic material) produces a potentially more realistic model for the ways many structural discontinuities function in rock masses.

Table 9.2 Comparison of LEFM and PYFM approaches

LEFM	PYFM
Linearly elastic material has infinite strength	**Observe:** Shallow near-tip displacement gradient
∴ Singular near-tip stress magnitudes	∴ Damage distributed away from fracture tip
Small-scale yielding	Yield strength comparable to stress magnitude
Stresses focused at fracture tip	Near-tip stresses comparable to remote stresses
Damage focused at fracture tip	K vs. K_c criterion not valid
Stresses ≫ remote stress magnitudes	Less displacement per unit fracture length
K vs. K_c criterion valid	End-zone length = f(yield strength, discontinuity length, and driving stress)
Steep displacement gradient near tip	
Angle between end structures ~140°	Angle between end structures <140°

9.4 Elastic–Plastic Fracture Growth

Previously we saw how the stress intensity factor, K, was computed from the fracture length and the applied stress state and then compared to a material parameter, K_c—the critical (dynamic) stress intensity factor or dynamic *fracture toughness*—to determine if the fracture would propagate (i.e., lengthen). This approach assumes that the fracture is trying to break an ideal elastic material, with no significant inelasticity or plasticity at or near the fracture tip. Many structural discontinuities in rocks are often associated with a swarm of small-scale deformation near their tips; this is called the *process zone* (e.g., Broberg, 1999, p. 5; Scholz, 2002; Gudmundsson, 2011). We also saw that the process zone may be large enough, especially for faults, compared to the parent fracture to make the elastic equations a less applicable approach to predicting fracture propagation in rock (see also Ingraffea, 1987). So if LEFM and $K = K_c$ don't work well in some geologic contexts, what can be used instead? Three approaches hold promise in this challenging and important area.

9.4.1 Apparent Dynamic Fracture Toughness

Although K by itself may have little meaning for elastic–plastic, post-yield fracture propagation (e.g., Rubin, 1993a), it turns out that we can still use it as a tool for assessing fracture propagation in rock. Here we'll see how to compare K against a modified form of dynamic fracture toughness—the *apparent dynamic fracture toughness*—to predict fracture growth in non-elastic rocks. A parallel approach was presented by Khazan and Fialko (1995). Kanninen and Popelar (1985) show how the *J*-integral (Rice, 1968) can be used to model the growth of elastic–plastic fractures.

Heald et al. (1972) compared the tendency for tall, 2-D tunnel cracks to propagate by deriving equations for the "apparent dynamic fracture toughness" of material surrounding non-LEFM fractures, assuming the same values for fracture length, driving stress, and end-zone strength. Heald et al.'s equation represents the **decrease in K** at the tip of a post-yield fracture due to toughening of the rock in its end zone. From this, an expression for the post-yield, apparent dynamic fracture toughness of the fractured near-tip rock bounding the fracture can be obtained. For a non-LEFM fracture, the **apparent dynamic fracture toughness** K_{ac} encountered by a post-yield fracture (having an end zone as in Section 9.4) can be given by

Apparent fracture toughness for post-yield fractures

$$K_{ac} = \left\{ 2\sigma_y \sqrt{\frac{c}{\pi}} \, \cos^{-1}\left[\exp\left(\frac{-\pi K_c^2}{8\sigma_y^2 Y^2 c} \right) \right] \right\}^{-1} \tag{9.7}$$

in which K_c is the LEFM dynamic fracture toughness, σ_y is the yield strength of the end zone, c is the fracture half-length (that excludes the end zone; see Fig. 9.6), and Y is the fracture's configuration parameter (Chapter 8). The strengthening effect of the end zone on non-LEFM fractures is shown in Fig. 9.15.

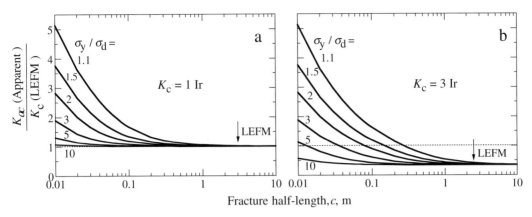

Fig. 9.15. The **strengthening** of rock ahead of a post-yield fracture is shown by the curves of apparent dynamic fracture toughness (equation (9.7)) normalized by the LEFM dynamic fracture toughness K_c. At small fracture lengths, it takes several times more stress to propagate a fracture with an end zone than an LEFM fracture that has a negligibly small plastic zone. Driving stress $\sigma_d = 1.0$ in all cases shown.

As one can see from equation (9.7), the apparent dynamic fracture toughness depends on several factors including *fracture length* and yield strength, along with Y and K_c. This means at once that $\boldsymbol{K_{ac}}$ **is not a material constant**, but instead depends on the geometry (i.e., size) of the fractured system. Thus, the resistance to fracture propagation in rocks (K_{ac}) varies with the applied stress or loading conditions (through the yield stress (Rubin, 1993a)) and with fracture length (e.g., Ingraffea, 1987; Rubin, 1993a). A dependence of apparent dynamic fracture toughness on confining pressure (for mode-I cracks) has been appreciated for some time (e.g., Perkins and Krech, 1966; Abou-Sayed, 1977; Schmidt and Huddle, 1977; Atkinson and Meredith, 1987b; Thallak et al., 1993).

Values of apparent dynamic fracture toughness K_{ac} (due to the fracture's end zone alone) are plotted in Fig. 9.15 against fracture half-length c for two values of dynamic fracture toughness K_c. Here, we've normalized the apparent fracture toughness K_{ac} by the LEFM dynamic fracture toughness K_c in order to show directly the *strengthening* of the rock at the fracture tipline (within the end zone) due to the elevated stress levels adjacent to the fracture tipline. The strengthening is related to the additional work required to deform the rock inelastically within the end zone (e.g., Li and Liang, 1986; Rubin, 1993a).

One can confirm immediately in Fig. 9.15 that K_{ac} now depends on fracture length—a new feature compared to the dynamic fracture toughness K_c that was independent of fracture length. When σ_y/σ_d is very large (i.e., >10), there is only a minimal increase above the LEFM value (K_c). In the limit of infinitely strong rock ($\sigma_y \rightarrow \infty$ for LEFM and small-scale yielding), $K_{ac} = K_c$. Also, as fracture lengths get very large, $K_{ac} \rightarrow K_c$, implying that small post-yield fractures can, to some extent, "grow out of" their damage zones to approximate LEFM fractures as they lengthen (e.g., Heald et al., 1972; Ingraffea, 1987). Fracture propagation would still be governed by σ_y, however (not K_c; Palmer and Rice, 1973; Cowie and Scholz, 1992b); one just can't see its effect as easily on plots like Fig. 9.15 for these sufficiently large fractures. As σ_y/σ_d becomes small (and perhaps more geologically reasonable), then the rock strengthens to perhaps five times its value from LEFM-based K_c estimates that one would find tabulated in books and journal articles (e.g., Atkinson and Meredith, 1987b; Chen et al., 2017). This means that, for geologically reasonable yield strengths ($2 < \sigma_y/\sigma_d < 3$), it takes **2–3 times more energy (or stress) to propagate a post-yield fracture** than would be implied by a simpler LEFM-based calculation.

9.4.2 The *J*-Integral

Simple fracture problems can often be adequately described by LEFM and small-scale yielding. More complex problems, however, require more complete treatments of the stress and strain fields, and related conditions, near a discontinuity tip. Some of these conditions include (Kanninen and Popelar, 1985, pp. 15–17):

- *Elevated temperatures*, leading to plasticity, thermal cracking, and dislocation movement in the end zone;
- *Reduced yield strength*, leading to cracking, faulting, and buckling in the end zone; and
- *Variations in loading rate*, with dynamic fracture toughness increasing with loading rate for post-yield fractures and decreasing with loading rate for fractures with smaller end zones.

For one or more of these three cases, the equations of LEFM begin to lose their validity in the strictest sense.

The *J*-integral (Rice, 1968; Begley and Landes, 1972, 1976) was developed to exploit the role of **strain energy** near a fracture tip in fracture propagation, as defined by the equations for nonlinear plasticity (e.g., Kanninen and Popelar, 1985, pp. 164–168; Li, 1987; Lawn, 1993, pp. 66–72). Here, one can include all the effects that are considered important in discontinuity propagation, as long as one can incorporate them mathematically, either through equations or numerically (e.g., finite elements). *J*-integral methods were applied to study deformation band propagation by Rudnicki and Sternlof (2005), Rudnicki (2007), Tembe et al. (2008), Torabi and Alikarami (2012), and Schultz and Soliva (2012). Essentially, *J* represents the sum (calculated as energies) of *driving forces* (that would encourage a fracture to propagate) and *dissipative forces* (that would resist or occur as a result of propagation; Li, 1987).

Let's see how the *J*-integral works for a post-yield fracture with an end zone of constant yield strength σ_y. We will use the propagation criterion

Criterion for propagation of an elastic–plastic (post-yield) fracture

$$J > J_c : \text{Growth} \tag{9.8a}$$

$$J < J_c : \text{No growth} \tag{9.8b}$$

Here, J_c is a material parameter analogous to the apparent dynamic fracture toughness K_a. In the limit of infinitely large yield strength and small-scale yielding (both appropriate to LEFM),

Equivalence of fracture parameters, LEFM and PYFM

$$J_{SSY} = \frac{K_i^2}{E(1-v^2)} = \frac{V(\sigma_d^2)\pi a}{E(1-v^2)} = \mathcal{G} \tag{9.9}$$

for plane-strain fractures. But in general, $J_c > J_{SSY}$; in either case, one can calculate values of J_c directly from the elastic–plastic end-zone equations (9.5). For typical post-yield discontinuities with an end zone,

$$J_c = \sigma_y \delta_c \tag{9.10}$$

in which σ_y is the constant yield strength in the end zone adjacent to the fracture's tipline and δ_c is the critical slip distance or fracture tip opening displacement (FTOD). Because values for δ_c are rarely known with precision for structural discontinuities in geologic materials one can rewrite this value (again using plane strain) as (Kanninen and Popelar, 1985, p. 283)

$$\delta_c = \frac{8}{\pi}\frac{\left(1-v^2\right)}{E}\sigma_y c \ln\left[\sec\left(\frac{\pi\,\sigma_d}{2\,\sigma_y}\right)\right] \tag{9.11}$$

in which all the variables are things we do know. From (9.10), the **critical value of J for fracture propagation** to occur is

J-integral for an elastic-plastic (post-yield) fracture

$$J_c = \frac{8}{\pi}\frac{\left(1-v^2\right)}{E}\sigma_y^2 c \ln\left[\sec\left(\frac{\pi\,\sigma_d}{2\,\sigma_y}\right)\right] \tag{9.12}$$

This equation (9.12) provides the right-hand side of the fracture criterion (equations (9.8)), or the left-hand side for particular cases of post-yield fractures (see Example Exercise 9.2 below). Notice that J_c is dependent on fracture length (e.g., Cowie and Scholz, 1992b) and the yield stress, just like apparent dynamic fracture toughness was in the previous section. It is thus a material parameter, not a constant material property like K_c. J_c has units of *energy per unit area* (Joules per square meter, or J m^{-2}), so one may consider it to be the critical energy flux in the tip region needed for a post-yield discontinuity to propagate. It also sounds remarkably similar to G_f, the strain energy release rate (per unit crack area) that we discussed in Chapter 8. The units for J imply that it is the **energy flux per unit fracture surface area**; it is *different* than strain energy near a fracture, as described in the following section.

What values should one expect for J_c? From equation (9.9), we can set typical values for K_c of 1–3 Ir (MPa m$^{1/2}$; see Table 9.2), $v = 0.25$, and $E = 50$–100 GPa (keeping units of MPa) and obtain $J_c = 1 \times 10^{-4}$ to 1×10^{-5} J m^{-2}, or 0.1–0.01 mJ m^{-2}. For more realistic cases for discontinuities in rock (i.e., non-LEFM conditions), the apparent dynamic toughness relations shown in Figure 9.14 suggest an increase of J_c by perhaps a value of 2–5. These values of J_c represent the amount of energy expended by the rock mass, and in turn, consumed by the discontinuity, when it propagates (i.e., creates new discontinuity surface area) under LEFM conditions. J_c is also known as the **work done by fracturing** and can include the creation of new fracture surface area, radiated seismic energy, frictional heating, and damage in the process zone (e.g., Li, 1987).

Larger fractures seem to be associated with larger values for J_c. Here's a sampling of approximate values of J_c (typically given as the fracture propagation energy, G_c, which corresponds here to J_c, given the post-yield nature of the systems studied) from the literature for different materials and conditions:

- **Single crystals (intact):** 0.2–12 J m^{-2} (Atkinson and Meredith, 1987a,b);
- **Rocks (intact):** 0.5–10^3 J m^{-2} (Atkinson and Meredith, 1987a,b);
- **Faulted surfaces (in experiments, meter scale):** 10–10^2 J m^{-2} (Okubo and Dieterich, 1984);
- **Faulted joints (in the field, meter scale):** 10^4 J m^{-2} (Martel and Pollard, 1989);
- **Earthquakes:** 10^2 (creeping) –10$^{5-7}$ (stick-slip) J m^{-2} (Li, 1987);
- **Faults:** 10^8 J m^{-2} (Cowie and Scholz, 1992b).

The values depend on lithology (weaker rocks like chalk have small values compared to stronger rocks like granite or sandstone), size, the nature of the surface (intact material vs. pre-existing fracture surfaces), and the *in situ* stress conditions.

Now let's calculate values for J that one would obtain from a particular geologic fracture; then we'll see if it will propagate according to equations (9.8).

Example Exercise 9.2

Using a critical value of J_c of 0.1 mJ m^{-2}, determine whether the following outcrop-scale fractures will be stable or will propagate.

(a) $a = 1.0$ m, $\sigma_y = 100$ MPa, $\sigma_d = 0.35$ MPa, $E = 10$ GPa, and $v = 0.25$.
(b) $a = 1.0$ m, $\sigma_y = 3.0$ MPa, $\sigma_d = 1.35$ MPa, $E = 10$ GPa, and $v = 0.25$.
(c) $a = 14.5$ m, $\sigma_y = 100$ MPa, $\sigma_d = 0.13$ MPa, $E = 10$ GPa, and $v = 0.25$.
(d) $a = 14.5$ m, $\sigma_y = 100$ MPa, $\sigma_d = 0.13$ MPa, $E = 5.5$ GPa, and $v = 0.25$.

Solution

We evaluate the inequalities of equations (9.8) to assess the stability of the fractures against J_c. We'll plug the values into equation (9.12) to calculate the left-hand side of the propagation criterion in (9.8). It's also important to keep the units consistent: meters and MPa throughout, and use the fracture length $2a$ that does not include any end zone (technically this is length $2c$ as shown in Fig. 9.6). We find

(a) 0.038 mJ m^{-2} < 0.1 mJ m^{-2} J/J_c = 38% No propagation
(b) 0.628 mJ m^{-2} > 0.1 mJ m^{-2} J/J_c = 628% *Propagation!*
(c) 0.077 mJ m^{-2} < 0.1 mJ m^{-2} J/J_c = 77% No propagation
(d) 0.140 mJ m^{-2} > 0.1 mJ m^{-2} J/J_c = 140% *Propagation!*

These results show how J might be interpreted. Propagation is difficult when the ratio σ_y/σ_d is large (as in part (a) and part (c)), and facilitated for smaller ratios (part (b) vs. part (a)), longer fractures (part (c) vs. part (a)), and/or more deformable rock masses (smaller modulus, E; part (d) vs. part (c)). The percentages represent the proximity to failure (akin to equation (8.35)) for each example.

Given a fracture with a certain length $L = 2a$ (excluding any end zone), what must happen for it to propagate? If the yield strength σ_y of the end zone rock is held constant, and the rock deformability properties (modulus, Poisson's ratio) also do not change, then your only remaining variable is the **driving stress**, σ_d. How might one increase the driving stress on a fracture to satisfy $J \geq J_c$ and therefore cause it to propagate? There are several ways, including mechanical interaction with other fractures or the Earth's surface (increasing the configuration parameter Y; see Chapter 8), increasing the remote stress or tectonic strain, elevating the internal pore-fluid pressure, decreasing the friction, or rotating the fracture into a different orientation. By using the J-integral approach instead of stress intensity and LEFM, one can now exploit the relationships between yield strength, driving stress, end-zone length, and apparent dynamic fracture toughness to better understand the conditions under which a geologic structural discontinuity might propagate.

9.4.3 Strain Energy Density

According to LEFM, faults cannot propagate in their own planes, but instead must propagate by cutting a new fracture (joint or fault) at some nonzero "take-off" angle to the parent fault's tip (e.g., Pollard and Segall, 1987; Petit and Barquins, 1988). This means that strike-slip faults, in map view, or dip-slip (normal or thrust) faults, in cross-section, cannot propagate in-plane but would resemble patterns such as those depicted schematically in Fig. 9.12.

However, field observations, laboratory experiments, and scale models clearly demonstrate that in-plane propagation actually does occur for faults and deformation bands (e.g., King and Yielding, 1984; Lockner et al., 1991; Du and Aydin, 1993, 1995; Cartwright et al., 1996; Fossen and Gabrielson, 1996; Cartwright and Mansfield, 1998; Morewood and Roberts, 2002; Buiter et al., 2006). Elevated confining pressures (e.g., depth) can inhibit dilatant crack growth and consequently promote in-plane growth of mode-II fractures (Melin, 1986). This effect can also be achieved by an increase of pore volume (compaction or soil-type consolidation) in porous rocks like sandstones (e.g., Aydin and Johnson, 1983) or tuffs (Wilson et al., 2003), promoting in-plane propagation of faults in these materials.

Several investigators have proposed or utilized criteria that would predict in-plane propagation of faults, given the field evidence that this phenomenon really does happen. Lin and Parmentier (1988) and Kemeny (2003) assumed that faults propagate in the direction of maximum shear stress. Reches (1978, 1983) and Wallace and Kemeny (1992) also assumed in-plane growth for normal faults viewed in cross section. Du and Aydin (1993, 1995) showed that distortional strain energy density predicts the propagation paths of strike-slip faults, in map view, into echelon stepovers (i.e., pull-apart basins and push-up ranges) surprisingly well; similar results for thrust faults in cross-section were demonstrated by Okubo and Schultz (2004). Schultz and Balasko (2003) and

Okubo and Schultz (2006) used distortional strain energy density to reproduce linked stepovers (also called "ladder structures" by Davis et al., 2000; see Chapter 7) between echelon deformation bands. Though not without controversy, criteria such as these are finding increased utility and applicability to fault and deformation band propagation studies.

Strain energy density has also been widely used in engineering fracture mechanics to predict the *propagation paths* of cracks having mixed-mode (I–II) displacements (e.g., Sih, 1974, 1991; Broek, 1986, pp. 377–380; Ingraffea, 1987; Lee and Olson, 2017). This criterion is a comparably good predictor of mixed-mode crack propagation paths (i.e., the "take-off" angle of the kink or wing crack) as the maximum tensile stress criterion that is perhaps more commonly used (see Sih, 1974, for discussion).

Strain energy density has units of energy per unit volume —J m⁻³

Strain energy density provides an alternative measure of the response of the rock to stress. Recalling the discussion in Chapter 2 and noting the change in symbol, both stress and strain energy density have the same units, either MPa or $J\ m^{-3}$ (joules per cubic meter). The units suggest immediately that strain energy density is different than what is represented by the *J*-integral (see previous section), which is energy flux per unit fracture area (i.e., $J\ m^{-2}$ (joules per square meter). Timoshenko and Goodier (1970) give a step-by-step derivation and clear discussion of strain energy density (pp. 244–249).

Because strain energy density is calculated from elastic stresses or strains, there is no *provision for failure* (i.e., fault, fracture, or deformation band propagation) within this concept. What we will do is look at the equations for strain energy density and then create a criterion—much like proximity to failure—that shows the predicted increase (or decrease) in strain energy density near a discontinuity tip relative to the strain energy density stored within the surrounding rock (far from the discontinuity) due just to the remote stress state. Strain energy density thus contrasts with a peak-strength failure criterion—like Coulomb or Hoek–Brown (see Chapter 3)—that requires an existing surface or discontinuity for failure to localize upon. Strain energy density, on the other hand, is pre-peak in nature (see the *Complete Stress–Strain Curve for Rock* in Chapter 2 and the discussion of yielding vs. failure in Chapter 7).

Strain energy density for pre-peak deformation, Coulomb or Hoek–Brown for post-peak cracking or frictional sliding

The strain energy density of a deformed material provides a convenient measure of its response to external forces (e.g., remote stresses or displacements). Cracked systems are the classical ones (e.g., Griffith, 1921, 1924; see Lawn, 1993, pp. 20–23; Scholz, 2002, pp. 4–9), although faulted systems have also been investigated (Reches, 1978, 1983; Cooke and Kameda, 2002; Cooke and Madden, 2014). Before splitting the strain energy density up into its two components—volumetric (for mode-I crack growth) and distortional (for mode-II fracture growth)—we first *subtract off the mean stress* to obtain **deviatoric stresses** (see Engelder, 1994, for discussion) for use in the equations near a discontinuity tip (see Jaeger, 1969, pp. 89–95). For example,

Deviatoric stresses from total stresses

$$s_{ii} = \sigma_{ii} - (\sigma_x + \sigma_y + \sigma_z)/3 \qquad (9.13)$$

in which s_{ii} are the deviatoric normal stresses, σ_{ii} are the two (for plane strain) or three total normal stresses, and the mean stress is the sum of the normal stresses (the term in parentheses). The principal stresses (σ_1, σ_2, and σ_3) can be used in (9.13) instead of their *xyz* components. Recall from the Mohr circle that shear stresses are not affected by changes in mean stress, so they are not recalculated here. The deviatoric (normal) stresses would then be denoted in the following equations by s_x (for example), rather than σ_x.

The total strain energy density SED_{total} is written using the Cartesian elastic (total) *stress* components as

Strain energy density from total stresses

$$SED_{\text{total}} = \frac{1}{2E}\left(\sigma_x^2 + \sigma_y^2 + \sigma_z^2\right) - \frac{v}{E}\left(\sigma_x\sigma_x + \sigma_y\sigma_z + \sigma_z\sigma_x\right) + \frac{1}{2G}\left(\tau_{xy}^2 + \tau_{yz}^2 + \tau_{xz}^2\right)$$

(9.14)

Here, E is Young's modulus, G is shear modulus, and v is Poisson's ratio. For 2-D stress states like plane strain, where we don't have a distinct value for the out-of-plane normal stress σ_z, one could use $\sigma_z = v\,(\sigma_x + \sigma_y)$ (see Chapter 2; Timoshenko and Goodier, 1970, p. 17). Written using the elastic *strains*, the total strain energy density is given by

Strain energy density from elastic strains

$$SED_{\text{total}} = \left[\frac{Ev\left(\varepsilon_x + \varepsilon_y + \varepsilon_z\right)}{(1+v)(1-2v)}\right] + G\left(\varepsilon_x^2 + \varepsilon_y^2 + \varepsilon_z^2\right) + \frac{G}{2}\left(\gamma_{xy}^2 + \gamma_{yz}^2 + \gamma_{xz}^2\right) \quad (9.15)$$

where ε_i are the normal strains and γ_{xy}, etc., the shear strains (Jaeger, 1969, p. 90). Either of these expressions (9.14) or (9.15) can be used depending only on which quantities you have available to plug into the right-hand side of the equations. The total strain energy density near a mixed-mode (I-II) LEFM crack tip can also be given as a function of the crack's *stress intensity factors* (Broek, 1986, p. 378).

There are two subclasses of strain energy density that are particularly useful. These both sum to give the total strain energy density (once the mean stress is re-introduced). The first, **volumetric strain energy density**, SED_v, states that discontinuity growth would occur in the direction that minimizes the total strain energy density (Sih, 1974, 1991; Broek, 1986, p. 377), or more clearly for mode-I cracks or dilational tabular discontinuities, either *maximizes the volumetric strain energy density* (that leads to dilation) or minimizes the distortional (i.e., shearing) strain energy. This quantity is given by (Timoshenko and Goodier, 1970, p. 247)

Volumetric strain energy density

$$SED_v = \frac{(1-2v)}{6E}(s_x + s_y + s_z)^2 \qquad (9.16)$$

Volumetric strain energy density also provides a criterion for the propagation of dilation or compaction bands (Okubo and Schultz, 2006) and perhaps stylolites.

The second subclass, which applies to the in-plane propagation of faults (Du and Aydin, 1993, 1995) and shear deformation bands (Schultz and Balasko, 2003; Okubo and Schultz, 2006), is the **distortional strain energy density**, SED_d. Here, discontinuity propagation would occur in the *direction* *that maximizes SED$_d$*; this is equivalent to stating that propagation occurs in areas where the shear stresses are greatest in absolute magnitude. This quantity is given by (Timoshenko and Goodier, 1970, p. 248)

Distortional strain energy density

$$SED_d = \frac{(1+v)}{6E}\left[\left(s_x - s_y\right)^2 + \left(s_y - s_z\right)^2 + \left(s_x - s_z\right)^2\right] + \frac{1}{2G}\left(\tau_{xy}^2 + \tau_{yz}^2 + \tau_{xz}^2\right) \quad (9.17)$$

In using strain energy density as a discontinuity propagation criterion, one can calculate it using either the local (deviatoric) stresses amplified near a discontinuity tip or the remote, far-field (deviatoric) stresses. A useful criterion for fault propagation is obtained by normalizing the distortional strain energy density calculated near the fault tip by that calculated by using the remote stresses (the background value, using deviatoric stresses; ratio > 1.0), so that

To propagate, or not to propagate

$$\left(\frac{SED_d^{\text{fault tip}}}{SED_d^{\text{remote}}}\right) < 1.0 : \text{Fault is stable}$$

$$\left(\frac{SED_d^{\text{fault tip}}}{SED_d^{\text{remote}}}\right) \geq 1.0 : \text{Fault may propagate} \quad (9.18)$$

One can also normalize the strain energy density near the tip of an interacting discontinuity to that for the isolated discontinuity, just as we did for echelon fractures in Fig. 5.7.

Figure 9.16 shows the distortional strain energy density near the tip of a fault (or a deformation band, also shearing in mode II). One can infer that the shearing discontinuity should propagate into the region of concentrated *SED$_d$*, and that **in-plane propagation of the mode-II fault is predicted**, rather than out-of-plane growth of secondary structures. Linkage, or secondary structures, can be predicted if the "bullseye" moves to either side of the discontinuity tip, as is possible in overlapped echelon configurations (e.g., Schultz and Balasko, 2003; Okubo and Schultz, 2006).

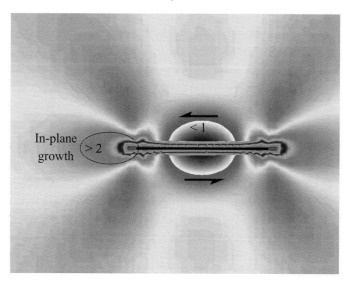

Fig. 9.16. Distortional strain energy density near a mode-II fracture or deformation band tip, showing the tendency for in-plane propagation (normalized following equations (9.18)).

9.5 Displacement–Length Scaling

Numerous studies have demonstrated that the maximum displacement D_{max} along joints, veins, faults with slip surfaces, and deformation bands scales with the horizontal (map or trace) length L (e.g., Muraoka and Kamata, 1983; Walsh and Watterson, 1988; Cowie and Scholz, 1992a; Dawers et al., 1993; Vermilye and Scholz, 1995; Clark and Cox, 1996; Schlische et al., 1996; Fossen and Hesthammer, 1997, 1998; Renshaw, 1997; Gudmundsson, 2000; Crider and Peacock, 2004; Kim and Sanderson, 2005; Xu et al., 2005; Schultz et al., 2006, 2008a, 2013; Fossen and Cavalcante, 2017). This scaling relationship is associated with the well-known proportionality between seismic moment and fault (or rupture) length (e.g., Scholz, 1982; Wells and Coppersmith, 1994). Displacements along a structural discontinuity increase from minimum values—close to zero—at the discontinuity tiplines to a maximum value near the discontinuity's midpoint. The gradient in displacement along the discontinuity is related to the strength and deformability of the rock mass immediately adjacent to the discontinuity tipline (Cowie and Scholz, 1992b; Schultz and Fossen, 2002). Smaller displacements are associated with larger values of modulus (either Young's or shear) and/or smaller strength (tensile or shear) of the surrounding rock mass for given values of driving stress.

Many attributes such as displacement magnitude and distribution, damage zone development and width, and slip rate vary systematically along a structural discontinuity

Gradients in slip (or shear offset) along a fault's trace (map) length have become well documented (discussed e.g. by Segall and Pollard, 1980; Muraoka and Kamata, 1983; Barnett et al., 1987; Walsh and Watterson, 1989; and Scholz, 1990). More detailed studies have focused on the sequence of fault slip, interaction, propagation, and linkage from these distributions (e.g., Peacock and Sanderson, 1991; Dawers et al., 1993; Davison, 1994; Cartwright et al., 1995; Willemse, 1997; Cartwright and Mansfield, 1998; Cowie and Shipton, 1998). As a result, a basic understanding of the mechanics of discontinuity segmentation and linkage has been established for a variety of simple configurations. One challenge in scaling analysis is to determine and quantify each of the responsible factors and their impact on the scaling relations.

Displacement magnitude varies along a fault

In this section we explore the displacement–length scaling relations for individual structural discontinuities from a mechanics perspective. These structures are either located sufficiently far away from nearby heterogeneities (such as other structures; the free surface; or stratigraphic, lithologic, or rheologic variations) that local stress perturbations on them are small relative to the average values of material properties and remote stress state. As structural discontinuities grow, they increasingly interact with their neighbors and with physical property changes (such as those noted above), leading to departures from the idealized relationships described in this section. These departures illuminate the rich diversity and complexity of processes by which faults, deformation bands, and other types of structural discontinuities grow and develop as part of natural geologic deformation that are so important to academic and applied geoscience. These applications or processes include active tectonics and seismic hazard (e.g., Main, 1995; Cowie and Roberts, 2001; Roberts et al., 2004; Soliva et al., 2008), fold-fault relationships (e.g., Moody and Hill, 1956; Suppe and Medwedeff, 1990; Schlische, 1995; McClay, 2011; Bernal et al., 2018), fault growth in sedimentary and petroleum basins (e.g., Morley, 1999; McLeod et al., 2000), and asynchronous growth and attendant scaling relations (e.g., Walsh et al., 2002; Nicol et al., 2016).

Identifying such departures however hinges on identifying the fundamental scaling relationships, without which discussions of displacement gradients near fault tips (related to mechanical interaction with nearby stress heterogeneities) and growth sequences (such as displacement increase on a reactivated or stratigraphically restricted fault) may not be very informative. Other useful and informative approaches to studying fault scaling include phenomenological (e.g., see Childs et al., 2017, for an overview), analog (clay or sandbox) models (e.g., McClay et al., 2005; Withjack et al., 2007), and purely statistical approaches (e.g., Cowie et al., 1995; Bonnet et al., 2001; Kolyukhin and Torabi, 2012); some of the key elements of these approaches will be touched on as needed but not emphasized in this section.

Scaling relations for individual structural discontinuities provide context for interpreting more complex geometries and systems

The **displacement distribution** along such an individual structural discontinuity is influenced by several physical factors, including the following.

- **Discontinuity Geometry:**
  - Trace *length* in map or (less commonly) cross-sectional view (e.g., Segall and Pollard, 1980; Muraoka and Kamata, 1983; Cowie and Scholz, 1992a).
  - Discontinuity *aspect ratio* (trace-length/down-dip height; Nicol et al., 1996; Willemse et al., 1996; Gudmundsson, 2000; Schultz and Fossen, 2002; Soliva et al., 2005).
  - Discontinuity *shape* (rectangular vs. elliptical: Sih, 1973; Scholz, 1982; Barnett et al., 1987; Willemse et al., 1996; Martel and Boger, 1998; Willemse and Pollard, 2000; Schultz and Fossen, 2002; Benedicto et al., 2003; Savalli and Engelder, 2005; Soliva et al., 2005).
- **Rock Strength at the Discontinuity Tip:**
  - Development, type, and extent of an *end zone* (Palmer and Rice, 1973; Li, 1987; Cowie and Scholz, 1992b; Rubin, 1993a; Cooke, 1997; Martel, 1997, 1990).
- **Remote Stress State:**
  - Configuration of the *far-field stresses* (Bürgmann et al., 1994; Crider and Pollard, 1998; Kattenhorn and Pollard, 1999; Schultz et al., 2006).
- **Local Stress State:**
  - "*Soft linkage*" with other discontinuities (e.g., Segall and Pollard, 1980; Peacock and Sanderson, 1991, 1995; Vermilye and Scholz, 1995; Willemse, 1997; Willemse and Pollard, 2000).
  - "*Hard linkage*" of discontinuity segments (Davison, 1994; Dawers and Anders, 1995; Willemse and Pollard, 2000).
  - *Proximity* to the free surface and other boundaries (e.g., Pollard and Holzhausen, 1979; Pollard and Aydin, 1988; Bruhn and Schultz, 1996; Crider and Pollard, 1998; Soliva and Schultz, 2008).
- **Rock Rheology and Large-Strain Kinematics:**
  - *Rheologic* (elastic, viscous) properties and variations in *lithology* displaced by the discontinuity (Martel, 1990; Scholz et al., 1993; Bürgmann et al., 1994; Rubin, 1995b; Schultz et al., 2006).

– Frictional, constitutive, and rate-and-state *properties* of the discontinuity (Lin and Parmentier, 1988; Aydin and Schultz, 1990; Bürgmann et al., 1994; Cooke, 1997; Martel, 1997).

– Inter-fracture *plate deformation* (King and Ellis, 1990; Gross et al., 1997; Moore and Schultz, 1999; Walsh and Schultz-Ela, 2003).

– *Time-dependent* rheologies of the host rock (Tse and Rice, 1986; Freed and Lin, 1998).

Because the displacement distributions that can be observed and measured on structural discontinuities, as well as the related displacement–length scaling relations, depend on one or more of the above factors, a systematic investigation of the likely factors, for each field example, is necessary for an understanding of discontinuity displacement profiles to be achieved.

Both displacement and slip rate vary along a fault

Cowie and Roberts (2001) demonstrated that **slip rate** along a segmented fault trace must vary with along-strike position, just as the displacement magnitude does. This fundamental insight, later supported by detailed studies (e.g., Roberts et al., 2004; Nicol et al., 2005), helps to reconcile different values of slip rate measured at various and often arbitrary positions along a fault with values obtained by methods such as GPS, trenching, geologic offsets, and geomorphology. As a result, the discussion of displacement profiles in this chapter also hints at how *displacement-rate profiles* for certain types of structural discontinuities grow and evolve.

Displacement–length scaling relations have been used extensively in many areas of rock mechanics, structural geology, and tectonics. Some of the more important or common of these include the following.

• The calculation of **tectonic strains** due to faulting (e.g., Scholz and Cowie, 1990; Cowie et al., 1993a; Scholz, 1997; Watters et al., 1998; Schultz, 2000a; Polit et al., 2009; Schultz et al., 2010c; Nahm and Schultz, 2011; Klimczak et al., 2015).

• Studies of **fault growth** from individual seismic slip events through linked fault arrays (e.g., Cowie et al., 1993b; Cartwright et al., 1995; Cladouhos and Marrett, 1996; Cowie, 1998a,b; Cowie and Shipton, 1998; Shaw et al., 2002; Nicol et al., 2005; Soliva and Schultz, 2008; Bergen and Shaw, 2010; Soliva et al., 2010; Kolyukhin and Torabi, 2012; Hughes and Shaw, 2014; Childs et al., 2017).

• Characterization of **background seismicity levels** and associated fault inventories for assessing the magnitude of induced seismicity potentially caused by wastewater injection, carbon dioxide sequestration, or similar subsurface activities (e.g., Zoback and Gorelick, 2012; Ellsworth, 2013; Petersen et al., 2015).

• Estimates of **mechanical interaction** between soft-linked fault strands (e.g., Dawers et al., 1993; Dawers and Anders, 1995; Gupta and Scholz, 2000a,b; Soliva et al., 2006, 2008).

• As a probe of **stratigraphic and rock properties** (e.g., Dawers et al., 1993; Bürgmann et al., 1994; Schultz and Fori, 1996; Cooke, 1997; Gross et al., 1997; Martel, 1997; Schultz, 1997, 2000a,b; Gudmundsson, 2000; Watters et al., 2000; Schultz and Fossen, 2002; Soliva et al., 2005; Schultz et al., 2006, 2008a).

- Bounding **values for fault lengths and geometries** such as linkage from displacements measured from seismic or other indirect methods (e.g., Barnett et al., 1987; Walsh and Watterson, 1991; Cartwright et al., 2003; Baudon and Cartwright, 2008; Bergen and Shaw, 2010).

This material provides some context for understanding how the outcrop patterns of faults, for example, can reveal growth histories (e.g., Crider and Peacock, 2004; Childs et al., 2017) and insights into stratigraphic properties and tectonics of fault-related deformation in three dimensions (e.g., Scholz, 1997; Soliva et al., 2005).

Geologic structural discontinuities are surfaces of finite extent (i.e., they terminate in the rock at their edges or "tiplines"). Most simply, the **magnitude of displacement** (or offset of stratigraphic units and formerly contiguous markers) accommodated along a fault or other type of structural discontinuity depends on (e.g., Pollard and Segall, 1987; Bürgmann et al., 1994; Wibberley et al., 1999, 2000; Gudmundsson, 2004):

- How much **stress** is applied (the "driving stress"),
- The **deformability** of the surroundings (given by Young's modulus, shear modulus, and Poisson's ratio),
- The **strength** of rock at the discontinuity's tipline,
- The **degree of interaction and linkage** of discontinuity segments, and
- The **size** of the discontinuity (given by its length L and height H).

Longer discontinuities can accommodate greater magnitudes of displacement along them, all other factors being equal (e.g., Watterson, 1986; Walsh and Watterson, 1988). This simple yet profound relationship tells us directly that a plot of the maximum displacement on a discontinuity (D_{max}) vs. its length L would show a strong positive correlation (e.g., Childs et al., 2017) and a positive constant of proportionality. The relationships revealed on such plots can also provide considerable physical insight into the mechanisms of strain localization and displacement accumulation in rock units deformed by brittle (and semi-brittle) structures. As we'll see, much of this hinges on how the discontinuities propagated as they grew in length and displacement. More specifically, these studies shed light on the interplay between near-tip stress concentration and rock strength.

Displacement-length scaling is not just for specialists but is useful for general *field mapping* as well. Let's suppose you're working in an area and you identify a partial exposure of faulted rocks. The stratigraphic offset there is found to be 1.8 m, but how long is the fault? How far should you try to walk it out, or should you just draw a line on your map and query or dot-in the unexposed ends? The D–L scaling relations can give you a general idea. Using the range for D_{max}/L in the literature (and in Fig. 1.19), $10^{-1} < D_{max}/L < 10^{-3}$, one could say that the fault should have a length of approximately 18–1800 m. The more astute geologist will also note that the fault cuts poorly cemented sedimentary rocks

*D–L scaling is now
a standard tool in
the arsenal of the
contemporary
structural geologist*

(sandstones and limestones with only weak diagenesis) that have a smaller value for Young's modulus than, say, basalt or granite. Using the relation given in Chapter 8 for the maximum displacement on a fracture (equation (8.10)), on might infer a value for fault length near the lower (shorter) end of the range (since a softer, more compliant rock will permit greater displacements along the fault; e.g., Wibberley et al., 1999). The field geologist will suspect that the fault has a map length of several tens of meters (say, 20–100 m) but not something in the kilometer range. This estimate may help to suggest other localities along-strike to investigate and thereby discover the true outcrop extent of the fault.

The next day, suppose our field geologist notices that a fault 1.5 km long has an offset of 570 m (D_{max} /L = 0.38). After locating one of the fault tips, the astute geologist identifies steep displacement gradients along with folding and pressure solution cleavage adjacent to the fault tip. Perhaps this might be a "eureka" moment. The fault may then be recognized as a *segment* in a longer, kinematically coherent echelon fault array, so the D_{max} /L ratio for the individual fault is increased over the more typical values (for widely separated, individual faults) given above. The segment behaves as though it is *longer* than it actually is (because of the influence of the nearby fault strands), leading to a larger value of displacement accommodated along it. The larger displacements occur as the fault segment interacts mechanically (see Chapter 7) with its closely spaced neighbors (Segall and Pollard, 1980) in the early stages of linkage and kinematic coherence (Willemse, 1997).

9.5.1 Two Basic Scaling Relationships

D–L scaling can be understood by using elastic–plastic fracture mechanics models (Dugdale, 1960; Barenblatt, 1962; Bilby et al., 1963; Heald et al., 1972; Turner, 1979; Cowie and Scholz, 1992b; Rubin, 1993a) that incorporate a specific yield strength at the discontinuity tips (Cooke, 1997; Martel, 1997; Schultz and Fossen, 2002; Yao, 2012). The yield strength of the end zone serves to regulate the compatibility between (large) displacements on the structural discontinuity and (negligible) displacements in its surroundings.

D–L scaling of structural discontinuities depends on the displacement magnitude and distribution that results from the remote stress state and the dimensions and boundary conditions of the discontinuity (in addition to the rock stiffness and any mechanical interactions). As has been noted previously, the degree of stress concentration at a fracture or deformation band tip is related to the ratio of the host-rock yield strength σ_y to the driving stress σ_d on the structural discontinuity. Referring to equations (8.10), (8.11), and (8.14), a 2-D scaling relation for either LEFM or post-yield structural discontinuities can be obtained. Setting $\theta = 0°$ and $\Omega = 1$ in equation (8.14), a more general relation for 2-D, tall discontinuities with length measured along the horizontal dimension becomes

$$D_{max} = \frac{2\left(1 - v^2\right)}{E}\left(\sigma_d - C\sigma_y\right)L \qquad (9.19)$$

in which E is Young's modulus, v is Poisson's ratio (both for the host rock), σ_d is the driving stress over the fully yielded discontinuity surface, σ_y is the host-rock yield strength, and C is a variable or function that specifies how the theoretical stress singularity at the discontinuity tip can be removed (Bürgmann et al., 1994; Schultz et al., 2006).

Equation (9.19) has the form

D–L scaling for 2-D structural discontinuities

$$D_{max} = \gamma L^n \qquad (9.20)$$

in which D_{max} is the maximum value of shear displacement along a discontinuity of horizontal (or map-view) length L, and γ is the proportionality constant, or *scaling factor*, between maximum displacement and length (e.g., Cowie and Scholz, 1992b). It is also considered to be a *characteristic shear strain* (Watterson, 1986) which must be attained at a fault tip for propagation to occur (Cowie and Scholz, 1992b; Scholz, 1997) under conditions of constant remote stress. The exponent n describes the rate of increase of the maximum displacement as discontinuity length increases. This scaling relationship is often compared to measurements taken from geologic structural discontinuities graphically, as in Figs. 1.19 and 9.19 (below).

The dependence of D_{max}/L scaling on the ratio of yield strength to driving stress, obtained by using equation (9.19), is shown in Fig. 9.17. LEFM conditions are implied for σ_y/σ_d greater than about 50, with **brittle conditions** associated with ratios greater than about 10, corresponding canonically to end-zone lengths less than about 10% of the fracture half-length a. Larger end zones, and smaller ratios, would imply **semi-brittle conditions**, with perfect plasticity (and no strain localization into discrete structural discontinuities such as fractures) would be implied by ratios of 1.0. Thus, the ratio of rock yield strength to driving stress informs the class of propagation criterion to be considered, which then drives the form of the displacement–length scaling relation.

Perfectly brittle, LEFM deformation:
$\sigma_y/\sigma_d >\sim 50$

Semi-brittle, post-yield deformation:
$\sigma_y/\sigma_d <\sim 10$

For the case of semi-brittle, post-yield structural discontinuities, C in (9.19) may be calculated by using

$$C = 1 - \cos\left(\frac{\pi}{2}\frac{\sigma_d}{\sigma_y}\right) \qquad (9.21a)$$

for the end-zone model (Schultz and Fossen, 2002), or

$$C = 1/\pi \qquad (9.21b)$$

for the linear displacement model (Bürgmann et al., 1994). The cosine term in equation (9.21a) defines the length of the end zone (s/a in eq. (7b) of Schultz and Fossen, 2002) adjoining the fully slipped central part of the fault. C in equation (9.21b) is obtained by setting $x/a = 0$ and $L = 2a$ in eq. (14) of Bürgmann et al. (1994), with their quantities S_r and S_g being interpreted as σ_d and σ_y, respectively (S. Martel, personal communication, 2004). The LEFM solution, with its inherent singularity in near-tip stress, constant value

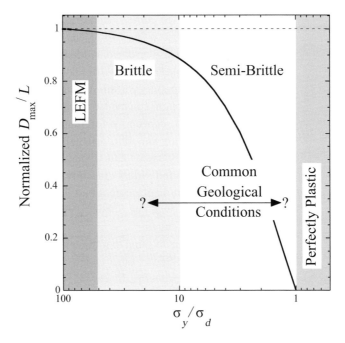

Fig. 9.17. Dependence of displacement–length ratio on the ratio of yield strength to driving stress (D_{max}/L is normalized by the value for $\sigma_y/\sigma_d = 100$; equation (9.19) is plotted.

of driving stress (or stress drop), and the associated elliptical displacement profile, is recovered by setting $C = 0$ in equations (9.21). Equations (9.21) are comparable with that obtained by Scholz (1997) in his discussion of end-zone models and D–L scaling relations. This figure shows that the maximum displacement is greatest for perfectly brittle conditions (LEFM) and that geologic structural discontinuities having $\sigma_y/\sigma_d < 10$ should have much smaller values per unit length, assuming that changes in displacement and length keep pace with each other (see also Schultz and Fossen, 2002).

Equation (9.19) implies that a yield strength of some type can replace the dynamic fracture toughness, under certain conditions, as the limiting parameter for propagation. If instead, C is a constant (i.e. $1/\pi$), then a nearly triangular displacement distribution is produced (Bürgmann et al., 1994). Thus, the displacement–length scaling relations depend upon the details of displacement distribution, the relative size of the end zone, and on the propagation criterion. Differences in the magnitudes of shear, tensile, or contractional yield strengths may also imply differences in the scaling relations between different types of geologic structural discontinuities.

The two physical D–L scaling relations (for 2-D cases of plane deformation) are first summarized and illustrated in Fig. 9.18 and then discussed in more detail in the following sections. Following the seminal work in the literature (e.g., Cowie and Scholz, 1992b; Olson, 2003), two basic controls on the propagation of geologic structural discontinuities are considered (see Fig. 9.5).

- Propagation under constant dynamic fracture toughness conditions; and
- Propagation under constant driving stress conditions.

The first of these conditions requires LEFM conditions, so that the tips of geologic structural discontinuities are characterized by small-scale yielding and indefinitely large stress concentrations (i.e., singular stress). A consequence of propagation under LEFM conditions is a sublinear scaling relation between maximum displacement and length, with the scaling exponent in the range of 0.5 (Scholz, 2002, p. 116; Olson, 2003). As previously noted, this case approximates perfectly brittle fracture.

The second case of constant driving stress conditions has been more widely recognized, especially for faults. In this case the near-tip stress concentration is not singular, so that LEFM does not apply, implying that stress intensity factor and dynamic fracture toughness don't either. This approach for individual structures predicts a linear scaling relation between maximum displacement and length, with the scaling exponent in the range of 1.0 (Cowie and Scholz, 1992b). This case applies to semi-brittle, post-yield structures.

It should be borne in mind that these basic scenarios can be, and likely are in many cases, modified by other factors that can alter the scaling relationships for certain structures over particular length-scales and/or time scales. Some of these modifying factors include variations in the rate of delivery of pore-fluid pressure to the structures, which will affect the magnitude of the driving stress; the remote loading conditions (fixed-grips or constant stress) (Olson, 2003); and changes in the mechanism of nucleation of the discontinuity (e.g., tensile fracture, deformation band) as a function of lithology and/or as it grows in size (e.g. Hatton et al., 1993; Wilkins and Gross, 2002; Crider and Peacock,

Fig. 9.18. Derivations of displacement–length scaling relations for propagation under conditions of (a) constant dynamic fracture toughness, and (b) constant remote stress loading conditions.

	D-L scaling for constant dynamic fracture toughness conditions (LEFM case)	*D-L* scaling for constant remote stress conditions (non-LEFM case)	
LEFM relationship	$D_{max} = \Delta\sigma \dfrac{2(1-v^2)}{E} L$	$D_{max} = \Delta\sigma \dfrac{2(1-v^2)}{E} L$	LEFM relationship
Propagation criterion for LEFM	$K_c = K_i = \Delta\sigma\sqrt{\pi a}$		Define driving stress for non-LEFM
Define driving stress for LEFM	$\Delta\sigma = \dfrac{K_i}{\sqrt{\pi a}}$	$\Delta\sigma = (\sigma_d - C\sigma_y)N$	
	$D_{max} = \dfrac{K_i}{\sqrt{\pi a}} \dfrac{2(1-v^2)}{E} L$	$D_{max} = (\sigma_d - C\sigma_y)N\dfrac{2(1-v^2)}{E}L$	**Scaling relation:** $\boldsymbol{D_{max} \propto L^1}$
Combine fracture length variables	$L = 2a$		
Collect terms	$D_{max} = \dfrac{K_i}{\sqrt{\pi a}} \dfrac{2(1-v^2)}{E} 2a$		
Scaling relation: $\boldsymbol{D_{max} \propto L^{1/2}}$	$D_{max} = K_i \dfrac{2(1-v^2)}{E} \dfrac{\sqrt{8}}{\sqrt{\pi}} \sqrt{L}$		

2004). The two basic physical scaling relationships may be thought of as an average or equilibrium about which excursions related to the many physical or observational factors described in this chapter and in the literature can occur.

Derivations of the two basic scaling relationships are given in Fig. 9.18. The side-by-side comparison shows how important it is to specify the propagation criterion; i.e., the dynamic fracture toughness (left-hand columns) or the host-rock yield strength (right-hand columns). Table 9.2 listed several approaches that can help inform which of these two scaling relations might best apply to a given geologic situation.

9.5.1.1 Perfectly Brittle LEFM Deformation

The relationship for propagation under conditions of constant (dynamic) fracture toughness, K_c, is derived first (left-hand columns on Fig. 9.18). This case assumes LEFM conditions for the structural discontinuity with displacements in any sense (i.e. opening or shear). Although one might have expected that this form of scaling relationship had been derived and utilized much earlier than it had, given its direct relevance to LEFM problems, it postdates the one applied to conditions of constant remote stress discussed below. Perhaps the delay could be related to the surge in fault studies, including displacement–length scaling, from the mid-1980s (e.g., Watterson, 1986; Barnett et al., 1987; Walsh and Watterson, 1988) that implicitly relied on fault propagation under conditions of constant remote stress.

Starting with the basic equation for maximum displacement along a fracture (equation (8.14)), the condition for propagation is given (equations (8.21)) and substituted, resulting in the final expression and a displacement-scaling relation having an exponent of 0.5 (i.e., $\sqrt{L}$).

For cases of opening-mode structural discontinuities such as cracks, joints, hydraulic fractures, veins, or igneous dikes, the displacement D_{max} is understood to correspond to the *kinematic* opening displacement (i.e. the mechanical aperture or physical separation of rock on either side of the fracture), rather than to a hydraulic aperture. The opening displacement relevant to this equation is also that which is not reduced by cement precipitation during diagenesis (e.g., Laubach et al., 2004; Olson et al., 2007). Consideration of these additional chemical and fluid effects, along with related changes in the stiffness and hydrologic properties of the host rock (since cement precipitation can occur not just in the fractures, but in the host rock; e.g. Lander and Laubach, 2015) would modify the scaling relations but likely not invalidate them from the point of view of fracture mechanics.

The case of structural discontinuities propagating under conditions of constant dynamic fracture toughness would imply that D_{max} would be proportional to the square root of length (solid curve in Fig. 9.5). This has been called **sublinear scaling** in the literature (e.g., Olson, 2003) and it applies regardless of whether the dynamic fracture toughness value used in the propagation criterion is critical or subcritical (e.g., Olson, 2003, 2004; Olson

Sublinear displacement–length scaling

et al., 2007) because the form of the final equation is unchanged by the value of this parameter. The D/L ratio for LEFM conditions depends on the host-rock properties, dynamic fracture toughness, and the square root of fracture length.

9.5.1.2 Semi-Brittle Post-Yield Deformation

The case of propagation of shearing structural discontinuities under conditions of constant remote stress, such as faults, is illustrated in the right-hand columns of Fig. 9.18. Starting with equation (9.19), the term for effective driving stress $\Delta\sigma$ (in this case, for a fault) is expanded following Cowie and Scholz (1992b) and Schultz et al. (2006, 2008a) so that σ_d is the shear driving stress, which is understood to include pore-fluid pressure (Cowie and Scholz, 1992b; Gupta and Scholz, 2000a; Schultz, 2003b), σ_y is the yield strength of rock, in shear, at the fault tip, N is the ratio of geologic offset to short-term slip, and C is a variable or function that specifies how the theoretical stress singularity at the fault tip is removed (Bürgmann et al., 1994; Schultz et al., 2006).

Sublinear displacement–length scaling

The D/L ratio depends explicitly (and **linearly**) on the driving stress, rock properties, and yield strength, with the scaling coefficient γ (in equation (9.20)) being equal to the right-hand side of the last equation on the right-hand side of Fig. 9.18. This expression can also be used for other types of non-LEFM, post-yield structural discontinuities by choosing appropriate values for N and C with driving stress appropriate to the structure of interest.

The driving stress term σ_d represents the difference between the remote shear stress resolved along (in this example) a fault before and after slippage (Fig. 9.6; Pollard and Segall, 1987, p. 290; Martel and Pollard 1989; Cooke, 1997; Wilkins and Schultz, 2005). As demonstrated in seismological studies of earthquake faults, the average stress drop (or shear driving stress) is relatively constant on faults in a wide range of rock types and tectonic settings (e.g. Scholz, 2002, p. 205), implying that these structures accumulate offsets and propagate under conditions of approximately constant (shear) driving stress (e.g. Cowie and Scholz, 1992b; Scholz, 1997). In particular, the shear driving stress acting on faults is related to the difference between the maximum ("static") friction μ_0 and the long-term, steady-state value μ_{ss} (with $\mu_0 > \mu_{ss}$; Marone, 1998a,b; Paterson and Wong, 2005, p. 261) times a constant N that, in a general way, scales individual slip events to cumulative geologic offset along the fault (Cowie and Scholz, 1992b; Wells and Coppersmith, 1994; Schultz, 2003b; Schultz et al., 2006).

Cowie and Scholz (1992b) used such a cumulative driving stress with a magnitude of 250 MPa (along with typical values of modulus and yield strength) in their models of fault scaling; corresponding values of N on the order of 10^2 appear to capture this aspect of the scaling changes between earthquakes and faults reasonably well (Gupta and Scholz, 2000a; Schultz, 2003b). For example, many compilations and site-specific studies in the literature suggest that the magnitude of D_{max} is commonly between one-tenth (1/10) and one-thousandth

(1/1000) the map length of an individual fault (e.g., Cowie and Scholz, 1992a; Clark and Cox, 1996; Schlische et al., 1996; Kim and Sanderson, 2005; Xu et al., 2005; Schultz et al., 2006, 2008a, 2013). In other words, D_{max}/L is commonly between $10^{-1} < \gamma < 10^{-3}$, corresponding to critical shear strains of host rock at propagating individual faults of approximately 0.1–10%. These values of D_{max}/L are about 100 times larger than those that relate the incremental seismic slip displacements to the length of the earthquake rupture (e.g., Wells and Coppersmith, 1994; Scholz, 1997; 10^{-4} to 10^{-5}), implying that some large number of seismic events (such as hundreds) may be associated with cumulative displacements on a geologic fault in the range noted in the literature.

Earthquakes:
$D_{max}/L = 10^{-4} < \gamma < 10^{-5}$

Faults:
$D_{max}/L = 10^{-1} < \gamma < 10^{-3}$

Values of maximum friction coefficient for rocks are typically in the range of 0.2–0.8 (e.g. Paterson and Wong, 2005, p. 167) and generally increase with decreasing clay content of the host rock (e.g. Meng et al., 2016, their figure 3). The steady-state friction value represents the residual frictional strength of a fault that is achieved over large displacements (e.g. near a fault's midpoint) and times, independent of the choice of rate-and-state formulations (Marone, 1998a,b; Paterson and Wong, 2005, p. 261). Values of steady-state friction depend on the time-dependence of friction and its evolution with sliding velocity and post-slip healing (e.g. Marone, 1998a,b).

For the case of constant remote stress, the "end-zone" model (e.g. Cowie and Scholz, 1992b; Bürgmann et al., 1994; Cooke, 1997; Martel, 1997; Willemse and Pollard, 1998; Schultz and Fossen, 2002; Wilkins and Schultz, 2005) and the "symmetric linear stress distribution" model (Bürgmann et al., 1994) can be applied. Such models are generally consistent with linear D–L scaling (Scholz, 2002, p. 116), finite rock (yield) strength at the tipline (Cowie and Scholz, 1992b; Scholz, 1997), and gentle near-tip displacement gradients (e.g. Cowie and Scholz, 1992b; Bürgmann et al., 1994; Moore and Schultz, 1999; Cooke, 1997; Cowie and Shipton, 1998). The "end-zone" model can apply to either an earthquake rupture (Wilkins and Schultz, 2005) or an individual fault having a central well-slipped portion bounded by frictionally stronger end zones. The continuously varying strength model assumes a linear increase in fault frictional strength, from the center to the tip (Bürgmann et al., 1994), predicting a nearly linear (or triangular) displacement distribution as has been demonstrated by measurements for many faults (e.g. Dawers et al., 1993; Cowie and Shipton, 1998; Manighetti et al., 2001; Soliva and Benedicto, 2004). Both of these approaches imply a non-constant driving stress, or stress drop, along a fault, in contrast to the constant value of driving stress (or stress drop) defined for LEFM fractures. They also specify a yield strength on the same order (e.g. MPa) as the driving stress, considerably smaller than that implicitly assumed in LEFM models of faults (e.g. Pollard and Segall, 1987). The particular choice of post-yield model is thus not critical to the scaling relations discussed in this chapter, and there may indeed be an evolution through these end-members from faults with end zones to faults with more continuously varying frictional strength with increasing displacement.

The post-yield models have several important and useful properties. The effective driving stress (or stress drop) on the fault ($\sigma_d - C\sigma_y$) is independent of fault length. Because that equation does not contain a term for the rock's dynamic fracture toughness, the propagation of faults and other structural discontinuities that are not characterized by singular near-tip stress states (i.e. non-LEFM conditions) can occur when the yield strength σ_y (not a dynamic fracture toughness) is exceeded at the discontinuity's tip (e.g. Bürgmann et al., 1994; Scholz and Lawler, 2004). As a result, this class of models provides a physical basis for D–L scaling relations of the form $D = \gamma L$ (e.g. Cowie and Scholz, 1992a; Clark and Cox, 1996; Schultz and Fossen, 2002; Scholz, 2002, p. 116; dashed curve in Fig. 9.5).

9.5.2 *D–L* Scaling for the Major Classes of Geologic Structural Discontinuities

Several representative datasets illustrate the ranges and characteristics of D–L scaling relationships for eight classes of geologic structural discontinuities. The datasets plotted in Fig. 9.17 include:

- 16 datasets for faults (normal, strike-slip, and/or thrust) (Schultz et al., 2008a; Polit et al., 2009);
- 10 datasets for dilatant fractures (Klimczak et al., 2010; Schultz et al., 2010a,b,c);
- 7 datasets for deformation bands, including:
  - 2 datasets for cataclastic shear deformation bands (Fossen and Hesthammer, 1997; Wibberley et al., 2000);
  - 3 datasets for disaggregation bands, in which no cataclasis or volumetric strain is apparent (Wibberley et al., 1999; Fossen, 2010a,b; Exner and Grassemann, 2010); and
  - 2 datasets for shear-enhanced compaction bands (Sternlof et al., 2005; Schultz, 2009);
- 3 descriptions of pure compaction bands (Mollema and Antonellini, 1996; Fossen et al., 2011; Liu et al., 2016);
- 1 dataset for stylolites (Benedicto and Schultz, 2009); and
- 2 datasets for shear zones (Ramsay and Allison, 1979; Pennacchioni, 2005).

All datasets are plotted in the figures using the same log–log scale to facilitate comparison. The scaling relations for geologic structural discontinuities define several separate groups (e.g., Schultz et al., 2008a, 2013; Schultz and Soliva 2012).

The first group has a power-law slope of approximately $n = 1$ and therefore a linear dependence of maximum displacement and discontinuity length ($D_{max} = \gamma L$). Structural discontinuities that scale in this fashion include individual faults (Fig. 9.19a) and isochoric disaggregation deformation bands (Fig.9.19f). By implication, faults and disaggregation bands propagate when a shear yield strength is exceeded at their tips and because singular stress magnitudes may not generally be produced or sustained during the growth of these structures.

Linear displacement–length

- ♦ *Scaling (n = 1.0): Faults*
- ♦ *Shear (disaggregation) deformation bands*

Fig. 9.19a. Compilation of displacement–length data for faults. Datasets: NF, normal faults: Solite quarry, Schlische et al. (1996); various, Walsh and Watterson (1988); Utah, Krantz (1988); UK, Peacock and Sanderson (1991); minor faults, Muraoka and Kamata (1983); Iceland, Opheim and Gudmundsson (1989); Bishop Tuff, Dawers and Anders (1995); Utah, Cartwright et al. (1995); Iceland, Gudmundsson and Bäckström (1991); Italy, Roberts et al. (2004); Central Greece, Roberts (2007); France, SSF, strike-slip faults: Villemin et al. (1995); UK, Peacock and Sanderson (1991). TF, thrust faults: British Columbia, Elliott (1976); Puentes Hills, Shaw et al. (2002); New Zealand, Davis et al. (2005).

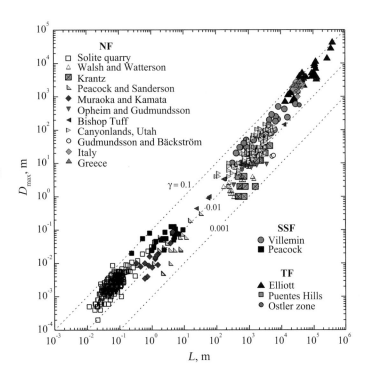

Fig. 9.19b. Compilation of displacement–length data for shear zones. Datasets: Switzerland, Ramsay and Allison (1979); Italy, Pennacchioni (2005).

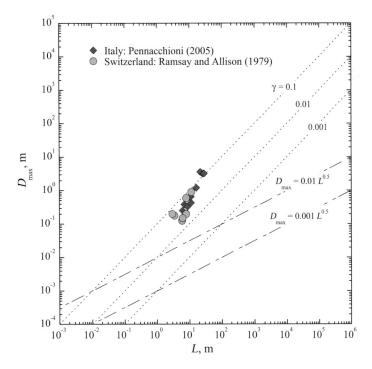

Fig. 9.19c. Compilation of displacement–length data for dilatant fractures. Datasets: Joints: Moros (1999); Candor, Schultz et al. (2010a). Veins: Florence Lake, Vermilye and Scholz (1995); Culpeper, Vermilye and Scholz (1995); Lodève, de Jossineau et al. (2005); Emerald Bay, Klimczak et al. (2010). Dikes: Ship Rock, Delaney and Pollard (1981); Donner Lake, Klimczak et al. (2010); Emerald Bay, Klimczak et al. (2010); Ethiopia, Schultz et al. (2008b).

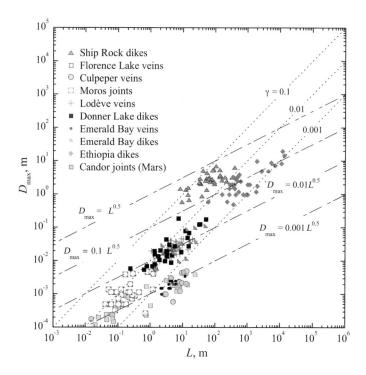

Fig. 9.19d. Compilation of displacement–length data for cataclastic compactional shear bands. Datasets: Utah, Fossen and Hesthammer (1997); Provence, Wibberley et al. (1999).

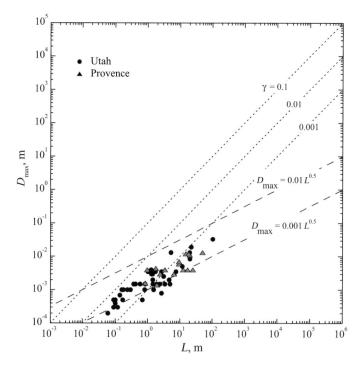

Fig. 9.19e. Compilation of displacement–length data for shear-enhanced compaction bands. Datasets: Valley of Fire, Sternlof et al. (2005); Buckskin Gulch, Schultz (2009).

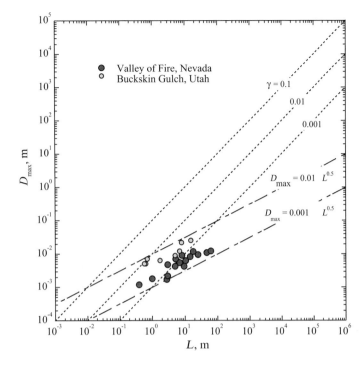

Fig. 9.19f. Compilation of displacement–length data for disaggregation bands. Datasets: Austria, Exner and Grassemann (2010); Arches, Fossen (2010a,b); Idni, Wibberley et al. (1999).

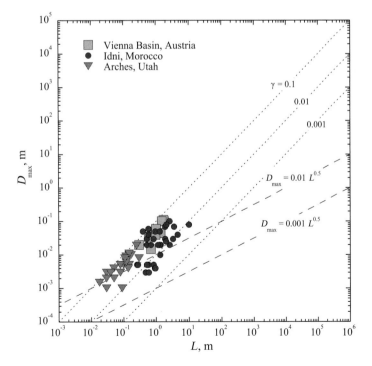

Fig. 9.19g. Compilation of displacement–length data for stylolites. Dataset: Italy, Benedicto and Schultz (2009).

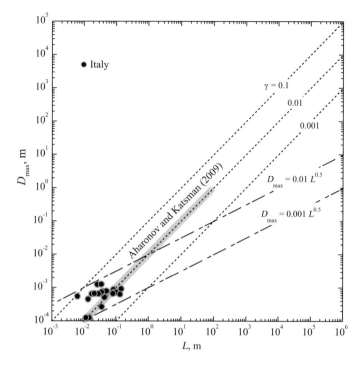

Fig. 9.19h. Plot of displacement–length descriptions for pure compaction bands. Datasets: Mollema and Antonellini (1996), Fossen et al. (2011), Liu et al. (2016).

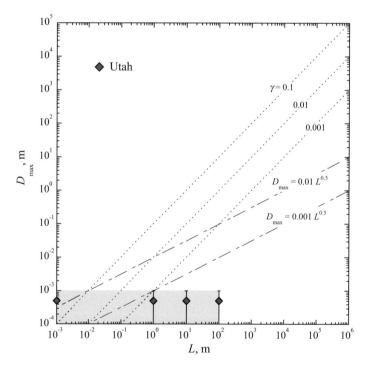

*Sublinear
displacement–length
scaling (n = 0.5):*

♦ *Opening-mode
  fractures*
♦ *Cataclastic
  (compactional shear)
  deformation bands*
♦ *Shear-enhanced
  compaction bands*

*Displacement–length
scaling (n > 1.0):*

♦ *Shear zones*

*Linear displacement–
length scaling (n = 1.0?):*

♦ *Stylolites*

The second group is characterized by a power-law slope of approximately $n = 0.5$ and therefore a square root dependence of maximum displacement and discontinuity length ($D_{max} = \alpha L^{0.5}$). Structural discontinuities within this group include dilatant cracks (i.e., individual joints, hydrofractures, veins, igneous dikes) (Fig. 9.19c), cataclastic (compactional shear) deformation bands (Fig. 9.19d), and shear-enhanced compaction bands (Fig. 9.19e), all of which accommodate significant changes in volume across them. Sublinear scaling with $n = 0.5$ implies that these structures approximate brittle fracture under LEFM conditions. Accordingly, their propagation is governed by mode-I dynamic fracture toughness (K_{Ic}) and attendant dilatant cracking of grains in small near-tip process zones subjected to large, nearly singular near-tip stress concentrations.

Shear zones don't appear to scale with $n = 1.0$ or $n = 0.5$ (Fig. 9.19b). Instead, both datasets define a steeper slope, perhaps approximately 1.5 (Fossen and Cavalcante, 2017). The steeper exponents imply that the rate of displacement accumulation along a shear zone is faster than that of length increase. Fossen and Cavalcante (2017) suggested that the slope for shear zones implies nucleation on weak zones or surfaces, similar to reactivated faults (Walsh et al., 2002; Nicol et al., 2005; Childs et al., 2017), but with different mechanisms operating to promote weakness in the host rock, and growth assisted by segment linkage.

Tectonic, non-sedimentary stylolites are recognized to thicken (i.e., exhibit larger topographic amplitude, or "teeth") from minima near their tips toward the central portions of the stylolite (e.g., Benedicto and Schultz, 2010; Nenna and Aydin, 2011; Toussaint et al., 2018). With the amplitude interpreted as a manifestation of the degree of dissolution and contractional strain accommodated across them (Koehn et al., 2007; Benedicto and Schultz, 2010), stylolites should scale in displacement and length (Toussaint et al., 2018). However, the growth process for stylolites is more complicated than simple mechanical fracturing and reorganization of grains (e.g., anticrack models: Cosgrove, 1976; Fletcher and Pollard, 1981; Meng and Pollard, 2014) as aqueous and chemical processes must also be explicitly included (Heald, 1959). Modeling work by Katsman et al. (2006a, b) and Aharonov and Katsman (2009) suggests that clay-enhanced dissolution along a stylolite is a necessary component, in addition to the far-field stress state, for both dissolution along the stylolite and propagation to occur. Consideration of reaction rates and electrochemical potentials of clay minerals leads to suggestions of self-similar, proportional growth in thickness (i.e., teeth amplitude) and length (i.e., propagation), with displacement/length achieving values on the order of 10^{-2} (Aharonov and Katsman, 2009). Given these results, it is possible that stylolites might follow displacement–length scaling relations in the form of $D_{max}/L = BC$, where B depends on rock, fluid, and chemical properties along with pressure, temperature, and related environmental factors and C is a proportionality constant, in the range of 0.011 or 0.086 (Aharonov and Katsman, 2009), depending on the mechanism of stylolite growth (Fig. 9.19g).

No displacement–length scaling (n = 0):

♦ *Pure compaction bands*

Pure compaction bands are the only class of geologic structural discontinuity that doesn't appear to change band thickness (or magnitude of accommodated displacement) as it grows in length. Field descriptions from sites in Nevada and Utah indicate band thicknesses of a few grain diameters along the entire length of the band (Mollema and Antonellini, 1996; Fossen et al., 2011; Liu et al., 2016). Numerical models of grain interactions by Katsman and Aharonov (2006) suggest that small increases in compressive stress near the tip of a pure compaction band can be sufficient to fracture grains there, leading to lateral propagation. However, the stresses elsewhere along the band are relaxed, so that grain fracturing or rearrangement would be inhibited there. Their work agrees with the field descriptions, implying an approximately constant and small magnitude of contractional displacement across a pure compaction band that would be independent of its length (Fig. 9.19h).

Role of rock brittleness on D–L scaling

As noted in Section 9.5.1 above, the scaling intercept and, correspondingly, the position of data on the *D–L* plot, increases with the degree of brittleness of the rock (Fig. 9.17). As a result, a smaller yield strength, corresponding to larger near-tip end zones and smaller near-tip displacement gradients, would promote smaller displacements for a given length. Structural discontinuities growing under more brittle (higher yield strength) conditions would plot higher on the *D–L* diagram than would discontinuities growing under semi-brittle conditions, all other factors being equal. As a result, studies of displacement–length scaling of geologic structural discontinuities need to consider the degree of brittleness (i.e., magnitude of yield strength relative to driving stress, size of near-tip end zone relative to discontinuity length) in addition to system stiffness to achieve a fuller understanding of the significance of the scaling relations.

9.5.3 *D–L* Scaling Relations for 3-D Structural Discontinuities

Many of the data collected from many (3-D) faults and other types of geologic structural discontinuities (Fig. 9.19a–h) have traditionally been understood by applying two-dimensional fracture mechanics models (e.g., Cowie and Scholz, 1992b; Kim and Sanderson, 2005; Kolyukhin and Torabi, 2012) that implicitly assume plane conditions (e.g., Irwin, 1962; Heald et al., 1972; Rudnicki, 1980; Bürgmann et al., 1994). This approach implies and requires that horizontal length is the smaller of the two discontinuity dimensions (e.g., Gudmundsson, 2000). In a 2-D approximation, discontinuity heights (in the down-dip direction) must be much greater than their map-view lengths (see Fig. 1.12) in order for displacements to be controlled purely (or primarily) by horizontal length dimension. Because discontinuity shapes can vary in their length/height (aspect) ratios (e.g., Nicol et al., 1996; Benedicto et al., 2003; Savalli and Engelder, 2005), 3-D models are useful for understanding displacement–length scaling relations of 3-D structural discontinuities (e.g., Soliva et al., 2005; Polit et al., 2009; Klimczak et al., 2013).

The inherent three-dimensionality of geologic structural discontinuities has been touched on earlier in this book as an important contributor to their shapes and patterns. In this section, scaling relations are described for 3-D "post-yield" structural discontinuities (i.e., those having non-negligible end zones). Several terms as illustrated in Figs. 1.12 and 5.21 will be used in this section.

Length – the horizontal dimension of a structural discontinuities ($L = 2a$), such as the trace of a fault in map view; the other two dimensions are the width and height.

Height – the vertical dimension of a structural discontinuities ($H = 2b$), such as the trace of a fault in cross-section; the other two dimensions are the length and width.

Width – the minimum dimension of a structural discontinuities, such as the dilation of a crack or the thickness of a fault zone; the other two dimensions are the discontinuity's length and height.

Aspect ratio – a measure of the shape of a structural discontinuity's surface, viewed normal to the surface, that relates the horizontal (strike) dimension, L, to the vertical dimension, H, so that $AR = L/H$. Using half-lengths, $AR = a/b$.

Length ratio – a measure of the shape of a surface-breaking structural discontinuity's surface, viewed normal to the surface, that relates the horizontal (strike) dimension, L, to the vertical dimension, $H/2$, so that $AR = 2 L/H$.

Long discontinuity – a 3-D structural discontinuity whose horizontal dimension (length) substantially exceeds its vertical dimension (height).

Tall discontinuity – a 3-D structural discontinuity whose vertical dimension (height) substantially exceeds its horizontal dimension (length).

Seismologists often use "width" to denote the vertical extent of a seismic event or a fault (e.g., Scholz, 1997, which is not adopted in this chapter). There has also some discussion in the literature of associating length with the shearing direction of faults (for example, the vertical dimension for dip-slip faults; e.g., Watterson, 1986; Torabi and Berg, 2011) although the physical or observational bases for making such a choice have not been convincingly demonstrated. However, differences in horizontal and vertical propagation rates, related in part to mechanical interaction with heterogeneities such as nearby structures or lithologic changes, can indeed contribute a major control on the shape of a geologic structural discontinuity (e.g., Nicol et al., 1996; Benedicto et al., 2003; Savalli and Engelder, 2005; Soliva et al., 2005).

There has been extensive work on the propagation and geometry of 3-D structural discontinuities, principally for mode-I fractures (e.g., Green and Sneddon, 1950; Irwin, 1962; Kassir and Sih, 1966; Chell, 1977; Kanninen and Popelar, 1985, p. 153; Lawn, 1993, p. 33; Olson, 1993, 2007; Dowling, 1999, pp. 307–309; Xue and Qu, 1999; Tada et al., 2000; Zhu et al., 2001; Savalli and Engelder, 2005; Wu and Olson, 2015). This work shows that the magnitude and distribution of displacement along the fracture, and especially that of D_{max} near its center, depends explicitly on **both length and height**. These results have been verified and applied to faults and deformation bands by many including Willemse et al. (1996), Martel and Boger (1998), Gudmundsson (2000), Schultz and Fossen (2002), Soliva et al. (2005, 2006), Polit et al. (2009), and Klimczak et al. (2013). In general, the mechanical interaction, linkage, and growth of 3-D structural discontinuities requires 3-D analyses to better understand what is seen in the field (e.g., Segall and Pollard, 1980; Rubin, 1992; Olson, 1993; Pollard et al., 1993; Crider et al., 1996; Willemse, 1997; Kattenhorn et al., 2000; Willemse and Pollard, 2000; Crider, 2001), and D–L scaling is no exception.

Building on equation (8.14) and the discussion in the previous section, the three-dimensional (3-D) displacement–length scaling relation for an elliptical post-yield fracture (or deformation band), measured in the horizontal plane, having horizontal length $L = 2a$ and down-dip height $H = 2b$ can be given by (Schultz and Fossen, 2002; Polit et al., 2009)

D–L scaling for 3-D structural discontinuities

$$D_{max} = \frac{2(1-v^2)}{E} \left\{ \frac{\left(\sigma_d - C\sigma_y\right)}{\sqrt{1 + 1.464\left(\dfrac{a}{b}\right)^{1.65}}} \right\} L \qquad (9.22)$$

in which E is Young's modulus and v is Poisson's ratio of the surrounding rock, σ_d is the driving stress, σ_y is the yield strength of the rock at the discontinuity tipline, a is the discontinuity's half-length (in map view), and b is its half-height (in the plane of the discontinuity; see Fig. 1.12b). For discontinuities that cut the Earth's ground (free) surface, $H = b$ (e.g., Segall and Pollard, 1980; Polit et al., 2009).

Willemse et al. (1996) compared the predicted values of D_{max} for 2-D and 3-D fractures, having negligibly small end zones, in elastic surroundings. These fractures met the small-scale yielding criterion for LEFM. They showed that D_{max} decreases systematically with increasing fracture aspect ratio (a/b, or L/H). This means that a "tall" fracture, say, 10 m long and 100 m high, can accommodate more displacement than a "long" fracture only 2 m high (with the same 10 m length and values of modulus and driving stress) can. Willemse et al. (1996), Gudmundsson (2000), and Schultz and Fossen (2002) showed (see Fig. 9.20) that a discontinuity that is 10 times greater in length than height ($L/H = 10$) can accommodate only about *one-tenth* of the displacement that one would measure along a tall one (with height 10 times its length; $L/H = 0.1$).

Tall discontinuities have more displacement per unit length than long ones

It can be seen from Fig. 9.20 that the ratio of maximum displacement to discontinuity length depends on the discontinuity's (i.e., fracture) aspect ratio. Relative to an indefinitely tall 2-D discontinuity ($b \gg a$), a penny discontinuity can accommodate only about 64% ($2/\pi$) of the displacement of the tall discontinuity (large dot on Fig. 9.20), whereas an indefinitely *long* 2-D discontinuity ($a \gg b$) can accommodate less than one-tenth of the displacement on the tall discontinuity (e.g., for $a/b > 8$ on Fig. 9.20). As a result, standard 2-D approaches

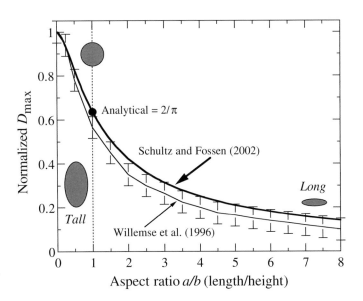

Fig. 9.20. Normalized maximum displacement (3-D value relative to standard 2-D value) decreases systematically with discontinuity aspect ratio a/b. Upper curve, 3-D scaling relationship (equation (9.22)); open symbols after Willemse et al. (1996) for $v = 0.25$. Shaded ellipses illustrate **tall, circular,** and **long discontinuities** that have the same values of length $2a$. Calculations assume $\sigma_y/\sigma_d = 50$, $\theta = 0°$, Poisson's ratio of 0.25, and shear modulus of 6.25 GPa (after Schultz and Fossen, 2002).

to displacement–length scaling that neglect the effect of discontinuity height ($2b$) may systematically *overpredict* the magnitude of D_{max} on a particular structural discontinuity by as much as an order of magnitude. For typical aspect ratios of $L/H = 2$–3 (e.g., Nicol et al., 1996; see also Savalli and Engelder, 2005), the growth into long, large aspect ratio discontinuities ($L/H > 10$) may yield a discrepancy in displacement of perhaps a factor of two or three. These errors will propagate through the population statistics if aspect ratio is not explicitly accounted for.

The relation for 3-D displacement–length scaling of a post-yield geologic structural discontinuity, equation (9.22), is plotted in Fig. 9.21. Curves representing values of maximum displacement for horizontal trace lengths L given different values of the scaling coefficient γ are plotted, illustrating contours of constant D_{max}/L ratio. Tall structural discontinuities ($L \ll 2b$) define a positive 45° (1:1) slope (left side of diagram, dashed lines) because D_{max} depends only on the length L, and not on the height, for given values of rock stiffness (modulus, Poisson's ratio). The lower set of curves shows the 3-D scaling relations for the same values of length (horizontal axis) but for varying values of discontinuity down-dip height. Inflection points in the curves occur approximately where $L = H$ (filled symbols on the figure). Slopes of the D_{max}/L curves flatten significantly beyond this point, indicating height-controlled displacement–length scaling (and $L > H$; Willemse et al., 1996; Gudmundsson, 2000; Schultz and Fossen, 2002).

Proportional vs. non-proportional growth

9.5.4 Stratigraphic Restriction and *D–L* Scaling

Proportional growth (straight lines on Fig. 9.21) occurs for structural discontinuities that grow in height and length at the same rate, leading to a constant aspect ratio. Proportional growth is also approximated for either tall discontinuities, with $H \gg L$, or 2-D growth models, because height does not influence the scaling relations in those cases.

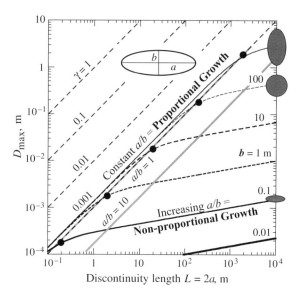

Fig. 9.21. Displacement–length scaling as a function of both horizontal map length ($L = 2a$) and vertical fault height ($H = 2b$) (after Schultz and Fossen, 2002). *Lines of constant aspect ratio* (AR = a/b; filled circles) require proportional fault growth on the *D–L* diagram.

In contrast, when height is considered and length becomes subequal in dimension to it (i.e., a penny fracture), the scaling relations change, leading to flatter curves for greater values of discontinuity length. As a discontinuity grows at constant height, its ability to accommodate displacements degrades with increasing aspect ratio, leading to slopes shallower than 1:1 on the *D–L* diagram. Growth of structural discontinuities with increasing values of aspect ratio is referred to as *non-proportional growth*.

A family of *D–L* curves (calculated by using equation (9.22)) describes how semi-brittle, non-LEFM 3-D structural discontinuities of arbitrary heights scale with their map lengths (Fig. 9.21). Each curve in the figure is defined by a characteristic thickness or vertical dimension (e.g., stratigraphic or mechanical layering) as labeled on the right-hand side (e.g., the vertical dimension ranges from 0.01–1000 m). "Small" discontinuities are those with heights $H = 2b$ and are much smaller than the characteristic vertical dimension shown; these scale linearly and proportionally in D_{max} and L. The point on each curve that corresponds to an aspect ratio $a/b = 1$ (circular discontinuity shape) defines the approximate transition from the more typical steep (proportional, 1:1) slope to shallow (non-proportional) slopes. "Large" discontinuities exhibit shallower, non-proportional slopes because their growth can occur only in length within the confined layer.

"Small" discontinuities are unrestricted

"Large" discontinuities are restricted

Field-based studies have demonstrated that many types of geologic structural discontinuities can become restricted to particular lithologic, rheologic, or stratigraphic layers (e.g., Gross, 1995; Gross et al., 1997; Rijken and Cooke, 2001; Wilkins and Gross, 2002; Soliva and Benedicto, 2004; Savalli and Engelder, 2005; Soliva et al., 2005; Roche et al., 2012; Ferrill et al., 2017, among many others). These observations have been complemented by theoretical and computational studies (e.g., Bai and Pollard, 2000a; Ackermann et al., 2001; Schultz and Fossen, 2002; Roche et al., 2013). Ductile layers such as shales or salt can exert a profound influence on the propagation of structural discontinuities, inhibiting propagation up-dip or down-dip into those layers. This process is referred to as *stratigraphic restriction* and it can sometimes be identified from displacement–length scaling relations. In such cases, the discontinuity's height becomes limited vertically by the stronger surroundings, leading to an *increasing aspect ratio* as the discontinuity lengthens and there is a concomitant decrease in the efficiency of displacement accumulation (Willemse et al., 1996; Fig. 9.20).

The switch from linear displacement–length scaling (i.e., proportional growth) to 3-D (non-proportional) growth is illustrated in Figs. 9.21 and 9.22. If a particular value of height is chosen, the scaling relations will change from linear (for $L \ll H$) to a flatter, nonlinear slope (for $L > 0.2\,H$) in a systematic fashion (Fig. 9.22). This characteristic was used by Soliva et al. (2005) to explain the change in displacement–length scaling for small faults in a layered stratigraphic sequence from northeast Spain (Fig. 1.3). There, the down-dip heights of faults in a bedded carbonate sequence were restricted vertically by intercalated shaley and coal-rich units, leading to non-proportional fault growth for lengths that attained and exceeded the bed thickness. Similar techniques were used by Polit et al. (2009) and Klimczak et al. (2013) to infer the thicknesses and characteristics of subsurface stratigraphic units from measurements of displacements and lengths of surface-breaking faults (Fig. 9.23). As will be seen later in this chapter, the growth style of geologic structural discontinuities (e.g., proportional or non-proportional) carries profound and practical implications for other characteristics of the discontinuity populations, such as spacing (Fig. 5.24) and strain accommodation.

Stratigraphic restriction of vertical discontinuity growth

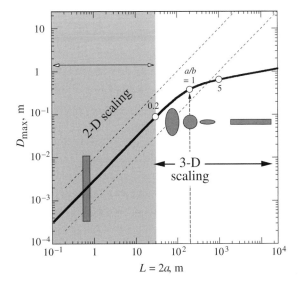

Fig. 9.22. Comparison between the 3-D scaling relations (for a given value of b and negligible end-zone length) and 2-D limiting cases (shaded area); equation (9.20) is plotted. The 3-D results converge toward the 2-D values (dotted diagonal lines, corresponding to values of γ) for tall discontinuities ($a/b \ll 0.2$). Curve shown calculated using $D_{max}/L = 0.006$, $b = 100$ m, $\sigma_y/\sigma_d = 5$, $\theta = 0°$, Poisson's ratio of 0.25, and shear modulus of 0.0625 GPa (after Schultz and Fossen, 2002).

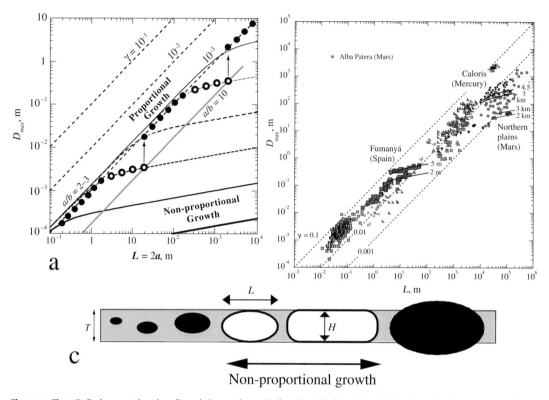

Fig. 9.23. The 3-D displacement–length scaling relations and growth of stratigraphically restricted faults (after Schultz et al., 2010c). (a) Discontinuity growth paths showing stair-step trajectory of alternating proportional (linear, filled symbols) and non-proportional (restricted, open symbols) growth. (b) Examples of vertically restricted normal fault populations from Fumayá, Spain (Soliva et al., 2005), northern plains, Mars (Polit et al., 2009), and Caloris basin, Mercury (Klimczak et al., 2013); thicknesses of restricting layers are shown. (c) Cross-sectional geometries shown schematically for the three parts of the growth sequence; filled and ellipses as in part (a).

Discontinuity growth within a rheologically stratified sequence can follow a stair-step pattern as shown in Fig. 9.23. A stair-step trajectory on the D–L diagram was identified for along-strike growth by segment linkage by Cartwright et al. (1995). Unrestricted growth of structural discontinuities occurs when they are small enough relative to the thickness of the layer in which they are growing to be unaffected by layer boundaries. Discontinuities grow at constant aspect ratio L/H (filled symbols in Fig. 9.23a,c), defining proportional growth and linear scaling on a D–L diagram. Proportional growth can be represented by 2-D scaling relations. Although data are scarce, a limiting value of about $L/H = 10$ may correspond to the longest a restricted discontinuity can attain before it propagates up- and/or down-dip (Soliva et al., 2005). Beyond this value the near-tip displacement gradients along the upper and lower tiplines of the discontinuity may be sufficient to generate stress concentrations that exceed the yield strength in the confining layers. Other growth paths have been recognized on the D–L diagram (e.g., Childs et al., 2017) that pertain, for example, to displacement accumulation along reactivated surfaces and faulting in active sedimentary basins.

In some cases, the surrounding material can be relatively weaker than the fractured layer. This can occur, for example, where a clastic sedimentary sequence is bounded below by salt, as in many oil and gas fields (e.g., Jackson, 1995) or when discontinuities such as joints or faults intersect the Earth's surface. The *D–L* data for the grabens in Canyonlands National Park, Utah (Cartwright et al., 1995), for example, show an *increase* in displacement for faults longer than about twice the 450 m layer thickness, in addition to increased scatter; the larger displacements were attributed to inter-block rotations by Walsh and Schultz-Ela (2003) due to the weaker (upper and lower) surroundings. Similarly, faults in the Afar region of north Africa, where the oceanic spreading center interacts onshore with the East African rift, show larger-than-expected displacements (Gupta and Scholz, 2000b) that again could be related to the relative strength of faulted upper lithosphere relative to its (upper and lower) surroundings. In the case of restricted fractures such as joints, veins, hydraulic fractures, or igneous dikes, opening displacements should tend to increase for fractures surrounded by relatively softer media (paralleling the increase in near-tip stress (e.g., Cook and Erdogan, 1972; Erdogan and Biricikoglu, 1973; Pollard and Aydin, 1988)), and decrease for those in layers surrounded by relatively stiffer materials.

In both cases of strength stratification, structural discontinuities can become vertically restricted, so that their aspect ratios increase with increasing length. Geologic structural discontinuities are sensitive to the degree of lithologic (mechanical, rheologic, petrophysical) stratification which exerts a first-order control on their geometries and patterns. Mechanical contrasts in layered sequences additionally control a wide and diverse variety of deformation-related processes, such as changes in the type of structural discontinuity in different layers (e.g., Gross et al., 1997; Fossen et al., 2007, 2017; Ferrill et al., 2014, 2017), first-order control of the seismic cycle (e.g., Tse and Rice, 1986), faulting vs. folding in a multi-layer sequence (e.g., Erickson, 1996), layer-parallel buckling (e.g., Currie et al., 1962; Johnson and Fletcher, 1994), and boudinage (e.g., Goscombe et al., 2004).

As may be evident, the displacement magnitude on geologic structural discontinuities depends on several important factors that may sometimes be either observable or measurable. This is useful for understanding the scaling relationships between displacement and discontinuity dimensions in different rock types, stratified vs. massive sequences, and as a geologic structural discontinuity set grows.

9.6 Populations of Geologic Structural Discontinuities

Few geologic structural discontinuities occur as solitary entities. Instead, they occur as members of a set, array, network, or population. In a population, discontinuities display wide variations in their primary characteristics, such as length, displacement, and spacing. However, these characteristics are not occurring at random. All of the discontinuity characteristics that are considered routinely depend on one another, sometimes directly and sometimes indirectly, so that knowledge of one or two key characteristics can provide insight into the values and relationships between the others.

Here are some definitions that are useful in discontinuity population studies (Fig. 9.24).

Population – a set or distribution of structural discontinuities whose measurable characteristics (number, size, displacements, lengths) can be used to define a mechanically integrated discontinuity network.

Array – a mechanically related, organized, collection of structural discontinuities, usually having a collinear and/or coplanar overall geometry (as in an echelon array); distinguished from a network or population by having a smaller number of discontinuities and more localized or focused character.

Set – a collection, array, network, or population of structural discontinuities.

Proportional growth – propagation of structural discontinuities in three dimensions that is accompanied by no change in the discontinuities' aspect ratio.

Non-proportional growth – propagation of structural discontinuities in three dimensions that is accompanied by a significant change in the discontinuities' aspect ratio.

Censoring – a qualitative filtering of discontinuity-population data that results in reliable statistics only for a certain length range, usually obtained by under-counting certain classes (too short, different type) of discontinuities.

Truncation bias – a reduction in slope on a cumulative-frequency vs. length diagram at progressively smaller lengths associated with under-counting of *shorter* discontinuities.

Curtailment bias – an increase in slope on a cumulative-frequency vs. length diagram at large lengths associated with underestimation of the actual lengths of the *longest* discontinuities.

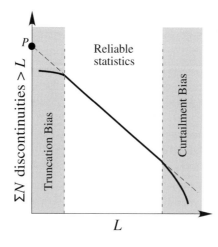

Fig. 9.24. Typical errors (truncation bias, curtailment bias) due to discontinuity sampling as represented on the cumulative length–frequency diagram. The intercept, extrapolated to the vertical axis (or to the minimum discontinuity length) is sometimes referred to as the productivity, *P*, of the discontinuity population.

Scaling dimension – a manifestation of the geometry of a discontinuity population ("small" or "large") relative to the volume or area of the deforming region.

The **proximity** of a structural discontinuity to its neighbors can control the lengths, spacings, displacement gradients, and—in short—the patterns of discontinuities in a population. As a discontinuity grows, it influences the state of stress in its vicinity. Any discontinuity within its "zone of influence" will be affected in some way—by either enhancing or impeding their growth in size and/or displacement magnitude. Earlier in this chapter it was shown how the near-tip environment can influence the propagation rate relative to other discontinuities that are also competing for driving stress. As individual discontinuities grow, the population evolves progressively with ongoing strain to produce an aggregate collection of inter-related discontinuities. The geologic structural discontinuity populations that one encounters are **snapshots** in this evolutionary sequence, and the task of a structural geologist is to figure out where the discontinuities are in the sequence, so one can both understand what has already occurred, and perhaps also predict what may happen next.

In this section we'll look at elements of discontinuity population statistics from a fracture mechanics point of view. The literature on this topic is vast, and entry can be gained from several sources.

- For treatments on the **data** and how to measure the discontinuity characteristics, see Muraoka and Kamata (1983), Watterson (1986), Walsh and Watterson (1988), Wojtal (1996), Fossen and Hesthammer (1997), Gross et al. (1997), Kim and Sanderson (2005), Twiss and Marrett (2010b), Torabi and Berg (2011), Peacock et al. (2016), and Lamarche et al. (2018).
- For **statistical analysis** of these basic measurements, Clark and Cox (1996), Clark et al. (1999), Borgos et al. (2000), Bonnet et al. (2001), and also Cowie and Scholz (1992b), Westaway (1994), Bonnet et al. (2001), Twiss and Marrett (2010a), Valliantos and Sammonds (2011), and Klimczak et al. (2013) provide a good start.
- **Simulations** of discontinuity population development are outlined by Gillespie et al. (1992, 1993), Kachanov (1992), Cowie et al. (1993a,b, 1995), Olson (1993), Renshaw and Pollard (1994a), Cladouhos and Marrett (1996).
- Discontinuity growth by **segment linkage**, a widely demonstrated and accepted mechanism for many rock types and discontinuity sets, was originally put on the map by Dawers and Anders (1995) and by Cartwright et al. (1995); see Gupta and Cowie (2000), Crider and Peacock (2004), and Childs et al. (2017).
- **Mechanical models** of discontinuity growth that take direct aim at population statistics include those of Segall and Pollard (1980), Olson

(1993, 2003, 2007), Bürgmann et al. (1994), Willemse et al. (1996), Gudmundsson (2000), Willemse and Pollard (2000), Olson et al. (2007), and Childs et al. (2017).

In much of the following, the emphasis will be on fault populations, since much of the work in this area was developed to better characterize and understand the growth, evolution, and tectonic significance of faults. However, the other types of structural discontinuity, such as cracks, dikes, anticracks, or deformation bands, can generally be considered in each of the following sections because the principles apply to any of these.

9.6.1 Discontinuity Length Statistics

Length-frequency distributions of faults can provide important clues to the geometrical and mechanical evolution of a fault array (e.g., Gillespie et al., 1993; Scholz et al., 1993; Cowie, 1998a,b; Bonnet et al., 2001) and, potentially, to parameters that influence the scaling relationships from small to large faults (e.g., Scholz, 1982, 1997; Shimazaki, 1986; Scholz et al., 1993; Cladouhos and Marrett, 1996). Length-frequency data (e.g., Jackson and Sanderson, 1992; Hatton et al., 1993; Knott et al., 1996; Marrett, 1996) augment displacement–length datasets that have been used to evaluate fault-growth and linkage models (e.g., Cowie and Scholz, 1992a; Cartwright et al., 1995; Dawers and Anders, 1995; Clark and Cox, 1996), displacement-length scaling relations (e.g., Cowie and Scholz, 1992a; Dawers et al., 1993; Marrett, 1996; Schlische et al., 1996; Kim and Sanderson, 2005; Torabi and Berg, 2011), and estimation of regional strains accommodated by fault arrays (Marrett and Allmendinger, 1990; Scholz and Cowie, 1990; Walsh et al., 1991; Westaway, 1994; Carter and Winter, 1995; Marrett, 1996; Clark et al., 1999; Bailey et al., 2005; Polun et al., 2018).

To illustrate how discontinuity statistical characteristics can be used to understand the associated population, we'll use the classic dataset for normal faults in the Bishop Tuff of California (Dawers et al., 1993; Dawers and Anders, 1995; Ferrill et al., 1999, 2016) as an illustrative example. The data are listed in Table 9.3. Notice that Dawers et al. report the field measurement (throw) so one would need to correct these values for fault dip angle(s) to obtain the maximum displacements (D_{max}). The correction for fault dip produces a small upward shift in the values of fault throw; shallower dips than the 75° used here would lead to larger increases in calculated values for D_{max} but these corrections would not be significant here. The data are sorted by length L, rather than by D_{max}. It is customary to use consistent units (such as meters or feet) for both parameters.

The D_{max}/L diagram for the Bishop Tuff normal faults, shown in Fig. 9.25, shows that the data define a consistent distribution, with few kinks and knees

Table 9.3 Displacement-length data from Bishop Tuff normal faults [a]

Throw T, m	D_{max}, m [b]	L, m
0.44	0.45	24
0.9	0.93	55
1.0	1.04	60
3.3	3.42	167
3.4	3.52	182
4.9	5.07	440
9.4	9.73	500
10.1	10.46	540
8.4	8.70	696
10.2	10.60	740
10.1	10.46	780
8.3	8.59	866
18.5	19.15	1620
17.3	17.91	1630
31.0	32.09	2210
145.0	150.12	12,360

[a] Data from Dawers et al. (1993)
[b] Assumes an average dip angle of $\delta = 75°$ (Dawers et al., 1993). $D_{max} = [T / \cos (90° - \delta)]$.

(bends in the slope). Some scatter is already evident in the values of D_{max}. Some workers also plot their data using linear axis scales to better resolve small differences in trend or position. Clearly, there are additional controls on D_{max} than just the fault length.

Dawers et al. (1993) showed that normal faults that cut the Bishop Tuff defined a linear D–L scaling relation (Fig. 9.25). However, they also suggested that the displacement profiles defined two basic groups: smaller faults having peaked or triangular displacement profiles and longer faults having more flat-topped, plateaued profiles. Faults longer than approximately twice the ~150-m thickness of the tuff assumed in their study (dashed line at $L = 300$ m in Fig. 9.25) showed plateaued profiles, suggesting, among other alternatives, their restriction to the tuff sequence.

The fault scarps are well preserved in the erosionally resistant, densely welded, 10-m-thick upper unit that forms the surface of the Tableland (Dawers et al., 1993; Pinter, 1995; Wilson and Hildreth, 1997; Hooper et al., 2003). Below the upper unit is a 40- to 150-m-thick lower unwelded unit (e.g., Evans and Bradbury, 2004). Below the lower unit is a 20- to 40-m-thick nonwelded basal airfall ash unit. Below the Tuff are unconsolidated alluvial deposits and

Fig. 9.25. D_{max}/L diagram for normal faults from Bishop Tuff, California; data from Dawers et al. (1993) and listed in Table 9.3. Vertical dashed line corresponds to estimated $L = H$ for tuff thickness 150 m.

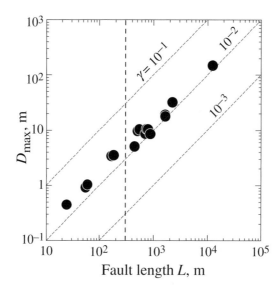

glacial Sherwin Till (Sharp, 1968). Granitic bedrock underlies the alluvium at depths of ~2 km (Mukhopadhyay, 2002). Although not yet demonstrated, it is likely that stiffness varies through the stratigraphic section.

The scaling of faults is well known to be sensitive to the stiffness of the faulted layers, so that displacements along faults in stiffer layers are smaller than those cutting softer layers. It follows that faults that cut a layered stratigraphic sequence having variations in layer stiffnesses could accumulate displacement magnitudes consistent with the average stiffness of the faulted layers. Because the fault displacement profiles also depend on the relative stiffness between faulted layers, a fault cutting down into a stiffer layer, for example, will develop a flatter displacement profile, with a correspondingly reduced value of maximum displacement, than would a fault of the same length cutting a layer of constant (lesser average) stiffness (Bürgmann et al., 1994). If the fault continued to grow in length and displacement, then its D_{max}/L ratio would decrease, but perhaps not sufficiently for restriction to occur or to become evident on the D–L diagram.

By using the methods outlined in Schultz and Fossen (2002), Schultz (2011a) suggested that the D–L data from Dawers et al. (1993) could be fitted as two sets: a shorter set, with lengths <300 m, and a longer set. Both sets have linear scaling and would therefore not be stratigraphically restricted. However, the longer set shows a systematic reduction in displacement relative to the shorter set. This can be modeled as either a fault of greater aspect (length to height) ratio (having a length ratio of 4), faulting of a stiffer layer, or some combination of both (Fig. 9.26). A fault propagating into a stiffer layer is known to change its shape to more flat-topped (Bürgmann et al., 1994), mirroring the throw profiles reported by Dawers et al. (1993). These results support an interpretation of faults in Bishop Tuff as not being stratigraphically restricted, but having reduced D–L ratios, and perhaps altered shapes, as they propagated downward into the underlying units.

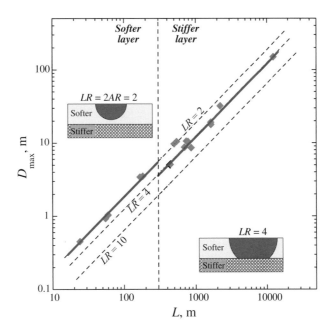

Fig. 9.26. Fits to normal fault data in the Bishop Tuff calculated by using 3-D scaling relations showing inferred increase in fault length ratio with depth (after Schultz, 2011b).

9.6.1.1 Cumulative frequency vs. length

A second, very informative, approach to analyzing discontinuity–length statistics is by using a plot of the *cumulative* frequency of discontinuities longer than a given value versus their lengths. Here again, the horizontal, map view lengths for the discontinuities will be used, although there is no reason why their height dimensions could not be used (i.e., from cross-sectional field exposures or from seismic sections). This approach provides information on the discontinuity population and its related attributes in cases where displacement magnitudes are not available (for example, a map or image that lacks topographic data). These diagrams are straightforward to create once the data are properly sorted.

The number, dimensions, and displacement magnitudes of discontinuities in a given area provide important inputs to calculate quantities such as discontinuity density and the magnitude of brittle strain accommodated by the population within that area. Other discontinuity attributes, such as spacing or especially displacement, have also been examined by plotting their cumulative frequencies (e.g., Bonnet et al., 2001; Hooker et al., 2014). Although length-frequency datasets can be compiled without keying them to the size of the associated area, treating the discontinuity population and the areal dimension as independent quantities may not be correct in general. By dividing the number of discontinuities by the area of the deformed region, the density of discontinuities can be obtained (e.g., Jackson and Sanderson, 1992; Westaway, 1994; Castaing et al., 1996; Schultz and Fori, 1996; Schultz, 2000a). Calculating cumulative number (Twiss and Marrett, 2010a) allows one to compare the discontinuity densities for differently sized regions.

Counting area for the cumulative-frequency vs. length diagram

In this section we work through the creation of a *cumulative-frequency vs. length diagram* by using the data for Bishop Tuff normal faults (Dawers et al., 1993). The data must first be sorted (Table 9.4) so that the lengths are listed in descending order (second column of Table 9.4). There is one fault of each length, so $N = 1$ for each entry. We normalize the number of faults of given length by the ratio of the standard (reference) area A_0 to that of the deformed area A (in this case, for the Bishop Tuff) using

$$N^* = N\left(\frac{A_0}{A}\right) \tag{9.23}$$

What this scaling operation does is to solve for the number of faults that we *should* find in an area with $A = A_0 = 10^4$ km². Suppose we are working with a small area ($A < A_0$). Letting $A = 1.5$ km², for example, and with one fault of length L in this small deformed area, then one might expect to find 6666.7 faults of length L in the (much larger) standard reference area.

Table 9.4 Cumulative-frequency vs. length data from Bishop Tuff normal faults [a]

N	L, m	N^* [b]	$\sum N^*$ [c]
1	12360	33.3	33.3
1	2210	33.3	66.6
1	1630	33.3	99.9
1	1620	33.3	133.2
1	866	33.3	166.5
1	780	33.3	199.8
1	740	33.3	233.1
1	696	33.3	266.4
1	540	33.3	299.7
1	500	33.3	333.0
1	440	33.3	366.3
1	182	33.3	399.6
1	167	33.3	432.9
1	60	33.3	466.2
1	55	33.3	499.5
1	24	33.3	532.8

[a] Data from Table 9.3.
[b] N^* is the normalized number of discontinuities of length L per reference area $A_0 = 10^4$ km² (Schultz, 2000a). A deformed area (Bishop Tuff) of $A = 15 \times 20$ km = 300 km² = 300 × 10⁶ m² (Dawers and Anders, 1995, their figure 2) is assumed. For $A < A_0$, $N^* = N \times (A_0 /A)$.
[c] $\sum N^*$ is the cumulative normalized number of discontinuities (N^*) longer than length L.

Conversely, if the deformed area is large, say, $A = 15,000,000$ km^2, then one might expect to find only 0.0007 faults of this length in the (smaller) reference area. Westaway (1994) emphasized the need to explicitly connect the number of faults in a deforming region to its dimensions (area or volume) and that recommendation has stood the test of time.

In this exercise, the discontinuity numbers are normalized to the final area of the *deformed* region. One might also restore the deformed area by undoing the discontinuity-related strain and then use that initial area in the cumulative-frequency vs. length plot. Depending on the discontinuity set you're working with, the difference in area from the initial to the final, deformed case may be on the order of several percent. On the other hand, the difference in size between two field areas (e.g., thin section, core, outcrop, or quadrangle) may be considerable! With this technique one can plot discontinuity population data from a thin section (millimeters), a hand sample (centimeters), a large-scale map of an outcrop (meters), and a satellite image (tens of kilometers) on the same diagram and obtain robust comparisons between them. The discrepancies between deformed and undeformed areas (of perhaps several percent) would likely be much less than the order-of-magnitude changes in the scales of the nested datasets. This allows one to plot data from widely varying scales on the same diagram (e.g., Scholz et al., 1993; Schultz and Fori, 1996) in a uniform and consistent manner (see also Westaway, 1994).

Scaling relationships for cumulative-frequency vs. length

Two important scaling relationships have been identified from discontinuity data: **power-law** and **exponential** (e.g., Cowie et al., 1993a; Bonnet et al., 2001). These relationships are given by

Power-law

$$N(L) = N_0 L^{-b} \tag{9.24a}$$

Exponential

$$N(L) = N_T e^{-\lambda L} \tag{9.24b}$$

in which N_0 is the intercept of the power-law distribution, b is the power-law exponent, N_T is the intercept of the power-law distribution (sometimes called the productivity, P; Fig. 9.24), and λ is a dimension term in the exponential distribution given by $\lambda = 1/L_{avg}$, where L_{avg} is the mean length from the population. Displacement or spacing can also be substituted for L in either of these relationships. Equations (9.24a,b) are plotted in Fig. 9.27 for various values of the parameters.

Many earthquake and fault populations exhibit power-law scaling of their lengths (e.g., Kakimi, 1980; Marrett and Allmendinger, 1990; Walsh et al., 1991; Scholz et al., 1993; Westaway, 1994; Cowie et al., 1995; Marrett, 1996; Bonnet et al., 2001) as can opening-mode fractures such as joints (e.g., Odling, 1997; Bertrand et al., 2015). **Power-law scaling** with negative exponents is associated with displacements and strains localized mainly along a small number of the largest discontinuities. Power-law exponents can range from approximately -3 to -0.5 (e.g., Bonnet et al., 2001; Twiss and Marrett, 2010b) with specific values

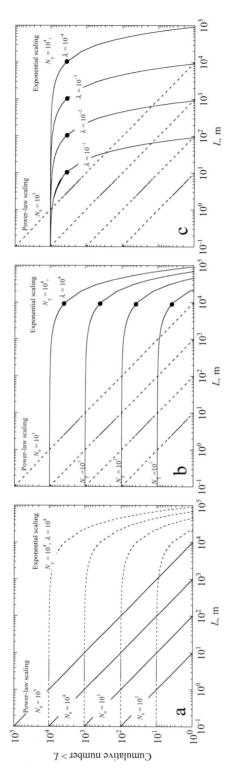

Fig. 9.27. Cumulative-frequency vs. length diagrams showing power-law (solid curve) and exponential (dashed curve) distributions calculated by using equations (9.24a) and (9.24b). Filled symbols in (b) and (c) show values of λ on curves.

depending on factors such as sampling dimensionality, spatial clustering, and degree of interaction and linkage. More importantly, a consistent power-law scaling relationship identified over a given range of dimension (i.e., length, aperture) can imply self-similar, fractal-like scaling of the population (e.g., Kakimi, 1980), which is a feature of many fault and structural discontinuity systems (e.g., Tchalenko, 1970; Cowie, 1998b). It is also consistent with length-dominated growth, interaction, and spacing of discontinuities that are small and unrestricted relative to the vertical dimension of the faulted layer or domain.

An **exponential scaling** relationship has been identified for length distributions of normal faults in oceanic crust adjacent to spreading centers (Cowie et al., 1994; Cowie, 1998b), faults in analog experiments (Ackermann et al., 2001), and for continental faults that are stratigraphically or otherwise vertically restricted (Soliva and Schultz, 2008). The parameter λ represents the slope of the exponential distribution on a semi-log plot (Fig. 9.28b). Exponential scaling can occur when discontinuity height, not length, controls the population. Confinement of structural discontinuities to a layer of given thickness can occur for superjacent and subjacent layers that have larger values of yield strength or dynamic fracture toughness so that up-dip or down-dip propagation is temporarily limited. This results in along-strike propagation within the layer that has the effect of limiting the perpendicular extent of strong mechanical interaction (e.g., Willemse et al., 1996; Willemse, 1997; Gudmundsson, 2000; Gupta and Scholz, 2000b; Soliva et al., 2006), leading to a population dominated by long discontinuities having relatively constant spacings (Figs. 1.3 and 5.24). The spacings scale with discontinuity height, and therefore with the

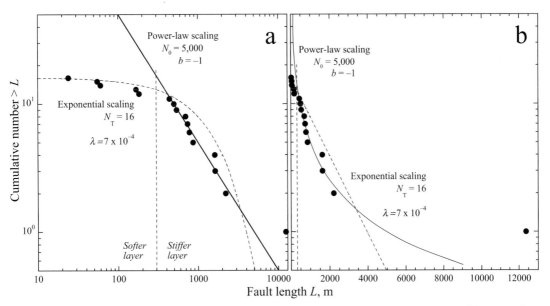

Fig. 9.28. Cumulative frequency vs. length data for normal faults from Bishop Tuff, California, plotted with power-law (solid curves) and exponential (dashed curves) distributions calculating by assuming $N_0 = 5000$, $b = -1.0$, $N_T = 16$, $L_{avg} = 1429.4$ m, $\lambda = 7 \times 10^{-4}$ m⁻¹. Vertical dashed lines at $L = 300$ m as in Fig. 9.25.

thickness of the confining layer, as demonstrated by vertically restricted joints and faults (e.g., Lachenbruch, 1961; Olson, 1993; Bai and Pollard, 2000a; Soliva et al., 2005, 2006; Schöpfer et al., 2011). In concert with spacing, the maximum or characteristic width of discontinuity linkages and stepovers, such as relays (Soliva et al., 2006), duplexes, pull-apart basins, push-up ranges, and similar structures can also be related to layer thickness for stratigraphically restricted populations. Either of these distributions arises because discontinuity length in a population depends on the interaction (or stress shielding) of nearby discontinuities; the interaction depends on the discontinuity spacing, which in turn depends on discontinuity displacement.

The cumulative-frequency vs. length data for the Bishop Tuff data are plotted in Fig. 9.28 along with power-law and exponential distributions. The data are generally consistent with a power-law scaling relation with a slope of −1 for faults longer than ~300 m, except for the longest (Fish Slough) fault that may not be part of the same population (i.e., contributing to rollover-related normal faulting of the tuff: Pinter, 1995; Ferrill et al., 2016). An exponential distribution (dashed curves in the figure) does not match the fault population well, except perhaps for faults shorter than ~300 m. A hypothesis that these shorter faults were vertically restricted would not be supported by their linear displacement–length scaling relation (Fig. 9.25) as a shallower, nonlinear trajectory on the D–L diagram would be expected (Fig. 9.21). Another explanation might be that faults in this length range were under-counted, leading to truncation bias in the fault-length distribution (e.g., Jackson and Sanderson, 1992; Fig. 9.24). Under-counting could be tested by re-examining the tuff for small faults and exploring reasons why they might be difficult to identify, such as erosion or poor scarp preservation in the upper welded tuff unit (e.g., Ferrill et al., 2016).

The data from normal faults in Bishop Tuff (Fig. 9.26) are generally consistent with a slope of −1. Values in this general range are commonly found from several independent datasets from various regions and settings from around the world (e.g., Soliva and Schultz, 2008; Twiss and Marrett, 2010b), Venus (e.g., Scholz, 1997), and Mars (Schultz, 2000a; Wilkins et al., 2002) to note just a few.

9.6.1.2 Relative frequency vs. length–the *R*-plot

One of the difficulties in interpreting traditional cumulative-length vs. frequency plots is that the process of summing faults can obscure details of the size–frequency distribution of the individual segments (e.g., Melosh, 1989, pp. 186–187; Westaway, 1994; Main, 1995). **Histograms of incremental length-frequency** can supplement cumulative plots and better reveal power-law distributions (e.g., Segall and Pollard, 1983a,b; Olson, 1993; Westaway, 1994). A standard technique was developed originally for power-law distributions of impact-crater diameters (Crater Analysis Techniques Working Group, 1979; Melosh, 1989; Strom et al., 2015) to better portray and interpret the length-frequency data. Called the *R*-plot, it has been adapted for use in fault population studies (Schultz, 2000a) and can be well suited to comparing discontinuity populations from regions of different areas and fault densities.

The *R*-plot is a histogram of the incremental, or differential, frequency of structural discontinuities per unit of dimension (i.e., area). Because the cumulative-length vs. frequency plot (Fig. 9.28) involves summing individual discontinuities with lengths greater than a given value, the power-law exponent from the cumulative plot is the integral of that from the incremental *R*-plot (Crater Analysis Techniques Working Group, 1979). This means that a cumulative slope of –1 corresponds to a differential slope of –2 (integration of L^{-2} reduces this differential power-law exponent to L^{-1}). However, the *R*-plot goes further than a simple histogram of incremental discontinuity lengths by normalizing the observed number of discontinuities with a given length to the number expected for that length from a specified power-law distribution. It thus directly compares the actual data to a particular model or expected

R-plot = Data/Model   distribution.

Building an *R*-plot is straightforward. First the length data are binned into √2-wide intervals, with upper bound L_b and lower bound L_a; this geometrically increasing progression of bin widths produces uniform point spacing on the resulting log–log plot (Melosh, 1989; Table 9.5). Next, the normalized or relative frequency *R* is calculated using (Crater Analysis Techniques Working Group, 1979; Strom et al., 2015)

Equation for the R-plot

$$R = \frac{\bar{L}N}{A\left(L_b - L_a\right)} \tag{9.25}$$

in which $\bar{L} = \sqrt{L_a L_b}$ is the geometric mean of the bin width, *n* is the absolute value of the power-law exponent for the reference distribution, *N* is the number of structural discontinuities contained within the given length bin, and *A* is the counting area of the region containing the discontinuity population (Schultz, 2000a).

Points on the *R*-plot are independent of those in adjacent bins, so details of the original discontinuity population (linked or unlinked) are displayed. For example, discontinuity arrays having a greater spatial density (number of discontinuities per unit of deformed area) plot higher on the diagram (larger values of *R*) than would discontinuity sets having lower spatial density. The *R*-plot was designed to highlight subtle variations in spatial density as compared to a single cumulative power-law slope.

In studies of impact cratering on which the *R*-plot is based, the best-fit slopes of cumulative size-frequency data (using crater diameters) are typically close to –2 (e.g., Melosh, 1989). A –2 slope causes the cumulative frequency to become dimensionless and, in turn, the scale dependence to vanish (Melosh, 1989, p. 187). The cumulative –2 slope is typically used in impact-cratering studies as a reference distribution to which actual datasets are compared (see also Chapman and McKinnon, 1986; Fassett, 2016). As

Table 9.5 Discontinuity length bins for R-plot

L_a, km	L_b, km	R, km	
0.0010	**0.0014**	**0.0012**	**1 m**
0.0014	0.0020	0.0017	
0.0020	0.0028	0.0024	
0.0028	0.0039	0.0033	
0.0039	0.0055	0.0046	
0.0055	0.0078	0.0065	
0.0078	**0.0110**	**0.0093**	**10 m**
0.0110	0.0156	0.0131	
0.0156	0.0221	0.0186	
0.0221	0.0313	0.0263	
0.0313	0.0443	0.0372	
0.0443	0.0626	0.0527	
0.0625	0.0884	0.0743	
0.0884	**0.1250**	**0.1051**	**100 m**
0.1250	0.1768	0.1487	
0.1768	0.2500	0.2102	
0.2500	0.3536	0.2973	
0.3536	0.5000	0.4204	
0.5000	0.7071	0.5946	
0.7071	1.0000	0.8409	
1.000	**1.414**	**1.189**	**1.0 km**
1.414	2.000	1.682	
2.000	2.828	2.378	
2.828	4.000	3.364	
4.000	5.657	4.757	
5.657	8.000	6.727	
8.000	**11.314**	**9.514**	**10 km**
11.314	16.000	13.454	
16.000	22.627	19.027	
22.627	32.000	26.909	
32.000	45.255	38.055	

Table 9.5 (Cont.)

L_a, km	L_b, km	R, km	
45.255	64.000	53.817	
64.000	90.510	76.109	
90.510	**128.000**	**107.635**	**100 km**
128.000	181.019	152.219	
181.019	256.000	215.269	
256.000	362.039	304.437	
362.039	512.000	430.539	
512.000	724.077	608.874	
724.077	**1024.000**	**861.078**	**1000 km**
1024.000	1448.155	1217.748	
1448.155	2048.000	1722.156	
2048.000	2896.309	2435.496	
2896.309	4096.000	3444.312	
4096.000	5792.619	4870.992	5000 km

a result, the R-plot was constructed to exploit this self-similar, fractal distribution of crater diameters by setting the differential exponent to –3 (Crater Analysis Techniques Working Group, 1979; Strom et al., 2015); this leads to R being dimensionless for this case. In general, however, R has units of $m^n\ m^{-3}$, where n is the absolute value of the power-law exponent used in the plot.

Best-fit slopes of cumulative-length vs. frequency data for many geologic structural discontinuities are commonly in the range of –0.5 to –3 (e.g., Cladouhos and Marrett, 1996; Marrett, 1996; Scholz, 1997; Cowie, 1998; Clark et al., 1999; Bonnet et al., 2001; Twiss and Marrett, 2010b). Integer values of n in equation (9.25) facilitate comparisons between datasets having differing numbers of discontinuities and counting areas.

Let's examine how the normal fault data from the Bishop Tuff look when displayed on an R-plot. First, the data are binned (Table 9.6) and the approximate reference slope of the cumulative distribution, in this example –1 (relative or differential slope of –2) from the cumulative-frequency vs. length diagram, was chosen. The binned data are plotted in Fig. 9.27 by using parameters $n = +1$, $N = 16$ faults, and counting area $A = 300 \times 10^6\ m^2$. The data are broadly consistent with a negative differential slope of about –2 (i.e., a horizontal distribution on the R-plot; inset to Fig. 9.29), consistent with the –1 cumulative power-law

Table 9.6 Binned fault length data for R-plot from Bishop Tuff normal faults[a]

$\bar{L}$, m[b]	N	R[c]
26.3	1	1.52×10^{-7}
37.2	0	–
52.7	2	3.07×10^{-7}
74.3	0	–
105.1	0	–
148.7	1	1.53×10^{-7}
210.2	1	1.53×10^{-7}
297.3	0	–
420.4	2	3.06×10^{-7}
594.6	2	3.06×10^{-7}
840.9	3	4.59×10^{-7}
1189	0	–
1682	2	3.06×10^{-7}
2378	1	1.53×10^{-7}
3364	0	–
4757	0	–
6727	0	–
9514	0	–
1345	1	1.53×10^{-7}

[a] Data from Table 9.4; lengths are in meters.
[b] By convention, a bin may commonly be set at $L_a = 1.0$ km
    (e.g., Melosh, 1989, p. 186).
[c] R is calculated by using equation (9.25).

slope. The spatial density is somewhat variable over the range of fault lengths, although the number of faults may be too few for compelling patterns to become apparent. The spatial density of normal faults for lengths $L > 300$ m is comparable to, if not somewhat greater than, that for smaller fault lengths, perhaps supporting an inference of truncation bias at those shorter lengths.

9.6.1.3 Comparison of the Diagrams

Each of the three types of discontinuity-population diagrams reveals a different component of the problem. The **D–L diagram** (Fig. 9.25) shows a linear trend for the normal fault data of Dawers et al. (1993). The exponent of the scaling relationship (equation (9.20)) can be taken to be $n = 1$. This is a key observation that reveals how the individual faults grew into a population (Cowie and Scholz, 1992b; Clark and Cox, 1996). Substantial departures from linear scaling, or non-uniform slopes on the D–L diagram, would suggest

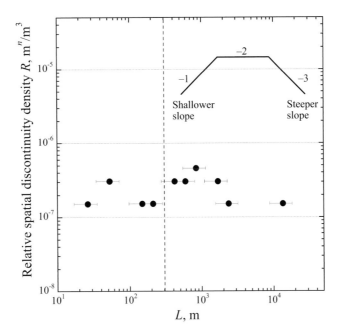

Fig. 9.29. *R*-plot diagram for normal faults from Bishop Tuff, California; data from Table 9.6. Inset shows expected orientations of three differential slopes; error bars illustrate bin widths.

instead that a more complex, or at least more comprehensive, physical model might be useful to describe the D_{max}/L relationship for that dataset.

Next, let's consider only the discontinuity lengths (in horizontal, map view). The **cumulative-frequency vs. length diagram** (Fig. 9.28) portrays how the fracture lengths scale across the deformed area. The format and interpretation are largely analogous to the *Gutenberg–Richter relationship* from seismology (e.g., Wesnousky, 1999) that defines a linear slope on a log–log plot of number of earthquakes vs. their size. In both cases, the negative slope indicates that proportionally fewer large events (or discontinuities) are represented than medium sized ones, with smaller events or discontinuities being most common (e.g., Gillespie et al., 1993; Westaway, 1994; Carter and Winter, 1995). Typical population slopes are between –0.5 and –3 (e.g., Twiss and Marrett, 2010b). The intercept of the data distribution (actually, of the best-fit linear slope) defines a useful parameter: the background value or "productivity" of a tectonically active region (based on seismicity or fault density; *P* in equation (9.24a) and Fig. 9.24).

Cumulative slopes of discontinuity populations tend to **decrease with increasing strain** (Cladouhos and Marrett, 1996) if the discontinuities grow by segment linkage. In this case, discontinuities lengthen by joining with, and thereby eliminating and reducing the number of, smaller ones; this decreases the value for the productivity (Schultz, 2000a). At the same time, discontinuities lengthen; both effects contribute to a reduction in slope as the discontinuities interact mechanically, link, and accommodate greater displacements and strains (e.g., Wojtal, 1986). These effects can be obscured by artifacts introduced by sampling or measurement techniques. Mapping scales (e.g.,

Cowie et al., 1993a,b), DEM resolution (e.g., Manighetti et al., 2001), how one defines and measures discontinuity length—all these (and more) can censor the data, complicating interpreting the physics of discontinuity-population development. Continuing developments in statistical analysis of discontinuity sets (e.g., Belfield, 1998; Clark et al., 1999; Main et al., 1999; Borgos et al., 2000; Twiss and Marrett, 2010b) show promise for separating artifact from physics.

The **R-plot** (Fig. 9.29) shows an explicit comparison between the data and a model. Departures from a horizontal distribution on the diagram reveal the discrepancies between an assumed distribution (the model) and the actual one (the data). Additionally, it may be possible in some cases to infer changes in the dimension of the deformed region (the "small" to "large" discontinuity transition) because of differences in the value of the cumulative power-law slope used for normalizing the data. In many cases, the R-plot can provide a clearer assessment of the linearity of slopes gleaned from the cumulative-frequency vs. length diagram, which can be used to predict how many more (or fewer) discontinuities of any given length match the model distribution.

These representations of discontinuity population characteristics can be used to separate out the important variables that represent the physics of discontinuity population evolution. As with other techniques, one should strive to adopt a standard set of conventions so that your diagrams look like everyone else's (e.g., Crater Analysis Techniques Working Group, 1979). This systematic approach facilitates the comparison of work between particular datasets, discontinuity types, and investigators.

9.6.2 Effect of Mechanical Interaction on the Propagation and Scaling of Structural Discontinuities

The physical mechanisms that control how and why structural discontinuities propagate are diverse and complex, depending in part on displacement mode (e.g., cracks vs. stylolites vs. faults) and in part on lithology (e.g., coalescence of microcracks in low-porosity compact rocks vs. cataclasis and dilatancy in porous sedimentary rocks). Two important processes are **in-plane propagation** and **growth by segment linkage**. First we'll look at these processes and see how they may be recorded in discontinuity displacement profiles. Then we'll examine their expression on D–L diagrams.

In-plane propagation results when the stress state at the discontinuity's tipline becomes sufficiently large to overcome the strength, or resistance to propagation, of the rock adjoining the discontinuity there. Under LEFM conditions (i.e., brittle rocks and small-scale yielding), the tendency to propagate in-plane is assessed by comparing the stress intensity factor, K_I, calculated at the discontinuity tip, to the rock's dynamic fracture toughness, K_c. This is commonly referred to in the literature as radial propagation (e.g., Childs et al., 2017).

Geologic structural discontinuities can also propagate in their own planes if the stress state near their tiplines equals the **yield strength** of the surrounding rock (e.g., Cowie and Scholz, 1992b; Rubin, 1993a). Du and Aydin (1993, 1995) suggested that (mode-II) deformation bands or strike-slip faults, in map view, can also propagate in-plane when the shear stresses achieve a sufficient value relative to the host rock's yield strength. Lin and Parmentier (1988) suggested the same for normal faults in cross-section (i.e., mode-II). The plaster models of Fossen and Gabrielsen (1996) clearly demonstrate the in-plane growth of faults (in this case, normal faults observed in cross-section, mode-II). In-plane propagation of dip-slip normal or thrust faults in map view (mode-III) has been well documented in the field (e.g., King and Yielding, 1984; Cartwright et al., 1996; Cartwright and Mansfield, 1998; Morewood and Roberts, 2002). Tectonic joints and igneous dikes (as mode-I cracks) also propagate by near-tip stress concentrations, leading to longer segments, echelon geometries, and populations of segmented fractures (e.g., Olson, 1993). Thus, any type of geologic structural discontinuity could potentially propagate in-plane, regardless of whether its displacement mode (mode I, II, or III) of the near-tip stress state meets or exceeds the value for the appropriate yield criterion there.

Structural discontinuities that grow by in-plane propagation will tend to keep pace, on average, between their maximum displacement (D_{max}) and length (L), as shown in Fig. 9.30. Changes in scaling can occur when either the near-tip stress state or the host rock's yield strength change; these changes can cause the scaling to diverge from linear. For example, **mechanical interaction** between

Fig. 9.30. Displacement profiles along isolated structural discontinuities of different lengths L_1 and L_2 showing how the distribution of displacements, as a function of position along the discontinuity trace, along with the value of maximum displacement D_{max}, grow in proportion with the discontinuity length. Note that, for interacting discontinuities (lower panel), the physical length (L_4) exceeds the effective, mechanical length (L_3) of the array. The same procedure can be done by assuming linear displacement distributions, rather than the elliptical ones shown here.

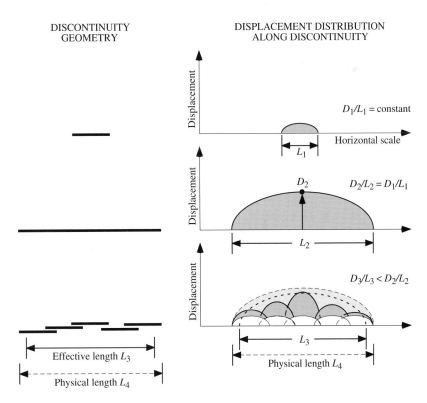

DISCONTINUITY GEOMETRY

DISPLACEMENT DISTRIBUTION ALONG DISCONTINUITY

closely spaced discontinuities (including *echelon* geometries; Fig. 5.6) has the effect of strengthening the rock adjacent to the tipline (Segall and Pollard, 1980; Gupta and Scholz, 2000a,b), leading to "anomalous" (i.e., larger) displacement values measured along interacting discontinuities (e.g., Willemse, 1997). These anomalous displacements have been identified and measured for interacting normal faults by many including Cartwright et al. (1995), Moore and Schultz (1998), and Morewood and Roberts (2002). The interaction, when sufficiently large, will "pin" the inner discontinuities' tips at a particular value of overlap and separation (Pollard and Aydin, 1988; Martel and Pollard, 1989; Aydin and Schultz, 1990)—effectively *freezing in* the echelon geometry. Any further displacement along the discontinuities will nucleate **linking structures** that will propagate out-of-plane—across the stepover.

Once they have grown into an overlapping echelon configuration, discontinuities grow to greater lengths by **segment linkage** (e.g., Segall and Pollard, 1980; Martel, 1990; Cartwright et al., 1995, 1996; Dawers and Anders, 1995; Crider and Pollard, 1998). A classic example of fault growth by segment linkage is given by the grabens of Canyonlands National Park, Utah (McGill and Stromquist, 1979; Cartwright et al., 1995; Moore and Schultz, 1998). Another clear example, in this case, of strike-slip faults nucleating on pre-existing joint arrays, was documented by Martel (1990).

The **efficiency of segment interaction and linkage** can be calculated as the ratio between the measured displacements (summed within the stepover; e.g., Dawers and Anders, 1995) and the measured or predicted maximum displacement of a single discontinuity of the same total length (see Fig. 9.31).

Fig. 9.31. Displacement profiles along two echelon discontinuities. The "excess" or anomalous displacement (dark shading), due to mechanical interaction between the discontinuities, produces a steeper displacement gradient at the inner discontinuity tips (within the stepover). The position of D_{max} also shifts from the discontinuity midpoint (vertical dashed lines) inward toward the stepover, as measured by the *skew*. The linkage efficiency is obtained by comparing the measured (and summed) profile to the theoretical profile (dashed line with D_0).

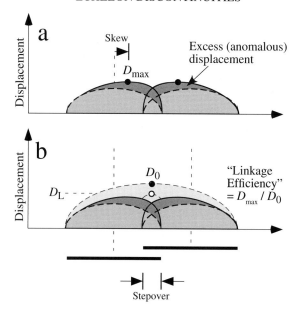

DISPLACEMENT DISTRIBUTION ALONG
ECHELON DISCONTINUITIES

This ratio provides a useful way to quantify the degree of mechanical inter-action (or, equivalently, the degree of "soft-linkage") between closely spaced discontinuities. However, regardless of the specific criterion for propagation (dynamic fracture toughness, yield strength, or shear stress), a reasonably good correspondence between D_{max} and L will be achieved and maintained for these discontinuities.

9.6.3 Discontinuity Growth Paths on the *D–L* Diagram

In this section we'll examine two common ways that geologic structural disconti-nuities can propagate: in-plane growth and growth by segment linkage. In doing so, we must also consider the three-dimensionality of the discontinuities (i.e., vertical restriction) when considering their growth paths on the *D–L* diagram.

The simplest case occurs for discontinuities that grow in-plane (i.e., radially, from the discontinuity tips) and whose aspect ratio *L/H* or length ratio 2 *L/H* remains constant during growth. In this case, the displacement–length ratio remains constant as the discontinuity grows in length (and, of course, in displacement and in height). This case of *proportional growth* is shown in Fig. 9.25 for normal faults in Bishop Tuff. The ratio between the maximum displacement and the discontinuity length remains a constant for these mem-bers of the discontinuity population.

Now let's see what happens for in-plane growth when the discontinuity aspect ratio changes. This important case arises for discontinuities that are vertically **restricted** (e.g., Nicol et al., 1996) to particular dimensions (such as down-dip height) by physical constraints like seismogenic thickness, rhe-ology, or stratigraphy. As a discontinuity grows in size, it initially is small enough not to feel the effects of the enclosing medium (so can be treated as an "isolated" discontinuity in LEFM). However, the discontinuity will, at some point, be limited—or restricted—in growing any larger in the vertical (height) dimension. Growth can then only occur in length, which requires that the aspect ratio (*L/H*) also must increase. A nonlinear growth path begins to become apparent for this case, as demonstrated for normal faults in the Spanish Pyrenees by Soliva et al. (2005). This case corresponds to *non-pro-portional growth*. For other discontinuity sets for which the change in aspect ratio is not as great (i.e., within the range of $2 < L/H < 3$ suggested by Nicol et al., 1996 to be typical of many faults), the degree of nonlinearity may be small, or indistinguishable within the scatter of the data.

In-plane, non-proportional growth on a *D–L* diagram is shown in Fig. 9.32. Initially, growth of discontinuities that are sufficiently small not to impinge on layer boundaries may be proportional (Fig. 9.32a, left-hand side, open dots). Trapping of the discontinuity within a layer leads to a vertically restricted discontinuity, increasing aspect ratio, non-proportional growth, and a shal-lowing slope on the *D–L* diagram (Fig. 9.30a, right-hand side, filled dots; Fossen and Hesthammer, 1997, 1998; Schultz and Fossen, 2002; Soliva et al.,

Fig. 9.32. Displacement–length scaling of discontinuities that grow faster in one direction (e.g., length) than the other (e.g., height; after Schultz and Fossen, 2002). (a) **Confinement of discontinuities by stratigraphy** leads to non-proportional growth (solid symbols) until the discontinuities break through bedding and cut up and/or down section. Discontinuities may grow non-proportionally again if confined to a thicker layer (arrow). (b) **Discontinuity linkage** produces composite discontinuities having different aspect ratios than those of the unlinked segments; trajectory on plot (arrows) depends on direction of linkage (along-strike vs. down-dip).

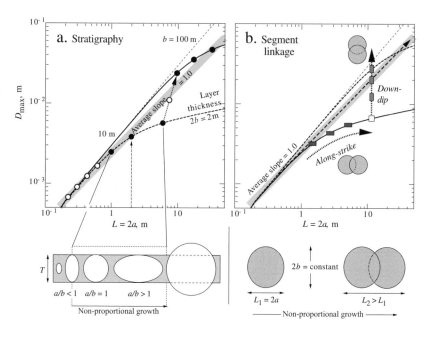

2005). Once the discontinuity breaks through the layer, however, it can follow a steeper trajectory on the $D–L$ diagram until it becomes confined within a thicker sequence (Fig. 9.32a). Fossen and Hesthammer (1997) suggested that a steeper growth path (ascending arrow in Fig. 9.32a; e.g., Nicol et al., 1996) probably connects deformation bands within the Entrada Sandstone to the larger, *faulted bands* studied by Krantz (1988).

A steeper growth path indicates that displacements accumulate faster than lengths, as can happen along reactivated discontinuities (Walsh et al., 2002) or during basin growth (e.g., Jackson et al., 2016). This latter scenario is referred to as a constant-length fault model. This scenario leads to a stair-stepped trajectory on the $D–L$ plot in which displacement increases for a given discontinuity length (Fig. 9.29), so that a discontinuity may be "over-displaced" relative to an isolated discontinuity of the same length. Eventually the discontinuity may propagate, rejoining the linear scaling relation. Ductile shear zones may grow in conjunction with this mechanism (Fossen and Cavalcante, 2017).

Next we examine discontinuity growth by **segment linkage**. Cartwright et al. (1995) demonstrated this mechanism in their studies of faulted joints in Canyonlands National Park (Fig. 9.33). They suggested that closely spaced, echelon normal faults there first linked up, leading to a rapid increase in (horizontal) length at a constant value of displacement. Then the "under-displaced" longer fault accumulates displacement, while remaining at the same new length, due to tip restriction, until the initial D/L ratio is again re-established. This growth mechanism and second, stair-step trajectory is now one of the fundamental

Fig. 9.33. Variable displacement gradients related to tip interaction are clearly expressed along the graben-bounding normal faults in Canyonlands National Park, Utah. Graben width (Devils Lane graben; view is to the north across S.O.B. Hill) is about 200 m. Photo by Matthew Soby.

canons in geologic fracture mechanics (e.g., Cowie and Roberts, 2001; Fossen and Gabrielson, 2005; Fossen, 2010a; Childs et al., 2017).

Linkage of discontinuities leads to trajectories on the _D–L_ diagram analogous to the effect of a confining stratigraphy (Fig. 9.32b) for the case in which horizontal (along-strike) linkage is faster than vertical (down-dip) linkage. Because along-strike (horizontal) linkage serves to _increase the aspect ratio_ of the composite (linked) discontinuity pair, linked arrays behave as "longer" elliptical discontinuities that can accommodate less displacement per unit map length than their 2-D ("tall") counterparts, as demonstrated in numerical simulations (Willemse and Pollard, 2000). Discontinuities and arrays that increase in length L by (along-strike) segment linkage follow a progressively shallowing growth path on the _D–L_ diagram (Fig. 9.32b). On the other hand, down-dip (vertical) linkage increases height instead of length, leading to a decrease in aspect ratio (to "taller" discontinuities) and an increase in the relative amount of displacement accommodated along the aggregate discontinuities. Down-dip linkage (by itself, with no increase in L) would be associated with vertical growth paths on the _D–L_ diagram (Fig. 9.32b).

Discontinuity growth by segment linkage in three dimensions would follow a unit slope if the rates of segment linkage were comparable, on average, in both directions (shaded bars in Fig. 9.32). Preferential propagation and/or linkage in either direction would lead to displacement discrepancies relative to 2-D expectations and consequent scatter on the _D–L_ diagram (Schultz and

Fossen, 2002). However, the "scatter" may indicate that particular discontinuity growth paths had not been identified in addition to the usual measurement uncertainties in discontinuity dimension and displacement magnitude.

9.6.4 Summary of Discontinuity Statistics and Scaling

As a geologic structural discontinuity grows, it accumulates displacements (normal or parallel, depending on its displacement mode). Its length L, measured in the horizontal plane, also increases, but at a rate depending on the degree of vertical or horizontal restriction (Fig. 9.34); this promotes changes in other statistical measures of the discontinuity population. One way to assess this is by quantifying its **aspect ratio** (length/height).

If the discontinuity grows in length and displacement at constant aspect ratio, then L/H becomes part of the constant of proportionality γ in $D_{max} = \gamma L$. **Proportional growth**—linear D–L scaling—occurs when all parameters that affect γ remain constant for discontinuities in the population as they grow. Mechanical interaction ("soft-linkage"), and eventual segment ("hard") linkage between closely spaced echelon discontinuities produce *transient departures* from linear D–L scaling that may smooth themselves out as the discontinuities continue to grow and accommodate displacements.

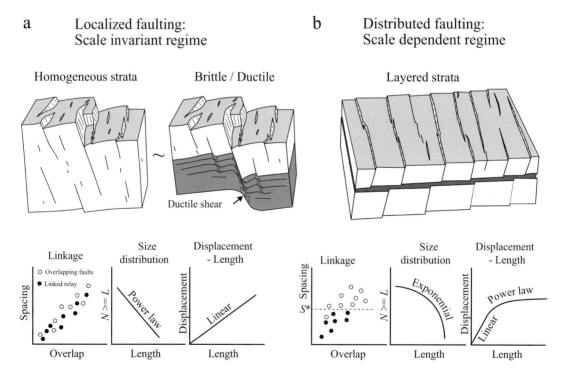

Fig. 9.34. Two end-member examples of discontinuity population statistics, after Soliva and Schultz (2008) and Schultz et al. (2010c). (a) Localized faulting along normal faults and length-dominated scaling relations. (b) Distributed faulting along faults and height-limited scaling relations. Lower panels show population statistics that would develop for each of the two examples.

Length–frequency statistics are characterized by **power-law** distributions, and the discontinuities are considered to be "small" and 2-D compared with the thickness of the fractured rock mass. Mechanical interaction and discontinuity spacing are modulated by the discontinuity aspect ratio.

Confinement of discontinuities to a mechanical or lithologic stratigraphy (like bedding) requires that their heights (in the vertical dimension) remain limited by the relevant thickness, whereas their lengths can continue to increase (e.g., Olson, 1993; Gudmundsson, 2000; Soliva and Benedicto, 2005). The restricted discontinuity then follows **non-proportional growth**. First, it becomes progressively harder to accumulate displacements as the discontinuity lengthens, leading to a reduced slope on the D–L diagram (for cases where the vertical discontinuity tips are impeded by stronger interfaces, rock types, and/or rheologic conditions). Because discontinuity spacing is related to the displacement magnitude, systematic reduction in relative displacement leads to a *stabilization of spacing* (Fig. 9.34). Second, the length–frequency statistics may change to an **exponential** distribution, reflecting the increased abundance of intermediate-length discontinuities in the deforming layer which may be related, in part, to a change from length scaling of the mechanical interaction to height scaling (e.g., Cowie et al., 1993a,b; Olson, 1993, 2003; Gudmundsson, 2000; Soliva et al., 2005). Thus, the 3-D mechanical interaction between discontinuities in a population largely can play a large role in the spatial organization and associated statistics of the aggregate population.

9.6.5 Brittle Strain

An extensive literature exists on the amount of strain accommodated by a population of discontinuities (see, e.g., Kostrov, 1974; Segall, 1984a,b; Marrett and Allmendinger, 1990; Scholz and Cowie, 1990; Westaway, 1994; Marrett, 1996; Scholz, 1997; and Borgos et al., 2000, for representative approaches). One would usually begin a calculation by doing a 1-D (line) traverse across a deformed region that displays faults, joints, deformation bands, or other type of geologic structural discontinuity (Fig. 9.35). The amount of displacement

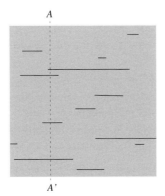

Fig. 9.35. A one-dimensional (1-D) sampling traverse (A–A′) across a sparse population of structural discontinuities.

across each discontinuity would be measured and then corrected to take into account only the component of displacement parallel to the traverse (i.e., to correct the values for the discontinuity's strike and dip; Peacock and Sanderson, 1993; Scholz, 1997). For very closely spaced discontinuities ("penetrative deformation") one could trace the offset of a passive marker from one side to the other, instead of measuring all the discontinuities' displacements (e.g., Pappalardo and Collins, 2005). In this section we'll see how brittle strain accommodated across a population of discontinuities can be calculated.

Here are some general principles.

- Longer discontinuities accommodate larger strains than shorter ones (all else held equal).
- A population of discontinuities, no matter how "penetrative" at a particular scale, contains discontinuities with a wide range of lengths.
- Discontinuities within a population may be either free to grow in any direction, or they may be restricted to a particular dimension by the deforming region.
- Discontinuity populations tend to be sparse; i.e., low density with wide spacings (e.g., Segall, 1984a; Barnett et al., 1987). In such sparse populations, it is possible to miss many discontinuities by using only a 1-D traverse. In addition, one may not be sampling each discontinuity at its value of maximum displacement. This sampling method works well for penetrative deformation, and methods for calculating average values of strain are available in any structural geology textbook. But for sparse systems, like most discontinuity populations, this method may not adequately characterize the population characteristics or brittle strain that it had accommodated.

Instead of a 1-D line traverse for calculating strain, an alternative method for obtaining discontinuity-related brittle strain comes to us from *seismology*, drawing on the close parallel between incremental displacements accumulated along rupture patches during an earthquake and the total (or cumulative) displacements accumulated along faults (e.g., Segall, 1984a; Scholz and Cowie, 1990; Scholz, 1997). The calculation of **brittle strain from discontinuity populations** is a customary and standard practice in structural geology and tectonics, and this section should help you to become better acquainted with it from the point of view of geologic fracture mechanics.

In this section, "brittle" is used to denote the strain distributed among, and accommodated by, a population of geologic structural discontinuities. This usage is in keeping with the relevant literature. As such, it is not meant to convey any mechanical implications such as brittle vs. ductile deformation or a comparison between near-tip stress state and the host rock's various yield strengths.

Recalling that any structural discontinuity has three key dimensions: *length* (taken to be the horizontal dimension), *maximum or average displacement* (usually measured near the discontinuity's midpoint), and *height* (measured normal to length), the procedure for calculating brittle strain is straightforward. Succinctly, one obtains the three key variables (L, D, and H) for each discontinuity in the population, sums them, and then divides them by the dimension of the deforming region.

First the amount of new volume created by displacement along the discontinuities within a region is calculated. This quantity is called the **geometric discontinuity moment** M_g which is given by

Geometric discontinuity moment

$$M_g = DLH \tag{9.26}$$

with *average* displacement D, discontinuity length L, down-dip discontinuity height H, and units of m³ (King, 1978; Scholz and Cowie, 1990; Ben-Zion, 2001). In this expression, D is the average offset along or across the discontinuity (*not* D_{max}), measured in the plane of the discontinuity and discussed below; it is not (yet) the component in a horizontal plane, as will be needed later for the calculation of the normal strains for extension or contraction.

The geometric discontinuity moment represents the volume of deformed rock associated with a discontinuity, and it applies to faults, joints (e.g., Segall, 1984a), or any other discontinuity type. As a discontinuity grows in size, its surface area increases; given that a discontinuity's displacement scales with L, the geometric moment M_g increases as a particular discontinuity grows in size and in displacement. The geometric discontinuity moment can be interpreted as:

- The volume increase across a population of opening-mode fractures, such as cracks;
- The volume of gouge and fault rock along a population of shearing mode-II or mode-II fractures (faults); or
- The volume decrease across a population of closing-mode discontinuities, such as anticracks.

The geometric discontinuity moment can be understood within the context of a 4 × 4 matrix of rock deformation dimensionality (Dershowitz, 1984; Dershowitz and Herda, 1992; Mauldon and Dershowitz, 2000). This matrix, shown in Fig. 9.36, underpins discrete-fracture network (DFN) approaches (e.g., Jing and Stephansson, 2007; Elmo et al., 2014). In the figure, the dimensionality of the sampling domain is given by the rows and first subscript of the matrix $\mathbf{P}_{ij}$, and the dimensionality of the discontinuity attribute is given by the columns and second subscript. For example, a linear traverse across a horizontal sampling plane (i.e., map view set of discontinuity traces; Fig. 9.35) would correspond to a 1-D sampling domain (second row in Fig. 9.36) and a number of discontinuity traces intersected along the traverse (column 0 in Fig. 9.36), leading to $\mathbf{P}_{10}$ and a discontinuity frequency or linear density.

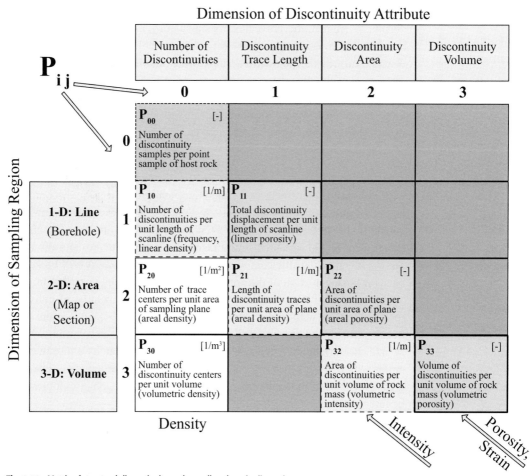

Fig. 9.36. Matrix of structural discontinuity and sampling domain dimensions.

The amount of deformation attributed to each discontinuity is given by a related scalar quantity, the **quasi-static discontinuity moment** M_f (after Pollard and Segall, 1987, p. 302):

Quasi-static discontinuity moment

$$M_f = GAD = GM_g = GDLH \qquad (9.27)$$

where G is the shear modulus of the surrounding rock mass, A is the surface area of the discontinuity as defined by its shape (length L times height H), and D is the average (relative) displacement across the discontinuity. M_f has units of MJ (joules $\times$ 10^6) for shear modulus in 10^6 Pa and L, H, and D in meters. An additional constant related to the shape of the discontinuity surface (circular, elliptical, rectangular) is sometimes added to this expression (e.g., Kostrov, 1974). The quasi-static discontinuity moment represents the energy consumed by the rock mass in producing the displacements along the discontinuity. The quantity is quasi-static because no dynamic terms are included in the formulation and the system is considered to be in static equilibrium. This quantity corresponds to the seismic moment of an earthquake (e.g., Twiss and Moores, 2007, p. 439).

The **total work done by discontinuity-related deformation** W_T is the sum of the quasi-static discontinuity moments for all discontinuities in a region:

Total (cumulative) work done by discontinuity-related deformation

$$W_T = \sum_{i=1}^{N} M_{fi} \qquad (9.28)$$

where M_{fi} is the quasi-static moment for each ith discontinuity and N is the total number of discontinuities in the deforming region. The work also has units of energy (MJ) or, equivalently, 10^6 N m. The quasi-static discontinuity moment represents the total energy consumed by the rock mass in producing the discontinuity displacements within the region. W_T does not explicitly depend on the *size* of the deforming region that contains the discontinuities, and it also omits the generally smaller contributions of processes such as discontinuity formation. Although there are some implicit relationships between the quantities in (9.27) and (9.28) and region size (e.g., A may be limited by stratal or crustal thickness (Scholz and Cowie, 1990; Westaway, 1994), and D and G may depend on scale and driving stress (Cowie and Scholz, 1992b; Schultz et al., 2006)), the total work done by discontinuity-related deformation represents an efficient way to quantify the role of structural discontinuities in brittle deformation.

Brittle strain is a tensor quantity having components such as normal and shear strain in various directions. For example, the strain accommodated by a population of **normal faults** will have a component of *extension*, perpendicular to the average strike of the faults (their "extension direction"), another component of extension parallel to the fault strike which will be zero for plane strain Anderson faults and commonly small for three-dimensional strain (Krantz, 1988), and a component of contraction in the vertical direction, corresponding to crustal thinning (e.g., Wilkins et al., 2002). Similarly, a **thrust fault** population will have a component of *contraction* perpendicular to the average strike of the faults (i.e., along their shortening or *vergence* direction), another, likely smaller, component of contraction parallel to the fault strike, and a component of extension in the vertical direction, corresponding to crustal thickening. These two cases of dip-slip faults are considered further in this section; the shear strain parallel to a population of strike-slip faults is also given (later) by this method. Comparable sets of length changes and strains can be calculated for populations of dipping shear deformation bands or, indeed, for any type of geologic structural discontinuity.

The required components of the strain tensor can be obtained by using either of two methods. First, all the information needed for equation (9.28)— the geometric discontinuity moment M_g—can be specified, along with fault dip, fault strike, and displacement rake for each fault. The component of interest can be obtained by solving using Kostrov's (1974) equation

Kostrov's formula for
brittle strain

$$\varepsilon_{kl} = \frac{1}{2V} \sum_{i=1}^{N} M_{gi} \tag{9.29}$$

as outlined for example by Aki and Richards (1980), pp. 117–118, and which has been used extensively in seismotectonics and structural geology (e.g., Molnar, 1983; Scholz and Cowie, 1990; Westaway, 1992; Scholz, 1997; Scholz, 2002, pp. 306–309; Wilkins et al., 2002; Schultz, 2003a,b; Dimitrova et al., 2006; Knapmeyer et al., 2006). Alternatively, measurements along a traverse can be taken, corrected explicitly for fault strike and dip angle, and then substituted into a set of strain equations that already have the dip correction incorporated into them (e.g., Scholz, 1997). Both methods will produce the same results.

The strike correction, illustrated here for normal or thrust faults, is the component of discontinuity displacement D in a particular horizontal direction (e.g., Priest, 1993, pp. 96–97; Peacock and Sanderson, 1993; Westaway, 1994). This is obtained by calculating the component of discontinuity displacement D_s along the direction of interest, such as a traverse line (such as one perpendicular to the average strike of a set of faults), by using

$$D_s = D_{max} \left| \cos\left(\Delta\psi\right) \right| \tag{9.30}$$

in which $\Delta\psi$ = (strike of fault ψ − strike of traverse ψ). The component of horizontal displacement along the traverse direction is given by

$$D = D_{max} \left| \cos\left(\Delta\psi\right) \right| \cos\delta \tag{9.31}$$

which includes the dip correction given by the last term in equation (9.31). A uniform dip angle δ, and a strike correction of $\Delta = 0°$, leads to the term $D = D_{max}(2 \sin\delta \cos\delta)$ when the direction cosine of the discontinuity plane is taken into account (e.g., Scholz, 1997).

The other correction that must be made to the displacement data is to reduce the value of D_{max} to an average value of displacement per unit discontinuity length. The **average displacement** D is used in fault-set inversions for paleo-stresses (e.g., Marrett and Allmendinger, 1990; Angelier, 1994), as well as for brittle strain (e.g., Scholz and Cowie, 1990; Scholz, 1997). The average displacement $D = \kappa D_{max}$, where κ is a fraction of the maximum displacement D_{max}, depending on the specific displacement distribution along the discontinuity. Representative values of κ are calculated and shown in Table 9.7 for different displacement distributions. A discontinuity having a linear displacement profile has $\kappa = 0.5$, whereas one with an ideal elliptical profile (assuming LEFM conditions) has $\kappa = 0.7854$. Dawers et al. (1993) obtained values of κ for small normal faults in Bishop Tuff of 0.61; Moore and Schultz (1999) found values of κ between 0.3 (for nearly linked echelon faults) and 0.7 for normal faults from Canyonlands National Park. Combined with the dip correction, this leads to $D = \kappa D_{max} (2 \sin\delta \cos\delta)$.

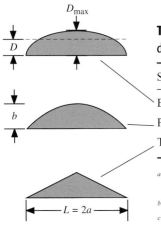

Table 9.7 Conversion factors from maximum displacement (D_{max}) to average displacement (D) [a]

Shape [b]	Equation [c]	D/D_{max}
Ellipse (LEFM)	$(1/2\ \pi ab)/2a = \pi/4b$	0.79
Parabolic slice [d]	$(4/3ab)/2a = 2/3b$	0.67
Triangular (linear)	$(ab)/2a = 1/2b$	0.50

[a] Average displacement = displacement area / discontinuity length (Dawers et al., 1993); $D_{max} = b$, discontinuity length $L = 2a$.
[b] Displacement distribution along discontinuity.
[c] Assumes $a = 1.0$, which drops out in third column when normalized by $L = 2a$.
[d] A nearly linear shape with vertex at discontinuity midpoint corresponding to D_{max} (Ginsberg and Genin, 1977, p. 418); similar to a bell-shaped profile from an end-zone model (e.g., Walsh and Watterson, 1987; Cowie and Scholz, 1992b).

Using these corrections to D_{max}, the horizontal strains due to a particular structural discontinuity can be calculated. The brittle normal strain ε is the horizontal component of the geometric discontinuity moment M_g normalized by the dimensions of the deformed region having thickness T, area A, and volume $V = TA$. For "small" discontinuities (e.g., Scholz and Cowie, 1990; Scholz, 1997), $H_i < T/\sin\delta_i$; for "large" discontinuities, $H_i = T/\sin\delta_i = H_0$, so the horizontal normal strain (assuming constant discontinuity dip angles) is obtained from Kostrov's (1974) equation (9.29) as (e.g., Scholz, 1997)

Brittle horizontal normal strain for dip-slip discontinuity populations

$$\varepsilon_n = \frac{\sin\delta\cos\delta}{V}\sum_{i=1}^{N}(D_iL_iH_i) \quad \text{for small discontinuities}$$

$$\varepsilon_n = \frac{\cos\delta}{A}\sum_{i=1}^{N}(D_iL_i) \quad\quad \text{for large discontinuities}$$

(9.32)

in which δ is fault dip angle and D is the average displacement on a particular discontinuity (using the correction for average displacement from D_{max}). The first of equations (9.32) is for small discontinuities, the second is for large ones. "Large" discontinuities in a population, as noted above, are considered to be vertically restricted; "small" discontinuities in a population are unrestricted (Fig. 9.21c). The sign of D must be specified for these equations, which is done by using

$D > 0$ *Normal sense*

$D < 0$ *Thrust sense*

Using this convention, extensional normal strain will be positive and contractional normal strain will be negative (Fig. 6.22).

The geometric moment corresponds to $\mathbf{P}_{33}$ once it is normalized by the volume V of the deforming domain. This implies that the diagonal terms $\mathbf{P}_{11}$, $\mathbf{P}_{22}$, and $\mathbf{P}_{33}$ on Fig. 9.34 (shaded, called "porosity" in the matrix) correspond to 1-D, 2-D, and 3-D normal strains, respectively.

In these equations, the geometric discontinuity moment M_g is calculated for the component of the complete moment tensor or the population in the horizontal plane and normal to discontinuity strike (e.g., Aki and Richards, 1980, pp. 117–118; Scholz, 1997). These equations are thus defined for discontinuities having pure normal or thrust offsets, with rakes of 90°, and provide the horizontal normal strain (e.g., extension for normal faults, contraction for thrust faults) accommodated by the discontinuity population perpendicular to its strike and in the horizontal plane. Analogous equations can be defined for the shear strain accommodated in the horizontal plane by a population of strike-slip faults or deformation bands.

The **vertical normal strains** associated with dip-slip discontinuities are given by (Aki and Richards, 1980, pp. 117–118)

Vertical strain for dip-slip discontinuity populations

$$\varepsilon_v = \frac{\sin\delta\cos\delta}{V} \sum_{i=1}^{N} (D_i L_i H_i) \quad \text{for small discontinuities}$$

$$\varepsilon_v = \frac{\cos\delta}{A} \sum_{i=1}^{N} (D_i L_i) \qquad \text{for large discontinuities}$$

(9.33)

The first of equations (9.33) is for small discontinuities, the second is for large ones. Note the sign change relative to the fault-normal (horizontal) strains: this is the "odd axis" strain of Krantz (1988, 1989; Fig. 6.22). For normal faulting, this strain component quantifies the amount of layer **thinning**; whereas for thrust faulting, it corresponds to the amount of **thickening** of the faulted section.

The *shear strain* parallel to strike in a population of **strike-slip discontinuities** is given by

Brittle tectonic shear strain for strike-slip discontinuity populations

$$\varepsilon_s = \frac{1}{2V} \sum_{i=1}^{N} (D_i L_i H_i) \quad \text{for small discontinuities}$$

$$\varepsilon_s = \frac{1}{2A} \sum_{i=1}^{N} (D_i L_i) \quad \text{for large discontinuities}$$

(9.34)

The extension (or contraction) component either perpendicular or parallel to the strike-slip discontinuities is equal to zero, indicating pure strike-slip deformation. This would change if either the rakes were different from 0° or if the discontinuity dips were different than 90°. One can also use equations (9.34) to calculate the shear strain parallel to a discontinuity array, such as in the dip direction of the Bishop Tuff normal faults.

Many structural discontinuities can be characterized as having $D_{max}/L = \gamma = $ constant (e.g., Scholz and Cowie, 1990; Cowie et al., 1993a,b; Clark and Cox, 1996), as discussed above. Using the displacement–length scaling relations, the equations for the horizontal normal strain component can be written without the explicit term for discontinuity displacement magnitude as

Brittle horizontal normal strain for dip-slip discontinuity populations using D_{max}/L scaling

$$\varepsilon_n = \frac{\kappa \gamma_0 \sin\delta \cos\delta}{V} \sum_{i=1}^{N} \left(L_i^2 H_i \right) \quad \text{for small discontinuities}$$

$$\varepsilon_n = \frac{\kappa \gamma_r \cos\delta}{A} \sum_{i=1}^{N} \left(L_i^2 \right) \qquad \text{for large discontinuities}$$

(9.35)

(see also Wilkins et al., 2002, and Schultz, 2003a,b).

In these equations γ is the D/L ratio for unrestricted discontinuities, and γ_r is the minimum D/L ratio for stratigraphically restricted ones (Nahm and Schultz, 2011). Values of γ_0 and γ_r are available from datasets from Earth (Fumanyá, near Barcelona, Spain; Soliva et al. 2005), the northern plains of Mars (Polit et al. 2009), and from the Caloris impact basin on Mercury (Klimczak et al., 2013). For Fumanyá, $\gamma_0/\gamma_r = 0.04/0.011 = 3.64$; for grabens on the Martian northern plains, $\gamma_0/\gamma_r = 0.001/0.0005 = 2.0$; and for grabens on Mercury's Caloris basin, $\gamma_0/\gamma_r = 0.01/0.003 = 3.3$. Although γ_r is not a constant, its minimum values provide an estimate of the difference between the two strain solutions.

Example Exercise 9.3

Using the population of normal faults from the Bishop Tuff of California (Dawers et al., 1993; Scholz et al., 1993), calculate the following:

(a) The work done by faulting.

(b) The tectonic strain.

Solution

(a) The first step is to calculate the *geometric discontinuity moments* for each normal fault in the population. Because it has been suggested that faults less than ~300 m long were considered to be "small" relative to the 150-m thickness of the tuff (Dawers et al., 1993; Scholz et al., 1993), larger faults will be considered as either small or large to illustrate the differences in moment and strain. Because the faults are surface-breaking, their geometries are best described as *half-ellipses*, their shapes are taken to be given by length ratios of $2a/b = 2 L/H = 2.0$. The maximum down-dip height for the small-fault case is $H_0 = T/\sin\delta$, using $\delta = 75°$ and $T = 150$ m, so that $H_0 = 155.3$ m. For a length ratio of $L/H_0 = 2.0$, $L = 310.6$ m for the longest "small" fault (or the shortest "large" fault). This value falls between those inferred independently by Dawers et al. (1993) from the shapes of the fault displacement profiles.

For the "large" faults, we will investigate the results if the normal faults were restricted to the 150-m thickness of the tuff as compared to if they were not. Large faults require that the aspect ratios of the faults increases with L, since $H = H_0$ because the fault height would be limited by the stratigraphic thickness.

Table 9.8 lists the results of the calculations for the Bishop Tuff fault population. The total geometric discontinuity moment is **3.15×10^8 m³**

Table 9.8 Moment calculations for Bishop Tuff normal faults [a]

D_{max}, m	L, m	Fault [b]	Aspect ratio	H, m	Geometric moment, m³		Horizontal normal strain, % [c]	
					Small + Large	Small [d]	Small + Large	Small
0.45	24	small	2.0	12.0	129.6	129.6	4.82×10^{-7}	4.82×10^{-7}
0.93	55	small	2.0	27.5	1406.6	1406.6	5.24×10^{-6}	5.24×10^{-6}
1.04	60	small	2.0	30.0	1872.0	1872.0	6.97×10^{-6}	6.97×10^{-6}
3.42	167	small	2.0	83.5	47,690.2	47,690.2	1.78×10^{-4}	1.78×10^{-4}
3.52	182	small	2.0	91.0	58,298.2	58,298.2	2.17×10^{-4}	2.17×10^{-4}
5.07	440	large	2.83 or 2.0	155.3 or 220.0	346,424.1	490,776.0	1.29×10^{-3}	1.83×10^{-3}
9.73	500	large	3.22 or 2.0	155.3 or 250.0	755,492.8	1,216,250.0	2.81×10^{-3}	4.53×10^{-3}
10.46	540	large	3.48 or 2.0	155.3 or 270.0	877,148.1	1,525,068.0	3.26×10^{-3}	5.68×10^{-3}
8.70	696	large	4.48 or 2.0	155.3 or 348.0	940,320.6	2,107,209.0	3.50×10^{-3}	7.84×10^{-3}
10.60	740	large	4.77 or 2.0	155.3 or 370.0	1,218,106.0	2,902,280.0	4.53×10^{-3}	1.08×10^{-2}
10.46	780	large	5.02 or 2.0	155.3 or 390.0	1,266,991.7	3,181,932.0	4.72×10^{-3}	1.18×10^{-2}
8.59	866	large	5.58 or 2.0	155.3 or 433.0	1,155,203.6	3,221,061.0	4.30×10^{-3}	1.20×10^{-2}
19.15	1620	large	10.4 or 2.0	155.3 or 810.0	4,817,605.9	25,128,630.0	1.79×10^{-2}	9.35×10^{-2}
17.91	1630	large	10.5 or 2.0	155.3 or 815.0	4,533,469.2	23,792,539.5	1.69×10^{-2}	8.86×10^{-2}
32.09	2210	large	14.2 or 2.0	155.3 or 1,105.0	11,013,097.2	78,365,384.5	4.10×10^{-2}	2.92×10^{-1}
150.12	12,360	large	79.6 or 2.0	155.3 or 6,180.0	288,140,634.0	11,466,886,176.0	1.07×10^{0}	4.27×10^{1}
				Totals:	**315,173,889.9**	**11,608,926,703**	**1.17%**	**43.2%**
				% small:	**0.035%**			
			Totals (without largest fault):		**27,033,255.9**	**142,040,527.3**	**0.10%**	**0.53%**
				% small:	**0.035%**			

[a] D_{max} and L from Dawers et al. (1993); Table 9.4.

[b] Relative to thickness of faulted tuff layer, using fault dip $\delta = 75°$ and thickness $T = 150$ m.

[c] Assumes shear modulus $G = 0.1$ GPa $= 100$ MPa for the tuff.

[d] Values for "small+large" faults assume constant aspect ratio for lengths < 5 km, then variable aspect ratios for the large-fault portion of the population. Values for "small" faults assume constant aspect ratio for all fault lengths.

(Fig. 9.37a). Using an approximate value of shear modulus of $G = 0.1$ GPa for the tuff (Schultz, 2011a,b), the total cumulative geometric moment, or the total work done by normal faulting, is calculated from equation (9.28) to be **3.15×10^4 MJ**, or 3.15×10^{10} N m. The geometric moments, both incremental and cumulative, are shown in Fig. 9.37; the work done by faulting is just these values times a constant (the shear modulus, G).

(b) Brittle strain is calculated from the geometric discontinuity moments for all the faults, as obtained in part a, along with the thickness of the faulted

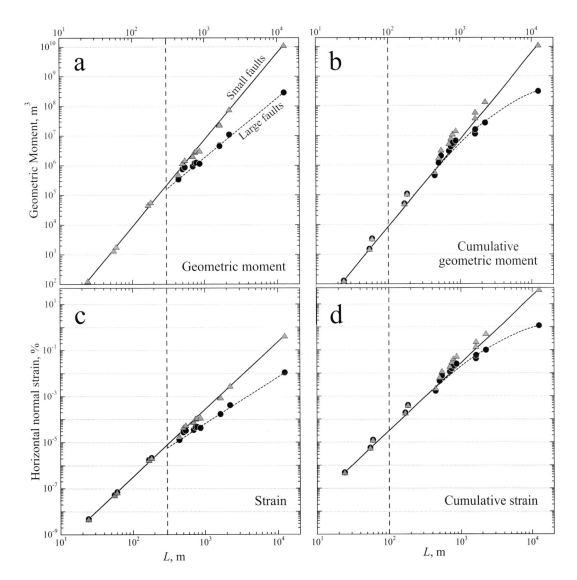

Fig. 9.37. Geometric discontinuity moments (a, b) and horizontal normal strain (c, d) calculated from normal fault population from Bishop Tuff, California. Dashed line at $L = 300$ m as in Fig. 9.27.

layer, T. Using a value for the area of the deformed tuff of ~300 km^2 with T = 150 m, the deforming rock has a volume of 45×10^8 m^3 (45 km^3). Using equation (9.32), the total horizontal extensional strain accommodated by the Bishop Tuff normal faults is either 1.17% for small and large faults or 43.7% for all small faults. The latter value seems large. Following the remarks on the cumulative-length distribution shown in Fig. 9.28, it may be plausible to consider that the longest fault is not a member of the Bishop Tuff normal fault population but helps to drive it, as discussed by Ferrill et al. (2016). If the Fish Slough fault were not included in the dataset, the horizontal normal strains would be 0.1% for the small-fault case and 0.53% for the small-and-large fault case. As implied on Fig. 9.37, stratigraphic restriction of fault heights (the large-fault case) reduces the geometric moments and associated horizontal normal strain accommodated by the normal-fault population. The cumulative horizontal normal strain obtained by excluding the longest (Fish Slough) fault, and assuming that all faults in the tuff were small and hence unrestricted at depth, would be 1.1%.

Calculation of the horizontal normal strains due to faulting in any dataset by using equations (9.32) or (9.35) can provide a robust measure of the extension (or contraction) and strain associated by a tectonic event. Similarly, the vertical strain (thinning or thickening) of a faulted section, or the shear strain in the plane of the faults (and in the shearing direction), can be obtained easily from an appropriate set of measurements of the fault population.

The magnitude of strain accommodated by a discontinuity population depends on factors such as the aspect ratio (L and H) of a given discontinuity, which in turn may be controlled by the degree of mechanical interaction, linkage, or stratigraphic confinement (restriction) of the discontinuity. The approach outlined in this section provides a means of extracting some of these factors out of the population statistics, especially when used in combination with the statistical diagrams discussed in this chapter.

9.7 Review of Important Concepts

1. Geologic fracture mechanics (GFM) can be thought of as an interdisciplinary field combining approaches from engineering, materials science, and geology. It includes Linear Elastic Fracture Mechanics (LEFM) but relaxes some of the assumptions that are required for LEFM to apply to geologic structural discontinuities.
2. Many characteristics of geologic structural discontinuities that imply non-LEFM conditions include: non-elliptical displacement profiles; non-parabolic near-tip displacement profiles; displacement–length scaling relations; discontinuity spacings; propagation at slow,

quasi-static rates rather than rapid, dynamic ones; propagation related to host-rock yield strength, rather than its dynamic fracture toughness; take-off angles of secondary structures; and in-plane propagation of shear discontinuities in mode II, such as deformation bands and strike-slip faults.

3. Environmental conditions within the Earth can exert a profound effect on the growth of geologic structural discontinuities. These conditions include temperature, pressure, (very low) strain rates, fluids, and chemistry.

4. Loading conditions influence the stability and rate of fracture growth. Stable fracture growth can occur under displacement-controlled, "fixed-grip" loading conditions that correspond to a stiff loading system. Conversely, unstable fracture can occur under stress-controlled, constant driving-stress, "dead-weight" loading conditions that correspond to a soft loading system.

5. An end zone is a region around and adjacent to the discontinuity tip (or tipline) that has partially broken (yielded) due to the elevated stress magnitudes generated away from the fracture tip by displacement sustained along the discontinuity. It is also referred to as a process or plastic zone. The size of the end zone is related to the near-tip stress distribution, discontinuity length, and host-rock strength. The location and orientation of secondary structures, such as wing cracks and splay faults, depend on the details of the end zone. Discontinuities that have end zones that are non-negligible in size imply semi-brittle conditions and are referred to as post-yield elastic–plastic structural discontinuities. Propagation of these structural discontinuities can be modeled by using an apparent dynamic fracture toughness, J-integral, or strain energy density methods.

6. Displacement–length scaling relations provide an observational foundation for assessing the main factors that influence the growth of geologic structural discontinuities. The value of maximum displacement (normal or shear) accommodated across a discontinuity can usually be related to its length through one or more systematic relationships, with important factors being the host-rock stiffness and brittleness, discontinuity dimensions and aspect ratio, and the mechanism of propagation. As structural discontinuities grow, they increasingly interact with their neighbors and with physical property changes, such as stratigraphic or rheological layering, leading to departures from the idealized relationships. These departures illuminate the rich diversity and complexity of processes by which faults, deformation bands, and other types of structural discontinuities grow and develop as part of natural geologic deformation. Scaling relations can be used to predict discontinuity spacings or length distributions in areas that may not be well sampled, such as in subseismic-scale applications.

7. Faults and disaggregation bands propagate when a shear yield strength is exceeded at their tips, leading to a linear dependence of maximum displacement and discontinuity length (unit slope). Dilatant cracks (joints, hydrofractures, veins, igneous dikes), cataclastic (compactional shear) deformation bands, and shear-enhanced compaction bands accommodate significant changes in volume across them, leading to sublinear scaling with a square root dependence of maximum displacement and discontinuity length. Accordingly, propagation of this latter type of structural discontinuity is governed by mode-I dynamic fracture toughness (K_{Ic}) and attendant dilatant cracking of grains in small near-tip process zones subjected to large, nearly singular near-tip stress concentrations. Shear zones are consistent with an exponent greater than one, implying that the rate of displacement accumulation along a shear zone is faster than that of length increase, consistent with nucleation on weak zones or surfaces, similar to reactivated faults, but with different mechanisms operating to promote weakness in the host rock, and growth assisted by segment linkage. Tectonic stylolites thicken (i.e., exhibit larger topographic amplitude, or "teeth") from minima near their tips toward their central portions; interpreting amplitude as a manifestation of the degree of dissolution and contractional strain accommodated across a stylolite, these discontinuities may scale linearly with the constant of proportionality depending on rock, fluid, and chemical properties along with pressure, temperature as well as the growth mechanism. Pure compaction bands exhibit an approximately constant and small magnitude of contractional displacement across a pure compaction band that appears to be largely independent of its length.

8. The population statistics of geologic structural discontinuities provide a means to explore inter-relationships between the discontinuity properties, such as length, displacement, spacing, and linkage. Proportional growth, and linear displacement–length scaling, occurs when all parameters remain constant for all discontinuities in the population. Mechanical interaction ("soft-linkage") and eventual segment ("hard") linkage between closely spaced echelon discontinuities, and vertical (stratigraphic or rheological) restriction can produce transient departures from linear displacement–length scaling that can be identified from various portrayals of the data. Length–frequency statistics that are fitted with power-law distributions imply "small" discontinuities relative to the thickness of the fractured rock mass; constant values of discontinuity aspect ratio; length-dependent spacing and stepover width; and strain dominated by the largest discontinuities. Exponential length–frequency distributions imply some mixture of small and large (restricted) discontinuities, variable (increasing) aspect ratios, scale-dependent spacing and stepover width; and strain distributed more evenly throughout the population of discontinuities.

9. The brittle strain accommodated by a population of structural discontinuities depends on their dimensions (length and height), displacement magnitude, attitude relative to a direction of interest, and the spatial dimensions of the deformed region. Strain magnitude increases as the slope of a power-law exponential distribution that describes the discontinuity population decreases, implying a greater degree of lengthening of discontinuities, and reduction of smaller discontinuities, in the population. Displacement–length scaling relations can be used to estimate brittle strains by extrapolating into domains having larger or smaller discontinuity lengths than represented in a particular dataset.

9.8 Exercises

1. Define subcritical fracture growth and discuss several geologic settings where you might encounter it.
2. Describe the sequence of physical mechanisms that control fracture propagation rates in geologic materials, using subcritical fracture growth as a guide.
3. What is the Charles Law and when is it applied in geology?
4. What is the Paris Law and when would you apply it in geology?
5. Define dynamic crack growth. What physical evidence would you look for to determine if a tectonic joint set propagated dynamically?
6. What are meant by stable and unstable fracture propagation?
7. Describe what is meant by the end zone and how its length can calculated. How would you assess whether a natural fracture conforms to LEFM or post-yield conditions?
8. Describe and sketch the shape of the displacement distribution along a post-yield structural discontinuity for different values of yield strength to driving stress ratio. Which one looks most reasonable to you, based on your own observations or reading about natural fractures in the field? How would you go about explaining any significant discrepancies?
9. Define discontinuity aspect ratio and discuss what controls it, and how it applies to discontinuity scaling and propagation. What controls discontinuity aspect ratio and how might this quantity change during lateral (tip) or vertical (stratigraphic) restriction?
10. Describe a test or set of measurements you could make in the field to determine how closely a fracture approximates LEFM conditions, using end (secondary) structures as your key. Be sure also to state the physical basis for each of your measurements.
11. Describe the effect that relaxing the condition of small-scale yielding has on the near-tip stresses and discontinuity-surface displacements. How might you use this in the field?
12. How could you differentiate between Riedel shears and splay faults located near the tips of a set of strike-slip faults?

13. Describe the degree of stress concentration developed near the tips of a post-yield structural discontinuity and what controls its magnitude. What values might you expect to encounter for natural examples, and why?

14. What is apparent dynamic fracture toughness and how is it related to small-scale yielding? What would you look for and measure in the field or in the laboratory to assess this quantity?

15. Define the J-integral and explain how it relates to the LEFM fracture parameters K_r, K_c, and G. How could you use it as a discontinuity propagation criterion?

16. Using values of $\sigma_x = 3.1$ MPa, $\sigma_y = 5.7$ MPa (both compression positive), $\tau_{xy} = -0.6$ MPa, $E = 10$ GPa, and $v = 0.3$, calculate:
    (a) the total strain energy density in the rock mass,
    (b) the volumetric strain energy component,
    (c) the distortional strain energy component.
    Hint: Remember to use total stress for part (a) and deviatoric normal stresses for parts (b) and (c).

17. Calculate the change in volumetric strain energy density near the tip of a mode-I crack. The crack is 1.2 m long and is loaded by a uniaxial remote compressive stress of –0.8 MPa with an internal pore-fluid pressure of –1.02 MPa (tension positive). Assume that LEFM applies and calculate the quantity near the crack tip using the near-tip stress equations at $r = 0.1$ m and $\theta = 0°$.
    Hint: use all three normal stress components (the shear stresses are zero here), but focus on the volumetric component for mode-I propagation. The values for background and near-tip use deviatoric stresses. The change is given by the strain energy density for the crack normalized by the background value.

18. Now calculate the change in distortional strain energy density near the tip of a fault (mode II). The fault has $L = 2.8$ m, $\mu_s = 0.6$, $\mu_r = 0.595$, with $\sigma_3 = 4.3$ MPa (compression positive), $\sigma_1 = 1.28$ MPa, and $\phi = 30°$. Assume again that LEFM applies and calculate the quantity near the crack tip using all six of the near-tip stress equations at $r = 0.1$ m and $\theta = 0°$. The values for background and near-tip again use deviatoric stresses. What do you conclude?

19. Describe the displacement–length scaling relations of:
    (a) a fault,
    (b) an igneous dike or hydraulic fracture,
    (c) a cataclastic deformation band.
    Describe what the main factors might be that influence the scaling relations. How would you be able to test your answer in the field?

20. Discuss the role of stratigraphy in the growth of geologic structural discontinuities, including joints, faults, and deformation bands, as portrayed on a D–L diagram. How might you expect the population statistics to change for the case of lateral or vertical restriction?

21. Calculate the linkage efficiency factor for segment interaction using the following field data: two echelon normal faults each with $L = 14$ m and $D_{max} = 0.283$ m whose inner tips overlap by 1.16 m. Assume that $\gamma = 3.1 \times 10^{-2}$ for isolated normal faults in this population.

22. Using the fault data for the Bishop Tuff, calculate:
    (a) the thinning of the tuff layer due to its extension,
    (b) the shear strain accommodated parallel to the normal faults (i.e., in the shearing direction, not the strike direction).

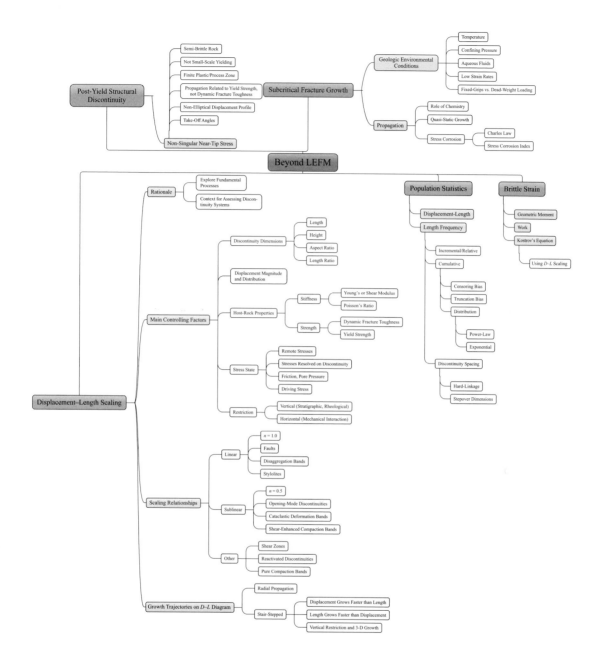

Glossary

This glossary provides definitions of the key terms used in this book. These terms may differ from some of the usages in other books or research articles, given the focus in this book on geologic fracture mechanics. The definitions may serve to review or clarify the approach and exposition in the text.

accommodation structure — one of several types of mappable structures, including faults, folds, and joints, developed within a stepover between two echelon faults; same as "relay-ramp."

accommodation zone — same as "relay-ramp" and "stepover" between echelon fault strands or sets of strands.

acute bisectrix — the angle between either conjugate set of faults and the greatest principal stress direction; it is commonly obtained by using statistical methods including stereonets or paleostress inversion techniques but must assume that both sets of discontinuities it relates are truly synchronous, shearing due to that applied stress state, and accommodating 2-D plane-strain conditions.

analytical solution — a set of one or more equations that solve a particular problem in mechanics by using specified parameters and variables; same as "closed-form solution," but see also "exact solution."

Anderson criterion — a physical explanation for the orientations and spatial arrangement of either single faults or fault pairs (conjugate faults) based on a 2-D plane-strain classification of principal stress orientations at the Earth's surface; may be supplemented by using a Coulomb criterion for fault slip to estimate stress magnitudes.

anisotropic — a description of properties of the stress state and/or the material that vary with orientation or direction.

anticrack — a sharp or tabular structural discontinuity having field evidence for predominantly closing displacements between the opposing walls.

anti-dilational — same as "contractional."

anti-plane shear — shear stress acting in the third direction, typically corresponding to vertical for crustal faults; same as "out-of-plane shear." The term comes from 2-D plane approximations of 3-D material behavior.

antithetic — a discontinuity, commonly a fault, whose dip direction is opposite to that of nearby discontinuities, usually with subequal strike direction; the term requires synchroneity to be applicable.

antithetic fault — in a graben, the bounding normal fault having smaller displacement than the opposing fault.

apparent dynamic fracture toughness — a material parameter related to the dynamic fracture toughness that depends on confining pressure, temperature, loading rate, and yield strength.

array — a mechanically related, organized, collection of discontinuities, usually having a collinear and/or coplanar overall geometry (as in an echelon array); distinguished from a network or population by having a smaller number of discontinuities and more localized or focused character.

arrest line — same as "rib mark."

aspect ratio — a measure of the relative dimensions of a discontinuity, viewed normal to it, that relates the vertical dimension, H, to the horizontal (strike) dimension, L, so that $AR = L/H$.

axial splitting — a macroscopic failure mode involving the formation of dilatant cracks parallel to the direction of maximum compressive stress; usually applied to small cylinders or discs of rock, it requires generation of absolute tension across flaws, resulting in mode-I or mixed-mode I-II crack growth.

Aydin's rule — a suggestion that "what you see in the field is completely determined by what you know."

balanced cross section — a method for assessing that geologic cross sections do not violate rules of basic geometry and kinematics by assuming constant line lengths (normal strains) or areas, along with plane-strain deformation normal to the section, in deformed strata; a first step in moving from an artistic rendition of deformation characteristics to a more rigorous, physically based model having testable and predictive capabilities.

band fault — the final stage in the evolutionary sequence of faulting in porous rocks such as sandstones involving instability in strain across a zone of deformation bands and formation of a slickensided fault surface; same as "faulted deformation band."

bifurcation of material properties — the onset of two sets of distinct material behaviors that occurs during deformation of an initially uniform, homogeneous rock. It marks a singularity, or abrupt transition, in the material behavior, such as zones of local strain softening (joints) or strain hardening (compactional deformation bands) in an otherwise less deformed and still homogeneous rock matrix. This condition describes the effect of strain localization on the overall material behavior.

Bingham plastic material — a composite ideal material having attributes of linear-viscous and rigid-plastic materials; a useful description of strong flows such as lava flows, mud and debris flows, and pyroclastic flows.

blade-like crack — a (2-D) crack whose horizontal dimension greatly exceeds its vertical dimension when viewed along the plume axis; analogous to "long discontinuity."

blind fault — a fault that does not break the Earth's surface; same as "bounded fault."

block tectonics — a conceptual model for rock-mass deformation that assumes rigid behavior of rock domains separated by frictionally slipping fault surfaces.

bookshelf tectonics — a term for the rotation, usually around a horizontal axis, of tabular blocks separated by unbounded faults, sometimes used as a model for progressive shear strain of faulted or fractured domains.

boundary condition — a constraint on the behavior of a geologic process; used mathematically to transform indefinite equations to definite ones with particular solutions that give answers to the problem being investigated. Specifying the remote stress and Coulomb criterion are examples of setting the boundary conditions for a frictional sliding problem.

bounded discontinuity — a discontinuity which has clear terminations (closed tiplines), in all directions, within the rock mass.

braided shear zone — same as "deformation band."

branch-line — the line of intersection where two fault surfaces meet.

branch-point — the point on the Earth's surface where two intersecting faults, at their branch line, crop out.

breakdown zone — same as "cohesive zone" located near the termination of a fracture.

brittle — a pressure-dependent deformation mechanism usually involving nucleation, growth, and coalescence of dilatant cracks. Also used to describe the behavior of rock near a discontinuity tip for which rock strength greatly exceeds stress magnitudes, promoting small-scale yielding conditions.

brittle–ductile transition — a widely used term to describe a change from faulting to flow in the crust, sometimes considered to be a planar mappable interface; because ductile deformation can occur using brittle mechanisms (cataclasis), the term has largely been replaced by "brittle–plastic transition" in tectonic applications. In rock deformation the brittle–ductile transition separates localized deformation (such as faulting) from distributed deformation (such as flow).

brittle–plastic transition — a change from brittle, pressure-dependent deformation mechanisms, such as cracking and faulting, to temperature-dependent deformation mechanisms such as creep; the transition is often gradual and dependent on several factors such as grain size and mineralogy, so that regimes marked by simultaneous brittle and plastic deformation are not uncommon.

brittle fault — a fault whose secondary structures and internal structures are characterized or dominated by dilatant crack arrays; usually synonymous with "fault."

bulk modulus — a parameter used to describe the ratio of hydrostatic pressure to the volumetric strain which it produces in an elastic material; also called the incompressibility.

Byerlee's rule — an experimentally determined proportionality between normal and shear stress at the onset of frictional sliding that is independent, to first order, of lithology and used in construction of brittle strength envelopes in the lithosphere; commonly referred to as "Byerlee's Law."

Cam cap — an ideal form of yield surface having an elliptical shape on the q–p diagram that marks the onset of compactional inelastic yielding of a rock.

Cartesian stress state — the normal and shear stress components referenced to an *xyz* coordinate system having three mutually perpendicular axes; contrasts with stress components referenced to radial (r, θ), spherical (r, θ, P), or cylindrical (r, θ, z) coordinate systems.

cataclastic deformation — fracturing and grain-size reduction within a zone or throughout the volume of a rock mass; can accommodate large ("ductile") strains by brittle mechanisms such as cracking and frictional sliding.

censoring — a qualitative filtering (or bias) of discontinuity-population data that results in reliable statistics only for a certain length range, commonly related to under-counting certain classes (too short, or different type) of discontinuities.

CHILE material — an acronym for "Continuous, Homogeneous, Isotropic, and Linearly Elastic" and representing the simplest idealization of a rock mass for calculation and modeling purposes.

cleavage — planar to irregular surfaces of mass loss, typically due to pressure solution processes; distinct from "cleavage" of single mineral crystals or rocks subjected to rapid dynamic loadings.

closed-form solution — same as "analytical solution."

coefficient of friction — same as "friction coefficient."

cohesion — the intercept of a linear criterion for frictional sliding, such as the Coulomb criterion, on the shear-stress axis of a Mohr diagram, corresponding to the value of shear stress at zero normal stress resolved on a surface of potential slip; compare to "instantaneous cohesion."

cohesive end zone — same as "cohesive zone."

cohesive zone — the region beyond a discontinuity tip characterized by localized cracking and/or frictional sliding and acting as a transition between a fully developed fracture and the rock beyond the tip; the zone serves to remove singular stress but represents a physically larger region ($>0.1a$) than in small-scale yielding.

collinear — discontinuities that are oriented parallel and also in line with each other; similar to "coplanar" in that 2-D case. Contrasts with "non-collinear."

compaction — the reduction and ultimate collapse of pore spaces within a porous rock due to compression and shearing.

compaction band — a tabular zone of decreased porosity in otherwise porous rock characterized by closure and reduction of volume and no shear displacement; see also "anticrack" and "weak discontinuity."

compaction deformation band — same as "compaction band."

Complete Stress–Strain Curve for Rock — a compact representation of the average properties of a rock, or rock mass, as they change progressively and systematically with increasing deformation.

compliance — a term used in classical elasticity theory for the reciprocal of the modulus; compare to "stiffness." The term arises from measuring elastic modulus (or, more precisely, the compliance) in the laboratory from $\varepsilon = E\sigma$, rather than from $\sigma = E\varepsilon$.

composite pull-apart veins — stepover structures along thrust-fault surfaces formed by localized dilation and vein-mineral deposition during slip; the term may supersede or qualify "shear fiber veins" and "bed-parallel veins," and may resemble dilational "crack-seal veins" but with slickensides along basal surfaces.

compressibility — an elastic property that is the inverse of the bulk modulus.

compression — a force, stress, or stress component which causes particles to move closer together; also used to denote the contractional side of a lithospheric strength envelope.

compressional quadrant — same as contractional quadrant.

concordant step — a term for a stepover in which distortion predominates over volumetric strain.

conjugate faults — synchronous fault pairs, rationalized using the 2-D Anderson classification, based on orientations of maximum shear stress in isotropic materials deforming in plane-strain.

conjugate shear planes — same as "conjugate faults."

consolidation — a coupled process in porous rocks and soils linking expulsion of pore-fluids during compaction to material behavior and deformation rate relative to fluid flow in closed (undrained, rapid loading) or open (drained, slow loading) systems.

continuum — a representation of a rock or other material whose imperfections are negligible at the scale of observation, resulting in a set of homogenous properties; compare to "effective continuum."

contraction — a strain defined by a decrease in line length.

contraction crack — a term for "dilatant crack" or "joint" that combines knowledge of the sign of the normal strain associated with crack dilation; compare to "tension crack."

contractional quadrant — the region of increased mean stress near the tip of a mode-II discontinuity.

contractional structure — a term for thrust faults, folds, pressure solution surfaces, and other strain indicators that have accommodated contractional strains.

coplanar — parallel discontinuities having the same strike and dip but not necessarily collinear; contrasts with "non-coplanar."

Coulomb criterion — a plastic yield criterion, originally defined for soils, relating the normal and shear stresses acting on a planar surface to the frictional resistance to sliding on that surface; the criterion is widely used to predict slip along faults or other fractures such as joints.

Coulomb failure function — a measure of the changes in driving stress for frictional slip resolved on surfaces due to slip or earthquakes along some surface (usually a fault).

crack — a sharp structural discontinuity having field evidence for discontinuous and predominantly opening displacements between the opposing walls in an opening-mode structural discontinuity, usually whose faces are open and stress-free; same as "mode-I fracture."

crack aficionado — a term occasionally applied to refer to a subject-matter expert in geologic fracture mechanics.

crack opening displacement — an engineering criterion (referred to as "COD") for crack propagation based on measuring the mode-I dilation across a crack and relating it to the stress intensity factor or to the dynamic fracture toughness.

crack tip opening displacement — an engineering criterion (referred to as "CTOD") for crack propagation based on measuring the shape of the crack tip (commonly expressed as radius of curvature) and relating it to the dynamic fracture toughness. The method is seldom used in LEFM given the microscopic size of typical flaws and the associated difficulty in observing and accurately measuring flaw shapes.

creep — slow deformation either localized along a surface ("fault creep") or distributed through a volume of rock.

critical stress intensity factor — same as "dynamic fracture toughness."

cross joint — a crack in a non-systematic joint set that intersects and terminates, usually at high angles, at or near cracks of the systematic set.

curtailment bias — an increase in slope on a cumulative-frequency vs. length diagram at large lengths associated with underestimation of the actual lengths of the longest discontinuities.

damage zone — a spatially heterogeneous zone of increased deformation density that is located around a discontinuity that formed at any stage in the evolution of the structure.

DB — a shortened form of "deformation band."

D_{max} — the value of maximum displacement (offset) along a discontinuity surface, commonly but not always located near its midpoint.

dead-weight loading — a remote loading condition on rock masses that applies an ever-increasing level of stress onto discontinuities, commonly leading to dynamic discontinuity propagation.

décollement — a sub-horizontal slip surface, usually having thrust fault displacement.

deformability — a measure of the ability of a rock to adjust to an applied stress state or displacement field, usually without incurring failure; related to the (Young's or deformation) modulus of the rock.

deformation — the response of a rock to stress, composed of rigid-body rotation and translation along with dilational (volume change) and distortional (strain) components.

deformation band — a tabular structural discontinuity having a continuous change in displacement, strength, or stiffness across a relatively narrow zone in porous rocks. See also "weak discontinuity."

deformation band shear zone — same as "deformation band."

deformation modulus — the local slope of the usually nonlinear stress–strain curve for a fractured rock mass, usually measured or back-calculated in the field; may include inelastic responses due to deformation of pre-existing fracture networks; usually smaller in value than Young's modulus for the intact rock component.

degenerate ellipse — a particular geometry in which the semi-minor axis equals zero, leading to a slit-like or crack-like shape; frequently used in a stress analysis to represent a geologic fracture mathematically.

detachment — a low angle structure, usually with normal-fault sense and having crustal-scale implications.

deviatoric stress — a part of the total stress state in a rock defined as the *in-situ* stress minus the mean (or lithostatic) stress; the associated Mohr circle would plot centered at the shear-stress axis and the terms "deviatoric tension" and "deviatoric compression" would apply. Commonly used in geophysics and seismology, it can be misapplied to structural geology if the total stress is ignored, leading to physically implausible concepts such as "tensional faults."

dextral — same as "right-lateral."

DIANE material — an acronym for "Discontinuous, Inhomogeneous, Anisotropic, Non-Elastic" and representing a more complicated idealization of a rock mass for calculation and modeling purposes.

differential stress — the difference between the values of the greatest and least principal stresses, and equal to twice the maximum shear stress. The actual value of differential stress that can be tolerated by a rock or rock mass is limited by that material's particular strength criterion, taking into account the relevant environmental conditions (pressure, temperature, strain rate, pore fluids).

dihedral angle — the angle that bisects two discontinuities or sets of discontinuities, and interpreted as diagnostic of paleo-stress trajectories associated with the discontinuities' formation and/or displacement.

dike — a sharp geologic structural discontinuity having field evidence for discontinuous and predominantly opening displacements between the opposing walls, oriented sub-vertically and/or normal to bedding, and containing liquid or solid magmatic materials (US English spelling).

dilatant crack — a fracture whose surfaces have been displaced perpendicular to the plane of the discontinuity, in opposite directions, and in an opening sense. See also "sharp discontinuity."

dilatation — the act of dilatating; same as "dilation" and sometimes used in seismology.

dilation — the opening displacement along a mode-I or mixed-mode discontinuity.

dilation angle — the ratio of plastic volume change to plastic shear strain; used in analyses of porous rock deformation to account for volumetric changes across a shearing zone.

dilation band — a thin tabular discontinuity having increased porosity within the band and opening displacements without shear. See also "weak discontinuity."

dilational quadrant — the region of reduced mean stress near the tip of a mode-II discontinuity; same as extensional quadrant.

dip — the angle between the horizontal plane (Earth's surface) and a fault or other planar structural discontinuity, usually expressed in degrees.

dip-slip fault — a surface on which the slip direction is subparallel to a vertical plane intersecting the fault surface; same as normal or thrust (reverse) fault.

discontinuity — a mechanical defect, flaw, tabular zone, or plane of weakness in a rock mass without regard to its origin or kinematics; see "structural discontinuity."

discordant step — a term for a stepover in which volumetric strain predominates over distortion.

displacement — the change in position of originally adjacent material points within a rock.

displacement-based classification of fractures — a method for identifying the mechanical or strain significance of a geologic structural discontinuity, and thereby assigning the appropriate name, based on direct observation whenever possible.

displacement discontinuity — a measure of the separation of formerly adjacent material points across a discontinuity surface; similar to the net slip across a fault or opening across a crack but evaluated at specific positions along the surfaces. See also "sharp discontinuity."

displacement–length scaling — a relationship or ratio between the maximum value of displacement and the length of the associated discontinuity.

displacement profile — a plot of the variation in the magnitude of displacement in relation to the position along the discontinuity; used for identifying the maximum value of displacement, and its location, along with the shape of the displacement gradient near the discontinuity terminations.

distension — same as "extension" when applied to a faulted region or terrane.

dog-leg pattern — a term sometimes used to describe the zig-zag geometry of linked dip-slip faults, having orthorhombic symmetry, slipping in a 3-D stress field.

domain — a local area within a larger region characterized by spatially homogeneous properties (e.g., discontinuity spacing or orientation) that may be deforming as a discrete unit within a larger, inhomogeneously deforming region.

drained conditions — an end-member response of fluid-saturated porous rock or rock mass in which the rate of fluid movement exceeds the deformation rate and for which the fluid phase thereby exerts a subset of effects on the deformation; also may characterize hydraulically open systems.

driving stress — the difference between the remote stress state resolved onto a discontinuity plane and the discontinuity's inherent resistance to displacement in opening, closing, and/or shearing senses.

D-shear — see "Riedel shear."

ductile — the capacity for a rock to sustain distributed flow or large deformations; the specific deformation mechanism through which this occurs (brittle cataclasis or plastic creep) is not specified in the term.

ductile fault — a fault that exhibits fabrics or relationships, such as drag, that imply both large strains and plastic deformation either along the fault or in the region surrounding it; essentially synonymous with "fault."

Dugdale–Barenblatt model — a fracture mechanics concept incorporating a cohesive end zone of non-negligible size relative to a discontinuity to explicitly eliminate the singular stresses generated by a perfectly sharp flaw in an ideal elastic material.

duplex — a term used to describe a stepover along parallel, non-collinear echelon discontinuities characterized by fully linked discontinuities; a "horse" in thrust-fault literature.

durability — an index property for the intact rock material that describes the resistance of the rock to mechanical or chemical degradation.

dyke — alternative spelling of "dike."

dynamic crack growth — rapid crack propagation velocities on the order of 0.3 times the sonic velocity of the host rock (km s^{-1}); splitting or branching of the tips of these "running" cracks is diagnostic of the high terminal velocity but rarely observed in tectonic joint sets.

dynamic fracture toughness — see "fracture toughness."

dynamic structural geology — a term for stress-based (as opposed to strain-based) methods in structural analysis; largely has been replaced by "mechanics" to avoid confusion with motion-based engineering analysis ("dynamics") or time-dependent hydrocarbon studies ("reservoir dynamics").

earthquake fault — a discontinuity surface having significant potential for accommodating future seismic slip events; same as "seismogenic fault."

echelon array — a set of closely spaced, parallel, non-collinear discontinuities; same as "en échelon" (from the original French).

edge crack — a discontinuity that intersects a free surface and appears as a semi-bounded fracture with only one termination.

effective continuum — a rock mass or other large mass of material that is fractured or polylithologic at some scale but considered to deform according to its average properties (as a continuum) at larger relative scales.

effective modulus — the value of Young's (or shear) modulus for a fractured rock that has been reduced by incorporating fractures (e.g., open cracks, closed yet sliding cracks) into the material.

elastic limit — the point on a stress–strain curve marking the departure from linear behavior and a decrease in slope and modulus, corresponding in general to the yield strength of a rock.

elastic material — an ideal rheology in which the strain is fully recoverable when the load is removed; a special case is a linearly elastic material where the relationship between stress and resulting strain is linear, with a constant value of Young's or shear modulus, and the strain is also fully recoverable; no provision for failure is implied in elastic behavior.

elastic modulus — same as "Young's modulus."

elastic–plastic fracture mechanics — a branch of fracture mechanics that includes the effects of plasticity, and materials that strain harden or strain soften in the discontinuity tip region, for which the assumptions of small-scale yielding may not strictly apply.

en échelon — same as "echelon."

end crack — a secondary fracture located at or near the termination of a larger parent discontinuity, that nucleates and propagates away as a result of either shearing or oblique dilation along the parent discontinuity; synonymous with "tail crack" and "wing crack."

end zone — same as "cohesive end zone" at a discontinuity periphery that is defined by localized inelastic deformation, usually coplanar with, and part of, the discontinuity.

exact solution — a particular type of analytical solution in which no approximations were assumed in its derivation.

exfoliation fractures — see "sheeting joints."

expansion joint — a geotechnical term for "dilatant crack" used when tensile far-field stresses are inferred.

extensile fracture — a term sometimes used in rock mechanics for a mode-I wing crack formed in a compressive-stress environment; same as "dilatant crack."

extension — a strain or displacement defined by an increase in line length.

extensional fault — a redundant term for "normal fault."

extensional shear fracture — same as "shear crack."

extensional stress — an incorrect term for "tensile stress."

eye structure — a stepover usually between deformation bands having inosculating segments.

factor of safety — the ratio of resisting to driving forces and commonly used in geological and geotechnical engineering design.

failure — in the engineering sense, the inability of a structure or component to fulfill its design function. Commonly identified with "peak strength."

failure criterion — a relationship between stress state and the deformation that occurs at peak stress levels; also referred to as a "peak yield criterion."

failure envelope — a contour or surface in stress space (constructed using either a Mohr diagram or the principal stresses) that represents the peak stresses that can be sustained by a rock or rock mass. "Failure" is suggested when the Mohr circle for stress becomes tangent to the failure envelope.

far-field stress — same as "remote stress" or "regional stress" where the stress components are applied sufficiently far from the point of interest that they are uniform or smoothly varying.

fatigue — one of the mechanisms that can lead to slow, time-dependent, subcritical fracture growth, usually under repetitive, cyclic loading.

fatigue crack — a crack whose slow growth contributes to weakening and eventual breakage of a material, usually at levels considerably less than that of the peak strength; as used in engineering design and testing, fatigue cracks grow during cyclic loading.

fault — a sharp structural discontinuity defined by slip planes (surfaces of discontinuous displacement) and related structures including fault core and damage zones that formed at any stage in the evolution of the structure.

fault-bend fold — a fold formed as a consequence of originally horizontal layers being flexed up and around geometric irregularities of the underlying fault surface, such as ramps that change the dip angle in the slip direction of a fault, as the layers are displaced along it.

fault bridge — same as "relay-ramp."

fault creep — the accumulation of offsets along a fault surface or fault zone by a macroscopically continuous and stable sequence of microseismic slip events.

fault-propagation fold — a geometric model of displacement transfer from an underlying fault to the overlying layered strata, manifested as folding localized above the blind fault tip.

fault reactivation — the recurrence of slip along a previously existing fault.

fault segment — one of several closely spaced bounded faults within a larger array or zone; the term emphasizes both the discontinuous geometry and the relationship to the larger fault array.

fault strain — same as "fracture strain" for shear displacements.

fault strand — same as "fault segment."

fault zone — a set of relatively closely spaced faults having similar strikes.

faulted joint — a slip surface that nucleated on a previously formed mode-I crack. The term implies a rotation of principal stresses about the crack and a change in sign of the resolved normal stress, from tensile to compressive, to permit frictional sliding along the joint; field recognition includes superposition of slickensides, steps, and risers onto the previously formed plumose structure.

feather fracture — same as "end crack" and "pinnate crack."

field quantity — a distribution of scalar or vector properties, within a rock mass with magnitudes and orientations that may be constant or vary smoothly or discontinuously with position or with time.

finite strain — a class of strains, usually exceeding 10% or so, common in geologic contexts and commonly involving non-coaxial principal stresses and strains.

fissure — a joint with large amounts of opening displacement, usually located at or near the surface; commonly used to describe ground cracks in volcanic or faulted areas, it is also used in rock engineering to label microcracks in a damaged sample.

fixed-grips loading — a remote loading condition that applies a remote strain or displacement to the rock masses and that modulates the level of stress on fractures, which can promote stable subcritical or quasi-static crack propagation.

flame structure — same as "hackle."

flat — the sub-horizontal surface of thrust-fault displacement, usually occurring on or close to a lithologic contact; same as "décollement."

flaw — a local inhomogeneity in rock properties, such as a crack, loadcast, or fossil, that serves to redistribute and concentrate the stresses and strains in that local vicinity.

flower structure — the cross-sectional geometry of stepovers along echelon strike-slip faults, usually described from seismic profiles; positive flower structures, with upward-diverging reverse faults, are contractional; negative flower structures, with upward-diverging normal faults, are extensional.

fluid-pressure ratio — a quantity relating the aqueous pore-fluid pressure at depth to the lithostatic stress.

footwall — the block located below the plane of a dipping fault.

footwall breach — see "upper ramp breach."

foot-wall cutoff — the line of intersection between the fault surface and a sub-horizontal reference plane (such as the Earth's surface) at the upper tipline of a normal or thrust fault.

fractal — a geometric description of a process, such as coastline erosion or fault-segment growth, that is independent of the scale of measurement or view; commonly taken to be synonymous with "self-similar" and "scale-independent."

fractography — a branch of fracture mechanics dealing with the morphology and significance of plumose structure and other decorations along mode-I crack surfaces.

fracture — A sharp structural discontinuity having a local reduction in strength and/or stiffness and an associated increase in fluid conductivity between the opposing pair of surfaces. Cracks, joints, and faults are all different types of fractures. See also "weak discontinuity" and "sharp discontinuity."

fracture front — the edge of a discontinuity; e.g., the ribs on a dilatant crack surface or the tipline of a fault.

fracture mechanics — the study of nucleation, displacement, and propagation of structural discontinuities in predominantly brittle materials; an interdisciplinary field drawing from engineering and materials science.

fracture or discontinuity population — a set or distribution of structural discontinuities whose measurable characteristics (number, size, displacements, lengths) can be used to define a mechanically integrated discontinuity network.

fracture porosity — a measure of the volume of discontinuities (usually cracks) to that of the host rock; may exceed the intrinsic porosity by several orders of magnitude.

fracture propagation energy — same as "strain energy release rate" applied to geological problems; the term emphasizes the units of energy, J m^{-2}, that relate this parameter to the original energy balance concept of Griffith.

fracture strain — a measure of the inhomogeneous, discontinuous deformation of a specified area or volume accommodated by structural discontinuities.

fracture strength — the breaking strength of an intact rock sample; sometimes also called rupture strength.

fracture toughness — a quantity (K_c) measured in the laboratory that characterizes the resistance to dynamic crack growth in a volume of rock; usually restricted to mode-I crack growth but sometimes used for other modes. It requires conditions of small-scale yielding to be a material constant; otherwise it is a material parameter (see "apparent fracture toughness").

free surface — an edge or boundary of a material, such as a cliff face or crack wall, that supports no shear or normal stresses; the absence of shear stresses makes this a principal plane.

friction angle — an alternative form for expressing the slope of the Coulomb criterion, particularly well suited for graphical interpretations and visual inferences.

friction coefficient — the local slope of the Mohr/Coulomb envelope for frictional sliding using a particular linear criterion (Coulomb, modified Griffith); compare to "instantaneous friction coefficient."

frictional breakdown zone — same as "cohesive zone" but usually applied to the parts of a fault surface near its periphery that exhibit elevated levels of frictional resistance and/or surface roughness.

frictional resistance — same as "frictional strength" and "shear strength," especially when applied to a fractured rock mass.

frictional sliding — the process of displacement accumulation by the mechanism of shear-stress dissipation along structural discontinuities.

frictional slip — same as "frictional sliding."

frictional strength — the "right-hand side" of the Coulomb criterion for frictional sliding along a pre-existing surface containing rock-surface properties (cohesion, friction angle or friction coefficient) and stress terms (resolved normal stress, internal pore-fluid pressure) that resist the action of resolved shear stress on the surface.

fringe joints — smaller crack segments forming echelon arrays, with constant sense of twist, near the edges of a parent joint; they are contiguous with hackle but have greater relief.

geodynamics — an interdisciplinary branch of planetary science and geophysics concerning the origin and development of global-scale and regional-scale processes of planets, usually involving elements of solid-earth geophysics, tectonics, mantle convection, core thermophysics and heat-flow, geomagnetism, gravity, and topography.

geologic fracture mechanics — the application of fracture mechanics to geological structural discontinuities.

geologic strength index — see "GSI."

geomechanics — a branch of materials science and geology that deals with the deformation of geological materials, including rock and soils; may be synonymous with "rock mechanics" depending on specific usage.

geotechnical engineering — a branch of civil engineering that applies principles of soil mechanics (primarily) to the design of structures such as bridges, dams, and railroad tunnels.

geotribology — the study of friction and wear of rocks, fault zones, gouge, and fracture surfaces.

graben — a pair of subparallel normal-fault arrays having opposing dip directions and common hanging walls.

grain crushing pressure — the value of the yield cap for a porous granular rock corresponding to collapse of porosity and onset of grain crushing at hydrostatic pressure (i.e., at negligible differential stress). It scales with the rock's porosity and average grain size, making it a convenient rock index property; also known as the "critical pressure."

granulation seam — same as "deformation band."

Green's function — a class of analytical solutions to problems in mechanics that can provide closed-form expressions for quantities such as stresses, strains, heat flow, or fluid flow from various partial derivatives of the function.

Griffith crack — same as "Griffith flaw."

Griffith criterion — a mathematical description of the strength of rocks in the tensile regime that is controlled by localized mode-I failures; usually written in principal-stress form and evaluated by using a Mohr diagram.

Griffith flaw — a generic term for small crack-shaped imperfections or inhomogeneities in a material that serve as stress concentrators that weaken the material relative to its ideal flawless strength.

growth — an increase in size or other related dimensions; same as "propagation" when applied to structural discontinuities.

GSI — an acronym for "geologic strength index" which is a generalized variant of the Rock Mass Rating (RMR) system and Hoek–Brown criterion, most commonly applied to weak and stratified rock masses.

hackle — subtle, systematic ridges and troughs on a mode-I crack surface that diverge away from the origin point of a propagating crack; they develop perpendicular to rib marks as tiny, incipient mixed-mode I-III (twist) breakdowns of the parent crack and are diagnostic of opening-mode displacement and strain.

hanging wall — the block located above the plane of a dipping fault.

hanging wall breach — see "lower ramp breach."

hanging-wall cutoff — the line of intersection between the fault surface and a sub-horizontal reference plane (such as the Earth's surface) at the lower tipline of a normal or thrust fault.

hard-linked faults — echelon faults whose stepovers have been bridged by secondary cross-discontinuities.

heave — the horizontal component of net slip across a fault.

height — the vertical dimension of a discontinuity; the other dimension is discontinuity length.

Hertzian fracture — a cuspate mode-I crack produced on a slip surface as a particle of similar hardness is dragged along it; the skipping produces impact fractures that cut into the parent plane, with the concave side of the

Hertzian fracture pointing in the slip direction. It differs from a "tool mark" formed from dragging of an indentor that is harder than the surface.

hesitation line — same as "rib mark."

Hoek–Brown criterion — an empirical approach to predicting the tensile and shear strengths of jointed rock masses; largely supersedes earlier attempts including Coulomb and Griffith criteria for large-scale rock exposures.

homogeneous properties — characteristics of a rock or rock mass that are statistically constant at a given relative scale.

homogeneous stress — an average stress state characterized by stress magnitudes and trajectories that do not change with position or location in the rock mass.

homologous temperature — a dimensionless ratio of the temperature of a rock under particular conditions to its melting point.

Hookean material — same as "elastic material."

Hooke's Law — the relationship between stress and strain in a material where the strain is fully recoverable when the stress is removed; an elastic material.

horse — a rhomb-shaped block of rock bounded on all sides by faults; a hard-linked echelon stepover between fault segments; same as "duplex."

horsetail cracks — same as "end cracks" so named for their braided appearance and location near the termination of the main slip surface.

horsetail solution surfaces — stylolites nucleated as secondary structures in the contractional quadrants of slipping faults, analogous (but with the opposite strain significance and a different formation mechanism) to end cracks in their kinematic interpretation.

host rock — the rock or rock mass that contains a structural discontinuity of interest.

hybrid extension-shear fracture — a seldom used term to designate a mixed-mode I-II (or less commonly, I-III) crack having oblique dilation of the crack walls.

hydrofracture — a mode-I crack whose opening and propagation are driven by a high fluid pressure within the crack; can be artificially or naturally generated. See also "sharp discontinuity."

hysteresis — a nonlinearity in stress–strain relations defined by unequal values of Young's modulus or accommodated strain for loading and unloading of a rock or rock mass.

inactive termination — a discontinuity tip isolated from further displacement accumulation by upper- or lower-ramp breaches (for dip-slip fault pairs), or secondary fractures (for strike-slip, opening-mode, or closing-mode fractures) across a stepover.

inelastic — a material that deforms permanently in response to a load or set of stresses.

infinite plate — a representation of a rock mass whose in-plane dimensions greatly exceed the size of a particular region or problem of interest, so that the boundaries of the rock mass do not influence the stresses or displacements anywhere in the plate.

infinitesimal strain — a class of strains, usually much less (orders of magnitude) than a few percent, common in geologic contexts (usually jointing) or engineering design and commonly involving coaxial principal stresses and strains.

Inglis/Kolosov Relation — a quantitative relationship between the maximum local stress at a cavity edge, the value of the remote stress, and the cavity geometry or shape. It is the starting point for LEFM.

inhomogeneous properties — characteristics of a rock or rock mass that are variable at a given relative scale; same as "heterogeneous."

inhomogeneous stress — a stress state with spatially varying magnitudes and/or trajectories; same as "heterogeneous."

inner boundary conditions — the parameters that describe tractions, stresses, displacements, strains, or material geometries located on or close to a discontinuity or other item of interest in a stress analysis.

inosculating fractures — a term used to describe the geometry of deformation bands where coplanar lens-shaped portions touch at a single common point, rather than along their edges as in a stepover.

in-plane direction — the direction in a thin sheet or plate that coincides with either of its longer (usually horizontal) axes.

in-plane stress —a stress component that acts in the direction perpendicular to the thickness of a thin plate.

in-situ stress — the total state of stress measured in the Earth by any of various methods; the vertical and horizontal components are almost always compressive except in rare circumstances.

instability — a significant change in conditions, either implied mathematically by singular quantities or physically by a shift to other conditions; localization and rapid growth of macrocracks near peak stress represents an instability in the deformation history of a rock.

instantaneous cohesion — the extrapolated intercept of the Mohr envelope for frictional sliding, valid for particular values of normal stress, using a nonlinear, pressure-dependent criterion (Griffith, Hoek–Brown).

instantaneous friction coefficient — the local slope of the Mohr envelope for frictional sliding, valid for particular values of normal stress, using a nonlinear, pressure-dependent criterion (Griffith, Hoek–Brown).

intact rock — an idealization of a rock or rock mass in which the properties at the scale of observation are considered to be uniform and homogenous, resulting in a continuum description of material behavior; best applied to small hand samples, and least appropriately to outcrops and rock masses.

intact rock material — the rock, usually intact but commonly weathered, that exists between fractures, bedding planes, faults, and other discontinuities in a rock mass; the term finds frequent use in rock engineering and in rock mass classification systems as a value of strength for the lithology being considered.

interface crack — a localized discontinuity along the contact between two geologic bodies or materials.

internal friction — the strength of the incipient shearing surface as represented by the curved envelope on a Mohr diagram.

Irwin — an informal unit of stress intensity or dynamic fracture toughness; 1 Ir = 1 MPa m$^{1/2}$.

isotropic — a property that does not vary with direction.

J-integral — a method to characterize the intensity of near-tip deformation at either LEFM or non-LEFM discontinuities as a more general criterion than stress intensity factor for discontinuity growth under conditions of plane strain, plane stress, or post-yielding (elastic plastic, high temperature, strain hardening or softening) conditions. Termed a "path-independent integral" for traction-free cracks, the end points for the closed integration contour must be specified for the *J*-integral to be used for discontinuity having arbitrary stress boundary conditions along their walls.

jog — a curve or bend in a fault trace, commonly found along linked echelon fault segments; can be described as dilational or contractional (same as "anti-dilational").

Johnson's rule — a reflexive maxim attributed to Professor Arvid Johnson from his classic 1970 book: "All geologic problems start in the field, and any (experimental, kinematic, or mathematical) solutions must fit what one sees in the field."

joint — a sharp structural discontinuity having field evidence for discontinuous and predominantly opening displacements between the opposing walls.

joint elimination direction — the direction associated with systematically increasing discontinuity length and spacing that is related mechanically to discontinuity interaction and that is diagnostic of the overall propagation direction of the discontinuity array.

joint saturation — see "saturation."

jointed fault — a mode-I crack that nucleated on a previously formed fault surface and recognized by superposition of plumose structure on the earlier fault surface textures such as slickensides.

joule — the SI unit for work or energy; 1 J = 1 N m.

K-dominant region — the volume surrounding a discontinuity tip defined where the near-tip stresses, elevated due to stress concentration about the discontinuity, exceed the level of the remote stress.

key-block theory — a "weakest-link" approach to designing in fractured rock masses that utilizes system geometry (statics) and related mechanics (frictional contact characteristics) to identify the most likely point for failure to initiate, usually by removal of a single, particular block.

kinematic coherence — the condition resulting from strong mechanical interaction ("soft-linkage") between closely spaced echelon discontinuity segments in which the discontinuity array functions as a single, composite discontinuity.

kinematics — the reconstruction of a structural problem that relies on geometric rules such as length or area balancing that may or may not be plausible on mechanical grounds.

Kirsch's problem — the state of stress in an infinite elastic plate, loaded by a uniaxial stress, and containing a circular cavity.

Kolosov's problem — a solution to the elasticity problem of the state of stress in an infinite plate, loaded by a uniaxial stress, and containing an elliptical cavity.

laccolith — a solidified igneous body that was emplaced initially as a sill, then lifted its overburden by bending into a mushroom-shaped body.

large discontinuity — a statistical description of discontinuities with a dimension (usually the down-dip height) that is comparable to the thickness of the deforming region or layer.

lateral ramp — a fault or fault zone bridging the stepover between two echelon thrust faults and oriented subparallel to the overall transport direction of the thrust sheet.

LEFM — an acronym for "linear elastic fracture mechanics."

left-oblique — a mixed-mode fault having both left-lateral strike-slip and dip-slip components of displacement.

length — the horizontal dimension of a discontinuity; the other dimension is discontinuity height.

length ratio — a measure of the relative dimensions of a surface-breaking discontinuity, viewed normal to it, that relates the vertical dimension, $H/2$, to the horizontal (strike) dimension, L, so that $LR = 2AR = 2L/H$.

ligament — a small area of intact rock carrying a large stress gradient; same as "rock bridge" or "relay" if located between two echelon discontinuities.

lineament — an alignment of points or textures that generates a feature, usually with the genetic interpretation of fracture or fault control.

linear elastic fracture mechanics — a subset of fracture mechanics that assumes perfectly elastic materials, sharp flaws, and small-scale yielding (LEFM), with propagation related to stress intensity factors and dynamic fracture toughness.

linear elastic material — an ideal rheology (Hookean material) that assumes that the stresses are linear combinations of the strains, which, in turn, requires that the material properties that relate stress and strain (Young's modulus, shear modulus, and Poisson's ratio) are constants. Interchangeable with "linearly elastic material."

linkage — a concluding stage in discontinuity interaction involving intersection of secondary end structures, nucleated near parent discontinuity tips, with adjacent discontinuities; synonymous with "hard linkage."

lithostatic stress — a state of stress in which all three principal stresses are defined by the dry weight of a column of rock in the crust.

load — a force or set of forces applied to a material; has units of force (N), not stress (Pa).

loading — a general term for a configuration of forces or stresses applied to a rock or engineering component.

local stress — the components of stress acting on a rock that are applied or generated sufficiently close to a point of interest that they may show variability in magnitude, orientation, and/or sign.

localization — the process of forming small zones having significantly different properties than those of the surroundings. The term is similar to "nucleation" which refers more specifically to the type of structure that is localized during the deformation.

long discontinuity — a structural discontinuity whose horizontal dimension (length) substantially exceeds its vertical dimension (height).

lower ramp breach — a secondary fault that links the two primary echelon faults across their stepover at the topographically lower end of the relay-ramp; also known as "hanging wall breach."

Lüders' bands — tabular zones of cataclasis and minor shearing produced experimentally in porous sandstone, limestone, and marble samples; they are the laboratory-scale equivalent of early deformation bands observed in the field.

macrocrack — a discontinuity visible with the naked eye in core or outcrop.

macroscale — a qualitative term for large size of discontinuity; larger than "mesoscale."

master fault — the bounding normal fault in a graben having larger displacement than the opposing fault.

maximum displacement — the largest value of displacement (offset) along a fault in a graben; it is commonly located near the discontinuity's midpoint but can be shifted off-center due to discontinuity interaction, stress gradients, or other factors. The location of maximum displacement does not necessarily correlate with the point of nucleation where the discontinuity initiated and grew.

mechanical interaction — the constructive or destructive interference between the local stresses generated along discontinuities, especially surrounding their terminations; synonymous with "interaction."

mechanics — a branch of engineering concerned with the application of forces, stresses, or displacements to a body, the constitutive response of the body, and the resulting stresses, displacements, and strains induced in the body by the first two sets of factors. Synonymous with "dynamics" in structural geology, mechanics properly encompasses both the static and dynamic aspects of rock deformation, with "kinematics" reserved for the geometric description (including retro-deformability) of a structural problem.

mechanism — the way in which a phenomenon occurs, usually given by a combination of boundary conditions and rheologic behavior; for example, the process of folding can be accomplished by several mechanisms including bending or buckling.

mesoscale — a qualitative term for an intermediate size of discontinuity.

microcrack — a discontinuity visible with a microscope or occurring at the grain scale of a rock; same as "fissure."

microfault — a qualitative term for a very small fault having very small displacement; compare to "minor fault" and "mesoscale fault."

microscale — a qualitative term for a tiny, commonly microscopic, scale of observation.

mirror zone — a small quasi-planar surface in fine-grained or amorphous materials that surrounds the crack origin and formed as the incipient crack front accelerates.

mixed-mode crack — a dilatant crack with some crack-parallel displacement of its crack walls; different than a fault, whose surfaces are in frictional contact and for which there is no significant opening displacement, a mixed-mode crack is a discontinuity with oblique dilation. See also "sharp discontinuity."

mode of fracture — an engineering classification of the sense of displacement accumulated along a discontinuity; may be distinct from nucleation and propagation.

mode-I discontinuity — same as "dilatant crack," in engineering and fracture mechanics; the opening mode of fracture.

mode-II discontinuity — a discontinuity characterized by the sliding mode of displacement; closely synonymous with "fault;" a strike-slip fault in map view.

mode-III discontinuity — a discontinuity characterized by the tearing mode of displacement; closely synonymous with "fault;" a normal or thrust fault in map view.

Modified Griffith criterion — a generalization of the plane Griffith criterion that assumes open, stress-free cracks even under compressive normal stresses, to include crack closure and frictional sliding; suitable for intact rock, the criterion is also widely used for fractured rock masses in lieu of the more comprehensive Hoek–Brown criterion.

modulus of rupture — a term for the tensile strength of a rock determined from a bending test and typically larger than the value obtained from a direct tensile test.

Mohr circle — a graphical technique used to visualize the variation of stress in a particular problem or to verify that the answers calculated from equations are in the correct range of magnitude and sign.

Mohr–Coulomb criterion — same as "Coulomb criterion" but preferred by some due to its implied connection to rock (as opposed to soil) deformation.

Mohr diagram — a plot of normal stress vs. shear stress, showing the variation of stress components in the local coordinate system of a discontinuity or other plane, used to graphically compare stress to a separate failure criterion such as that for Coulomb slip.

N (Newton) — the SI unit for force; $1 \text{ N} = 1 \text{ kg m s}^{-2}$.

neoformed fracture — an informal term for a discontinuity of any displacement mode that forms in rock as an original structure and is thereby related directly to the stress state and constitutive relations at the time of its formation; contrasts with, for example, "faulted joints" and other sequentially displaced discontinuity types.

net slip — the magnitude and direction (i.e., the slip vector) of a piercing point that is displaced, or offset, across a fault; also called the "displacement across the fault."

non-proportional growth — propagation of structural discontinuities that is accompanied by a significant change in the discontinuities' aspect ratio.

non-systematic joint set — an array of discontinuous parallel cracks usually associated with, but oriented at high angles to, a systematic joint set; see also "cross-joint."

normal stress — a stress component that acts perpendicular to the associated surface.

nucleation — the initiation and earliest growth of a discontinuity, such as a slip patch, that is localized during the deformation of a rock or rock mass.

oblique extension fracture — same as "shear crack."

Oertel faults — an informal designation referring to the set of four synchronous faults (usually normal or reverse) that together are necessary to accommodate 3-D, plane-stress deformation at the Earth's surface; contrasts with "Andersonian faults" for plane-strain and also occasionally called "bi-conjugate" faults.

offset — the net slip across a structural discontinuity.

origin — the point at which a mode-I crack begins to dilate and propagate, marking the location from which hackle and rib marks radiate out along the crack wall.

orthorhombic fault sets — same as "Oertel faults" but used as the preferred term for the geometry.

outer boundary conditions — the parameters that describe stresses, displacements, strains, or material geometries located very far away from a discontinuity or other item of interest in a stress analysis.

outer-normal vector — a directional indicator that points away from a reference surface into the surroundings.

overlap — a measure of the length of the stepover along the strike of echelon discontinuities.

overstep — same as "stepover."

Pa (Pascal) — the SI unit for pressure or stress; $1 \text{ Pa} = 1 \text{ N m}^{-2}$; in rock mechanics kPa (10^3 Pa) are commonly used for values of cohesion, MPa (10^6 Pa) for stress, and GPa (10^9 Pa) for modulus. Because $1 \text{ Pa} = 1 \text{ J m}^{-3}$, stress is equivalent to strain-energy density or work per unit volume.

Paris Law — in fatigue cracking, a well-known power-law relationship between the change in crack length and the stress intensity factor raised to a power that depends on (cyclic) load frequency and environmental effects.

peak strength — the maximum level of stress on the Complete Stress–Strain Curve for Rock associated with microcrack growth, the beginnings of macrocrack growth and linkage, and strain softening in brittle compact rocks.

peak strength criterion — an estimate of the maximum value of (usually differential) stress withstood by a rock or rock mass before an instability in deformation rate or behavior results.

peak stress — the maximum value of differential stress applied to a rock sample or rock mass, corresponding to the change from a positive to a negative slope on the Complete Stress–Strain Curve for Rock.

penny crack — a fully bounded crack having a circular geometric shape when viewed normal to the crack plane, parallel to the opening-displacement vector.

permeability — a measurement of the ease of fluid flow through a rock or rock mass, expressed in units of area (cm^2 or m^2).

pervasively fractured rock — a concept used by geophysicists to describe a rock mass having discontinuities oriented optimally for frictional sliding and negligible cohesive and unconfined compressive strengths; used to apply the Coulomb criterion in brittle strength envelopes of the Earth's crust.

pinnate joints — same as "end cracks" but sometimes applied to outcrop-scale examples. They nucleate in the tip regions of discontinuities such as faulted joints, leaving steps on the fault surface that also indicate the paleostress direction and sense of slip along the parent slip surface.

plane strain — a 2-D representation of 3-D stress states in which the normal strain in the anti-plane direction (z) is negligibly small, with nonzero normal stress in that direction, compared to the strain and stress in the other two in-plane directions; the thick-plate approximation.

plane stress — a 2-D representation of 3-D stress states in which the normal stress in the anti-plane direction (z) is negligibly small, with nonzero normal strain in that direction, compared to the stress and strain in the other two in-plane directions; the thin-plate approximation.

planetary science — an interdisciplinary study of the heliocentric objects of a solar system and their satellites.

planetary structural geology — a branch of planetary science that applies the methods of structural geology, rock mechanics, and tectonics to investigate deformation processes and specific location-dependent problems in the deformation of planetary lithospheres. The term is applied to both heliocentric bodies (planets) or planetocentric bodies (satellites) regardless of size or composition (silicate or hydrous ice).

plastic (material) — an ideal rheology involving two-fold behavior under stress: no deformation if the stress is less than a specified level, the "yield strength," and permanent deformation for greater values of stress. Important subtypes include elastic–plastic and visco-plastic materials.

plastic (rheology) — a deformation mechanism involving temperature-dependent processes such as dislocation movement, twinning, and creep.

plastic zone — a mathematical expression of a rock's limited tensile or shear strength in a small region surrounding a discontinuity termination; it is used to remove the theoretically infinite level of stress at the termination and its extent (or radius) can reasonably well approximate the extent of localized deformation in process zones observed in natural rock examples.

plate — a representation of rock geometry whose dimensions are much greater in two directions (x and y) than in the third (z).

plumose structure — the collection of origin, hackle, and rib marks on a preserved crack wall that demonstrate the geometry and sequence of nucleation and propagation of the crack.

Poisson's ratio — the absolute value of the ratio between the smaller and larger values of strain in an elastic material measured in orthogonal directions.

polygonal fault — a particular dip-slip fault of an orthorhombic set that developed under conditions of 3-D strain.

poorly sorted — a granular rock having a wide range of grain sizes.

poroelastic material — a material type consisting of an elastic framework surrounding void space filled by rheologically distinct phases (typically fluids); commonly used to model soils or fractured rock masses deformed under drained or usually undrained conditions.

porosity — an index property of a rock or rock mass defined as the ratio of void volume to host-rock volume; typically expressed as a percentage.

post-peak deformation — permanent inelastic strains associated with discontinuity growth, linkage, and faulting that is large compared to rock dimensions, leading to either strain-softening or (more rarely) strain hardening response.

post-yield fracture mechanics — a branch of fracture mechanics that relaxes perfectly elastic behavior beyond a negligibly small plastic zone at the crack tip to incorporate any of several aspects: slow subcritical crack growth, plane stress dynamic fracture toughness, non-uniqueness of strain energy release rate G as a material constant, and limitation of near-tip stress magnitudes to yield strength values for the material. The Dugdale and Barenblatt "strip models" are approaches to this class of fracture problems.

precursory substrate — the pre-faulting zone of (usually) weakness that formed during the same deformation event, and under the same far-field stress state, as the later slip surface.

pre-existing substrate — the pre-faulting zone of (usually) weakness that formed prior to, and under a different far-field stress state from, the later slip surface.

pressure — a scalar quantity having units of force/unit area (N m^{-2}); used without reference to particular planes or force directions (e.g., atmospheric or pore-water pressure).

pressure solution surface — same as "stylolite."

process — a series of actions or conditions that lead to a geologically recognizable phenomenon, such as faulting.

process zone — a local zone of increased microcrack (or discontinuity) density located ahead of a discontinuity termination; usually related conceptually to the plastic zone concept for sharp flaws.

propagation — an increase in length of a discontinuity; typical growth criteria involve a comparison of near-tip quantities (stress, displacement, or energy) with discontinuity-scale strength properties (dynamic fracture toughness, microcrack coalescence, frictional resistance, or yield strength).

proportional growth — propagation of structural discontinuities that is accompanied by no significant change in the discontinuities' aspect ratio

P-shear — see "Riedel shear."

pull-apart basin — a region of horizontal extensional strain between echelon strike-slip faults involving subsidence and normal faulting within their stepover.

push-up range — a region of horizontal contractional strain between echelon strike-slip faults involving uplift, folding, and reverse faulting within their stepover.

q–p diagram — a plot of mean stress vs. differential stresses, or the first and second invariants of the 3-D stress state, related to yielding and failure of a soil or high-porosity rock.

Q-system — one of two main rock mass classification systems currently in use in rock engineering; the Q(uality) system was designed originally for estimating the need for roof support in underground tunnels in compact rocks.

quasi-continuum — a representation of a rock mass whose imperfections including discontinuity networks and faults are negligible at the scale of observation, resulting in a set of homogenous properties.

quasi-static — the relatively slow growth velocity of structural discontinuities in which acceleration and dynamic inertial terms are unimportant in the near-tip stress fields.

rake — a measure of the slip vector on a fault plane.

ramp — a panel of sub-horizontal rock located near the termination of a normal fault that is flexed gently downward as part of the hanging wall; also the inclined segment of a nonplanar thrust fault. See also "relay-ramp" and "stepover."

rate-and-state friction — a class of approaches that relate the frictional resistance of rock surfaces to changes in rates of loading and sliding and the properties (state) of the surface.

R curve — called "crack growth resistance curve" in engineering, it is a technique for assessing the non-LEFM growth of discontinuities under plane stress configurations or others for which the strain energy release rate G is not a material constant.

relative scale — the formal comparison or ratio between the scale of observation and scale of structure (joint spacing or block size) used to define the degree of homogeneity of a rock or rock mass.

relay — the region located within a stepover between echelon normal faults.

relay-ramp — a region of mainly vertical distortion and strain between echelon dip-slip faults, usually involving folding and cross faulting; also known as a "fault bridge."

releasing bend — a linkage zone involving horizontal dilational strain in the stepover between two echelon strike-slip faults.

remote stress — the components of stress acting on a rock that are applied sufficiently far away from a point of interest that they can be considered to be applied uniformly.

restraining bend — a linkage zone involving horizontal contractional strain in the stepover between two echelon strike-slip faults.

restricted discontinuity — a discontinuity whose propagation is impeded by local stress gradients or physical boundaries, leading to increased near-tip displacement gradients.

reverse fault — same as "thrust fault" but with steeper dips.

rheology — the study of deformation mechanisms within rocks and other materials that lead to macroscopic, or overall, classes of behavior including elastic, viscous, and plastic.

rib mark — subtle to pronounced topographic changes on a fracture wall that are concave back to the crack origin and perpendicular to hackle; also called "arrest lines" or "hesitation lines."

Riedel shear — a term for echelon fault strands formed early in a progressive deformation sequence, later linked by P-shears across the stepovers to form a continuous slip surface, called a principal-displacement surface or D-shear.

right-oblique — a mixed-mode fault having both right-lateral strike-slip and dip-slip components of displacement.

rille — a generic term used primarily in the Apollo era of the 1960s for lunar grabens (straight rilles) or lava-carved channels (sinuous rilles).

riser — a rare stair-step-like topography preserved on some fault surfaces, in association with steps, that indicates the sense and direction of slip. Steps represent broken contractional stepovers along secondary or initial echelon fault segments found along the aggregate fault surface. The slip direction is given by the rough direction (as opposed to the smooth direction) as one rubs one's hand over the surface, noting that the risers are beveled and chipped by the fault displacement, whereas the steps are cleanly broken at the fault surface.

river pattern — a term for "hackle."

RMR — an acronym for "rock mass rating" whose value is calculated from a systematic assessment of intact-rock strength, discontinuity characteristics and spacing, and ground water conditions.

rock bridge — a small panel of rock located within the stepover between two echelon discontinuities; same as "ligament."

rockburst — sudden dynamic crack propagation in underground settings, such as tunnels or chambers, caused by removal of lateral confining pressure at a free face and resulting in an explosive release of rock material into the cavity.

rock engineering — a branch of engineering that uses the principles of rock mechanics and engineering mechanics to design structures in, or utilizing, rock masses.

rock mass — the aggregate material consisting of both the intact rock and the associated joints, faults, bedding planes, blast fractures, solution surfaces, and other discontinuities, and including ground water conditions.

rock mass classification system — an empirically-based template for identifying and recording characteristics of a field-scale rock exposure that are relevant to the overall strength and deformability properties of the rock mass.

rock mass rating — see "RMR."

rock mechanics — an interdisciplinary branch of materials science and geology dealing with the physical processes of rock deformation, based on laboratory experiments and theory.

RQD — an acronym for "rock quality designation" and used in the Rock Mass Rating system (RMR) as a measure of discontinuity spacing and concomitant strength reduction in the rock mass.

R-shear — same as "echelon fault strand;" see "Riedel shear."

running crack — a crack propagating at dynamic velocities of perhaps 30–40% the sonic velocity of the rock.

rupture — a term for "failure" and to denote either tensile crack growth or frictional sliding; the term finds widespread use in seismology (e.g., "earthquake rupture").

rupture patch — same as "slip patch."

rupture strength — a term for the peak strength of a rock sample; synonymous with, and largely supplanted by, "tensile strength" or "shear strength."

rupture surface — same as "slip surface" but usually reserved for seismic slip along a fault.

saturation — close spacing between parallel discontinuities (usually within a layer) attributed to growth of new discontinuities within the shadow zones of stress relief associated with displacement along the discontinuities whose dimensions span the layer thickness.

scalar — a quantity from physics represented by its magnitude, such as mass, time, or temperature.

scale of observation — a measure of the external dimensions of a rock or rock mass, whether in thin section, hand sample, outcrop, seismic section, or satellite view.

scale of structure — a measure of the internal dimensions of a rock mass such as discontinuity spacing or block size that determines the response of the rock mass to natural or engineered configurations such as free surfaces or imposed states of stress.

scale-independent — same as "self-similar."

scaling dimension — a manifestation of the geometry of a discontinuity population ("small" or "large") relative to the volume or area of the deforming region.

scan-line survey — a common method for assessing the spacing of discontinuities by counting the number of discontinuities that cross a measuring tape placed along the exposure.

secant modulus — the ratio of the cumulative values of stress and strain in a rock; measured from the origin of the stress–strain curve and identical to the tangent modulus for an elastically deforming rock, a comparison between tangent and secant moduli provides a measure of the amount of inelastic deformation sustained by the rock as it was loaded.

second-order fault — a term to describe smaller faults that formed nearby, and probably in response to, a larger discortinuity; now referred to as "secondary faults."

secondary fracture — same as "end crack."

segmentation — growth of discontinuities into an echelon array; a process that arises from lateral restriction of discontinuity lengths.

seismogenic fault — same as "earthquake fault."

self-similar — a process that is independent of scale.

semi-bounded discontinuity — a discontinuity that has clear terminations within the rock mass along only part of its periphery.

semi-major axis — one-half of the long dimension of a discontinuity usually denoted by *a*.

semi-minor axis — one-half of the short dimension of a discontinuity usually denoted by *b*.

sense of slip — the direction of displacement of rocks along a fault with reference to an independent coordinate system such as a horizontal or vertical plane.

sense of step — a description of stepover geometry in which an observer standing on one echelon discontinuity tip must physically step to the right or left to stand on the adjacent echelon segment.

sense of strain — the type of normal strain (extensional or contractional) characteristic of rocks within a discontinuity stepover.

separation — a measure of the perpendicular width of the stepover between echelon discontinuities; also the component of net slip in a particular direction (e.g., horizontal or vertical).

set — a collection, array, network, or population of discontinuities.

shadow zone — same as "stress shadow."

sharp discontinuity — a discontinuity characterized by a discontinuous gradient, or jump, in displacement across it; the normal and shear stresses may also be discontinuous. A sharp discontinuity affects its surroundings strongly, with the effect increasing as the displacement magnitude across the discontinuity increases. See also "crack" and "fault."

shear band — same as "deformation band" in which shear displacement of discontinuity walls is accompanied by no change in porosity or volume within the band.

shear crack — a term for a discontinuity that forms during shear deformation in the plane (or orientation) of the maximum shear stress.

shear-enhanced compaction — the reduction in pore volume in a granular rock caused by grain translations and rotations during remote shear loading.

shear fault — a redundant term for "fault."

shear fracture — same as "shear crack."

shear modulus — the local slope of the shear stress vs. shear strain curve for a rock; can also be back-calculated for an elastic material from Young's modulus and Poisson's ratio.

shear strength — a measure of the resistance of a rock or discontinuity surface to frictional sliding; typically dependent on resolved normal stress, pore-fluid pressure, strength and geometry of the material, loading rate, gouge properties, and temperature.

shear stress — a stress component that acts parallel to the associated discontinuity surface.

shear zone — a tabular structural discontinuity having a continuous change in strength or stiffness across a relatively narrow zone of shearing; shear and volumetric strains are continuous across the zone, and large or continuous (linked) slip surfaces are rare or absent. See also "weak discontinuity."

sheeting joint — a curved crack that forms close to, and subparallel to, a topographic surface beneath domes, ridges, and saddles in hard compact rock; thought to result from a combination of large compressive stress parallel to a curved topographic surface, rather than from uplift, reduction of overburden stress, weathering, pluton solidification, buckling, axial splitting, or residual stresses.

SI units — the set of "mks" (meter, kilogram, second) metric measurement units adopted as the international standard for scientific work; Système Internationale d'Unités (SI) units replace the previous metric "cgs" (centimeter, gram, second) and British (inch, pound, second) units.

sill — a sharp geologic structural discontinuity having field evidence for discontinuous and predominantly opening displacements between the opposing walls, oriented sub-horizontally and/or parallel to bedding, and containing liquid or solid magmatic materials.

simple stiffness — an informal rock mass parameter related to the bulk modulus or incompressibility defined as the modulus (Young's or shear) divided by a function of Poisson's ratio. It is used in fracture mechanics and differs from the stiffness of the loading system (see "system stiffness") and from the stiffness of a rock layer or discontinuity-filling material, which is defined as the slope of the stress–displacement curve, with units of MPa m^{-1}.

singularity — an occurrence of infinite values in an analysis, usually attributed to dividing by a variable that becomes zero for particular cases or geometries.

singular stress — a mathematically permitted, but physically inadmissible, value of stress state or a stress component that is infinite in magnitude at some particular point or points.

sinistral — same as "left-lateral."

sinuous rille — a term used primarily in the Apollo era of the 1960s for lunar lava tubes and channels; once used in connection with lunar grabens in early mapping studies.

Skempton's coefficient — a parameter that quantifies the relative volume of fluid (in pore spaces) and solid in a soil; values range from 0 (dry soil) to 1.0 (fully saturated).

Skempton's coefficient for rock — a parameter analogous to Skempton's coefficient (for soils) that characterizes the relative volume of fluid (in pore spaces) and solid in a rock; values typically range from 0.47 (Indiana Limestone) to ~1.0 (fully saturated rock).

slickenlines — grooves or striations inscribed on a fault surface that demonstrate the slip direction.

slickensides — fibrous mineral growths (usually calcite or serpentine) located in stepovers or other dilational geometric irregularities on a fault surface that suggest the slip direction on that surface.

slip — the magnitude of displacement or offset along a fault.

slip direction — the net slip vector that describes the direction of relative movement on a fault surface of the opposing walls; also known as "sense of slip" of the fault when referred to an independent coordinate system.

slip patch — an area of localized frictional sliding along a larger discontinuity surface.

slip surface — A displacement discontinuity with a shearing sense parallel to the plane of the discontinuity.

slip weakening — the behavior of fault zones and rupture patches characterized by a progressive reduction in frictional resistance with increasing amounts of slip or displacement; similar to "strain softening" of rocks and rock masses.

slit — same as "degenerate ellipse;" a discontinuity having atomistically sharp tips.

small discontinuity — a description of discontinuities whose dimensions (length or down-dip height) are less than those of the deforming volume.

small-scale yielding — a region commonly less than ~0.1a in radius around a discontinuity tip in an elastic material that removes the theoretically infinite stress levels by localized plastic deformation; used along with stress intensity factor in crack propagation and interaction criteria; it is defined where the near-tip stress exceeds the appropriate rock (yield) strength.

soft-linked faults — echelon faults whose stepovers have not been bridged by secondary cross-discontinuities, but where mechanical interaction between the fault strands is inferred to be significant.

solution surface — same as "pressure solution surface" or "stylolite."

sorting — a description of the variability of grain sizes within a granular rock.

splay cracks — same as "end cracks."

SSY — an acronym for "small-scale yielding."

stability — a term used to describe a process or simulation (mathematical or experimental) that is well behaved or smoothly varying; also, a term that describes the resistance of a discontinuity to processes such as frictional sliding or propagation.

stable discontinuity growth — quasi-static or subcritical discontinuity growth that can occur when the rate of increase of driving stresses is less than the rate of discontinuity propagation; the major process associated with the growth of many tectonic joint sets.

stable sliding — frictional sliding that is velocity strengthening, with a negative stress drop or driving stress; same as creep along faults.

static friction coefficient — the maximum value of shear stress/normal stress ratio (assuming negligibly small cohesion) required for frictional sliding to occur along a discontinuity; also called the "maximum static friction."

step — a stair-step-like topography preserved on fault surfaces produced at the intersection between secondary (end or pinnate) discontinuities and the fault and indicating the sense and direction of slip. Steps may also represent broken pull-aparts along an aggregate fault surface that are inhabited by fibrous mineral growths and slickensides.

stepover — the region bounded by linking strands between the tips of echelon discontinuities; commonly rhombohedral in shape.

stiffness — a term that describes the deformability of a rock or rock mass before failure occurs; see also "simple stiffness."

straight rille — a term used primarily in the Apollo era of the 1960s for lunar grabens.

strain — a measure of the change in volume and/or shape of a homogeneous material; normal strain is the change in line length, shear strain is the change in angle within a sample.

strain ellipse — a graphical representation of the principal strains due to homogeneous deformation, commonly used to predict the orientations of faults, folds, and joints during regional shearing but lacking the capability to describe the spatial organization, interaction, inter-relationships, or relative timing of such structures.

strain energy — the product of force times the resulting displacement, with units of N m or joules (J); also known as "work."

strain energy density — a criterion for discontinuity propagation based on finding the location about the discontinuity tip corresponding to the minimum energy needed for growth. Physically defined as the area under a stress–strain curve, the quantity has units of either MPa or J m^{-3}.

strain energy release rate — a criterion for discontinuity propagation (G) that compares the stress intensity factors K from LEFM to the dynamic fracture toughness K_c; also known as "fracture propagation energy" given its units of J m^{-2}. In general, G is not dependent on a determination of K at discontinuity tips but represents a global measure of the energy from all sources that resists discontinuity propagation.

strain hardening — a decrease in the slope of the stress–strain curve for rocks related to closure of discontinuities, mechanical interaction between them, and/or broadening of deformation, leading to more stable strain; same as "work hardening."

strain localization — a set of processes by which inhomogeneous strains develop in materials.

strain softening — a reduction in the capacity of a rock in the post-peak region to support additional stress usually attributable to localization of strain into potentially unstable discontinuities; same as "work softening."

strength — the amount of stress withstood by a rock measured at the initiation of an instability in deformation, such as rapid discontinuity growth; also a measure of the relative responses of a discontinuity and its surroundings.

strength envelope — a peak-strength criterion that relates differential stress to depth and used for placing limits on the stress levels in terrestrial and planetary lithospheres.

strength of singularity — a measure of how rapidly a value such as a stress component increases to an infinite value; stronger singularities have a steeper slope.

stress — a calculated quantity defined by the amount of force per given unit of area; a tensor quantity of second rank involving both the force directions and orientation of the plane on which it acts.

stress analysis — same as "mechanical analysis" in which stresses, displacements, or strains are used to calculate the stresses, displacements, or strains in a body described by a given rheological behavior.

stress concentration — a local increase in the magnitude of stress components, usually near a flaw or other inhomogeneity.

stress concentrator — same as "flaw."

stress corrosion — a mechanism for slow, quasi-static fracture propagation that is modulated by chemically assisted weakening of atomic bonds in the near-tip region, leading to degradation in strength over time by the action of chemically reactive pore fluids.

stress function — a class of analytical solutions to problems in mechanics that can provide closed-form expressions for stresses, strains, and displacements from various partial derivatives of the stress function.

stress intensity factor — a parameter in linear elastic fracture mechanics that measures the intensity of the stress concentration near the tip of a sharp flaw; it scales with the applied stress and discontinuity length, along with secondary factors related to the geometry of the fractured material.

stress shadow — a region of locally decreased stress components adjacent to and partly surrounding a discontinuity that can reduce the driving stress acting on nearby discontinuities.

stress trajectory — a set of imaginary lines or curves that trace out the local directions of the principal stresses in a rock mass.

strip model — a mathematical simplification to plastic deformation near a discontinuity tip that collapses the volumetric deformation onto a single planar zone coplanar with and adjacent to the discontinuity.

structural discontinuity — a localized curviplanar change in strength or stiffness caused by deformation of a rock that is characterized by two opposing surfaces that are bounded in extent, approximately planar compared to the longest dimension, and having a displacement of originally adjacent points on the opposing walls that is small relative to the longest dimension.

St. Venant material — same as "plastic material."

St. Venant's principle — a law of mechanics stating that the greatest degree of stress gradient, magnitude, and change is located nearest the area of application of a load; related to near-tip stress concentration, stress redistribution, and displacement accumulation along discontinuities in rock.

stylolite — a discontinuity marked by insoluble clay residue associated with localized pressure solution and mass transfer in soluble rocks such as carbonates; same as "pressure solution surface" or "selvage seam."

subcritical fracture growth — propagation of discontinuities of any mode at very slow rates which are controlled by chemical (stress corrosion) and other processes near the fracture's tipline.

substrate — the planar or tabular zone having different mechanical properties than the surrounding host rock that serves as a nucleation site for frictional sliding.

synthetic — a discontinuity whose dip direction is subparallel to that of nearby ones, usually with subequal strike direction; the term usually implies, but may not require, synchroneity to be applicable.

systematic joint set — an array of parallel, coplanar, kinematically coherent cracks having aggregate lengths much greater than their spacings.

system stiffness — the role of the loading system or confining medium on the material properties and deformation response of a rock or rock mass.

tabular discontinuity — a structural discontinuity having a continuous change in strength or stiffness that occurs across a relatively thin band.

tail crack — same as "end crack."

tall discontinuity — a structural discontinuity whose vertical dimension (height) substantially exceeds its horizontal dimension (length).

tangent modulus — the value of slope of a stress–strain curve for rock at a particular level of stress; same as "Young's modulus" in the elastic range.

teeth — irregularities on a stylolite surface due to differential dissolution of the host rock and crudely aligned with the dirction of shortening.

tension — a force, stress, or stress component which causes particles to move apart.

tension crack — a term for "dilatant crack" or "joint" that combines knowledge of the sign of the normal strain associated with crack dilation; compare to "contraction crack."

tensional fault — a term for "normal fault" that obscures the requirement for contact and frictional sliding between adjacent fault walls.

tensional quadrant — same as dilational quadrant.

tensor — a mathematical quantity involving magnitude, direction, and another component such as orientation; stress is a second-rank tensor, having combined elements of force (magnitude and direction) and the plane on which it acts (orientation).

termination — the end of a discontinuity where displacements decrease toward zero, often associated with arrays of unlinked or partially linked microstructures; synonymous with "tip."

through-the-thickness crack — same as "tunnel crack."

throw — the vertical component of net slip across a fault.

thrust anticline — the surface expression of a fold developed above a blind thrust fault; same as "forced fold," "fault-bend fold," "fault-propagation fold," and "wrinkle ridge."

tip displacement criterion (TDC) — a propagation criterion for discontinuities that relies on comparing the displacement at the discontinuity tip's to the appropriate yield strength there; also known as "crack opening displacement" (COD) or "crack tip opening displacement" (CTOD).

tipline — the edge of a bounded or semi-bounded fault in the subsurface; the surface-breaking equivalent is the "fault trace."

tool mark — a linear groove formed on a slipping interface from dragging of an indentor that is harder than the surface, either from its material properties or due to stress concentration related to the indentor's shape.

torsion — shear stress imparted by either twisting or anti-plane shearing.

traction — a term for stress components resolved onto a particular plane or orientation (such as a discontinuity wall); a vector rather than a tensor, traction is sometimes referred to as the "stress vector."

transcurrent fault — a term for "strike-slip fault."

transfer fault — a secondary discontinuity that links echelon faults; same as "accommodation structure" and "hard-linkage."

transfer zone — a stepover between two echelon dip-slip faults.

transform fault — a special class of strike-slip faults that connects opposing oceanic spreading centers and is unbounded in map view; its sense of slip is opposite to that of strike-slip faults.

transformational faulting — a mechanism for deep-focus (410–670 km depth) earthquakes in the Earth's mantle in which localized volume reductions due to mineral phase changes may nucleate anticracks that serve as slip surfaces and rupture source regions.

transitional tensile joint — a term for a crack having oblique dilation (opening) of the fracture walls and obtained from a classical interpretation of the Mohr diagram and a Griffith failure criterion; now largely replaced by "mixed-mode crack."

triangle dike — a dike whose opening-mode displacements decrease linearly from the maximum value toward minima at its terminations; the term describes the dike's shape in cross section.

tribology — the study of friction and wear of engineering components.

truncation bias — a reduction in slope on a cumulative-frequency vs. length diagram at progressively smaller lengths associated with undercounting of shorter discontinuities.

tunnel crack — a 2-D representation of a discontinuity that extends into the material, parallel to the plume axis, much farther than its in-plane horizontal length.

twist — the angle measured between the centers of echelon discontinuities and their local strikes.

twist hackle — synonymous with "hackle" as used to describe the plume-like textural markings on a mode-I crack surface.

ultimate strength — same as "peak strength" but normally used for steel and other engineering materials.

unbounded discontinuity — a discontinuity that has no terminations within the rock mass; its longitudinal dimensions equal those of the material.

undrained conditions — an end-member response of fluid-saturated porous rock or rock mass in which the rate of fluid movement is significantly less than the deformation rate and for which the fluid phase thereby exerts non-negligible effects on the deformation; also may characterize hydraulically closed systems.

unrestricted discontinuity — a discontinuity whose propagation is not impeded by local stress gradients or physical boundaries, leading to near-tip displacement gradients typical of discontinuities in infinite rock masses loaded by uniform remote stresses.

unstable crack growth — same as "dynamic growth of a running crack;" can occur when the rate of increase of crack-driving stresses exceeds the rate of crack propagation that serves to absorb, or dissipate, the driving stress.

unstable sliding — frictional slip along a surface that is sudden or episodic; same as "stick-slip."

upper ramp breach — a secondary fault that links the two primary echelon dip-slip faults across their stepover at the topographically higher end of the relay-ramp; also known as "footwall breach."

vector — a quantity from physics having both magnitude and direction, such as velocity, displacement, and force.

vein — a sharp geologic structural discontinuity having field evidence for discontinuous and predominantly opening displacements between the opposing walls and containing hydrothermal minerals such as calcite, silica, or epidote.

vergence — the overall direction of displacement of a linked fault system or thrust sheet.

viscosity — the slope of the stress–strain rate relation for a viscous material.

viscous material — an ideal rheology in which an applied stress leads to a particular strain rate and flow; a special case is a linear viscous material with a constant value of viscosity. Usually synonymous with "viscous fluid."

volumetric strain — the sum of the normal strains in all three coordinate directions.

weak discontinuity — a discontinuity characterized by a continuous gradient in displacement across its thickness; the normal and shear stresses may also be continuous. A weak discontinuity affects its surroundings weakly, with the effect increasing as the displacement gradient across the discontinuity increases. See also "deformation band."

well sorted — a granular rock having a uniform grain size.

wing crack — same as "end crack."

work — in mechanics, the product of force times displacement (with units of joules), used as a measure of deformation. Work is also a measure of the area under a force- (or load-) displacement curve; same as "strain energy."

work hardening — same as "strain hardening."

wrench fault — a term for strike-slip fault, commonly encountered in petroleum applications.

wrench tectonics — strike-slip tectonics; a classical term used primarily by the petroleum industry.

wrinkle ridge — a blind-thrust anticline (or forced fold) typically cut by one or more backthrust faults, commonly found exposed on planetary surfaces such as Mercury, Venus, the Moon, and Mars.

yield cap — the part of the yield surface for a porous granular rock at high values of mean stress associated with compactional strain with or without shearing.

yield criterion — a mathematical relationship between stress state and deformation; can be the preferred term for "failure criterion," or "strength criterion" and used to describe either localized or distributed behavior, often for pre-peak conditions.

yield strength — a transitional characteristic of a geologic material given by the value of stress state at which significant inelastic deformation occurs; although yielding may carry connotations of plastic or ductile flow, the general term can also include brittle (pressure-dependent) processes and either localized (e.g., crack growth) or distributed (e.g., cataclasis) types of deformation.

yield stress — the value of applied stress at which yielding in a rock begins or becomes significant.

yield surface — a locus of stress states that separate elastic behavior from permanent, inelastic, plastic yielding of a rock or rock mass.

Young's modulus — the local slope of the stress–strain curve for rock loaded within the elastic range, measured in a laboratory sample at 50% of the peak stress; same as "tangent modulus."

References

Aadnøy, B. S. and Looyeh, R. (2010). *Petroleum Rock Mechanics: Drilling Operations and Well Design*, Elsevier, Amsterdam.

Abercrombie, R. W. and Ekström, G. (2001). Earthquake slip on oceanic transform faults. *Nature* **410**:74–76.

Abou-Sayed, A. S. (1977). Fracture toughness, K_{1c}, of triaxially-loaded Indiana limestone. *Proc. US Symp. Rock Mech.* **18**:2A3/1–2A3/8.

Ackermann, R. V., Schlische, R. W. and Withjack, M. O. (2001). The geometric and statistical evolution of normal fault systems: an experimental study of the effects of mechanical layer thickness on scaling laws. *J. Struct. Geol.* **23**:1803–1819.

Acocella, V., Gudmundsson, A. and Funiciello, R. (2000). Interaction and linkage of extension fractures and normal faults: examples from the rift zone of Iceland. *J. Struct. Geol.* **22**:1233–1246.

Aharonov, E. and Katsman, R. (2009). Interaction between pressure solution and clays in stylolite development: insights from modeling. *Am. J. Sci.* **309**:607–632.

Ahlgren, S. G. (2001). The nucleation and evolution of Riedel shear zones as deformation bands in porous sandstone. *J. Struct. Geol.* **23**:1203–1214.

Aki, K. and Richards, P. G. (1980). *Quantitative Seismology: Theory and Methods*, Vol. I, W. H. Freeman, San Francisco, California.

Albert, R. A., Phillips, R. J., Dombard, A. J. and Brown, C. D. (2002). A test of the validity of yield strength envelopes with an elastoviscoplastic finite element method. *Geophys. J. Int.* **140**:399–409.

Al-Chalabi, M. and Huang, C. L. (1974). Stress distribution within circular cylinders in compression. *Int. J. Rock Mech. Min. Sci.* **11**:45–56.

Aliha, M. R. M., Ayatollahi, M. R., Smith, D. J. and Pavier, M. J. (2010). Geometry and size effects on fracture trajectory in a limestone rock under mixed mode loading. *Eng. Fracture Mech.* **77**:2200–2212.

Allmendinger, R. W. (1998). Inverse and forward numerical modeling of trishear fault-propagation folds. *Tectonics* **17**:640–656.

Altman, S. J., Aminzadeh, B., Balhoff, M. T., et al. (2014). Chemical and hydrodynamic mechanisms for long-term geological carbon storage. *J. Phys. Chem.* **118**:15,103–15,113.

Ameen, M. S. (1995). Fractography and fracture characterization in the Permo-Triassic sandstones and the Lower Palaeozoic Basement, West Cumbria, UK. In *Fractography: Fracture Topography as a Tool in Fracture Mechanics and Stress Analysis* (ed. M. S. Ameen), pp. 97–147, Geol. Soc. Spec. Publ. 92.

Anders, M. H. and Schlische, R. W. (1994). Overlapping faults, intrabasin highs, and the growth of normal faults. *J. Geol.* **102**:165–179.

Anderson, E. M. (1938). The dynamics of sheet intrusion. *Proc. Royal Soc. Edinburgh* **58**:242–251.

Anderson, E. M. (1951). *The Dynamics of Faulting and Dyke Formation, with Applications to Britain*, Oliver & Boyd, Edinburgh.

Anderson, O. L. and Grew, P. C. (1977). Stress corrosion theory of crack propagation with applications to geophysics. *Rev. Geophys.* **15**:77–104.

Anderson, R. E. (1973). Large-magnitude late Tertiary strike-slip faulting north of Lake Mead, Nevada. *US Geol. Surv. Prof. Pap.* 794, 18 pp.

Anderson, T. L. (1995). *Fracture Mechanics: Fundamentals and Applications* (2nd edn), CRC Press, Boca Raton, Florida.

Anderson, T. L. (2017). *Fracture Mechanics: Fundamentals and Applications* (4th edn), CRC Press, Boca Raton, Florida, 688 pp.

Andrade, J. E., Avila, C. F., Hall, S. A., Lenoir, N. and Viggiani, G. (2011). Multiscale modeling and characterization of granular matter: from grain kinematics to continuum mechanics. *J. Mech. Phys. Solids* **59**:237–250.

Andrews-Hanna, J. C., Zuber, M. T. and Hauck, S. A., II (2008). Strike-slip faults on Mars: Observations and implications for global tectonics and geodynamics. *J. Geophys. Res.* **113**:E08002; doi:10.1029/2007JE002980.

Angelier, J. (1989). From orientation to magnitudes in paleostress determinations using fault slip data. *J. Struct. Geol.* **11**:37–50.

Angelier, J. (1994). Fault slip analysis and paleostress reconstruction. In *Continental Deformation* (ed. P. L. Hancock), pp. 53–100, Pergamon, New York.

Angelier, J., Colletta, B. and Anderson, R. E. (1985). Neogene paleostress changes in the Basin and Range: a case study at Hoover Dam, Nevada-Arizona. *Geol. Soc. Am. Bull.* **96**:347–361.

Antonellini, M. and Aydin, A. (1994). Effect of faulting on fluid flow in porous sandstones: geometry and spatial distribution. *Am. Assoc. Petrol. Geol. Bull.* **78**:355–377.

Antonellini, M. and Pollard, D. D. (1995). Distinct element modeling of deformation bands in sandstone. *J. Struct. Geol.* **17**:1165–1182.

Antonellini, M., Aydin, A. and Pollard, D. D. (1994). Microstructure of deformation bands in porous sandstones at Arches National Park, Utah. *J. Struct. Geol.* **16**:941–959.

Aplin, A. C. and Macquaker, J. H. S. (2011). Mudstone diversity: origin and implications for source, seal, and reservoir properties in petroleum systems. *Am. Assoc. Petrol. Geol. Bull.* **95**:2031–2059.

Argon, A. S., Andrews, R. D., Godrick, J. A. and Whitney, W. (1968). Plastic deformation bands in glassy polystyrene. *J. Appl. Physics* **39**:1899–1906.

Arthur, J., Dunstan, T., Al-Ani, Q. and Assadi, A. (1977). Plastic deformation and failure in granular media. *Géotechnique* **27**:53–74.

Ashby, M. F. (2016). *Materials Selection in Mechanical Design* (5th edn), Butterworth-Heinemann, 660 pp.

Ashby, M. F. and Sammis, C. G. (1990). The damage mechanics of brittle solids in compression. *Pure Appl. Geophys.* **133**:489–521.

Atkinson, B. K. (1984). Subcritical crack growth in geologic materials. *J. Geophys. Res.* **89**:4077–4114.

Atkinson, B. K. (1987). Introduction to fracture mechanics and its geophysical applications. In *Fracture Mechanics of Rock* (ed. B. K. Atkinson), pp. 1–26, Academic Press, New York.

Atkinson, B. K. and Meredith, P. G. (1987a). The theory of subcritical crack growth with applications to minerals and rocks. In *Fracture Mechanics of Rock* (ed. B. K. Atkinson), pp. 111–166, Academic Press, New York.

Atkinson, B. K. and Meredith, P. G. (1987b). Experimental fracture mechanics data for rocks and minerals. In *Fracture Mechanics of Rock* (ed. B. K. Atkinson), pp. 477–525, Academic Press, New York.

Atkinson, G. M., Eaton, D. W., Ghofrani, H., et al. (2016). Hydraulic fracturing and seismicity in the Western Canada Sedimentary Basin. *Seismol. Res. Lett.* **87**:631–647.

Attewell, P. B. and Woodman, J. P. (1971). Stability of discontinuous rock masses under polyaxial stress systems. In *13th Symp. on Rock Mech., Stability of Rock Slopes*, pp. 665–683, Am. Soc. Civil Eng., New York.

Attewell, P. B. and Farmer, I. W. (1973). Fatigue behaviour of rocks. *Int. J. Rock Mech. Min. Sci.* **10**:1–9.

Atwater, T. (1970). Implications of plate tectonics for the Cenozoic tectonic evolution of western North America. *Geol. Soc. Am. Bull.* **81**:3513–3536.

Awdal, A., Healy, D. and Alsop, G. I. (2014). Geometrical analysis of deformation band lozenges and their scaling relationships to fault lenses. *J. Struct. Geol.* **66**:11–23.

Axen, G. J. (1992). Pore pressure, stress increase, and fault weakening in low-angle normal faults. *J. Geophys. Res.* **97**:8979–8991.

Aydin, A. (1978). Small faults formed as deformation bands in sandstone. *Pure Appl. Geophys.* **116**:913–930.

Aydin, A. (1988). Discontinuities along thrust faults and the cleavage duplexes. In *Geometries and Mechanisms of Thrusting, with Special Reference to the Appalachians* (eds. G. Mitra and S. Wojtal), Geol. Soc. Am. Spec. Pap. 222:223–232.

Aydin, A. (2000). Fractures, faults, and hydrocarbon entrapment, migration and flow. *Marine Petrol. Geol.* **17**:797–814.

Aydin, A. (2014). Failure modes of shales and their implications for natural and man-made fracture assemblages. *Am. Assoc. Petrol. Geol. Bull.* **98**:2391–2409.

Aydin, A. and Ahmadov, R. (2009). Bed-parallel compaction bands in aeolian sandstone: their identification, characterization and implications. *Tectonophysics* **479**:277–284.

Aydin, A. and Berryman, J. G. (2010). Analysis of the growth of strike-slip faults using effective medium theory. *J. Struct. Geol.* **32**:1629–1642.

Aydin, A. and DeGraff, J. M. (1988). Evolution of polygonal fracture patterns in lava flows. *Science* **239**:471–476.

Aydin, A. and Johnson, A. (1978). Development of faults as zones of deformation bands and as slip surfaces in sandstone. *Pure Appl. Geophys.* **116**:931–942.

Aydin, A. and Johnson, A. (1983). Analysis of faulting in porous sandstones. *J. Struct. Geol.* **5**:19–31.

Aydin, A. and Nur, A. (1982). Evolution of pull-apart basins and their scale independence. *Tectonics* **1**:11–21.

Aydin, A. and Nur, A. (1985). The types and role of stepovers in strike-slip tectonics. In *Strike-Slip Deformation, Basin Formation, and Sedimentation* (eds. K. T. Biddle and N. Christie-Blick), pp. 35–44, Soc. Econ. Paleon. Miner., Spec. Publ. 37.

Aydin, A. and Page, B. M. (1984). Diverse Pliocene–Quaternary tectonics in a transform environment, San Francisco Bay region, California. *Geol. Soc. Am. Bull.* **95**:1303–1317.

Aydin, A. and Reches, Z. (1982). Number and orientation of fault sets in the field and in experiments. *Geology* **10**:107–112.

Aydin, A. and Schultz, R. A. (1990). Effect of mechanical interaction on the development of strike-slip faults with echelon patterns. *J. Struct. Geol.* **12**:123–129.

Aydin, A., Schultz, R. A. and Campagna, D. (1990). Fault-normal dilation in pull-apart basins: implications for the relationship between strike-slip faults and volcanic activity. *Annales Tectonicae* **4**:45–52.

Aydin, A., Borja, R. I. and Eichhubl, P. (2006). Geological and mathematical framework for failure modes in granular rock. *J. Struct. Geol.* **28**:83–98.

Bahat, D. (1979). Theoretical considerations on mechanical parameters of joint surfaces based on studies on ceramics. *Geol. Mag.* **116**:81–92.

Bahat, D. (1991). *Tectonofractography*, Springer-Verlag, Heidelberg.

Bahat, D. and Engelder, T. (1984). Surface morphology on cross-fold joints of the Appalachian plateau, New York and Pennsylvania. *Tectonophysics* **104**:299–313.

Bahat, D., Bankwitz, P. and Bankwitz, E. (2003). Preuplift joints in granites: evidence for subcritical and postcritical fracture growth. *Geol. Soc. Am. Bull.* **115**:148–165.

Bai, T. and Gross, M. R. (1999). Theoretical analysis of cross-joint geometries and their classification. *J. Geophys. Res.* **104**:1163–1177.

Bai, T. and Pollard, D. D. (1997). Experimental study of joint surface morphology and fracture propagation rate (abstract). *Eos, Trans. Am. Geophys. Union* **78**:F711.

Bai, T. and Pollard, D. D. (2000a). Fracture spacing in layered rocks: a new explanation based on the stress transition. *J. Struct. Geol.* **22**:43–57.

Bai, T. and Pollard, D. D. (2000b). Closely spaced fractures in layered rocks: initiation mechanism and propagation kinematics. *J. Struct. Geol.* **22**:1409–1425.

Bai, T., Maerten, L., Gross, M. R. and Aydin, A. (2002). Orthogonal cross joints: do they imply a regional stress rotation? *J. Struct. Geol.* **24**:77–88.

Baig, A. M., Urbancic, T. and Viegas, G. (2012). Do hydraulic fractures induce events large enough to be felt on surface? *Can. Soc. Explor. Geophys. Recorder*, October 2012: 40–46.

Bailey, W. R., Walsh, J. J. and Manzocchi, T. (2005). Fault populations, strain distribution and basement fault reactivation in the East Pennines Coalfield, UK. *J. Struct. Geol.* **27**:913–928.

Ballas, G., Soliva, R., Sizun, J.-P., et al. (2012). The importance of the degree of cataclasis in shear bands for fluid flow in porous sandstone, Provence, France. *Am. Assoc. Petrol. Geol. Bull.* **96**:2167–2186.

Ballas, G., Soliva, R., Sizun, J.-P., et al. (2013). Shear-enhanced compaction bands formed at shallow burial conditions; implications for fluid flow (Provence, France). *J. Struct. Geol.* **47**:3–15.

Ballas, G., Soliva, R., Benedicto, A. and Sizun, J.-P. (2014). Control of tectonic setting and large-scale faults on the basin-scale distribution of deformation bands in porous sandstone (Provence, France). *Marine Petrol. Geol.* **55**:142–159.

Ballas, G., Fossen, H. and Soliva, R. (2015). Factors controlling permeability of cataclastic deformation bands and faults in porous sandstone reservoirs. *J. Struct. Geol.* **76**:1–21.

Banks, C. J. and Warburton, J. (1986). 'Passive-roof' duplex geometry in the frontal structures of the Kirthar and Sulaiman mountain belts, Pakistan. *J. Struct. Geol.* **8**:229–237.

Barenblatt, G. I. (1962). The mathematical theory of equilibrium cracks in brittle fracture. *Adv. Appl. Mech.* **7**:55–129.

Barka, A. A. and Kadinsky-Cade, K. (1988). Strike-slip fault geometry in Turkey and its influence on earthquake activity. *Tectonics* **7**:663–684.

Barnett, J. A. M., Mortimer, J., Rippon, J. H., Walsh, J. J. and Watterson, J. (1987). Displacement geometry in the volume containing a single normal fault. *Am. Assoc. Petrol. Geol. Bull.* **71**:925–937.

Barree, R. D., Gilbert, J. V. and Conway, M. W. (2009). Stress and rock property profiling for unconventional reservoir stimulation. Paper SPE 118703 presented at the 2009 SPE Hydraulic Fracturing Technology Conference, The Woodlands, Texas, 19–21 January 2009.

Barrell, D. J. A., Litchfield, N. J., Townsend, D. B., et al. (2011). Strike-slip ground-surface rupture (Greendale Fault) associated with the 4 September 2010 Darfield earthquake, Canterbury, New Zealand. *Quart. J. Eng. Geol. Hydrogeol.*, **44**:283–291; doi:10.1144/1470-9236/11-034.

Bartlett, W. L., Friedman, M. and Logan, J. M. (1981). Experimental folding and faulting of rocks under confining pressure, Part IX. Wrench faults in limestone layers. *Tectonophysics* **79**:255–277.

Barton, C. A. and Zoback, M. D. (1994). Stress perturbations associated with active faults penetrated by boreholes: possible evidence for near-complete stress drop and a new technique for stress magnitude measurement. *J. Geophys. Res.* **99**:9373–9390.

Barton, C. A., Zoback, M. D. and Moos, D. (1995). Fluid flow along potentially active faults in crystalline rock. *Geology* **23**:683–686.

Barton, C. C. (1983). Systematic jointing in the Cardium Sandstone along the Bow River, Alberta, Canada. PhD thesis, Yale University, New Haven, CT.

Barton, N. R. (1976). The shear strength of rock and rock joints. *Int. J. Rock Mech. Min. Sci. Geomech. Abs.* **13**:255–279.

Barton, N. R. (1990). Scale effects or sampling bias? In *Scale Effects in Rock Masses* (ed. A. P. Cunha), pp. 31–55. Balkema, Rotterdam.

Barton, N. R. (2013). Shear strength criteria for rock, rock joints, rockfill and rock masses: problems and some solutions. *J. Rock Mech. Geotech. Eng.* **5**:249–261.

Barton, N. R. and Bandis, S. C. (1990). Review of predictive capabilities of JRC-JCS model in engineering practice. In *Proc. Int. Symp. Rock Joints* (eds. N. Barton and O. Stephansson), pp. 603–610, Loen, Norway.

Barton, N. R. and Choubey, V. (1977). The shear strength of rock joints in theory and practice. *Rock Mech.* **10**:1–54.

Barton, N. R., Lien, R. and Lunde, J. (1974). Engineering classification of rock masses for the design of tunnel support. *Rock Mech.* **6**:189–236.

Baud, P., Klein, E. and Wong, T.-f. (2004). Compaction localization in porous sandstones: spatial evolution of damage and acoustic emission activity. *J. Struct. Geol.* **26**:603–624.

Baud, P., Vajdova, V. and Wong, T.-f. (2006). Shear-enhanced compaction and strain localization: inelastic deformation and constitutive modeling of four porous sandstones. *J. Geophys. Res.* **111**, B12401; doi:10.1029/2005JB004101.

Baudon, C. and Cartwright, J. A. (2008). 3D seismic characterization of an array of linked normal faults in the Levant Basin, Eastern Mediterranean. *J. Struct. Geol.* **30**:746–760.

Baxevanis, T., Papamichos, E., Flornes, O. and Larsen, I. (2006). Compaction bands and induced permeability reduction in Tuffeau de Maastricht calcarenite. *Acta Geotechnica* **1**:123–135.

Bayly, B. (1992). *Mechanics in Structural Geology*, Springer-Verlag, New York.

Bazant, Z. P. (1976). Instability, ductility and size effect in strain-softening concrete. *J. Engng. Mech.* **102**:331–344.

Bazant, Z. P. and Hubler, M. H. (2014). Theory of cyclic creep of concrete based on Paris law for fatigue growth of subcritical microcracks. *J. Mech. Phys. Solids* **63**:187–200.

Beach, A. (1975). The geometry of en echelon vein arrays. *Tectonophysics* **28**:245–263.

Becker, S. P., Eichhubl, P., Laubach, S. E., et al. (2010). A 48 m.y. history of fracture opening, temperature, and fluid pressure: Cretaceous Travis Peak Formation, East Texas basin. *Geol. Soc. Am. Bull.* **122**:1081–1093.

Beeler, N. M. and Lockner, D. A. (2003). Why earthquakes correlate weakly with the solid earth tides: effects of periodic stress on the rate and probability of earthquake occurrence. *J. Geophys. Res.* **108**:2391; doi:10.1029/2001JB001518.

Beeler, N. M., Tullis, T. E. and Weeks, J. D. (1996). Frictional behavior of large displacement experimental faults. *J. Geophys. Res.* **101**:8697–8715.

Beeler, N. M., Simpson, R. W., Hickman, S. H. and Lockner, D. A. (2000). Pore fluid pressure, apparent friction, and Coulomb failure. *J. Geophys. Res.* **105**:25,533–25,542.

Beer, F. P., Johnston, E. R., Jr. and DeWolf, J. T. (2004). *Mechanics of Materials (in SI Units)* (3rd edn), McGraw-Hill, New York.

Begley, J. A. and Landes, J. D. (1972). The *J*-integral as a fracture criterion. In *Fracture Toughness*, pp. 1–20, Am. Soc. Testing Mat., ASTM STP 514.

Begley, J. A. and Landes, J. D. (1976). Serendipity and the *J*-integral. *Int. J. Fracture* **12**:764–766.

Behn, M. D., Lin, J. and Zuber, M. T. (2002a). Mechanisms of normal fault development at mid-ocean ridges. *J. Geophys. Res.* **107**; doi:10.1029/2001JB000503.

Behn, M. D., Lin, J. and Zuber, M. T. (2002b). Evidence for weak transform faults. *Geophys. Res. Lett.* **29**:2207; doi:10.1029/2002GL015612.

Belfield, W. C. (1998). Incorporating spatial distribution into stochastic modelling of fractures: multifractals and Lévy-stable statistics. *J. Struct. Geol.* **20**:473–486.

Bell, F. G. (1993). *Engineering Geology*, Blackwell, London.

Benedicto, A. and Schultz, R. A. (2010). Stylolites in limestone: magnitude of contractional strain accommodated and scaling relationships. *J. Struct. Geol.* **32**:1250–1256.

Benedicto, A., Schultz, R. and Soliva, R. (2003). Layer thickness and the shape of faults. *Geophys. Res. Lett.* **30**:2076, doi:10.1029/2003GL018237.

Bense, V. F., Gleeson, T., Loveless, S. E., Bour, O. and Scibek, J. (2013). Fault zone hydrology. *Earth-Sci. Rev.* **127**:171–192.

Ben-Zion, Y. (2001). On quantification of the earthquake source. *Seismol. Res. Lett.* **72**:151–152.

Bergen, K., and Shaw, J. H. (2010). Displacement profiles and displacement-length scaling relationships of thrust faults constrained by seismic reflection data. *Geol. Soc. Am. Bull.* **122**:1209–1219; doi:10.1130/B26373.1.

Bergsaker, A. S., Røyne, A., Ougier-Simonin, A. and Renard, F. (2016). The effect of fluid composition, salinity, and acidity on subcritical crack growth in calcite crystals. *J. Geophys. Res.* **121**:1631–1651; doi:10.1002/2015JB012723.

Bernal, A., Hardy, S. and Gawthorpe, R. L. (2018). Three-dimensional growth of flexural slip fault-bend and fault-propagation folds and their geomorphic expression. *Geosciences* **8**: 110; doi:10.3390/geosciences8040110.

Bertrand, L., Géraud, Y., Le Garzic, E., et al. (2015). A multiscale analysis of a fracture pattern in granite: a case study of the Tamariu granite, Catalunya, Spain. *J. Struct. Geol.* **78**:52–66.

Bessenger, B. A., Liu, Z., Cook, N. G. W. and Myer, L. R. (1997). A new fracturing mechanism for granular media. *Geophys. Res. Lett.* **24**:2605–2608.

Bésuelle, P. (2001a). Compacting and dilating shear bands in porous rock: theoretical and experimental conditions. *J. Geophys. Res.* **106**:13,435–13,442.

Bésuelle, P. (2001b). Evolution of strain localisation with stress in a sandstone: brittle and semi-brittle regimes. *Phys. Chem. Earth* **A26**:101–106.

Bésuelle, P. and Rudnicki, J. W. (2004). Localization: shear bands and compaction bands. In *Mechanics of Fluid-Saturated Rocks* (eds. Y. Guéguen and M. Boutéca), pp. 219–321, Elsevier, Amsterdam.

Biddle, K. T. and Christie-Blick, N. (1985). Glossary—Strike-slip deformation, basin formation, and sedimentation. In *Strike-Slip Deformation, Basin Formation, and Sedimentation* (eds. K. T. Biddle and N. Christie-Blick), pp. 375–386, Soc. Econ. Paleon. Miner., Spec. Publ. 37.

Biegel, R. L., Sammis, C. G. and Dieterich, J. H. (1989). The frictional properties of a simulated gouge having a fractal particle distribution. *J. Struct. Geol.* **11**:827–846.

Bieniawski, Z. T. (1968). The effect of specimen size on compressive strength of coal. *Int. J. Rock Mech. Min. Sci.* **5**:325–335.

Bieniawski, Z. T. (1974). Estimating the strength of rock materials. *J. S. Afr. Inst. Min. Metall.* **74**:312–320.

Bieniawski, Z. T. (1978). Determining rock mass deformability—experience from case histories. *Int. J. Rock Mech. Min. Sci.* **15**:237–247.

Bieniawski, Z. T. (1989). *Engineering Rock Mass Classifications: A Complete Manual for Engineers and Geologists in Mining, Civil, and Petroleum Engineering*, Wiley, New York.

Bieniawski, Z. T. and Van Heerden, W. L. (1975). The significance of in-situ tests on large rock specimens. *Int. J. Rock Mech. Min. Sci.* **12**:101–113.

Biggs, J., Bastow, I. D., Keir, D. and Lewi, E. (2011). Pulses of deformation reveal frequently recurring shallow magmatic activity beneath the Main Ethiopian Rift. *Geochem. Geophys. Geosystems* **12**:Q0AB10; doi:10.1029/2011GC003662.

Bilby, B. A., Cotterell, A. H. and Swindon, K. H. (1963). The spread of plastic yield from a notch. *Proc. Royal Soc. London* A, **272**:304–314.

Bilham, R. and King, G. (1989). The morphology of strike-slip faults: examples from the San Andreas Fault, California. *J. Geophys. Res.*, **94**:10,204–10,216.

Billi, A., Salvini, F. and Storti, F. (2003). The damage zone-fault core transition in carbonate rocks: implications for fault growth, structure and permeability. *J. Struct. Geol.* **25**:1779–1794.

Billings, M. P. (1972). *Structural Geology* (3rd edn), Prentice-Hall, Englewood Cliffs, New Jersey.

Bilotti, F. and Suppe, J. (1999). The global distribution of wrinkle ridges on Venus. *Icarus* **139**:137–159.

Bishop, D. G. (1968). The geometric relationships of structural features associated with major strike-slip faults in New Zealand. *N.Z. J. Geol. Geophys.* **11**:405–417.

Boettcher, M. S. and Marone, C. (2004). Effects of normal stress variations on the strength and stability of creeping faults. *J. Geophys. Res.* **109**:3406; doi:10.1029/2003JB002824.

Bolton, M. (1986). The strength and dilatancy of sands. *Géotechnique* **36**:65–78.

Bonnet, E., Bour, O., Odling, N. E., et al. (2001). Scaling of fracture systems in geological media. *Rev. Geophys.* **39**:347–383.

Bons, P. D., Elburg, M. A. and Gomez-Rivas, E. (2012). A review of the formation of tectonic veins and their microstructures. *J. Struct. Geol.* **43**:33–62.

Borgia, A., Burr, J., Montero, W., Morales, L. D. and Alvarado, G. A. (1990). Fault propagation folds induced by gravitational failure and slumping of the central Costa Rica volcanic range: implications for large terrestrial and Martian volcanic edifices. *J. Geophys. Res.* **95**:14,357–14,382.

Borgos, H. G., Cowie, P. A. and Dawers, N. H. (2000). Practicalities of extrapolating one-dimensional fault and fracture size–frequency distributions to higher-dimensional samples. *J. Geophys. Res.* **105**:28,377–28,391.

Borja, R. I. (2002). Bifurcation of elastoplastic solids to shear band mode at finite strain. *Comput. Methods Appl. Mech. Eng.* **191**:5287–5314.

Borja, R. I. (2004). Computational modeling of deformation bands in granular media. II. Numerical simulations. *Comput. Methods Appl. Mech. Eng.* **193**:2699–2718.

Borja, R. I. and Aydin, A. (2004). Computational modeling of deformation bands in granular media. I. Geological and mathematical framework. *Comput. Methods Appl. Mech. Eng.* **193**:2667–2698.

Bosworth, W. (1985). Geometry of propagating continental rifts. *Nature* **316**:625–627.

Bott, M. H. P. (1959). The mechanics of oblique slip faulting. *Geol. Mag.* **96**:109–117.

Bowden, P. B. and Raha, S. (1970). The formation of micro shear bands in polystyrene and polymethylmethacrylate. *Philosoph. Mag.* **22**:463–482.

Boyer, S. E. and Elliott, D. (1982). Thrust systems. *Am. Assoc. Petrol. Geol. Bull.* **66**:1196–1230.

Boyer, S. E. and Mitra, G. (1988). Relations between deformation of crystalline basement and sedimentary cover at the basement/cover transition zone of the Appalachian Blue Ridge province. In *Geometries and Mechanisms of Thrusting, with Special Reference to the Appalachians* (eds. G. Mitra and S. Wojtal), Geol. Soc. Am. Spec. Pap. **222**:119–136.

Brace, W. F. (1960). An extension of the Griffith theory of fracture to rocks. *J. Geophys. Res.* **65**:3477–3480.

Brace, W. F. (1961). Dependence of the fracture strength of rocks on grain size. *Penn. State Univ. Min. Ind. Bull.* **76**:99–103.

Brace, W. F. and Bombolakis, E. G. (1963). A note on brittle crack growth in compression. *J. Geophys. Res.* **68**:3709–3713.

Brace, W. F. and Byerlee, J. D. (1966). Stick-slip as a mechanism for earthquakes. *Science* **153**:990–992.

Brace, W. F. and Kohlstedt, D. L. (1980). Limits on lithospheric stress imposed by laboratory experiments. *J. Geophys. Res.* **85**:6248–6252.

Brace, W. F., Paulding, B. W. and Scholz, C. H. (1966). Dilatancy in the fracture of crystalline rocks. *J. Geophys. Res.* **71**:3939–3953.

Brady, B. T. (1971a). An exact solution to the radially end-constrained circular cylinder under triaxial loading. *Int. J. Rock Mech. Min. Sci.* **8**:165–178.

Brady, B. T. (1971b). The effect of confining pressure on the elastic stress distribution in a radially end-constrained circular cylinder. *Int. J. Rock Mech. Min. Sci.* **8**:153–164.

Brady, B. H. G. and Brown, E. T. (1993). *Rock Mechanics for Underground Mining*, Chapman and Hall, London.

Brandes, C. and Tanner, D. C. (2012). Three-dimensional geometry and fabric of shear deformation-bands in unconsolidated Pleistocene sediments. *Tectonophysics* **518–521**:84–92.

Brantut, N., Baud, P., Heap, H. J. and Meredith, P. G. (2012). Micromechanics of brittle creep in rocks. *J. Geophys. Res.* **117**, B08412; doi:10.1029/2012JB009299.

Brantut, N., Heap, H. J., Meredith, P. G. and Baud, P. (2013). Time-dependent cracking and brittle creep in crustal rocks: a review. *J. Struct. Geol.* **52**:17–43.

Breckels, I. M. and van Eekelen, H. A. M. (1982). Relationship between horizontal stress and depth in sedimentary basins. *J. Petrol. Tech.* **34**:2191–2199.

Bridgman, P. W. (1936). Shearing phenomena at high pressure of possible importance to geology. *J. Geol.* **44**:653–669.

Broberg, K. B. (1999). *Cracks and Fracture*, Academic, San Diego.

Broch, E. and Franklin, J. A. (1972). The point-load strength test. *Int. J. Rock Mech. Min. Sci.* **9**:669–697.

Brodsky, E. E., Roeloffs, E., Woodcock, D., Gall, I. and Manga, M. (2003). A mechanism for sustained groundwater pressure changes induced by distant earthquakes. *J. Geophys. Res.* **108**, 2390; doi:1029/2002/JB002321.

Broek, D. (1986). *Elementary Engineering Fracture Mechanics* (4th edn), Martinus Nijhoff, Boston.

Brown, C. D. and Phillips, R. J. (1999). Flexural rift flank uplift at the Rio Grande rift, New Mexico. *Tectonics* **18**:1275–1291.

Brown, E. and Hoek, E. (1978). Trends in relationships between measured in situ stress and depth. *Int. J. Rock Mech. Min. Sci. & Geomech. Abs*. **15**:211–215.

Brown, E. and Hoek, E. (1988). Determination of shear failure envelope in rock masses. *J. Geotech. Eng. Div. Am. Soc. Civ. Engrs*. **114**:371–376.

Bruhn, R. L. and Schultz, R. A. (1996). Geometry and slip distribution in normal fault systems: implications for mechanics and fault-related hazards. *J. Geophys. Res*. **101**:3401–3412.

Bruhn, R. L., Yonkee, W. A. and Parry, W. T. (1990). Structural and fluid-chemical properties of seismogenic normal faults. *Tectonophysics* **175**:130–157.

Brzesowsky, R. H., Hangx, S. J. T., Brantut, N. and Spiers, C. J. (2014). Compaction creep of sands due to time-dependent grain failure: effects of chemical environment, applied stress, and grain size. *J. Geophys. Res*. **119**:7521–7541.

Bucher, W. H. (1920). The mechanical interpretation of joints. *J. Geol*. **28**:707–730.

Budkewitsch, P. and Robin, P.-Y. (1994). Modelling the evolution of columnar joints. *J. Volcanol. Geotherm. Res*. **59**:219–239.

Buiter, S. J. H., Babeyko, A. Y., Ellis, S., et al. (2006). The numerical sandbox: comparison of model results for a shortening and an extension experiment. In *Analogue and Numerical Modeling of Crustal-Scale Processes* (eds. S. J. H. Buiter and G. Schreurs), pp. 29–64, Geol. Soc. London Spec. Publ. 253.

Bunger, A. P. (2008). A rigorous tool for evaluating the importance of viscous dissipation in sill formation: it's in the tip. In *Dynamics of Crustal Magma Transfer, Storage and Differentiation* (eds. C. Annen and G. F. Zellmer), pp. 71–81, Geol. Soc. London Spec. Publ. 304.

Burdekin, F. M. and Stone, D. E. W. (1966). The crack opening displacement approach to fracture mechanics in yielding materials. *J. Strain Anal*. **1**:145–153.

Burchfiel, B. C. and Stewart, J. H. (1966). "Pull-apart" origin of the central segment of Death Valley, California. *Geol. Soc. Am. Bull*. **77**:439–442.

Bürgmann, R. and Pollard, D. D. (1992). Influence of the state of stress on the brittle-ductile transition in granitic rock: evidence from fault steps in the Sierra Nevada, California. *Geology* **20**:645–648.

Bürgmann, R. and Pollard, D. D. (1994). Strain accommodation about strike-slip fault discontinuities in granitic rock under brittle-to-ductile conditions. *J. Struct. Geol*. **16**:1655–1674.

Bürgmann, R., Pollard, D. D. and Martel, S. J. (1994). Slip distributions on faults: effects of stress gradients, inelastic deformation, heterogeneous host-rock stiffness, and fault interaction. *J. Struct. Geol*. **16**:1675–1690.

Burnley, P. C., Green, H. W., II and Prior, D. (1991). Faulting associated with the olivine to spinel transformation in Mg_2GeO_4 and its implications for deep-focus earthquakes. *J. Geophys. Res*. **96**:425–443.

Busetti, S., Mish, K., Hennings, P. and Reches, Z. (2012). Damage and plastic deformation of reservoir rocks: part 2. Propagation of a hydraulic fracture. *Amer. Assoc. Petrol. Geol. Bull.* **96**:1711–1732.

Butler, R. W. H. (1982). The terminology of structures in thrust belts. *J. Struct. Geol.* **4**:239–245.

Byerlee, J. D. (1970). The mechanics of stick-slip. *Tectonophysics* **9**:475–486.

Byerlee, J. D. (1978). Friction of rocks. *Pure Appl. Geophys.* **116**:615–626.

Byerlee, J. D. and Brace, W. F. (1968). Stick-slip, stable sliding, and earthquakes: effect of rock type, pressure, strain rate, and stiffness. *J. Geophys. Res.* **73**:6031–6037.

Caine, J. S., Evans, J. P. and Forster, C. B. (1996). Fault zone architecture and permeability structure. *Geology* **24**:1025–1028.

Calais, E., d'Oreye, N., Albaric, J., et al. (2008). Aseismic strain accommodation by slow slip and dyking in a youthful continental rift, East Africa. *Nature* **456**; doi:10.1038/nature07478.

Callister, W. D., Jr. (2000). *Materials Science and Engineering: An Introduction* (5th edn), Wiley, New York.

Campagna, D. J. and Aydin, A. (1991). Tertiary uplift and shortening in the Basin and Range; the Echo Hills, southeastern Nevada. *Geology* **19**:485–488.

Campagna, D. J. and Levandowski, D. W. (1991). The recognition of strike-slip fault systems using imagery, gravity, and topographic data sets. *Photog. Engng. Rem. Sens.* **57**:1195–1201.

Candela, T. and Renard, F. (2012). Segment linkage process at the origin of slip surface roughness: evidence from the Dixie Valley fault. *J. Struct. Geol.* **45**:87–100.

Carbotte, S. and Macdonald, K. (1994). Comparison of seafloor tectonic fabric at intermediate, fast, and super fast spreading ridges: influence of spreading rate, plate motions, and ridge segmentation on fault patterns. *J. Geophys. Res.* **99**:13,609–13,631.

Carr, M. H. (1974). Tectonism and volcanism of the Tharsis region of Mars. *J. Geophys. Res.* **79**:3943–3949.

Carter, B. J., Scott Duncan, E. J. and Lajtai, E. Z. (1991). Fitting strength criteria to intact rock. *Geotech. Geol. Eng.* **9**:73–81.

Carter, K. E. and Winter, C. L. (1995). Fractal nature and scaling of normal faults in the Española Basin, Rio Grande rift, New Mexico: implications for fault growth and brittle strain. *J. Struct. Geol.* **17**:863–873.

Cartwright, J. A. and Lonergan, L. (1996). Volumetric contraction during the compaction of mudrocks: a mechanism for the development of regional-scale polygonal fault systems. *Basin Res.* **8**:183–193.

Cartwright, J. A. and Mansfield, J. S. (1998). Lateral tip geometry and displacement gradients on normal faults in the Canyonlands National Park, Utah. *J. Struct. Geol.* **20**:3–19.

Cartwright, J. A., Trudgill, B. D. and Mansfield, C. S. (1995). Fault growth by segment linkage: an explanation for scatter in maximum displacement and trace length data from the Canyonlands Grabens of SE Utah. *J. Struct. Geol.* **17**:1319–1326.

Cartwright, J. A., Mansfield, C. and Trudgill, B. (1996). The growth of normal faults by segment linkage. In *Modern Developments in Structural Interpretation, Validation and Modelling* (eds. P. G. Buchanan and D. A. Nieuwland), pp. 163–177, Geol. Soc. Spec. Publ. 99.

Cartwright, J. A., Bolton, A. J. and James, D. M. D. (2003). The genesis of polygonal fault systems: a review. In *Subsurface Sediment Mobilization* (eds. P. Van Rensbergen et al.), pp. 223–243, Geol. Soc. Lond. Spec. Publ. 216; doi:10.1144/GSL.SP.2003.216.01.15.

Cartwright, J. A., Huuse, M. and Aplin, A. (2007). Seal bypass systems. *Amer. Assoc. Petrol. Geol. Bull.* **91**:1141–1166.

Casagrande, A. (1936). The determination of the pre-consolidation load and its practical significance. *Proc. 1st Int. Soil Mech. and Found. Eng. Conf.*, June 22–26, 1936, Cambridge, Massachusetts, pp. 60–64.

Cashman, S., and Cashman, K. (2000). Cataclasis and deformation-band formation in unconsolidated marine terrace sand, Humboldt County, California. *Geology* **28**:111–114.

Castaing, C., Halawani, M. A., Gervais, F., et al. (1996). Scaling relationships in intraplate fracture systems related to Red Sea rifting. *Tectonophysics* **261**:291–314.

Cates, M. E., Wittmer, J. P., Bouchaud, J.-P. and Claudin, P. (1998). Jamming, force chains, and fragile matter. *Phys. Rev. Lett.* **81**:1841–1844.

Chadwick, W. W., Jr. and Embley, R. W. (1998). Graben formation associated with recent dike intrusions and volcanic eruptions on the mid-ocean ridge. *J. Geophys. Res.* **103**:9807–9825.

Challa, V. and Issen, K. A. (2004). Conditions for compaction band formation in porous rock using a two yield surface model. *J. Eng. Mech.* **130**:1089–1097.

Chambon, G., Schmittbuhl, J., Corfdir, A., et al. (2006). The thickness of faults: from laboratory experiments to field scale observations. *Tectonophysics* **426**:77–94.

Champion, J. A., Tate, A., Mueller, K. J. and Guccione, M. (2001). Geometry, numerical modeling and revised slip rate for the Reelfoot blind thrust and trishear fault-propagation fold, New Madrid seismic zone. *Eng. Geol.* **62**:31–49.

Chang, M. F. (1991). Interpretation of overconsolidation ratio from in situ tests in Recent clay deposits in Singapore and Malaysia. *Can. Geotech. J.* **28**:210–225.

Chapman, C. R. and McKinnon, W. B. (1986). Cratering of planetary satellites. In *Satellites*, University of Arizona Press, pp. 492–580.

Charles, R. J. (1958). Dynamic fatigue of glass. *J. Appl. Phys.* **29**:1657–1662.

Chell, G. G. (1977). The application of post-yield fracture mechanics to penny-shaped and semi-circular cracks. *Engng. Fract. Mech.* **9**:55–63.

Chemenda, A. I. (2009). The formation of tabular compaction-band arrays: theoretical and numerical analysis. *J. Mech. Phys. Solids* **57**:851–868.

Chemenda, A. I. (2011). Origin of compaction bands: anti-cracking or constitutive instability? *Tectonophysics* **499**:156–164.

Chemenda, A., Deverchere, J. and Calais, E. (2002). Three-dimensional laboratory modeling of rifting: application to the Baikal rift, Russia. *Tectonophysics* **356**:253–273.

Chemenda, A. I., Nguyen, S.-H., Petit, J.-P. and Ambre, J. (2011). Mode I cracking versus dilatancy banding: experimental constraints on the mechanisms of extension fracturing. *J. Geophys. Res.* **116**:B04401; doi:10.1029/2010JB008104.

Chemenda, A. I., Ballas, G. and Soliva, R. (2014). Impact of a multilayer structure on initiation and evolution of strain localization in porous rocks: field observations and numerical modeling. *Tectonophysics* **631**:29–36.

Chemenda, A. I., Cavalié, O., Vergnolle, M., Bouissou, S. and Delouis, B. (2016). Numerical modeling of formation of a 3-D strike-slip fault system. *Comptus Rendus Geoscience* **348**:61–69.

Chen, X., Eichhubl, P. and Olson, J. E. (2017). Effect of water on critical and subcritical fracture properties of Woodford shale. *J. Geophys. Res.* **122**:2736–2750; doi:10.1002/2016JB013708.

Chester, F. M., Evans, J. P. and Biegel, R. L. (1993). Internal structure and weakening mechanisms of the San Andreas Fault. *J. Geophys. Res.* **98**:771–786.

Chester, J. S., Logan, J. M. and Spang, J. H. (1991). Influence of layering and boundary conditions on fault-bend and fault-propagation folding. *Geol. Soc. Am. Bull.* **103**:1059–1072.

Cheung, C. S. N., Baud, P. and Wong, T.-f. (2012). Effect of grain size distribution on the development of compaction localization in porous sandstone. *Geophys. Res. Lett.* **39**; doi:10.1029/2012GL053739.

Childs, C., Nicol, A., Walsh, J. J., and Watterson, J. (1996). Growth of vertically segmented normal faults. *J. Struct. Geol.* **18**:1389–1397.

Childs, C., Walsh, J. J. and Watterson, J. (1997). Complexity in fault zone structure and implications for fault seal prediction. In *Hydrocarbon Seals: Importance for Exploration and Production.* Norwegian Petroleum Society (eds. P. Møller-Pedersen and A. G. Koestler), Special Publication 7 (Elsevier), pp. 61–72.

Childs, C., Holdsworth, R. E., Jackson, C. A.-L., et al. (2017). Introduction to the geometry and growth of normal faults. In *The Geometry and Growth of Normal Faults* (eds. C. Childs, R. E. Holdsworth, C. A.-L. Jackson, et al.), Geol. Soc. London Spec. Publ. 439; https://doi.org/10.1144/SP439.24.

Chinnery, M. A. (1961). The deformation of the ground around surface faults. *Seismol. Soc. Amer. Bull.* **51**:355–372.

Chinnery, M. A. (1963). The stress changes that accompany strike-slip faulting. *Seismol. Soc. Amer. Bull.* **53**:921–932.

Chinnery, M. A. (1965). The vertical displacements associated with transcurrent faulting. *J. Geophys. Res.* **70**:4627–4632.

Chinnery, M. A. and Petrak, J. A. (1967). The dislocation fault model with a variable discontinuity. *Tectonophysics* **5**:513–529.

Choi, J.-H., Edwards, P., Ko, K. and Kim, Y.-S. (2016). Definition and classification of fault damage zones: a review and a new methodological approach. *Earth-Sci. Rev.* **152**:70–87.

Christensen, R. M. (2013). *The Theory of Materials Failure*, Oxford University Press.

Christie, M. A. (1996). Upscaling for reservoir simulation. Paper SPE 37324, in *J. Petrol. Technol.* **48**:1004–1010.

Christie-Blick, N. and Biddle, K. T. (1985). Deformation and basin formation along strike-slip faults. In *Strike-Slip Deformation, Basin Formation, and Sedimentation* (eds. K. T. Biddle and N. Christie-Blick), pp. 187–196, Soc. Econ. Paleon. Miner. Spec. Publ. 37.

Churchill, R. V. and Brown, J. W. (1984). *Complex Variables and Applications* (4th edn), McGraw-Hill, New York.

Cladouhos, T. T. and Marrett, R. (1996). Are fault growth and linkage models consistent with power-law distributions of fault lengths? *J. Struct. Geol.* **18**:281–293.

Clark, D., McPherson, A. and Collins, C. (2011). Australia's seismogenic neotectonic record: a case for heterogeneous intraplate deformation. *Geosci. Australia Record 2011/11*, Canberra, 95 pp.

Clark, R. M. and Cox, S. J. D. (1996). A modern regression approach to determining fault displacement–length scaling relationships. *J. Struct. Geol.* **18**:147–152.

Clark, R. M., Cox, S. J. D. and Laslett, G. M. (1999). Generalizations of power-law distributions applicable to sampled fault-trace lengths: model choice, parameter estimation and caveats. *Geophys. J. Int.* **136**:357–372.

Clark, T. A., Gordon, D., Himwich, W. E., et al. (1987). Determination of relative site motions in the western United States using Mark III very long baseline interferometry. *J. Geophys. Res.* **92**:12,741–12,750.

Clayton, L. (1966). Tectonic depressions along the Hope Fault, a transcurrent fault in North Canterbury, New Zealand. *N. Z. J. Geol. Geophys.* **9**:95–104.

Cleary, M. P. (1976). Continuously distributed dislocation model for shear-bands in softening materials. *Int. J. Num. Methods Engng.* **10**:679–702.

Cleary, M. P., Wright, C. A. and Wright, T. B. (1991). Experimental and modeling evidence for major changes in hydraulic fracturing design and field procedures. Paper presented at SPE Gas Technology Symposium, Soc. of Pet. Eng., Houston, Texas, paper SPE 21494.

Clifton, A. E. and Schlische, R. W. (2001). Nucleation, growth, and linkage of faults in oblique rift zones: results from experimental clay models and implications for maximum fault size. *Geology* **29**:455–458.

Cohen, S. C. (1999). Numerical models of crustal deformation in seismic fault zones. *Adv. Geophys.*, **41**:133–231.

Colmenares, L. B. and Zoback, M. D. (2002). A statistical evaluation of rock failure criteria constrained by polyaxial test data for five different rocks. *Int. J. Rock Mech. Min. Sci.* **39**:695–729.

Committee on Fracture Characterization and Fluid Flow (1996). *Rock Fractures and Fluid Flow: Contemporary Understanding and Applications*, National Academy Press, Washington, DC.

Connolly, P. and Cosgrove, J. (1999). Prediction of fracture-induced permeability and fluid flow in the crust using experimental stress data. *Am. Assoc. Petrol. Geol. Bull.* **83**:757–777.

Cook, J., Frederiksen, R. A., Hasbo, K., et al. (2007). Rocks matter: ground truth in geomechanics. *Oilfield Rev.*, Autumn 2007, 36–55.

Cook, N. G. W. (1992). Jaeger memorial dedication lecture—Natural joints in rock: mechanical, hydraulic and seismic behaviour and properties under normal stress. *Int. J. Rock Mech. Min. Sci. Geomech. Abstr.* **29**:198–223.

Cook, T. S. and Erdogan, F. (1972). Stresses in bonded materials with a crack perpendicular to the interface. *Int. J. Eng. Sci.* **10**:677–697.

Cooke, M. L. (1997). Fracture localization along faults with spatially varying friction. *J. Geophys. Res.* **102**:22,425–22,434.

Cooke, M. L. and Kameda, A. (2002). Mechanical fault interaction within the Los Angeles Basin: a two-dimensional analysis using mechanical efficiency. *J. Geophys. Res.* **107**; doi:10.1029/2001JB000542.

Cooke, M. L. and Madden, E. H. (2014). Is the Earth lazy? A review of work minimization in fault evolution. *J. Struct. Geol.* **66**:334–346.

Cooke, M. L. and Pollard, D. D. (1996). Fracture propagation paths under mixed mode loading within rectangular blocks of polymethyl methacrylate. *J. Geophys. Res.* **101**:3387–3400.

Cooke, M. L. and Pollard, D. D. (1997). Bedding-plane slip in initial stages of fault-related folding. *J. Struct. Geol.* **19**:567–581.

Cooke, M. L. and Underwood, C. A. (2001). Fracture termination and step-over at bedding interfaces due to frictional slip and interface opening. *J. Struct. Geol.* **23**:223–238.

Cooke, M. L., Mollema, P. N., Aydin, A. and Pollard, D. D. (2000). Interlayer slip and joint localization in the East Kaibab Monocline, Utah: field evidence and results from numerical modeling. In *Forced Folds and Fractures* (eds. J. W. Cosgrove and M. S. Ameen), pp. 23–49, Geol. Soc. London Spec. Publ. 169.

Cosgrove, J. W. (1976). The formation of crenulation cleavage. *J. Geol. Soc. London* **132**:155–178.

Cosgrove, J. W. (1995). The expression of hydraulic fracturing in rocks and sediments. In *Fractography: Fracture Topography as a Tool in Fracture Mechanics and Stress Analysis* (ed. M. S. Ameen), pp. 97–147, Geol. Soc. Spec. Publ. 92.

Costin, L. S. (1987). Time-dependent deformation and failure. In *Fracture Mechanics of Rock* (ed. B. K. Atkinson), pp. 167–215, Academic Press, New York.

Cotterell, B. (1966). Notes on the paths and stability of cracks. *Int. J. Fracture Mech.* **2**:526–533.

Cotterell, B. and Rice, J. R. (1980). Slightly curved or kinked cracks. *Int. J. Fracture* **16**:155–169.

Cowie, P. A. (1998a). A healing-reloading feedback control on the growth rate of seismogenic faults. *J. Struct. Geol.* **20**:1075–1087.

Cowie, P. A. (1998b). Normal fault growth in three-dimensions in continental and oceanic crust. In *Faulting and Magmatism at Mid-Ocean Ridges* (eds. W. R. Buck, P. T. Delaney, J. A. Karson and Y. Lagabrielle), pp. 325–348, American Geophysical Union Geophys. Mon. 106.

Cowie, P. A. and Roberts, G. P. (2001). Constraining slip rates and spacings for active normal faults. *J. Struct. Geol.* **23**:1901–1915.

Cowie, P. A. and Scholz, C. H. (1992a). Displacement–length scaling relationship for faults: data synthesis and analysis. *J. Struct. Geol.* **14**:1149–1156.

Cowie, P. A. and Scholz, C. H. (1992b). Physical explanation for the displacement–length relationship of faults using a post-yield fracture mechanics model. *J. Struct. Geol.* **14**:1133–1148.

Cowie, P. A. and Scholz, C. H. (1992c). Growth of faults by accumulation of seismic slip. *J. Geophys. Res.* **97**:11,085–11,095.

Cowie, P. A. and Shipton, Z. K. (1998). Fault tip displacement gradients and process zone dimensions. *J. Struct. Geol.* **20**:983–997.

Cowie, P. A., Scholz, C. H., Edwards, M. and Malinverno, A. (1993a). Fault strain and seismic coupling on mid-ocean ridges. *J. Geophys. Res.* **98**:17,911–17,920.

Cowie, P. A., Sornette, D. and Vanneste, C. (1993b). Statistical physical model for the spatio-temporal evolution of faults. *J. Geophys. Res.* **98**:21,809–21,821.

Cowie, P. A., Malinverno, A., Ryan, W. B. F. and Edwards, M. H. (1994). Quantitative fault studies on the East Pacific Rise: a comparison of sonar imaging techniques. *J. Geophys. Res.* **99**:15,205–15,218.

Cowie, P. A., Sornette, D. and Vanneste, C. (1995). Multifractal scaling properties of a growing fault population. *Geophys. J. Int.* **122**:457–469.

Cowie, P. A., Roberts, G. P. and Mortimer, E. (2007). Strain localization within fault arrays over timescales of 10^0–10^7 years. In *Tectonic Faults: Agents of Change on a Dynamic Earth* (eds. M. R. Handy, G. Hirth, and N. Hovius), pp. 47–77, Dahlem Workshop Report 95, MIT Press, Cambridge, MA.

Cox, S. J. D. and Meredith, P. G. (1993). Microcrack formation and material softening in rock measured by monitoring acoustic emissions. *Int. J. Rock Mech. Min. Sci. Geomech. Abstr.* **30**:11–24.

Cox, S. J. D. and Scholz, C. H. (1988a). Rupture initiation in shear fracture of rocks: an experimental study. *J. Geophys. Res.* **93**:3307–3320.

Cox, S. J. D. and Scholz, C. H. (1988b). On the formation and growth of faults. *J. Struct. Geol.* **10**:413–430.

Crater Analysis Techniques Working Group (1979). Standard techniques for presentation and analysis of crater size-frequency data. *Icarus* **37**:467–474.

Crawford, B. R. (1998). Experimental fault sealing: shear band permeability dependency on cataclastic fault gouge characteristics. In *Structural Geology in Reservoir Characterization* (eds. M. P. Coward, T. S. Daltaban and H. Johnson), pp. 27–47, Geol. Soc. London Spec. Publ. 127.

Crider, J. G. (2001). Oblique slip and the geometry of normal-fault linkage: mechanics and a case study from the Basin and Range in Oregon. *J. Struct. Geol.* **23**:1997–2009.

Crider, J. G. and Peacock, D. C. P. (2004). Initiation of brittle faults in the upper crust: a review of field observations. *J. Struct. Geol.* **26**:691–707.

Crider, J. G. and Pollard, D. D. (1998). Fault linkage: three-dimensional mechanical interaction between echelon normal faults. *J. Geophys. Res.* **103**:24,373–24,391.

Crider, J. G., Cooke, M. L., Willemse, E. J. M. and Arrowsmith, J. R. (1996). Linear-elastic crack models of jointing and faulting. In *Structural Geology and Personal Computers* (ed. D. G. De Paor), pp. 359–388.

Crouch, S. L. (1970). Experimental determination of volumetric strains in failed rock. *Int. J. Rock Mech. Min. Sci. Geomech. Abs.* **7**:589–603.

Crouch, S. L. (1976). Analysis of stresses and displacements around underground excavations: an application of the displacement discontinuity method. Geomechanics report to the National Science Foundation, University of Minnesota, Minneapolis, 268 pp.

Crouch, S. L. and Starfield, A. M. (1983). *Boundary Element Methods in Solid Mechanics*. George Allen & Unwin, London.

Crowell, J. C. (1962). Displacement along the San Andreas fault, California. *Geol. Soc. Am. Spec. Pap.* **71**, 61 pp.

Cruikshank, K. M. and Aydin, A. (1994). Role of fracture localization in arch formation, Arches National Park, Utah. *Geol. Soc. Am. Bull.* **106**:879–891.

Cruikshank, K. M. and Aydin, A. (1995). Unweaving the joints in Entrada Sandstone, Arches National Park, Utah, U.S.A. *J. Struct. Geol.* **17**:409–421.

Cruikshank, K. M., Zhao, G. and Johnson, A. M. (1991a). Analysis of minor fractures associated with joints and faulted joints. *J. Struct. Geol.* **13**: 865–886.

Cruikshank, K. M., Zhao, G. and Johnson, A. M. (1991b). Duplex structures connecting fault segments in Entrada sandstone. *J. Struct. Geol.* **13**:1185–1196.

Currie, J. B., Patnode, H. W. and Trump, R. P. (1962). Development of folds in sedimentary strata. *Geol. Soc. Am. Bull.* **73**: 655–674.

Currie, K. L. and Ferguson, J. (1970). The mechanism of intrusion of lamprophyre dikes indicated by "offsetting" of dikes. *Tectonophysics* **9**:525–535.

Cuss, R. J., Rutter, E. H. and Holloway, R. F. (2003). The application of critical state soil mechanics to the mechanical behaviour of porous sandstones. *Int. J. Rock Mech. Min. Sci.* **40**:847–862.

Dahlen, F. A. (1990). Critical taper model of fold-and-thrust belts and accretionary wedges. *Annu. Rev. Earth Planet. Sci.* **18**:55–89.

Dahlen, F. A., Suppe, J. and Davis, D. (1984). Mechanics of fold-and-thrust belts and accretionary wedges: cohesive Coulomb theory. *J. Geophys. Res.* **89**:10,087–10,101.

Dahlstrom, C. D. A. (1970). Structural geology in the eastern margin of the Canadian Rocky Mountains. *Bull. Can. Petrol. Geol.* **18**:332–406.

d' Alessio, M. A. and Martel, S. J. (2004). Fault terminations and barriers to fault growth. *J. Struct. Geol.* **26**:1885–1896.

Das, A., Nguyen, G. D. and Einav, I. (2011). Compaction bands due to grain crushing in porous rocks: a theoretical approach based on breakage mechanics. *J. Geophys. Res.* **116**:B08203; doi:10.1029/2011JB008265.

Das, B. M. (1983). *Advanced Soil Mechanics*, McGraw-Hill, New York.

Davatzes, N. C. and Aydin, A. (2003). Overprinting faulting mechanisms in high porosity sandstones of SE Utah. *J. Struct. Geol.* **25**:1795–1813.

Davatzes, N. C., Aydin, A. and Eichhubl, P. (2003). Overprinting faulting mechanisms during the development of multiple fault sets, Chimney Rock fault array, Utah. *Tectonophysics* **363**:1–18.

Davatzes, N. C., Eichubl, P. and Aydin, A. (2005). Structural evolution of fault zones in sandstone by multiple deformation mechanisms: Moab fault, southeast Utah. *Geol. Soc. Am. Bull.* **117**:135–148.

David, C., Menendez, B., Zhu, W. and Wong, T.-f. (2001). Mechanical compaction, microstructures and permeability evolution in sandstones. *Phys. Chem. Earth (A)* **26**:45–51.

Davies, R. K. and Pollard, D. D. (1986). Relations between left-lateral strike-slip faults and right-lateral monoclinal kink bands in granodiorite, Mt. Abbot Quadrangle, Sierra Nevada, California. *Pure Appl. Geophys.* **124**:177–201.

Davies, R. K., Crawford, M., Dula, W. F., Jr., Cole, M. J. and Dorn, G. A. (1997). Outcrop interpretation of seismic-scale normal faults in southern Oregon: description of structural styles and evaluation of subsurface interpretation methods. *Leading Edge* **16**:1135–1141.

Davis, D. M. and Engelder, T. (1985). The role of salt in fold-and-thrust belts. In *Collision Tectonics: Deformation of Continental Lithosphere* (eds. N. L. Carter and S. Uyeda), *Tectonophysics* **119**:67–88.

Davis, D., Suppe, J. and Dahlen, F. A. (1983). Mechanics of fold-and-thrust belts and accretionary wedges. *J. Geophys. Res.* **88**:1153–1172.

Davis, G. A. and Burchfiel, B. C. (1973). Garlock fault: an intracontinental transform structure, southern California. *Geol. Soc. Am. Bull.* **84**:1407–1422.

Davis, G. H. (1997). Field guide to geologic structures in the Bryce Canyon region, Utah. In Am. Assoc. Petrol. Geol., Hedberg Research Conference on "Reservoir-Scale Deformation – Characterization and Prediction" (W. G. Higgs and C. F. Kluth, convenors), June 22–28, 1997, 119 pp.

Davis, G. H. (1998). Fault-fin landscape. *Geol. Mag.* **135**:283–286.

Davis, G. H. (1999). Structural geology of the Colorado Plateau region of southern Utah, with special emphasis on deformation bands. Geol. Soc. Am. Spec. Pap. 342, 157 pp.

Davis, G. H., Bump, A. P., García, P. E. and Ahlgren, S. G. (2000). Conjugate Riedel deformation band shear zones. *J. Struct. Geol.* **22**:169–190.

Davis, G. H. and Reynolds, S. J. (1996). *Structural Geology of Rocks and Regions* (2nd edn), Wiley, New York.

Davis, K., Burbank, D. W., Fisher, D., Wallace, S. and Nobes, D. (2005). Thrust-fault growth and segment linkage in the active Ostler fault zone, New Zealand. *J. Struct. Geol.* **27**:1528–1546.

Davis, R. O. and Selvadurai, A. P. S. (1996). *Elasticity and Geomechanics*. Cambridge University Press, New York, 201 pp.

Davis, R. O. and Selvadurai, A. P. S. (2002). *Plasticity and Geomechanics*. Cambridge University Press, New York, 287 pp.

Davison, I. (1994). Linked fault systems; extensional, strike-slip and contractional. In *Continental Deformation* (ed. P. L. Hancock), pp. 121–142, Pergamon, New York.

Dawers, N. H. and Anders, M. H. (1995). Displacement-length scaling and fault linkage. *J. Struct. Geol.* **17**:607–614.

Dawers, N. H., Anders, M. H. and Scholz, C. H. (1993). Growth of normal faults: displacement–length scaling. *Geology* **21**:1107–1110.

DeGraff, J. M. and Aydin, A. (1987). Surface morphology of columnar joints and its significance to mechanics and direction of joint growth. *Geol. Soc. Am. Bull.* **99**:605–617.

DeGraff, J. M. and Aydin, A. (1993). Effect of thermal regime on growth increment and spacing of contraction joints in basaltic lava. *J. Geophys. Res.* **98**:6411–6430.

Deere, D. U. (1963). Technical description of cores for engineering purposes. *Rock Mech. Eng. Geol.* **1**:16–22.

Deere, D. U. and Miller, R. P. (1966). Engineering classification and index properties for intact rock. Tech. Rept. No. AFWL–TR–65–116, Air Force Weapons Laboratory, Kirtland Air Force Base, New Mexico.

de Joussineau, G. and Aydin, A. (2009). Segmentation along strike-slip faults revisited. *Pure Appl. Geophys.* **166**:1575–1594.

de Joussineau, G., Bazalgette, L., Petit, J.-P., and Lopez, M. (2005). Morphology, intersections, and syn/late-diagenetic origin of vein networks in pelites of the Lodève Permian Basin, Southern France. *J. Struct. Geol.* **27**:67–87.

Delaney, P. T. (1982). Rapid intrusion of magma into wet rock: groundwater flow due to pore-pressure increases. *J. Geophys. Res.* **87**:7739–7756.

Delaney, P. T. and Pollard, D. D. (1981). Deformation of host rocks and flow of magma during growth of minette dikes and breccia-bearing intrusions near Ship Rock, New Mexico. US Geol. Surv. Prof. Pap. 1202, 61 pp.

Delaney, P. T., Pollard, D. D., Ziony, J. I. and McKee, E. H. (1986). Field relations between dikes and joints: emplacement processes and paleostress analysis. *J. Geophys. Res.* **91**:4920–4938.

DeMets, C., Gordon, R. G. and Argus, D. F. (2010). Geologically current plate motions. *Geophys. J. Int.* **181**:1–80(erratum: *Geophys. J. Int.* **187**:538).

Deng, S. and Aydin, A. (2012). Distribution of compaction bands in 3D in an aeolian sandstone: the role of cross-bed orientation. *Tectonophysics* **574–575**:204–218.

Deng, S. and Aydin, A. (2015). The strength anisotropy of localized compaction: a model for the role of the nature and orientation of cross-beds on the orientation and distribution of compaction bands in 3-D. *J. Geophys. Res.* **120**:1523–1542; doi:10.1002/2014JB011689.

Deng, S., Zuo, L., Aydin, A., Dvorkin, J. and Mukerji, T. (2015). Permeability characterization of natural compaction bands using core flooding experiments and three-dimensional image-based analysis: comparing and contrasting the results from two different methods. *Am. Assoc. Petrol. Geol. Bull.* **99**:27–49.

Denkhaus, H. G. (2003). Brittleness and drillability. *J. S. Afr. Inst. Min. Metall.* **103**:523–524.

Dershowitz, W. S. (1984). Rock joint systems. Unpublished. PhD dissertation, Massachusetts Inst. Tech., 987 pp.

Dershowitz, W. S. and Herda, H. H. (1992). Interpretation of fracture spacing and intensity. In *Proc. 33rd US Symp. Rock Mech.* (eds. N. Tillerson and W. Wawersik), pp. 757–766, Balkema, Rotterdam.

Desai, C. S. and Siriwardane, H. J. (1984). *Constitutive Laws for Engineering Materials with Emphasis on Geologic Materials*. Prentice-Hall, Englewood Cliffs, NJ.

Desroches, J. and Bratton, T. (2000). Formation characterization: well logs. In *Reservoir Stimulation* (eds. M. J. Economides and K. G. Nolte), pp. 4.1–4.26, Wiley, New York.

Dewey, J. F., Holdsworth, R. E. and Trachan, R. A. (1998). Transpression and transtension zones. In *Continental Transpressional and Transtensional Tectonics* (eds. R.E. Holdsworth, R. E. Strachan and J. F. Dewey), pp. 1–14, Geol. Soc. London Spec. Publ. 135.

Dholakia, S. K., Aydin, A., Pollard, D. D. and Zoback, M. D. (1998). Fault-controlled hydrocarbon pathways in the Monterey Formation, California. *Am. Assoc. Petrol. Geol. Bull.* **82**:1551–1574.

Dibblee, T. W., Jr. (1977). Strike-slip tectonics of the San Andreas fault and its role in Cenozoic basin evolvement. In *Late Mesozoic and Cenozoic Sedimentation and Tectonics in California*, San Joaquin Geol. Soc. short course, pp. 26–38.

Dieterich, J. H. (1972). Time-dependent friction in rocks. *J. Geophys. Res.* **77**:3690–3697.

Dieterich, J. H. (1978). Time-dependent friction and the mechanics of stick-slip. *Pure Appl. Geophys.* **116**:790–806.

Dieterich, J. H. and Linker, M. F. (1992). Fault stability under conditions of variable normal stress. *Geophys. Res. Lett.* **19**:1691–1694.

Dieterich, J. H., Richards-Dinger, K. B. and Kroll, K. A. (2015). Modeling injection-induced seismicity with the physics-based earthquake simulator RSQSim. *Seismol. Res. Lett.* **86**; doi 10.1785/0220150057.

DiGiovanni, A. A., Fredrich, J. T., Holcomb, D. J. and Olsson, W. A. (2007). Microscale damage evolution in compacting sandstone. In: *The Relationship Between Damage and Localization* (eds. H. Lewis and G. D. Couples), pp. 89–103, Geol. Soc. Spec. Publ. 289.

Dimitrova, L. L., Holt, W. E., Haines, A. J. and Schultz, R. A. (2006). Towards understanding the history and mechanisms of Martian faulting: the contribution of gravitational potential energy. *Geophys. Res. Lett.* **33**:L08202; doi: 10.1029/2005GL025307

Dmowska, R. and Rice, J. R. (1986). Fracture theory and its seismological applications. In *Continuum Theories in Solid Earth Physics* (ed. R. Teisseyre), pp. 187–255, PWN-Polish Scientific Publishers, Warsaw.

Dokka, R. K. and Travis, C. J. (1990). Late Cenozoic strike-slip faulting in the Mojave Desert, California. *Tectonics* **9**:311–340.

Dong, P. and Pan, J. (1991). Elastic-plastic analysis of cracks in pressure-sensitive materials. *Int. J. Solids Structures* **28**:1113–1127.

Douglas, K. J. (2002). The shear strength of rock masses. Unpublished PhD dissertation, University of New South Wales, Sydney, Australia.

Dowling, N. E. (1999). *Mechanical Behavior of Materials: Engineering Methods for Deformation, Fracture, and Fatigue* (2nd edn), Prentice-Hall, Upper Saddle River, New Jersey.

Dowling, N. E. and Thangjitham, S. (2000). An overview and discussion of basic methodology for fatigue. In *Fatigue and Fracture Mechanics: 31st Volume* (eds. G. R. Halford and J. P. Gallagher), pp. 3–36, Amer. Soc. Test. Mat. Spec. Publ. 1389, West Conshohocken, Penn.

Downey, M. W. (1984). Evaluating seals for hydrocarbon accumulations. *Amer. Assoc. Petrol. Geol. Bull.* **68**:1752–1763.

Du, Y. and Aydin, A. (1991). Interaction of multiple cracks and formation of echelon crack arrays. *Int. J. Num. Anal. Methods Geomech.* **15**:205–218.

Du, Y. and Aydin, A. (1993). The maximum distortional strain energy density criterion for shear fracture propagation with applications to the growth paths of *en échelon* faults. *Geophys. Res. Lett.* **20**:1091–1094.

Du, Y. and Aydin, A. (1995). Shear fracture patterns and connectivity at geometric complexities along strike-slip faults. *J. Geophys. Res.* **100**:18,093–18,102.

Du Bernard, X., Labame, P., Darcel, C., Davy, P. and Bour, O. (2002a). Cataclastic slip band distribution in normal fault damage zones, Nubian sandstones, Suez rift. *J. Geophys. Res.* **107**:2141; doi: 10.129/2001JB000493.

Du Bernard, X., Eichhubl, P. and Aydin, A. (2002b). Dilation bands: a new form of localized failure in granular media. *Geophys. Res. Lett.* **29**: 2176; doi: 10.1029/2002GL015966.

Dugdale, D. S. (1960). Yielding in steel sheets containing slits. *J. Mech. Phys. Solids* **8**:100–104.

Dunand, D. C., Schuh, C. and Goldsby, D. L. (2001). Pressure-induced transformation plasticity of H_2O ice. *Phys. Rev. Lett.* **86**:668.

Dunn, D. E., LaFountain, L. J. and Jackson, R. E. (1973). Porosity dependence and mechanism of brittle fracture in sandstones. *J. Geophys. Res.* **78**:2403–2417.

Dunne, W. M. and Ferrill, D. A. (1988). Blind thrust systems. *Geology* **16**:33–36.

Dunne, W. M. and Hancock, P. L. (1994). Paleostress analysis of small-scale brittle structures. In *Continental Deformation* (ed. P. L. Hancock), pp. 101–120, Pergamon, New York.

Durney, D. W. (1974). The influence of stress concentrations on the lateral propagation of pressure solution zones and surfaces. *Geol. Soc. Austrail. Tecton. Struct. Newslett.* **3**:19.

Durney, D. W. and Kisch, H. (1994). A field classification and intensity scale for first generation cleavages. *J. Aust. Geol. Geophys.* **15**:257–295.

Dyer, R. (1988). Using joint interactions to estimate paleostress ratios. *J. Struct. Geol.* **10**:685–699.

Ebinger, C. J. (1989). Geometric and kinematic development of border faults and accommodation zones, Kivu-Rusizi rift, Africa. *Tectonics* **8**:117–133.

Edwards, M., Fornari, D., Malinverno, A. and Ryan, W. (1991). The regional tectonic fabric of the East Pacific Rise from 12°50'N to 15°10'N. *J. Geophys. Res.* **96**:7995–8017.

Eftis, J. and Liebowitz, H. (1972). On the modified Westergaard equations for certain plane crack problems. *Int. J. Fracture Mech.* **8**:383–392.

Eichhubl, P. (2004). Growth of ductile opening-mode fractures in geomaterials. In *The Initiation, Propagation, and Arrest of Joints and Other Fractures: Interpretations Based on Field Observations* (eds. J. W. Cosgrove and T. Engelder), pp. 11–24. Geol. Soc. London Spec. Publ. 231.

Eichhubl, P. and Boles, J. R. (2000). Focused fluid flow along faults in the Monterey Formation, coastal California. *Geol. Soc. Am. Bull.* **112**:1667–1679.

Eichhubl, P., Davatzes, N. C. and Becker, S. P. (2009). Structural and diagenetic control of fluid migration along the Moab fault, Utah. *Amer. Assoc. Petrol. Geol. Bull.* **93**:653–681.

Eichhubl, P., Hooker, J. N. and Laubach, S. E. (2010). Pure and shear-enhanced compaction bands in Aztec sandstone. *J. Struct. Geol.* **32**:1873–1886.

Eisenstadt, G. and DePaor, D. G. (1987). Alternative model of thrust-fault propagation. *Geology* **15**:630–633.

Elfgren, L. (editor) (1989). *Fracture Mechanics of Concrete Structures*. Chapman and Hall, London.

Elliott, D. (1976). The energy balance and deformation mechanism of thrust sheets. *Phil. Trans. Royal Soc. London* A, **283**: 289–312.

Elliott, S. J., Eichhubl, P. and Landry, C. J. (2014). Effects of coupled structural and diagenetic processes on deformation localization and fluid flow properties in sandstone reservoirs of the southwestern United States (abstract). *Eos (Trans. AGU)*, MR23A–4336.

Ellis, S., Schreurs, G. and Panien, M. (2004). Comparisons between analogue and numerical models of thrust wedge development. *J. Struct. Geol.* **26**:1659–1675.

Ellsworth, W. L. (2013). Injection-induced earthquakes. *Science*, **341**, no. 6142; doi:10.1126/science.1225942.

Elmo, D., Rogers, S., Stead, D. and Eberhardt, E. (2014). Discrete Fracture Network approach to characterise rock mass fragmentation and implications for geomechanical upscaling. *Mining Technol.* **123**:149–161.

Elyasi, A. and Goshtasbi, K. (2016). Using different failure criteria in wellbore stability analysis. *Geomech. Energy Environ.* **2**:15–21.

Engelder, T. (1974). Cataclasis and the generation of fault gouge. *Geol. Soc. Am. Bull.* **85**:1515–1522.

Engelder, T. (1985). Loading paths to joint propagation during a tectonic cycle: an example from the Appalachian Plateau, USA. *J. Struct. Geol.* **7**:459–476.

Engelder, T. (1987). Joints and shear fractures in rock. In *Fracture Mechanics of Rock* (ed. B. K. Atkinson), pp. 27–69, Academic Press, New York.

Engelder, T. (1989). The analysis of pinnate joints in the Mount Desert Island Granite: Implications for post-intrusion kinematics in the coastal volcanic belt, Maine. *Geology* **17**:564–567.

Engelder, T. (1993). *Stress Regimes in the Lithosphere*. Princeton University Press, Princeton, New Jersey.

Engelder, T. (1994). Deviatoric stressitis: A virus infecting the Earth science community. *Eos (Trans. Am. Geophys. Un.)* **75**:209–212.

Engelder, T. (1999). Transitional-tensile fracture propagation: A status report. *J. Struct. Geol.* **21**:1049–1055.

Engelder, T. (2007). Propagation velocity of joints: a debate over stable vs. unstable growth of cracks in the Earth. In *Fractography of Glasses and Ceramics V* (eds. G. D. Quinn, J. R. Varner and M. Wightman), pp. 500–525, American Ceramic Society, Westerville, OH.

Engelder, T. and Fischer, M. P. (1996). Loading configurations and driving mechanisms for joints based on the Griffith energy-balance concept. *Tectonophysics* **256**:253–277.

Engelder, T. and Geiser, P. (1980). On the use of regional joint sets as trajectories of paleostress fields during the development of the Appalachian plateau, New York. *J. Geophys. Res.* **85**:6319–6341.

Engelder, T. and Gross, M. R. (1993). Curving cross joints and the lithospheric stress field in eastern North America. *Geology* **21**:817–820.

Engelder, T. and Lacazette, A. (1990). Natural hydraulic fracturing. In *Rock Joints* (eds. N. Barton and O. Stephansson), pp. 35–43, Balkema, Rotterdam.

Engelder, T. and Marshak, S. (1985). Disjunctive cleavage formed at shallow depths in sedimentary rocks. *J. Struct. Geol.* **7**:327–343.

Engelder, T., Fischer, M. P. and Gross, M. R. (1993). Geological aspects of fracture mechanics. Geol. Soc. Am. Short Course Notes.

Engelder, T., Lash, G. G. and Uzcátegui, R. S. (2009). Joint sets that enhance production from Middle and Upper Devonian gas shales of the Appalachian Basin. *Am. Assoc. Petrol. Geol. Bull.* **93**:857–889.

Erdogan, F. and Biricikoglu, V. (1973). Two bonded half planes with a crack going through the interface. *Int. J. Eng. Sci.* **11**:745–766.

Erdogan, F. and Sih, G. C. (1963). On the crack extension in plates under plane loading and transverse shear. *J. Basic Eng.* **85**:519–527.

Erickson, S. G. (1996). Influence of mechanical stratigraphy on folding vs. faulting. *J. Struct. Geol.* **18**:443–450.

Erslev, E. A. (1991). Trishear fault-propagation folding. *Geology* **19**:617–620.

Evans, A. G. and Wiederhorn, S. M. (1974). Crack propagation and failure prediction in silicon nitride at elevated temperatures. *J. Mat. Sci.* **9**:270–278.

Evans, B. and Kohlstedt, D. L. (1995). Rheology of rocks. In *Rock Physics and Phase Relations—A Handbook of Physical Constants* (ed. T. J. Ahrens), American Geophysical Union Reference Shelf 3, pp. 148–165.

Evans, B., Fredrich, J. T. and Wong, T.-f. (1990). The brittle–ductile transition in rocks: recent experimental and theoretical progress. In *The Brittle–Ductile Transition in Rocks* (eds. A. G. Duba, W. B. Durham, J. W. Handin and H. F. Wang), American Geophysical Union Geophys. Monog. 56, pp. 1–20.

Evans, J. P. and Bradbury, K. K. (2004). Faulting and fracturing of nonwelded Bishop Tuff, eastern California: deformation mechanisms in very porous materials in the vadose zone. *Vadose Zone J.* **3**:602–623.

Exner, U. and Grassemann, B. (2010). Deformation bands in gravels: displacement gradients and heterogeneous strain. *J. Geol. Soc. London* **167**:905–913.

Exner, U. and Tschegg, C. (2012). Preferential cataclastic grain size reduction of feldspar in deformation bands in poorly consolidated arkosic sands. *J. Struct. Geol.* **43**:63–72.

Exner, U., Kaiser, J. and Gier, S. (2013). Deformation bands evolving from dilation to cementation bands in a hydrocarbon reservoir (Vienna Basin, Austria). *Marine Petrol. Geol.* **43**:504–515.

Fall, A., Eichhubl, P., Cumella, S. P., et al. (2012). Testing the basin-centered gas accumulation model using fluid inclusion observations: southern Piceance Basin, Colorado. *Am. Assoc. Petrol. Geol. Bull.* **96**:2297–2318.

Farmer, I. W. (1983). *Engineering Behaviour of Rocks* (2nd edn), Chapman and Hall, London.

Fassett, C. I. (2016). Analysis of impact crater populations and the geochronology of planetary surfaces in the inner solar system. *J. Geophys. Res.* **121**:1900–1926.

Faulds, J. E. and Varga, R. J. (1998). The role of accommodation zones and transfer zones in the regional segmentation of extended terranes. In *Accommodation Zones and Transfer Zones: The Regional Segmentation of the Basin and Range Province* (eds. J. E. Faulds and J. H. Stewart), pp. 1–45, Geol. Soc. Am. Spec. Pap. 323.

Faulkner, D. R., Jackson, C. A. I., Lunn, R. J.. et al. (2010). A review of recent developments concerning the structure, mechanics and fluid flow properties of fault zones. *J. Struct. Geol.* **32**:1557–1575.

Felbeck, D. K. and Atkins, A. G. (1984). *Strength and Fracture of Engineering Solids*, Prentice-Hall, Englewood Cliffs, NJ.

Fender, M., Lechenault, F., and Daniels, K. E. (2010). Universal shapes formed by interacting cracks. *Phys. Rev. Lett.* **105**:125505.

Ferrill, D. A. and Morris, A. P. (2003). Dilational normal faults. *J. Struct. Geol.* **25**:183–196.

Ferrill, D. A., Stamatakos, J. A. and Sims, D. (1999). Normal fault corrugation: implications for growth and seismicity of active normal faults. *J. Struct. Geol.* **21**:1027–1038.

Ferrill, D. A., McGinnis, R. N., Morris, A. P., et al. (2014). Control of mechanical stratigraphy on bed-restricted jointing and normal faulting: Eagle Ford Formation, south-central Texas. *Am. Assoc. Petrol. Geol. Bull.* **98**:2477–2506.

Ferrill, D. A., Morris, A. P., McGinnis, R. N., et al. (2016). Observations on normal-fault scarp morphology and fault system evolution of the Bishop Tuff in the Volcanic Tableland, Owens Valley, California, USA. *Lithosphere* **8**:238–253.

Ferrill, D. A., Morris, A. P., McGinnis, R. N., Smart, K. J. and Wigginton, S. S. (2017). Mechanical stratigraphy and normal faulting. *J. Struct. Geol.* **94**:275–302.

Fialko, Y. I. and Rubin, A. M. (1998). Thermodynamics of lateral dike propagation: implications for crustal accretion at slow spreading mid-ocean ridges. *J. Geophys. Res.* **103**:2501–2514.

Fidan, M. Nielsen, H., Anderson, N., et al. (2012). Characterization of overburden anisotropy improves wellbore stability in North Sea field. *World Oil*, May 2012:1–3.

Field, J. E. (1971). Brittle fracture: its study and application. *Contemp. Phys.* **12**:1–31.

Fields, R. J. and Ashby, M. F. (1976). Finger-like crack growth in solids and liquids. *Philos. Mag.* **33**:33–48.

Fink, J. H. (1985). Geometry of silicic dikes beneath the Inyo Domes, California. *J. Geophys. Res.* **90**:11,127–11,133.

Finkbeiner, T., Zoback, M., Stump, B. B. and Flemings, P. B. (1998). *In situ* stress, pore pressure, and hydrocarbon migration in the South Eugene Island field, Gulf of Mexico. In *Overpressures in Petroleum Exploration* (eds. A. Mitchell and D. Grauls), pp. 103–110, Proc. Workshop Pau, France, April 1998, Bull. Centre Rech. Elf Explor. Prod., Mém. 22.

Finkbeiner, T., Zoback, M., Flemings, P. and Stump, B. (2001). Stress, pore pressure, and dynamically constrained hydrocarbon columns in the South Eugene Island 330 field, northern Gulf of Mexico. *Bull. Am. Assoc. Petrol. Geol.* **85**:1007–1031.

Fischer, G. J. and Paterson, M. S. (1989). Dilatancy during rock deformation at high temperatures and pressures. *J. Geophys. Res.* **94**:17,607–17,617.

Fischer, M. P., Gross, M. R., Engelder, T. and Greenfield, R. J. (1995). Finite-element analysis of the stress distribution around a pressurized crack in a layered elastic medium: implications for the spacing of fluid-driven joints in bedded sedimentary rock. *Tectonophysics* **247**:49–64.

Fletcher, R. C. (1982). Coupling of diffusional mass transport and deformation in a tight rock. *Tectonophysics* **83**:275–291.

Fletcher, R. C. and Pollard, D. D. (1981). Anticrack model for pressure solution surfaces. *Geology* **9**:419–424.

Fletcher, R. C. and Pollard, D. D. (1999). Can we understand structural and tectonic processes and their products without appeal to a complete mechanics? *J. Struct. Geol.* **21**:1071–1088.

Flodin, E. and Aydin, A. (2004a). Faults with asymmetric damage zones in sandstone, Valley of Fire State Park, southern Nevada. *J. Struct. Geol.* **26**:983–988.

Flodin, E. and Aydin, A. (2004b). Evolution of a strike-slip fault network, Valley of Fire State Park, southern Nevada. *Geol. Soc. Am. Bull.* **116**:42–59.

Folger, P. and Tiemann, M. (2016). Human-induced earthquakes from deep-well injection: a brief overview. Congressional Research Service Report for Members of Congress 7-5700, R43836, 29 pp.

Fookes, P. G. and Parrish, D. G. (1969). Observations on small-scale structural discontinuities in the London Clay and their relationship to regional geology. *Quart. J. Eng. Geol.* **1**:217–240.

Fortin, J., Stanchits, S., Dresen, G. and Gueguen, Y. (2009). Acoustic emissions monitoring during inelastic deformation of porous sandstone: comparison of three modes of deformation. *Pure Appl. Geophys.* **166**:823–841.

Fossen, H. (2010a). *Structural Geology*, Cambridge University Press, 463 pp.

Fossen, H. (2010b). Deformation bands formed during soft-sediment deformation: observations from SE Utah. *Marine Petrol. Geol.* **27**:215–222.

Fossen, H. and Bale, A. (2007). Deformation bands and their influence on fluid flow. *Amer. Assoc. Petrol. Geol. Bull.* **91**:1685–1700.

Fossen, H. and Cavalcante, G. C. G. (2017). Shear zones: a review. *Earth-Sci. Rev.* **171**:434–455.

Fossen, H. and Gabrielsen, R. H. (1996). Experimental modeling of extensional fault systems by use of plaster. *J. Struct. Geol.* **18**:673–687.

Fossen, H. and Gabrielsen, R. H. (2005). *Strukturgeologi*, Fagbokforlaget (in Norwegian), Bergen, Norway, 375 pp.

Fossen, H. and Hesthammer, J. (1997). Geometric analysis and scaling relations of deformation bands in porous sandstone. *J. Struct. Geol.* **19**:1479–1493.

Fossen, H. and Hesthammer, J. (1998). Deformation bands and their significance in porous sandstone reservoirs. *First Break* **16**:21–25.

Fossen, H. and Hesthammer, J. (2000). Possible absence of small faults in the Gullfaks Field, northern North Sea: implications for downscaling of faults in some porous sandstones. *J. Struct. Geol.* **22**:851–863.

Fossen, H. and Rotevatn, A. (2016). Fault linkage and relay structures in extensional settings—a review. *Earth-Sci. Rev.* **154**:14–28.

Fossen, H. and Tikoff, B. (1998). Extended models of transpression and transtension, and application to tectonic settings. In *Continental Transpressional and Transtensional Tectonics* (eds. R. E. Holdsworth, R. E. Strachan and J. F. Dewey), pp. 15–33, Geol. Soc. London Spec. Publ. 135.

Fossen, H., Johansen, T. E. S., Hesthammer, J. and Rotevatn, A. (2005). Fault interaction in porous sandstone and implications for reservoir management: examples from southern Utah. *Amer. Assoc. Petrol. Geol. Bull.* **89**:1593–1606.

Fossen, H., Schultz, R. A., Shipton, Z. K. and Mair, K. (2007). Deformation bands in sandstone: a review. *J. Geol. Soc. London* **164**:755–769.

Fossen, H., Schultz, R. A., Rundhovde, E., Rotevatn, A. and Buckley, S. J. (2010). Fault linkage and graben stepovers in Canyonlands (Utah) and the North Sea Viking Graben, with implications for hydrocarbon migration and accumulation. *Amer. Assoc. Petrol. Geol. Bull.* **94**:597–613.

Fossen, H., Schultz, R. A. and Torabi, A. (2011). Conditions and implications for compaction band formation in Navajo Sandstone, Utah. *J. Struct. Geol.* **33**:1477–1490.

Fossen, H., Zuluaga, L. F., Ballas, G., Soliva, R. and Rotevatn, A. (2015). Contractional deformation of porous sandstone: insights from the Aztec Sandstone, SE Nevada, USA. *J. Struct. Geol.* **74**:172–184.

Fossen, H., Soliva, R., Ballas, G., et al. (2017). A review of deformation bands in reservoir sandstones: geometries, mechanisms and distribution. In *Subseismic-Scale Reservoir Deformation* (eds. M. Ashton, S. J. Dee and O. P. Wennberg), Geol. Soc. London Spec. Publ. 459; doi:10.1144/SP459.4.

Fox, P. J. and Gallo, D. G. (1984). A tectonic model for ridge-transform-ridge plate boundaries: implications for the structure of oceanic lithosphere. *Tectonophysics* **104**:205–242.

Franklin, J. A. (1993). Empirical design and rock mass characterization. In *Comprehensive Rock Engineering* (ed. J. A. Hudson), vol. 2 (ed. C. Fairhurst), pp. 795–806, Pergamon, New York.

Frankowicz, E. and McClay, K. R. (2010). Extensional fault segmentation and linkages, Bonaparte Basin, outer North West Shelf, Australia. *Am. Assoc. Petrol. Geol. Bull.* **94**:977–1010.

Fréchette, V. D. (1972). The fractography of glass. In *Introduction to Glass Science* (ed. D. L. Pye), pp. 432–450, Plenum Press, New York.

Fréchette, V. D. (1990). *Failure Analysis of Brittle Materials. Advances in Ceramics, vol. 28*, Am. Ceram. Soc., Westerville, Ohio.

Fredrich, J. T., Evans, B. and Wong, T.-f. (1989). Micromechanics of the brittle to plastic transition in Carrara marble. *J. Geophys. Res.* **94**:4129–4145.

Freed, A. M. (2005). Earthquake triggering by static, dynamic, and postseismic stress transfer. *Annu. Rev. Earth Planet. Sci.* **33**:335–367.

Freed, A. M. and Lin, J. (1998). Time-dependent changes in failure stress following thrust earthquakes. *J. Geophys. Res.* **103**:24,393–24,409.

Freeze, A. R. and Cherry, J. A. (1979). *Groundwater*, Prentice-Hall, New Jersey.

Freund, L. B. (1990). *Dynamic Fracture Mechanics*, Cambridge University Press, New York, 563 pp.

Freund, R. (1970). Rotation of strike-slip faults in Sistan, southeast Iran. *J. Geol.* **78**:188–200.

Freund, R. (1974). Kinematics of transform and transcurrent faults. *Tectonophysics* **21**:93–134.

Friedman, M. and Logan, J. M. (1973). Lüders' bands in experimentally deformed sandstone and limestone. *Geol. Soc. Am. Bull.* **84**:1465–1476.

Frohlich, C. (2012). Two-year survey comparing earthquake activity and injection-well locations in the Barnett shale, Texas. *Proc. US Nat. Acad. Sci.* **109**:13,934–13,938.

Fueten, F. and Robin, P.-Y. F. (1992). Finite element modeling of the propagation of a pressure solution cleavage seam. *J. Struct. Geol.* **14**:953–962.

Gabrielsen, R. H. and Kløvjan, O. S. (1997). Late Jurassic–early Cretaceous caprocks of the southwestern Barents Sea: fracture systems and rock mechanical properties. In *Hydrocarbon Seals: Importance for Exploration and Production* (eds. P. Møller-Pedersen and A. G. Koestler), pp. 73–89, Norwegian Petrol. Soc. Spec. Publ. 7.

Gabrielsen, R. H. and Koestler, A. G. (1987). Description and structural implications of fractures in the late Jurassic sandstones of the Troll Field, northern North Sea. *Norsk Geologisk Tidsskrift* **67**:371–381.

Gabrielsen, R. H., Aarland, R.-K. and Alsaker, E. (1998). Identification and spatial distribution of fractures in porous, silisiclastic sediments. In *Structural Geology in Reservoir Characterization* (eds. M. P. Coward, T. S. Daltaban and H. Johnson), pp. 49–64, Geol. Soc. London Spec. Publ. 127.

Gale, J. F. W., Laubach, S. E., Olson, J. E., Eichhubl, P. and Fall, A. (2014). Natural fractures in shale—a review and new observations. *Am. Assoc. Petrol. Geol. Bull.* **98**:2165–2216.

Gallagher, J. J., Friedman, M., Handin, J. and Sowers, G. M. (1974). Experimental studies relating to microfracture in sandstone. *Tectonophysics* **21**:203–247.

Gamond, J. F. (1983). Displacement features associated with fault zones: a comparison between observed examples and experimental models. *J. Struct. Geol.* **5**:33–45.

Gamond, J. F. (1987). Bridge structures as sense of displacement criteria on brittle faults. *J. Struct. Geol.* **9**:609–620.

Garfunkel, Z. (1981). Internal structure of the Dead Sea leaky transform (rift) in relation to plate tectonics. *Tectonophysics* **80**:81–108.

Gawthorpe, R. L. and Leeder, M. R. (2000). Tectono-sedimentary evolution of active extensional basins. *Basin Res.* **12**:195–218.

Geiser, P. A. (1988). The role of kinematics in the construction and analysis of geological cross sections in deformed terranes. In *Geometries and Mechanisms of Thrusting, with Special Reference to the Appalachians* (eds. G. Mitra and S. Wojtal), Geol. Soc. Am. Spec. Pap. 222, pp. 47–76.

Geiser, P. A. and Sansone, S. (1981). Joints, microfractures, and the formation of solution cleavage in limestone. *Geology* **9**:280–285.

Gercek, H. (2007). Poisson's ratio for rocks. *Int. J. Rock Mech. Min. Sci.* **44**:1–13.

Germanovich, L. N. and Cherepanov, G. P. (1995). On some general properties of strength criteria. *Int. J. Fracture* **71**:37–56.

Germanovich, L. N., Salganik, R. L., Dyskin, A. V. and Lee, K. K. (1994). Mechanisms of brittle fracture of rock with pre-existing cracks in compression. *Pure Appl. Geophys.* **143**:117–149.

Gerya, T. (2012). Origin and models of oceanic transform faults. *Tectonophysics* **522–523**:34–54.

Giba, M., Walsh, J. J. and Nicol, A. (2012). Segmentation and growth of an obliquely reactivated normal fault. *J. Struct. Geol.* **39**:253–267.

Gibbs, A. D. (1984). Structural evolution of extensional basin margins. *J. Geol. Soc. London* **141**:609–620.

Gibbs, A. D. (1990). Linked fault families in basin formation. *J. Struct. Geol.* **12**:795–803.

Gibson, R. G. (1998). Physical character and fluid-flow properties of sandstone-derived fault zones. In *Structural Geology in Reservoir Characterization* (eds. M. P. Coward, T. S. Daltaban and H. Johnson), pp. 83–97, Geol. Soc. London Spec. Publ. 127.

Gillespie, P. A., Walsh, J. J. and Watterson, J. (1992). Limitations of dimension and displacement data from single faults and the consequences for data analysis and interpretation. *J. Struct. Geol.* **14**:1157–1172.

Gillespie, P. A., Howard, C. B., Walsh, J. J. and Watterson, J. (1993). Measurement and characterisation of spatial distributions of fractures. *Tectonophysics* **226**:113–141.

Ginsberg, J. H. and Genin, J. (1977). *Statics*, Wiley, New York.

Goehring, L., Mahadevan, L. and Morris, S. W. (2009). Nonequilibrium scale selection mechanism for columnar jointing. *Proc. US Nat. Acad. Sci.* **106**:387–392.

Goetze, C. and Evans, B. (1979). Stress and temperature in the bending lithosphere as constrained by experimental rock mechanics. *Geophys. J. Royal Astron. Soc.* **59**:463–478.

Gokceoglu, C., Somnez, H. and Kayabasi, A. (2003). Predicting the deformation moduli of rock masses. *Int. J. Rock Mech. Min. Sci.* **40**:701–710.

Golombek, M. P. and Phillips, R. J. (2010). Mars tectonics. In *Planetary Tectonics* (eds. T. R. Watters and R. A. Schultz), pp. 183–232, Cambridge University Press.

Golombek, M. P., Anderson, F. S. and Zuber, M. T. (2001). Martian wrinkle ridge topography: evidence for subsurface faults from MOLA. *J. Geophys. Res.* **106**:23,811–23,821.

Gomberg, J., Blanpied, M. L. and Beeler, N. M. (1997). Transient triggering of near and distant earthquakes. *Bull. Seismol. Soc. Am.* **87**:294–309.

Goodier, J. N. (1968). Mathematical theory of equilibrium cracks. *Fracture, An Advanced Treatise* **2**:1–66.

Goodman, R. E. (1976). *Methods of Engineering in Discontinuous Rocks*, West, St. Paul.

Goodman, R. E. (1989). *Introduction to Rock Mechanics* (2nd edn), Wiley, New York.

Goodman, R. E. and Shi, G. (1985). *Block Theory and its Application to Rock Engineering*, Prentice-Hall, New Jersey.

Gordon, F. R. and Lewis, J. D. (1980). The Meckering and Calingiri earthquakes October 1968 and March 1970. *West. Aust. Geol. Surv. Bull.* **126**, 229 pp.

Goscombe, B. D., Passchier, C. W. and Hand, M. (2004). Boudinage classification: end-member boudin types and modified boudin structures. *J. Struct. Geol.* **26**:739–763.

Goudy, C. L., and Schultz, R. A. (2005). Dike intrusions beneath grabens south of Arsia Mons, Mars. *Geophys. Res. Lett.* **32**:5, doi: 10.1029/2004GL021977.

Goudy, C. L., Schultz, R. A., and Gregg, T. K. P. (2005). Coulomb stress changes in Hesperia Planum, Mars, reveal regional thrust fault reactivation. *J. Geophys. Res.* **110**:E10005; doi:10.1029/2004JE002293.

Gowd, T. N. and Rummel, F. (1980). Effect of confining pressure on fracture behavior of a porous rock. *Int. J. Rock Mech. Min. Sci.* **17**:225–229.

Grady, D. E. and Kipp, M. E. (1987). Dynamic rock fragmentation. In *Fracture Mechanics of Rock* (ed. B. K. Atkinson), pp. 429–475, Academic Press, New York.

Graham, B., Antonellini, M. and Aydin, A. (2003). Formation and growth of normal faults in carbonates within a compressive environment. *Geology* **31**:11–14.

Grasemann, B., Martel, S. J. and Passchier, C. (2005). Reverse and normal drag along a single dip-slip fault. *J. Struct. Geol.* **27**:999–1010.

Grasso, J. R., and Wittlinger, G. (1990). 10 years of seismic monitoring over a gas field area. *Seismol. Soc. Am. Bull.* **80**:450–473.

Graveleau, F., Malavieille, J. and Dominguez, S. (2012). Experimental modelling of orogenic wedges: a review. *Tectonophysics* **538–540**:1–66.

Gray, D., Anderson, P., Logel, J., et al. (2012). Estimation of stress and geomechanical properties using 3D seismic data. *First Break* **30**:59–68.

Green, A. E. and Sneddon, I. N. (1950). The distribution of stresses in the neighborhood of a flat elliptical crack in an elastic solid. *Proc. Cambridge Phil. Soc.* **46**:159–163.

Green, H. W., II (1984). How and why does olivine transform to spinel? *Geophys. Res. Lett.* **11**:817–820.

Green, H. W., II and Burnley, P. C. (1989). A new, self-organizing, mechanism for deep-focus earthquakes. *Nature* **341**:733–737.

Green, H. W., II, Young, T. E., Walker, D. and Scholz, C. H. (1990). Anticrack-associated faulting at very high pressure in natural olivine. *Nature* **348**:720–722.

Green, H. W., II and Houston, H. (1995). The mechanics of deep earthquakes. *Ann. Rev. Earth Planet. Sci.* **23**:169–213.

Green, H. W., II and Marone, C. (2002). Instability of deformation. In *Plastic Deformation of Minerals and Rocks* (eds. S.-i. Karato and H.-R. Wenk), pp. 181–199, Mineral. Soc. Am. Reviews in Mineral. and Geochem. v. 51.

Grieve, R. A. F. (1987). Terrestrial impact structures. *Ann. Rev. Earth Planet. Sci.* **15**:245–270.

Griffith, A. A. (1921). The phenomena of rupture and flow in solids. *Phil. Trans. Royal Soc. London* **A221**:163–198.

Griffith, A. A. (1924). The theory of rupture. In *Proc. 1st Int. Congress Appl. Mech.* (eds. C. B. Biezeno and J. M. Burgers), pp. 55–63.

Griggs, D. T. (1936). Deformation of rocks under high confining pressures. *J. Geol.* **44**:541–577.

Griggs, D. T. and Handin, J. (1960). Observations on fracture and a hypothesis of earthquakes. In *Rock Deformation* (eds. D. Griggs and J. Handin), pp. 347–373, Geol. Soc. Am. Memoir 79.

Griggs, D. T., Turner, F. J. and Heard, H. C. (1960). Deformation of rocks at 500° to 800° C. In *Rock Deformation* (eds. D. Griggs and J. Handin), pp. 39–104, Geol. Soc. Am. Memoir 79.

Grimm, R. E. and Phillips, R. J. (1991). Gravity anomalies, compensation mechanisms, and the geodynamics of western Ishtar Terra, Venus. *J. Geophys. Res.* **96**:8305–8324.

Grosfils, E. B., Schultz, R. A. and Kroeger, G. (2003). Geophysical exploration within northern Devils Lane graben, Canyonlands National Park, Utah: implications for sediment thickness and tectonic evolution. *J. Struct. Geol.* **25**:455–467.

Groshong, R. H. Jr. (1988). Low-temperature deformation mechanisms and their interpretation. *Geol. Soc. Am. Bull.* **100**:1329–1360.

Groshong, R. H. Jr. (1989). Half-graben structures: balanced models of extensional fault-bend folds. *Geol. Soc. Am. Bull.* **101**:96–105.

Groshong, R. H. Jr. and Usdansky, S. I. (1988). Kinematic models of plane-roofed duplex styles. In *Geometries and Mechanisms of Thrusting, with Special Reference to the Appalachians* (eds. G. Mitra and S. Wojtal), pp. 197–206, Geol. Soc. Am. Spec. Pap. 222.

Gross, M. R. (1993). The origin and spacing of cross joints: examples from the Monterey Formation, Santa Barbara coastline, California. *J. Struct. Geol.* **15**:737–751.

Gross, M. R. (1995). Fracture partitioning: failure mode as a function of lithology in the Monterey Formation of coastal California. *Geol. Soc. Am. Bull.* **107**:779–792.

Gross, M. R., Gutierrez-Alonzo, G., Bai, T., et al. (1997). Influence of mechanical stratigraphy and kinematics on fault scaling relationships. *J. Struct. Geol.* **19**:171–183.

Grueschow, E. and Rudnicki, J. W. (2005). Elliptic yield cap constitutive modeling for high porosity sandstone. *Int. J. Solids Structures* **42**:4574–4587.

Gu, Y. and Wong, T.-f. (1994). Development of shear localization in simulated quartz gouge: effects of cumulative slip and gouge particle size. *Pure Appl. Geophys.* **143**:387–423.

Gudmundsson, A. (1992). Formation and growth of normal faults at the divergent plate boundary in Iceland. *Terra Nova* **4**:464–471.

Gudmundsson, A. (1995). Stress fields associated with oceanic transform faults. *Earth Planet. Sci. Lett.* **136**:603–614.

Gudmundsson, A. (2000). Fracture dimensions, displacements and fluid transport. *J. Struct. Geol.* **22**:1221–1231.

Gudmundsson, A. (2004). Effects of Young's modulus on fault displacement. *Comptes Rendus Geoscience* **336**:85–92.

Gudmundsson, A. (2011). *Rock Fractures in Geological Processes*, Cambridge University Press, Cambridge, 560 pp.

Gudmundsson, A. and Bäckström, K. (1991). Structure and development of the Sveeinagja graben, Northeast Iceland. *Tectonophysics* **200**:111–125.

Gupta, A. and Scholz, C. H. (2000a). A model of normal fault interaction based on observations and theory. *J. Struct. Geol.* **22**:865–879.

Gupta, A. and Scholz, C. H. (2000b). Brittle strain regime transition in the Afar depression: implications for fault growth and seafloor spreading. *Geology* **28**:1078–1090.

Gupta, S. and Cowie, P. (2000). Processes and controls in the stratigraphic development of extensional basins. *Basin Res.* **12**:185–194.

Gupta, S., Cowie, P. A., Dawers, N. H. and Underhill, J. R. (1998). A mechanism to explain rift-basin subsidence and stratigraphic patterns through fault-array evolution. *Geology* **26**:595–598.

Gürbüz, A. (2014). Geometric characteristics of pull-apart basins. *Lithosphere* **2**:199–206.

Gutierrez, M. and Homand, S. (1998). Formulation of a basic chalk constitutive model. In *Chalk V – Chalk Geomechanics*. Report prepared for Joint Chalk Research Phase V, Norwegian Geotechnical Institute, Oslo, 50 pp.

Haddad, M. and Sepehrnoori, K. (2014). Cohesive fracture analysis to model multiple-stage fracturing in quasibrittle shale formations. 2014 SIMULIA Conference, www.3ds.com/simulia, 15 pp.

Haddad, M. and Sepehrnoori, K. (2015). Simulation of hydraulic fracturing in quasi-brittle shale formations using characterized cohesive layer: stimulation controlling factors. *J. Unconv. Oil Gas Res.* **9**:65–83.

Haimson, B. C. (2001). Fracture-like borehole breakouts in high-porosity sandstone: are they caused by compaction bands? *Physics Chem. Earth* **26**:15–20.

Haimson, B. C. and Fairhurst, C. (1967). Initiation and extension of hydraulic fractures in rocks. *Soc. Petrol. Eng. J.* **7**:310–318.

Haimson, B. C. and Kim, R. Y. (1971). Mechanical behavior of rock under cyclic failure. *Proc. US Rock Mech. Symp.* **13**:845–862.

Haimson, B. C. and Rummel, F. (1982). Hydrofracturing stress measurements in the Iceland research drilling project drill hole at Reydarfjördur, Iceland. *J. Geophys. Res.* **87**:6631–6649.

Hancock, P. L. (1972). The analysis of en echelon veins. *Geol. Mag.* **109**:269–276.

Hancock, P. L. (1985). Brittle microtectonics: principles and practice. *J. Struct. Geol.* **7**:437–457.

Handin, J. (1969). On the Coulomb–Mohr failure criterion. *J. Geophys. Res.* **74**:5343–5348.

Handin, J., Hager, R. V. Jr., Friedman, M. and Feather, J. N. (1963). Experimental deformation of sedimentary rocks under confining pressure: pore pressure tests. *Am. Assoc. Petrol. Geol. Bull.* **47**:717–755.

Hansen, B. (1958). Line ruptures regarded as narrow rupture zones: basic equations based on kinematic considerations. *Proc. Brussels Conf. 58 on Earth Pressure Problems* **1**:39–48.

Hansen, F. D., Hardin, E. L., Rechard, R. P., et al. (2010). Shale disposal of U.S. high-level radioactive waste. Report SAND2010–2843, Sandia National Laboratories, Albuquerque, New Mexico, 148 pp.

Harding, T. P. (1974). Petroleum traps associated with wrench faults. *Am. Assoc. Petrol. Geol. Bull.* **58**:1290–1304.

Harding, T. P. and Lowell, J. D. (1979). Structural styles, their plate-tectonic habitats, and hydrocarbon traps in petroleum provinces. *Am. Assoc. Petrol. Geol. Bull.* **63**:1016–1058.

Harding, T. P., Vierbuchen, R. C. and Christie-Blick, N. (1985). Structural styles, plate-tectonic settings, and hydrocarbon traps of divergent (transtensional) wrench faults. In *Strike-Slip Deformation, Basin Formation, and Sedimentation* (eds. K. T. Biddle and N. Christie-Blick), pp. 51–77, Soc. Econ. Paleon. Miner., Spec. Publ. 37.

Hardy, S. and Ford, M. (1997). Numerical modeling of trishear fault-propagation folding. *Tectonics* **16**:841–854.

Hardy, S. and Allmendinger, R.W. (2011). Trishear: a review of kinematics, mechanics, and applications. In *Thrust Fault-Related Folding* (eds. K. R. McClay, J. H. Shaw and J. Suppe), pp. 95–119, Amer. Assoc. Petrol. Geol. Mem. 94.

Harland, W. B. (1957). Exfoliation joints and ice action. *J. Glaciol.* **3**:8–10.

Harland, W. B. (1971). Tectonic transpression in Caledonian Spitzbergen. *Geol. Mag.* **108**:27–42.

Harris, R. A. (1998). Introduction to special section: Stress triggers, stress shadows, and implications for seismic hazard. *J. Geophys. Res.* **103**:24,347–24,358.

Harris, R. A. and Day, S. M. (1993). Dynamics of fault interaction: parallel strike-slip faults. *J. Geophys. Res.* **98**:4461–4472.

Harris, R. A. and Simpson, R. W. (1998). Suppression of large earthquakes by stress shadows: a comparison of Coulomb and rate-and-state failure. *J. Geophys. Res.* **103**:24,439–24,451.

Hashida, T., Oghikubo, H., Takahashi, H. and Shoji, T. (1993). Numerical simulation with experimental verification of the fracture behavior in granite under confining pressures based on the tension-softening model. *Int. J. Fracture* **59**:227–244.

Hatheway, A. W. (1996). Fractures; discontinuities that control your project (Perspective No. 28). *AEG News* 39/4:19–22.

Hatton, C. G., Main, I. G. and Meredith, P. G. (1993). A comparison of seismic and structural measurements of scaling exponents during tensile subcritical crack growth. *J. Struct. Geol.* **15**:1485–1495.

Hawkes, C. D., Bachu, S. and Mclellan, P. J. (2005). Geomechanical factors affecting geologic storage of CO_2 in depleted oil and gas reservoirs. *J. Can. Petrol. Technol.* **44**:52–61.

Hawkes, I. and Mellor, M. (1970). Uniaxial testing in rock mechanics laboratories. *Eng. Geol.* **4**:177–285.

Hayward, N. and Ebinger, C. J. (1996). Variations in the along-axis segmentation of the Afar rift system. *Tectonics* **15**:244–257.

Heald, M. T. (1956). Cementation of Simpson and St. Peter sandstones in parts of Oklahoma, Arkansas, and Missouri. *J. Geol.* **64**:16–30.

Heald, M. T. (1959). Significance of stylolites in permeable sandstones. *J. Sedimentary Res.* **29**:251–253.

Heald, P. T., Spink, G. M. and Worthington, P. J. (1972). Post yield fracture mechanics. *Mat. Sci. Engng.* **10**:129–138.

Healy, D., Blenkinsop, T. G., Timms, N. E., et al. (2015). Polymodal faulting: time for a new angle on shear failure. *J. Struct. Geol.* **80**:57–71.

Heidbach, O., Tingay, M., Barth, A., et al. (2010). Global crustal stress pattern based on the World Stress Map database release 2008. *Tectonophysics* **482**:2–15.

Helgeson, D. E. and Aydin, A. (1991). Characteristics of joint propagation across layer interfaces in sedimentary rocks. *J. Struct. Geol.* **13**:897–911.

Hennings, P. H., Olson, J. E. and Thompson, L. B. (2000). Combining outcrop data and three-dimensional structural models to characterize fractured reservoirs: an example from Wyoming. *Amer. Assoc. Petrol. Geol. Bull.* **84**:830–849.

Hennings, P. H., Allwardt, P., Paul, P., et al. (2012). Relationship between fractures, fault zones, stress, and reservoir productivity in the Suban gas field, Sumatra, Indonesia. *Amer. Assoc. Petrol. Geol. Bull.* **96**:753–772.

Hergarten, S. and Kenkmann, T. (2015). The number of impact craters on Earth: any room for further discoveries? *Earth Planet. Sci. Lett.* **425**:187–192.

Hesthammer, J., Johansen, T. E. S. and Watts, L. (2000). Spatial relationships within fault damage zones in sandstone. *Marine Petrol. Geol.* **17**:873–893.

Heuzé, F. E. (1980). Scale effects in the determination of rock mass strength and deformability. *Rock Mech.* **12**:167–192.

Hickman, R. J. (2004). Formulation and implementation of a constitutive model for soft rock, PhD dissertation, Virginia Polytechnic Institute and State University, Blacksburg, Virginia.

Hickman, R. J., and Gutierrez, M. S. (2007). Formulation of a three-dimensional rate-dependent constitutive model for chalk and porous rocks. *Int. J. Numer. Anal. Meth. Geomech.* **31**: 583–605.

Higgins, R. I. and Harris, L. B. (1997). The effect of cover composition on extensional faulting above re-activated basement faults; results from analog modeling. *J. Struct. Geol.* **19**:89–98.

Hill, D. P. and Prejean, S. (2007). Dynamic triggering. In *Earthquake Seismology, V. 4* (ed. H. Kanamori), pp. 258–288, Treatise on Geophysics (G. Schubert, ed. in chief), Elsevier, Amsterdam.

Hill, M. L. (1984). Earthquakes and folding, Coalinga, California. *Geology* **12**:711–712.

Hill, R. (1963). Elastic properties of reinforced solids: some theoretical principles. *J. Mech. Phys. Solids* **11**:357–372.

Hill, R. E. (1989). Analysis of deformation bands in the Aztec Sandstone, Valley of Fire, Nevada, MS Thesis, Geosciences Department, University of Nevada, Las Vegas.

Hillerborg, A. (1991). Application of the ficticious crack model to different types of materials. *Int. J. Fracture* **51**:95–102.

Hillerborg, A., Modéer, M. and Petersson, P.-E. (1976). Analysis of crack formation and crack growth in concrete by means of fracture mechanics and finite elements. *Cement Concrete Res.* **6**:773–782.

Hobbs, B. E., Means, W. D. and Williams, P. F. (1976). *An Outline of Structural Geology*, Wiley, New York.

Hodgkinson, K. M., Stein, R. S. and King, G. C. P. (1996). The 1954 Rainbow Mountain–Fairview Peak–Dixie Valley earthquakes: a triggered normal faulting sequence. *J. Geophys. Res.* **101**:25,459–25,471.

Hodgson, R. A. (1961). Classification of structures on joint surfaces. *Am. J. Sci.* **259**:493–502.

Hoek, E. (1983). Strength of jointed rock masses. *Géotechnique* **33**:187–223.

Hoek, E. (1990). Estimating Mohr–Coulomb friction and cohesion from the Hoek-Brown failure criterion. *Int. J. Rock Mech. Min. Sci. Geomech. Abstr.* **27**:227–229.

Hoek, E. (2002). A brief history of the development of the Hoek–Brown failure criterion. Online document, www.rocscience.com, 7 pages.

Hoek, E. (2007). *Practical Rock Engineering*. RocScience, available online at www.rocscience.com.

Hoek, E. and Bieniawski, Z. T. (1965). Brittle fracture propagation in rock under compression. *Int. J. Fracture Mech.* **1**:137–155.

Hoek, E. and Brown, E. T. (1980). Empirical strength criterion for rock masses. *J. Geotech. Engng. Div. Am. Soc. Civ. Engrs.* **106**:1013–1035.

Hoek, E. and Brown, E. T. (1997). Practical estimates of rock mass strength. *Int. J. Rock Mech. Min. Sci. Geomech. Abstr.* **34**:1165–1186.

Hoek, E. and Diederichs, M. (2006). Empirical estimates of rock mass modulus. *Int. J. Rock Mech. Min. Sci.* **43**:203–215.

Hoek, E., Kaiser, P. K. and Bawden, W. F. (1995). *Support of Underground Excavations in Hard Rock*, Balkema, Rotterdam.

Hoek, E., Carranza-Torres, C. and Corkum, B. (2002). Hoek–Brown failure criterion—2002 edition, 5th N. Am. Rock Mech. Symp. and 17th Tunnel. Assoc. Canada Conf., NARMS-TAC, 267–273.

Hoek, E., Carter, T. G. and Diederichs, M. S. (2013). Quantification of the Geological Strength Index chart. Paper presented at the 47th US Rock Mech./Geomech. Symp., ARMA, 13–672.

Holcomb, D., Rudnicki, J. W., Issen, K. A. and Sternlof, K. (2007). Compaction localization in the Earth and the laboratory: state of the research and research directions. *Acta Geotech.* **2**:1–15.

Holder, J., Olson, J. E. and Philip, Z. (2001). Experimental determination of subcritical crack growth parameters in sedimentary rock. *Geophys. Res. Lett.* **28**:599–602.

Holt, R. M., Fjaer, E., Nes, O.-M. and Alassi, H. T. (2011). A shaley look at brittleness. Paper ARMA 11–366 presented at the 45th US Rock Mechanics/Geomechanics Symposium, San Francisco, California, 26–29 June 2011.

Holt, R. M., Fjaer, E., Stenebråten, J. F. and Nes, O.-M. (2015). Brittleness of shales: relevance to borehole collapse and hydraulic fracturing. *J. Petrol. Sci. Eng.* **131**:200–209.

Holzhausen, G. R. and Johnson, A. M. (1979). Analyses of longitudinal splitting of uniaxially compressed rock cylinders. *Int. J. Rock Mech. Min. Sci. Geomech. Abstr.* **16**:163–177.

Hook, J. R. (2003). An introduction to porosity. *Petrophysics* **44**:205–212.

Hooker, J. N., Laubach, S. E. and Marrett, R. (2014). A universal power-law scaling exponent for fracture apertures in sandstones. *Geol. Soc. Am. Bull.* **126**:1340–1362.

Hooper, D. M., Bursik, M. I. and Webb, F. H. (2003). Application of high-resolution, interferometric DEMs to geomorphic studies of fault scarps, Fish Lake Valley, Nevada-California, USA. *Rem. Sens. Environ.* **84**:255–267.

Hoppa, G. V., Tufts, B. R., Greenberg, R. and Geissler, P. E. (1999). Formation of cycloid features on Europa. *Science* **285**:1899–1902.

Hoppin, R. A. (1961). Precambrian rocks and their relationship to Laramide structure along the east flank of the Bighorn Mountains near Buffalo, Wyoming. *Geol. Soc. Am. Bull.* **72**:351–368.

Horii, H. and Nemat-Nasser, S. (1985). Compression-induced microcrack growth in brittle solids: axial splitting and shear failure. *J. Geophys. Res.* **90**:3105–3125.

Hornbach, M. J., DeShon, H. R., Ellsworth, W. L., et al. (2015). Causal factors for seismicity near Azle, Texas. *Nature Comm.* **6**: 6728; doi:10.1038/ncomms7728.

Hubbert, M. K. and Rubey, W. W. (1959). Role of fluid pressure in mechanics of overthrust faulting. Pts. I & II. *Geol. Soc. Am. Bull.* **70**:115–205.

Hubbert, M. K. and Willis, D. G. (1957). Mechanics of hydraulic fracturing. *AIME Petrol. Trans.* **210**:153–163.

Hucka, V. and Das, B. (1974). Brittleness determination of rocks by different methods. *Int. J. Rock Mech. Min. Sci. Geomech. Abstr.* **11**:389–392.

Hudleston, P. (1999). Strain compatibility and shear zones: is there a problem? *J. Struct. Geol.* **21**:923–932.

Hudson, J. A., Brown, E. T. and Fairhurst, C. (1971). Shape of the complete stress-strain curve for rock. *Proc. US Rock Mech. Symp.* **13**:773–795.

Hudson, J. A., Crouch, S. L. and Fairhurst, C. (1972). Soft, stiff, and servo-controlled testing machines: a review with reference to rock failure. *Eng. Geol.* **6**:155–189.

Hudson, J. A. and Harrison, J. P. (1997). *Engineering Rock Mechanics: An Introduction to the Principles*, Pergamon, New York.

Hughes, A. N. and Shaw, J. H. (2014). Fault displacement-distance relationships as indicators of contractional fault-related folding style. *Am. Assoc. Petrol. Geol. Bull.* **98**:227–251.

Ikari, M. J., Marone, C. and Saffer, D. M. (2011). On the relation between fault strength and frictional stability. *Geology* **39**:83–86.

Inglis, C. E. (1913). Stresses in a plate due to the presence of cracks and sharp corners. *Royal Inst. Naval Architec. Trans.* **55**:219–241.

Ingraffea, A. R. (1987). Theory of crack initiation and propagation in rock. In *Fracture Mechanics of Rock* (ed. B. K. Atkinson), pp. 71–110, Academic Press, New York.

Ingram, G. M. and Urai, J. L. (1999). Top-seal leakage through faults and fractures: the role of mudrock properties. In *Muds and Mudstones—Physical and Fluid Flow Properties* (eds. A. C. Aplin, A. J. Fleet and J. H. S. Macquaker), pp. 125–135, Geol. Soc. London Spec. Publ. 158.

Ingram, G. M., Urai, J. L. and Naylor, M. A. (1997). Sealing processes and top seal assessment. In *Hydrocarbon Seals: Importance for Exploration and Production* (eds. P. Møller-Pedersen and A. G. Koestler), pp. 165–174, Elsevier, Amsterdam.

International Society for Rock Mechanics, Commission on Standardization of Laboratory and Field Tests (1978). Suggested methods for the quantitative description of discontinuities in rock masses. *Int. J. Rock Mech. Min. Sci. Geomech. Abs.* **15**:319–368.

Irwin, G. R. (1957). Analysis of stresses and strains near the end of a crack traversing a plate. *J. Appl. Mech.* **24**:361–364.

Irwin, G. R. (1960). Fracture mode transition for a crack traversing a plate. *J. Basic Engng.* **82**:417–425.

Irwin, G. R. (1962). The crack extension force for a part-through crack in a plate. *J. Appl. Mech.* **29**:651–654.

Irwin, G. R., Kies, J. A. and Smith, H. L. (1958). Fracture strengths relative to onset and arrest of crack propagation. *Proc. Amer. Soc. Test. Mater.* **58**:640–657.

Ishii, E. (2012). Microstructure and origin of faults in siliceous mudstone at the Horonobe Underground Research laboratory site, Japan. *J. Struct. Geol.* **34**:20–29.

Ishii, E., Sanada, H., Funaki, J., Sugita, Y. and Kurikami, H. (2011). The relationships among brittleness, deformation behavior, and transport properties in mudstones: an example from the Horonobe Underground Research Laboratory, Japan. *J. Geophys. Res.* **116**: B09206; doi 10.1029/2011JB008279.

Issen, K. A. (2002). The influence of constitutive models on localization conditions for porous rock. *Eng. Fracture Mech.* **69**:1891–1906.

Issen, K. A. and Challa, V. (2003). Conditions for dilation band formation in granular materials. In *Proc. 16th ASCE Eng. Mech. Conf.* 16th, pp. 1–4.

Issen, K. A. and Rudnicki, J. W. (2000). Conditions for compaction bands in porous rock. *J. Geophys. Res.* **105**:21,529–21,536.

Issen, K. A. and Rudnicki, J. W. (2001). Theory of compaction bands in porous rock. *Phys. Chem. Earth* **A26**:95–100.

Jackson, C. A.-L., Bell, R. E., Rotevatn, A. and Tvedt, A. B. M. (2016). Techniques to determine the kinematics of synsedimentary normal faults and implications for fault growth models. In *The Geometry and Growth of Normal Faults* (eds. C. Childs, R. E. Holdsworth, C. A.-L. Jackson et al.), Geol. Soc. London Spec. Publ. 439; doi:10.1144/SP429.22.

Jackson, J. A. and White, N. J. (1989). Normal faulting in the upper continental crust: observations from regions of active extension. *J. Struct. Geol.* **11**:15–36.

Jackson, M. D. and Pollard, D. D. (1988). The laccolith-stock controversy: new results from the southern Henry Mountains, Utah. *Geol. Soc. Am. Bull.* **100**:117–139.

Jackson, M. D. and Pollard, D. D. (1990). Flexure and faulting of sedimentary host rocks during growth of igneous domes, Henry Mountains, Utah. *J. Struct. Geol.* **12**:185–206.

Jackson, M. P. A. (1995). Retrospective salt tectonics. In *Salt Tectonics* (eds. M. P. A. Jackson, D. G. Roberts and S. Snelson), pp. 1–28, AAPG Memoir 65, Tulsa, Oklahoma.

Jackson, M. P. A. and Vendeville, B. C. (1994). Regional extension as a geologic trigger for diapirism. *Geol. Soc. Am. Bull.* **106**:57–73.

Jackson, P. and Sanderson, D. J. (1992). Scaling of fault displacement from the Badajoz-Córdoba shear zone, SW Spain. *Tectonophysics* **210**:179–190.

Jaeger, J. C. (1969). *Elasticity, Fracture and Flow, with Engineering and Geological Applications*, Chapman and Hall, London, 268 pp.

Jaeger, J. C. (1971). Friction of rocks and stability of rock slopes. Rankine lecture. *Géotechnique* **21**:97–134.

Jaeger, J. C. and Cook, N. G. W. (1979). *Fundamentals of Rock Mechanics* (3rd edn), Chapman and Hall, New York, 593 pp.

Jaeger, J. C., Cook, N. G. W. and Zimmerman, R. W. (2007). *Fundamentals of Rock Mechanics* (4th edn), Blackwell, Oxford, 475 pp.

Jamison, W. R. (1989). Fault-fracture strain in Wingate Sandstone. *J. Struct. Geol.* **11**:959–974.

Jamison, W. R. and Stearns, D. W. (1982). Tectonic deformation of Wingate Sandstone, Colorado National Monument. *Amer. Assoc. Petrol. Geol. Bull.* **66**:2584–2608.

Jaroszewski, W. (1984). *Fault and Fold Tectonics*, Wiley, New York, 565 pp.

Jarrard, R. D. (1986). Terrane motion by strike-slip faulting of forearc slivers. *Geology* **14**:780–783.

Jing, L. and Stephansson, O. (2007). 10 – Discrete fracture network (DFN) method. *Dev. Geotech. Eng.* **85**:365–398.

Johansen, T. E. S. and Fossen, H. (2008). Internal geometry of fault damage zones in interbedded siliciclastic sediments. In *The Internal Structure of Fault Zones: Implications for Mechanical and Fluid-Flow Properties* (eds. C. A. J. Wibberley, W. Kurz, J. Imber, R. E. Holdsworth and C. Collettini), pp. 35–56, Geol. Soc. London Spec. Publ. 299.

Johnson, A. M. (1970). *Physical Processes in Geology: a Method for Interpretation of Natural Phenomena—Intrusions in Igneous Rocks, Fractures and Folds, Flow of Debris and Ice*, Freeman, Cooper, and Co., San Francisco, California, 577 pp.

Johnson, A. M. (1980). Folding and faulting of strain-hardening sedimentary rocks. *Tectonophysics* **62**:251–278.

Johnson, A. M. (1995). Orientations of faults determined by premonitory shear zones. *Tectonophysics* **247**:161–238.

Johnson, A. M. (2001). Propagation of deformation bands in porous sandstones. Unpubl. manuscript, Purdue University, West Lafayette, Indiana, 73 pp.

Johnson, A. M. and Fletcher, R. C. (1994). *Folding of Viscous Layers—Mechanical Analysis and Interpretation of Structures in Deformed Rock*, Columbia University Press, New York, 461 pp.

Johnson, A. M. and Pollard, D. D. (1973). Mechanics of growth of some laccolithic intrusions in the Henry Mountains, Utah, I. *Tectonophysics* **18**:261–309.

Johnson, E. and Cleary, M. P. (1991). Implications of recent laboratory experimental results for hydraulic fractures. Paper presented at SPE Joint Rocky Mountain Regional Meeting and Low Permeability Reservoir Symposium, Soc. of Pet. Eng., Denver, Colorado, paper SPE 21846.

Johri, M., Zoback, M. D. and Hennings, P. (2014). A scaling law to characterize fault-damage zones at reservoir depths. *Amer. Assoc. Petrol. Geol. Bull.* **98**:2057–2079.

Justo, J. L., Justo, E., Azañón, J. M., Durand, P. and Morales, A. (2010). The use of rock mass classification systems to estimate the modulus and strength of jointed rock. *Rock Mech. Rock Eng.* **43**:287–304.

Kachanov, M. (1992). Effective elastic properties of cracked solids: critical review of some basic concepts. *Appl. Mech. Rev.* **45**:304–335.

Kakimi, T. (1980). Magnitude–frequency relation for displacement of minor faults and its significance in crustal deformation. *Bull. Geol. Survey Japan* **31**:467–487.

Kamb, W. B. (1959). Petrofabric observations from Blue Glacier, Washington, in relation to theory and experiment. *J. Geophys. Res.* **64**:1891–1909.

Kalthoff, J. F. (1971). On the characteristic angle or crack branching in ductile materials. *Int. J. Fracture Mech.* **7**:478–480.

Kanninen, M. F. and Popelar, C. H. (1985). *Advanced Fracture Mechanics*, Oxford University Press, New York, 563 pp.

Kaplan, M. F. (1961). Crack propagation and the fracture of concrete. *Am. Concrete Inst. J.* **58**:591–610.

Kaproth, B. M., Cashman, S. M. and Marone, C. (2010). Deformation band formation and strength evolution in unlithified sand: the role of grain breakage. *J. Geophys. Res.* **115**:B12103; doi:10.1029/2010JB007406.

Karato, S.-I. (2008). *Deformation of Earth Materials: an Introduction to the Rheology of Solid Earth*, Cambridge University Press, New York, 463 pp.

Karcz, Z. and Scholz, C. H. (2003). The fractal geometry of some stylolites from the Calcare Massiccio Formation, Italy. *J. Struct. Geol.* **25**:1301–1316.

Karner, S. L. (2006). An extension of rate and state theory to poromechanics. *Geophys. Res. Lett.* **33**:L03308, 10.1029/2005GL024934.

Karner, S. L., Marone, C. and Evans, B. (1997). Laboratory study of fault healing and lithification in simulated fault gouge under hydrothermal conditions. *Tectonophysics* **277**:41–55.

Karner, S. L., Chester, F. M., Kronenberg, A. K. and Chester, J. S. (2003). Subcritical compaction and yielding of granular quartz sand. *Tectonophysics* **377**:357–381.

Karner, S. L., Chester, J. S., Chester, F. M., Kronenberg, A. K. and Hajash, A. Jr. (2005). Laboratory deformation of granular quartz sand: implications for the burial of clastic rocks. *Am. Assoc. Petrol. Geol. Bull.* **89**:603–625.

Karson, J. A. and Dick, H. (1983). Tectonics of ridge-transform intersections at the Kane fracture zone. *Mar. Geophys. Res.* **6**:51–98.

Karson, K. A. (2002). Geologic structure of the uppermost oceanic crust created at fast- to intermediate-rate spreading centers. *Ann. Rev. Earth Planet. Sci.* **30**:347–384.

Kassir, M. K. and Sih, G. C. (1966). Three-dimensional stress distribution around an elliptical crack under arbitrary loadings. *J. Appl. Mech.* **33**:601–611.

Kastens, K. A. (1987). A compendium of causes and effects of processes at transform faults and fracture zones. *Rev. Geophys.* **25**:1554–1562.

Katsman, R. and Aharonov, E. (2006). A study of compaction bands originating from cracks, notches, and compacted defects. *J. Struct. Geol.* **28**:508–518.

Katsman, R., Aharonov, E. and Scher, H. (2004). Numerical simulation of compaction bands in high-porosity sedimentary rock. *Mech. Mater.* **37**:371–390.

Katsman, R., Aharonov, E. and Scher, H. (2006a). A numerical study on localized volume reduction in elastic media: some insights on the mechanics of anticracks. *J. Geophys. Res.* **111**:B03204; doi:10.1029/2004JB003607.

Katsman, R., Aharonov, E. and Scher, H. (2006b). Localized compaction in rocks: Eshelby's inclusion and the spring network model. *Geophys. Res. Lett.* **33**:L10311; doi:10.1029/2005GL025628.

Katsube, T. J. and Williamson, M. A. (1998). Shale petrophysical characteristics: permeability history of subsiding shales. In *Shales and Mudstones*, vol. II (eds. J. Schieber, W. Zimmerle and P. Sethi), pp. 69–91, E. Schweizerbart Science Publishers, Stuttgart, Germany.

Kattenhorn, S. A. and Marshall, S. T. (2006). Fault-induced stress fields and associated tensile and compressive deformation at fault tips in the ice shell of Europa: implications for fault mechanics. *J. Struct. Geol.* **28**:2204–2221.

Kattenhorn, S. A. and Pollard, D. D. (1999). Is lithostatic loading important for the slip behavior and evolution of normal faults in the Earth's crust? *J. Geophys. Res.* **104**:28,879–28,898.

Kattenhorn, S. A., Aydin, A. and Pollard, D. D. (2000). Joints at high angles to normal fault strike: an explanation using 3-D numerical models of fault-perturbed stress fields. *J. Struct. Geol.* **22**:1–23.

Kattenhorn, S. A. and Watkeys, M. K. (1995). Blunt-ended dyke segments. *J. Struct. Geol.* **17**:1535–1542.

Katz, Y., Weinberger, R. and Aydin, A. (2004). Geometry and kinematic evolution of Riedel shear structures, Capitol Reef National Park, Utah. *J. Struct. Geol.* **26**:491–501.

Katzman, R., ten Brink, U. S. and Lin, J. (1995). Three-dimensional modeling of pull-apart basins: implications for the tectonics of the Dead Sea basin. *J. Geophys. Res.* **100**:6295–6312.

Kemeny, J. W. (1991). A model for non-linear rock deformation under compression due to sub-critical crack growth. *Int. J. Rock Mech. Min. Sci. Geomech. Abstr.* **28**:459–467.

Kemeny, J. (2003). The time-dependent reduction of sliding cohesion due to rock bridges along discontinuities: a fracture mechanics approach. *Rock Mech. Rock Eng.* **36**:27–38.

Kemeny, J. and Cook, N. G. W. (1986). Effective moduli, non-linear deformation and strength of a cracked elastic solid. *Int. J. Rock Mech. Min. Sci. Geomech. Abstr.* **23**:107–118.

Kemeny, J. M. and Cook, N. G. W. (1987). Crack models for the failure of rock under compression. *Proc. 2nd. Int. Conf. Constitutive Laws for Engng. Materials* **2**:879–887.

Kenkmann, T. (2002). Folding within seconds. *Geology* **30**:231–234.

Kenkmann, T. (2003). Dike formation, cataclastic flow, and rock fluidization during impact cratering: an example from the Upheaval Dome structure, Utah. *Earth Planet. Sci. Lett.* **214**:43–58.

Keranen, K. M., Savage, H. M., Abers, G. A. and Cochran, E. S. (2013). Potentially induced earthquakes in Oklahoma, USA—links between wastewater injection and the 2011 Mw 5.7 earthquake sequence. *Geology* **41**; doi:10.1130/G34045.1.

Key, W. R. O. and Schultz, R. A. (2011). Fault formation in porous rocks at high strain rates: first results from the Upheaval Dome impact crater, Utah, USA. *Geol. Soc. Am. Bull.* **123**:1161–1170.

Khabbazi, A., Ghafoori, M., Lashkaripour, G. R. and Cheshomi, A. (2012). Estimation of the rock mass deformation modulus using a rock classification system. *Geomech. Geoeng.*; doi:10.1080/17486025.2012.695089.

Khan, A. S., Xiang, Y. and Huang, S. (1991). Behavior of Berea Sandstone under confining pressure part 1: Yield and failure surfaces, and nonlinear elastic response. *Int. J. Plasticity* **7**:607–624.

Khazan, Y. M. and Fialko, Y. A. (1995). Fracture criteria at the tip of fluid-driven cracks in the earth. *Geophys. Res. Lett.* **22**:2541–2544.

Kidambi, T. and Kumar, G. S. (2016). Mechanical Earth Modeling for a vertical well drilled in a naturally fractured tight carbonate gas reservoir in the Persian Gulf. *J. Petrol. Sci. Eng.* **141**:38–51.

Kies, J. A., Krafft, J. M., Sanford, R. J., Smith, H. L. and Sullivan, A. M. (1975). Historical note on the development of fracture mechanics by G. R. Irwin. In *Linear Fracture Mechanics: Historical Developments and Applications of Linear Fracture Mechanics Theory* (eds. G. C. Sih, R. P. Wei and F. Erdogan), pp. 1–27, Envo Publishing Company, Lehigh, Pennsylvania.

Kilburn, C. R. J. and Voight, B. (1998). Slow rock fracture as eruption precursor at Soufriere Hills volcano, Montserrat. *Geophys. Res. Lett.* **25**:3665–3668.

Kim, W.-Y. (2013). Induced seismicity associated with fluid injection into a deep well in Youngstown, Ohio. *J. Geophys. Res.* **118**:3506–3518.

Kim, Y.-S. and Sanderson, D. J. (2005). The relationship between displacement and length of faults: a review. *Earth-Sci. Rev.* **68**:317–334.

Kim, Y.-S., Andrews, J. R. and Sanderson, D. J. (2003). Reactivated strike-slip faults: examples from north Cornwall, UK. *Tectonophysics* **340**:173–194.

Kim, Y.-S., Peacock, D. C. P. and Sanderson, D. J. (2004). Fault damage zones. *J. Struct. Geol.* **26**:503–517.

King, G. C. P. (1978). Geological faulting: fracture, creep and strain. *Phil. Trans. Royal Soc. London A*, **288**:197–212.

King, G. C. P. (1983). The accommodation of large strains in the upper lithosphere of the Earth and other solids by self-similar fault systems—the geometrical origin of *b*-value. *Pure Appl. Geophys.* **121**:762–815.

King, G. and Ellis, M. (1990). The origin of large local uplift in extensional regions. *Nature* **348**:689–693.

King, G. C. P. and Nábêlek, J. L. (1985). The role of fault bends in the initiation and termination of earthquake rupture. *Science* **228**:984–987.

King, G. and Yielding, G. (1984). The evolution of a thrust fault system: processes of rupture initiation, propagation and termination in the 1980 El Asnam (Algeria) earthquake. *Geophys. J. Royal Astron. Soc.* **77**:915–933.

King, G. C. P., Stein, R. S. and Lin, J. (1994). Static stress changes and the triggering of earthquakes. *Bull. Seismol. Soc. Am.* **84**:935–953.

Kingma, J. T. (1958). Possible origin of piercement structures, local unconformities, and secondary basins in the Eastern Geosyncline, New Zealand. *N. Z. J. Geol. Geophys.* **1**:269–274.

Kirby, S. H. and Kronenberg, A. K. (1987). Rheology of the lithosphere: selected topics. *Rev. Geophys.* **25**:1219–1244.

Kirby, S. H., Durham, W. B. and Stern, L. (1991). Mantle phase changes and deep earthquake faulting in subducting lithosphere. *Science* **252**:216–225.

Kirby, S. H., Durham, W. B. and Stern, L. (1992). The ice I–II transformation: mechanisms and kinetics under hydrostatic and nonhydrostatic conditions. In *Physics and Chemistry of Ice* (eds. N. Maeno and T. Hondoh), pp. 456–463, Hokkaido University Press, Sapporo, Japan.

Klimczak, C. (2014). Geomorphology of lunar grabens requires igneous dikes at depth. *Geology* **42**:963–966.

Klimczak, C. and Schultz, R. A. (2013a). Shear-enhanced compaction in dilating granular materials. *Int. J. Rock Mech. Min. Sci.* **64**:139–147.

Klimczak, C. and Schultz, R. A. (2013b). Fault damage zone origin of the Teufelsmauer, Subhercynian Cretaceous Basin, Germany. *Int. J. Earth Sci. (Geologische Rundschau)* **102**:121–138; doi:10.1007/s00531-012-0794-z.

Klimczak, C., Schultz, R. A. Parashar, R. and Reeves, D. M. (2010). Cubic law with aperture-length correlation: implications for network scale fluid flow. *Hydrogeol. J.* **18**:851–862; doi:10.1007/s10040–009–0572–6.

Klimczak, C., Soliva, R., Schultz, R. A. and Chéry, J. (2011). Growth of deformation bands in a multilayer sequence. *J. Geophys. Res.* **116**, B09209; doi:10.1029/2011JB008365.

Klimczak, C., Ernst, C. M., Byrne, P. K., et al. (2013). Insights into the subsurface structure of the Caloris basin, Mercury, from assessments of mechanical layering and changes in long-wavelength topography. *J. Geophys. Res.* **118**:2030–2044; doi:10.1002/JGRE.20157.

Klimczak, C., Byrne, P. K. and Solomon, S. C. (2015). A rock-mechanical assessment of Mercury's global tectonic fabric. *Earth Planet. Sci. Lett.* **416**:82–90.

Knapmeyer, M., Oberst, J., Hauber, E., et al. (2006). Working models for spatial distribution and level of Mars' seismicity. *J. Geophys. Res*. **111**:E11006, doi:10.1029/2006JE002708.

Knipe, R. J. (1989). Deformation mechanisms—recognition from natural tectonites. *J. Struct. Geol.* **11**:127–146.

Knott, S. D. (1993). Fault seal analysis in the North Sea. *Am. Assoc. Petrol. Geol. Bull.* **77**:778–792.

Knott, S. D., Beach, A., Brockbank, P. J., et al. (1996). Spatial and mechanical controls on normal fault populations. *J. Struct. Geol.* **18**:359–372.

Ko, T. Y. and Kemeny, J. (2011). Subcritical crack growth in rocks under shear loading. *J. Geophys. Res.* **116**:B01407; doi:10.1029/2010JB000846.

Koehn, D., Renard, F., Toussaint, R. and Passchier, C. W. (2007). Growth of stylolite teeth patterns depending on normal stress and finite compaction. *Earth Planet. Sci. Lett.* **257**:582–595.

Koenig, E. and Aydin, A. (1998). Evidence for large-scale strike-slip faulting on Venus. *Geology* **26**:551–554.

Kohlstedt, D. L. and Mackwell, S. J. (2010). Strength and deformation of planetary lithospheres. In *Planetary Tectonics* (eds. T. R. Watters and R. A. Schultz), pp. 397–456, Cambridge University Press.

Kohlstedt, D. L., Evans, B. and Mackwell, S. J. (1995). Strength of the lithosphere: constraints imposed by laboratory experiments. *J. Geophys. Res.* **100**:17,587–17,602.

Kolosov, G. V. (1935). Application of a complex variable to the theory of elasticity. *Objed. Nauchno-tekhn. Izd.*, Moscow-Leningrad (in Russian).

Kolyukhin, D. and Torabi, A. (2012). Statistical analysis of the relationships between faults attributes. *J. Geophys. Res.* **117**, B05406; doi:10.1029/2011JB008880.

Kostrov, B. (1974). Seismic moment and energy of earthquakes, and seismic flow of rock. *Izvestiya, Phys. Solid Earth* **13**:13–21.

Kramer, E. J. (1974). The stress–strain curve of shear-banding polystyrene. *J. Macromol. Sci.* **B10**:191–202.

Krantz, R. L. (1983). Microcracks in rocks: a review. *Tectonophysics* **100**:449–480.

Krantz, R. W. (1988). Multiple fault sets and three-dimensional strain: theory and application. *J. Struct. Geol.* **10**:225–237.

Krantz, R. W. (1989). Orthorhombic fault patterns: the odd axis model and slip vector orientations. *Tectonics* **8**:483–495.

Krantz, R. W. (1995). The transpressional strain model applied to strike-slip, oblique-convergent and oblique-divergent deformation. *J. Struct. Geol.* **17**:1125–1137.

Kreemer, C., Blewett, G. and Klein, E. C. (2014). A geodetic plate motion and global strain rate model. *Geochem. Geophys. Geosyst.* **15**:3849–3889.

Kulander, B. R. and Dean, S. L. (1985). Hackle plume geometry and joint propagation dynamics. In *Proceedings of the International Symposium on Fundamentals of Rock Joints* (ed. O. Stephansson), pp. 85–94, Centek.

Kulander, B. R. and Dean, S. L. (1995). Observations on fractography with laboratory experiments for geologists. In *Fractography: Fracture Topography as a Tool in Fracture Mechanics and Stress Analysis* (ed. M. S. Ameen), pp. 97–147, Geol. Soc. Spec. Publ. 92.

Kulander, B. R., Barton, C. C. and Dean, S. L. (1979). The application of fractography to core and outcrop fracture investigations. US Department of Energy, METC/SP–79/3, National Technical Information Service, US Department of Commerce, Springfield, VA.

Kulhawy, F. H. (1975). Stress deformation properties of rock and rock discontinuities. *Eng. Geol.* **9**:327–350.

Labuz, J. F., Zeng, F., Makhnenko, R. and Li, Y. (2018). Brittle failure of rock: a review and general linear criterion. *J. Struct. Geol.* **112**:7–28.

Lacazette, A. and Engelder, T. (1992). Fluid-driven cyclic propagation of a joint in the Ithaca Siltstone, Appalachian Basin, New York. In *Fault Mechanics and Transport Properties of Rocks* (eds. B. Evans and T.-f. Wong), pp. 297–324, Academic Press, New York.

Lachenbruch, A. H. (1961). Depth and spacing of tension cracks. *J. Geophys. Res.* **66**:4273–4292.

Lahee, F. H. (1961). *Field Geology* (6th edn), McGraw-Hill, New York.

Lajtai, E. Z. (1969). Mechanics of second order faults and tension gashes. *Geol. Soc. Am. Bull.* **80**:2253–2272.

Lajtai, E. Z. (1974). Brittle fracture in compression. *Int. J. Fracture* **10**:525–536.

Lajtai, E. Z. (1991). Time dependent behaviour of the rock mass. *Geotech. Geol. Eng.* **9**:109–124.

Lamarche, J., Chabani, A. and Gauthier, B. D. M. (2018). Dimensional threshold for fracture linkage and hooking. *J. Struct. Geol.* **108**:171–179.

Lander, R. H. and Laubach, S. E. (2015). Insight into rates of fracture growth and sealing from a model for quartz cementation in fractured sandstones. *Geol. Soc. Am. Bull.* **127**:516–538.

Landes, J. D. (2000). The contributions of George Irwin to elastic-plastic fracture mechanics development. In *Fatigue and Fracture Mechanics: 31st Volume* (eds. G. R. Halford and J. P. Gallagher), pp. 54–63, Am. Soc. Test. Mat. Spec. Publ. 1389, West Conshohocken, Penn.

Laney, R. T. and Van Schmus, W. R. (1978). A structural study of the Kentland, Indiana, impact site. In *Proc. 9th Lunar Planet. Sci. Conf.* pp. 2609–2632.

Langford, J. C. and Diederichs, M. S. (2015). Quantifying uncertainty in Hoek–Brown intact strength envelopes. *Int. J. Rock Mech. Min. Sci.* **74**:91–102.

Larsen, P. H. (1988). Relay structures in a Lower Permian basement-involved extension system, East Greenland. *J. Struct. Geol.* **10**:3–8.

Latzko, D. G. H. (editor) (1979). *Post-Yield Fracture Mechanics*, Elsevier Science Ltd, 364 pp.

Laubach, S. E., Reed, R. M., Olson, J. E., Lander, R. H. and Bonnell, L. M. (2004). Coevolution of crack-seal texture and fracture porosity in sedimentary rocks: cathodoluminescence observations of regional fractures. *J. Struct. Geol.* **26**:967–982.

Laubach, S. E., Olson, J. E. and Gross, M. R. (2009). Mechanical and fracture stratigraphy. *Am. Assoc. Petrol. Geol. Bull.* **93**:1413–1426.

Laubach, S. E., Eichhubl, P., Hilgers, C. and Lander, R. H. (2010). Structural diagenesis. *J. Struct. Geol.* **32**:1866–1872.

Lawn, B. (1993). *Fracture of Brittle Solids* (2nd edn), Cambridge University Press, 378 pp.

Leblond, J.-B., Karma, A. and Lazarus, V. (2011). Theoretical analysis of crack front instability in mode I+III. *J. Mech. Phys. Solids* **59**:1872–1887.

Lecampion, B., Bunger, A. and Zhang, X. (2018). Numerical methods for hydraulic fracture propagation: a review of recent trends. *Journal of Natural Gas Science and Engineering* **49**:66–83; doi:10.1016/j.jngse.2017.10.012.

Lee, H. P. and Olson, J. E. (2017). The effect of remote and internal crack stresses on mixed-mode brittle fracture propagation of open cracks under compressive loading. *Int. J. Fracture* **207**; doi:10.1007/s10704-017-0231-1.

Lemaitre, J. and Desmorat, R. (2005). *Engineering Damage Mechanics: Ductile, Creep, Fatigue and Brittle Failures*, Springer, Berlin.

Lensen, G. J. (1958). A method of horst and graben formation. *J. Geol.* **66**:579–587.

Li, A. J., Merifield, R. S. and Lyamin, A. V. (2008). Stability charts for rock slopes based on the Hoek–Brown failure criterion. *Int. J. Rock Mech. Min. Sci.* **45**:689–700.

Li, V. C. (1987). Mechanics of shear rupture applied to earthquake zones. In *Fracture Mechanics of Rock* (ed. B. K. Atkinson), pp. 351–428, Academic Press, New York.

Li, V. C. and Liang, E. (1986). Fracture processes in concrete and fiber reinforced cementitious composites. *J. Eng. Mech.* **112**:566–586.

Lin, B., Mear, M. E. and Ravi-Chandar, K. (2010). Criterion for initiation of cracks under mixed-mode I + III loading. *Int. J. Fracture* **165**:175–188.

Lin, J. and Parmentier, E. M. (1988). Quasistatic propagation of a normal fault: a fracture mechanics model. *J. Struct. Geol.* **10**:249–262.

Lin, J. and Stein, R. S. (1989). Coseismic folding, earthquake recurrence, and the 1987 source mechanism at Whittier Narrows, Los Angeles basin, California. *J. Geophys. Res.* **94**:9614–9632.

Lin, M., Hardy, M. P., Agapito, J. F. T. et al. (1993). Rock mass mechanical property estimations for the Yucca Mountain site characterization project. Sandia Nat. Lab. Rept. SAND92–0450, Albquerque, New Mexico.

Linker, M. F. and Dieterich, J. H. (1992). Effects of variable normal stress on rock friction: observations and constitutive equations. *J. Geophys. Res.* **97**:4923–4940.

Lisle, R. J. and Leyshon, P. R. (2004). *Stereographic Projection Techniques for Geologists and Civil Engineers* (2nd edn), Cambridge University Press, Cambridge.

Liu, C., Pollard, D. D., Deng, S. and Aydin, A. (2016). Mechanism of formation of wiggly compaction bands in porous sandstone: 1. observations and conceptual model. *J. Geophys. Res.* **120**:8138–8152; doi:10.1002/2015JB012372.

Lockner, D. (1993). The role of acoustic emission in the study of rock fracture. *Int. J. Rock Mech. Min. Sci. Geomech. Abstr.* **30**:883–899.

Lockner, D. A. (1995). Rock failure. In *Rock Physics and Phase Relations—A Handbook of Physical Constants* (ed. T. J. Ahrens), pp. 127–147, American Geophysical Union Reference Shelf 3.

Lockner, D. A. (1998). A generalized law for brittle deformation of Westerly granite. *J. Geophys. Res.* **103**:5107–5123.

Lockner, D. A. and Beeler, N. M. (1999). Premonitory slip and tidal triggering of earthquakes. *J. Geophys. Res.* **104**:20,133–20,151.

Lockner, D. A., Byerlee, J. D., Kuksenko, V., Ponomarev, A. and Sidorin, A. (1991). Quasi-static fault growth and shear fracture energy in granite. *Nature* **350**:39–42.

Loizzo, M., Lecampion, B. and Mogilevskaya, S. (2017). The role of geological barriers in achieving robust well integrity. *Energy Procedia* **114**:5193–5205.

Lonergan, L. and Cartwright, J. A. (1999). Polygonal faults and their influence on deep-water sandstone reservoir geometries, Alba field, United Kingdom central North Sea. *Am. Assoc. Petrol. Geol. Bull.* **83**:410–432.

Long, J. J. and Imber, J. (2011). Geological controls on fault relay zone scaling. *J. Struct. Geol.* **33**:1790–1800.

Lorenz, J. C. and Cooper, S. P. (2017) Deformation bands. In *Atlas of Natural and Induced Fractures in Core*, John Wiley & Sons, Ltd, Chichester, UK; doi:10.1002/9781119160014.ch11.

Lorenz, J. C., Sterling, J. L., Schechter, D. S., Whigham, C. L. and Jensen, J. J. (2002). Natural fractures in the Sprayberry Formation, Midland Basin, Texas: the effects of mechanical stratigraphy on fracture variability and reservoir behavior. *Am. Assoc. Petrol. Geol. Bull.*, **86**:505–524.

Lucchitta, B. K. (1976). Mare ridges and related highland scarps—results of vertical tectonism? *Proc. Lunar Sci. Conf.* **7**:2761–2782.

Lucchitta, B. K. (1977). Topography, structure, and mare ridges in southern Mare Imbrium and northern Oceanus Procellarum. *Proc. Lunar Sci. Conf.* **8**:2691–2703.

Luyendyk, B. P., Kamerling, M. J. and Terres, R. (1980). Geometric model for Neogene crustal rotations in southern California. *Geol. Soc. Am. Bull.* **91**:211–217.

Lyakhovsky, V., Zhu, W. and Shalev, E. (2015). Visco-poroelastic damage model for brittle-ductile failure of porous rocks. *J. Geophys. Res.* **120**; doi:10.1002/2014JB011805.

Lyzenga, G. A., Wallace, K. S., Faneslow, J. L., Raefsky, A. and Groth, P. M. (1986). Tectonic motions in California inferred from VLBI observations, 1980–1984. *J. Geophys. Res.* **91**:9473–9487.

Ma, X. Q. and Kusznir, N. J. (1993). Modelling of near-field subsurface displacements for generalized faults and fault arrays. *J. Struct. Geol.* **15**:1471–1484.

Macdonald, G. A. (1957). Faults and monoclines on Kilauea Volcano, Hawaii. *Geol. Soc. Am. Bull.* **68**:269–271.

Macdonald, K. (1982). Mid-ocean ridges: fine scale tectonic, volcanic, and hydrothermal processes within the plate boundary zone. *Annu. Rev. Earth Planet. Sci.* **10**:155–190.

Macdonald, K. C., Fox, P. J., Perram, L. J., et al. (1988). A new view of the mid-ocean ridge from the behavior of ridge-axis discontinuities. *Nature* **335**:217–225.

Mack, G. H. and Seager, W. R. (1995). Transfer zones in the southern Rio Grande rift. *J. Geol. Soc. London* **152**:551–560.

Maerten, L., Willemse, E. J. M., Pollard, D. D. and Rawnsley, K. (1999). Slip distributions on intersecting normal faults. *J. Struct. Geol.* **21**:259–272.

Main, I. G. (1991). A modified Griffith criterion for the evolution of damage with a fractal distribution of crack lengths: application to seismic event rates and b-values. *Geophys. J. Int.* **107**:353–362.

Main, I. G. (1995). Earthquakes as critical phenomena: implications for probabilistic seismic hazard analysis. *Bull. Seismol. Soc. Am.* **85**:1299–1308.

Main, I. G. (2000). A damage mechanics model for power-law creep and earthquake aftershock and foreshock sequences. *Geophys. J. Int.* **142**:151–161.

Main, I. G. and Meredith, P. G. (1991). Stress corrosion constitutive laws as a possible mechanism of intermediate-term and short-term seismic quiescence. *Geophys. J. Int.* **107**:363–372.

Main, I. G., Sammonds, P. R. and Meredith, P. G. (1993). Application of a modified Griffith criterion to the evolution of fractal damage during compressional rock failure. *Geophys. J. Int.* **115**:367–380.

Main, I. G., Leonard, T., Papasouliotis, O., Hatton, C. G. and Meredith, P. G. (1999). One slope or two? Detecting statistically significant breaks of slope in geophysical data, with applications to fracture scaling relationships. *Geophys. Res. Lett.* **26**:2801–2804.

Mair, K., Main, I. and Elphick, S. (2000). Sequential growth of deformation bands in the laboratory. *J. Struct. Geol.* **22**:25–42.

Mair, K., Frye, K. M. and Marone, C. (2002). Influence of grain characteristics on the friction of granular shear zones. *J. Geophys. Res.* **107**; doi:10.1029/2001JB000516, 2219.

Maiti, S. K. and Smith, T. A. (1983). Comparison of the criteria for mixed mode brittle failure based on the preinstability stress-strain field—part I: slit and elliptical cracks under uniaxial tensile loading. *Int. J. Fracture* **23**:281–295.

Maiti, S. K. and Smith, T. A. (1984). Comparison of the criteria for mixed mode brittle failure based on the preinstability stress-strain field—part II: pure shear and uniaxial compressive loading. *Int. J. Fracture* **24**:5–22.

Majer, E. L., Baria, R., Stark, M., et al. (2007). Induced seismicity associated with Enhanced Geothermal Systems. *Geothermics* **36**:185–222.

Malik, J. N., Shah, A. A., Sahoo, A. K., et al. (2010). Active fault, fault growth and segment linkage along the Janauri anticline (frontal foreland fold), NW Himalaya, India. *Tectonophysics* **483**:327–343.

Maltman, A. J. (1984). On the term 'soft-sediment deformation.' *J. Struct. Geol.* **6**:589–592.

Maltman, A. J. (1988). The importance of shear zones in naturally deformed wet sediments. *Tectonophysics* **145**:163–175.

Maltman, A. J. (1994). Prelithification deformation. In *Continental Deformation* (ed. P. L. Hancock), pp. 143–158, Pergamon, New York.

Mandl, G. (1987a). Discontinuous fault zones. *J. Struct. Geol.* **9**:105–110.

Mandl, G. (1987b). Tectonic deformation by rotating parallel faults: the "bookshelf" mechanism. *Tectonophysics* **141**:277–316.

Mandl, G. (1988). *Mechanics of Tectonic Faulting—Models and Basic Concepts,* Elsevier, Amsterdam.

Mandl, G. (2000). *Faulting in Brittle Rocks: an Introduction to the Mechanics of Tectonic Faults,* Springer-Verlag, Heidelberg.

Mandl, G. (2005). *Rock Joints: the Mechanical Genesis,* Springer-Verlag, Heidelberg.

Manighetti, I., King, G. C. P., Gaudemer, Y., Scholz, C. H. and Doubre, C. (2001). Slip accumulation and lateral propagation of active normal faults in Afar. *J. Geophys. Res.* **106**:13,667–13,696.

Manighetti, I., King, G. and Sammis, C. G. (2004). The role of off-fault damage in the evolution of normal faults. *Earth Planet. Sci. Lett.* **217**:399–408.

Mangold, N., Allemand, P. and Thomas, P. G. (1998). Wrinkle ridges of Mars: structural analysis and evidence for shallow deformation controlled by ice-rich décollements. *Planet. Space Sci.* **46**:345–356.

Mann, P., Hempton, M. R., Bradley, D. C. and Burke, K. C. (1983). Development of pull-apart basins. *J. Geol.* **91**:529–554.

Mann, P., Draper, G. and Burke, K. (1985). Neotectonics of a strike-slip restraining bend system, Jamaica. In *Strike-Slip Deformation, Basin Formation, and Sedimentation* (eds. K. T. Biddle and N. Christie-Blick), pp. 211–226, Soc. Econ. Paleon. Miner., Spec. Publ. 37.

Mansinha, L. and Smylie, D. E. (1971). The displacement fields of inclined faults. *Bull. Seismol. Soc. Am.* **61**:1433–1440.

Marder, M. and Fineberg, J. (1996). How things break. *Phys. Today* **49**:24–29.

Marinos, P. and Hoek, E. (2001). Estimating the geotechnical properties of heterogeneous rock masses such as flysch. *Bull. Eng. Geol. Env.* **60**:85–92.

Marinos, V., Marinos, P. and Hoek, E. (2005). The geologic strength index: applications and limitations. *Bull. Eng. Geol. Env.* **64**:55–65.

Marketos, G. and Bolton, M. D. (2009). Compaction bands simulated in discrete element models. *J. Struct. Geol.* **31**:479–490.

Marone, C. (1995). Fault zone strength and failure criteria. *Geophys. Res. Lett.* **22**:723–726.

Marone, C. (1998a). The effect of loading rate on static friction and the rate of fault healing during the earthquake cycle. *Nature* **391**:69–72.

Marone, C. (1998b). Laboratory-derived friction laws and their application to seismic faulting. *Annu. Rev. Earth Planet. Sci.* **26**:643–696.

Marone, C. and Scholz, C. H. (1988). The depth of seismic faulting and the transition from stable to unstable slip regimes. *Geophys. Res. Lett.* **15**:621–624.

Marone, C., Scholz, C. H. and Bilham, R. (1991). On the mechanics of earthquake afterslip. *J. Geophys. Res.* **96**:8441–8452.

Marrett, R. (1996). Aggregate properties of fracture populations. *J. Struct. Geol.* **18**:169–178.

Marrett, R. and Allmendinger, R. W. (1990). Kinematic analysis of fault slip data. *J. Struct. Geol.* **12**:973–986.

Marrett, R. and Peacock, D. C. P. (1999). Strain and stress. *J. Struct. Geol.* **21**:1057–1063.

Marshak, S. and Mitra, G. (eds.) (1988). *Basic Methods of Structural Geology*, Prentice-Hall, New York.

Martel, S. J. (1990). Formation of compound strike-slip fault zones, Mount Abbot quadrangle, California. *J. Struct. Geol.* **12**:869–882.

Martel, S. J. (1997). Effects of cohesive zones on small faults and implications for secondary fracturing and fault trace geometry. *J. Struct. Geol.* **19**:835–847.

Martel, S. J. (2004). Mechanics of landslide initiation as a shear fracture phenomenon. *Marine Geol.* **203**:319–339.

Martel, S. J. (2017). Progress in understanding sheeting joints over the past two centuries. *J. Struct. Geol.* **94**:68–86.

Martel, S. J. and Boger, W. A. (1998). Geometry and mechanics of secondary fracturing around small three-dimensional faults in granitic rock. *J. Geophys. Res.* **103**:21,299–21,314.

Martel, S. J. and Pollard, D. D. (1989). Mechanics of slip and fracture along small faults and simple strike-slip fault zones in granitic rock. *J. Geophys. Res.* **94**:9417–9428.

Martel, S. J., Pollard, D. D. and Segall, P. (1988). Development of simple strike-slip fault zones, Mount Abbot quadrangle, Sierra Nevada, California. *Geol. Soc. Am. Bull.* **100**:1451–1465.

Mastin, L. G. and Pollard, D. D. (1988). Surface deformation and shallow dike intrusion processes at Inyo Craters, Long Valley, California. *J. Geophys. Res.* **93**:13,221–13,235.

Masuda, K. (2001). Effects of water on rock strength in a brittle regime. *J. Struct. Geol.* **23**:1653–1657.

Matthäi, S., Aydin, A., Pollard, D. D. and Roberts, S. (1998). Numerical simulation of departures from radial drawdown in a faulted sandstone reservoir with joints and deformation bands, pp. 157–191, Geol. Soc. London Spec. Publ. 147.

Mazars, J. and Pijaudier-Cabot, G. (1996). From damage to fracture mechanics and conversely: a combined approach. *Int. J. Solids Structures* **33**:3327–3342.

Mauldon, M. and Dershowitz, W. (2000). A multi-dimensional system of fracture abundance measures (abstract). Paper presented at the annual meeting, Geol. Soc. Amer., Reno, Nevada, November 2000.

McCaffrey, R. (1992). Oblique plate convergence, slip vectors, and forearc deformation. *J. Geophys. Res.* **97**:8905–8915.

McClay, K. R. (1992). Glossary of thrust tectonics terms. In *Thrust Tectonics* (ed. K. R. McClay), pp. 419–433, Chapman and Hall, London.

McClay, K. R. (2011). Introduction to thrust fault-related folding. In *Thrust Fault-Related Folding* (eds. K. R. McClay, J. H. Shaw and J. Suppe), pp. 1–19, Amer. Assoc. Petrol. Geol. Mem. 94.

McClay, K. R., Dooley, T., Whitehouse, P. S. and Anadon-Ruiz, S. (2005). 4D analogue models of extensional fault systems in asymmetric rifts: 3D visualizations and comparisons with natural examples. In *Petroleum Geology: North-West Europe and Global Perspectives—Proc. 6th Petrol. Geol. Conf.* (eds. A. G. Doré and B. A. Vining), pp. 1543–1556, Geological Society, London.

McClintock, F. A. and Walsh, J. B. (1962). Friction on Griffith cracks in rocks under pressure. Proc., 4th US National Congress of Appl. Mech., pp. 1015–1021.

McConaughy, D. T. and Engelder, T. (2001). Joint initiation in bedded clastic rocks. *J. Struct. Geol.* **23**:203–221.

McCoss, A. M. (1986). Simple constructions for deformation in transpression/transtension zones. *J. Struct. Geol.* **8**:715–718.

McGarr, A. (1988). On the state of lithospheric stress in the absence of applied tectonic forces. *J. Geophys. Res.* **93**:13,609–13,617.

McGarr, A. (2014). Maximum magnitude earthquakes induced by fluid injection. *J. Geophys. Res.* **119**:1008–1019.

McGarr, A. and Gay, N. C. (1978). State of stress in the Earth's crust. *Ann. Rev. Earth Planet. Sci.* **6**:405–436.

McGarr, A., Pollard, D. D., Gay, N. C. and Ortlepp, W. D. (1979). Observations and analysis of structures in exhumed mine-induced faults. US Geological Survey Open-File Report, 79(1239), pp.101–120.

McGarr, A., Simpson, D. and Seeber, L. (2002). Case histories of induced and triggered seismicity. In *International Handbook of Earthquake and Engineering Seismology*, vol. 81A, pp. 647–661, Academic Press, San Francisco.

McGill, G. E. (1993). Wrinkle ridges, stress domains and kinematics of Venusian plains. *Geophys. Res. Lett.* **20**:2407–2410.

McGill, G. E. and Stromquist, A. W. (1979). The grabens of Canyonlands National Park, Utah: geometry, mechanics, and kinematics. *J. Geophys. Res.* **84**:4547–4563.

McGill, G. E., Stofan, E. R. and Smrekar, S. E. (2010). Venus tectonics. In *Planetary Tectonics* (eds. T. R. Watters and R. A. Schultz), pp. 81–120, Cambridge University Press.

McGovern, P. J. and Solomon, S. C. (1993). State of stress, faulting, and eruption characteristics of large volcanoes on Mars. *J. Geophys. Res.* **98**:23,553–23,579.

McGrath, A. G. and Davison, I. (1995). Damage zone geometry around fault tips. *J. Struct. Geol.* **17**:1011–1024.

McKinstry, H. E. (1948). *Mining Geology*, Prentice-Hall, New York, 680 pp.

McLeod, A. E., Dawers, N. H. and Underhill, J. R. (2000). The propagation and linkage of normal faults: insights from the Strathspey-Brent-Statfjord fault array, northern North Sea. *Basin Res.* **12**:263–284.

Means, W. D. (1976). *Stress and Strain—Basic Concepts of Continuum Mechanics for Geologists*, Springer-Verlag, New York.

Medwedeff, D. A. (1992). Geometry and kinematics of an active, laterally propagating wedge thrust, Wheeler Ridge, California. In *Structural Geology of Fold and Thrust Belts* (ed. S. Mitra and G. W. Fisher), pp. 3–28, Johns Hopkins University Press, Baltimore, Maryland.

Melin, S. (1982). Why do cracks avoid each other? *Int. J. Fracture* **23**:37–45.

Melin, S. (1986). When does a crack grow under mode II conditions? *Int. J. Fracture* **30**:103–114.

Melosh, H. J. (1989). *Impact Cratering: A Geologic Process*, Oxford University Press, New York.

Melosh, H. J. and Williams, C. A., Jr. (1989). Mechanics of graben formation in crustal rocks: a finite element analysis. *J. Geophys. Res.* **94**:13,961–13,973.

Menéndez, B., Zhu, W. and Wong, T.-f. (1996). Micromechanics of brittle faulting and cataclastic flow in Berea sandstone. *J. Struct. Geol.* **18**:1–16.

Meng, C. and Pollard, D. D. (2014). Eshelby's solution for ellipsoidal inhomogeneous inclusions with applications to compaction bands. *J. Struct. Geol.* **67**:1–19.

Meng, L., Fu, X., Lv, Y., et al. (2016). Risking fault reactivation induced by gas injection into depleted reservoirs based on the heterogeneity of geomechanical properties of fault zones. *Petrol. Geosci.* **23**:29–38.

Mercier, E., Outtani, F. and Frizon de Lamotte, D. (1997). Late-stage evolution of fault-propagation folds: principles and example. *J. Struct. Geol.* **19**:185–193.

Meredith, P. G. and Atkinson, B. K. (1983). Stress corrosion and acoustic emission during tensile crack propagation in Whin Sill dolerite and other basic rocks. *Geophys. J. R. Astron. Soc.* **75**:1–21.

Meredith P. G., Main I. G. and Jones C. (1990). Temporal variations in seismicity during quasi-static and dynamic rock failure. *Tectonophysics* **175**:249–268.

Mériaux, C., Lister, J. R., Lyakhovsky, V. and Agnon, A. (1999). Dyke propagation with distributed damage of the host rock. *Earth Planet. Sci. Lett.* **165**:177–185.

Michalske, T. A., Singh, M. and Fréchette, V. D. (1981). Experimental observation of crack velocity and crack front shape effects in double-torsion fracture mechanics tests. In *Fracture Mechanics for Ceramics, Rocks, and Concrete* (eds. S. W. Freiman and E. R. Fuller, Jr.), pp. 3–12, Am. Soc. Test. Mat. Spec. Publ. 745, Philadelphia, Penn.

Micklethwaite, S., Ford, A., Witt, W. and Sheldon, H. A. (2014). The where and how of faults, fluids and permeability: insights from fault stepovers, scaling properties and gold mineralisation. *Geofluids*; doi 10.1111/gfl.12102.

Mitra, G. (1979). Ductile deformation zones in Blue Ridge basement rocks and estimation of finite strain. *Geol. Soc. Am. Bull.* **90**:935–951.

Mitra, G. and Sussman, A. J. (1997). Structural evolution of connecting splay duplexes and their implications for critical taper: an example based on geometry and kinematics of the Canyon Range culmination, Sevier Belt, central Utah. *J. Struct. Geol.* **19**:503–521.

Mitra, S. (1986). Duplex structures and imbricate thrust systems: geometry, structural position, and hydrocarbon potential. *Am. Assoc. Petrol. Geol. Bull.* **70**:1087–1112.

Mitra, S. (1990). Fault-propagation folds: geometry, kinematic evolution, and hydrocarbon traps. *Am. Assoc. Petrol. Geol. Bull.* **74**:921–945.

Mitra, S. and Mount, V. S. (1998). Foreland basement-involved structures. *Am. Assoc. Petrol. Geol. Bull.* **82**:70–109. (Errata, *Am. Assoc. Petrol. Geol. Bull.* **83**:2024–2027, 1999.)

Mitra, S. and Mount, V. S. (1999). Foreland basement-involved structures: Reply. *Am. Assoc. Petrol. Geol. Bull.* **83**:2017–2023.

Moeck, I. and Backers, T. (2011). Fault reactivation potential as a critical factor during reservoir stimulation. *First Break* **29**:73–80.

Moeck, I., Kwiatek, G. and Zimmermann, G. (2009). Slip tendency analysis, fault reactivation potential and induced seismicity in a deep geothermal reservoir. *J. Struct. Geol.* **31**:1174–1182.

Mollema, P. N. and Antonellini, M. A. (1996). Compaction bands: a structural analog for anti-mode I cracks in aeolian sandstone. *Tectonophysics* **267**:209–228.

Mollema, P. N. and Antonellini, M. A. (1999). Development of strike-slip faults in the dolomites of the Sella Group, northern Italy. *J. Struct. Geol.* **21**:273–292.

Molnar, P. (1983). Average regional strain due to slip on numerous faults of different orientations. *J. Geophys. Res.* **88**:6430–6432.

Molnar, P. (1988). Continental tectonics in the aftermath of plate tectonics. *Nature* **335**:131–137.

Molnar, P. and Tapponnier, P. (1975). Cenozoic tectonics of Asia: effects of a continental collision. *Science* **189**:419–426.

Moody, J. D. and Hill, M. L. (1956). Wrench-fault tectonics. *Geol. Soc. Am. Bull.* **67**:1207–1246.

Moore, D. E. and Lockner, D. A. (1995). The role of microcracking in shear-fracture propagation in granite. *J. Struct. Geol.* **17**:95–114.

Moore, J. M. and Schultz, R. A. (1999). Processes of faulting in jointed rocks of Canyonlands National Park, Utah. *Geol. Soc. Am. Bull.* **111**:808–822.

Moos, D. and Zoback, M. D. (1990). Utilization of observations of well bore failure to constrain the orientation and magnitude of crustal stresses: application to continental Deep Sea Drilling Project and Ocean Drilling Program boreholes. *J. Geophys. Res.* **95**:9305–9325.

Morewood, N. C. and Roberts, G. P. (2002). Surface observations of active normal fault propagation: implications for growth. *J. Geol. Soc. London* **159**:263–272.

Morgan, J. K. and Boettcher, M. S. (1999). Numerical simulations of granular shear zones using the distinct element method, 1. Shear zone kinematics and the micromechanics of localization. *J. Geophys. Res.* **104**:2703–2719.

Morgan, J. K. and McGovern, P. J. (2005). Discrete element simulations of gravitational volcanic deformation: 2. Mechanical analysis. *J. Geophys. Res.* **110**:B05403; doi:10.1029/2004JB003253.

Morgan, W. J. (1968). Rises, trenches, great faults, and crustal blocks. *J. Geophys. Res.* **73**:1959–1982.

Morley, C. K., Nelson, R. A., Patton, T. L. and Munn, S. G. (1990). Transfer zones in the East African rift system and their relevance to hydrocarbon exploration in rifts. *Am. Assoc. Petrol. Geol. Bull.* **74**:1234–1253.

Morley, C. K. (1999). How successful are analogue models in addressing the influence of pre-existing fabrics on rift structure? *J. Struct. Geol.* **21**:1267–1274.

Moros, J. G. (1999). Relationship between fracture aperture and length in sedimentary rocks. Unpublished MS thesis, The University of Texas, Austin.

Moustafa, A. R. (1997). Controls on the development and evolution of transfer zones: the influence of basement structure and sedimentary thickness in the Suez rift and Red Sea. *J. Struct. Geol.* **19**:755–768.

Mount, V. S. and Suppe, J. (1987). State of stress near the San Andreas fault: implications for wrench tectonics. *Geology* **15**:1143–1146.

Mount, V. S., Suppe, J. and Hook, S. C. (1990). A forward modeling strategy for balancing cross sections. *Am. Assoc. Petrol. Geol. Bull.* **74**:521–531.

Muehlberger, W. R. (1961). Conjugate joint sets of small dihedral angle. *J. Geol.* **69**:211–218.

Mueller, K. and Golombek, M. P. (2004). Compressional structures on Mars. *Ann. Rev. Earth Planet. Sci.* **32**:435–464.

Mueller, K. and Talling, P. (1997). Geomorphic evidence for tear faults accommodating lateral propagation of an active fault-bend fold, Wheeler Ridge, California. *J. Struct. Geol.* **19**:397–411.

Mühlhaus, H. B. and Vardoulakis, I. (1988). The thickness of shear bands in granular materials. *Géotechnique* **38**:271–284.

Mukhopadhyay, B. (2002). Water-rock interactions in the basement beneath Long Valley caldera: an oxygen isotope study of the Long Valley Exploratory Well drill cores. *J. Volcanol. Geotherm. Res.* **116**:325–359.

Muir Wood, D. (1990). *Soil Behaviour and Critical State Soil Mechanics*, Cambridge University Press, New York, 462 pp.

Mullen, M., Roundtree, R., Barree, B. and Turk, G. (2007). A composite determination of mechanical rock properties for stimulation design (what to do when you don't have a sonic log). Paper SPE 108139 presented at the 2007 SPE Rocky Mountain Oil and Gas Technology Symposium, Denver, Colorado, 16–18 April 2007.

Muller, J. R. and Martel, S. J. (2000). Numerical models of translational landslide rupture surface growth. *Pure Appl. Geophys.* **157**:1009–1038.

Murgatroyd, J. B. (1942). The significance of surface marks on fractured glass. *J. Soc. Glass Technol.* **26**:155–171.

Muraoka, H. and Kamata, H. (1983). Displacement distribution along minor fault traces. *J. Struct. Geol.* **5**:483–495.

Murrell, S. A. F. (1958). The strength of coal under triaxial compression. In *Mechanical Properties of Non-Metallic Materials* (ed. W. Walton), pp. 123–153, Butterworths.

Murrell, S. A. F. (1964). The theory of the propagation of elliptical Griffith cracks under various conditions of plane strain or plane stress, Pts. II, III. *Br. J. Appl. Phys.* **15**:1211–1223.

Muskhelishvili, N. I. (1963). *Some Basic Problems of the Mathematical Theory of Elasticity* (4th edn), Noordhoff, Groningen, Holland.

Myers, R. and Aydin, A. (2004). The evolution of faults formed by shearing across joint zones in sandstone. *J. Struct. Geol.* **26**:947–966.

Nadai, A. (1950). *Theory of Flow and Fracture of Solids* (Vol. 1), McGraw-Hill, New York.

Nahm, A. K. and Schultz, R. A. (2007). Outcrop-scale properties of Burns Formation at Meridiana Planum, Mars. *Geophys. Res. Lett.* **34**, L20203; doi:10.1029/2007GL031005.

Nahm, A. L., and Schultz, R. A. (2010). Evaluation of the orogenic belt hypothesis for the origin of the Thaumasia Highlands, Mars. *J. Geophys. Res.* **115**:E04008; doi:10.1029/2009JE003327.

Nahm, A. L. and Schultz, R. A. (2011). Magnitude of global contraction on Mars from analysis of surface faults: implications for Martian thermal history. *Icarus* **211**:389–400; doi:10.1016/j.icarus.2010.11.003.

Nakamura, K. (1977). Volcanoes as possible indicators of tectonic stress orientation: principle and proposal. *J. Volcanol. Geotherm. Res.* **2**:1–16.

Nara, Y., Yamanaka, H., Oe, Y. and Kaneko, K. (2013). Influence of temperature and water on subcritical crack growth parameters and long-term strength for igneous rocks. *Geophys. J. Int.* **193**:47–60.

Narr, W. and Suppe, J. (1991). Joint spacing in sedimentary rocks. *J. Struct. Geol.* **13**:1037–1048.

National Academy of Sciences (1996). *Rock Fractures and Fluid Flow: Contemporary Understanding and Applications*, National Academy Press, Washington, DC, 551 pp.

Naylor, M. A., Mandl, G. and Sijpesteijn, C. H. K. (1986). Fault geometries in basement-induced wrench faulting under different initial stress states. *J. Struct. Geol.* **8**:737–752.

Naylor, M., Sinclair, H. D., Willett, S. and Cowie, P. A. (2005). A discrete element model for orogenesis and accretionary wedge growth. *J. Geophys. Res.* **110**:B12403; doi:10.1029/2003JB002940.

Nehari, Z. (1952). *Conformal Mapping*, Dover Publications, New York.

Nemat-Nasser, S. and Horii, H. (1982). Compression-induced nonplanar crack extension with application to splitting, exfoliation, and rockburst. *J. Geophys. Res.* **87**:6805–6821.

Nemčok, M., Schamel, S. and Gayer, R. (2005). *Thrustbelts: Structural Architecture, Thermal Regimes, and Petroleum Systems*, Cambridge University Press, 541 pp.

Nenna, F. and Aydin, A. (2011). The formation and growth of pressure solution seams in clastic rocks: a field and analytical study. *J. Struct. Geol.* **33**:633–643.

Neuffer, D. P., Schultz, R. A. and Watters, R. J. (2006). Mechanisms of slope failure on Pyramid Mountain, a subglacial volcano in Wells Gray Provincial Park, British Columbia. *Can. J. Earth Sci.* **43**:147–155.

Neuffer, D. P. and Schultz, R. A. (2006). Mechanisms of slope failure in Valles Marineris, Mars. *Quart. J. Eng. Geol. Hydrogeol.* **39**:227–240.

Neuendorf, K. K. E., Mehl, J. P., Jr. and Jackson, J. A. (2005). *Glossary of Geology* (5th edn), Am. Geol. Inst., Alexandria, VA.

Newman, J. C., Jr. (2000). Irwin's stress intensity factor—A historical perspective. In *Fatigue and Fracture Mechanics: 31st Volume* (eds. G. R. Halford and J. P. Gallagher), pp. 39–53, Am. Soc. Test. Mat. Spec. Publ. 1389, West Conshohocken, Penn.

Nguyen, G. D., Nguyen, C. T., Bui, H. H. and Nguyen, V. P. (2016). Constitutive modelling of compaction localization in porous sandstones. *Int. J. Rock Mech. Min. Sci.* **83**:57–72.

Nguyen, O., Repetto, E. A., Ortiz, M. and Radovitzky, R. A. (2001). A cohesive model of fatigue crack growth. *Int. J. Fracture* **110**:351–369.

Nicol, A., Walsh, J. J., Watterson, J. and Bretan, P. G. (1995). Three-dimensional geometry and growth of conjugate normal faults. *J. Struct. Geol.* **17**:847–862.

Nicol, A., Watterson, J., Walsh, J. J. and Childs, C. (1996). The shapes, major axis orientations and displacement patterns of fault surfaces. *J. Struct. Geol.* **18**:235–248.

Nicol, A., Gillespie, P. A., Childs, C. and Walsh, J. J. (2002). Relay zones between mesoscopic thrust fault in layered sedimentary sequences. *J. Struct. Geol.* **24**:709–727.

Nicol, A., Walsh, J., Berryman, K. and Nodder, S. (2005). Growth of a normal fault by the accumulation of slip over millions of years. *J. Struct. Geol.* **27**:327–342.

Nicol, A., Childs, C., Walsh, J. J. and Schafer, K. W. (2013). A geometric model for the formation of deformation band clusters. *J. Struct. Geol.* **55**:21–33.

Nicol, A., Childs, C., Walsh, J. J., Manzocchi, T. and Schöpfer, M. P. J. (2016). Interactions and growth of faults in an outcrop-scale system. In *The Geometry and Growth of Normal Faults* (eds. C. Childs, R. E. Holdsworth, C. A.-L. Jackson et al.), Geol. Soc. London Spec. Publ. 439. (First published online March 10, 2016; https://doi.org/10.1144/SP439.9.)

Nicolas, A. (1995). *The Mid-Oceanic Ridges: Mountains Below Sea Level*, Springer-Verlag, Berlin.

Nicholson, C., Seeber, L., Williams, P. and Sykes, L. R. (1986). Seismic evidence for conjugate slip and block rotation within the San Andreas fault system, southern California. *Tectonics* **5**:629–648.

Nicholson, R. and Pollard, D. D. (1985). Dilation and linkage of echelon cracks. *J. Struct. Geol.* **7**:583–590.

Nickelsen, R. P. and Hough, V. N. D. (1967). Jointing in the Appalachian plateau of Pennsylvania. *Geol. Soc. Am. Bull.* **78**:609–630.

Niño, F., Philip, H. and Chéry, J. (1998). The role of bed-parallel slip in the formation of blind thrust faults. *J. Struct. Geol.* **20**:503–516.

Nova, R. (2005). A simple elastoplastic model for soils and soft rocks. In *Soil Constitutive Models: Evaluation, Selection, and Calibration* (eds. J. A. Yamamuro and V. N. Kaliakin), pp. 380–399, Am. Soc. Civ. Eng. Geotech. Spec. Publ. 128.

Nova, R. and Lagioia, R. (2000). Soft rocks: behaviour and modeling. In *Eurock '96, Prediction and Performance in Rock Mechanics and Rock Engineering* (ed. G. Barla), pp. 1521–1540, Proc. ISRM Symp., vol. 3, Balkema, Rotterdam.

Nur, A., Ron, H. and Scotti, O. (1986). Fault mechanics and the kinematics of block rotations. *Geology* **14**:746–749.

Nygård, R., Gutierrez, M., Bratli, R. K. and Høeg, K. (2006). Brittle-ductile transition, shear failure and leakage in shales and mudrocks. *Marine Petrol. Geol.* **23**:201–212.

Odé, H. (1957). Mechanical analysis of the dike pattern of the Spanish Peaks area, Colorado. *Geol. Soc. Am. Bull.* **68**:567–576.

Odling, N. E. (1997). Scaling and connectivity of joint systems in sandstones from western Norway. *J. Struct. Geol.* **19**:1257–1271.

Odonne, F. and Massonnat, G. (1992). Volume loss and deformation around conjugate fractures: comparison between a natural example and analogue experiments. *J. Struct. Geol.* **14**:963–972.

Oertel, G. (1965). The mechanism of faulting in clay experiments. *Tectonophysics* **2**:343–393.

Oertel, G. (1996). *Stress and Deformation: A Handbook on Tensors in Geology*, Oxford University Press.

Ogilvie, S. R. and Glover, P. W. J. (2001). The petrophysical properties of deformation bands in relation to their microstructure. *Earth Planet. Sci. Lett.* **193**:129–142.

Ohlmacher, G. C. and Aydin, A. (1995). Progressive deformation and fracture patterns during foreland thrusting in the southern Appalachians. *Am. J. Sci.* **295**:943–987.

Ohlmacher, G. C. and Aydin, A. (1997). Mechanics of vein, fault and solution surface formation in the Appalachian Valley and Ridge, northeastern Tennessee, USA: implications for fault friction, state of stress and fluid pressure. *J. Struct. Geol.* **19**:927–944.

Okal, E. A. and Langenhorst, A. R. (2000). Seismic properties of the Etanin Transform System, South Pacific. *Phys. Earth Planet. Inter.* **119**:185–208.

Okubo, C. H. (2007). Strength and deformability of light-toned layered deposits observed by MER Opportunity: Eagle to Erebus craters, Mars. *Geophys. Res. Lett.* **34**:L20205; doi:10.1029/2007GL031327.

Okubo, C. H. and Martel, S. J. (1998). Pit crater formation on Kilauea volcano, Hawaii. *J. Volcanol. Geotherm. Res.* **86**:1–18.

Okubo, C. H. and Schultz, R. A. (2003). Thrust fault vergence directions on Mars: a foundation for investigating global-scale Tharsis-driven tectonics. *Geophys. Res. Lett.* **30**:2154, 10.1029/2003GL018664.

Okubo, C. H. and Schultz, R. A. (2004). Mechanical stratigraphy in the western equatorial region of Mars based on thrust fault-related fold topography and implications for near-surface volatile reservoirs. *Geol. Soc. Am. Bull.* **116**:594–605.

Okubo, C. H. and Schultz, R. A. (2005). Evolution of damage zone geometry and intensity in porous sandstone: insight from strain energy density. *J. Geol. Soc. London* **162**:939–949.

Okubo, C. H. and Schultz, R. A. (2006). Near-tip stress rotation and the development of deformation band stepover geometries in mode II. *Geol. Soc. Am. Bull.* **118**:343–348.

Okubo, C. H. and Schultz, R. A. (2007). Compactional deformation bands in Wingate Sandstone; additional evidence of an impact origin for Upheaval Dome, Utah. *Earth Planet. Sci. Lett.* **256**:169–181.

Okubo, C. H., Schultz, R. A., Chan, M. A., Komatsu, G., and the High-Resolution Imaging Science Experiment (HiRISE) Team (2009). Deformation band clusters on Mars and implications for subsurface fluid flow. *Geol. Soc. Am. Bull.* **121**:474–482.

Okubo, P. G. and Dieterich, J. H. (1984). Effects of fault properties on frictional instabilities produced on simulated faults. *J. Geophys. Res.* **89**:5817–5827.

Okubo, P. G. and Dieterich, J. H. (1986). State variable fault constitutive relations for dynamic slip. In *Earthquake Source Mechanics* (eds. S. Das, J. Boatwright and C. H. Scholz), pp. 25–35, Am. Geophys. Un. Geophys. Monog. 37.

Oldenburg, D. W. and Brune, J. N. (1975). An explanation for the orthogonality of ocean ridges and transform faults. *J. Geophys. Res.* **80**:2575–2585.

Olson, E. L. and Cooke, M. L. (2005). Application of three fault growth criteria to the Puente Hills thrust system, Los Angeles, California, USA. *J. Struct. Geol.* **27**:1765–1777.

Olson, J. E. (1993). Joint pattern development: effects of subcritical crack growth and mechanical crack interaction. *J. Geophys. Res.* **98**:12,251–12,265.

Olson, J. E. (2003). Sublinear scaling of fracture aperture versus length: an exception or the rule? *J. Geophys. Res.* **108**:2413; doi:10.1029/2001JB000419.

Olson, J. E. (2004). Predicting fracture swarms—the influence of subcritical crack growth and the crack-tip process zone on joint spacing in rock. In *The Initiation, Propagation, and Arrest of Joints and Other Fractures: Interpretations Based on Field Observations* (eds. J. W. Cosgrove and T. Engelder), pp. 73–87, Geol. Soc. London Spec. Publ. 231.

Olson, J. E. (2007). Fracture aperture, length and pattern geometry development under biaxial loading: a numerical study with applications to natural, cross-jointed systems. In *Fracture-Like Damage and Localisation* (eds. G. Couples and H. Lewis), pp. 123–142, Geol. Soc. London Spec. Publ. 289.

Olson, J. and Pollard, D. D. (1989). Inferring paleostresses from natural fracture patterns: a new method. *Geology* **17**:345–348.

Olson, J. and Pollard, D. D. (1991). The initiation and growth of en échelon veins. *J. Struct. Geol.* **13**:595–608.

Olson, J. E. and Schultz, R. A. (2011). Comment on "A note on the scaling relations for opening mode fractures in rock" by C.H. Scholz. *J. Struct. Geol.* **33**:1523–1524.

Olson, J. E., Hennings, P. H. and Laubach, S. E. (1998). Integrating wellbore data and geomechanical modeling for effective characterization of naturally fractured reservoirs. SPE/ISRM Eurock Conference, Trondheim, Norway, July 8–10.

Olson, J. E., Laubach, S. E., and Lander, R. H. (2007). Combining diagenesis and mechanics to quantify fracture aperture distributions and fracture pattern permeability. In *Fractured Reservoirs* (eds. L. Lonergan, R. J. H. Jolly, K. Rawnsley and D. J. Sanderson), pp. 101–116, Geol. Soc. London Spec. Publ. 270.

Olsson, W. A. (1999). Theoretical and experimental investigation of compaction bands in porous rock. *J. Geophys. Res.* **104**:7219–7228.

Olsson, W. A. (2000). Origin of Lüders' bands in deformed rock. *J. Geophys. Res.* **105**:5931–5938.

Olsson, W. A. (2001). Quasistatic propagation of compaction fronts in porous rock. *Mech. Mat.* **33**:659–668.

Olsson, W. A. and Holcomb, D. J. (2000). Compaction localization in porous rock. *Geophys. Res. Lett.* **27**:3537–3540.

Omdal, E., Madland, M. V., Kristiansen, T. G., et al. (2010). Deformation behavior of chalk studied close to in situ reservoir conditions. *Rock Mech. Rock Eng.* **43**:557–580.

Onasch, C. M., Farver, J. R. and Dunne, W. M. (2010). The role of dilation and cementation in the formation of cataclasite in low temperature deformation of well cemented quartz-rich rocks. *J. Struct. Geol.* **32**:1912–1922.

Opheim, J. A. and Gudmundsson, A. (1989). Formation and geometry of fractures, and related volcanism, of the Krafla fissure swarm, Northeast Iceland. *Geol. Soc. Am. Bull.* **101**:1608–1622.

Pachell, M. A. and Evans, J. P. (2002). Growth, linkage, and termination processes of a 10-km-long strike-slip fault in jointed granite: the Gemini fault zone, Sierra Nevada, California. *J. Struct. Geol.* **24**:1903–1924.

Palmer, A. C. and Rice, J. R. (1973). The growth of slip surfaces in the progressive failure of over-consolidated clay. *Proc. Royal Soc. London A*, **332**:527–548.

Papamichos, E. and Vardoulakis, I. (1995). Shear band formation in sand according to non-coaxial plasticity model. *Géotechnique* **45**:649–661.

Papamichos, E., Brignoli, M. and Santerelli, F. J. (1997). An experimental and theoretical study of partially saturated collapsible rocks. *Mech. Cohesive-Frictional Mat.* **2**:251–278.

Pappalardo, R. T. and Collins, G. C. (2005). Strained craters on Ganymede. *J. Struct. Geol.* **27**:827–838.

Pappalardo, R. T., Head, J. W., Collins, G. C., et al. (1998). Grooved terrain on Ganymede: first results from Galileo high-resolution imaging. *Icarus* **135**:276–302.

Paris, P., Gomez, M. and Anderson, W. (1961). A rational analytic theory of fatigue. *The Trend in Eng.* **13**:9–14.

Paris, P. and Erdogan, F. (1963). A critical analysis of crack propagation laws. *J. Basic Eng.* **85**:528–534.

Paris, P. C. and Sih, G. C. (1965). Stress analysis of cracks. In *Fracture Toughness Testing and its Applications*. Amer. Soc. Testing Mater., Spec. Tech. Publ. 381:30–81.

Park, K. and Paulino, G. H. (2011). Cohesive zone models: a critical review of traction-separation relationships across fracture surfaces. *Appl. Mech. Rev.* **64**:061002; doi:10.1115/1.4023110.

Park, R. G. (1989). *Foundations of Structural Geology* (2nd edn), Chapman and Hall, New York, 148 pp.

Parker, J. M. (1942). Regional systematic jointing in slightly deformed sedimentary rocks. *Geol. Soc. Am. Bull.* **53**:381–408.

Parnell, J. (2010). Potential of palaeofluid analysis for understanding oil charge history. *Geofluids* **10**:73–82.

Parnell, J., Watt, G. R., Middleton, D., Kelley, J. and Baron, M. (2004). Deformation band control on hydrocarbon migration. *J. Sed. Res.* **74**:552–560.

Parry, W. T., Chan, M. A. and Beitler, B. (2004). Chemical weathering indicates episodes of fluid flow in deformation bands in sandstone. *Am. Assoc. Petrol. Geol. Bull.* **88**:175–191.

Passchier, C. W. and Trouw, R. A. J. (1996). *Microtectonics*, Springer-Verlag, Berlin.

Paterson, M. S. (1978). *Experimental Rock Deformation—the Brittle Field*, Springer-Verlag, Heidelberg.

Paterson, M. S. and Wong, T.-f. (2005). *Experimental Rock Deformation—the Brittle Field* (2nd edn), Springer-Verlag, Heidelberg.

Paul, B. (1961). A modification of the Coulomb–Mohr theory of fracture. *J. Appl. Mech.* **28**:259–268.

Pawar, R. J., Bromhal, G. S., Carey, J. W., et al. (2015). Recent advances in risk assessment and risk management of geologic CO_2 storage. *Int. J. Greenhouse Gas Control* **40**:292–311.

Peacock, D. C. P. (2002). Propagation, interaction and linkage in normal fault systems. *Earth Sci. Rev.* **58**:121–142.

Peacock, D. C. P. and Azzam, I. N. (2006). Development and scaling relationships of a stylolite population. *J. Struct. Geol.* **28**:1883–1889.

Peacock, D. C. P. and Sanderson, D. J. (1991). Displacements, segment linkage and relay ramps in normal fault systems. *J. Struct. Geol.* **13**:721–733.

Peacock, D. C. P. and Sanderson, D. J. (1992). Effects of layering and anisotropy on fault geometry. *J. Geol. Soc. London* **149**:793–802.

Peacock, D. C. P. and Sanderson, D. J. (1993). Estimating strain from fault slip using a line sample. *J. Struct. Geol.* **15**: 1513–1516.

Peacock, D. C. P. and Sanderson, D. J. (1994). Geometry and development of relay ramps in normal fault systems. *Am. Assoc. Petrol. Geol. Bull.* **78**:147–165.

Peacock, D. C. P. and Sanderson, D. J. (1995). Strike-slip relay ramps. *J. Struct. Geol.* **17**:1351–1360.

Peacock, D. C. P., Jones, G., Knipe, F. J. and McAllister, E. (1998). Large lateral ramps in the Eocene Valkyr shear zone: extensional ductile faulting controlled by plutonism in southern British Columbia: Discussion. *J. Struct. Geol.* **20**:487–488.

Peacock, D. C. P., Knipe, R. J. and Sanderson, D. J. (2000). Glossary of normal faults. *J. Struct. Geol.* **22**:291–305.

Peacock, D. C. P., Nixon, C. W., Rotevatn, A., Sanderson, D. J. and Zuluaga, L. F. (2016). Glossary of fault and other fracture networks. *J. Struct. Geol.* **92**:12–29.

Peacock, D. C. P., Nixon, C. W., Rotevatn, A., Sanderson, D. J. and Zuluaga, L. F. (2017). Interacting faults. *J. Struct. Geol.* **97**:1–22.

Peacock, D. C. P., Sanderson, D. J. and Rotevatn, A. (2018). Relationships between fractures. *J. Struct. Geol.* **106**:41–53.

Peng, S. D. (1971). Stresses within elastic circular cylinders loaded uniaxially and triaxially. *Int. J. Rock Mech. Min. Sci. & Geomech. Abstr.* **8**:399–432.

Peng, S. and Johnson, A. M. (1972). Crack growth and faulting in cylindrical specimens of Chelmsford granite. *Int. J. Rock Mech. Min. Sci.* **9**:37–86.

Pennacchioni, G. (2005). Control of the geometry of precursor brittle structures on the type of ductile shear zone in the Adamello tonalites, Southern Alps (Italy). *J. Struct. Geol.* **27**:627–644.

Perfettini, H., Schmittbul, J. and Cochard, A. (2003). Shear and normal load perturbations on a two-dimensional continuous fault: 2. Dynamic triggering. *J. Geophys. Res.* **108**:2409; doi:10.1029/2002JB001805.

Perkins, T. K. and Krech, W. W. (1966). Effect of cleavage rate and stress level on apparent surface energies of rocks. *Soc. Petrol. Eng. J.* **6**:308–314.

Peters, J. F., Muthuswamy, M., Wibowo, J. and Tordesillas, A. (2005). Characterization of force chains in granular material. *Phys. Rev. E* **72**:041307.

Petersen, M. D., Mueller, C. S., Moschetti, M. P., et al. (2015). Incorporating induced seismicity in the 2014 United States National Seismic Hazard Model—Results of 2014 workshop and sensitivity studies: US Geological Survey Open-File Report 2015–1070, 69 pp.; http://dx.doi.org/10.3133/ofr20151070.

Petit, J.-P. (1987). Criteria for the sense of movement on fault surfaces in brittle rocks. *J. Struct. Geol.* **9**:597–608.

Petit, J.-P. and Barquins, M. (1988). Can natural faults propagate under mode-II conditions? *Tectonics* **7**:1243–1256.

Petit, J.-P. and Mattauer, M. (1995). Palaeostress superimposition deduced from mesoscale structures in limestone: the Matelles exposure, Languedoc, France. *J. Struct. Geol.* **17**:245–256.

Petrie, E. S., Evans, J. P. and Bauer, S. J. (2014). Failure of cap-rock seals as determined from mechanical stratigraphy, stress history, and tensile-failure analysis of exhumed analogs. *Am. Assoc. Petrol. Geol. Bull.* **98**:2365–2389.

Philip, H. and Meghraoui, M. (1983). Structural analysis and interpretation of the surface deformations of the El Asnam earthquake of October 10, 1980. *Tectonics* **2**:17–49.

Phipps Morgan, J. and Parmentier, E. M. (1985). Causes and rate limiting mechanisms of ridge propagation: a fracture mechanics model. *J. Geophys. Res.* **90**:8603–8612.

Philit, S. (2017). Elaboration of a structural, petrophysical and mechanical model of faults in porous sandstones; implication for migration and fluid entrapment, PhD thesis, University of Montpellier, France, 298 pp.

Philit, S., Soliva, R., Ballas, G. and Fossen, H. (2017). Grain deformation processes in porous quartz sandstones: insight from the clusters of cataclastic deformation bands. *EPJ Web of Conferences* 140:07002, Powders and Grains 2017; doi:10.1051/epjconf/201714007002.

Philit, S., Soliva, R., Castilla, R., Ballas, G. and Taillefer, A. (2018). Clusters of deformation bands in porous sandstones. *J. Struct. Geol.* **114**:235–250.

Pinter, N. (1995). Faulting on the volcanic tableland, Owens Valley, California. *J. Geol.* **103**:73–83.

Pittman, E. D. (1981). Effect of fault related granulation on porosity and permeability of quartz sandstones, Simpson Group (Ordovician), Oklahoma. *Am. Assoc. Petrol. Geol. Bull.* **65**:2381–2387.

Plescia, J. B. and Golombek, M. P. (1986). Origin of planetary wrinkle ridges based on the study of terrestrial analogs. *Geol. Soc. Am. Bull.* **97**:1289–1299.

Plumb, R. A. (1994). Variations of the least horizontal stress magnitude in sedimentary basins. In *Rock Mechanics: Models and Measurements, Challenges from Industry* (eds. P. Nelson and S. E. Laubach), pp. 71–77, Balkema, Rotterdam.

Polit, A. T., Schultz, R. A. and Soliva, R. (2009). Geometry, displacement-length scaling, and extensional strain of normal faults on Mars with inferences on mechanical stratigraphy of the Martian crust. *J. Struct. Geol.* **31**:662–673.

Pollard, D. D. (1987). Elementary fracture mechanics applied to the structural interpretation of dykes. In *Mafic Dyke Swarms* (eds. H. C. Halls and W. F. Fahrig), pp. 5–24, Geol. Assoc. Canada Spec. Paper 34.

Pollard, D. D. and Aydin, A. (1984). Propagation and linkage of oceanic ridge segments. *J. Geophys. Res.* **89**:10,017–10,128.

Pollard, D. D. and Aydin, A. (1988). Progress in understanding jointing over the past century. *Geol. Soc. Am. Bull.* **100**:1181–1204.

Pollard, D. D. and Fletcher, R. C. (2005). *Fundamentals of Structural Geology*, Cambridge University Press, Cambridge.

Pollard, D. D. and Holzhausen, G. (1979). On the mechanical interaction between a fluid-filled fracture and the Earth's surface. *Tectonophysics* **53**:27–57.

Pollard, D. D. and Johnson, A. M. (1973). Mechanics of growth of some laccolithic intrusions in the Henry Mountains, Utah II: Bending and failure of overburden layers and sill formation. *Tectonophysics* **18**:311–354.

Pollard, D. D. and Muller, O. H. (1976). The effect of regional gradients in stress and magma pressure on the form of sheet intrusions in cross section. *J. Geophys. Res.* **81**:975–984.

Pollard, D. D. and Segall, P. (1987). Theoretical displacements and stresses near fractures in rock: with applications to faults, joints, dikes, and solution surfaces. In *Fracture Mechanics of Rock* (ed. B. K. Atkinson), pp. 277–349, Academic Press, New York.

Pollard, D. D., Muller, O. H. and Dockstader, D. R. (1975). The form and growth of fingered sheet intrusions. *Geol. Soc. Am. Bull.* **86**:351–363.

Pollard, D. D., Segall, P. and Delaney, P. T. (1982). Formation and interpretation of dilatant echelon cracks. *Geol. Soc. Am. Bull.* **93**:1291–1303.

Pollard, D. D., Delaney, P. T., Duffield, W. A., Endo, E. T. and Okamura, A. T. (1983). Surface deformation in volcanic rift zones. *Tectonophysics* **94**:541–584.

Pollard, D. D., Saltzer, S. D. and Rubin, A. M. (1993). Stress inversion methods: are they based on faulty assumptions? *J. Struct. Geol.* **15**:1045–1054.

Polun, S. G., Gomez, F. and Tesfaye, S. (2018). Scaling properties of normal faults in the central Afar, Ethiopia and Djibouti: implications for strain partitioning during the final stages of continental breakup. *J. Struct. Geol.*, **155**:178–189.

Poncelet, E. F. (1958). The markings on fracture surfaces. *J. Soc. Glass Tech.* **42**:279 T–288 T.

Pook, L. P. (1971). The effect of crack angle on fracture toughness. *Eng. Frac. Mech.* **3**:205–218.

Preston, F. W. (1926). A study of the rupture of glass. *J. Soc. Glass Technol.* **10**:234–269.

Price, N. J. (1966). *Fault and Joint Development in Brittle and Semi-Brittle Rock*, Pergamon Press, Oxford.

Price, N. J. and Cosgrove, J. W. (1990). *Analysis of Geological Structures*, Cambridge University Press.

Priest, S. D. (1993). *Discontinuity Analysis for Rock Engineering*, Chapman and Hall, London.

Priest, S. D. and Hudson, J. A. (1976). Discontinuity spacings in rock. *Int. J. Rock Mech. Min. Sci. Geomech. Abstr.* **13**:135–148.

Priest, S. D. and Hudson, J. A. (1981). Estimation of discontinuity spacing and trace length using scanline surveys. *Int. J. Rock Mech. Min. Sci. Geomech. Abstr.* **18**:183–197.

Proffett, J. M., Jr. (1977). Cenozoic geology of the Yerington district, Nevada, and implications for the nature and origin of Basin and Range faulting. *Geol. Soc. Am. Bull.* **88**:247–266.

Prost, G. L. and Newsome, J. (2015). Caprock integrity determination at the Christina Lake thermal recovery project, Alberta.: *Canadian Soc. Petrol. Geol. Bull.* **64**:309–323.

Pugno, N., Ciavarella, M., Cornetti, P. and Carpinteri, A. (2006). A generalized Paris' law for fatigue crack growth. *J. Mech. Phys. Solids* **54**:1333–1349.

Qu, D. and Tveranger, J. (2016). Incorporation of deformation band fault damage zones in reservoir models. *Am. Assoc. Petrol. Geol. Bull.* **100**:423–443.

Quennell, A. M. (1958). The structural and geomorphic evolution of the Dead Sea rift. *J. Geol. Soc. Lond.* **114**:1–24.

Quigley, M., Van Dissen, R., Litchfield, N., et al. (2012). Surface rupture during the 2010 M_w 7.1 Darfield (Canterbury) earthquake: implications for fault rupture dynamics and seismic-hazard analysis. *Geology* **40**:55–58.

Rabinovitch, A. and Bahat, D. (1999). Model of joint spacing distribution based on shadow compliance. *J. Geophys. Res.* **104**:4877–4886.

Rabinowicz, E. (1958). The intrinsic variables affecting the stick-slip process. *Proc. Phys. Soc.* **71**:668–675.

Raduha, S., Butler, D., Mozley, P. S., et al. (2016). Potential seal bypass and caprock storage produced by deformation-band-to-opening-mode-fracture transition at the reservoir/caprock interface. *Geofluids*; doi:10.1111/gfl.12177.

Ragan, D. M. (2009). *Structural Geology: an Introduction to Geometrical Techniques* (4th edn), Cambridge University Press, 602 pp.

Ramulu, M. and Kobayashi, A. S. (1983). Dynamic crack curving: a photoelastic evaluation. *Exp. Mech.* **23**:1–9.

Ramsay, J. G. and Allison, I. (1979). Structural analysis of shear zones in an alpinised Hercynian granite (Maggia Lappen, Pennine zone, central Alps). *Schweitz. mineral. petrogr. Mitt.* **59**:251–279.

Ramsay, J. G. and Huber, M. I. (1987). *The Techniques of Modern Structural Geology—Volume 2: Folds and Fractures*, Academic, San Diego, 700 pp.

Ramsay, J. G. and Lisle, R. J. (2000). *The Techniques of Modern Structural Geology—Volume 3: Applications of Continuum Mechanics in Structural Geology*, Academic, San Diego, 1061 pp.

Ramsey, J. M. and Chester, F. M. (2004). Hybrid fracture and the transition from extension fracture to shear fracture. *Nature* **428**:63–66.

Rawnsley, K. D., Rives, T., Petit, J.-P., Henscher, S. R. and Lumsden, A. C. (1992). Joint development in perturbed stress fields near faults. *J. Struct. Geol.* **14**:939–951.

Raynaud, S. and Carrio-Schaffhauser, E. (1992). Rock matrix structures in a zone influenced by a stylolite. *J. Struct. Geol.* **14**:973–980.

Reading, H. G. (1980). Characteristics and recognition of strike-slip systems. In *Sedimentation in Oblique-Slip Mobile Zones* (eds. P. F. Ballance and H. G. Reading) pp. 7–26, Int. Assoc. Sedimentol. Spec. Publ. 4.

Reasenberg, P. A. and Simpson, R. W. (1992). Response of regional seismicity to the static stress change produced by the Loma Prieta earthquake. *Science* **255**:1687–1690.

Reches, Z. (1978). Analysis of faulting in three-dimensional strain field. *Tectonophysics* **47**:109–129.

Reches, Z. (1983). Faulting of rocks in three-dimensional strain fields II. Theoretical analysis. *Tectonophysics* **95**:133–156.

Reches, Z. (1987). Mechanical aspects of pull-apart basins and push-up swells with applications to the Dead Sea transform. *Tectonophysics* **141**:75–88.

Reches, Z. and Dieterich, J. H. (1983). Faulting of rocks in three-dimensional strain fields I. Failure of rocks in polyaxial, servo-control experiments. *Tectonophysics* **95**:111–132.

Reches, Z. and Lockner, D. A. (1994). Nucleation and growth of faults in brittle rocks. *J. Geophys. Res.* **99**:18,159–18,173.

Reid, H. F., Davis, W. M., Lawson, A. C. and Ransome, F. L. (1913). Report of the committee on the nomenclature of faults. *Geol. Soc. Am. Bull.* **24**:163–186.

Rekach, V. G. (1979). *Manual of the Theory of Elasticity*. Mir Publishers, Moscow (trans. M. Konyaeva).

Renard, F., Gratier, J.-P. and Jamtveit, B. (2000). Kinetics of crack-sealing, intergranular pressure solution, and compaction around active faults. *J. Struct. Geol.* **22**:1395–1407.

Renard, F., Schmittbuhl, J., Gratier, J.-P., Meakin, P. and Merino, E. (2004). Three-dimensional roughness of stylolites in limestones. *J. Geophys. Res.* **109**:B03209.

Renshaw, C. E. (1996). Influence of subcritical fracture growth on the connectivity of fracture networks. *Water Resour. Res.* **32**:1519–1530.

Renshaw, C. E. (1997). Mechanical controls on the spatial density of opening-mode fracture networks. *Geology* **25**:923–926.

Renshaw, C. E. and Pollard, D. D. (1994a). Are large differential stresses required for straight fracture propagation? *J. Struct. Geol.* **16**:817–822.

Renshaw, C. E. and Pollard, D. D. (1994b). Numerical simulation of fracture set formation: a fracture mechanics model consistent with experimental observations. *J. Geophys. Res.* **99**:9359–9372.

Renshaw, C. E. and Pollard, D. D. (1995). An experimentally verified criterion for propagation across unbonded frictional interfaces. *Int. J. Rock Mech. Min. Sci. Geomech. Abstr.* **32**:237–249.

Reynolds, O. (1885). On the dilatancy of media composed of rigid particles in contact. *Philosophical Mag.* **5**:469.

Rice, J. R. (1968). A path-independent integral and the approximate analysis of strain concentration by notches and cracks. *J. Appl. Mech.* **35**:379–386.

Rice, J. R. (1978). Thermodynamics of the quasi-static growth of Griffith cracks. *J. Mech. Phys. Solids* **26**:61–78.

Rice, J. R. (1979). Theory of precursory processes in the inception of earthquake rupture. *Beitr. Geophys.* **88**:91–127.

Rice, J. R. (1980). The mechanics of earthquake rupture. In *Physics of the Earth's Interior* (eds. A. M. Dziewonski and E. Boschi), pp. 555–649, Proc. Int. School of Phys. "Enrico Fermi" Course 78, Italian Physical Society, North Holland Publ. Co.

Rice, J. R. (1983). Constitutive relations for fault slip and earthquake instabilities. *Pure Appl. Geophys.* **121**:443–475.

Rice, J. R. (1992). Fault stress states, pore-pressure distributions, and the weakness of the San Andreas fault. In *Fault Mechanics and Transport Properties of Rocks* (eds. B. Evans and T.-f. Wong), pp. 435–459, Academic, New York.

Richards, L. R. and Read, S. A. L. (2013). Estimation of the Hoek–Brown parameter m_i using Brazilian tensile test. Paper presented at the 47th US Rock Mech./Geomech. Symp., ARMA, 13–465.

Rickman, R., Mullen, M., Petre, E., Grieser, B. and Kundert, D. (2008). A practical use of shale petrophysics for stimulation design optimization: all shale plays are not clones of the Barnett Shale. Paper SPE 115258 presented at the 2008 SPE Annual Technical Conference and Exhibition, Denver, Colorado, 21–24 September 2008.

Rickman, R., Mullen, M., Petre, E., Grieser, B. and Kundert, D. (2009). Petrophysics key in stimulating shales. American Oil and Gas Reporter, March 2009.

Riedel, J. J. and Labuz, J. F. (2006). Propagation of a shear band in sandstone. *Int. J. Numer. Analyt. Methods Geomech.* **31**:1281–1299.

Riedel, W. (1929). Zur Mechanik geologischer Brucherscheinungen ein Beitrag zum Problem der "Fiederspalten." *Centralbl. f. Mineral. Geol. u. Paläont. Abt.* **1929B**:354–368.

Riggs, E. M. and Green, H. W., II (2001). Shear localization in transformation-induced faulting: first order similarities to brittle shear failure. *Tectonophysics* **340**:95–107.

Rijken, M. C. M. (2005). Modeling naturally fractured reservoirs—From experimental rock mechanics to flow simulation, PhD dissertation, The University of Texas at Austin, 239 pp.

Rijken, P. and Cooke, M. L. (2001). Role of shale thickness on vertical connectivity of fractures: application of crack-bridging theory to the Austin Chalk, Texas. *Tectonophysics* **337**:117–133.

Risnes, R. (2001). Deformation and yield in high porosity outcrop chalk. *Phys. Chem. Earth* **26**:53–57.

Rispoli, R. (1981). Stress fields about strike-slip faults inferred from stylolites and tension gashes. *Tectonophysics* **75**:T29–T36.

Ritchie, R. O. (2011). The conflicts between strength and toughness. *Nature Mater.* **10**:817–822.

Rivero, C. and Shaw, J. H. (2011). Active folding and blind thrust faulting induced by basin inversion processes, inner California borderlands. In *Thrust Fault-Related Folding* (eds. K. R. McClay, J. H. Shaw and J. Suppe), pp. 187–214, Amer. Assoc. Petrol. Geol. Mem. 94.

Robert, R., Robion, P., Souloumiac, P., David, C. and Saillet, E. (2018). Deformation bands, early markers of tectonic activity in front of a fold-and-thrust belt: example from the Tremp-Graus basin, southern Pyrenees, Spain. *J. Struct. Geol.* **110**:65–85.

Roberts, A. and Yielding, G. (1994). Continental extensional tectonics. In *Continental Deformation* (ed. P. L. Hancock), pp. 223–250, Pergamon, New York.

Roberts, G. P. (2007). Fault orientation variations along the strike of active normal fault systems in Italy and Greece: implications for predicting the orientations of subseismic-resolution faults in hydrocarbon reservoirs. *Am. Assoc. Petrol. Geol. Bull.* **91**:1–20.

Roberts, G. P., Cowie, P., Papanikoulaou, I. and Michetti, A. M. (2004). Fault scaling relationships, deformation rates and seismic hazards: An example from the Lazio-Abruzzo Apennines, central Italy. *J. Struct. Geol.* **26**:377–398.

Roberts, J. C. (1961). Feather-fracture, and the mechanics of rock-jointing. *Amer. J. Sci.* **259**:481–492.

Roche, V., Homberg, C. and Rocher, M. (2012). Architecture and growth of normal fault zones in multilayer systems: a 3D field analysis in the South-Eastern Basin, France. *J. Struct. Geol.* **37**:19–35.

Roche, V., Homberg, C. and Rocher, M. (2013). Fault nucleation, restriction, and aspect ratio in layered sections: quantification of the strength and stiffness roles using numerical modeling. *J. Geophys. Res.* **118**:4446–4460; doi:10.1002/JGRB.50279.

Rodgers, D. A. (1980). Analysis of pull-apart basin development produced by *en echelon* strike-slip faults. In *Sedimentation in Oblique-Slip Mobile Zones* (eds. P. F. Ballance and H. G. Reading), pp. 27–41, Int. Assoc. Sedimentol. Spec. Publ. 4.

Rögnvaldsson, S. T., Gudmundsson, A. and Slunga, R. (1998). Seismotectonic analysis of the Tjörnes Fracture Zone, an active transform fault in north Iceland. *J. Geophys. Res.* **103**:30,117–30,129.

Roering, C. (1968). The geometrical significance of natural en echelon crack arrays. *Tectonophysics* **5**:107–123.

Roering, J. J., Cooke, M. L. and Pollard, D. D. (1997). Why blind thrust faults do not propagate to the Earth's surface: numerical modeling of coseismic deformation associated with thrust-related anticlines. *J. Geophys. Res.* **102**:11,901–11,912.

Rogers, R. D. and Bird, D. K. (1987). Fracture propagation associated with dike emplacement at the Skaergaard intrusion, East Greenland. *J. Struct. Geol.* **9**:71–86.

Rogers, S. F. (2003). Critical stress-related permeability in fractured rocks. In *Fracture and in-situ Stress Characterization of Hydrocarbon Reservoirs* (ed. M. Ameen), pp. 7–16, Geol. Soc. London Spec. Publ. 209.

Rogers, S. F., Elmo, D., Webb, G. and Moreno, C. G. (2016). DFN modelling of major structural instabilities in a large open pit for end of life planning purposes. Paper ARMA 16–882 presented at the 50th US Rock Mechanics/Geomechanics Symposium, Houston, Texas, June 26–29, 2016.

Romijn, R. and Groenenboom, J. (1997). Acoustic monitoring of hydraulic fracture growth. *First Break* **15**:295–303.

Ron, H., Aydin, A. and Nur, A. (1986). Strike-slip faulting and block rotation in the Lake Mead fault system. *Geology* **14**:1020–1023.

Ron, H., Freund, R., Garfunkel, Z. and Nur, A. (1984). Block rotation by strike-slip faulting: structural and paleomagnetic evidence. *J. Geophys. Res.* **89**:6256–6270.

Ron, H., Nur, A. and Aydin, A. (1993). Rotation of stress and blocks in the Lake Mead, Nevada, fault system. *Geophys. Res. Lett.* **20**:1703–1706.

Roscoe, K. H. (1970). Influence of strains in soil mechanics. *Géotechnique* **20**:129–170.

Roscoe, K. H. and Burland, J. B. (1968). On the generalized stress-strain behaviour of 'wet' clay. In *Engineering Plasticity* (eds. J. Heyman and F. A. Leckie), pp. 535–609, Cambridge University Press.

Roscoe, K. H. and Poorooshasb, H. B. (1963). A theoretical and experimental study of strains in triaxial tests on normally consolidated clays. *Géotechnique* **13**:12–38.

Roscoe, K. H., Schofield, A. N. and Wroth, C. P. (1958). On the yielding of soils. *Géotechnique* **8**:22–53.

Roscoe, K. H., Schofield, A. N. and Thurairajah, A. (1963). Yielding of clays in states wetter than critical. *Géotechnique* **13**:211–240.

Rosendahl, B. R. (1987). Architecture of continental rifts with special reference to East Africa. *Annu. Rev. Earth Planet. Sci.* **15**:445–503.

Rotevatn, A., Tveranger, J., Howell, J. and Fossen, H. (2009). Dynamic investigation of the effect of a relay ramp on simulated fluid flow: geocellular modeling of the Delicate Arch Ramp, Utah. *Petrol. Geoscience* **15**:45–58.

Rowe, P. W. (1962). The stress-dilatancy relation for static equilibrium of an assembly of particles in contact. *Proc. Royal Soc. London A*, **269**:500–527.

Rowland, S. M. and Duebendorfer, E. M. (1994). *Structural Analysis and Synthesis: A Laboratory Course in Structural Geology* (2nd edn), Blackwell Science, Cambridge, Mass.

Royden, L. H. (1985). The Vienna Basin: a thin-skinned pull-apart basin. In *Strike-Slip Deformation, Basin Formation, and Sedimentation* (eds. K. T. Biddle and N. Christie-Blick), pp. 319–338, Soc. Econ. Paleon. Miner. Spec. Publ. 37.

Roznovsky, T. A. and Aydin, A. (2001). Concentration of shearing deformation related to changes in strike of monoclinal fold axes: the Waterpocket monocline, Utah. *J. Struct. Geol.* **23**:1567–1579.

Rubin, A. M. (1990). A comparison of rift-zone tectonics in Iceland and Hawaii. *Bull. Volcanol.* **52**:302–319.

Rubin, A. M. (1992). Dike-induced faulting and graben subsidence in volcanic rift zones. *J. Geophys. Res.* **97**:1839–1858.

Rubin, A. M. (1993a). Tensile fracture of rock at high confining pressure: implications for dike propagation. *J. Geophys. Res.* **98**:15,919–15,935.

Rubin, A.M. (1993b). Dikes vs. diapirs in viscoelastic rock. *Earth Planet. Sci. Lett.* **119**:641–569.

Rubin, A. M. (1995a). Getting granitic dikes out of the source region. *J. Geophys. Res.* **100**:5911–5929.

Rubin, A. M. (1995b). Propagation of magma-filled cracks. *Ann. Rev. Earth Planet. Sci.* **23**:287–336.

Rubin, A. M. (1998). Dike ascent in partially molten rock. *J. Geophys. Res.* **103**:20,901–20,919.

Rubin, A. M. and Pollard, D. D. (1987). Origins of blade-like dikes in volcanic rift zones. In *Volcanism in Hawaii* (eds. R. W. Decker, T. L. Wright and P. H. Stauffer), pp. 1449–1470, US Geol. Surv. Prof. Pap. 1350.

Rubin, A. M. and Pollard, D. D. (1988). Dike-induced faulting in rift zones of Iceland and Afar. *Geology* **16**:413–417.

Rudnicki, J. W. (1977). The inception of faulting in a rock mass with a weakened zone. *J. Geophys. Res.* **82**:844–854 (correction, *J. Geophys. Res.* **82**:3437).

Rudnicki, J. W. (1980). Fracture mechanics applied to the Earth's crust. *Ann. Rev. Earth Planet. Sci.* **8**:489–525.

Rudnicki, J. W. (2004). Shear and compaction band formation on an elliptic yield cap. *J. Geophys. Res.* **109**:B03402, doi:10.1029/2003JB002633.

Rudnicki, J. W. (2007). Models for compaction band propagation. In *Rock Physics and Geomechanics in the Study of Reservoirs and Repositories* (eds. C. David and M. Le Ravalec-Dupin), pp. 107–125, Geol. Soc. London Spec. Publ. 284.

Rudnicki, J. W. and Rice, J. R. (1975). Conditions for the localization of deformation in pressure-sensitive dilatant materials. *J. Mech. Phys. Solids* **23**:371–394.

Rudnicki, J. W. and Sternlof, K. R. (2005). Energy release model of compaction band propagation. *Geophys. Res. Lett.* **32**:L16303; doi 10.1029/2005GL023602.

Ruegg, J. C., Kasser, M., Tarantola, A., Lepine, L. C. and Chouikrat, B. (1982). Deformations associated with the El Asnam earthquake of October 1980: geodetic determination of vertical and horizontal movements. *Bull. Seismol. Soc. Am.* **72**:2227–2244.

Ruina, A. L. (1983). Slip instability and state variable friction laws. *J. Geophys. Res.* **88**:10,359–10,370.

Ruf, J. C., Rust, K. A. and Engelder, T. (1998). Investigating the effect of mechanical discontinuities on joint spacing. *Tectonophysics* **295**:245–257.

Rummel, F. (1987). Fracture mechanics approach to hydraulic fracturing stress measurements. In *Fracture Mechanics of Rock* (ed. B. K. Atkinson), pp. 217–240, Academic Press, New York.

Rutqvist, J. (2012). The geomechanics of CO_2 storage in deep sedimentary formations. *Geotech. Geol. Eng.* **30**:525–551.

Rutqvist, J., Tsang, C. F. and Stephansson, O. (2000). Uncertainty in the maximum principal stress estimated from hydraulic fracturing measurements due to the presence of the induced fracture. *Int. J. Rock Mech. Min. Sci.* **37**:107–120.

Rutter, E. H. (1983). Pressure solution in nature, theory and experiment. *J. Geol. Soc. London* **140**:725–740.

Rutter, E. H. (1986). On the nomenclature of mode of failure transitions in rocks. *Tectonophysics* **122**:381–387.

Rutter, E. H. and Brodie, K. H. (1991). Lithosphere rheology—a note of caution. *J. Struct. Geol.* **13**:363–367.

Rutter, E. H. and Glover, C. T. (2012). The deformation of porous sandstones; are Byerlee friction and the critical state line equivalent? *J. Struct. Geol.* **44**:129–140.

Rutter, E. H., Maddock, R. H., Hall, S. H. and White, S. H. (1986). Comparative microstructures of natural and experimentally produced clay-bearing fault gouges. *Pure Appl. Geophys.* **124**:3–30.

Ryan, M. P. and Sammis, C. G. (1978). Cyclic fracture mechanisms in cooling basalt. *Geol. Soc. Am. Bull.* **89**:1295–1308.

Rybacki, E., Meier, T. and Dresen G. (2016). What controls mechanical properties of rocks? —part II: brittleness. *J. Petrol. Sci. Eng.* **144**:39–58.

Saada, A. S., Liang, L., Figueroa, J. L. and Cope, C. T. (1999). Bifurcation and shear band propagation in sands. *Géotechnique* **49**:367–385.

Saade, A., Abou-Jaoude, G. and Wartman, J. (2016). Regional-scale co-seismic landslide assessment using limit equilibrium analysis. *Eng. Geol.* **204**:53–64.

Safaricz, M. and Davison, I. (2005). Pressure solution in chalk. *Amer. Assoc. Petrol. Geol. Bull.* **89**:383–401.

Saffer, D. M., Frye, K. M., Marone, C. and Mair, K. (2001). Laboratory results indicating complex and potentially unstable frictional behavior of smectite clay. *Geophys. Res. Lett.* **28**:2297–2300.

Sagy, A., Reches, Z. and Roman, I. (2001). Dynamic fracturing: field and experimental observations. *J. Struct. Geol.* **23**:1223–1239.

Saillet, E. and Wibberley, C. A. J. (2013). Permeability and flow impacts of faults and deformation bands in high-porosity sand reservoirs: Southeast Basin, France, analog. *Am. Assoc. Petrol. Geol. Bull.* **97**:437–464.

Sammis, C., King, G. and Biegel, R. (1987). The kinematics of gouge deformation. *Pure Appl. Geophys.* **125**:777–812.

Sammonds, P. R., Meredith, P. G. and Main, I. G. (1992). Role of pore fluids in the generation of seismic precursors to shear fracture. *Nature* **359**:228–230.

Sample, J. C., Woods, S., Bender, E. and Loveall, M. (2006). Relationship between deformation bands and petroleum migration in an exhumed reservoir rock, Los Angeles Basin, California, USA. *Geofluids* **6**:105–112.

Sanderson, D. J. and Marchini, W. R. D. (1984). Transpression. *J. Struct. Geol.* **6**:449–458.

Sandwell, D. (1986). Thermal stress and the spacings of transform faults. *J. Geophys. Res.* **91**:6405–6417.

Sauber, J., Thatcher, W. and Solomon, S. (1986). Geodetic measurement of deformation in the central Mojave Desert, California. *J. Geophys. Res.* **91**:12,683–12,693.

Savage, J. C., and Hastie, L. M. (1966). Surface deformation associated with dip-slip faulting. *J. Geophys. Res.* **71**:4897–4904.

Savage, J. C., and Burford, R. O. (1973). Geodetic determination of relative plate motion in central California. *J. Geophys. Res.* **78**:832–845.

Savage, J. C., Byerlee, J. D. and Lockner, D. A. (1996). Is internal friction friction? *Geophys. Res. Lett.* **23**:487–490.

Savalli, L. and Engelder, T. (2005). Mechanisms controlling rupture shape during subcritical growth of joints in layered rock. *Geol. Soc. Am. Bull.* **117**:436–449.

Schardin, E. B. (1959). Velocity effects in fracture. In *Proceedings, International Conference on Atomic Mechanisms of Fracture* (eds. B. L. Averback, D. K. Felbeck, G. T. Hahn and D. A. Thomas), Wiley, New York.

Schenk, P. M. and McKinnon, W. B. (1989). Fault offsets and lateral crustal movement on Europa: evidence for a mobile ice shell. *Icarus* **79**:75–100.

Schlische, R. W. (1995). Geometry and origin of fault-related folds in extensional settings. *Am. Assoc. Petrol. Geol. Bull.* **79**:1661–1678.

Schlische, R. W., Young, S. S., Ackermann, R. V. and Gupta, A. (1996). Geometry and scaling relations of a population of very small rift-related normal faults. *Geology* **24**:683–686.

Schmidt, R. A. (1980). A microcrack model and its significance to hydraulic fracturing and fracture toughness testing. Paper ARMA 80–581 presented at the *21st US Symposium on Rock Mechanics*, 27–30 May 1980, Rolla, Missouri.

Schmidt, R. A. and Huddle, C. W. (1977). Effect of confining pressure on fracture toughness of Indiana limestone. *Int. J. Rock Mech. Min. Sci. Geomech. Abstr.* **14**:289–293.

Schmitt, D. R., Currie, C. A. and Zhang, L. (2012). Crustal stress determination from boreholes and rock cores: fundamental principles. *Tectonophysics* **580**:1–26 (corrigendum, *Tectonophysics* **586**:206–207, 2013).

Schofield, A. (2005). *Disturbed Soil Properties and Geotechnical Design*, Thomas Telford, London.

Schofield, A. and Wroth, P. (1968). *Critical State Soil Mechanics*, McGraw-Hill, New York, 310 pp.

Scholz, C. H. (1968a). Microcracking and the inelastic deformation of rock in compression. *J. Geophys. Res.* **73**:1417–1432.

Scholz, C. H. (1968b). Experimental study of the fracturing process in brittle rock. *J. Geophys. Res.* **73**:1447–1454.

Scholz, C. H. (1982). Scaling laws for large earthquakes: consequences for physical models. *Bull. Seismol. Soc. Amer.* **72**:1–14.

Scholz, C. H. (1990). *The Mechanics of Earthquakes and Faulting*, Cambridge University Press, 439 pp.

Scholz, C. H. (1997). Earthquake and fault populations and the calculation of brittle strain. *Geowissenschaften* **15**:124–130.

Scholz, C. H. (1998). Earthquakes and friction laws. *Nature* **391**:37–42.

Scholz, C. H. (2002). *The Mechanics of Earthquakes and Faulting* (2nd edn), Cambridge University Press, 471 pp.

Scholz, C. H. (2003). Good tidings. *Nature* **425**:670–671.

Scholz, C. H. and Cowie, P. A. (1990). Determination of total strain from faulting using slip measurements. *Nature* **346**:837–838.

Scholz, C. H. and Lawler, T. M. (2004). Slip tapers at the tips of faults and earthquake ruptures. *Geophys. Res. Lett.* **31**:L21609; doi:10.1029/2004GL021030.

Scholz, C. H., Dawers, N. H., Yu, J.-Z. and Anders, M. H. (1993). Fault growth and fault scaling laws: preliminary results. *J. Geophys. Res.* **98**:21,951–21,961.

Schöpfer, M. P. J. and Childs, C. (2013). The orientation and dilatancy of shear bands in a bonded particle model for rock. *Int. J. Rock Mech. Min. Sci.* **57**:75–88.

Schöpfer, M. P. J., Childs, C. and Walsh, J. J. (2006). Localisation of normal faults in multilayer sequences. *J. Struct. Geol.* **28**:816–823.

Schöpfer, M. P. J., Arslan, A., Walsh, J. J. and Childs, C. (2011). Reconciliation of contrasting theories for fracture spacing in layered rocks. *J. Struct. Geol.* **33**:551–565.

Schreurs, G. and Colletta, B. (1998). Analogue modelling of faulting in zones of continental transpression and transtension. In *Continental Transpressional and Transtensional Tectonics* (eds. R. E. Holdsworth, R. E. Strachan and J. F. Dewey), pp. 59–79, Geol. Soc. London Spec. Publ. 135.

Schreurs, G., Buiter, S .J. H., Boutelier, D., et al. (2006). Analogue benchmarks of shortening and extension experiments. In *Analogue and Numerical Modeling of Crustal-Scale Processes* (eds. S. J. H. Buiter and G. Schreurs), pp. 1–27, Geol. Soc. London Spec. Publ. 253.

Schueller, S., Braathen, A., Fossen, H. and Tveranger, J. (2013). Spatial distribution of deformation bands in damage zones of extensional faults in porous sandstones: statistical analysis of field data. *J. Struct. Geol.* **52**:148–162.

Schultz, R. A. (1989). Strike-slip faulting of ridged plains near Valles Marineris, Mars. *Nature* **341**:424–426.

Schultz, R. A. (1992). Mechanics of curved slip surfaces in rock. *Eng. Anal. Bound. Elements* **10**:147–154.

Schultz, R. A. (1993). Brittle strength of basaltic rock masses with applications to Venus. *J. Geophys. Res.* **98**:10,883–10,895.

Schultz, R. A. (1995a). Limits on strength and deformation properties of jointed basaltic rock masses. *Rock Mech. Rock Eng.* **28**:1–15.

Schultz, R. A. (1995b). Gradients in extension and strain at Valles Marineris, Mars. *Planet. Space Sci.* **43**:1561–1566.

Schultz, R. A. (1996). Relative scale and the strength and deformability of rock masses. *J. Struct. Geol.* **18**:1139–1149.

Schultz, R. A. (1997). Displacement-length scaling for terrestrial and Martian faults: implications for Valles Marineris and shallow planetary grabens. *J. Geophys. Res.* **102**:12,009–12,015.

Schultz, R. A. (1998). Integrating rock mechanics into traditional geoscience curricula. *J. Geosci. Ed.* **46**:141–145.

Schultz, R. A. (1999). Understanding the process of faulting: selected challenges and opportunities at the edge of the 21st century. *J. Struct. Geol.* **21**:985–993.

Schultz, R. A. (2000a). Fault-population statistics at the Valles Marineris Extensional Province, Mars: implications for segment linkage, crustal strains, and its geodynamical development. *Tectonophysics* **316**:169–193.

Schultz, R. A. (2000b). Growth of geologic fractures into large-strain populations: nomenclature, subcritical crack growth, and implications for rock engineering. *Int. J. Rock Mech. Min. Sci.* **37**:403–411.

Schultz, R. A. (2000c). Localization of bedding plane slip and backthrust faults above blind thrust faults: keys to wrinkle ridge structure. *J. Geophys. Res.* **105**:12,035–12,052.

Schultz, R. A. (2002). Stability of rock slopes in Valles Marineris, Mars. *Geophys. Res. Lett.* **30**:1932; doi:10.1029/2002GL015728.

Schultz, R. A. (2003a). Seismotectonics of the Amenthes Rupes thrust fault population, Mars. *Geophys. Res. Lett.* **30**:1303, 10.1029/2002GL016475.

Schultz, R. A. (2003b). A method to relate initial elastic stress to fault population strains. *Geophys. Res. Lett.* **30**:1593, 10.1029/2002GL016681.

Schultz, R. A. (2009). Scaling and paleodepth of compaction bands, Nevada and Utah. *J. Geophys. Res.* **114**:B03407; doi:10.1029/2008JB005876.

Schultz, R. A. (2011a). Relationship of compaction bands in Utah to Laramide fault-related folding. *Earth Planet. Sci. Lett.* **304**, 29–35.

Schultz, R. A. (2011b). The influence of near-surface stratigraphy on the growth and scaling of normal faults in Bishop Tuff, California, with planetary implications (abstr.), in Lunar and Planetary Science XLII, #1471.

Schultz, R. A. and Aydin, A. (1990). Formation of interior basins associated with curved faults in Alaska. *Tectonics* **9**:1387–1407.

Schultz, R. A. and Balasko, C. M. (2003). Growth of deformation bands into echelon and ladder geometries. *Geophys. Res. Lett.* **30**:2033; doi:10.1029/2003GL018449.

Schultz, R. A. and Fori, A. N. (1996). Fault-length statistics and implications of graben sets at Candor Mensa, Mars. *J. Struct. Geol.* **18**:373–383.

Schultz, R. A. and Fossen, H. (2002). Displacement-length scaling in three dimensions: the importance of aspect ratio and application to deformation bands. *J. Struct. Geol.* **24**:1389–1411.

Schultz, R. A. and Fossen, H. (2008). Terminology for structural discontinuities. *Am. Assoc. Petrol. Geol. Bull.* **92**:853–867.

Schultz, R. A. and Li, Q. (1995). Uniaxial strength testing of non-welded Calico Hills tuff, Yucca Mountain, Nevada. *Eng. Geol.* **40**:287–299.

Schultz, R. A. and Moore, J. M. (1996). New observations of grabens from the Needles District, Canyonlands National Park, Utah. In *Geology and Resources of the Paradox Basin* (eds. A. C. Huffman, Jr., W. R. Lund and L. H. Godwin), pp. 295–302, Utah Geological Association and Four Corners Geological Society, Salt Lake City.

Schultz, R. A. and Siddharthan, R. (2005). A general framework for the occurrence and faulting of deformation bands in porous granular rocks. *Tectonophysics* **411**:1–18.

Schultz, R. A. and Soliva, R. (2012). Propagation energies inferred from deformation bands in sandstone. *Int. J. Fracture* **176**:135–149.

Schultz, R. A. and Watters, T. R. (1995). Elastic buckling of fractured basalt on the Columbia Plateau, Washington State. In *Rock Mechanics: Proceedings of the 35th US Symposium* (eds. J. J. K. Daemen and R. A. Schultz), pp. 855–860, Balkema, Rotterdam.

Schultz, R. A. and Watters, T. R. (2001). Forward mechanical modeling of the Amenthes Rupes thrust fault on Mars. *Geophys. Res. Lett.* 28:4659–4662.

Schultz, R. A. and Zuber, M. T. (1994). Observations, models, and mechanisms of failure of surface rocks surrounding planetary surface loads. *J. Geophys. Res.* **99**:14,691–14,702.

Schultz, R. A., Ward, K. A. and Hibbard, M. J. (1993). Dike emplacement in the Donner Summit pluton, northern Sierra Nevada, California. In *Crustal Evolution of the Great Basin and Sierra Nevada* (eds. M. M. Lahren, J. H. Trexler, Jr. and C. Spinosa), pp. 277–284, Geol. Soc. Am. Guidebook, Cordilleran/Rocky Mountains section, Reno, Nevada.

Schultz, R. A., Jensen, M. and Bradt, R. C. (1994). Single-crystal cleavage of brittle materials. *Int. J. Fracture* **65**:291–312.

Schultz, R. A., Okubo, C. H., Goudy, C. L. and Wilkins, S. J. (2004). Igneous dikes on Mars revealed by MOLA topography. *Geology* **32**:889–892.

Schultz, R. A., Okubo, C. H. and Wilkins, S. J. (2006). Displacement-length scaling relations for faults on the terrestrial planets. *J. Struct. Geol.* **28**:2182–2193.

Schultz, R. A., Soliva, R., Fossen, H., Okubo, C. H. and Reeves, D. M. (2008a). Dependence of displacement–length scaling relations for fractures and deformation bands on the volumetric changes across them. *J. Struct. Geol.* **30**:1405–1411.

Schultz, R. A., Mège, D. and Diot, H. (2008b). Emplacement conditions of igneous dikes in Ethiopian Traps. *J. Volcanol. Geotherm. Res.* **178**:673–692.

Schultz, R. A., Hauber, E., Kattenhorn, S., Okubo, C. H. and Watters, T. R. (2010a). Interpretation and analysis of planetary structures. *J. Struct. Geol.* **32**:855–875.

Schultz, R. A., Okubo, C. H. and Fossen, H. (2010b). Porosity and grain size controls on compaction band formation in Jurassic Navajo Sandstone. *Geophys. Res. Lett.* **37**:L22306; doi:2010GL044909.

Schultz, R. A., Soliva, R., Okubo, C. H. and Mège, D. (2010c). Fault populations. In *Planetary Tectonics* (eds. T. R. Watters and R. A. Schultz), pp. 457–510, Cambridge University Press.

Schultz, R. A., Klimczak, C., Fossen, H., et al. (2013). Statistical tests of scaling relationships for geologic structures. *J. Struct. Geol.* **48**:85–94.

Schultz, R. A., Tong, X., Soofi, K. A., Sandwell, D. A. and Hennings, P. H. (2014). Using InSAR to detect active deformation associated with faults in Suban field, South Sumatra Basin, Indonesia. *The Leading Edge* 33:882–888.

Schultz-Ela, D. D. and Walsh, P. (2002). Modeling of grabens extending above evaporites in Canyonlands National Park, Utah. *J. Struct. Geol.* **24**:247–275.

Searle, R. and Laughton, A. (1981). Fine-scale sonar study of tectonics and volcanism on the Reykjanes Ridge. *Oceanol. Acta* **4**(suppl.):5–13.

Secor, D. T. (1965). Role of fluid pressure in jointing. *Am. J. Sci.* **263**:633–646.

Secor, D. T. and Pollard, D. D. (1975). On the stability of open hydraulic fractures in the Earth's crust. *Geophys. Res. Lett.* **2**:510–513.

Segall, P. (1984a). Formation and growth of extensional fracture sets. *Geol. Soc. Am. Bull.* **95**:454–462.

Segall, P. (1984b). Rate-dependent extensional deformation resulting from crack growth in rock. *J. Geophys. Res.* **89**:4185–4195.

Segall, P. (1989). Earthquakes triggered by fluid extraction. *Geology* **17**: 942–946.

Segall, P. (2010). *Earthquake and Volcano Deformation*, Princeton University Press, 432 pp.

Segall, P. and Fitzgerald, S. D. (1998). A note on induced stress changes in hydrocarbon and geothermal reservoirs. *Tectonophysics* **289**:117–128.

Segall, P. and Pollard, D. D. (1980). Mechanics of discontinuous faults. *J. Geophys. Res.* **85**:4337–4350.

Segall, P. and Pollard, D. D. (1983a). Joint formation in granitic rock of the Sierra Nevada. *Geol. Soc. Am. Bull.* **94**:563–575.

Segall, P., and Pollard, D. D. (1983b). Nucleation and growth of strike-slip faults in granite. *J. Geophys. Res.* **88**:555–568.

Segall, P. and Pollard, D. D. (1983c). From joints and faults to photolineaments. *Proc. Intern. Conf. Basement Tectonics* **4**: 11–20.

Segall, P. and Simpson, C. (1986). Nucleation of ductile shear zones on dilatant fractures. *Geology* **14**:56–59.

Segall, P., Grasso, J. R., and Mossop, A. (1994). Poroelastic stressing and induced seismicity near the Laq gas field, southwestern France. *J. Geophys. Res.* **99**:15,423–15,438.

Sempere, J.-C. and Macdonald, K. C. (1986). Overlapping spreading centers: implications from crack growth simulation by the displacement discontinuity method. *Tectonics* **5**:151–163.

Serafim, J. L. and Pereira, J. P. (1983). Considerations on the geomechanical classification of Bieniawski. Proc. Int. Symp. on Eng. Geol. and Underground Construction, vol. I, Lisbon, Portugal, pp. 33–44.

Shah, R. C. and Kobayashi, A. S. (1973). Stress intensity factors for an elliptical crack approaching the surface of a semi-infinite solid. *Int. J. Fracture* **9**:133–146.

Shand, E. B. (1959). Breaking strength of glass determined from dimensions of fracture mirrors. *J. Am. Ceramic Soc.* **42**:474–477.

Sharon, E., Gross, S. P. and Fineberg, J. (1995). Local crack branching as a mechanism for instability in dynamic fracture. *Phys. Rev. Lett.* **74**, 5096–5099.

Sharp, R. P. (1968). Sherwin till-Bishop tuff geological relationships, Sierra Nevada, California. *Geol. Soc. Amer. Bull.* **79**, 351–364.

Shaw, J. H., Bilotti, F. and Brennan, P. A. (1999). Patterns of imbricate thrusting. *Geol. Soc. Am. Bull.* **111**:1140–1154.

Shaw, J. H., Plesch, A., Dolan, J. F., Pratt, T. L. and Fiore, P. (2002). Puente Hills blind-thrust system, Los Angeles, California. *Seismol. Soc. Am. Bull.* **92**:2946–2960.

Shaw, P. R. and Lin, J. (1993). Causes and consequences of variations in faulting style at the Mid-Atlantic Ridge. *J. Geophys. Res.* **98**:21,839–21,851.

Sheldon, H. A., Barnicoat, A. C. and Ord, A. (2006). Numerical modelling of faulting and fluid flow in porous rocks: an approach based on critical state soil mechanics. *J. Struct. Geol.* **28**:1468–1482.

Sheriff, R. E. and Geldart, L. P. (1995). *Exploration Seismology*, Cambridge University Press, New York.

Shimazaki, K. (1986). Small and large earthquakes: the effects of the thickness of seismogenic layer and the free surface. *Earthquake Source Mech.* **37**:209–216.

Shin, H., Santamarina, J. C. and Cartwright, J. A. (2010). Displacement field in contraction-driven faults. *J. Geophys. Res.* **115**, B07408; doi:10.1029/2009JB006572.

Shipton, Z. K. and Cowie, P. A. (2001). Damage zone and slip-surface evolution over μm to km scales in high-porosity Navajo Sandstone, Utah. *J. Struct. Geol.* **23**:1825–1844.

Shipton, Z. K. and Cowie, P. A. (2003). A conceptual model for the origin of fault damage zone structures in high-porosity sandstone. *J. Struct. Geol.* **25**: 333–344 (erratum, *J. Struct. Geol.* **25**:1343–1345).

Shipton, Z. K., Evans, J. P., Robeson, K., Forster, C. B. and Snelgrove, S. (2002). Structural heterogeneity and permeability in faulted eolian sandstone: implications for subsurface modelling of faults. *Am. Assoc. Petrol. Geol. Bull.* **86**:863–883.

Shipton, Z. K., Evans, J. P. and Thompson, L. B. (2005). The geometry and thickness of deformation-band fault core and its influence on sealing characteristics of deformation-band fault zones. In *Faults, Fluid Flow, and Petroleum Traps* (eds. R. Sorkhabi and Y. Tsuji), pp. 181–195, Am. Assoc. Petrol. Geol. Mem. 85.

Sibson, R. H. (1974). Frictional constraints on thrust, wrench and normal faults. *Nature* **249**:542–544.

Sibson, R. H. (1984). Roughness at the base of the seismogenic zone: contributing factors. *J. Geophys. Res.* **89**:5791–5799.

Sibson, R. H. (1985). A note on fault reactivation. *J. Struct. Geol.* **7**:751–754.

Sibson, R. H. (1986a). Earthquakes and rock deformation in crustal fault zones. *Annu. Rev. Earth Planet. Sci.* **14**:149–175.

Sibson, R. H. (1986b). Brecciation processes in fault zones: inferences from earthquake rupturing. *Pure Appl. Geophys.* **124**:159–175.

Sibson, R. H. (1986c). Rupture interaction with fault jogs. In *Earthquake Source Mechanics* (eds. S. Das, J. Boatwright and C.H. Scholz), pp. 157–167, Am. Geophys. Union. Geophys. Monog. 37.

Sibson, R. H. (1989). Earthquake faulting as a structural process. *J. Struct. Geol.* **11**:1–14.

Sibson, R. H. (1994). An assessment of field evidence for 'Byerlee' friction. *Pure Appl. Geophys.* **142**:645–662.

Sibson, R. H. (1998). Brittle failure mode plots for compressional and extensional tectonic regimes. *J. Struct. Geol.* **20**:655–660.

Siddans, A. (1972). Slaty cleavage—a review of research since 1815. *Earth Sci. Rev.* **8**:205–232.

Sih, G. C. (1973). Some basic problems in fracture mechanics and new concepts. *J. Eng. Frac. Mech.* **5**:365–377.

Sih, G. C. (1974). Strain energy density factor applied to mixed mode crack problems. *Int. J. Fracture* **10**:305–322.

Sih, G. C. (1991). *Mechanics of Fracture Initiation and Propagation—Surface and Volume Energy Density Applied as Failure Criterion,* Kluwer Academic, Boston.

Sih, G. C., Paris, P. C. and Erdogan, F. (1962). Crack-tip, stress-intensity factors for plane extension and plate bending problems. *J. Appl. Mech.* **29**:306–312.

Sih, G. C. and Liebowitz, H. (1968). Mathematical theories of brittle fracture. In *Fracture: An Advanced Treatise* (ed. H. Liebowitz), vol. II, pp. 67–190, Academic, New York.

Sih, G. C. and Macdonald, B. (1974). Fracture mechanics applied to engineering problems: strain energy density fracture criterion. *Eng. Fract. Mech.* **6**:361–386.

Simony, P. S. and Carr, S. D. (1998). Large lateral ramps in the Eocene Valkyr shear zone: extensional ductile faulting controlled by plutonism in southern British Columbia: Reply. *J. Struct. Geol.* **20**:489–490.

Simpson, R. W. (1997). Quantifying Anderson's fault types. *J. Geophys. Res.* **102**:17,909–17,919.

Singh, R. N. and Sun, G. (1990). Applications of fracture mechanics to some mining engineering problems. *Mining Sci. Technol.* **10**:53–60.

Skempton, A. W. (1954). The pore-pressure coefficients *A* and *B*. *Géotechnique* **4**:143–147.

Skempton, A. W. (1966). Some observations on tectonic shear zones. Paper presented at the 1st ISRM Congress, 25 September–1 October 1966, Lisbon, Portugal.

Skurtveit, E., Torabi, A., Gabrielsen, R. H. and Zoback, M. D. (2013). Experimental investigation of deformation mechanisms during shear-enhanced compaction in poorly lithified sandstone and sand. *J. Geophys. Res.* **118**:4083–4100; doi:10.1002/jgrb.50342.

Smith, D. J., Ayatollahi, M. R. and Pavier, M. J. (2001). The role of T-stress in brittle fracture for linear elastic materials under mixed-mode loading. *Fatigue Fracture Eng. Mater. Struct.* **24**:137–150.

Smith, G. A. (1983). Porosity dependence of deformation bands in the Entrada Sandstone, La Plata County, Colorado. *Mountain Geologist* **20**:82–85.

Sneddon, I. N. (1946). The distribution of stress in the neighborhood of a crack in an elastic solid. *Proc. Royal Soc. London A* **187**:229–260.

Soliva, R. and Benedicto, A. (2004). A linkage criterion for segmented normal faults. *J. Struct. Geol.* **26**:2251–2267.

Soliva, R. and Benedicto, A. (2005). Geometry, scaling relations and spacing of vertically restricted normal faults. *J. Struct. Geol.* **27**:317–325.

Soliva, R. and Schultz, R. A. (2008). Distributed and localized faulting in extensional settings: insight from the North Ethiopian Rift – Afar transition area. *Tectonics* **27**:TC2003; doi:10.1029/2007TC002148.

Soliva, R., Schultz, R. A. and Benedicto, A. (2005). Three-dimensional displacement-length scaling and maximum dimension of normal faults in layered rocks. *Geophys. Res. Lett.* **32**:L16302, 10.1029/2005GL023007.

Soliva, R., Benedicto, A. and Maerten, L. (2006). Spacing and linkage of confined normal faults: importance of mechanical thickness. *J. Geophys. Res.* **111**:B01402, 10.1029/2004JB003507.

Soliva, R., Benedicto, A., Schultz, R. A., Maerten, L. and Micarelli, L. (2008). Displacement and interaction of normal fault segments branched at depth: implications for fault growth and potential earthquake rupture size. *J. Struct. Geol.* **30**:1288–1299.

Soliva, R., Maerten, F., Petit, J.-P. and Auzias, V. (2010). Field evidences for the role of static friction on fracture orientation in extensional relays along strike-slip faults: comparison with photoelasticity and 3-D numerical modeling. *J. Struct. Geol.* **32**:1721–1731.

Soliva, R., Schultz, R. A., Ballas, G., et al. (2013). A model of strain localization in porous sandstone as a function of tectonic setting, burial and material properties; new insight from Provence (southern France). *J. Struct. Geol.* **49**:50–63.

Soliva, R., Ballas, G., Fossen, H. and Philit, S. (2016). Tectonic regime controls clustering of deformation bands in porous sandstone. *Geology* **44**:423–426.

Solomon, S., Huang, P. and Meinke, L. (1988). The seismic moment budget of slowly spreading ridges. *Nature* **334**:58–60.

Solomon, S. C., McNutt, R. L., Jr., Watters, T. R., et al. (2008). Return to Mercury: a global perspective on MESSENGER's first Mercury flyby. *Science* **321**:59–62.

Solum, J. G., Brandenburg, J. P., Naruk, S. J., et al. (2010). Characterization of deformation bands associated with normal and reverse stress states in Navajo Sandstone, Utah. *Amer. Assoc. Petrol. Geol. Bull.* **94**:1453–1474.

Sommer, E. (1969). Formation of fracture lances in glass. *Eng. Fracture Mech.* **1**:539–546.

Sone, H. and Zoback, M. D. (2014). Time-dependent deformation of shale gas reservoir rocks and its long-term effect on the in situ state of stress. *Int. J. Rock Mech. Min. Sci.* **69**:120–132.

Sonmez, H., Ulusay, R. and Gokceoglu, C. (1998). A practical procedure for the back analysis of slope failures in closely jointed rock masses. *Int. J. Rock Mech. Min. Sci.* **35**:219–233.

Sonmez, H., Gokceoglu, C., Nefeslioglu, H. A., and Kayabasi, A. (2006). Estimation of rock modulus: for intact rocks with an artificial neural network and for rock masses with a new empirical equation. *Int. J. Rock Mech. Min. Sci.* **43**:224–235.

Stanchits, S., Fortin, J., Gueguen, Y. and Dresen, G. (2009). Initiation and propagation of compaction bands in dry and wet Bentheim Sandstone. *Pure Appl. Geophys.* **166**:843–868.

Steefel, C. I., DePaolo, D. J. and Lichtner, P. C. (2005). Reactive transport modeling: an essential tool and a new research approach for the Earth sciences. *Earth Planet. Sci. Lett.* **240**:539–558.

Stefanov, Y. P. and Bakeev, R. A. (2014). Deformation and fracture structures in strike-slip faulting. *Eng. Fract. Mech.* **129**:102–111.

Stefanov, Y. P., Bakeev, R. A., Rebetsky, Y. L. and Kontorovich, V. A. (2014). Structure and formation stages of a fault zone in a geomedium layer in strike-slip displacement of the basement. *Physical Mesomechanics* **17**:1–12; translated from original Russian text in *Fizicheskaya Mezomekhanika* **16**:41–52, 2013.

Stein, R. S. (1999). The role of stress transfer in earthquake occurrence. *Nature* **402**:605–609.

Stein, R. S. and King, G. C. P. (1984). Seismic potential revealed by folding: 1983 Coalinga, California, earthquake. *Science* **224**:869–872.

Stein, R. S., King, G. C. P. and Lin, J. (1992). Change in failure stress on the southern San Andreas fault system caused by the magnitude = 7.4 Landers earthquake. *Science* **258**:1328–1332.

Stein, R. S., King, G. C. P. and Lin, J. (1994). Stress triggering of the 1994 M = 6.7 Northridge, California, earthquake by its predecessors. *Science* **165**:1432–1435.

Sternlof, K. R., Rudnicki, J. W. and Pollard, D. D. (2005). Anticrack inclusion model for compaction bands in sandstone. *J. Geophys. Res.* **110**:B11403; doi:10.1029/2005JB003764.

Sternlof, K. R., Karimi-Fard, M., Pollard, D. D. and Durlofsky, L. J. (2006). Flow and transport effects of compaction bands in sandstone at scales relevant to aquifer and reservoir management. *Water Resour. Res.* **42**:W07425; doi:10.1029/2005WR004664.

Stesky, R. M., Brace, W. F., Riley, D. K. and Robin, P.-Y. F. (1975). Friction in faulted rock at high temperature and pressure. *Tectonophysics* **23**:177–203.

Stevens, B. (1911). The laws of intrusion. *Bull. Am. Inst. Min. Eng.* **49**:1–23.

Stewart, S. A., Harvey, M. J., Otto, S. C. and Weston, P. J. (1996). Influence of salt on fault geometry: examples from the UK salt basins. In *Salt Tectonics* (eds. G. I. Alsop, D. J. Blundell and I. Davison), pp. 175–202, Geol. Soc. Spec. Publ. 100.

Stirling, M. W., Wesnousky, S. G. and Shimazaki, K. (1996). Fault trace complexity, cumulative slip, and the shape of the magnitude-frequency distribution for strike-slip faults: a global survey. *Geophys. J. Int.* **124**:833–868.

Stockdale, P. B. (1922). Stylolites: their nature and origin. *Indiana Univ. Stud.* **9**:1–97.

Stone, D. S. (1999). Foreland basement-involved structures: discussion. *Am. Assoc. Petrol. Geol. Bull.* **83**:2006–2016.

Storti, F., Holdsworth, R. E. and Salvini, F. (2003). Intraplate strike-slip deformation belts. In *Intraplate Strike-Slip Deformation Belts* (eds. F. Storti, R. E. Holdsworth and F. Salvini), pp. 1–14, Geol. Soc. London Spec. Publ. 210.

Strokova, L. (2013). Effect of the overconsolidation ratio of soils in surface settlements due to tunneling. *Sciences in Cold and Actic Regions* **5**:637–643.

Strom, R. G., Trask, N. J. and Guest, J. E. (1975). Tectonism and volcanism on Mercury. *J. Geophys. Res.* **80**:2478–2507.

Strom, R. G., Malhotra, R., Xiao, Z., et al. (2015). The inner solar system cratering record and the evolution of impactor populations. *Res. Astron. Astrophys.* **15**:407.

Sumy, D. F., Cochran, E. S., Keranen, K. M., Wei, M. and Abers, G. A. (2014). Observations of static Coulomb stress triggering of the November 2011 M5.7 Oklahoma earthquake sequence. *J. Geophys. Res.* **119**; doi:10.1002/2013JB010612.

Suppe, J. (1983). Geometry and kinematics of fault-bend folding. *Am. J. Sci.* **283**:648–721.

Suppe, J. (1985). *Principles of Structural Geology*, Prentice-Hall, Englewood Cliffs, New Jersey.

Suppe, J. (2007). Absolute fault and crustal strength from wedge tapers. *Geology* **35**:1127–1130.

Suppe, J. and Connors, C. (1992). Critical taper wedge mechanics of fold-and-thrust belts on Venus: initial results from Magellan. *J. Geophys. Res.* **97**:13,545–13,561.

Suppe, J. and Medwedeff, D. (1990). Geometry and kinematics of fault propagation folding. *Eclogae Geol. Helv.* **83**:409–454.

Swain, M. V. and Hagan, J. T. (1978). Some observations of overlapping interacting cracks. *Eng. Fracture Mech.* **10**:299–304.

Swain, M. V., Lawn, B. R. and Burns, S. J. (1974). Cleavage step deformation. *J. Mat. Sci.* **9**:175–183.

Swanson, P. L. (1984). Subcritical crack growth and other time- and environment-dependent behavior in crustal rocks. *J. Geophys. Res.* **89**:4137–4152.

Swanson, P. L. (1987). Tensile fracture resistance mechanisms in brittle polycrystals: an ultrasonics and in situ microscopy investigation. *J. Geophys. Res.* **92**:8015–8036.

Sykes, L. R. (1967). Mechanism of earthquakes and nature of faulting on the Mid-Atlantic Ridge. *J. Geophys. Res.* **72**:2131–2153.

Sylvester, A. G. (1988). Strike-slip faults. *Geol. Soc. Am. Bull.* **100**:1666–1703.

Sylvester, A. G. and Smith, R. R. (1976). Tectonic transpression and basement-controlled deformation in San Andreas fault zone, Salton trough, California. *Am. Assoc. Petrol. Geol. Bull.* **60**:2081–2102.

Taboada, A., Bousquet, J. C. and Philip, H. (1993). Coseismic elastic models of folds above blind thrusts in the Betic Cordilleras (Spain) and evaluation of seismic hazard. *Tectonophysics* **220**:223–241.

Tada, H., Paris, P. C. and Irwin, G. R. (2000). *The Stress Analysis of Cracks Handbook* (3rd edn), Am Soc. Mat. Eng., New York.

Tanaka, K. L., Anderson, R., Dohm, J. M., et al. (2010). Planetary structural mapping. In *Planetary Tectonics* (eds. T. R. Watters and R. A. Schultz), pp. 351–396, Cambridge University Press.

Tapp, B. and Cook, J. (1988). Pressure solution zone propagation in naturally deformed carbonate rocks. *Geology* **16**:182–185.

Tapponier, P. and Brace, W. F. (1976). Development of stress-induced microcracks in Westerly granite. *Int. J. Rock Mech. Min. Sci. Geomech. Abstr.* **13**:103–112.

Tavani, S. and Storti, F. (2011). Layer-parallel shortening templates associated with double-edge fault-propagation folding. In *Thrust Fault-Related Folding* (eds. K. R. McClay, J. H. Shaw and J. Suppe), pp. 121–135, Amer. Assoc. Petrol. Geol. Mem. 94.

Taylor, B., Goodliffe, A. and Martinez, F. (2009). Initiation of transform faults at rifted continental margins. *C. R. Geosci.* **341**:428–438.

Taylor, W. L., Pollard, D. D. and Aydin, A. (1999). Fluid flow in discrete joint sets: field observations and numerical simulations. *J. Geophys. Res.* **104**:28,983–29,006.

Tchalenko, J. S. (1970). Similarities between shear zones of different magnitudes. *Geol. Soc. Am. Bull.* **81**:1625–1640.

Tchalenko, J. S. and Ambrasays, N. N. (1970). Structural analyses of the Dasht-e Bayaz (Iran) earthquake fractures. *Geol. Soc. Am. Bull.* **81**:41–60.

Tembe, S., Vajdova, V., Wong, T.-f. and Zhu, W. (2006). Initiation and propagation of strain localization in circumferentially notched samples of two porous sandstones. *J. Geophys. Res.* **111**:B02409; doi 10.1029/2005JB003611.

Tembe, S., Baud, P. and Wong, T.-f. (2008). Stress conditions for the propagation of discrete compaction bands in porous sandstone. *J. Geophys. Res.* **113**:B09409; doi:10.1029/2007JB005439.

ten Brink, U. S., Katzman, R. and Lin, J. (1996). Three-dimensional models of deformation near strike-slip faults. *J. Geophys. Res.* **101**:16,205–16,220.

Terzaghi, K. (1946). Rock defects and loads on tunnel support. In *Rock Tunneling with Steel Supports* (eds. R. V. Proctor and T. White), pp. 15–99, Commercial Shearing Co., Youngstown, Ohio.

Terzaghi, K. (1943). *Theoretical Soil Mechanics*, Wiley, New York, 265 pp.

Terzaghi, K., Peck, R. B. and Mesri, G. (1996). *Soil Mechanics in Engineering Practice* (3rd edn), Wiley, New York, 565 pp.

Teufel, L. W., Rhett, D. W. and Farrell, H. E. (1991). Effect of reservoir depletion and pore pressure drawdown on in situ stress and deformation in the Ekofisk Field, North Sea. In *Rock Mechanics as a Multidisciplinary Science* (ed. J. C. Roegiers), pp. 63–72, Balkema, Rotterdam.

Thallak, S., Holder, J. and Gray, K. E. (1993). The pressure dependence of apparent hydrofracture toughness. *Int. J. Rock Mech. Min. Sci. Geomech. Abstr.* **30**:831–835.

Tharp, T. M. and Coffin, D. T. (1985). Field application of fracture mechanics analysis to small rock slopes. In *Proc. 26th US Symp. Rock Mech.*, pp. 667–674.

Thomas, A. L. and Pollard, D. D. (1993). The geometry of echelon fractures in rock: implications from laboratory and numerical experiments. *J. Struct. Geol.* **15**:323–334.

Timoshenko, S. P. and Goodier, J. N. (1970). *Theory of Elasticity*, McGraw-Hill, New York.

Tirosh, J. and Catz, E. (1981). Mixed-mode fracture angle and fracture locus of materials subjected to compressive loading. *Eng. Frac. Mech.* **14**:27–38.

Tondi, E., Antonellini, M., Aydin, A., Marchegiani, L. and Cello, G. (2006). The role of deformation bands, stylolites and sheared stylolites in fault development in carbonate grainstones of Majella Mountain, Italy. *J. Struct. Geol.* **28**:376–391.

Torabi, A. and Alikarami, R. (2012). Heterogeneity within deformation bands in sandstone reservoirs. Paper ARMA-2012-347 presented at the *46th US Rock Mechanics/Geomechanics Symposium*, 24–27 June, Chicago, Illinois.

Torabi, A. and Berg, S. S. (2011). Scaling of fault attributes: a review. *Mar. Petrol. Geol.* **28**:1444–1460.

Torabi, A. and Fossen, H. (2009). Spatial variation of microstructure and petrophysical properties along deformation bands in reservoir sandstones. *Am. Assoc. Petrol. Geol. Bull.* **93**:919–938.

Toussaint, R., Aharonov, E., Koehn, D., et al. (2018). Stylolites: a review. *J. Struct. Geol.* **114**:163–195; doi:10.1016/j.jsg.2018.05.003.

Townend, E., Thompson, B. D., Benson, P. M., et al. (2008). Imaging compaction band propagation in Diemelstadt sandstone using acoustic emission locations. *Geophys. Res. Lett.* **35**, L15301; doi:10.1029/2008GL034723.

Townend, J. and Zoback, M. D. (2000). How faulting keeps the crust strong. *Geology* **28**:399–402.

Treagus, S. H. and Lisle, R. J. (1997). Do principal surfaces of stress and strain always exist? *J. Struct. Geol.* **19**:997–1010.

Trudgill, B. and Cartwright, J. (1994). Relay-ramp forms and normal-fault linkages, Canyonlands National Park, Utah. *Geol. Soc. Am. Bull.* **106**:1143–1157.

Tse, S. T. and Rice, J. R. (1986). Crustal earthquake instability in relation to the variation of frictional slip properties. *J. Geophys. Res.*, **91**:9452–9472.

Tucholke, B. E. and Lin, J. (1994). A geological model for the structure of ridge segments in slow-spreading ocean crust. *J. Geophys. Res.* **99**:11,937–11,958.

Tucholke, B. E., Lin, J., Kleinrock, M. C., et al. (1997). Segmentation and crustal structure of the western Mid-Atlantic Ridge flank, 25°25'–27°10'N and 0–29 m.y. *J. Geophys. Res.* **102**:10,203–10,223.

Tucholke, B. E., Lin, J. and Kleinrock, M. C. (1998). Megamullions and mullion structure defining oceanic metamorphic core complexes on the Mid-Atlantic Ridge. *J. Geophys. Res.* **103**:9857–9866.

Tufts, B. R., Greenberg R., Hoppa, G. and Geissler, P. (1999). Astypalaea Linea: a large-scale strike-slip fault on Europa. *Icarus* **141**:53–64.

Turcotte, D. L. and Schubert, G. (1982). *Geodynamics: Applications of Continuum Physics to Geological Problems*, Wiley, New York, 450 pp.

Turner, C. E. (1979). Methods for post-yield fracture safety assessment. In *Post-Yield Fracture Mechanics* (ed. D. G. H. Latzko), pp. 23–210, Applied Science Publishers, London.

Twidale, C. R. (1973). On the origin of sheet jointing. *Rock Mech.* **5**:163–187.

Twiss, R. J. and Marrett, R. (2010a). Determining brittle extension and shear strain using fault length and displacement systematics: part I: theory. *J. Struct. Geol.* **32**:1960–1977.

Twiss, R. J. and Marrett, R. (2010b). Determining brittle extension and shear strain using fault length and displacement systematics: part II: data evaluation and test of the theory. *J. Struct. Geol.* **32**:1978–1995.

Twiss, R. J. and Moores, E. M. (1992). *Structural Geology*, Freeman and Co., New York.

Twiss, R. J. and Moores, E. M. (2007). *Structural Geology* (2nd edn), Freeman and Co., New York.

Tworzydlo, W. W. and Hamzeh, O. N. (1997). On the importance of normal vibrations in modeling of stick slip in rock sliding. *J. Geophys. Res.* **102**:15,091–15,103.

Ucar, R. (1986). Determination of shear failure envelope in rock masses. *J. Geotech. Eng. Div. Am. Soc. Civ. Engrs.* **112**: 303–315.

Unger, D. J. (1995). *Analytical Fracture Mechanics*, Academic, New York.

Vajdova, V. and Wong, T.-f. (2003). Incremental propagation of discrete compaction bands: acoustic emission and microstructural observations on circumferentially notched samples of Bentheim sandstone. *Geophys. Res. Lett.* **30**:1775; doi:10.1029/2003GL017750.

Vajdova, V., Baud, P. and Wong, T.-f. (2004a). Compaction, dilatancy, and failure in porous carbonate rocks. *J. Geophys. Res.* **109**:B05204; doi:10.1029/2003JB002508.

Vajdova, V., Baud, P. and Wong, T.-f. (2004b). Permeability evolution during localized deformation in Bentheim sandstone. *J. Geophys. Res.* **109**:B10406; doi:10.1029/2003JB002942.

Valkó, P. and Economides, M. J. (1995). *Hydraulic Fracture Mechanics*, Wiley, New York, 298 pp.

Vallianatos, F. and Sammonds, P. (2011). A non-extensive statistics of the fault-population at the Valles Marineris extensional province, Mars. *Tectonophysics* **509**:50–54.

van der Pluijm, B. A. and Marshak, S. (1997). *Earth Structure: an Introduction to Structural Geology and Tectonics*, McGraw-Hill.

Vermeer, P. A. (1990). The orientation of shear bands in biaxial tests. *Géotechnique* **40**:223–236.

Vermeer, P. A. and de Borst, R. (1984). Non-associated plasticity for soils, concrete, and rock. *Heron* **29**:3–64.

Vermilye, J. M. and Scholz, C. H. (1995). Relation between vein length and aperture. *J. Struct. Geol.* **17**:423–434.

Vermilye, J. M. and Scholz, C. H. (1998). The process zone: a microstructural view of fault growth. *J. Geophys. Res.* **103**:12,223–12,237.

Vidale, J. E., Agnew, D. C., Johnson, M. J. S. and Oppenheimer, D. H. (1998). Absence of earthquake correlation with Earth tides: an indication of high preseismic fault stress rate. *J. Geophys. Res.* **103**:24,567–24,572.

Villemin, T., Angelier, J. and Sunwoo, C. (1995). Fractal distribution of fault length and offsets: implications of brittle deformation evaluation – the Lorrain coal basin. In: *Fractals in the Earth Sciences* (eds. C. C. Barton and P. R. LaPointe), pp. 205–225, Plenum Press, New York.

Wachtman, J. B. (1996). *Mechanical Properties of Ceramics*, Wiley, New York.

Wallace, M. H. and Kemeny, J. (1992). Nucleation and growth of dip-slip faults in a stable craton. *J. Geophys. Res.* **97**: 7145–7157.

Wallner, H. (1939). Linienstrukturen an Bruchflachen. *Zeitschrift für Physik* **114**:368–378.

Walsh, F. R., III, and Zoback, M. D. (2015). Oklahoma's recent earthquakes and saltwater disposal. *Sci. Adv.* **1**:e1500195; doi:10.1126/sciadv.1500195.

Walsh, J. B. (1965). The effect of cracks on the uniaxial compression of rocks. *J. Geophys. Res.* **70**:399–411.

Walsh, J. J. and Watterson, J. (1987). Distributions of cumulative displacement and seismic slip on a single normal fault surface. *J. Struct. Geol.* **9**:1039–1046.

Walsh, J. J. and Watterson, J. (1988). Analysis of the relationship between displacements and dimensions of faults. *J. Struct. Geol.* **10**:239–247.

Walsh, J. J. and Watterson, J. (1989). Displacement gradients on fault surfaces. *J. Struct. Geol.* **11**:307–316.

Walsh, J. J. and Watterson, J. (1991). Geometric and kinematic coherence and scale effects in normal fault systems. In *The Geometry of Normal Faults* (eds. A. M. Roberts, G. Yielding and B. Freeman), pp. 193–203, Geol. Soc. London Spec. Publ. 56.

Walsh, J. J., Watterson, J. and Yielding, G. (1991). The importance of small-scale faulting in regional extension. *Nature* **351**:391–393.

Walsh, J. J., Watterson, J., Bailey, W. R. and Childs, C. (1999). Fault relays, bends and branch lines. *J. Struct. Geol.* **21**:1019–1026.

Walsh, J. J., Nicol, A. and Childs, C. (2002). An alternative model for the growth of faults. *J. Struct. Geol.* **24**:1669–1675.

Walsh, J. J., Bailey, W. R., Childs, C., Nicol, A. and Bonson, C. G. (2003). Formation of segmented normal faults: a 3-D perspective. *J. Struct. Geol.* **25**:1251–1262.

Walsh, P. and Schultz-Ela, D. D. (2003). Mechanics of graben evolution in Canyonlands National Park, Utah. *Geol. Soc. Am. Bull.* **115**:259–270.

Wang, H., Marongiu-Porcu, M. and Economides, M. J. (2016). Poroelastic and poroplastic modeling of hydraulic fracturing in brittle and ductile formations. *SPE Production and Operations* **31**:47–59.

Wang, R. and Kemeny, J. M. (1995). A new empirical failure criterion for rock under polyaxial compressive stresses. In *Rock Mechanics: Proceedings of the 35th US Symposium* (eds. J. J. K. Daemen and R. A. Schultz), pp. 453–458, Balkema, Rotterdam.

Warpinski, N. R. (1985). Measurement of width and pressure in a propagating hydraulic fracture. *Soc. Petrol. Eng. J.* **25**:46–54.

Watters, T. R. (1988). Wrinkle ridge assemblages on the terrestrial planets. *J. Geophys. Res.* **93**:10,236–10,254.

Watters, T. R. (1991). Origin of periodically spaced wrinkle ridges on the Tharsis Plateau of Mars. *J. Geophys. Res.* **96**:15,599–15,616.

Watters, T. R. (1993). Compressional tectonism on Mars. *J. Geophys. Res.* **98**:17049–17060.

Watters, T. R. and Johnson, C. L. (2010). Lunar tectonics. In *Planetary Tectonics* (eds. T. R. Watters and R. A. Schultz), pp. 121–182, Cambridge University Press.

Watters, T. R. and Maxwell, T. A. (1983). Cross-cutting relations and relative age of ridges and faults in the Tharsis region of Mars. *Icarus* **56**:278–298.

Watters, T. R. and Maxwell, T. A. (1986). Orientation, relative age, and extent of the Tharsis Plateau ridge system. *J. Geophys. Res.* **91**:8113–8125.

Watters, T. R. and Nimmo, F. (2010). The tectonics of Mercury. In *Planetary Tectonics* (eds. T. R. Watters and R. A. Schultz), pp. 15–80, Cambridge University Press.

Watters, T. R., Robinson, M. S. and Cook, A. C. (1998). Topography of lobate scarps on Mercury: new constraints on the planet's contraction. *Geology* **26**:991–994.

Watters, T. R., Schultz, R. A. and Robinson, M. S. (2000). Displacement-length relations of thrust faults associated with lobate scarps on Mercury and Mars: Comparison with terrestrial faults. *Geophys. Res. Lett.* **27**:3659–3662.

Watterson, J. (1986). Fault dimensions, displacements and growth. *Pure Appl. Geophys.* **124**:365–373.

Watterson, J., Nicol, A., Walsh, J. J. and Meier, D. (1998). Strains at the intersections of synchronous conjugate normal faults. *J. Struct. Geol.* **20**:363–370.

Wawersik, W. R. and Brace, W. F. (1971). Post-failure behavior of a granite and diabase. *Rock Mech.* **3**:61–85.

Wawersik, W. R. and Fairhurst, C. (1970). A study of brittle rock fracture in laboratory compression experiments. *Int. J. Rock Mech. Min. Sci.* **7**:561–575.

Weijermars, R. (1997). *Principles of Rock Mechanics*, Alboran Science Publishing, Amsterdam.

Weinberger, R., Baer, G., Shamir, G. and Agnon, A. (1995). Deformation bands associated with dyke propagation in porous sandstone, Makhtesh Ramon, Israel. In *Physics and Chemistry of Dikes* (eds. G. Baer and A. Heimann), pp. 95–112, Balkema, Rotterdam.

Weissel, J. K. and Karner, G. D. (1989). Flexural uplift of rift flanks due to mechanical unloading of the lithosphere during extension. *J. Geophys. Res.* **94**:13,919–13,950.

Welch, M. J., Knipe, R. J., Souque, C. and Davis, R. K. (2009). A quadshear kinematic model for folding and clay smear development in fault zones. *Tectonophysics* **471**:186–202.

Wells, A. A. (1961). Unstable crack propagation in metals: cleavage and fast fracture. Proc. Crack Prop. Symp. *1*, paper 84, Cranfield, UK.

Wells, A. A. (1963). Application of fracture mechanics at and beyond general yielding. *British Welding J.* **10**:563–570.

Wells, D. L. and Coppersmith, K. J. (1994). New empirical relationships among magnitude, rupture length, rupture width, rupture area, and surface displacement. *Bull. Seismol. Soc. Am.* **84**:974–1002.

Wennberg, O. P., Casani, G., Jahanpanah, A., et al. (2013). Deformation bands in chalk, examples from the Shetland Group of the Oseberg Field, North Sea, Norway. *J. Struct. Geol.* **56**:103–117.

Wernicke, B. (1981). Low-angle normal faults in the Basin and Range province: nappe tectonics in an extending orogen. *Nature* **291**:645–648.

Wernicke, B. (1992). Cenozoic extensional tectonics of the U.S. Cordillera. In *The Cordilleran Orogen: Conterminous U.S.* (eds. B. C. Burchfiel, P. W. Lipman and M. L. Zoback) pp. 553–581, Geol. Soc. Am., The Geology of North America, v. G–3.

Wesnousky, S. G. (1988). Seismological and structural evolution of strike-slip faults. *Nature* **335**:340–343.

Wesnousky, S. G. (1999). Crustal deformation and the stability of the Gutenberg-Richter relationship. *Bull. Seismol. Soc. Am.* **89**:1131–1137.

Wesnousky, S. G. (2005). The San Andreas and Walker Lane fault systems, western North America: transpression, transtension, cumulative slip and the structural evolution of a major transform plate boundary. *J. Struct. Geol.* **27**:1505–1512.

Wesnousky, S. G. (2008). Displacement and geometrical characteristics of earthquake surface ruptures: issues and implications for seismic hazard analysis and the earthquake rupture process. *Bull. Seismol. Soc. Am.* **98**:1609–1632.

Westaway, R. (1992). Seismic moment summation for historical earthquakes in Italy: tectonic implications. *J. Geophys. Res.* **97**:15,437–15,464.

Westaway, R. (1994). Quantitative analysis of populations of small faults. *J. Struct. Geol.* **16**:1259–1273.

Westergaard, H. M. (1939). Bearing pressures and cracks. *J. Appl. Mech.* **61**:A49–A53.

Whittaker, B. N., Singh, R. N. and Sun, G. (1992). *Rock Fracture Mechanics: Principles, Design and Applications*, Elsevier, New York, 570 pp.

Wibberley, C. A. J., Petit, J.-P. and Rives, T. (1999). Mechanics of high displacement gradient faulting prior to lithification. *J. Struct. Geol.* **21**:251–257.

Wibberley, C. A. J., Petit, J.-P. and Rives, T. (2000). Mechanics of cataclastic 'deformation band' faulting in high-porosity sandstone, Provence. *Comptes Rendus Acad. Sci., Paris* **331**:419–425.

Wibberley, C. A. J., Petit, J.-P. and Rives, T. (2007). The mechanics of fault distribution and localization in high-porosity sands, Provence, France. In *The Relationship Between Damage and Localization* (eds. H. Lewis and G. D. Couples), pp. 19–46, J. Geol. Soc. London Spec. Publ. 289.

Wilcox, R. E., Harding, T. P. and Seely, D. R. (1973). Basic wrench tectonics. *Am. Assoc. Petrol. Geol. Bull.* **57**:74–96.

Wilkins, S. J. (2002). Mechanical and statistical aspects of faulting: from coseismic rupture to cumulative deformation. Unpublished PhD dissertation, University of Nevada, Reno, 181 pp.

Wilkins, S .J. and Gross, M. R. (2002). Normal fault growth in layered rocks at Split Mountain, Utah: influence of mechanical stratigraphy on dip linkage, fault restriction and fault scaling. *J. Struct. Geol.* **24**:1413–1429 (erratum, *J. Struct. Geol.* **24**, 2007).

Wilkins, S. J. and Naruk, S. J. (2007). Quantitative analysis of slip-induced dilation with application to fault seal. *Am. Assoc. Petrol. Geol. Bull.* **91**:97–113.

Wilkins, S. J. and Schultz, R. A. (2003). Cross faults in extensional settings: stress triggering, displacement localization, and implications for the origin of blunt troughs at Valles Marineris, Mars. *J. Geophys. Res.* **108**:5056, 10.1029/2002JE001968.

Wilkins, S. J. and Schultz, R. A. (2005). 3-D cohesive end-zone model for source scaling of strike-slip interplate earthquakes. *Bull. Seismol. Soc. Am.* **95**:2232–2258.

Wilkins, S. J., Gross, M. R., Wacker, M., Eyal, Y. and Engelder, T. (2001). Faulted joints: kinematics, displacement–length scaling relations and criteria for their identification. *J. Struct. Geol.* **23**:315–327.

Wilkins, S. J., Schultz, R. A., Anderson, R. C., Dohm, J. M., and Dawers, N. C. (2002). Deformation rates from faulting at the Tempe Terra extensional province, Mars. *Geophys. Res. Lett.* **29**:1884, 10.1029/2002GL015391.

Willemse, E. J. M. (1997). Segmented normal faults: correspondence between three-dimensional mechanical models and field data. *J. Geophys. Res.* **102**:675–692.

Willemse, E. J. M. and Pollard, D. D. (1998). On the orientation and pattern of wing cracks and solution surfaces at the tips of a sliding flaw or fault. *J. Geophys. Res.* **103**:2427–2438.

Willemse, E. J. M. and Pollard, D. D. (2000). Normal fault growth: evolution of tipline shapes and slip distributions. In *Aspects of Tectonic Faulting* (eds. F. K. Lehner and J. L. Urai), pp. 193–226, Springer, New York.

Willemse, E. J. M., Pollard, D. D. and Aydin, A. (1996). Three-dimensional analyses of slip distributions on normal fault arrays with consequences for fault scaling. *J. Struct. Geol.* **18**:295–309.

Willemse, E. J. M., Peacock, D. C. P. and Aydin, A. (1997). Nucleation and growth of strike-slip faults in limestones from Somerset, UK. *J. Struct. Geol.* **19**:1461–1477.

Williams, A. (1958). Oblique-slip faults and rotated stress systems. *Geol. Mag.* **95**:207–218.

Williams, C. A., Connors, C., Dahlen, F. A., Price, E. J. and Suppe, J. (1994). Effect of the brittle-ductile transition on the topography of compressive mountain belts on Earth and Venus. *J. Geophys. Res.* **99**:19,947–19,974.

Williams, J. G. and Ewing, P. D. (1972). Fracture under complex stress: the angled crack problem. *Int. J. Fracture Mech.* **8**: 441–446.

Williams, M. L. (1957). On the stress distribution at the base of a stationary crack. *J. Appl. Mech.* **24**:109–114.

Willis, B. and Willis, R. (1934). *Geologic Structures*, McGraw-Hill, New York.

Wilson, C. J. N. and Hildreth, W. (1997). The Bishop Tuff: new insights from eruptive stratigraphy. *J. Geol.* **105**:407–439; doi:10.1086/515937.

Wilson, J. E., Goodwin, L. B. and Lewis, C. J. (2003). Deformation bands in nonwelded ignimbrites: petrophysical controls on fault-zone deformation and evidence of preferential fluid flow. *Geology* **31**:837–840.

Wilson, J. T. (1965). A new class of faults and their bearing on continental drift. *Nature* **207**:343–347.

Wiprut, D. and Zoback, M. D. (2002). Fault reactivation, leakage potential, and hydrocarbon column heights in the North Sea. In *Hydrocarbon Seal Quantification* (eds. A. G. Koestler and R. Hunsdale), pp. 203–219, Norwegian Petroleum Society Spec. Publ. 11.

Withjack, M. O., Schlische, R. W. and Henza, A. A. (2007). Scaled experimental models of extension: dry sand vs. wet clay. *Bull. Houston Geol. Soc.* **49**:31–49.

Wojtal, S. (1986). Deformation within foreland thrust sheets by populations of minor faults. *J. Struct. Geol.* **8**:341–360.

Wojtal, S. (1989). Measuring displacement gradients and strains in faulted rocks. *J. Struct. Geol.* **11**:669–678.

Wojtal, S. F. (1996). Changes in fault displacement populations correlated to linkage between faults. *J. Struct. Geol.* **18**:265–279.

Wojtal, S. and Mitra, G. (1988). Nature of deformation in some fault rocks from Appalachian thrusts. In *Geometries and Mechanisms of Thrusting, with Special Reference to the Appalachians* (eds. G. Mitra and S. Wojtal), pp. 17–33, Geol. Soc. Am. Spec. Pap. 222.

Wolf, H., König, D. and Triantafyllidis, T. (2003). Experimental investigation of shear band patterns in granular material. *J. Struct. Geol.* **25**:1229–1240.

Wood, D. S. (1974). Current views on the development of slaty cleavage. *Rev. Earth Planet. Sci.* **2**:1–37.

Woodcock, N. H. and Fischer, M. (1986). Strike-slip duplexes. *J. Struct. Geol.* **8**:725–735.

Woodcock, N. H. and Schubert, C. (1994). Continental strike-slip tectonics. In *Continental Deformation* (ed. P. L. Hancock), pp. 251–263, Pergamon, New York.

Woodworth, J. B. (1896). On the fracture system of joints, with remarks on certain great fractures. *Boston Soc. Natural History Proc.* **27**:163–183.

Wong, T.-f. (1982). Shear fracture energy of Westerly granite from post-failure behavior. *J. Geophys. Res.* **87**:990–1000.

Wong, T.-f. and Baud, P. (1999). Mechanical compaction of porous sandstone. *Oil & Gas Sci. Technol. – Rev. IFP* **54**:715–727.

Wong, T.-f. and Baud, P. (2012). The brittle–ductile transition in rock: a review. *J. Struct. Geol.* **44**:25–53.

Wong, T.-f., Szeto, H. and Zhang, J. (1992). Effect of loading path and porosity on the failure mode of porous rocks. In *Micromechanical Modelling of Quasi-Brittle Materials Behavior* (ed. V. C. Li), *Appl. Mech. Rev.* **45**:281–293.

Wong, T.-f., David, C. and Zhu, W. (1997). The transition from brittle faulting to cataclastic flow in porous sandstones: mechanical deformation. *J. Geophys. Res.* **102**:3009–3025.

Wong, T.-f., Baud, P. and Klein, E. (2001). Localized failure modes in a compactant porous rock. *Geophys. Res. Lett.* **28**: 2521–2524.

Wong, T.-f., David, C. and Menéndez, B. (2004). Mechanical compaction. In *Mechanics of Fluid-Saturated Rocks* (eds. Y. Guéguen and M. Boutéca), pp. 55–114, Elsevier, Amsterdam.

Worrall, D. M. and Snelson, S. (1989). Evolution of the northern Gulf of Mexico, with emphasis on Cenozoic growth faulting and the role of salt. In *The Geology of North America, vol. A, The Geology of North America: an Overview* (ed. A.W. Bally), pp. 97–137, Geol. Soc. Am., Boulder, CO.

Wu, J. E. and McClay, K. R. (2011). Two-dimensional analog modeling of fold and thrust belts: dynamic interactions with syncontractional sedimentation and erosion. In *Thrust Fault-Related Folding* (eds. K. R. McClay, J. H. Shaw and J. Suppe), pp. 301–333, Amer. Assoc. Petrol. Geol. Mem. 94.

Wu, K. and Olson, J. E. (2015). Simultaneous multifracture treatments: fully coupled fluid flow and fracture mechanics for horizontal wells. *Soc. Petrol. Eng. J.* **20**:337–346.

Wyllie, D. C. and Norrish, N. I. (1996). Rock strength properties and their measurement. In *Landslides: Investigation and Mitigation* (eds. A. K. Turner and R. L. Schuster), pp. 372–390, National Research Council, Transportation Research Board Special Report 247, National Academy Press, Washington, DC.

Xie, S. Y. and Shao, J. F. (2006). Elastoplastic deformation of a porous rock and water interaction. *Int. J. Plasticity* **22**: 2195–2225.

Xu, S.-S., Nieto-Samaniego, A. F., Alaniz-Álvarez, S. A. and Velasquillo-Martínez, L. G. (2005). Effect of sampling and linkage on fault length and length–displacement relationship. *Int. J. Earth Sci.* **95**:841–853.

Xue, Y. and Qu, J. (1999). Mixed-mode fracture mechanics parameters of elliptical interface cracks in anisotropic bimaterials. In *Mixed-Mode Crack Behavior* (eds. K. J. Miller, and D. L. McDowell), pp. 143–159, ASTM STP 1359, Am. Soc. Test. Mat., West Conshohocken, Penn.

Yang, Y., Sone, H., Hows, A. and Zoback, M. D. (2013). Comparison of brittleness indices in organic-rich shale formations. Paper ARMA 13–403 presented at the *47th US Symposium on Rock Mechanics/Geomechanics*, San Francisco, California, 23–26 June 2013.

Yao, Y. (2012). Linear elastic and cohesive fracture analysis to model hydraulic fracture in brittle and ductile rocks. *Rock Mech. Rock Eng.* **45**:375–387; doi:10.1007/s00603-011-0211-0.

Yasuhara, H., Elsworth, D. and Polak, A. (2004). Evolution of permeability in a natural fracture: significant role of pressure solution. *J. Geophys. Res.* **109**:3204; doi:10.1029/2003JB002663.

Yeats, R. S., Sieh, K. and Allen, C. R. (1997). *The Geology of Earthquakes*, Cambridge University Press, 568 pp.

Yerkes, R. F. and Castle, R. O. (1976). Seismicity and faulting attributable to fluid extraction. *Eng. Geol.* **10**: 151–167.

Yoffe, E. H. (1951). The moving Griffith crack. *Philosoph. Mag.* **42**:739–750.

Younes, A. I. and Engelder, T. (1999). Fringe cracks: key structures for the interpretation of the progressive Alleghanian deformation of the Appalachian Plateau. *Geol. Soc. Am. Bull.* **111**:219–239.

Zakharova, N. V. and Goldberg, D. S. (2014). In situ stress analysis in the northern Newark Basin: implications for induced seismicity from CO_2 injection. *J. Geophys. Res.* **119**; doi:10.1002/2013JB010492.

Zhang, D., Ranjith, P. G. and Perera, M. S. A. (2016). The brittleness indices used in rock mechanics and their application to shale hydraulic fracturing: a review. *J. Petrol. Sci. Eng.* **143**:158–170.

Zhang, J., Wong, T.-f. and Davis, D. M. (1990). Micromechanics of pressure-induced grain crushing in porous rocks. *J. Geophys. Res.* **95**:341–352.

Zhang, P., Burchfiel, B. C., Chen, S. and Deng, Q. (1989). Extinction of pull-apart basins. *Geology* **17**:814–817.

Zhao, G. and Johnson, A. M. (1991). Sequential and incremental formation of conjugate sets of faults. *J. Struct. Geol.* **13**: 887–895.

Zhao, G. and Johnson, A. M. (1992). Sequence of deformation recorded in joints and faults, Arches National Park, Utah. *J. Struct. Geol.* **14**:225–236.

Zhao, X. G. and Cai, M. (2010). A mobilized dilation angle model for rocks. *Int. J. Rock Mech. Min. Sci.* **47**:368–384.

Zhou, X. and Aydin, A. (2010). Mechanics of pressure solution seam growth and evolution. *J. Geophys. Res.* **115**:B12207; doi:12.1029/2010JB007614.

Zhou, X. and Aydin, A. (2012). Mechanics of the formation of orthogonal sets of solution seams, and solution seams and veins and parallel solution seams and veins. *Tectonophysics* **532–535**:242–257.

Zhu, W. and Wong, T.-f. (1997). The transition from brittle faulting to cataclastic flow: permeability evolution. *J. Geophys. Res.* **102**:3027–3041.

Zhu, X. K., Liu, G. T. and Chao, Y. J. (2001). Three-dimensional stress and displacement fields near an elliptical crack front. *Int. J. Fracture* **109**:383–401.

Ziv, A., Rubin, A. M. and Agnon, A. (2000). Stability of dike intrusion along preexisting fractures. *J. Geophys. Res.* **105**: 5947–5961.

Zoback, M. D. (2007). *Reservoir Geomechanics*, Cambridge University Press, New York.

Zoback, M. D. and Gorelick, S. M. (2012). Earthquake triggering and large-scale geologic storage of carbon dioxide. *Proc. US Nat. Acad. Sci.* **109**:10,164–10,168.

Zoback, M. D. and Healy, J. H. (1984). Friction, faulting, and "in situ" stress. *Ann. Geophys.* **2**:689–698.

Zoback, M. D., and Zinke, J. C. (2002). Production-induced normal faulting in the Valhall and Ekofisk oil fields. *Pure Appl. Geophys.* **159**:403–420.

Zoback, M. D., Zoback, M. L., Mount, V., et al. (1987). New evidence on the state of stress of the San Andreas fault system. *Science* **238**:1105–1111.

Zoback, M. D., Apel, R., Baumgärtner, J., et al. (1993). Upper-crustal strength inferred from stress measurements to 6 km depth in the KTB borehole. *Nature* **365**:633–635.

Zoback, M. D., Townend, J. and Grollimund, B. (2002). Steady-state failure equilibrium and deformation of intraplate lithosphere. *Inter. Geol. Rev.* **44**:383–401.

Zoback, M. D., Barton, C. A., Brudy, M., et al. (2003). Determination of stress orientation and magnitude in deep wells. *Int. J. Rock Mech. Min Sci.* **40**:1049–1076.

Zoback, M. D., Kohli, A., Das, I. and McClure, M. (2012). The importance of slow slip on faults during hydraulic fracturing stimulation of shale gas reservoirs. Paper SPE–155476 presented at the Americas Unconventional Resources Conference, Pittsburg, Pennsylvania, 5–7 June 2012.

Zoback, M. L. (1992). First and second order patterns of tectonic stress: the World Stress Map Project. *J. Geophys. Res.* **97**:11,703–11,728.

Index